	Concept Illustrated	Description	Applet Activity
...e from a ...ation	Assesses how well a sample represents the population and the role that sample size plays in the process.	Produces random sample from population from specified sample size and population distribution shape. Reports mean, median, and standard deviation; applet creates plot of sample.	**4.4**, 186; **4.6**, 200
...pling distributions	Compares means and standard deviations of distributions; assesses effect of sample size; illustrates undbiasedness.	Simulates repeatedly choosing samples of a fixed size n from a population with specified sample size, number of samples, and shape of population distribution. Applet reports means, medians, and standard deviations; creates plots for both.	**4.7**, 227; **4.8**, 228
...om numbers	Uses a random number generator to determine the experimental units to be included in a sample.	Generates random numbers from a range of integers specified by the user.	**1.1**, 18; **1.2**, 18; **3.6**, 154; **4.1**, 174
...-run probability demonstrations illustrate the concept that theoretical probabilities are long-run experimental probabilities.			
...ulating probability ...olling a 6	Investigates relationship between theoretical and experimental probabilities of rolling 6 as number of die rolls increases.	Reports and creates frequency histogram for each outcome of each simulated roll of a fair die. Students specify number of rolls; applet calculates and plots proportion of 6s.	**3.1**, 118; **3.3**, 129; **3.4**, 130; **3.5**, 144
Simulating probability of rolling a 3 or 4	Investigates relationship between theoretical and experimental probabilities of rolling 3 or 4 as number of die rolls increases.	Reports outcome of each simulated roll of a fair die; creates frequency histogram for outcomes. Students specify number of rolls; applet calculates and plots proportion of 3s and 4s.	**3.3**, 129; **3.4**, 130
Simulating the probability of heads: fair coin	Investigates relationship between theoretical and experimental probabilities of getting heads as number of fair coin flips increases.	Reports outcome of each fair coin flip and creates a bar graph for outcomes. Students specify number of flips; applet calculates and plots proportion of heads.	**3.2**, 118; **4.2**, 174
Simulating probability of heads: unfair coin $(P(H) = .2)$	Investigates relationship between theoretical and experimental probabilities of getting heads as number of unfair coin flips increases.	Reports outcome of each flip for a coin where heads is less likely to occur than tails and creates a bar graph for outcomes. Students specify number of flips; applet calculates and plots the proportion of heads.	**4.3**, 186
Simulating probability of heads: unfair coin $(P(H) = .8)$	Investigates relationship between theoretical and experimental probabilities of getting heads as number of unfair coin flips increases.	Reports outcome of each flip for a coin where heads is more likely to occur than tails and creates a bar graph for outcomes. Students specify number of flips; applet calculates and plots the proportion of heads.	**4.3**, 186
...imulating the stock ...rket	Theoretical probabilities are long run experimental probabilities.	Simulates stock market fluctuation. Students specify number of days; applet reports whether stock market goes up or down daily and creates a bar graph for outcomes. Calculates and plots proportion of simulated days stock market go...	**4.5**, 186
... versus median	Investigates how skewedness and outliers affect measures of central tendency.	Students visualize rel... and median by addin... points; applet autom... and median.	2.3, 57

(*Continued*)

Applet	Concept Illustrated	Description	Applet Activity
Standard deviation	Investigates how distribution shape and spread affect standard deviation.	Students visualize relationship between mean and standard deviation by adding and deleting data points; applet updates mean and standard deviation.	**2.4**, 64; **2.5**, 64; **2.6**, 65; **2.**
Confidence intervals for a proportion	Not all confidence intervals contain the population proportion. Investigates the meaning of 95% and 99% confidence.	Simulates selecting 100 random samples from the population and finds the 95% and 99% confidence intervals for each. Students specify population proportion and sample size; applet plots confidence intervals and reports number and proportion containing true proportion.	**5.5**, 269; **5.6**, 269
Confidence intervals for a mean (the impact of confidence level)	Not all confidence intervals contain the population mean. Investigates the meaning of 95% and 99% confidence.	Simulates selecting 100 random samples from population; finds 95% and 99% confidence intervals for each. Students specify sample size, distribution shape, and population mean and standard deviation; applet plots confidence intervals and reports number and proportion containing true mean.	**5.1**, 250; **5.2**, 251
Confidence intervals for a mean (not knowing standard deviation)	Confidence intervals obtained using the sample standard deviation are different from those obtained using the population standard deviation. Investigates effect of not knowing the population standard deviation.	Simulates selecting 100 random samples from the population and finds the 95% z-interval and 95% t-interval for each. Students specify sample size, distribution shape, and population mean and standard deviation; applet plots confidence intervals and reports number and proportion containing true mean.	**5.3**, 260; **5.4**, 260
Hypothesis tests for a proportion	Not all tests of hypotheses lead correctly to either rejecting or failing to reject the null hypothesis. Investigates the relationship between the level of confidence and the probabilities of making Type I and Type II errors.	Simulates selecting 100 random samples from population; calculates and plots z-statistic and P-value for each. Students specify population proportion, sample size, and null and alternative hypotheses; applet reports number and proportion of times null hypothesis is rejected at 0.05 and 0.01 levels.	**6.5**, 329; **6.6**, 329
Hypothesis tests for a mean	Not all tests of hypotheses lead correctly to either rejecting or failing to reject the null hypothesis. Investigates the relationship between the level of confidence and the probabilities of making Type I and Type II errors.	Simulates selecting 100 random samples from population; calculates and plots t statistic and P-value for each. Students specify population distribution shape, mean, and standard deviation; sample size, and null and alternative hypotheses; applet reports number and proportion of times null hypothesis is rejected at both 0.05 and 0.01 levels.	**6.1**, 304; **6.2**, 308; **6.3**, 308; **6.4**, 308
Correlation by eye	Correlation coefficient measures strength of linear relationship between two variables. Teaches user how to assess strength of a linear relationship from a scattergram.	Computes correlation coefficient r for a set of bivariate data plotted on a scattergram. Students add or delete points and guess value of r; applet compares guess to calculated value.	**9.2**, 510
Regression by eye	The least squares regression line has a smaller SSE than any other line that might approximate a set of bivariate data. Teaches students how to approximate the location of a regression line on a scattergram.	Computes least squares regression line for a set of bivariate data plotted on a scattergram. Students add or delete points and guess location of regression line by manipulating a line provided on the scattergram; applet plots least squares line and displays the equations and the SSEs for both lines.	**9.1**, 486

A FIRST COURSE IN

STATISTICS

A FIRST COURSE IN

STATISTICS

ELEVENTH EDITION

James T. McClave

Info Tech, Inc.

University of Florida

Terry Sincich

University of South Florida

PEARSON

Boston Columbus Indianapolis New York San Francisco Upper Saddle River
Amsterdam Cape Town Dubai London Madrid Milan Munich Paris Montréal Toronto
Delhi Mexico City São Paulo Sydney Hong Kong Seoul Singapore Taipei Tokyo

Editor in Chief: Deirdre Lynch
Acquisitions Editor: Marianne Stepanian
Associate Content Editor: Dana Bettez
Editorial Assistant: Sonia Ashraf
Senior Managing Editor: Karen Wernholm
Senior Production Project Manager: Tracy Patruno
Associate Director of Design, USHE North and West: Andrea Nix
Text and Cover Designer: Barbara T. Atkinson
Digital Assets Manager: Marianne Groth
Production Coordinator: Katherine Roz
Media Producer: Jean Choe
Software Developers: Mary Durnwald and Bob Carroll
Marketing Manager: Erin Lane
Marketing Assistant: Kathleen DeChavez
Senior Author Support/Technology Specialist: Joe Vetere
Rights and Permissions Advisor: Michael Joyce
Image Manager: Rachel Youdelman
Procurement Manager: Evelyn Beaton
Procurement Specialist: Linda Cox
Senior Media Procurement Sepcialist: Ginny Michaud
Production Coordination, Composition, Illustrations: Integra
Cover Image: Crowd of small symbolic 3d figures linked by lines, ©Higyou/Shutterstock

Credits appear on page 587, which constitutes a continuation of the copyright page.

Library of Congress Cataloging-in-Publication Data

McClave, James T.
  A first course in statistics / James T. McClave, Terry Sincich.—11th ed.
    p. cm.
  ISBN 0-321-75595-2
1. Statistics.   I. Sincich, Terry.   II. Title.
  QA276.M378 2013
  519.5—dc23

                                        2011031869

2 3 4 5 6 7 8 9 10—V011—15 14

www.pearsonhighered.com

ISBN 10: 0-321-75595-2
ISBN 13: 978-0-321-75595-7

Contents

Appendices

Preface

A First Course in Statistics is an introductory text designed for one-semester courses that emphasizes inference and sound decision-making through extensive coverage of data collection and analysis. As in earlier editions, the eleventh edition text stresses the development of statistical thinking, the assessment of credibility, and value of the inferences made from data, both by those who consume and those who produce them. It assumes a mathematical background of basic algebra.

The text incorporates the following strategies, developed from the American Statistical Association's (ASA) Guidelines for Assessment and Instruction in Statistics Education (GAISE) Project:

- Emphasize statistical literacy and develop statistical thinking
- Use real data in applications
- Use technology for developing conceptual understanding and analyzing data
- Foster active learning in the classroom
- Stress conceptual understanding rather than mere knowledge of procedures

New in the Eleventh Edition

- **More than 1,200 exercises, with revisions and updates to 20%.** Many new and updated exercises, based on contemporary studies and real data, have been added. Most of these exercises foster and promote critical thinking skills.
- **Updated technology.** All printouts from statistical software (SAS®, IBM® SPSS®, MINITAB®, and the TI-83/84 Plus Graphing Calculator) and corresponding instructions for use have been revised to reflect the latest versions of the software.
- **New and Revised Statistics in Action Cases.** More than one-third of the *Statistics in Action* cases are new or revised, each based on real data from a recent study.
- **Redesigned end–of–chapter summaries.** Summaries at the end of each chapter have been redesigned to make them better study aids for students. Important points are reinforced through flow graphs (which aid in selecting the appropriate statistical method) and boxed notes with key ideas, terms, symbols/notation, and formulas. A supplementary exercise section and a hands-on activity are also included at the end of each chapter.
- **Emphasis on ethics.** Where appropriate, boxes have been added emphasizing the importance of ethical behavior when collecting, analyzing, and interpreting data.
- **Learning objectives.** Chapter opening *Where We're Going* bullet points include section numbers that correspond to where that concept is discussed in the chapter.

Content–Specific Changes to This Edition

- **Chapter 5 (Confidence Intervals).** The methodology for finding a confidence interval for a population mean is developed based on using either a normal (z) statistic (Section 5.2) or a Student's t-statistic (Section 5.3). Also, we added an optional section (Section 5.6) on estimating a population variance.
- **Chapter 6 (Tests of Hypothesis).** A new section emphasizing the formulation of the null and alternative hypotheses (Section 6.2) has been added.
- **Chapter 8 (Comparing Population Proportions).** A subsection on contingency tables with fixed marginals as been added to Section 8.4.

Hallmark Strengths

We have maintained or strengthened the pedagogical features that make *A First Course in Statistics* unique among introductory statistics texts. These features, which assist the student in achieving an overview of statistics and an understanding of its relevance in the world and everyday life, are as follows:

- **Use of examples as a teaching device.**   Almost all new ideas are introduced and illustrated by data-based applications and examples. We believe that students better understand definitions, generalizations, and theoretical concepts *after* seeing an application. All examples have three components: (1) Problem, (2) Solution, and (3) Look Back (or Look Ahead). This step-by-step process provides students with a defined structure by which to approach problems and enhances their problem-solving skills. The Look Back feature often gives helpful hints to solving the problem and/or provides a further reflection or insight into the concept or procedure that is covered.

- **Now Work.**   A *Now Work* exercise suggestion follows each example. The *Now Work* exercise (marked with the icon [NW] in exercise sets) is similar in style and concept to the text example. This provides students with an opportunity to immediately test and confirm their understanding.

- **Statistics in Action.**   Each chapter begins with a case study based on an actual contemporary, controversial or high-profile issue. Relevant research questions and data from the study are presented and the proper analysis demonstrated in short *Statistics in Action Revisited* sections throughout the chapter. These motivate students to critically evaluate the findings and think through the statistical issues involved.

- **Applet Exercises.**   The text is accompanied by a resource CD containing applets (short JAVA computer programs). These point-and-click applets allow students to easily run simulations that visually demonstrate some of the more difficult statistical concepts (e.g., sampling distributions and confidence intervals.) Each chapter contains several optional applet exercises in the exercise sets. They are denoted with the following icon:

- **Real data-driven exercises.**   The text includes more than 1,000 exercises based on a wide variety of applications in various disciplines and research areas. Nearly all of the applied exercises use current, real data extracted from newspapers, magazines, current journals, and the Internet. Some students have difficulty learning the mechanics of statistical techniques when all problems are couched in terms of realistic applications. For this reason, all exercise sections are divided into four parts:

   Learning the Mechanics.   Designed as straightforward applications of new concepts, these exercises allow students to test their ability to comprehend a mathematical concept or a definition.

   Applying the Concepts—Basic.   Based on applications taken from a wide variety of journals, newspapers, and other sources, these short exercises help students begin developing the skills necessary to diagnose and analyze real-world problems.

   Applying the Concepts—Intermediate.   Based on more detailed real-world applications, these exercises require students to apply their knowledge of the technique presented in the section.

   Applying the Concepts—Advanced.   These more difficult real-data exercises require students to utilize their critical thinking skills.

- **Critical Thinking Challenges.**   Placed at the end of the Supplementary Exercises section only, students are asked to apply their critical thinking skills to solve one or two challenging real-life problems. These exercises expose students to real-world problems with solutions that are derived from careful, logical thought and selection of the appropriate statistical analysis tool.

- **Exploring data with statistical computer software and the graphing calculator.** We demonstrate each statistical analysis method presented using output from three leading statistical software packages: SAS, SPSS, and MINITAB. These outputs appear throughout the text in examples and exercises, exposing students to the output they will encounter in today's high-tech world. In addition, we provide output and keystroke instructions for the TI-83/84 Plus Graphing Calculator in the *Using Technology* section at the end of appropriate chapters.

- **Using Technology tutorials.** At the end of each chapter we've included statistical software tutorials with instructions and screen shots for MINITAB and, where appropriate, the TI-83/84 Plus Graphing Calculator. These step-by-step tutorials are easily located and show students how to best use statistical software.

- **Biographies.** Brief descriptions of famous statisticians and his/her achievements are presented in-text and in marginal boxes. With these profiles, students will develop an appreciation for the statistician's efforts and the discipline of statistics as a whole.

- **CD-ROM.** New copies of the text are accompanied by a resource CD that contains files for all of the text examples, exercises, Statistics in Action, and Real-World case data sets marked with a 💿. Data sets are provided in multiple formats. The CD also contains a set of applets that illustrate statistical concepts.

Supplements

Student Resources

Student's Solutions Manual, by Nancy Boudreau (Bowling Green State University), includes complete worked out solutions to all odd-numbered text exercises (ISBN-13: 978-0-321-78343-1; ISBN-10: 0-321-78343-3).

Excel® Manual (download only), by Mark Dummeldinger (University of South Florida). Available for download from www.pearsonhighered.com/mathstatsresources.

MINITAB® Manual (download only), by Keith Bower, available for download from www.pearsonhighered.com/mathstatsresources.

Graphing Calculator Manual (download only), by Susan Herring (Sonoma State University), available for download from www.pearsonhighered.com/mathstatsresources.

Study Cards for Statistics Software. This series of study cards, available for Excel, MINITAB, JMP®, SPSS, R, StatCrunch®, and TI-83/84 Plus Graphing Calculators provides students with easy step-by-step guides to the most common statistics software. Visit myPearsonstore.com for more information.

Instructor Resources

Annotated Instructor's Edition contains answers to text exercises. Annotated marginal notes include Teaching Tips, suggested exercises to reinforce the statistical concepts discussed in the text, and short answers to exercises and examples (ISBN-13: 978-0-321-78304-2; ISBN-10: 0-321-78304-2).

Instructor's Solutions Manual (download only), by Nancy Boudreau (Bowling Green State University), includes complete worked-out solutions to all even-numbered text exercises Careful attention has been paid to ensure that all methods of solution and notation are consistent with those used in the core text. Available for download from Pearson's online catalog at www.pearsonhighered.com/irc.

PowerPoint® Lecture Slides include figures and tables from the textbook. Available for download from Pearson's online catalog at www.pearsonhighered.com/irc.

TestGen® (www.pearsoned.com/testgen) enables instructors to build, edit, print, and administer tests using a computerized bank of questions developed to cover all the objectives of the text. TestGen is algorithmically based, allowing instructors to create multiple but equivalent versions of the same question or test with the click of a button. Instructors can also modify test bank questions or add new questions. The software and test bank are available for download from Pearson Education's online catalog.

Online Test Bank is a test bank derived from TestGen®. It includes multiple-choice and short-answer questions for each section of the text, along with the answer keys. Available for download from Pearson's online catalog at www.pearsonhighered.com/irc.

The Pearson Math Adjunct Support Center (http://www.pearsontutorservices.com/math-adjunct.html) is staffed by qualified instructors with more than 100 years of combined experience at both the community college and university levels. Assistance is provided for faculty in the following areas:

- Suggested syllabus consultation
- Tips on using materials packed with your book
- Book-specific content assistance
- Teaching suggestions, including advice on classroom strategies

Technology Resources

A companion **CD-ROM** is bound in new copies of *A First Course in Statistics.* The CD holds a number of support materials, including:

- **Data sets** formatted as .csv, .txt, and TI files
- **Applets** (short JAVA computer programs) that allow students to run simulations that visually demonstrate statistical concepts

Data sets are also available for download from www.pearsonhighered.com/mathstatsresources.

MathXL® for Statistics Online Course (access code required) MathXL® is the homework and assessment engine that runs MyStatLab. (MyStatLab is MathXL plus a learning management system.) With MathXL for Statistics, instructors can:

- Create, edit, and assign online homework and tests using algorithmically generated exercises correlated at the objective level to the textbook.
- Create and assign their own online exercises and import TestGen tests for added flexibility.
- Maintain records of all student work, tracked in MathXL's online gradebook.

With MathXL for Statistics, students can:

- Take chapter tests in MathXL and receive personalized study plans and/or personalized homework assignments based on their test results.
- Use the study plan and/or the homework to link directly to tutorial exercises for the objectives they need to study.

- Students can also access supplemental animations and video clips directly from selected exercises.
- Knowing that students often use external statistical software, we make it easy to copy our data sets, both from the ebook and the MyStatLab questions, into software like StatCrunch, MINITAB, Excel, and more.

MathXL for Statistics is available to qualified adopters. For more information, visit our website at www.mathxl.com, or contact your Pearson representative.

MyStatLab™ Online Course (access code required)

MyStatLab is a course management system that delivers **proven results** in helping individual students succeed.

- MyStatLab can be successfully implemented in any environment—lab-based, hybrid, fully online, traditional—and demonstrates the quantifiable difference that integrated usage has on student retention, subsequent success, and overall achievement.
- MyStatLab's comprehensive online gradebook automatically tracks students' results on tests, quizzes, homework, and in the study plan. Instructors can use the gradebook to intervene if students have trouble or to provide positive feedback. Data can be easily exported to a variety of spreadsheet programs, such as Microsoft Excel.

MyStatLab provides **engaging experiences** that personalize, stimulate, and measure learning for each student.

- **Tutorial Exercises with Multimedia Learning Aids.** The homework and practice exercises in MyStatLab align with the exercises in the textbook, and they regenerate algorithmically to give students unlimited opportunity for practice and mastery. Exercises offer immediate helpful feedback, guided solutions, sample problems, animations, videos, and eText clips for extra help at point-of-use.
- **Getting Ready for Statistics.** A library of questions now appears within the MyStatLab assessment manager to offer the developmental math topics students need for the course. These can be assigned as a prerequisite to other assignments, if desired.
- **Conceptual Question Library.** In addition to algorithmically regenerated questions that are aligned with your textbook, there is a library of 1,000 Conceptual Questions available in the assessment managers that require students to apply their statistical understanding.
- **StatCrunch.** MyStatLab includes a web-based statistical software, StatCrunch, within the online assessment platform so that students can easily analyze data sets from exercises and the text. In addition, MyStatLab includes access to www.statcrunch.com, a web site where users can access more than 13,000 shared data

sets, conduct online surveys, perform complex analyses using the powerful statistical software, and generate compelling reports.

- **Integration of Statistical Software.** Knowing that students often use external statistical software, we make it easy to copy our data sets, both from the ebook and MyStatLab questions, into software like StatCrunch, MINITAB, Excel and more. Students have access to a variety of support—Technology Instruction Videos, Technology Study Cards, and Manuals—to learn how to effectively use statistical software.
- **Expert Tutoring.** Although many students describe the whole of MyStatLab as "like having your own personal tutor," students also have access to live tutoring from Pearson. Qualified statistics instructors provide tutoring sessions for students via MyStatLab.

And, MyStatLab comes from a **trusted partner** with educational expertise and an eye on the future.

Knowing that you are using a Pearson product means knowing that you are using quality content. That means that our eTexts are accurate, that our assessment tools work, and that our questions are error-free. And whether you are just getting started with MyStatLab, or have a question along the way, we're here to help you learn about our technologies and how to incorporate them into your course.

To learn more about how MyStatLab combines proven learning applications with powerful assessment, visit www.mystatlab.com or contact your Pearson representative.

StatCrunch

StatCrunch is powerful web-based statistical software that allows users to perform complex analyses, share data sets, and generate compelling reports of their data. The vibrant online community offers more than 13,000 data sets for students to analyze.

- **Collect.** Users can upload their own data to StatCrunch or search a large library of publicly shared data sets, spanning almost any topic of interest. Also, an online survey tool allows users to quickly collect data via web-based surveys.
- **Crunch.** A full range of numerical and graphical methods allow users to analyze and gain insights from any data set. Interactive graphics help users understand statistical concepts, and are available for export to enrich reports with visual representations of data.
- **Communicate.** Reporting options help users create a wide variety of visually-appealing representations of their data.

Full access to StatCrunch is available with a MyStatLab kit, and StatCrunch is available by itself to qualified adopters. For more information, visit our website at www.statcrunch.com, or contact your Pearson representative.

The Student Edition of MINITAB is a condensed edition of the professional release of MINITAB statistical software. It offers the full range of statistical methods and graphical capabilities, along with worksheets that can include up to 10,000 data points. Individual copies of the software can be bundled with the text (ISBN-13: 978-0-321-11313-9; ISBN-10: 0-321-11313-6).

JMP Student Edition is an easy-to-use, streamlined version of JMP desktop statistical discovery software from SAS Institute Inc. and is available for bundling with the text (ISBN-13: 978-0-321-67212-4; ISBN-10: 0-321-67212-7).

Acknowledgments

This book reflects the efforts of a great many people over a number of years. First, we would like to thank the following professors, whose reviews and comments on this and prior editions have contributed to the 11th edition:

Reviewers Involved with the Eleventh Edition of *A First Course in Statistics*

Ali Arab, *Georgetown University*
Jen Case, *Jacksonville State University*
Maggie McBride, *Montana State University—Billings*
Surajit Ray, *Boston University*
JR Schott, *University of Central Florida*
Susan Schott, *University of Central Florida*
Lewis Shoemaker, *Millersville University*
Engin Sungur, *University of Minnesota—Morris*
Sherwin Toribio, *Universitiy of Wisconsin—La Crosse*
Michael Zwilling, *Mt. Union College*

Reviewers of Previous Editions

Bill Adamson, South Dakota State; Ibrahim Ahmad, Northern Illinois University; Roddy Akbari, Guilford Technical Community College; David Atkinson, Olivet Nazarene University; Mary Sue Beersman, Northeast Missouri State University; William H. Beyer, University of Akron; Marvin Bishop, Manhattan College; Patricia M. Buchanan, Pennsylvania State University; Dean S. Burbank, Gulf Coast Community College; Ann Cascarelle, St. Petersburg College; Kathryn Chaloner, University of Minnesota; Hanfeng Chen, Bowling Green State University; Gerardo Chin-Leo, The Everygreen State College; Linda Brant Collins, Iowa State University; Brant Deppa, Winona State University; John Dirkse, California State University—Bakersfield; N. B. Ebrahimi, Northern Illinois University; John Egenolf, University of Alaska—Anchorage; Dale Everson, University of Idaho; Christine Franklin, University of Georgia; Khadiga Gamgoum, Northern Virginia CC; Rudy Gideon, University of Montana; Victoria Marie Gribshaw, Seton Hill College; Larry Griffey, Florida Community College; David Groggel, Miami University at Oxford; Sneh Gulati, Florida International University; John E. Groves, California Polytechnic State University—San Luis Obispo; Dale K. Hathaway, Olivet Nazarene University; Shu-ping Hodgson, Central Michigan University; Jean L. Holton, Virginia Commonwealth University; Soon Hong, Grand Valley; Ina Parks S. Howell, Florida International University; Gary Itzkowitz, Rowan College of New Jersey; John H. Kellermeier, State University College at Plattsburgh; Golan Kibria, Florida International University; Timothy J. Killeen, University of Connecticut; William G. Koellner, Montclair State University; James R. Lackritz, San Diego State University; Diane Lambert, AT&T/Bell Laboratories; Edwin G. Landauer, Clackamas Community College; James Lang, Valencia Junior College; Glenn Larson, University of Regina; John J. Lefante, Jr., University of South Alabama; Pi-Erh Lin, Florida

State University; R. Bruce Lind, University of Puget Sound; Rhonda Magel, North Dakota State University; Linda C. Malone, University of Central Florida; Allen E. Martin, California State University—Los Angeles; Rick Martinez, Foothill College; Brenda Masters, Oklahoma State University; Leslie Matekaitis, Cal Genetics; E. Donice McCune, Stephen F. Austin, State University; Mark M. Meerschaert, University of Nevada—Reno; Greg Miller, Steven F. Austin State University; Satya Narayan Mishra, University of South Alabama; Kazemi Mohammed, UNC–Charlotte; Christopher Morrell, Loyola College in Maryland; Mir Mortazavi, Eastern New Mexico University; A. Mukherjea, University of South Florida; Steve Nimmo, Morningside College (Iowa); Susan Nolan, Seton Hall University; Thomas O'Gorman, Northern Illinois University; Bernard Ostle, University of Central Florida; William B. Owen, Central Washington University; Won J. Park, Wright State University; John J. Peterson, Smith Kline & French Laboratories; Ronald Pierce, Eastern Kentucky University; Betty Rehfuss, North Dakota State University—Bottineau; Andrew Rosalsky, University of Florida; C. Bradley Russell, Clemson University; Rita Schillaber, University of Alberta; James R. Schott, University of Central Florida; Susan C. Schott, University of Central Florida; George Schultz, St. Petersburg Junior College; Carl James Schwarz, University of Manitoba; Mike Seyfried, Shippensburg University; Arvind K. Shah, University of South Alabama; Lewis Shoemaker, Millersville University; Sean Simpson, Westchester CC; Charles W. Sinclair, Portland State University; Robert K. Smidt, California Polytechnic State University—San Luis Obispo; Vasanth B. Solomon, Drake University; W. Robert Stephenson, Iowa State University; Thaddeus Tarpey, Wright State University; Kathy Taylor, Clackamas Community College; Barbara Treadwell, Western Michigan University; Dan Voss, Wright State University; Augustin Vukov, University of Toronto; Dennis D. Wackerly, University of Florida; Barbara Wainwright, Salisbury University; Matthew Wood, University of Missouri—Columbia.

Other Contributors

Special thanks are due to our ancillary authors, Nancy Boudreau, Mark Dummeldinger, Keith Bower, and Susan Herring. Thank you to our accuracy checkers, Engin Sungur and Cathleen Zucco-Teveloff, who helped to insure a highly accurate, clean text. Finally, the Pearson staff of Deirdre Lynch, Marianne Stepanian, Tracy Patruno, Dana Bettez, Sonia Ashraf, Barbara Atkinson, Jean Choe, Kathleen DeChavez, Roxanne McCarley, and Erin Lane, and Integra-Chicago's Amanda Zagnoli all helped greatly with various stages of the book and media.

Applications Index

STATISTICS

1 Statistics, Data, and Statistical Thinking

CONTENTS

Where We're Going

- Introduce the field of statistics (1.1)
- Demonstrate how statistics applies to real-world problems (1.2)
- Introduce the language of statistics and the key elements to any statistical problem. (1.3)
- Differentiate between population and sample data (1.3)
- Differentiate between descriptive and inferential statistics (1.3)
- Identify the different types of data and data collection methods (1.4–1.5)
- Discover how critical thinking through statistics can help improve our quantitative literacy (1.6)

Statistics in Action Social Media Networks and the Millennial Generation

The Pew Research Center, a nonpartisan organization funded by a Philadelphia-based charity, has conducted over 100 surveys on Internet usage in the United States as part of the Pew Internet & American Life Project (PIALP). The PIALP has recently published a series of reports on teens and adults from ages 18 to 29 years—called the "Millennial Generation." In a 2010 report titled "Social Media & Mobile Internet Use," the PIALP examined the Millennial Generation's attitudes and behavior towards online social networks (e.g., Facebook, MySpace, Twitter). According to Wikipedia (the free online encyclopedia), *social media* are "media for social interaction, using highly accessible and scalable publishing techniques" such as Weblogs, Internet forums, Twitter, and social networking sites such as Facebook and MySpace.

Results from several of the survey questions asked of the teens are provided here:

- *Internet Use*
  When asked how often they use the Internet, teens responded:

Several times a day	36%
About once a day	27%
3–5 days a week	14%
1–2 days a week	12%
Every few weeks	7%
Less often	4%

- *Social Networking*
  When asked if they use social network sites like Facebook or MySpace, teens responded:

Yes	73%
No	27%

- *Twitter*
  When asked if they use Twitter, teens responded:

Yes	8%
No	91%

- *Text Messaging*
  When asked how often they send text messages on their cell phones, teens responded:

Every day	54%
Several times per week	10%
At least once a week	5%
Less than once a week	3%
Never	28%

- *Average Number of Phone Calls*
  On an average day, teens make and receive 10.7 phone calls on their cell phones.

- *Average Number of Text Messages*
  On an average day, teens send and receive 112.4 text messages on their cell phones.

In the following Statistics in Action Revisited sections, we discuss several key statistical concepts covered in this chapter that are relevant to the Pew Internet & American Life Project survey.

- Identifying the Population, Sample, and Inference (p. 9)
- Identifying the Data Collection Method and Data Type (p. 14)
- Critically Assessing the Ethics of a Statistical Study (p. 17)

Based on Pew Research Center for the People & the Press, "Social Media & Mobile Internet Use" report. © 2010 PIALF.

1.1 The Science of Statistics

What does statistics mean to you? Does it bring to mind batting averages, Gallup polls, unemployment figures, or numerical distortions of facts (lying with statistics!)? Or is it simply a college requirement you have to complete? We hope to persuade you that statistics is a meaningful, useful science whose broad scope of applications to business, government, and the physical and social sciences is almost limitless. We also want to show that statistics can lie only when they are misapplied. Finally, we wish to demonstrate the key role statistics plays in critical thinking—whether in the classroom, on the job, or in everyday life. Our objective is to leave you with the impression that the time you spend studying this subject will repay you in many ways.

The *Random House College Dictionary* defines **statistics** as "the science that deals with the collection, classification, analysis, and interpretation of information or data." Thus, a statistician isn't just someone who calculates batting averages at baseball games or tabulates the results of a Gallup poll. Professional statisticians are trained in *statistical science*. That is, they are trained in collecting information in the form of **data**, evaluating the information, and drawing conclusions from it. Furthermore, statisticians determine what information is relevant in a given problem and whether the conclusions drawn from a study are to be trusted.

> **Statistics** is the science of data. This involves collecting, classifying, summarizing, organizing, analyzing, presenting, and interpreting numerical information.

In the next section, you'll see several real-life examples of statistical applications that involve making decisions and drawing conclusions.

1.2 Types of Statistical Applications

"Statistics" means "numerical descriptions" to most people. Monthly housing starts, the failure rate of liver transplants, and the proportion of African-Americans who feel brutalized by local police all represent statistical descriptions of large sets of data collected on some phenomenon. (Later, in Section 1.4, we learn that not all data is numerical in nature.) Often the data are selected from some larger set of data whose characteristics we wish to estimate. We call this selection process *sampling*. For example, you might collect the ages of a sample of customers who shop for a particular product online to estimate the average age of *all* customers who shop online for the product. Then you could use your estimate to target the Web site's advertisements to the appropriate age group. Notice that statistics involves two different processes: (1) describing sets of data and (2) drawing conclusions (making estimates, decisions, predictions, etc.) about the sets of data on the basis of sampling. So, the applications of statistics can be divided into two broad areas: **descriptive statistics** and **inferential statistics.**

> **Descriptive statistics** utilizes numerical and graphical methods to look for patterns in a data set, to summarize the information revealed in a data set, and to present that information in a convenient form.

> **Inferential statistics** utilizes sample data to make estimates, decisions, predictions, or other generalizations about a larger set of data.

BIOGRAPHY FLORENCE NIGHTINGALE (1820–1910)

The Passionate Statistician

In Victorian England, the "Lady of the Lamp" had a mission to improve the squalid field hospital conditions of the British army during the Crimean War. Today, most historians consider Florence Nightingale to be the founder of the nursing profession. To convince members of the British Parliament of the need for supplying nursing and medical care to soldiers in the field, Nightingale compiled massive amounts of data from army files. Through a remarkable series of graphs (which included the first pie chart), she demonstrated that most of the deaths in the war either were due to illnesses contracted outside the battlefield or occurred long after battle action from wounds that went untreated. Florence Nightingale's compassion and self-sacrificing nature, coupled with her ability to collect, arrange, and present large amounts of data, led some to call her the Passionate Statistician. ■

Although we'll discuss both descriptive and inferential Statistics in the chapters that follow, the primary theme of the text is **inference**.

Let's begin by examining some studies that illustrate applications of statistics.

Study 1.1   "Best-Selling Girl Scout Cookies" (*www.girlscouts.org*)

Since 1917, the Girl Scouts of America have been selling boxes of cookies. Currently, there are eight varieties for sale: Thin Mints, Samoas, Caramel DeLites, Tagalongs,

Peanut Butter Patties, Do-si-dos, Peanut Butter Sandwiches, and Trefoils. Each of the approximately 150 million boxes of Girl Scout cookies sold each year is classified by variety. The results are summarized in Figure 1.1. From the graph, you can clearly see that the best-selling variety is Thin Mints (25%), followed by Samoas (19%) and Tagalongs (13%). Since the figure describes the various categories of boxes of Girl Scout cookies sold, the graphic is an example of descriptive statistics.

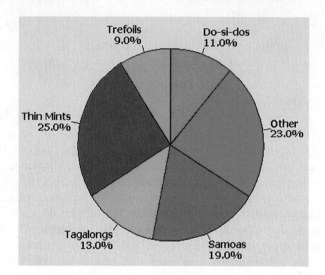

Figure 1.1

MINITAB graph of best-selling Girl Scout cookies
Based on www.girlscouts.org.

Study 1.2 "Does Playing Video Games Make for Better Visual Attention Skills?" (*Journal of Articles in Support of the Null Hypothesis*, Vol. 6, No. 1, 2009)

Researchers at Griffin University (Australia) conducted a study to determine whether video game players have superior visual attention skills compared to non–video game players. Each in a sample of 65 male psychology students was classified as a video game player or a nonplayer. The two groups were then subjected to a series of visual attention tasks that included the "attentional blink" test, the "field of view" test, and the "repetition blindness" test. Except for attentional blink, no differences in the performance of the two groups were found. From this analysis, the researchers inferred "a limited role for video game playing in the modification of visual attention." Thus, inferential statistics was applied to arrive at this conclusion.

Study 1.3 "Animal Assisted Therapy…[for] Hospitalized Heart Failure Patients" (*American Heart Association Conference*, November 2005)

A team from the UCLA Medical Center and School of Nursing, led by RN Kathie Cole, conducted a study to gauge whether animal-assisted therapy can improve the physiological responses of heart failure patients. Cole and her colleagues studied 76 heart patients, randomly divided into three groups. Each person in one group of patients was visited by a human volunteer accompanied by a trained dog, each person in another group was visited by a volunteer only, and the third group was not visited at all. The researchers measured patients' physiological responses (levels of anxiety, stress, and blood pressure) before and after the visits. An analysis of the data revealed that those patients with animal-assisted therapy had significantly greater drops in levels of anxiety, stress, and blood pressure. Thus, the researchers concluded that "pet therapy has the potential to be an effective treatment…for patients hospitalized with heart failure." Like Study 1.2, this study is an example of the use of inferential statistics. The medical researchers used data from 76 patients to make inferences about the effectiveness of animal-assisted therapy for all heart failure patients.

These studies provide three real-life examples of the uses of statistics. Notice that each involves an analysis of data, either for the purpose of describing the data set (Study 1.1) or for making inferences about a data set (Studies 1.2 and 1.3).

1.3 Fundamental Elements of Statistics

Statistical methods are particularly useful for studying, analyzing, and learning about **populations** of **experimental units.**

> An **experimental** (or **observational**) **unit** is an object (e.g., person, thing, transaction, or event) about which we collect data.

> A **population** is a set of units (usually people, objects, transactions, or events) that we are interested in studying.

For example, populations may include (1) *all* employed workers in the United States, (2) *all* registered voters in California, (3) *everyone* who is afflicted with AIDS, (4) *all* the cars produced last year by a particular assembly line, (5) the *entire* stock of spare parts available at United Airlines' maintenance facility, (6) *all* sales made at the drive-in window of a McDonald's restaurant during a given year, or (7) the set of *all* accidents occurring on a particular stretch of interstate highway during a holiday period. Notice that the first three population examples (1–3) are sets (groups) of people, the next two (4–5) are sets of objects, the next (6) is a set of transactions, and the last (7) is a set of events. Notice also that *each set includes all the units in the population.*

In studying a population, we focus on one or more characteristics or properties of the units in the population. We call such characteristics **variables.** For example, we may be interested in the variables age, gender, and number of years of education of the people currently unemployed in the United States.

> A **variable** is a characteristic or property of an individual experimental (or observational) unit in the population.

The name *variable* is derived from the fact that any particular characteristic may vary among the units in a population.

In studying a particular variable, it is helpful to be able to obtain a numerical representation for it. Often, however, numerical representations are not readily available, so measurement plays an important supporting role in statistical studies. **Measurement** is the process we use to assign numbers to variables of individual population units. We might, for instance, measure the performance of the president by asking a registered voter to rate it on a scale from 1 to 10. Or we might measure the age of the U.S. workforce simply by asking each worker, "How old are you?" In other cases, measurement involves the use of instruments such as stopwatches, scales, and calipers.

If the population you wish to study is small, it is possible to measure a variable for every unit in the population. For example, if you are measuring the GPA for all incoming first-year students at your university, it is at least feasible to obtain every GPA. When we measure a variable for every unit of a population, it is called a **census** of the population. Typically, however, the populations of interest in most applications are much larger, involving perhaps many thousands, or even an infinite number, of units. Examples of large populations are those following the definition of population above, as well as all graduates of your university or college, all potential buyers of a new iPhone, and all pieces of first-class mail handled by the U.S. Post Office. For such populations, conducting a census would be prohibitively time consuming or costly. A reasonable alternative would be to select and study a *subset* (or portion) of the units in the population.

> A **sample** is a subset of the units of a population.

For example, instead of polling all 145 million registered voters in the United States during a presidential election year, a pollster might select and question a sample of just 1,500 voters. (See Figure 1.2.) If he is interested in the variable "presidential preference," he would record (measure) the preference of each vote sampled.

Figure 1.2

A sample of voter registration cards for all registered voters

After the variables of interest for every unit in the sample (or population) are measured, the data are analyzed, either by descriptive or inferential statistical methods. The pollster, for example, may be interested only in *describing* the voting patterns of the sample of 1,500 voters. More likely, however, he will want to use the information in the sample to make inferences about the population of all 145 million voters.

> A **statistical inference** is an estimate, prediction, or some other generalization about a population based on information contained in a sample.

That is, *we use the information contained in the smaller sample to learn about the larger population.** Thus, from the sample of 1,500 voters, the pollster may estimate the percentage of all the voters who would vote for each presidential candidate if the election were held on the day the poll was conducted, or he might use the results to predict the outcome on election day.

*The terms *population* and *sample* are often used to refer to the sets of measurements themselves, as well as to the units on which the measurements are made. When a single variable of interest is being measured, this usage causes little confusion. But when the terminology is ambiguous, we'll refer to the measurements as *population data sets* and *sample data sets*, respectively.

Example 1.1
Key Elements of a Statistical Problem— Ages of TV Viewers

Problem According to *Variety* (Aug. 27, 2009), the average age of viewers of live television programs broadcast on CBS, NBC, and ABC is 51 years. Suppose a rival network (e.g., Fox) executive hypothesizes that the average age of Fox viewers is less than 51. To test her hypothesis, she samples 200 Fox viewers and determines the age of each.

a. Describe the population.

b. Describe the variable of interest.

c. Describe the sample.

d. Describe the inference.

Solution

a. The population is the set of units of interest to the TV executive, which is the set of all Fox viewers.

b. The age (in years) of each viewer is the variable of interest.

c. The sample must be a subset of the population. In this case, it is the 200 Fox viewers selected by the executive.

d. The inference of interest involves the *generalization* of the information contained in the sample of 200 viewers to the population of all Fox viewers. In particular, the executive wants to *estimate* the average age of the viewers in order to determine whether it is less than 51 years. She might accomplish this by calculating the average age of the sample and using that average to estimate the average age of the population.

Look Back A key to diagnosing a statistical problem is to identify the data set collected (in this example, the ages of the 200 Fox viewers) as a population or a sample.

Example 1.2
Key Elements of a Statistical Problem— Pepsi vs. Coca-Cola

Problem "Cola wars" is the popular term for the intense competition between Coca-Cola and Pepsi displayed in their marketing campaigns, which have featured movie and television stars, rock videos, athletic endorsements, and claims of consumer preference based on taste tests. Suppose, as part of a Pepsi marketing campaign, 1,000 cola consumers are given a blind taste test (i.e., a taste test in which the two brand names are disguised). Each consumer is asked to state a preference for brand A or brand B.

a. Describe the population.

b. Describe the variable of interest.

c. Describe the sample.

d. Describe the inference.

Solution

a. Since we are interested in the responses of cola consumers in a taste test, a cola consumer is the experimental unit. Thus, the population of interest is the collection or set of all cola consumers.

b. The characteristic that Pepsi wants to measure is the consumer's cola preference, as revealed under the conditions of a blind taste test, so *cola preference* is the variable of interest.

c. The sample is the 1,000 cola consumers selected from the population of all cola consumers.

d. The inference of interest is the *generalization* of the cola preferences of the 1,000 sampled consumers to the population of all cola consumers. In particular, the preferences of the consumers in the sample can be used to *estimate* the percentages of cola consumers who prefer each brand.

Look Back In determining whether the study is inferential or descriptive, we assess whether Pepsi is interested in the responses of only the 1,000 sampled customers (descriptive statistics) or in the responses of the entire population of consumers (inferential statistics).

Now Work Exercise 1.14b

The preceding definitions and examples identify four of the five elements of an inferential statistical problem: a population, one or more variables of interest, a sample, and an inference. But making the inference is only part of the story; we also need to know its **reliability**—that is, how good the inference is. The only way we can be certain that an inference about a population is correct is to include the entire population in our sample. However, because of *resource constraints* (i.e., insufficient time or money), we usually can't work with whole populations, so we base our inferences on just a portion of the population (a sample). Thus, we introduce an element of *uncertainty* into our inferences. Consequently, whenever possible, it is important to determine and report the reliability of each inference made. Reliability, then, is the fifth element of inferential statistical problems.

The **measure of reliability** that accompanies an inference separates the science of statistics from the art of fortune-telling. A palm reader, like a statistician, may examine a sample (your hand) and make inferences about the population (your life). However, unlike statistical inferences, the palm reader's inferences include no measure of reliability.

Suppose, like the TV executive in Example 1.1, we are interested in the *error of estimation* (i.e., the difference between the average age of a population of TV viewers and the average age of a sample of viewers). Using statistical methods, we can determine a *bound on the estimation error*. This bound is simply a number that our estimation error (the difference between the average age of the sample and the average age of the population) is not likely to exceed. We'll see in later chapters that this bound is a measure of the uncertainty of our inference. The reliability of statistical inferences is discussed throughout this text. For now, we simply want you to realize that an inference is incomplete without a measure of its reliability.

A **measure of reliability** is a statement (usually quantitative) about the degree of uncertainty associated with a statistical inference.

Let's conclude this section with a summary of the elements of descriptive and of inferential statistical problems and an example to illustrate a measure of reliability.

Four Elements of Descriptive Statistical Problems

1. The population or sample of interest
2. One or more variables (characteristics of the population or sample units) that are to be investigated
3. Tables, graphs, or numerical summary tools
4. Identification of patterns in the data

Five Elements of Inferential Statistical Problems

1. The population of interest
2. One or more variables (characteristics of the population units) that are to be investigated
3. The sample of population units
4. The inference about the population based on information contained in the sample
5. A measure of the reliability of the inference

Example 1.3
Reliability of an Inference—Pepsi vs. Coca-Cola

Problem Refer to Example 1.2, in which the preferences of 1,000 cola consumers were indicated in a taste test. Describe how the reliability of an inference concerning the preferences of all cola consumers in the Pepsi bottler's marketing region could be measured.

Solution When the preferences of 1,000 consumers are used to estimate those of all consumers in a region, the estimate will not exactly mirror the preferences of the population. For example, if the taste test shows that 56% of the 1,000 cola consumers preferred Pepsi, it does not follow (nor is it likely) that exactly 56% of all cola drinkers in the region prefer Pepsi. Nevertheless, we can use sound statistical reasoning (which we'll explore later in the text) to ensure that the sampling procedure will generate estimates that are almost certainly within a specified limit of the true percentage of all cola consumers who prefer Pepsi. For example, such reasoning might assure us that the estimate of the preference for Pepsi is almost certainly within 5% of the preference of the population. The implication is that the actual preference for Pepsi is between 51% [i.e., $(56 - 5)\%$] and 61% [i.e., $(56 + 5)\%$]—that is, $(56 \pm 5)\%$. This interval represents a measure of the reliability of the inference.

Look Ahead The interval 56 ± 5 is called a *confidence interval*, since we are confident that the true percentage of cola consumers who prefer Pepsi in a taste test falls into the range $(51, 61)$. In Chapter 5, we learn how to assess the degree of confidence (e.g., a 90% or 95% level of confidence) in the interval.

Statistics ɪɴ Action | Revisited

Identifying the Population, Sample, and Inference

Consider the 2010 Pew Internet & American Life Project survey on social networking and cell phone use by teenagers. In particular, consider the survey results on social networking sites like Facebook or MySpace. The experimental unit for the study is a teenager (the person answering the question), and the variable measured is the response ("yes" or "no") to the question.

The Pew Research Center reported that approximately 800 teens participated in the study. Obviously, that number is not all of the teenagers in the United States. Consequently, the 800 responses represent a sample selected from the much larger population of all American teenagers.

Earlier surveys found that 55% of American teenagers used an online social networking site in 2006, and 65%

in 2008. These are descriptive statistics that provide information on the popularity of social networking in past years. Since 73% of the surveyed teens in 2010 used an online social networking site, the Pew Research Center inferred that teens' usage of social networking cites continues its upward trend, with more and more teens getting online each year. That is, the researchers used the descriptive statistics from the sample to make an inference about the current population of American teenagers' use of social networking.

1.4 Types of Data

You have learned that statistics is the science of data and that data are obtained by measuring the values of one or more variables on the units in the sample (or population). All data (and hence the variables we measure) can be classified as one of two general types: **quantitative data** and **qualitative data.**

Quantitative data are data that are measured on a naturally occurring numerical scale.* The following are examples of quantitative data:

1. The temperature (in degrees Celsius) at which each piece in a sample of 20 pieces of heat-resistant plastic begins to melt

*Quantitative data can be subclassified as either *interval data* or *ratio data*. For ratio data, the origin (i.e., the value 0) is a meaningful number. But the origin has no meaning with interval data. Consequently, we can add and subtract interval data, but we can't multiply and divide them. Of the four quantitative data sets listed as examples, (1) and (3) are interval data while (2) and (4) are ratio data.

2. The current unemployment rate (measured as a percentage) in each of the 50 states

3. The scores of a sample of 150 law school applicants on the LSAT, a standardized law school entrance exam administered nationwide

4. The number of convicted murderers who receive the death penalty each year over a 10-year period

> **Quantitative data** are measurements that are recorded on a naturally occurring numerical scale.

In contrast, qualitative data cannot be measured on a natural numerical scale; they can only be classified into categories.* (For this reason, this type of data is also called *categorical data*.) Examples of qualitative data include the following:

1. The political party affiliation (Democrat, Republican, or Independent) in a sample of 50 voters

2. The defective status (defective or not) of each of 100 computer chips manufactured by Intel

3. The size of a car (subcompact, compact, midsize, or full size) rented by each of a sample of 30 business travelers

4. A taste tester's ranking (best, worst, etc.) of four brands of barbecue sauce for a panel of 10 testers

Often, we assign arbitrary numerical values to qualitative data for ease of computer entry and analysis. But these assigned numerical values are simply codes: They cannot be meaningfully added, subtracted, multiplied, or divided. For example, we might code Democrat = 1, Republican = 2, and Independent = 3. Similarly, a taste tester might rank the barbecue sauces from 1 (best) to 4 (worst). These are simply arbitrarily selected numerical codes for the categories and have no utility beyond that.

> **Qualitative** (or **categorical**) **data** are measurements that cannot be measured on a natural numerical scale; they can only be classified into one of a group of categories.

Example 1.4
Data Types—Army Corps of Engineers Study of a Contaminated River

Problem Chemical and manufacturing plants often discharge toxic-waste materials such as DDT into nearby rivers and streams. These toxins can adversely affect the plants and animals inhabiting the river and the riverbank. The U.S. Army Corps of Engineers conducted a study of fish in the Tennessee River (in Alabama) and its three tributary creeks: Flint Creek, Limestone Creek, and Spring Creek. A total of 144 fish were captured, and the following variables were measured for each:

1. River/creek where each fish was captured

2. Species (channel catfish, largemouth bass, or smallmouth buffalo fish)

3. Length (centimeters)

4. Weight (grams)

5. DDT concentration (parts per million)

*Qualitative data can be subclassified as either *nominal data* or *ordinal data*. The categories of an ordinal data set can be ranked or meaningfully ordered, but the categories of a nominal data set can't be ordered. Of the four qualitative data sets listed as examples, (1) and (2) are nominal and (3) and (4) are ordinal.

(For future analyses, these data are saved in the **FISHDDT** file.) Classify each of the five variables measured as quantitative or qualitative.

Solution The variables length, weight, and DDT concentration are quantitative because each is measured on a numerical scale: length in centimeters, weight in grams, and DDT in parts per million. In contrast, river/creek and species cannot be measured quantitatively: They can only be classified into categories (e.g., channel catfish, largemouth bass, and smallmouth buffalo fish for species). Consequently, data on river/creek and species are qualitative.

Look Ahead It is essential that you understand whether the data you are interested in are quantitative or qualitative, since the statistical method appropriate for describing, reporting, and analyzing the data depends on the data type (quantitative or qualitative).

Now Work Exercise 1.12

We demonstrate many useful methods for analyzing quantitative and qualitative data in the remaining chapters of the text. But first, we discuss some important ideas on data collection in the next section.

1.5 Collecting Data

Once you decide on the type of data—quantitative or qualitative—appropriate for the problem at hand, you'll need to collect the data. Generally, you can obtain data in three different ways:

1. From a *published source*
2. From a *designed experiment*
3. From an *observational study* (e.g., a *survey*)

Sometimes, the data set of interest has already been collected for you and is available in a **published source**, such as a book, journal, or newspaper. For example, you may want to examine and summarize the divorce rates (i.e., number of divorces per 1,000 population) in the 50 states of the United States. You can find this data set (as well as numerous other data sets) at your library in the *Statistical Abstract of the United States*, published annually by the U.S. government. Similarly, someone who is interested in monthly mortgage applications for new home construction would find this data set in the *Survey of Current Business*, another government publication. Other examples of published data sources include *The Wall Street Journal* (financial data) and *The Sporting News* (sports information). The Internet (World Wide Web) now provides a medium by which data from published sources are readily obtained.*

A second method of collecting data involves conducting a **designed experiment**, in which the researcher exerts strict control over the units (people, objects, or things) in the study. For example, an often-cited medical study investigated the potential of aspirin in preventing heart attacks. Volunteer physicians were divided into two groups: the *treatment* group and the *control* group. Each physician in the treatment group took one aspirin tablet a day for one year, while each physician in the control group took an aspirin-free placebo made to look like an aspirin tablet. The researchers—not the physicians under study—controlled who received the aspirin (the treatment) and who received the placebo. A properly designed experiment allows you to extract more information from the data than is possible with an uncontrolled study.

Finally, observational studies can be employed to collect data. In an **observational study**, the researcher observes the experimental units in their natural setting and records the variable(s) of interest. For example, a child psychologist might observe and record the

---

*With published data, we often make a distinction between the *primary source* and a *secondary source*. If the publisher is the original collector of the data, the source is primary. Otherwise, the data is secondary-source data.

level of aggressive behavior of a sample of fifth graders playing on a school playground. Similarly, a zoologist may observe and measure the weights of newborn elephants born in captivity. Unlike a designed experiment, an observational study is a study in which the researcher makes no attempt to control any aspect of the experimental units.

The most common type of observational study is a **survey**, where the researcher samples a group of people, asks one or more questions, and records the responses. Probably the most familiar type of survey is the political poll, conducted by any one of a number of organizations (e.g., Harris, Gallup, Roper, and CNN) and designed to predict the outcome of a political election. Another familiar survey is the Nielsen survey, which provides the major networks with information on the most-watched programs on television. Surveys can be conducted through the mail, with telephone interviews, or with in-person interviews. Although in-person surveys are more expensive than mail or telephone surveys, they may be necessary when complex information is to be collected.

> A **designed experiment** is a data collection method where the researcher exerts full control over the characteristics of the experimental units sampled. These experiments typically involve a group of experimental units that are assigned the *treatment* and an untreated (or *control*) group.

> An **observational study** is a data collection method where the experimental units sampled are observed in their natural setting. No attempt is made to control the characteristics of the experimental units sampled. (Examples include *opinion polls* and *surveys*.)

Regardless of which data collection method is employed, it is likely that the data will be a sample from some population. And if we wish to apply inferential statistics, we must obtain a **representative sample.**

> A **representative sample** exhibits characteristics typical of those possessed by the target population.

For example, consider a political poll conducted during a presidential election year. Assume that the pollster wants to estimate the percentage of all 145 million registered voters in the United States who favor the incumbent president. The pollster would be unwise to base the estimate on survey data collected for a sample of voters from the incumbent's own state. Such an estimate would almost certainly be *biased* high; consequently, it would not be very reliable.

The most common way to satisfy the representative sample requirement is to select a random sample. A **random sample** ensures that every subset of fixed size in the population has the same chance of being included in the sample. If the pollster samples 1,500 of the 145 million voters in the population so that every subset of 1,500 voters has an equal chance of being selected, she has devised a random sample. The procedure for selecting a random sample is discussed in Chapter 3. Here, we look at two examples involving actual sampling studies.

> A **random sample** of *n* experimental units is a sample selected from the population in such a way that every different sample of size *n* has an equal chance of selection. (See Section 3.7 for how to generate a random sample.)*

---

*In statistical sampling theory, this sample is better known as a *simple random sample*.

Example 1.5
Method of Data Collection—Internet Addiction Study

Problem What percentage of Web users are addicted to the Internet? To find out, a psychologist designed a series of 10 questions based on a widely used set of criteria for gambling addiction and distributed them through the Web site *ABCNews.com*. (A sample question: "Do you use the Internet to escape problems?") A total of 17,251 Web users responded to the questionnaire. If participants answered "yes" to at least half of the questions, they were viewed as addicted. The findings, released at an annual meeting of the American Psychological Association, revealed that 990 respondents, or 5.7%, are addicted to the Internet.

a. Identify the data collection method.

b. Identify the target population.

c. Are the sample data representative of the population?

Solution

a. The data collection method is a survey: 17,251 Internet users responded to the questions posed at the *ABCNews.com* Web site.

b. Since the Web site can be accessed by anyone surfing the Internet, presumably the target population is *all* Internet users.

c. Because the 17,251 respondents clearly make up a subset of the target population, they do form a sample. Whether or not the sample is representative is unclear, since we are given no information on the 17,251 respondents. However, a survey like this one in which the respondents are *self-selected* (i.e., each Internet user who saw the survey chose whether or not to respond to it) often suffers from *nonresponse bias*. It is possible that many Internet users who chose not to respond (or who never saw the survey) would have answered the questions differently, leading to a higher (or lower) percentage of affirmative answers.

Look Back Any inferences based on survey samples that employ self-selection are suspect due to potential nonresponse bias.

Example 1.6
Method of Data Collection—Study of Susceptibility to Hypnosis

Problem Psychologists at the University of Tennessee carried out a study of the susceptibility of people to hypnosis (*Psychological Assessment*, Mar. 1995). In a random sample of 130 undergraduate psychology students at the university, each experienced both traditional hypnosis and computer-assisted hypnosis. Approximately half were randomly assigned to undergo the traditional procedure first, followed by the computer-assisted procedure. The other half were randomly assigned to experience computer-assisted hypnosis first, then traditional hypnosis. Following the hypnosis episodes, all students filled out questionnaires designed to measure a student's susceptibility to hypnosis. The susceptibility scores of the two groups of students were compared.

a. Identify the data collection method.

b. Is the sample data representative of the target population?

Solution

a. Here, the experimental units are the psychology students. Since the researchers controlled which type of hypnosis—traditional or computer assisted—the students experienced first (through random assignment), a designed experiment was used to collect the data.

b. The sample of 130 psychology students was randomly selected from all psychology students at the University of Tennessee. If the target population is *all University of Tennessee psychology students*, it is likely that the sample is representative. However, the researchers warn that the sample data should not be used to make inferences about other, more general, populations.

Look Ahead By using randomization in a designed experiment, the researcher is attempting to eliminate different types of bias, including self-selection bias. (We discuss this type of bias in the next section.)

Now Work Exercise 1.15

Statistics IN Action **Revisited** **Identifying the Data Collection Method and Data Type**

In the Pew Internet & American Life Project report, American teenagers are asked to respond to a variety of questions about Internet, cell phone, and social networking site usage. According to the report, the data were obtained through phone interviews in the continental United States of 800 teenagers who were 12 to 17 years old at the time. Consequently, the data collection method is a survey (observational study).

Both quantitative and qualitative data were collected in the survey. For example, the survey question asking teens if they use social networking sites is phrased to elicit a "yes" or "no" response. Since the responses produced for this question are categorical in nature, these data are qualitative. However, the question asking teens for the number of text messages they send and receive per day will give meaningful numerical responses, such as 5, 10, 12, etc. Thus, these data are quantitative.

1.6 The Role of Statistics in Critical Thinking and Ethics

According to H. G. Wells, author of such science-fiction classics as *The War of the Worlds* and *The Time Machine,* "Statistical thinking will one day be as necessary for efficient citizenship as the ability to read and write." Written more than a hundred years ago, Wells's prediction is proving true today.

The growth in data collection associated with scientific phenomena, business operations, and government activities (quality control, statistical auditing, forecasting, etc.) has been remarkable in the past several decades. Every day the media present us with published results of political, economic, and social surveys. In increasing government emphasis on drug and product testing, for example, we see vivid evidence of the need for *quantitative literacy* (i.e., the ability to evaluate data intelligently). Consequently, each of us has to develop a discerning sense—an ability to use rational thought to interpret and understand the meaning of data. Quantitative literacy can help you make intelligent decisions, inferences, and generalizations; that is, it helps you *think critically* using statistics.

BIOGRAPHY **H. G. WELLS (1866–1946)**
Writer and Novelist

English-born Herbert George Wells published his first novel, *The Time Machine*, in 1895 as a parody of the English class division and as a satirical warning that human progress is inevitable. Although most famous as a science-fiction novelist, Wells was a prolific writer as a journalist, sociologist, historian, and philosopher. Wells's prediction about statistical thinking is just one of a plethora of observations he made about life on this world. Here are a few more of H. G. Wells's more famous quotes:

"Advertising is legalized lying."

"Crude classification and false generalizations are the curse of organized life."

"The crisis of today is the joke of tomorrow."

"Fools make researchers and wise men exploit them."

"The only true measure of success is the ratio between what we might have done and what we might have been on the one hand, and the thing we have made and the things we have made of ourselves on the other." ∎

> **Statistical thinking** involves applying rational thought and the science of statistics to critically assess data and inferences. Fundamental to the thought process is that variation exists in populations of data.

To gain some insight into the role statistics plays in critical thinking, we present two examples of some misleading or faulty surveys.

Example 1.7
Biased Sample—Motorcyclists and Helmets

Problem An article in the *New York Times* considered the question of whether motorcyclists should be required by law to wear helmets. In supporting his argument for no helmets, the editor of a magazine for Harley-Davidson bikers presented the results of one study that claimed "nine states without helmet laws had a lower fatality rate (3.05 deaths per 10,000 motorcycles) than those that mandated helmets (3.38)" and a survey that found "of 2,500 bikers at a rally, 98% of the respondents opposed such laws." Based on this information, do you think it is safer to ride a motorcycle without a helmet? What further statistical information would you like?

Solution You can use "statistical thinking" to help you critically evaluate the study. For example, before you can evaluate the validity of the 98% estimate, you would want to know how the data were collected. If a survey was, in fact, conducted, it's possible that the 2,500 bikers in the sample were not selected at random from the target population of all bikers, but rather were "self-selected." (Remember, they were all attending a rally—a rally, likely, for bikers who oppose the law.) If the respondents were likely to have strong opinions regarding the helmet law (e.g., to strongly oppose the law), the resulting estimate is probably biased high. Also, if the biased sample was intentional, with the sole purpose to mislead the public, the researchers would be guilty of **unethical statistical practice**.

You would also want more information about the study comparing the motorcycle fatality rate of the nine states without a helmet law to those states that mandate helmets. Were the data obtained from a published source? Were all 50 states included in the study, or were only certain states selected? That is, are you seeing sample data or population data? Furthermore, do the helmet laws vary among states? If so, can you really compare the fatality rates?

Look Back Questions such as these led a group of mathematics and statistics teachers attending an American Statistical Association course to discover a scientific and statistically sound study on helmets. The study reported a dramatic *decline* in motorcycle crash deaths after California passed its helmet law.

Now Work Exercise 1.31c

Example 1.8
Manipulative or Ambiguous Survey Questions—Howard Stern on Sirius Radio

Problem Talk-show host, Howard Stern moved his controversial radio program from free, over-the-air (AM/FM) radio to Sirius satellite radio. The move was perceived in the industry to boost satellite radio subscriptions. This led American Media Services, a developer of AM/FM radio properties, to solicit a January 2006 nationwide random-digit dialing phone survey of 1,008 people. The purpose of the survey was to determine how much interest Americans really have in buying satellite radio service. After providing some background on Howard Stern's controversial radio program, one of the questions asked, "How likely are you to purchase a subscription to satellite radio after Howard Stern's move to Sirius?" The result: Eighty-six percent of the respondents stated that they weren't likely to buy satellite radio because of Stern's move. Consequently, American Media Services concluded that "the Howard Stern Factor is overrated" and that "few Americans expect to purchase satellite radio"—claims that made the headlines of news reports and Weblogs. Do you agree?

Solution First, we need to recognize that American Media Services has a vested interest in the outcome of the survey—the company makes its money from over-the-air broadcast radio stations. Second, although the phone survey was conducted using random-digit dialing, there is no information provided on the response rate. It's possible that nonrespondents (people who were not home or refused to answer the survey questions) tend to be people who use cell phones more than their landline phone, and, consequently, are more likely to use the latest in electronic technology, including satellite radio. Finally, the survey question itself is ambiguous. Do the respondents have negative feelings about satellite radio, Howard Stern, or both? If not for Howard Stern's program, would the respondents be more likely to buy satellite radio? To the critical thinker, it's unclear what the results of the survey imply.

Look Back Examining the survey results from the perspective of satellite radio providers, 14% of the respondents indicated that they would be likely to purchase satellite radio. Projecting the 14% back to the population of all American adults, this figure represents about 50 million people; what is interpreted as "few Americans" by American Media Services could be music to the ears of satellite radio providers.

Now Work Exercise 1.29b

Ethics IN Statistics

Intentionally selecting a biased sample in order to produce misleading statistics is considered *unethical statistical practice*.

As with many statistical studies, both the motorcycle helmet study and the satellite radio study are based on survey data. Most of the problems with these surveys result from the use of *nonrandom* samples. These samples are subject to errors such as **selection bias, nonresponse bias** (recall Example 1.5), and **measurement error.** Researchers who are aware of these problems yet continue to use the sample data to make inferences are practicing unethical statistics.

Selection bias results when a subset of the experimental units in the population is excluded so that these units have no chance of being selected in the sample.

Nonresponse bias results when the researchers conducting a survey or study are unable to obtain data on all experimental units selected for the sample.

Measurement error refers to inaccuracies in the values of the data recorded. In surveys, this kind of error may be due to ambiguous or leading questions and the inter-viewer's effect on the respondent.

Statistics IN Action | Revisited | Critically Assessing the Ethics of a Statistical Study

The results from the Pew Internet & American Life Project (PIALP) survey on social networking and cell phone use by teenagers led the research center to make conclusions such as "teens continue to be avid users of social networking Web sites" and that the percentage "continues to climb upwards." The survey included a sample of 800 teens. In order to assess the validity of this inference, a critical thinker would consider several issues.

First, is the sample representative of all American teenagers? In the PIALP report, the claim is made that the survey results are based on "telephone interviews with a nationally representative sample of teenagers." Two samples of teenagers (and their parents) were selected. The first sample consisted of landline telephone numbers randomly selected (using a random number generator) from all landline numbers listed in the continental United States. The second sample consisted of cellular telephone numbers randomly selected from all U.S. cell numbers that were not associated with directory-listed landline numbers. As many as seven attempts were made to contact and interview the teenager at each number. These contact calls were staggered over times of day and days of the week in an attempt to reduce nonresponse bias. Clearly, there was no attemtp to bias the results of the study by intentionally selecting a sample with a certain point of view. In fact, just the opposite occurred. Consequently, the sample of 800 teenagers appears to be a truly random sample of all teenagers in the United States;

therefore, the sample is very likely to be representative of the population of interest.

Second, is the sample of 800 teenagers large enough to draw a reliable inference? Although there are over 25 million teenagers living in the continental United States, we will learn in later chapters that a random sample of size 800 will allow the researchers to draw a reliable inference. In fact, the PIALP report states that the margin of sampling error for the study is about 4%. This means that anywhere from 69% to 77% (73% ± 4%) of all the teens in the United States regularly visit a social networking site.

Third, is the wording of the survey questions ambiguous or leading? Such questions are often crafted to elicit an answer that supports a specific point of view. For example, consider a hypothetical question about Twitter: "Twitter is becoming more and more popular with affluent teenagers. Have you ever used or desired to use Twitter in order to keep up with your friends?" The question is meant to elicit a "yes" answer and bias the results of the study. If you examine the six questions in the actual study (see p. 2), you will conclude that all are very straightforward and clear. Consequently, the survey results appear to be both ethical and valid.

In the remaining chapters of the text, you'll become familiar with the tools essential for building a firm foundation in statistics and statistical thinking.

CHAPTER NOTES

Key Terms

Census 5
Data 2
Descriptive statistics 3
Designed experiment 12
Experimental (or observational) unit 5
Inference 3
Inferential statistics 3
Measure of reliability 8
Measurement 5
Measurement error 16
Nonresponse bias 16
Observational study 12
Population 5

Published source 11
Qualitative (or categorical) data 10
Quantitative data 10
Random sample 12
Reliability (of an inference) 8
Representative sample 12
Sample 5
Selection bias 16
Statistical inference 6
Statistical thinking 15
Statistics 3
Survey 12
Unethical statistical practice 15
Variable 5

Key Ideas

Types of Statistical Applications

Descriptive
1. Identify **population** or **sample** (collection of **experimental units**)
2. Identify **variable(s)**
3. Collect **data**
4. **Describe** data

Inferential
1. Identify **population** (collection of *all* **experimental units**)
2. Identify **variable(s)**

3. Collect **sample** data (*subset* of population)
4. **Inference** about population based on sample
5. **Measure of reliability** of inference

Types of Data

1. **Quantitative** (numerical in nature)
2. **Qualitative** (categorical in nature)

Data Collection Methods

1. **Published source**
2. **Observational** (e.g., survey)
3. **Designed experiment**

Problems with Nonrandom Samples

1. **Selection bias**
2. **Nonresponse bias**
3. **Measurement error**

Exercises 1.1–1.33

Understanding the Principles

1.1 What is statistics?

1.2 Explain the difference between descriptive and inferential statistics.

1.3 List and define the five elements of an inferential statistical analysis.

1.4 List the four major methods of collecting data, and explain their differences.

1.5 Explain the difference between quantitative and qualitative data.

1.6 Explain how populations and variables differ.

1.7 Explain how populations and samples differ.

1.8 What is a representative sample? What is its value?

1.9 Why would a statistician consider an inference incomplete without an accompanying measure of its reliability?

1.10 Define statistical thinking.

1.11 Suppose you're given a data set that classifies each sample unit into one of four categories: A, B, C, or D. You plan to create a computer database consisting of these data, and you decide to code the data as $A = 1, B = 2, C = 3,$ and $D = 4$. Are the data consisting of the classifications A, B, C, and D qualitative or quantitative? After the data are input as 1, 2, 3, or 4, are they qualitative or quantitative? Explain your answers.

Applet Exercise 1.1

The *Random Numbers* applet generates a list of n random numbers from 1 to N, where n is the size of the sample and N is the size of the population. The list generated often contains repetitions of one or more numbers.

a. Using the applet *Random Numbers*, enter 1 for the minimum value, 10 for the maximum value, and 10 for the number of samples. Then click on *Sample*. Look at the results, and list any numbers that are repeated and the number of times each of these numbers occurs.

b. Repeat part (a), changing the maximum value to 20 and keeping the size of the sample fixed at 10. If you still have repetitions, repeat the process, increasing the maximum value by 10 each time but keeping the size of the sample fixed. What is the smallest maximum value for which you had no repetitions?

c. Describe the relationship between the population size (maximum value) and the number of repetitions in the list of random numbers as the population size increases and the sample size remains the same. What

can you conclude about using a random number generator to choose a relatively small sample from a large population?

Applet Exercise 1.2

The *Random Numbers* applet can be used to select a random sample from a population, but can it be used to simulate data? In parts (a) and (b), you will use the applet to create data sets. Then you will explore whether those data sets are realistic.

a. Consider the number of customers waiting in line to order at a fast-food outlet. Use the *Random Numbers* applet to simulate this data set by setting the minimum value equal to 0, the maximum value equal to 99, and the sample size equal to 30. Explain what the numbers in the list produced by the applet represent in the context of the problem. Do the numbers produced by the applet seem reasonable? Explain.

b. Use the *Random Numbers* applet to simulate grades on a statistics test by setting the minimum value equal to 0, the maximum value equal to 100, and the sample size equal to 30. Explain what the numbers in the list produced by the applet represent in this context. Do the numbers produced by the applet seem reasonable? Explain.

c. Referring to parts (a) and (b), why do the randomly generated data seem more reasonable in one situation than in the other? Comment on the usefulness of using a random-number generator to produce data.

Applying the Concepts—Basic

1.12 **College application.** Colleges and universities are requiring an increasing amount of information about applicants before making acceptance and financial aid decisions. Classify each of the following types of data required on a college application as quantitative or qualitative.

  a. High school GPA
  b. High school class rank
  c. Applicant's score on the SAT or ACT
  d. Gender of applicant
  e. Parents' income
  f. Age of applicant

1.13 **Ground motion of earthquakes.** In the *Journal of Earthquake Engineering* (Nov. 2004), a team of civil and environmental engineers studied the ground motion characteristics of 15 earthquakes that occurred around the world since 1940. Three (of many) variables measured on each earthquake were the type of ground motion (short, long, or forward directive), the magnitude of the earthquake (on the Richter scale), and peak ground acceleration (feet per second). One of the goals of the study was to estimate the inelastic spectra of any ground motion cycle.

a. Identify the experimental units for this study.

b. Do the data for the 15 earthquakes represent a population or a sample? Explain.

c. Define the variables measured and classify them as quantitative or qualitative.

1.14 **Sprint speed training.** *The Sport Journal* (Winter 2004) reported on a study of a speed-training program for high school football players. Each participant was timed in a 40-yard sprint both before and after training. The researchers measured two variables: (1) the difference between the before and after sprint times (in seconds), and (2) the category of improvement ("improved," "no change," and "worse") for each player.

a. Identify the type (quantitative or qualitative) of each variable measured.

[NW] **b.** A total of 14 high school football players participated in the speed-training program. Does the data set collected represent a population or a sample? Explain.

1.15 **Going online for health information.** *A cyberchondriac* is defined as a person who regularly searches the Web for health care information. Each year a Harris Poll is conducted to determine the number of cyberchondriacs in the United States. In 2008, the Harris Poll surveyed 1,010 U.S. adults by telephone and asked the following questions:

1. Have you ever gone online to look for health care information?

2. How many times per month do you look for health care information online?

3. In the past year, have you ever discussed with your doctor the information you found online?

a. For each question, determine whether the type of data collected is quantitative or qualitative.

b. Do the data collected for the 1,010 adults represent a sample or a population? Explain.

1.16 **Student GPAs.** Consider the set of all students enrolled in your statistics course this term. Suppose you're interested in learning about the current grade point averages (GPAs) of this group.

a. Define the population and variable of interest.

b. Is the variable qualitative or quantitative?

c. Suppose you determine the GPA of every member of the class. Would this determination represent a census or a sample?

d. Suppose you determine the GPA of 10 members of the class. Would this determination represent a census or a sample?

e. If you determine the GPA of every member of the class and then calculate the average, how much reliability does your calculation have as an "estimate" of the class average GPA?

f. If you determine the GPA of 10 members of the class and then calculate the average, will the number you get necessarily be the same as the average GPA for the whole class? On what factors would you expect the reliability of the estimate to depend?

g. What must be true in order for the sample of 10 students you select from your class to be considered a random sample?

1.17 **Treasury deficit prior to the Civil War.** In *Civil War History* (June 2009), historian Jane Flaherty researched the condition of the U.S. Treasury on the eve of the Civil War in 1861. Between 1854 and 1857 (under President Franklin Pierce), the annual surplus/deficit was +18.8, +6.7, +5.3, and +1.3 million dollars, respectively. In contrast, between 1858 and 1861 (under President James Buchanan), the annual surplus/deficit was −27.3, −16.2, −7.2, and −25.2 million dollars, respectively. Flaherty used these data to aid in portraying the exhausted condition of the U.S. Treasury when Abraham Lincoln took office in 1861. Does this study represent a descriptive or inferential statistical study? Explain.

1.18 **Extinct birds.** Biologists at the University of California (Riverside) are studying the patterns of extinction in the New Zealand bird population. (*Evolutionary Ecology Research*, July 2003.) At the time of the Maori colonization of New Zealand (prior to European contact), the following variables were measured for each bird species:

a. Flight capability (volant or flightless)

b. Type of habitat (aquatic, ground terrestrial, or aerial terrestrial)

c. Nesting site (ground, cavity within ground, tree, cavity above ground)

d. Nest density (high or low)

e. Diet (fish, vertebrates, vegetables, or invertebrates)

f. Body mass (grams)

g. Egg length (millimeters)

h. Extinct status (extinct, absent from island, present)

Identify each variable as quantitative or qualitative.

1.19 **Study of quality of drinking water.** *Disasters* (Vol. 28, 2004) published a study of the effects of a tropical cyclone on the quality of drinking water on a remote Pacific island. Water samples (size 500 milliliters) were collected approximately four weeks after Cyclone Ami hit the island. The following variables were recorded for each water sample:

a. Town where sample was collected

b. Type of water supply (river intake, stream, or borehole)

c. Acidic level (pH scale, 1 to 14)

d. Turbidity level (nephalometric turbidity units = NTUs)

e. Temperature (degrees centigrade)

f. Number of fecal coliforms per 100 milliliters

g. Free-chlorine residual (milligrams per liter)

h. Presence of hydrogen sulphide (yes or no)

Identify each variable as quantitative or qualitative.

Applying the Concepts—Intermediate

1.20 **Herbal medicines.** *The American Association of Nurse Anesthetists Journal* (Feb. 2000) published the results of a study on the use of herbal medicines before surgery. Of 500 surgical patients randomly selected for the study, 51% used herbal or alternative medicines (e.g., garlic, ginkgo, kava, fish oil) against their doctor's advice prior to surgery.

a. Do the 500 surgical patients represent a population or a sample? Explain.

b. If your answer was "sample" in part **a**, is the sample representative of the population? If you answered "population" in part **a**, explain how to obtain a representative sample from the population.

c. For each patient, what variable is measured? Are the data collected quantitative or qualitative?

1.21 **National Bridge Inventory.** All highway bridges in the United States are inspected periodically for structural deficiency by the Federal Highway Administration (FHWA). Data from the FHWA inspections are compiled into the National Bridge Inventory (NBI). Several of the nearly 100 variables maintained by the NBI are listed next. Classify each variable as quantitative or qualitative.

a. Length of maximum span (feet)
b. Number of vehicle lanes
c. Toll bridge (yes or no)
d. Average daily traffic
e. Condition of deck (good, fair, or poor)
f. Bypass or detour length (miles)
g. Type of route (interstate, U.S., state, county, or city)

1.22 **Annual survey of computer crimes.** The Computer Security Institute (CSI) conducts an annual survey of computer crime committed at U.S. businesses. CSI sends survey questionnaires to computer security personnel at all U.S. corporations and government agencies. In 2006, 616 organizations responded to the CSI survey. Fifty-two percent of the respondents admitted unauthorized use of computer systems at their firms during the year. (*Computer Security Issues & Trends*, Spring 2006.)

a. Identify the population of interest to CSI.
b. Identify the data collection method used by CSI. Are there any potential biases in the method used?
c. Describe the variable measured in the CSI survey. Is it quantitative or qualitative?
d. What inference can be made from the result of the study?

1.23 **CT scanning for lung cancer.** According to the American Lung Association, lung cancer accounts for 28% of all cancer deaths in the United States. A new type of screening for lung cancer, computed tomography (CT), has been developed. Medical researchers believe that CT scans are more sensitive than regular X-rays in pinpointing small tumors. The H. Lee Moffitt Cancer Center at the University of South Florida is currently conducting a clinical trial of 50,000 smokers nationwide to compare the effectiveness of CT scans with X-rays for detecting lung cancer. (*Todays' Tomorrows*, Fall 2002.) Each participating smoker is randomly assigned to one of two screening methods, CT or chest X-ray, and his or her progress tracked over time. The age at which the scanning method first detects a tumor is the variable of interest.

a. Identify the data collection method used by the cancer researchers.
b. Identify the experimental units of the study.
c. Identify the type (quantitative or qualitative) of the variable measured.
d. Identify the population and sample.
e. What is the inference that will ultimately be drawn from the clinical trial?

1.24 **Satellite radio in cars.** A recent survey conducted for the National Association of Broadcasters investigated satellite radio subscriber service and usage. The June 2007 survey, conducted by Wilson Research Strategies, consisted of a random sample of 501 satellite radio subscribers. One of the questions of interest was, "Do you have a satellite radio receiver in your car?" The survey found that 396 subscribers did, in fact, have a satellite receiver in their car.

a. Identify the population of interest to the National Association of Broadcasters.
b. Based on the survey question, what is the variable of interest?
c. Does the variable produce quantitative or qualitative data?
d. Describe that sample of interest.
e. What inference can be made from the survey results?

1.25 **Massage vs. rest in boxing.** Does a massage enable the muscles of tired athletes to recover from exertion faster than usual? To answer this question, researchers recruited eight amateur boxers to participate in an experiment. (*British Journal of Sports Medicine*, April 2000.) After a 10-minute workout in which each boxer threw 400 punches, half the boxers were given a 20-minute massage and half just rested for 20 minutes. Before they returned to the ring for a second workout, the heart rate (beats per minute) and blood lactate level (micromoles) were recorded for each boxer. The researchers found no difference in the means of the two groups of boxers for either variable.

a. Identify the data collection method used by the researchers.
b. Identify the experimental units of the study.
c. Identify the variables measured and their type (quantitative or qualitative).
d. What is the inference drawn from the analysis?
e. Comment on whether this inference can be made about all athletes.

1.26 **Insomnia and education.** Is insomnia related to education status? Researchers at the Universities of Memphis, Alabama at Birmingham, and Tennessee investigated this question in the *Journal of Abnormal Psychology* (Feb. 2005). Adults living in Tennessee were selected to participate in the study, which used a random-digit telephone dialing procedure. Two of the many variables measured for each of the 575 study participants were number of years of education and insomnia status (normal sleeper or chronic insomniac). The researchers discovered that the fewer the years of education, the more likely the person was to have chronic insomnia.

a. Identify the population and sample of interest to the researchers.
b. Identify the data collection method. Are there any potential biases in the method used?
c. Describe the variables measured in the study as quantitative or qualitative.
d. What inference did the researchers make?

1.27 **Guilt in decision making.** The effect of the emotion of guilt on how a decision maker focuses on a problem was investigated in the *Journal of Behavioral Decision Making* (January 2007). A total of 155 volunteer students participated in the experiement, where each was randomly assigned to one of three emotional states (guilt, anger, or neutral) through a reading/writing task. Immediately after the task, the students were presented with a decision problem (e.g., whether or not to spend money on repairing a very old car). The researchers found that a higher proportion of students in the guilty-state group chose not to repair the car than those in the neutral-state and anger-state groups.

a. Identify the population, sample, and variables measured for this study.
b. Identify the data collection method used.
c. What inference was made by the researcher?

d. In later chapters you will learn that the reliability of an inference is related to the size of the sample used. In addition to sample size, what factors might affect the reliability of the inference drawn in this study?

Applying the Concepts—Advanced

1.28 **Object recall study.** Are men or women more adept at remembering where they leave misplaced items (such as car keys)? According to University of Florida psychology professor Robin West, women show greater competence in actually finding these objects (*Explore*, Fall 1998). Approximately 300 men and women from Gainesville, Florida, participated in a study in which each person placed 20 common objects in a 12-room "virtual" house represented on a computer screen. Thirty minutes later, the subjects were asked to recall where they put each of the objects. For each object, a recall variable was measured as "yes" or "no."

a. Identify the population of interest to the psychology professor.

b. Identify the sample.

c. Does the study involve descriptive or inferential statistics? Explain.

d. Are the variables measured in the study quantitative or qualitative?

1.29 **Dating and disclosure.** As an adolescent, did you voluntarily disclose information about dating and romantic relationships to your parents? This was the research question of interest in the *Journal of Adolescence* (Apr. 2010). A sample of 222 high school students was recruited to participate in the study. Some of the many variables measured on each student were age (years), gender, dating experience (number of dates), and the extent to which the student was willing to tell his/her parent (without being asked) about a dating issue (e.g., how late the daters stayed out). The responses for the last variable were categorized as "never tell," "rarely tell," "sometimes tell," "almost always tell," and "always tell."

a. Identify the data type for each variable.

`NW` **b.** The study was unclear on exactly how the sample of students was selected, stating only that "participants were recruited from health or government classes in a primarily European American middle-class school district." Based on this information, what are the potential caveats to using the sample to make inferences on dating and disclosure to parents for all high school students?

1.30 **Success/failure of software reuse.** The PROMISE Software Engineering Repository, hosted by the University of Ottawa, is a collection of publicly available data sets to serve researchers in building prediction software models. A PROMISE data set on software reuse, saved in the **SWREUSE** file, provides information on the success or failure of reusing previously developed software for each in a sample of 24 new software development projects. (*Data source: IEEE Transactions on Software Engineering*, Vol. 28, 2002.) Of the 24 projects, 9 were judged failures and 15 were successfully implemented.

a. Identify the experimental units for this study.

b. Describe the population from which the sample is selected.

c. What is the variable of interest in the study? Is it quantitative or qualitative?

d. Critically evaluate the statement "Since $15/24 = .625$, it follows that 62.5% of all new software development projects will be successfully implemented."

Critical Thinking Challenges

1.31 **Your choice for a mom.** After running stories on current First Lady Michelle Obama and former vice presidential candidate Sarah Palin on consecutive weeks, *USA Weekend* magazine asked its readers on Mother's Day, "Who would you rather have as your mom, Sarah Palin or Michelle Obama?" Readers were asked to vote online at www.usaweekend.com. Based on over 34,000 votes cast, the results were: Obama—71%, Palin—29%. (*USA Weekend Magazine* press release, May 11, 2010.)

a. What type of data collection method is used in this study?

b. Is the data collected quantitative or qualitative? Explain.

`NW` **c.** Discuss the validity of the study results. What are the potential problems with running a poll where voting is done online?

1.32 **Poll on alien spacecraft.** "Have you ever seen anything that you believe was a spacecraft from another planet?" This was the question put to 1,500 American adults in a national poll conducted by ABC News and *The Washington Post*. The pollsters used random-digit telephone dialing to contact adult Americans until 1,500 responded. Ten percent (i.e., 150) of the respondents answered that they had, in fact, seen an alien spacecraft. (*Chance*, Summer 1997.) No information was provided on how many adults were called and, for one reason or another, did not answer the question.

a. Identify the data collection method.

b. Identify the target population.

c. Comment on the validity of the survey results.

1.33 ***20/20* survey exposés.** The popular prime-time ABC television program *20/20* presented several misleading (and possibly unethical) surveys in a segment titled "Facts or Fiction? Exposés of So-Called Surveys" (March 31, 1995). The information reported from four of these surveys, conducted by businesses or special-interest groups with specific objectives in mind, are given. (Actual survey facts are provided in parentheses.)

Quaker Oats Study: Eating oat bran is a cheap and easy way to reduce your cholesterol count. (*Fact:* Diet must consist of nothing but oat bran to achieve a slightly lower cholesterol count.)

March of Dimes Report: Domestic violence causes more birth defects than all medical issues combined. (*Fact:* No study—false report.)

American Association of University Women (AAUW) study: Only 29% of high school girls are happy with themselves, compared with 66% of elementary school girls. (*Fact:* Of 3,000 high school girls, 29% responded "Always true" to the statement, "I am happy the way I am." Most answered, "Sort of true" and "Sometimes true.")

Food Research and Action Center study: One in four American children under age 12 is hungry or at risk of hunger. (*Fact:* Survey results are based on responses to the questions "Do you ever cut the size of meals?," "Do you ever eat less than you feel you should?" and "Did you ever rely on limited numbers of foods to feed your children because you were running out of money to buy food for a meal?")

a. Refer to the Quaker Oats study relating oat bran to cholesterol levels. Discuss why it is unethical to report the results as stated.

b. Consider the false March of Dimes report on domestic violence and birth defects. Discuss the type of data required to investigate the impact of domestic violence

on birth defects. What data collection method would you recommend?

c. Refer to the AAUW study of the self-esteem of high school girls. Explain why the results of the study are likely to be misleading. What data might be appropriate for assessing the self-esteem of high school girls?

d. Refer to the Food Research and Action Center study of hunger in America. Explain why the results of the study are likely to be misleading. What data would provide insight into the proportion of hungry American children?

Activity   Data in the News

Scan your daily newspaper or weekly news magazine, or search the Internet for articles that contain data. The data might be a summary of the results of a public opinion poll, the results of a vote by the U.S. Senate, or a list of crime rates, birth or death rates, etc. For each article you find, answer the following questions:

a. Do the data constitute a sample or an entire population? If a sample has been taken, clearly identify both the sample and the population; otherwise, identify the population.

b. What type of data (quantitative or qualitative) has been collected?

c. What is the source of the data?

d. If a sample has been observed, is it likely to be representative of the population?

e. If a sample has been observed, does the article present an explicit (or implied) inference about the population of interest? If so, state the inference made in the article.

f. If an inference has been made, has a measure of reliability been included? What is it?

g. Use your answers to questions d–f to critically evaluate the article. Comment on the ethics of the study.

References

What Is a Survey? American Statistical Association (F. Scheuren, editor), 2nd. ed. 2005. (*www.amstat.org*)

Careers in Statistics, American Statistical Association, Biometric Society, Institute of Mathematical Statistics and Statistical Society of Canada, 2008. (*www.amstat.org*)

Cochran, W. G. *Sampling Techniques*, 3rd ed. New York: Wiley, 1977.

Deming, W. E. *Sample Design in Business Research*. New York: Wiley, 1963.

Ethical Guidelines for Statistical Practice. American Statistical Association, 1999. (*www.amstat.org*)

Hansen, M. H., Hurwitz, W. N., and Madow, W. G. *Sample Survey Methods and Theory*, Vol. 1. New York: Wiley, 1953.

Hoerl, R., and Snee, R. *Statistical Thinking: Improving Business Performance.* Boston: Duxbury, 2002.

Kirk, R. E., ed. *Statistical Issues: A Reader for the Behavioral Sciences.* Monterey, CA: Brooks/Cole, 1972.

Kish, L. *Survey Sampling.* New York: Wiley, 1965 (paperback: 1995).

Peck, R., Casella, G., Cobb, G., Hoerl, R., Nolan, D., Starbuck, R., and Stern, H. *Statisics: A Guide to the Unknown*, 4th ed. Monterey, CA: Brooks/Cole, 2006.

Scheaffer, R., Mendenhall, W., and Ott, R. L. *Elementary Survey Sampling*, 6th ed. Boston: Duxbury, 2005.

USING TECHNOLOGY

MINITAB: Accessing and Listing Data

When you start a MINITAB session, you will see a screen similar to Figure 1.M.1. The bottom portion of the screen is an empty spreadsheet—called a MINITAB worksheet—with columns representing variables and rows representing observations (or cases). The very top of the screen is the MINITAB main menu bar, with buttons for the different functions and procedures available in MINITAB. Once you have entered

data into the spreadsheet, you can analyze the data by clicking the appropriate menu buttons. The results will appear in the Session window.

Entering Data

Create a MINITAB data file by entering data directly into the worksheet. Figure 1.M.2 shows data entered for a variable called "RATIO." Name the variables (columns) by typing in the name of each variable in the box below the column number.

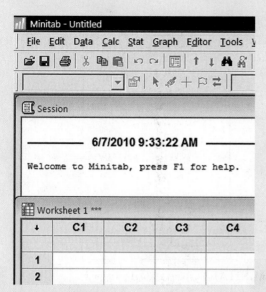

Figure 1.M.1 Initial screen viewed by the MINITAB user

Figure 1.M.2 Data entered into the MINITAB worksheet

Accessing External Data from a File

Step 1 Click the "File" button on the menu bar, and then click "Open Worksheet" as shown in Figure 1.M.3. A dialog box similar to Figure 1.M.4 will appear.

![Minitab File menu]

Figure 1.M.3 MINITAB options for reading data from an external file

Step 2 Specify the disk drive and folder that contains the external data file and the file type, and then click on the file name, as shown in Figure 1.M.4.

Step 3 If the data set contains qualitative data or data with special characters, click on the "Options" button as shown in Figure 1.M.4. The Options dialog box, shown in Figure 1.M.5, will appear.

Figure 1.M.4 Selecting the external data file in MINITAB

Figure 1.M.5 Selecting the MINITAB data input options

Step 4 Specify the appropriate options for the data set, and then click "OK" to return to the "Open Worksheet" dialog box (Figure 1.M.4).

Step 5 Click "Open" and the MINITAB worksheet will appear with the data from the external data file, as shown in Figure 1.M.6.

↓	C1	C2-T	C3-T
1	3.1	Male	Extract
2	42.0	Male	Control
3	6.2	Male	Extract

Figure 1.M.6 MINITAB worksheet with the imported data

Reminder: The variables (columns) can be named by typing in the name of each variable in the box under the column number.

To access a previously saved MINITAB worksheet, click "File," then "Open Worksheet," then select the MINITAB file.

Listing (Printing) Data

Step 1 Click on the "Data" button on the MINITAB main menu bar, and then click on "Display Data." The resulting menu, or dialog box, appears as in Figure 1.M.7.

Step 2 Enter the names of the variables you want to print in the "Columns, constants, and matrices to display" box (you can do this by simply double clicking on the variables), and then click "OK." The printout will show up on your MINITAB session screen.

Figure 1.M.7 MINITAB Display Data dialog box

2 Methods for Describing Sets of Data

CONTENTS

Where We've Been

- Examined the difference between inferential and descriptive statistics
- Described the key elements of a statistical problem
- Learned about the two types of data: quantitative and qualitative
- Discussed the role of statistical thinking in managerial decision making

Where We're Going

- Describe qualitative data with graphs (2.1)
- Describe quantitative data with graphs (2.2)
- Describe quantitative data with numerical measures (2.3–2.8)
- Describe the relationship between two quantitative variables with a graph (2.9)
- Detecting descriptive methods which distort the truth (2.10)

Statistics in Action   Body Image Dissatisfaction: Real or Imagined?

"Everything has beauty, but not everyone sees it" — ancient Chinese sage Confucius

"Body dissatisfaction can produce extreme body-shaping behaviors, such as eating disorders. Women and girls can't help being exposed to ultra-thin models in advertising whose body size is unrealistic and unhealthy. There is good evidence already that exposure to these unhealthy models leads a large proportion of women to feel dissatisfied with their own bodies." — Helga Dittmar, University of Sussex researcher

"Action figures (like G.I. Joe) present subtle messages of unrealistic role models of well-sculpted, heavily muscled, 'perfect' bodies that little boys see as their role models." — Sondra Kronberg, director of Eating Disorder Associates Treatment & Referral Centers

"By age 13, 53% of American girls are unhappy with their bodies; that figure grows to 78% by the time girls reach 17. In another study on fifth graders, 10-year-old girls and boys told researchers they were dissatisfied with their own bodies after watching a music video by Britney Spears or a clip from the TV show Friends. And adolescent girls who viewed commercials depicting unrealistically thin models felt 'less confident, more angry, and more dissatisfied with their weight and appearance.'" — statistics posted by the National Institute on Media and the Family

Are you dissatisfied with your physical appearance? Do you have a negative image of your own body? In today's media-driven society, many of us would answer yes to these questions (as the statistics in the previous quote show). Much research has been conducted on the body images of normal adolescents and adults. However, there is a lack of information on how patients with a body image disorder evaluate their own appearance, health, and fitness. To fill this gap, researchers from the Department of Psychiatry and Human Behavior at Brown University conducted and published a study in *Body Image: An International Journal of Research*, January 2010.

Data were collected on 92 patients diagnosed with *body dysmorphic disorder* (BDD). This disorder "is characterized by a distressing or impairing preoccupation with an imagined or slight defect in appearance that causes clinically significant distress or functional impairment." (Patients were also diagnosed with additional mental disorders, such as major depression or social phobia, called a comorbid disorder.) Each patient completed the Multidimensional Body-Self Relations

Questionnaire (MBSRQ). The questionnaire elicits responses that assess how satisfied one is with his/her appearance, health, fitness, and weight. In this Statistics in Action, our focus is on the appearance evaluations (e.g., "How satisfied are you with your physical attractiveness and looks?") of the BDD patients. The scores for each of seven appearance items (questions) were recorded on 5-point Likert scales, where the possible responses are 1 = definitely dissatisfied, 2 = somewhat dissatisfied, 3 = neutral, 4 = somewhat satisfied, and 5 = definitely satisfied. These scores were summed and a total appearance score (ranging from 7 to 35 points) was recorded for each patient.

The data for the study (simulated on the basis of summary statistics presented in the journal article) are provided in the **BDD** file. For each of the 92 patients in the experiment, the following variables were measured:

1. *Total Appearance Score* (points)
2. *Gender* (M or F)
3. *Comorbid Disorder* (Major Depression, Social Phobia, Obsessive Compulsive, or Anorexia/Bulimia Nervosa)
4. *Dissatisfied with Looks* (Yes or No)

For this application, we are interested in evaluating the data collected on the BDD patients. Specifically, we want to know if (1) BDD females tend to be more dissatisfied with their looks than BDD males, (2) certain comorbid disorders lead to a higher level of dissatisfaction with body appearance, and (3) BDD patients have lower appearance scores than normal people. We apply the graphical and numerical descriptive techniques of this chapter to the BDD data to answer these questions in the following Statistics in Action Revisited sections:

Statistics in Action Revisited

- Interpreting Pie Charts for the Body Image Data (p. 31)
- Interpreting Histograms for the Body Image Data (p. 44)
- Interpreting Descriptive Statistics for the Body Image Data (p. 70)
- Detecting Outliers in the Body Image Data (p. 84)

Data Set: BDD

Suppose you wish to evaluate the mathematical capabilities of a class of 1,000 first-year college students, based on their quantitative Scholastic Aptitude Test (SAT) scores. How would you describe these 1,000 measurements? Characteristics of interest include the typical, or most frequent, SAT score; the variability in the scores; the highest and lowest scores; the "shape" of the data; and whether the data set contains any unusual scores. Extracting this information isn't easy. The 1,000 scores provide too many bits of information for our minds to comprehend. Clearly, we need some method for summarizing and characterizing the information in such a data set. Methods for describing data sets are also essential for statistical inference. Most populations make for large data sets. Consequently, we need methods for describing a data set that let us make inferences about a population on the basis of information contained in a sample.

Two methods for describing data are presented in this chapter, one *graphical* and the other *numerical*. Both play an important role in statistics. Section 2.1 presents both graphical and numerical methods for describing qualitative data. Graphical methods for describing quantitative data are illustrated in Sections 2.2, 2.8, and 2.9; numerical descriptive methods for quantitative data are presented in Sections 2.3–2.7. We end the chapter with a section on the *misuse* of descriptive techniques.

2.1 Describing Qualitative Data

Consider a study of aphasia published in the *Journal of Communication Disorders* (Mar. 1995). Aphasia is the "impairment or loss of the faculty of using or understanding spoken or written language." Three types of aphasia have been identified by researchers: Broca's, conduction, and anomic. The researchers wanted to determine whether one type of aphasia occurs more often than any other and, if so, how often. Consequently, they measured the type of aphasia for a sample of 22 adult aphasics. Table 2.1 gives the type of aphasia diagnosed for each aphasic in the sample.

For this study, the variable of interest, type of aphasia, is qualitative in nature. Qualitative data are nonnumerical in nature; thus, the value of a qualitative variable can only be classified into categories called *classes*. The possible types of aphasia—Broca's, conduction, and anomic—represent the classes for this qualitative variable. We can summarize such data numerically in two ways: (1) by computing the *class frequency*—the number of observations in the data set that fall into each class—or (2) by computing the *class relative frequency*—the proportion of the total number of observations falling into each class.

> A **class** is one of the categories into which qualitative data can be classified.

> The **class frequency** is the number of observations in the data set that fall into a particular class.

> The **class relative frequency** is the class frequency divided by the total number of observations in the data set; that is,
>
> $$\text{class relative frequency} = \frac{\text{class frequency}}{n}$$

Table 2.1	Data on 22 Adult Aphasias		
Subject	Type of Aphasia	Subject	Type of Aphasia
1	Broca's	12	Broca's
2	Anomic	13	Anomic
3	Anomic	14	Broca's
4	Conduction	15	Anomic
5	Broca's	16	Anomic
6	Conduction	17	Anomic
7	Conduction	18	Conduction
8	Anomic	19	Broca's
9	Conduction	20	Anomic
10	Anomic	21	Conduction
11	Conduction	22	Anomic

Based on Li, E. C., Williams, S. E., and Volpe, R. D. "The effects of topic and listener familiarity of discourse variables in procedural and narrative discourse tasks." *The Journal of Communication Disorders*, Vol. 28, No. 1, Mar. 1995, p. 44 (Table 1).

Data Set: APHASIA

> The **class percentage** is the class relative frequency multiplied by 100; that is,
>
> class percentage = (class relative frequency) × 100

Examining Table 2.1, we observe that 10 aphasics in the study were diagnosed as suffering from anomic aphasia, 5 from Broca's aphasia, and 7 from conduction aphasia. These numbers—10, 5, and 7—represent the class frequencies for the three classes and are shown in the summary table, Figure 2.1, produced with SPSS.

TYPE

		Frequency	Percent	Valid Percent	Cumulative Percent
Valid	Anomic	10	45.5	45.5	45.5
	Brocas	5	22.7	22.7	68.2
	Conduction	7	31.8	31.8	100.0
	Total	22	100.0	100.0	

Figure 2.1

SPSS summary table for types of aphasia

Figure 2.1 also gives the relative frequency of each of the three aphasia classes. From the class relative frequency definition, we calculate the relative frequency by dividing the class frequency by the total number of observations in the data set. Thus, the relative frequencies for the three types of aphasia are

$$\text{Anomic:} \frac{10}{22} = .455$$

$$\text{Broca's:} \frac{5}{22} = .227$$

$$\text{Conduction:} \frac{7}{22} = .318$$

These values, expressed as a percent, are shown in the SPSS summary table of Figure 2.1. From these relative frequencies, we observe that nearly half (45.5%) of the 22 subjects in the study are suffering from anomic aphasia.

Although the summary table of Figure 2.1 adequately describes the data of Table 2.1, we often want a graphical presentation as well. Figures 2.2 and 2.3 show two of the most widely used graphical methods for describing qualitative data: bar graphs and pie charts. Figure 2.2 shows the frequencies of the three types of aphasia in a **bar graph** produced with SAS. Note that the height of the rectangle, or "bar," over each class is equal to the class frequency. (Optionally, the bar heights can be proportional to class relative frequencies.)

Figure 2.2

SAS bar graph for type of aphasia

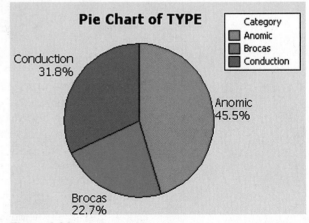

Figure 2.3

MINITAB pie chart for type of aphasia

In contrast, Figure 2.3 shows the relative frequencies of the three types of aphasia in a **pie chart** generated with MINITAB. Note that the pie is a circle (spanning 360°) and the size (angle) of the "pie slice" assigned to each class is proportional to the class relative frequency. For example, the slice assigned to anomic aphasia is 45.5% of 360°, or $(.455)(360°) = 163.8°$.

Before leaving the data set in Table 2.1, consider the bar graph shown in Figure 2.4, produced with SPSS. Note that the bars for the types of aphasia are arranged in descending order of height, from left to right across the horizontal axis. That is, the tallest bar (Anomic) is positioned at the far left and the shortest bar (Broca's) is at the far right. This rearrangement of the bars in a bar graph is called a **Pareto diagram**. One goal of a Pareto diagram (named for the Italian economist Vilfredo Pareto) is to make it easy to locate the "most important" categories—those with the largest frequencies.

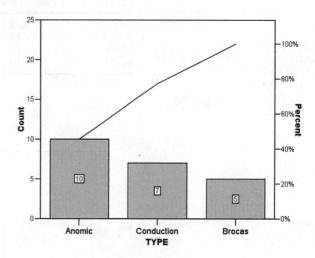

Figure 2.4

SPSS Pareto diagram for type of aphasia

Summary of Graphical Descriptive Methods for Qualitative Data

Bar Graph: The categories (classes) of the qualitative variable are represented by bars, where the height of each bar is either the class frequency, class relative frequency, or class percentage.

Pie Chart: The categories (classes) of the qualitative variable are represented by slices of a pie (circle). The size of each slice is proportional to the class relative frequency.

Pareto Diagram: A bar graph with the categories (classes) of the qualitative variable (i.e., the bars) arranged by height in descending order from left to right.

Now Work Exercise 2.6

Let's look at a practical example that requires interpretation of the graphical results.

Example 2.1

Graphing and Summarizing Qualitative Data— Drug Designed to Reduce Blood Loss

Problem A group of cardiac physicians in southwest Florida has been studying a new drug designed to reduce blood loss in coronary bypass operations. Blood loss data for 114 cor-onary bypass patients (some who received a dosage of the drug and others who did not) are saved in the **BLOODLOSS** file. Although the drug shows promise in reducing blood loss, the physicians are concerned about possible side effects and complications. So their data set includes not only the qualitative variable DRUG, which indicates whether or not the patient received the drug, but also the qualitative variable COMP, which specifies the type (if any) of complication experienced by the patient. The four values of COMP are (1) redo surgery, (2) post-op infection, (3) both, or (4) none.

a. Figure 2.5, generated by SAS, shows summary tables for the two qualitative variables, DRUG and COMP. Interpret the results.

b. Interpret the MINITAB and SPSS printouts shown in Figures 2.6 and 2.7, respectively.

The FREQ Procedure

DRUG	Frequency	Percent	Cumulative Frequency	Cumulative Percent
NO	57	50.00	57	50.00
YES	57	50.00	114	100.00

COMP	Frequency	Percent	Cumulative Frequency	Cumulative Percent
BOTH	6	5.26	6	5.26
INFECT	15	13.16	21	18.42
NONE	79	69.30	100	87.72
REDO	14	12.28	114	100.00

Figure 2.5

SAS summary tables for DRUG and COMP

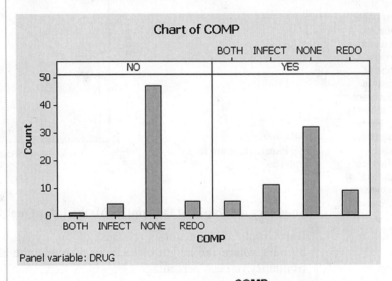

Figure 2.6

MINITAB side-by-side bar graphs for COMP by value of DRUG

COMP

DRUG			Frequency	Percent	Valid Percent	Cumulative Percent
NO	Valid	BOTH	1	1.8	1.8	1.8
		INFECT	4	7.0	7.0	8.8
		NONE	47	82.5	82.5	91.2
		REDO	5	8.8	8.8	100.0
		Total	57	100.0	100.0	
YES	Valid	BOTH	5	8.8	8.8	8.8
		INFECT	11	19.3	19.3	28.1
		NONE	32	56.1	56.1	84.2
		REDO	9	15.8	15.8	100.0
		Total	57	100.0	100.0	

Figure 2.7

SPSS summary tables for COMP by value of drug

Solution

a. The top table in Figure 2.5 is a summary frequency table for DRUG. Note that exactly half (57) of the 114 coronary bypass patients received the drug and half did not. The bottom table in Figure 2.5 is a summary frequency table for COMP. The class percentages are given in the Percent column. We see that about 69% of the 114 patients had no complications, leaving about 31% who experienced either a redo surgery, a post-op infection, or both.

b. Figure 2.6 is a MINITAB side-by-side bar graph of the data. The four bars in the left-side graph represent the frequencies of COMP for the 57 patients who did not receive the drug; the four bars in the right-side graph represent the frequencies of COMP for the 57 patients who did receive a dosage of the drug. The graph clearly shows that patients who did not receive the drug suffered fewer complications. The exact percentages are displayed in the SPSS summary tables of Figure 2.7. Over 56% of the patients who got the drug had no complications, compared with about 83% for the patients who got no drug.

Look Back Although these results show that the drug may be effective in reducing blood loss, Figures 2.6 and 2.7 imply that patients on the drug may have a higher risk of incurring complications. But before using this information to make a decision about the drug, the physicians will need to provide a measure of reliability for the inference. That is, the physicians will want to know whether the difference between the percentages of patients with complications observed in this sample of 114 patients is generalizable to the population of all coronary bypass patients.

Now Work Exercise 2.15

Statistics in Action | Revisited | Interpreting Pie Charts for the Body Image Data

In the *Body Image: An International Journal of Research* (Jan. 2010) study, Brown University researchers measured several qualitative (categorical) variables for each of 92 body dysmorphic disorder (BDD) patients: *Gender* (M or F), *Comorbid Disorder* (Major Depression, Social Phobia, Obsessive Compulsive Disorder—OCD, or Anorexia/Bulimia Nervosa), and *Dissatisfied with Looks* (Yes or No). [*Note:* "Yes" values for Dissatisfied with Looks correspond to total appearance evaluation responses of 20 points (out of 35) or less. "No" values correspond to totals of 21 points or more.] Pie charts and bar graphs can be used to summarize and describe the responses for these variables. Recall that the data are saved in the **BDD** file. We used MINITAB to create pie charts for these variables.

Figure SIA2.1 shows individual pie charts for each of the three qualitative variables, Gender, Disorder, and Dissatisfied with Looks. First, notice that of the 92 BDD patients, 65% are females and 35% are males. Of these patients, the most common comorbid disorder is major depression (42%), followed by social phobia (35%) and OCD (20%). The third pie chart shows that 77% of the BDD patients are dissatisfied in some way with their body.

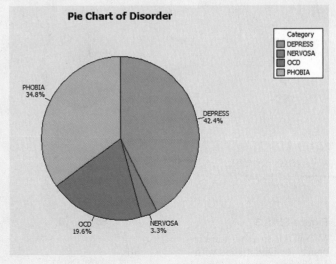

Figure SIA2.1
MINITAB pie charts for Gender, Disorder, and Dissatisfied with Looks

(continued)

Figure SIA2.1
(*continued*)

Figure SIA2.2
MINITAB side-by-side pie charts for Dissatisfied with Looks by Gender

Figure SIA2.3
MINITAB side-by-side Pie charts for Dissatisfied with Looks by Comorbid Disorder

Statistics IN Action
(continued)

Of interest in the study is whether BDD females tend to be more dissatisfied with their looks than BDD males. We can gain insight into this question by forming side-by-side pie charts of the Dissatisfied with Looks variable, one chart for females and one for males. These pie charts are shown in Figure SIA2.2. You can see that about 82% of the females are dissatisfied in some way with their body as compared to about 69% of the males. Thus, it does appear that BDD females tend to be more dissatisfied with their looks than males, at least for this sample of patients.

Also of interest is whether certain comorbid disorders lead to a higher level of dissatisfaction with body appearance. Figure SIA2.3 is a set of side-by-side pie charts for Dissatisfied with Looks, one chart for each of four comorbid disorders. The graphs show that the percentage of BDD patients who are dissatisfied in some way with their body image range from about 67% for those with a nervosa disorder to about 81% for those diagnosed with a social phobia.

 Data Set: BBD

⚠ **CAUTION** The information produced in these pie charts should be limited to describing the sample of 92 BDD patients. If one is interested in making inferences about the population of all BDD patients (as were the Brown University researchers), inferential statistical methods need to be applied to the data. These methods are the topics of later chapters. ▲

Exercises 2.1–2.21

Understanding the Principles

2.1 Explain the difference between class frequency, class relative frequency, and class percentage for a qualitative variable.

2.2 Explain the difference between a bar graph and a pie chart.

2.3 Explain the difference between a bar graph and a Pareto diagram.

Learning the Mechanics

2.4 Complete the following table:

Grade on Statistics Exam	Frequency	Relative Frequency
A: 90–100		.08
B: 80–89	36	
C: 65–79	90	
D: 50–64	30	
F: Below 50	28	
Total	200	1.00

2.5 A qualitative variable with three classes (X, Y, and Z) is measured for each of 20 units randomly sampled from a target population. The data (observed class for each unit) are as follows:

```
Y  X  X  Z  X  Y  Y  Y  Y  X  X  Z  X
Y  Y  X  Z  Y  Y  Y  X
```

a. Compute the frequency for each of the three classes.
b. Compute the relative frequency for each of the three classes.
c. Display the results from part **a** in a frequency bar graph.
d. Display the results from part **b** in a pie chart.

Applying the Concepts—Basic

2.6 **Study of importance of libraries.** In *The Canadian Journal of Information and Library Science* (Vol. 33, 2009), researchers from the University of Western Ontario reported on mall shoppers' opinions about libraries and their importance to today's society. Each in a sample of over 200 mall shoppers was asked the following question: "In today's world, with Internet access and online and large book sellers like Amazon, do you think libraries have become more, less, or the same in importance to their community?" The accompanying graphic summarizes the mall shoppers' responses.

a. What type of graph is shown?
b. Identify the qualitative variable described in the graph.
c. From the graph, identify the most common response.
d. Convert the graph into a Pareto diagram. Interpret the results.

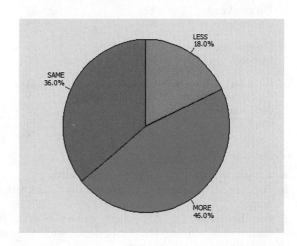

2.7 **Cheek teeth of extinct primates.** The characteristics of cheek teeth (e.g., molars) can provide anthropologists with information on the dietary habits of extinct mammals. The cheek teeth of an extinct primate species was the subject of research reported in the *American Journal of Physical Anthropology* (Vol. 142, 2010). These data are saved in the **CHEEKTEETH** file. A total of 18 cheek teeth extracted from skulls discovered in western Wyoming were analyzed. Each tooth was classified according to degree of wear (unworn, slight, light-moderate, moderate, moderate-heavy, or heavy). The 18 measurements are listed here.

Data on Degree of Wear	
Unknown	Slight
Unknown	Slight
Unknown	Heavy
Moderate	Unworn
Slight	Light-moderate
Unknown	Light-moderate
Moderate-heavy	Moderate
Moderate	Unworn
Slight	Unknown

a. Identify the variable measured in the study and its type (quantitative or qualitative).

b. Count the number of cheek teeth in each wear category.

c. Calculate the relative frequency for each wear category.

d. Construct a relative frequency bar graph for the data.

e. Construct a Pareto diagram for the data.

f. Identify the degree of wear category that occurred most often in the sample of 18 cheek teeth.

2.8 **Humane education and classroom pets.** In grade school, did your teacher have a pet in the classroom? Many teachers do so in order to promote humane education. The *Journal of Moral Education* (Mar. 2010) published a study designed to examine elementary school teachers' experiences with humane education and animals in the classroom. Based on a survey of 75 elementary school teachers, the following results were reported:

Survey Question	Response	Number of Teachers
Q1: Do you keep classroom pets?	Yes	61
	No	14
Q2: Do you allow visits by pets?	Yes	35
	No	40
Q3: What formal humane education program does your school employ?	Character counts	21
	Roots of empathy	9
	Teacher designed	3
	Other	14
	None	28

Based on *Journal of Moral Education*, March 2010.

a. Explain why the data collected is qualitative.

b. For each survey question, summarize the results with a graph.

c. For this sample of 75 teachers, write two or three sentences describing their experience with humane education and classroom pets.

2.9 **Japanese reading levels.** University of Hawaii language professors C. Hitosugi and R. Day incorporated a 10-week extensive reading program into a second-semester Japanese language course in an effort to improve students' Japanese reading comprehension. (*Reading in a Foreign Language*, Apr. 2004.) The professors collected 266 books originally written for Japanese children and required their students to read at least 40 of them as part of the grade in the course. The books were categorized into reading levels (color coded for easy selection) according to length and complexity. The reading levels for the 266 books are summarized in the following table:

Reading Level	Number
Level 1 (Red)	39
Level 2 (Blue)	76
Level 3 (Yellow)	50
Level 4 (Pink)	87
Level 5 (Orange)	11
Level 6 (Green)	3
Total	266

Source: Hitosugi, C. I., and Day, R. R. "Extensive reading in Japanese." *Reading in a Foreign Language*, Vol. 16, No. 1. Apr. 2004 (Table 2). Reprinted with permission from the National Foreign Language Resource Center, University of Hawaii.

a. Calculate the proportion of books at reading level 1 (red).

b. Repeat part a for each of the remaining reading levels.

c. Verify that the proportions in parts **a** and **b** sum to 1.

d. Use the previous results to form a bar graph for the reading levels.

e. Construct a Pareto diagram for the data. Use the diagram to identify the reading level that occurs most often.

2.10 **PIN pad shipments.** Personal identification number (PIN) pads are devices that connect to point-of-sale electronic cash registers for debit and credit card purchases. The PIN pad allows the customer card to be accessed and the PIN encrypted before it is sent to the transaction manager. *The Nilson Report* (Oct. 2008) listed the volume of PIN pad shipments by manufacturers worldwide in 2007. For the 12 manufacturers listed in the table, a total of 334,039 PIN pads were shipped in 2007 (These data are saved in the **PINPADS** file.)

Manufacturer	Number Shipped (units)
Bitel	13,500
CyberNet	16,200
Fujian Landi	119,000
Glintt (ParaRede)	5,990
Intelligent	4,562
KwangWoo	42,000
Omron	20,000
Pax Tech.	10,072
ProvencoCadmus	20,000
SZZT Electronics	67,300
Toshiba TEC	12,415
Urmet	3,000

Based on *The Nilson Report*, No. 913, Oct. 2008 (p. 9).

a. One of the 334,039 PIN pads is selected and the manufacturer of the pad is determined. What type of data (quantitative or qualitative) is measured?

MINITAB Output for Exercise 2.12

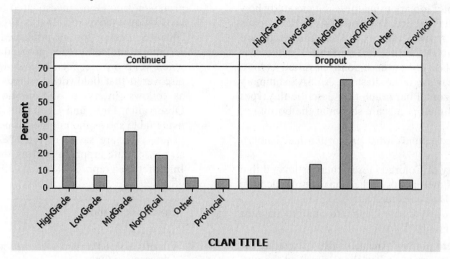

b. Construct a frequency bar chart for the data summarized in the table.

c. Convert the frequency bar chart, part **b,** into a Pareto diagram. Interpret the results.

2.11 Estimating the rhino population. The International Rhino Federation estimates that there are 17,800 rhinoceroses living in the wild in Africa and Asia. A breakdown of the number of rhinos of each species is reported in the accompanying table:

Rhino Species	Population Estimate
African Black	3,610
African White	11,330
(Asian) Sumatran	300
(Asian) Javan	60
(Asian) Indian	2,500
Total	17,800

Based on International Rhino Foundation, Mar. 2007.

a. Construct a relative frequency table for the data.

b. Display the relative frequencies in a bar graph.

c. What proportion of the 17,800 rhinos are African rhinos? Asian?

Applying the Concepts—Intermediate

2.12 Genealogy research. Writing in the *Journal of Family History* (Jan. 2010), Hong Kong University historian Sangkuk Lee investigated the genealogy of a certain Korean clan. Of interest to Lee was whether or not a family name continued into the next generation in the clan genealogy (called a *continued line*) or dropped out (called a *dropout line*). Lee used side-by-side pie charts to describe the rate at which certain occupational titles of clan individuals occurred within each line type. Similarly constructed MINITAB bar charts are shown above.

a. Identify the two qualitative variables graphed in the side-by-side pie charts.

b. Give a full interpretation of the charts. Identify the major differences (if any) between the two line groups.

Based on Sangkuk Lee, *Journal of Family History* (Jan 2010).

2.13 Characteristics of ice melt ponds. The National Snow and Ice Data Center (NSIDC) collects data on the albedo, depth, and physical characteristics of ice melt ponds in

SAS output for Exercise 2.13

The FREQ Procedure

ICETYPE	Frequency	Percent	Cumulative Frequency	Cumulative Percent
First-year	88	17.46	88	17.46
Landfast	196	38.89	284	56.35
Multi-year	220	43.65	504	100.00

the Canadian Arctic. Environmental engineers at the University of Colorado are using these data to study how climate affects the sea ice. Data on 504 ice melt ponds located in the Barrow Strait in the Canadian Arctic are saved in the **PONDICE** file. One variable of interest is the type of ice observed for each pond, classified as first-year ice, multiyear ice, or landfast ice. An SAS summary table and a horizontal bar graph that describe the types of ice of the 504 melt ponds are shown at the bottom of page 35.

a. Of the 504 melt ponds, what proportion had landfast ice?

b. The University of Colorado researchers estimated that about 17% of melt ponds in the Canadian Arctic have first-year ice. Do you agree?

c. Convert the horizontal bar graph into a Pareto diagram. Interpret the graph.

2.14 **Excavating ancient pottery.** Archaeologists excavating the ancient Greek settlement at Phylakopi classified the pottery found in trenches. (*Chance*, Fall 2000.) The accompanying table describes the collection of 837 pottery pieces uncovered in a particular layer at the excavation site. Construct and interpret a graph that will aid the archaeologists in understanding the distribution of the types of pottery found at the site.

Pot Category	Number Found
Burnished	133
Monochrome	460
Slipped	55
Painted in curvilinear decoration	14
Painted in geometric decoration	165
Painted in naturalistic decoration	4
Cycladic white clay	4
Conical cup clay	2
Total	837

Based on Berg, I., and Bliedon, S. "The pots of Phylakopi: Applying statistical techniques to archaeology." *Chance*, Vol. 13, No. 4, Fall 2000.

2.15 **Satellites in orbit.** According to the Union of Concerned Scientists (www.ucsusa.org), as of April 1, 2009, there were 442 low earth orbit (LEO) and 366 geosynchronous orbit (GEO) satellites in space. Each satellite is owned by an entity in either the government, military, commercial, or civil sector. A breakdown of the number of satellites in orbit for each sector is displayed in the accompanying table. Use this information to construct a pair of graphs that compare the ownership sectors of LEO and GEO satellites in orbit. What observations do you have about the data?

	LEO Satellites	GEO Satellites
Government	193	45
Military	88	67
Commercial	130	253
Civil	31	1
Total	442	366

2.16 **Sociology fieldwork methods.** University of New Mexico professor Jane Hood investigated the fieldwork methods used by qualitative sociologists. (*Teaching Sociology*, July 2006.) Searching for all published journal articles, dissertations, and conference proceedings over the previous seven years in the *Sociological Abstracts* database, she discovered that fieldwork methods could be categorized as follows: Interview, Observation plus Participation, Observation Only, and Grounded Theory. The accompanying table shows the number of papers in each category. (These data are saved in the **FIELDWORK** file.) Use an appropriate graph to portray the results of the study. Interpret the graph.

Fieldwork Method	Number of Papers
Interview	5,079
Observation + Participation	1,042
Observation Only	848
Grounded Theory	537

Based on Hood, J. C. "Teaching against the text: The case of qualitative methods." *Teaching Sociology*, Vol. 34, Issue 3, p. 207 (Exhibit 2).

2.17 **Best-paid CEOs.** *Forbes* magazine conducts an annual survey of the salaries of chief executive officers. In addition to salary information, *Forbes* collects and reports personal data on the CEOs, including their level of education and age. Do most CEOs have advanced degrees, such as master's degrees or doctorates? The data in the accompanying table (saved in the **FORBES40** file) represent the highest degree earned by each of the top 40 best-paid CEOs of 2010. Use a graphical method to summarize the highest degree earned by these CEOs. What is your opinion about whether most CEOs have advanced degrees?

MBA	PhD	Bachelor's	Bachelor's
PhD	Bachelor's	Bachelor's	MBA
None	Law	Bachelor's	Bachelor's
Bachelor's	MBA	None	Master's
Bachelor's	MBA	MBA	MBA
PhD	Bachelor's	None	None
Bachelor's	MBA	MBA	MBA
Masters	Bachelor's	Law	Bachelor's
MBA	MBA	None	Bachelor's
Bachelor's	None	Law	Master's

Based on "40 highest-paid CEOs of 2010" from "What the boss makes: CEO compensation." *Forbes*, April 28, 2010.

2.18 **Do you believe in the Bible?** In its annual General Social Survey (GSS), the National Opinion Research Center (NORC) elicits the opinions of Americans on a wide variety of social topics. One question in the survey asked about a person's belief in the Bible. Data for the approximately 2,800 Americans who participated in the 2004 GSS are saved in the **BIBLE** file. Each respondent selected from one of the following answers: (1) The Bible is the actual word of God and is to be taken literally; (2) the Bible is the inspired word of God, but not everything is to be taken literally; (3) the Bible is an ancient book of fables; and (4) the Bible has some other origin, but is recorded by men. The variable "Bible1" contains the responses.

(*Note:* A response value of 8 represents "Don't Know"; a value of 9 represents a missing value.)

a. Summarize the responses to the Bible question in the form of a relative frequency table.

b. Summarize the responses to the Bible question in a pie chart.

c. Write a few sentences that give a practical interpretation of the results shown in the summary table and graph.

Applying the Concepts—Advanced

2.19 Museum management. What criteria do museums use to evaluate their performance? In a worldwide survey reported in *Museum Management and Curatorship* (June 2010), managers of 30 leading museums of contemporary art were asked to provide the performance measure used most often. A summary of the results is provided in the table. The researcher concluded that "there is a large amount of variation within the museum community with regard to… performance measurement and evaluation." Do you agree? Use a graph to support your conclusion.

Performance Measure	Number of Museums
Total visitors	8
Paying visitors	5
Big shows	6
Funds raised	7
Members	4

2.20 Extinct New Zealand birds. Refer to the *Evolutionary Ecology Research* (July 2003) study of the patterns of extinction in the New Zealand bird population, Exercise 1.18 (p. 19). Data on flight capability (volant or flightless), habitat (aquatic, ground terrestrial, or aerial terrestrial), nesting site (ground, cavity within ground, tree, or cavity above ground), nest density (high or low), diet (fish, vertebrates, vegetables, or invertebrates), body mass (grams), egg length (millimeters), and extinct status (extinct, absent from island, or present) for 132 bird species that existed at the time of the Maori colonization of New Zealand are saved in the **NZBIRDS** file. Use a graphical method to investigate the theory that extinct status is related to flight capability, habitat, and nest density.

2.21 Groundwater contamination in wells. In New Hampshire, about half the counties mandate the use of reformulated gasoline. This has led to an increase in the contamination of groundwater with methyl *tert*-butyl ether (MTBE). *Environmental Science & Technology* (Jan. 2005) reported on the factors related to MTBE contamination in public and private New Hamsphire wells. Data were collected on a sample of 223 wells. These data are saved in the **MTBE** file. Three of the variables are qualitative in nature: well class (public or private), aquifer (bedrock or unconsolidated), and detectable level of MTBE (below limit or detect). [*Note:* A detectable level of MTBE occurs if the MTBE value exceeds .2 microgram per liter.] The data on 11 selected wells are shown in the accompanying table.

(11 selected observations from 223)

Well Class	Aquifer	Detect MTBE?
Private	Bedrock	Below Limit
Private	Bedrock	Below Limit
Public	Unconsolidated	Detect
Public	Unconsolidated	Below Limit
Public	Unconsolidated	Below Limit
Public	Unconsolidated	Below Limit
Public	Unconsolidated	Detect
Public	Unconsolidated	Below Limit
Public	Unconsolidated	Below Limit
Public	Bedrock	Detect
Public	Bedrock	Detect

Based on Ayotte, J. D., Argue, D. M., and McGarry, F. J. "Methyl *tert*-Butyl ether occurrence and related factors in public and private wells in southeast New Hampshire." *Environmental Science & Technology*, Vol. 39, No. 1, Jan. 2005, pp. 9–16.

a. Use graphical methods to describe each of the three qualitative variables for all 223 wells.

b. Use side-by-side bar charts to compare the proportions of contaminated wells for private and public well classes.

c. Use side-by-side bar charts to compare the proportions of contaminated wells for bedrock and unconsolidated aquifers.

d. What inferences can be made from the bar charts of parts **a–c**?

2.2 Graphical Methods for Describing Quantitative Data

Recall from Section 1.4 that quantitative data sets consist of data that are recorded on a meaningful numerical scale. To describe, summarize, and detect patterns in such data, we can use three graphical methods: *dot plots, stem-and-leaf displays,* and *histograms*. Since most statistical software packages can be used to construct these displays, we'll focus here on their interpretation rather than their construction.

For example, the Environmental Protection Agency (EPA) performs extensive tests on all new car models to determine their mileage ratings. Suppose that the 100 measurements in Table 2.2 represent the results of such tests on a certain new car model. How can we summarize the information in this rather large sample?

Table 2.2	EPA Mileage Ratings on 100 Cars			
36.3	41.0	36.9	37.1	44.9
32.7	37.3	41.2	36.6	32.9
40.5	36.5	37.6	33.9	40.2
36.2	37.9	36.0	37.9	35.9
38.5	39.0	35.5	34.8	38.6
36.3	36.8	32.5	36.4	40.5
41.0	31.8	37.3	33.1	37.0
37.0	37.2	40.7	37.4	37.1
37.1	40.3	36.7	37.0	33.9
39.9	36.9	32.9	33.8	39.8
36.8	30.0	37.2	42.1	36.7
36.5	33.2	37.4	37.5	33.6
36.4	37.7	37.7	40.0	34.2
38.2	38.3	35.7	35.6	35.1
39.4	35.3	34.4	38.8	39.7
36.6	36.1	38.2	38.4	39.3
37.6	37.0	38.7	39.0	35.8
37.8	35.9	35.6	36.7	34.5
40.1	38.0	35.2	34.8	39.5
34.0	36.8	35.0	38.1	36.9

⊙ *Data Set: EPAGAS*

A visual inspection of the data indicates some obvious facts. For example, most of the mileages are in the 30s, with a smaller fraction in the 40s. But it is difficult to provide much additional information on the 100 mileage ratings without resorting to some method of summarizing the data. One such method is a dot plot.

Dot Plots

A MINITAB **dot plot** for the 100 EPA mileage ratings is shown in Figure 2.8. The horizontal axis of the figure is a scale for the quantitative variable in miles per gallon (mpg). The rounded (to the nearest half gallon) numerical value of each measurement in the data set is located on the horizontal scale by a dot. When data values repeat, the dots are placed above one another, forming a pile at that particular numerical location. As you can see, this dot plot verifies that almost all of the mileage ratings are in the 30s, with most falling between 35 and 40 miles per gallon.

Figure 2.8
MINITAB dot plot for 100 EPA mileage ratings

Stem-and-Leaf Display

Another graphical representation of these same data, a MINITAB **stem-and-leaf display**, is shown in Figure 2.9. In this display, the *stem* is the portion of the measurement (mpg) to the left of the decimal point, while the remaining portion, to the right of the decimal point, is the *leaf*.

In Figure 2.9, the stems for the data set are listed in the second column, from the smallest (30) to the largest (44). Then the leaf for each observation is listed to the right, in the row of the display corresponding to the observation's stem.* For example, the leaf 3 of the first observation (36.3) in Table 2.2 appears in the row corresponding to the stem 36. Similarly, the leaf 7 for the second observation (32.7) in Table 2.2 appears in the row corresponding to the stem 32, while the leaf 5 for the third observation (40.5) appears in the row corresponding to the stem 40. (The stems and leaves for these first three observations are highlighted in Figure 2.9.) Typically, the leaves in each row are ordered as shown in the MINITAB stem-and-leaf display.

Stem-and-Leaf Display: MPG

```
Stem-and-leaf of MPG  N  = 100
Leaf Unit = 0.10


   1    30  0
   2    31  8
   6    32  5799
  12    33  126899
  18    34  024588
  29    35  01235667899
  49    36  01233445566777888999
 (21)   37  000011122334456677899
  30    38  0122345678
  20    39  00345789
  12    40  0123557
   5    41  002
   2    42  1
   1    43
   1    44  9
```

Figure 2.9

MINITAB stem-and-leaf display for 100 mileage ratings

The stem-and-leaf display presents another compact picture of the data set. You can see at a glance that the 100 mileage readings were distributed between 30.0 and 44.9, with most of them falling in stem rows 35 to 39. The 6 leaves in stem row 34 indicate that 6 of the 100 readings were at least 34.0, but less than 35.0. Similarly, the 11 leaves in stem row 35 indicate that 11 of the 100 readings were at least 35.0, but less than 36.0. Only five cars had readings equal to 41 or larger, and only one was as low as 30.

The definitions of the stem and leaf for a data set can be modified to alter the graphical description. For example, suppose we had defined the stem as the tens digit for

BIOGRAPHY

JOHN TUKEY (1915–2000)

The Picasso of Statistics

Like the legendary artist Pablo Picasso, who mastered and revolutionized a variety of art forms during his lifetime, John Tukey is recognized for his contributions to many subfields of statistics. Born in Massachusetts, Tukey was home schooled, graduated with his bachelor's and master's degrees in chemistry from Brown University, and received his Ph.D. in mathematics from Princeton University. While at Bell Telephone Laboratories in the 1960s and early 1970s, Tukey developed exploratory data analysis, a set of graphical descriptive methods for summarizing and presenting huge amounts of data. Many of these tools, including the stem-and-leaf display and the box plot, are now standard features of modern statistical software packages. (In fact, it was Tukey himself who coined the term *software* for computer programs.) ∎

*The first column of the MINITAB stem-and-leaf display represents the cumulative number of measurements from the class interval to the nearest extreme class interval.

the gas mileage data, rather than the ones and tens digits. With this definition, the stems and leaves corresponding to the measurements 36.3 and 32.7 would be as follows:

Stem	Leaf
3	6

Stem	Leaf
3	2

Note that the decimal portion of the numbers has been dropped. Generally, only one digit is displayed in the leaf.

If you look at the data, you'll see why we didn't define the stem this way. All the mileage measurements fall into the 30s and 40s, so all the leaves would fall into just two stem rows in this display. The resulting picture would not be nearly as informative as Figure 2.9.

Now Work Exercise 2.27

Histograms

An SPSS **histogram** for the 100 EPA mileage readings given in Table 2.2 is shown in Figure 2.10. The horizontal axis of the figure, which gives the miles per gallon for a given automobile, is divided into **class intervals**, commencing with the interval from 30–31 and proceeding in intervals of equal size to 44–45 mpg. The vertical axis gives the number (or *frequency*) of the 100 readings that fall into each interval. It appears that about 21 of the 100 cars, or 21%, attained a mileage between 37 and 38 mpg. This class interval contains the highest frequency, and the intervals tend to contain a smaller number of the measurements as the mileages get smaller or larger.

Histograms can be used to display either the frequency or relative frequency of the measurements falling into the class intervals. The class intervals, frequencies, and relative frequencies for the EPA car mileage data are shown in the summary table, Table 2.3.*

By summing the relative frequencies in the intervals 35–36, 36–37, 37–38, and 38–39, you find that 65% of the mileages are between 35 and 39. Similarly, only 2% of

Figure 2.10
SPSS histogram for 100 EPA gas mileage ratings

*SPSS, like many software packages, will classify an observation that falls on the borderline of a class interval into the next-highest interval. For example, the gas mileage of 37.0, which falls on the border between the class intervals 36–37 and 37–38, is classified into the 37–38 class. The frequencies in Table 2.3 reflect this convention.

Table 2.3	Class Intervals, Frequencies, and Relative Frequencies for the Gas Mileage Data	
Class Interval	Frequency	Relative Frequency
30–31	1	0.01
31–32	1	0.01
32–33	4	0.04
33–34	6	0.06
34–35	6	0.06
35–36	11	0.11
36–37	20	0.20
37–38	21	0.21
38–39	10	0.10
39–40	8	0.08
40–41	7	0.07
41–42	3	0.03
42–43	1	0.01
43–44	0	0.00
44–45	1	0.01
Totals	100	1.00

the cars obtained a mileage rating over 42.0. Many other summary statements can be made by further examining the histogram and accompanying summary table. Note that the sum of all class frequencies will always equal the sample size n.

In interpreting a histogram, consider two important facts. First, the proportion of the total area under the histogram that falls above a particular interval on the x-axis is equal to the relative frequency of measurements falling into that interval. For example, the relative frequency for the class interval 37–38 in Figure 2.10 is .20. Consequently, the rectangle above the interval contains .20 of the total area under the histogram.

Second, imagine the appearance of the relative frequency histogram for a very large set of data (representing, say, a population). As the number of measurements in a data set is increased, you can obtain a better description of the data by decreasing the width of the class intervals. When the class intervals become small enough, a relative frequency histogram will (for all practical purposes) appear as a smooth curve. (See Figure 2.11.)

Figure 2.11
The effect of the size of a data set on the outline of a histogram

a. Small data set   b. Larger data set   c. Very large data set

Some recommendations for selecting the number of intervals in a histogram for smaller data sets are given in the following box:

Determining the Number of Classes in a Histogram

Number of Observations in Data Set	Number of Classes
Fewer than 25	5–6
25–50	7–14
More than 50	15–20

While histograms provide good visual descriptions of data sets—particularly very large ones—they do not let us identify individual measurements. In contrast, each of the original measurements is visible to some extent in a dot plot and is clearly visible in a stem-and-leaf display. The stem-and-leaf display arranges the data in ascending order, so it's easy to locate the individual measurements. For example, in Figure 2.9 we can easily see that two of the gas mileage measurements are equal to 36.3, but we can't see that fact by inspecting the histogram in Figure 2.10. However, stem-and-leaf displays can become unwieldy for very large data sets. A very large number of stems and leaves causes the vertical and horizontal dimensions of the display to become cumbersome, diminishing the usefulness of the visual display.

Example 2.2

Graphing a Quantitative Variable—The "Water-Level Task"

Problem Over 60 years ago, famous child psychologist Jean Piaget devised a test of basic perceptual and conceptual skills dubbed the "water-level task." Subjects were shown a drawing of a glass being held at a 45° angle and asked to draw a line representing the true surface of the water. Today, research psychologists continue to use the task to test the perception of both adults and children. In one study, the water-level task was given to several groups that included 20 male bartenders and 20 female waitresses (*Psychological Science*, March 1995). For each participant, the researchers measured the deviation (in angle degrees) of the judged line from the true line. These deviations (simulated on the basis of summary results presented in the journal article) are shown in Table 2.4. [*Note:* Deviations can be negative if the judged angle is smaller than the angle of the true line.]

Table 2.4	Water-Level Task Deviations (angle degrees)

| Bartenders: | −9 | 6 | 10 | 6 | 10 | −3 | 7 | 8 | 6 | 14 | 7 | 8 | −5 | 2 | −1 | 0 | 2 | 3 | 0 | 2 |
| Waitresses: | 7 | 10 | 25 | 8 | 10 | 8 | 12 | 9 | 35 | 10 | 12 | 11 | 7 | 10 | 21 | −1 | 4 | 0 | 16 | −1 |

Data Set: WLTASK

a. Use a statistical software package to create a frequency histogram for the combined data in Table 2.4. Then, shade the area under the histogram that corresponds to deviations recorded for waitresses. Interpret the result.

b. Use a statistical software package to create a create a stem-and-leaf display for these combined data. Again, shade each leaf of the display that corresponds to a deviation recorded for a waitress. Interpret the result.

Solution:

a. We used SPSS to generate the frequency histogram, shown in Figure 2.12. Note that SPSS formed 20 classes, with class intervals −20 to −10, −10 to −5, ..., 30 to 35, and 35 to 40. This histogram clearly shows the clustering of the deviation angles between 0° and 15°, with a few deviations in the upper end of the distribution (greater than 20°). SPSS used green bars to shade the areas of the histogram that correspond to the measurements for waitresses. The graph clearly shows that waitresses tend to have greater (positive) deviations than do bartenders, and fewer deviations near 0° relative to bartenders.

b. We used MINITAB to produce the stem-and-leaf display in Figure 2.13. Note that the stem (the second column on the printout) represents the first digit (including 0) in the deviation angle measurement while the leaf (the third column on the printout) represents the second digit. Thus, the leaf 5 in the stem 2 row represents the deviation angle of 25°. The shaded leaves represent deviations recorded for waitresses. As with the histogram, the stem-and-leaf display shows that deviations for waitresses tend to appear in the upper tail of the distribution. Together, the graphs imply that waitresses tend to overestimate the angle of the true line relative to bartenders.

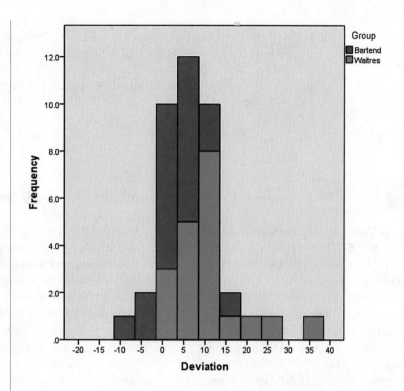

Figure 2.12

SPSS histogram for task deviations

Look Back As is usually the case with data sets that are not too large (say, fewer than 100 measurements), the stem-and-leaf display provides more detail than the histogram without being unwieldy. For instance, the stem-and-leaf display in Figure 2.13 clearly indicates the values of the individual measurements in the data set. For example, the largest deviation angle (representing the measurement 35°) is shown in the last stem row. By contrast, histograms are most useful for displaying very large data sets when the overall shape of the distribution of measurements is more important than the identification of individual measurements.

Stem-and-Leaf Display: Deviation

```
Stem-and-leaf of Deviation  N  = 40
Leaf Unit = 1.0


  2   -0  95
  7   -0  31110
 14    0  0022234
(12)   0  666777788889
 14    1  0000001224
  4    1  6
  3    2  1
  2    2  5
  1    3
  1    3  5
```

Figure 2.13

MINITAB stem-and-leaf display for task deviations

Now Work Exercise 2.31

Most statistical software packages can be used to generate histograms, stem-and-leaf displays, and dot plots. All three are useful tools for graphically describing data sets. We recommend that you generate and compare the displays whenever you can.

Summary of Graphical Descriptive Methods for Quantitative Data

Dot Plot: The numerical value of each quantitative measurement in the data set is represented by a dot on a horizontal scale. When data values repeat, the dots are placed above one another vertically.

Stem-and-Leaf Display: The numerical value of the quantitative variable is partitioned into a "stem" and a "leaf." The possible stems are listed in order in a column. The leaf for each quantitative measurement in the data set is placed in the corresponding stem row. Leaves for observations with the same stem value are listed in increasing order horizontally.

(continued)

> **Histogram:** The possible numerical values of the quantitative variable are partitioned into class intervals, each of which has the same width. These intervals form the scale of the horizontal axis. The frequency or relative frequency of observations in each class interval is determined. A vertical bar is placed over each class interval, with the height of the bar equal to either the class frequency or class relative frequency.

Statistics IN Action | **Revisited**

Interpreting Histograms for the Body Image Data

In the *Body Image: An International Journal of Research* (Jan. 2010) study of 92 BDD patients, the researchers asked each patient to respond to a series of questions on body image (e.g., "How satisfied are you with your physical attractiveness and looks?"). Recall that the scores were summed to yield an Appearance Evaluation score which ranged from 7 to 35 points. This score represents a quantitative variable. Consequently, to graphically investigate whether BDD females tend to be more dissatisfied with their looks than BDD males, we can form side-by-side histograms for the total score, one histogram for females and one for males. These histograms are shown in Figure SIA2.4.

Like the pie charts in the previous Statistics in Action Revisited section, the histograms tend to support the theory. For females, the histogram for appearance evaluation score is centered at about 17 points, while for males the histogram is centered higher, at about 20 points. Also from the histograms you can see that about 55% of the female patients had a score of less than 20, compared to only about 45% of the males. Again, the histograms seem to indicate that BDD females tend to be more dissatisfied with their looks than males. In later chapters, we'll learn how to attach a measure of reliability to such an inference.

 Data Set: BBD

Figure SIA2.4
MINITAB side-by-side histograms for Appearance Evaluation by Gender

Exercises 2.22–2.43

Understanding the Principles

2.22 Explain the difference between a bar graph and a histogram.

2.23 Explain the difference between a dot plot and a stem-and-leaf display.

2.24 Explain the difference between the stem and the leaf in a stem-and-leaf display.

2.25 In a histogram, what are the class intervals?

2.26 How many classes are recommended in a histogram of a data set with more than 50 observations?

Learning the Mechanics

2.27 Consider the stem-and-leaf display shown here:
NW

Stem	Leaf
5	1
4	457
3	00036
2	1134599
1	2248
0	012

   a. How many observations were in the original data set?

   b. In the bottom row of the stem-and-leaf display, identify the stem, the leaves, and the numbers in the original data set represented by this stem and its leaves.

   c. Re-create all the numbers in the data set, and construct a dot plot.

2.28 Graph the relative frequency histogram for the 500 measurements summarized in the accompanying relative frequency table.

Class Interval	Relative Frequency
.5–2.5	.10
2.5–4.5	.15
4.5–6.5	.25
6.5–8.5	.20
8.5–10.5	.05
10.5–12.5	.10
12.5–14.5	.10
14.5–16.5	.05

2.29 Refer to Exercise 2.28. Calculate the number of the 500 measurements falling into each of the measurement classes. Then graph a frequency histogram of these data.

2.30 Consider the MINITAB histogram shown below.

   a. Is this a frequency histogram or a relative frequency histogram? Explain.

   b. How many class intervals were used in the construction of this histogram?

   c. How many measurements are there in the data set described by this histogram?

Applying the Concepts—Basic

2.31 **Body length of armadillos.** A group of environmentalists reported the results of a study of giant armadillos inhabiting the southeastern region of Venezuela. A sample of 80 armadillos was captured and the body length (not including the tail) of each was measured (in centimeters). A MINITAB graph summarizing the data is shown below.

   a. What type of graph is employed?

   b. How many armadillos have body lengths between 87 and 91 centimeters?

   c. What proportion of the armadillos have body lengths between 87 and 91 centimeters?

   d. The dotted vertical lines on the graph show the minimum and maximum sizes of giant armadillos that can be legally captured for commercial purposes. What proportion of the captured armadillos are illegal?

2.32 **Is honey a cough remedy?** Does a teaspoon of honey before bed really calm a child's cough? To test the folk remedy, pediatric researchers at Pennsylvania State University carried out a designed study involving a sample of 105 children who were ill with an upper respiratory tract infection (*Archives of Pediatrics and Adolescent Medicine*, Dec. 2007). On the first night, parents rated their children's cough symptoms on a scale from 0 (no problems at all) to 30 (extremely severe). On the second night, the parents were instructed to give their sick child a dosage of liquid "medicine" prior to bedtime. Unknown to the parents, some were given a dosage of dextromethorphan (DM)—an over-the-counter cough medicine—while others were given a similar dose of honey. Also, a third group of parents (the control group) gave their sick children no dosage at all. Again, the parents rated their children's cough symptoms, and the improvement in total cough symptoms score was determined for each child. The data (improvement scores) for the study are shown in the table (p. 46) and saved in the **HONEYCOUGH** file.

   a. Construct a dot plot for the coughing improvement scores for the 35 children in the honey dosage group.

   b. Refer to part **a.** What coughing improvement score occurred most often in the honey dosage group?

MINITAB dot plot for Exercise 2.32

Data for Exercise 2.32

Honey Dosage:	12	11	15	11	10	13	10	4	15	16	9
	14	10	6	10	8	11	12	12	8	12	9
	11	15	10	15	9	13	8	12	10	8	9
	5	12									

DM Dosage:	4	6	9	4	7	7	7	9	12	10	11	6
	3	4	9	12	7	6	8	12	12	4	12	13
	7	10	13	9	4	4	10	15	9			

No Dosage (Control):	5	8	6	1	0	8	12	8	7	7	1	6	7	7
	12	7	9	7	9	5	11	9	5	6	8	8	6	7
	10	9	4	8	7	3	1	4	3					

Based on Paul, I. M., et al. "Effect of honey, dextromethorphan, and no treatment on nocturnal cough and sleep quality for coughing children and their parents." *Archives of Pediatrics and Adolescent Medicine*, Vol. 161, No. 12, Dec. 2007 (data simulated).

c. A MINITAB dot plot for the improvement scores of all three groups is shown above. Note that the green dots represent the children who received a dose of honey, the red dots represent those who got the DM dosage, and the black dots represent the children in the control group. What conclusions can pediatric researchers draw from the graph? Do you agree with the statement (extracted from the article), "Honey may be a preferable treatment for the cough and sleep difficulty associated with childhood upper respiratory tract infection"?

2.33 **Reading Japanese books.** Refer to the *Reading in a Foreign Language* (Apr. 2004) experiment to improve the Japanese reading comprehension levels of University of Hawaii students, Exercise 2.9 (p. 34). Fourteen students participated in a 10-week extensive reading program in a second-semester Japanese language course. The number of books read by each student and the student's grade in the course are listed in the next table and saved in the **JAPANESE** file.

a. Construct a stem-and-leaf display for the number of books read by the students.

b. Highlight (or circle) the leaves in the display that correspond to students who earned an A grade in the course. What inference can you make about these students?

Number of Books	Course Grade
53	A
42	A
40	A
40	B
39	A
34	A
34	A
30	A
28	B
24	A
22	C
21	B
20	B
16	B

Source: Hitosugi, C. I., and Day, R. R. "Extensive reading in Japanese." *Reading in a Foreign Language,* Vol. 16, No. 1, Apr. 2004 (Table 4). Reprinted with permission from the National Foreign Language Resource Center, University of Hawaii.

2.34 **Cheek teeth of extinct primates.** Refer to the *American Journal of Physical Anthropology* (Vol. 142, 2010) study of the characteristics of cheek teeth (e.g., molars) in an extinct primate species, Exercise 2.7 (p. 34). In addition to degree of wear, the researchers recorded the dentary depth of molars (in millimeters) for 18 cheek teeth extracted from skulls. These depth measurements are listed in the accompanying table and saved in the **CHEEKTEETH** file. Summarize the data graphically with a stem-and-leaf display. Is there a particular molar depth that occurs more frequently in the sample?

Data on Dentary Depth (mm) of Molars	
18.12	16.55
19.48	15.70
19.36	17.83
15.94	13.25
15.83	16.12
19.70	18.13
15.76	14.02
17.00	14.04
13.96	16.20

Based on Boyer, D. M., Evans, A. R., and Jernvall, J. "Evidence of dietary differentiation among late Paleocene–Early Eocene Plesiadapids (Mammalia, Primates)." *American Journal of Physical Anthropology*, Vol. 142, © 2010 (Table A3).

2.35 Sanitation inspection of cruise ships. To minimize the potential for gastrointestinal disease outbreaks, all passenger cruise ships arriving at U.S. ports are subject to unannounced sanitation inspections. Ships are rated on a 100-point scale by the Centers for Disease Control and Prevention. A score of 86 or higher indicates that the ship is providing an accepted standard of sanitation. The latest (as of Jan. 2010) sanitation scores for 186 cruise ships are saved in the **SHIPSANIT** file. The first five and last five observations in the data set are listed in the following table:

Ship Name	Sanitation Score
Adventure of the Seas	98
AID Aaura	100
Albatross	69
Amadea	84
Amsterdam	99
⋮	⋮
Westerdam	99
Wind Spirit	97
Wind Surf	95
Zaandam	99
Zuiderdam	95

Based on National Center for Environmental Health, Centers for Disease Control and Prevention, Jan. 6, 2010.

a. Generate a stem-and-leaf display of the data. Identify the stems and leaves of the graph.

b. Use the stem-and-leaf display to estimate the proportion of ships that have an accepted sanitation standard.

c. Locate the inspection score of 69 (*Albatrass*) on the stem-and-leaf display.

Applying the Concepts—Intermediate

2.36 Radioactive lichen. Lichen has a high absorbance capacity for radiation fallout from nuclear accidents. Since lichen is a major food source for Alaskan caribou, and caribou are, in turn, a major food source for many Alaskan villagers, it is important to monitor the level of radioactivity in lichen. Researchers at the University of Alaska, Fairbanks, collected data on nine lichen specimens at various locations for this purpose. The amount of the radioactive element cesium-137 was measured (in microcuries per milliliter) for each specimen. These data are saved in the **LICHEN** file. The data values, converted to logarithms, are given in the following table (note that the closer the value is to zero, the greater is the amount of cesium in the specimen).

Location			
Bethel	−5.50	−5.00	
Eagle Summit	−4.15	−4.85	
Moose Pass	−6.05		
Turnagain Pass	−5.00		
Wickersham Dome	−4.10	−4.50	−4.60

Based on Lichen Radionuclide Baseline Research Project, 2003, p. 25. *Orion,* University of Alaska–Fairbanks.

a. Construct a dot plot for the nine measurements.

b. Construct a stem-and-leaf display for the nine measurements.

c. Construct a histogram plot of the nine measurements.

d. Which of the three graphs in parts **a–c,** respectively, is most informative?

e. What proportion of the measurements has a radioactivity level of −5.00 or lower?

2.37 Sound waves from a basketball. An experiment was conducted to characterize sound waves in a spherical cavity (*American Journal of Physics,* June 2010). A fully inflated basketball, hanging from rubber bands, was struck with a metal rod, producing a series of metallic-sounding pings. Of particular interest were the frequencies of sound waves resulting from the first 24 resonances (echoes). A mathematical formula, well known in physics, was used to compute the theoretical frequencies. The data are saved in the **BBALL** file. These frequencies (measured in hertz) are listed in the table. Use a graphical method to describe the distribution of sound frequencies for the first 24 resonances.

Resonance	Frequency
1	979
2	1572
3	2113
4	2122
5	2659
6	2795
7	3181
8	3431
9	3638
10	3694
11	4038
12	4203
13	4334
14	4631
15	4711
16	4993
17	5130
18	5210
19	5214
20	5633
21	5779
22	5836
23	6259
24	6339

Based on Russell, D. A. "Basketballs as spherical acoustic cavities." *American Journal of Physics,* Vol. 78, No. 6, June 2010 (Table I).

2.38 Crab spiders hiding on flowers. Crab spiders use camouflage to hide on flowers while lying in wait to prey on other insects. Ecologists theorize that this natural camouflage also enables the spiders to hide from their own predators, such as birds and lizards. Researchers at the French Museum of Natural History conducted a field test of this theory and published the results in *Behavioral Ecology* (Jan. 2005). They collected a sample of 10 adult female crab spiders, each sitting on the yellow central part of a daisy. The chromatic contrast between each spider and the flower it was sitting on was measured numerically with a spectroradiometer, on which higher values indicate a greater contrast (and, presumably, easier detection by predators). The data for the 10 crab spiders are shown in the following table and saved in the **SPIDER** file.

57	75	116	37	96	61	56	2	43	32

Based on Thery, M., et al. "Specific color sensitivities of prey and predator explain camouflage in different visual systems." *Behavioral Ecology,* Vol. 16, No. 1, Jan. 2005 (Table 1).

a. Summarize the chromatic contrast measurements for the 10 spiders with a stem-and-leaf display.

b. For birds, the detection threshold is 70. (A contrast of 70 or greater allows the bird to see the spider.) Locate the spiders that can be seen by bird predators by circling their respective contrast values on the stem-and-leaf display.

c. Use the result of part **b** to make an inference about the likelihood of a bird detecting a crab spider sitting on the yellow central part of a daisy.

2.39 **Research on brain specimens.** The *postmortem interval* (PMI) is defined as the time elapsed (in days) between death and an autopsy. Knowledge of the PMI is considered essential to conducting medical research on human cadavers. The data in the accompanying table are the PMIs of 22 human brain specimens obtained at autopsy in a recent study (*Brain and Language*, June 1995). Describe the PMI data graphically with a dot plot. On the basis of the plot, make a summary statement about the PMIs of the 22 human brain specimens. (The data are saved in the **BRAINPMI** file.)

Postmortem Intervals for 22 Human Brain Specimens							
5.5	14.5	6.0	5.5	5.3	5.8	11.0	6.1
7.0	14.5	10.4	4.6	4.3	7.2	10.5	6.5
3.3	7.0	4.1	6.2	10.4	4.9		

Based on Hayes, T. L., and Lewis, D. A. "Anatomical specialization of the anterior motor speech area: Hemispheric differences in magnopyramidal neurons." *Brain and Language,* Vol. 49, No. 3, June 1995, p. 292 (Table 1).

2.40 **Research on eating disorders.** Data from a psychology experiment were reported and analyzed in *The American Statistician* (May 2001). Two samples of female students participated in the experiment. One sample consisted of 11 students known to suffer from the eating disorder bulimia; the other sample consisted of 14 students with normal eating habits. Each student completed a questionnaire from which a "fear of negative evaluation" (FNE) score was produced. (The higher the score, the greater was the fear of negative evaluation.) The data are saved in the **BULIMIA** file and displayed in the following table:

Bulimic
students: 21 13 10 20 25 19 16 21 24 13 14

Normal
students: 13 6 16 13 8 19 23 18 11 19 7 10 15 20

Based on Randles, R. H. "On neutral responses (zeros) in the sign test and ties in the Wilcoxon–Mann–Whitney test." *The American Statistician,* Vol. 55, No. 2, May 2001 (Figure 3).

a. Construct a dot plot or stem-and-leaf display for the FNE scores of all 25 female students.

b. Highlight the bulimic students on the graph you made in part **a**. Does it appear that bulimics tend to have a greater fear of negative evaluation? Explain.

c. Why is it important to attach a measure of reliability to the inference made in part **b**?

Applying the Concepts—Advanced

2.41 **College protests of labor exploitation.** The United Students Against Sweatshops (USAS) was formed by students on U.S. and Canadian college campuses in 1999 to protest labor exploitation in the apparel industry. Clark University sociologist Robert Ross analyzed the USAS movement in the *Journal of World-Systems Research* (Winter 2004). Between 1999 and 2000, there were 18 student "sit-ins" for a "sweat-free campus" organized at several universities. These data are saved in the **SITIN** file. The following table gives the duration (in days) of each sit-in, as well as the number of student arrests. Do the data support the theory that sit-ins of longer duratian are more likely to lead to arrests? Support your answer with a graph.

Sit-in	Year	University	Duration (days)	Number of Arrests	Tier Ranking
1	1999	Duke	1	0	1st
2	1999	Georgetown	4	0	1st
3	1999	Wisconsin	1	0	1st
4	1999	Michigan	1	0	1st
5	1999	Fairfield	1	0	1st
6	1999	North Carolina	1	0	1st
7	1999	Arizona	10	0	1st
8	2000	Toronto	11	0	1st
9	2000	Pennsylvania	9	0	1st
10	2000	Macalester	2	0	1st
11	2000	Michigan	3	0	1st
12	2000	Wisconsin	4	54	1st
13	2000	Tulane	12	0	1st
14	2000	SUNY Albany	1	11	2nd
15	2000	Oregon	3	14	2nd
16	2000	Purdue	12	0	2nd
17	2000	Iowa	4	16	2nd
18	2000	Kentucky	1	12	2nd

Based on Ross, R. J. S. "From antisweatshop to global justice to antiwar: How the new new left is the same and different from the old new left." *Journal of World-Systems Research,* Vol. X, No. 1, Winter 2004 (Table 1 and 3).

2.42 **Comparing SAT scores.** Educators are constantly evaluating the efficacy of public schools in the education and training of U.S. students. One quantitative assessment of change over time is the difference in scores on the SAT, which has been used for decades by colleges and universities as one criterion for admission. The **SATSCORES** file contains the average SAT scores for each of the 50 states and the District of Columbia for 2009 and 2005. Selected observations are shown in the following table:

State	2009	2005
Alabama	1,109	1,126
Alaska	1,036	1,042
Arizona	1,037	1,056
Arkansas	1,144	1,115
California	1,013	1,026
⋮	⋮	⋮
Wisconsin	1,202	1,191
Wyoming	1,135	1,087

Based on College Entrance Examination Board, 2010.

a. Use graphs to display the two SAT score distributions. How have the distributions of state scores changed from 1990 to 2005?

b. As another method of comparing the 1990 and 2005 average SAT scores, compute the **paired difference** by subtracting the 1990 score from the 2005 score for each state. Summarize these differences with a graph.

c. Interpret the graph you made in part **b**. How do your conclusions compare with those of part **a**?

d. Identify the state with the largest improvement in the SAT score between 2005 and 2009.

2.43 **Phishing attacks to e-mail accounts.** *Phishing* is the term used to describe an attempt to extract personal/financial information (e.g., PIN numbers, credit card information, bank account numbers) from unsuspecting people through fraudulent e-mail. An article in *Chance* (Summer 2007) demonstrates how statistics can help identify phishing attempts and make e-commerce safer. Data from an actual phishing attack against an organization were used to determine whether the attack may have been an "inside job" that originated within the company. The company set up a publicized e-mail account—called a "fraud box"—which enabled employees to notify it if they suspected an e-mail phishing attack. The interarrival times, i.e., the time differences (in seconds), for 267 fraud box e-mail notifications were recorded. Researchers showed that if there is minimal or no collaboration or collusion from within the company, the interarrival times would have a frequency distribution similar to the one shown in the accompanying figure. The 267 interarrival times are saved in the **PHISHING** file. Construct a frequency histogram for the interarrival times. Give your opinion on whether the phishing attack against the organization was an "inside job."

2.3 Summation Notation

Now that we've examined some graphical techniques for summarizing and describing quantitative data sets, we turn to numerical methods for accomplishing that objective. Before giving the formulas for calculating numerical descriptive measures, let's look at some shorthand notation that will simplify our calculation instructions. Remember that such notation is used for one reason only: to avoid repeating the same verbal descriptions over and over. If you mentally substitute the verbal definition of a symbol each time you read it, you'll soon get used to the symbol.

We denote the measurements of a quantitative data set as

$$x_1, x_2, x_3, \ldots, x_n$$

where x_1 is the first measurement in the data set, x_2 is the second measurement in the data set, x_3 is the third measurement in the data set, $\ldots$, and x_n is the nth (and last) measurement in the data set. Thus, if we have five measurements in a set of data, we will write x_1, x_2, x_3, x_4, x_5 to represent the measurements. If the actual numbers are 5, 3, 8, 5, and 4, we have $x_1 = 5, x_2 = 3, x_3 = 8, x_4 = 5$, and $x_5 = 4$.

Most of the formulas we use require a summation of numbers. For example, one sum we'll need to obtain is the sum of all the measurements in the data set, or $x_1 + x_2 + x_3 + \cdots + x_n$. To shorten the notation, we use the symbol $\sum$ for the summation. That is, $x_1 + x_2 + x_3 + \cdots + x_n = \sum_{i=1}^{n} x_i$. Verbally translate $\sum_{i=1}^{n} x_i$ as follows:

"The sum of the measurements whose typical member is x_i, beginning with the member x_1 and ending with the member x_n."

Suppose, as in our earlier example, $x_1 = 5, x_2 = 3, x_3 = 8, x_4 = 5,$ *and* $x_5 = 4$.

Then the sum of the five measurements, denoted $\sum_{i=1}^{5} x_i$, is obtained as follows:

$$\sum_{i=1}^{5} x_i = x_1 + x_2 + x_3 + x_4 + x_5$$

$$= 5 + 3 + 8 + 5 + 4 = 25$$

Another important calculation requires that we square each measurement and then sum the squares. The notation for this sum is $\sum_{i=1}^{n} x_i^2$. For the five measurements, we have

$$\sum_{i=1}^{5} x_i^2 = x_1^2 + x_2^2 + x_3^2 + x_4^2 + x_5^2$$

$$= 5^2 + 3^2 + 8^2 + 5^2 + 4^2$$

$$= 25 + 9 + 64 + 25 + 16 = 139$$

In general, the symbol following the summation sign $\sum$ represents the variable (or function of the variable) that is to be summed.

The Meaning of Summation Notation $\sum_{i=1}^{n} x_i$

Sum the measurements of the variable that appears to the right of the summation symbol, beginning with the first measurement and ending with the nth measurement.

Exercises 2.44–2.47

Learning the Mechanics

Note: In all exercises, $\sum$ represents $\sum_{i=1}^{n}$.

2.44 A data set contains the observations 5, 1, 3, 2, 1. Find

a. $\sum x$ **b.** $\sum x^2$ **c.** $\sum (x-1)$

d. $(\sum x - 1)^2$ **e.** $(\sum x)^2$

2.45 Suppose a data set contains the observations 3, 8, 4, 5, 3, 4, 6. Find

a. $\sum x$ **b.** $\sum x^2$ **c.** $\sum (x-5)^2$

d. $\sum (x-2)^2$ **e.** $(\sum x)^2$

2.46 Refer to Exercise 2.44. Find

a. $\sum x^2 - \dfrac{(\sum x)^2}{5}$ **b.** $\sum (x-2)^2$ **c.** $\sum x^2 - 10$

2.47 A data set contains the observations 6, 0, −2, −1, 3. Find

a. $\sum x$ **b.** $\sum x^2$ **c.** $\sum x^2 - \dfrac{(\sum x)^2}{5}$

2.4 Numerical Measures of Central Tendency

When we speak of a data set, we refer to either a sample or a population. If statistical inference is our goal, we'll ultimately wish to use sample **numerical descriptive measures** to make inferences about the corresponding measures for a population.

As you'll see, a large number of numerical methods are available to describe quantitative data sets. Most of these methods measure one of two data characteristics:

1. The **central tendency** of the set of measurements—that is, the tendency of the data to cluster, or center, about certain numerical values. (See Figure 2.14a.)

2. The **variability** of the set of measurements—that is, the spread of the data. (See Figure 2.14b.)

In this section, we concentrate on **measures of central tendency.** In the next section, we discuss measures of variability.

The most popular and best understood measure of central tendency for a quantitative data set is the *arithmetic mean* (or simply the mean) of the data set.

> The **mean** of a set of quantitative data is the sum of the measurements, divided by the number of measurements contained in the data set.

In everyday terms, the mean is the average value of the data set and is often used to represent a "typical" value. We denote the **mean of a sample** of measurements by $\bar{x}$ (read "x-bar"), and represent the formula for its calculation as shown in the following box:

Figure 2.14
Numerical descriptive measures

Formula for a Sample Mean

$$\bar{x} = \frac{\sum_{i=1}^{n} x_i}{n}$$

Example 2.3

Computing the Sample Mean

Problem Calculate the mean of the following five sample measurements: 5, 3, 8, 5, 6.

Solution Using the definition of sample mean and the summation notation, we find that

$$\bar{x} = \frac{\sum_{i=1}^{5} x_i}{5} = \frac{5 + 3 + 8 + 5 + 6}{5} = \frac{27}{5} = 5.4$$

Thus, the mean of this sample is 5.4.

Look Back There is no specific rule for rounding when calculating $\bar{x}$, because $\bar{x}$ is specifically defined to be the sum of all measurements, divided by n; that is, it is a specific fraction. When $\bar{x}$ is used for descriptive purposes, it is often convenient to round the calculated value of $\bar{x}$ to the number of significant figures used for the original measurements. When $\bar{x}$ is to be used in other calculations, however, it may be necessary to retain more significant figures.

Now Work Exercise 2.57

Example 2.4

Finding the Mean on a Printout— Mean Gas Mileage

Problem Calculate the sample mean for the 100 EPA mileages given in Table 2.2.

Solution The mean gas mileage for the 100 cars is denoted

$$\bar{x} = \frac{\sum_{i=1}^{100} x_i}{100}$$

Rather than compute $\bar{x}$ by hand (or even with a calculator), we employed SAS to compute the mean. The SAS printout is shown in Figure 2.15. The sample mean, highlighted on the printout, is $\bar{x} = 36.9940$.

Figure 2.15

SAS numerical descriptive measures for 100 EPA gas mileages

The MEANS Procedure

Analysis Variable : MPG

Mean	Std Dev	Variance	N	Minimum	Maximum	Median
36.9940000	2.4178971	5.8462263	100	30.0000000	44.9000000	37.0000000

Look Back Given this information, you can visualize a distribution of gas mileage readings centered in the vicinity of $\bar{x} \approx 37$. An examination of the relative frequency histogram (Figure 2.10) confirms that $\bar{x}$ does in fact fall near the center of the distribution.

The sample mean $\bar{x}$ will play an important role in accomplishing our objective of making inferences about populations on the basis of information about the sample. For this reason, we need to use a different symbol for the *mean of a population*—the mean of the set of measurements on every unit in the population. We use the Greek letter μ (mu) for the population mean.

Symbols for the Sample Mean and the Population Mean

In this text, we adopt a general policy of using Greek letters to represent numerical descriptive measures of the population and Roman letters to represent corresponding descriptive measures of the sample. The symbols for the mean are

$$\bar{x} = \text{Sample mean} \qquad \mu = \text{Population mean}$$

We'll often use the sample mean $\bar{x}$ to estimate (make an inference about) the population mean μ. For example, the EPA mileages for the population consisting of *all* cars has a mean equal to some value μ. Our sample of 100 cars yielded mileages with a mean of $\bar{x} = 36.9940$. If, as is usually the case, we don't have access to the measurements for the entire population, we could use $\bar{x}$ as an estimator or approximator for μ. Then we'd need to know something about the reliability of our inference. That is, we'd need to know how accurately we might expect $\bar{x}$ to estimate μ. In Chapter 5, we'll find that this accuracy depends on two factors:

1. The *size of the sample*. The larger the sample, the more accurate the estimate will tend to be.

2. The *variability, or spread, of the data*. All other factors remaining constant, the more variable the data, the less accurate is the estimate.

Another important measure of central tendency is the **median.**

The **median** of a quantitative data set is the middle number when the measurements are arranged in ascending (or descending) order.

The median is of most value in describing large data sets. If a data set is characterized by a relative frequency histogram (Figure 2.16), the median is the point on the x-axis such that half the area under the histogram lies above the median and half lies below. [*Note*: In Section 2.2, we observed that the relative frequency associated with a particular interval on the x-axis is proportional to the amount of area under the histogram that lies above the interval.] We denote the *median* of a *sample* by M.

Figure 2.16
Location of the median

Calculating a Sample Median M

Arrange the n measurements from the smallest to the largest.

1. If n is odd, M is the middle number.

2. If n is even, M is the mean of the middle two numbers.

Example 2.5

Computing the Median

Problem Consider the following sample of $n = 7$ measurements: 5, 7, 4, 5, 20, 6, 2.

a. Calculate the median M of this sample.

b. Eliminate the last measurement (the 2), and calculate the median of the remaining $n = 6$ measurements.

Solution

a. The seven measurements in the sample are ranked in ascending order: 2, 4, 5, 5, 6, 7, 20. Because the number of measurements is odd, the median is the middle measurement. Thus, the median of this sample is $M = 5$.

b. After removing the 2 from the set of measurements, we rank the sample measurements in ascending order as follows: 4, 5, 5, 6, 7, 20. Now the number of measurements is even, so we average the middle two measurements. The median is $M = (5 + 6)/2 = 5.5$.

Look Back When the sample size n is even (as in part **b**), exactly half of the measurements will fall below the calculated median M. However, when n is odd (as in part **a**), the percentage of measurements that fall below M is approximately 50%. The approximation improves as n increases.

Now Work Exercise 2.54

In certain situations, the median may be a better measure of central tendency than the mean. In particular, the median is less sensitive than the mean to extremely large or small measurements. Note, for instance, that all but one of the measurements in part **a** of Example 2.5 are close to $x = 5$. The single relatively large measurement, $x = 20$, does not affect the value of the median, 5, but it causes the mean, $\bar{x} = 7$, to lie to the right of most of the measurements.

As another example of data for which the central tendency is better described by the median than the mean, consider the household incomes of a community being studied by a sociologist. The presence of just a few households with very high incomes will affect the mean more than the median. Thus, the median will provide a more accurate picture of the typical income for the community. The mean could exceed the vast majority of the sample measurements (household incomes), making it a misleading measure of central tendency.

Example 2.6
Finding the Median on a Printout—Median Gas Mileage

Problem Calculate the median for the 100 EPA mileages given in Table 2.2. Compare the median with the mean computed in Example 2.4.

Solution For this large data set, we again resort to a computer analysis. The median is highlighted on the SAS printout displayed in Figure 2.15 (p. 51). You can see that the median is 37.0. Thus, half of the 100 mileages in the data set fall below 37.0 and half lie above 37.0. Note that the median, 37.0, and the mean, 36.9940, are almost equal, a relationship that indicates a lack of **skewness** in the data. In other words, the data exhibit a tendency to have as many measurements in the left tail of the distribution as in the right tail. (Recall the histogram of Figure 2.10.)

Look Back In general, extreme values (large or small) affect the mean more than the median, since these values are used explicitly in the calculation of the mean. The median is not affected directly by extreme measurements, since only the middle measurement (or two middle measurements) is explicitly used to calculate the median. Consequently, if measurements are pulled toward one end of the distribution, the mean will shift toward that tail more than the median will.

A data set is said to be **skewed** if one tail of the distribution has more extreme observations than the other tail.

A comparison of the mean and the median gives us a general method for detecting skewness in data sets, as shown in the next box. With *rightward skewed* data, the right tail (high end) of the distribution has more extreme observations. These few, but large, measurements pull the mean away from the median towards the right; that is, rightward skewness indicates that the mean is greater than the median. Conversely, with *leftward skewed* data, the left tail (low end) of the distribution has more extreme observations. These few, but small, measurements also pull the mean away from the median, but towards the left; consequently, leftward skewness implies that the mean is smaller than the median.

Detecting Skewness by Comparing the Mean and the Median

If the data set is skewed to the right, then the median is less than the mean.

If the data set is symmetric, then the mean equals the median.

If the data set is skewed to the left, then the mean is less than the median.

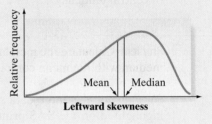

Now Work Exercise 2.53

A third measure of central tendency is the **mode** of a set of measurements.

> The **mode** is the measurement that occurs most frequently in the data set.

Therefore, the mode shows where the data tend to concentrate.

Example 2.7
Finding the Mode

Problem Each of 10 taste testers rated a new brand of barbecue sauce on a 10-point scale, where 1 = awful and 10 = excellent. Find the mode for the following 10 ratings:

$$8 \quad 7 \quad 9 \quad 6 \quad 8 \quad 10 \quad 9 \quad 9 \quad 5 \quad 7$$

Solution Since 9 occurs most often (three times), the mode of the ten taste ratings is 9.

Look Back Note that the data are actually qualitative in nature (e.g., "awful," "excellent"). The mode is particularly useful for describing qualitative data. The modal category is simply the category (or class) that occurs most often.

Now Work Exercise 2.55

Because it emphasizes data concentration, the mode is also used with quantitative data sets to locate the region in which much of the data is concentrated. A retailer of men's clothing would be interested in the modal neck size and sleeve length of potential customers. The modal income class of the laborers in the United States is of interest to the U.S. Department of Labor.

For some quantitative data sets, the mode may not be very meaningful. For example, consider the EPA mileage ratings in Table 2.2. A reexamination of the data reveals that the gas mileage of 37.0 occurs most often (four times). However, the mode of 37.0 is not particularly useful as a measure of central tendency.

A more meaningful measure can be obtained from a relative frequency histogram for quantitative data. The measurement class containing the largest relative frequency is called the **modal class.** Several definitions exist for locating the position of the mode within a modal class, but the simplest is to define the mode as the midpoint of the modal class. For example, examine the frequency histogram for the EPA mileage ratings in Figure 2.10 (p. 40). You can see that the modal class is the interval 37–38. The mode (the midpoint) is 37.5 This modal class (and the mode itself) identifies the area in which the data are most concentrated and, in that sense, is a measure of central tendency. However, for most applications involving quantitative data, the mean and median provide more descriptive information than the mode.

Example 2.8
Comparing the Mean, Median, and Mode—Earthquake Aftershocks

Problem Seismologists use the term "aftershock" to describe the smaller earthquakes that follow a main earthquake. Following the Northridge earthquake in 1994, the Los Angeles area experienced a record 2,929 aftershocks in a three-week period. The magnitudes (measured on the Richter scale) of these aftershocks as well as their interarrival times (in minutes) were recorded by the U.S. Geological Survey. (The data are saved in the **EARTHQUAKE** file.) Today seismologists continue to use these data to model future earthquake characteristics. Find and interpret the mean, median, and mode for both of these variables. Which measure of central tendency is better for describing the magnitude distribution? The distribution of interarrival times?

Solution Measures of central tendency for the two variables, magnitude and interarrival time, were produced using MINITAB. The means, medians, and modes are displayed in Figure 2.17.

Figure 2.17
MINITAB descriptive statistics for earthquake data

Descriptive Statistics: MAGNITUDE, INT-TIME

Variable	N	Mean	Median	Mode	N for Mode
MAGNITUDE	2929	2.1197	2.0000	1.8	298
INT-TIME	2928	9.771	6.000	2	354

For magnitude, the mean, median, and mode are 2.12, 2.00, and 1.8, respectively, on the Richter scale. The average magnitude is 2.12; half the magnitudes fall below 2.0; and the most commonly occurring magnitude is 1.8. These values are nearly identical, with the mean slightly larger than the median. This implies a slight rightward skewness in the data, which is shown graphically in the MINITAB histogram for magnitude displayed in Figure 2.18a. Because the distribution is nearly symmetric, any of the three measures would be adequate for describing the "center" of the earthquake aftershock magnitude distribution.

The mean, median, and mode of the interarrival times of the aftershocks are 9.77, 6.0, and 2.0 minutes, respectively. On average, the aftershocks arrive 9.77 minutes apart; half the aftershocks have interarrival times below 6.0 minutes; and the most commonly occurring interarrival time is 2.0 minutes. Note that the mean is much larger than the median, implying that the distribution of interarrival times is highly skewed to the right. This extreme rightward skewness is shown graphically in the histogram, in Figure 2.18b. The skewness is due to several exceptionally large interarrival times. Consequently, we would probably want to use the median of 6.0 minutes as the "typical" interarrival time for the aftershocks. You can see that the mode of 2.0 minutes is not very descriptive of the "center" of the interarrival time distribution.

Figure 2.18a
MINITAB Histogram for
Magnitudes of Aftershocks

Figure 2.18b
MINITAB Histogram for
Inter-Arrival Times of
Aftershocks

Look Back The choice of which measure of central tendency to use will depend on the properties of the data set analyzed and the application of interest. Consequently, it is vital that you understand how the mean, median, and mode are computed.

Now Work Exercise 2.68

Exercises 2.48–2.70

Understanding the Principles

2.48 Explain the difference between a measure of central tendency and a measure of variability.

2.49 Give three different measures of central tendency.

2.50 What is the symbol used to represent the sample mean? The population mean?

2.51 What two factors affect the accuracy of the sample mean as an estimate of the population mean?

2.52 Explain the concept of a skewed distribution.

2.53 Describe how the mean compares with the median for a
[NW] distribution as follows:
 a. Skewed to the left
 b. Skewed to the right
 c. Symmetric

Learning the Mechanics

2.54 Calculate the mean and median of the following grade
[NW] point averages:

3.2   2.5   2.1   3.7   2.8   2.0

2.55 Calculate the mode, mean, and median of the following data:
[NW]

    18   10   15   13   17   15   12   15   18   16   11

2.56 Construct one data set consisting of five measurements, and another consisting of six measurements, for which the medians are equal.

2.57 Calculate the mean for samples for which
[NW]
   a. $n = 10, \sum x = 85$

   b. $n = 16, \sum x = 400$

   c. $n = 45, \sum x = 35$

   d. $n = 18, \sum x = 242$

2.58 Calculate the mean, median, and mode for each of the following samples:
   a. 7, −2, 3, 3, 0, 4
   b. 2, 3, 5, 3, 2, 3, 4, 3, 5, 1, 2, 3, 4
   c. 51, 50, 47, 50, 48, 41, 59, 68, 45, 37

Applet Exercise 2.1

Use the applet entitled *Mean versus Median* to find the mean and median of each of the three data sets presented in Exercise 2.58. For each data set, set the lower limit to a number less than all of the data, set the upper limit to a number greater than all of the data, and then click on *Update*. Click on the approximate location of each data item on the number line. You can get rid of a point by dragging it to the trash can. To clear the graph between data sets, simply click on the trash can.

   a. Compare the means and medians generated by the applet with those you calculated by hand in Exercise 2.58. If there are differences, explain why the applet might give values slightly different from the hand-calculated values.
   b. Despite providing only approximate values of the mean and median of a data set, describe some advantages of using the applet to find those values.

Applet Exercise 2.2

Use the applet *Mean versus Median* to illustrate your descriptions in Exercise 2.53. For each part **a, b,** and **c,** create a data set with 10 items that has the given property. Using the applet, verify that the mean and median have the relationship you described in Exercise 2.53.

Applet Exercise 2.3

Use the applet *Mean versus Median* to study the effect that an extreme value has on the difference between the mean and median. Begin by setting appropriate limits and plotting the following data on the number line provided in the applet:

      0 6 7 7 8 8 8 9 9 10

   a. Describe the shape of the distribution and record the value of the mean and median. On the basis of the shape of the distribution, do the mean and median have the relationship that you would expect?
   b. Replace the extreme value of 0 with 2, then 4, and then 6. Record the mean and median each time. Describe what is happening to the mean as 0 is replaced, in turn, by the higher numbers stated. What is happening to the

median? How is the difference between the mean and the median changing?
   c. Now replace 0 with 8. What values does the applet give you for the mean and the median? Explain why the mean and the median should now be the same.

Applying the Concepts—Basic

2.59 **Characteristics of a rock fall.** In *Environmental Geology* (Vol. 58, 2009) computer simulation was employed to estimate how far a block from a collapsing rock wall will bounce—called *rebound length*—down a soil slope. Based on the depth, location, and angle of block-soil impact marks left on the slope from an actual rockfall, the following 13 rebound lengths (in meters) were estimated. The data are saved in the **ROCKFALL** file. Compute the mean and median of the rebound lengths and interpret these values.

10.94 13.71 11.38 7.26 17.83 11.92 11.87 5.44 13.35 4.90 5.85 5.10 6.77

Based on Paronuzzi, P. "Rockfall-induced block propagation on a soil slope, northern Italy," *Environmental Geology*, Vol. 58, 2009 (Table 2).

2.60 **Most powerful businesswomen in America.** *Fortune* (Oct. 16, 2008) published a list of the 50 most powerful women in business in the United States. The data on age (in years) and title of each of these 50 women are stored in the **WPOWER50** file. The first five and last five observations of the data set are listed in the table below. Descriptive statistics for the ages are displayed on the MINITAB printouts (p. 58).

Rank	Name	Age	Company	Title
1	Indra Nooyi	52	PepsiCo	CEO/Chairman
2	Irene Rosenfeld	55	Kraft Foods	CEO/Chairman
3	Pat Woertz	55	Archer Daniels Midland	CEO/Chairman
4	Anne Mulcahy	55	Xerox	CEO/Chairman
5	Angela Braley	47	Wellpoint	CEO/President
⋮	⋮	⋮	⋮	⋮
46	Lorrie Norrington	48	eBay	CEO
47	Terri Dial	58	Citigroup	CEO
48	Lynn Elsenhans	52	Sunoco	CEO/President
49	Cathie Black	64	Hearst Magazines	President
50	Marissa Mayer	33	Google	Vice president

Based on "50 most powerful businesswomen," *Fortune,* October 16, 2008.

   a. Find the mean, median, and modal age of these 50 women on the printout. Interpret these values.
   b. What do the mean and median indicate about the skewness of the age distribution?
   c. Find the modal age class on the histogram

2.61 **Reading Japanese books.** Refer to the *Reading in a Foreign Language* (Apr. 2004) experiment to improve the Japanese reading comprehension levels of 14 University of Hawaii students, Exercise 2.33 (p. 46). The data are saved in the **JAPANESE** file. The number of books read by each student and the student's course grade are repeated in the table on the next page.

MINITAB Output for Exercise 2.60

Descriptive Statistics: AGE

Variable	N	Mean	StDev	Minimum	Median	Maximum	Mode	N for Mode
AGE	50	50.020	6.444	28.000	51.000	64.000	54	7

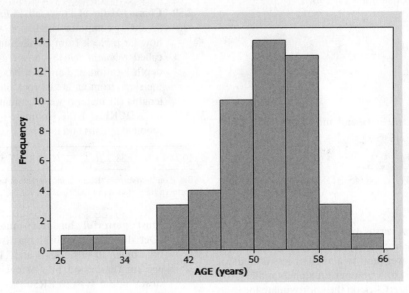

Data for Exercise 2.61

Number of Books	Course Grade	Number of Books	Course Grade
53	A	30	A
42	A	28	B
40	A	24	A
40	B	22	C
39	A	21	B
34	A	20	B
34	A	16	B

Source: Hitosugi, C. I., and Day, R. R., "Extensive Reading in Japanese." *Reading in a Foreign Language,* Vol. 16, No. 1, Apr. 2004 (Table 4). Reprinted with permission from the National Foreign Language Resource Center, University of Hawaii.

a. Find the mean, median, and mode of the number of books read. Interpret these values.

b. What do the mean and median indicate about the skewness of the distribution of the data?

2.62 **Is honey a cough remedy?** Refer to the *Archives of Pediatrics and Adolescent Medicine* (Dec. 2007) study of honey as a remedy for coughing, Exercise 2.32 (p. 45). Recall that the 105 ill children in the sample were randomly divided into three groups: those who received a dosage of an over-the-counter cough medicine (DM), those who received a dosage of honey (H), and those who received no dosage (control group). The coughing improvement scores (saved in the **HONEYCOUGH** file) for the patients are reproduced in the next table.

a. Find the median improvement score for the honey dosage group.

b. Find the median improvement score for the DM dosage group.

c. Find the median improvement score for the control group.

d. Based on the results, parts **a–c,** what conclusions can pediatric researchers draw? (We show how to support

these conclusions with a measure of reliability in subsequent chapters.)

Honey Dosage:	12	11	15	11	10	13	10	4	15	16	9
	14	10	6	10	8	11	12	12	8	12	9
	11	15	10	15	9	13	8	12	10	8	9
	5	12									
DM Dosage:	4	6	9	4	7	7	7	9	12	10	11
	6	3	4	9	12	7	6	8	12	12	4
	12	13	7	10	13	9	4	4	10	15	9
No Dosage (Control):	5	8	6	1	0	8	12	8	7	7	1
	6	7	7	12	7	9	7	9	5	11	9
	5	6	8	8	6	7	10	9	4	8	7
	3	1	4	3							

Based on Paul, I. M., et al. "Effect of honey, dextromethorphan, and no treatment on nocturnal cough and sleep quality for coughing children and their parents." *Archives of Pediatrics and Adolescent Medicine,* Vol. 161, No. 12, Dec. 2007 (data simulated).

2.63 **Radioactive lichen.** Refer to the University of Alaska study to monitor the level of radioactivity in lichen, presented in Exercise 2.36 (p. 47). The amount of the radioactive element cesium-137 (measured in microcuries per milliliter) for each of nine lichen specimens is repeated in the following table and saved in the **LICHEN** file:

Location			
Bethel	− 5.50	− 5.00	
Eagle Summit	− 4.15	− 4.85	
Moose Pass	− 6.05		
Turnagain Pass	− 5.00		
Wickersham Dome	− 4.10	− 4.50	− 4.60

Based on Lichen Radionuclide Baseline Research Project, 2003, p. 25. *Orion,* University of Alaska-Fairbanks.

a. Find the mean, median, and mode of the radioactivity levels.

b. Interpret the value of each measure of central tendency calculated in part **a.**

Applying the Concepts—Intermediate

2.64 **Ranking driving performance of professional golfers.** A group of Northeastern University researchers developed a new method for ranking the total driving performance of golfers on the Professional Golf Association (PGA) tour (*The Sport Journal*, Winter 2007). The method requires knowing a golfer's average driving distance (yards) and driving accuracy (percent of drives that land in the fairway). The values of these two variables are used to compute a driving performance index. Data for the top 40 PGA golfers (ranked by the new method) are saved in the **PGADRIVER** file. The first five and last five observations are listed in the accompanying table.

Rank	Player	Driving Distance (yards)	Driving Accuracy (%)	Driving Performance Index
1	Woods	316.1	54.6	3.58
2	Perry	304.7	63.4	3.48
3	Gutschewski	310.5	57.9	3.27
4	Wetterich	311.7	56.6	3.18
5	Hearn	295.2	68.5	2.82
⋮	⋮	⋮	⋮	⋮
36	Senden	291	66	1.31
37	Mickelson	300	58.7	1.30
38	Watney	298.9	59.4	1.26
39	Trahan	295.8	61.8	1.23
40	Pappas	309.4	50.6	1.17

Based on Wiseman, F., Habibullah, M., and Yilmaz, M. "Ranking driving performance on the PGA Tour," *Sports Journal*, Vol. 10, No. 1, Winter 2007 (Table 2).

a. Find the mean, median, and mode for the 40 driving performance index values.

b. Interpret each of the measures of central tendency calculated in part **a**.

c. Use the results from part **a** to make a statement about the type of skewness in the distribution of driving performance indexes. Support your statement with a graph.

2.65 **Training zoo animals.** "The Training Game" is an activity used in psychology in which one person shapes an arbitrary behavior by selectively reinforcing the movements of another person. A group of 15 psychology students at the Georgia Institute of Technology played the game at Zoo Atlanta while participating in an experimental psychology laboratory in which they assisted in the training of animals (*Teaching of Psychology*, May 1998). At the end of the session, each student was asked to rate the statement: "'The Training Game is a great way for students to understand the animal's perspective during training." Responses were recorded on a 7-point scale ranging from 1 (strongly disagree) to 7 (strongly agree). The 15 responses were summarized as follows:

$$\text{mean} = 5.87, \text{mode} = 6.$$

a. Interpret the measures of central tendency in the words of the problem.

b. What type of skewness (if any) is likely to be present in the distribution of student responses? Explain.

2.66 **Symmetric or skewed?** Would you expect the data sets that follow to possess relative frequency distributions that are symmetric, skewed to the right, or skewed to the left? Explain.

a. The salaries of all persons employed by a large university

b. The grades on an easy test

c. The grades on a difficult test

d. The amounts of time students in your class studied last week

e. The ages of automobiles on a used-car lot

f. The amounts of time spent by students on a difficult examination (maximum time is 50 minutes)

2.67 **Cheek teeth of extinct primates.** Refer to the *American Journal of Physical Anthropology* (Vol. 142, 2010) study of the characteristics of cheek teeth (e.g., molars) in an extinct primate species, Exercise 2.34 (p. 46). The data on dentary depth of molars (in millimeters) for 18 cheek teeth extracted from skulls are reproduced below and saved in the **CHEEKTEETH** file.

Data on Dentary Depth (mm) of Molars	
18.12	16.55
19.48	15.70
19.36	17.83
15.94	13.25
15.83	16.12
19.70	18.13
15.76	14.02
17.00	14.04
13.96	16.20

Based on Boyer, D. M., Evans, A. R., and Jernvall, J. "Evidence of dietary differentiation among Late Paleocene–Early Eocene Plesiadapids (Mammalia, Primates)." *American Journal of Physical Anthropology*, Vol. 142, © 2010 (Table A3).

a. Find and interpret the mean of the data set. If the largest depth measurement in the sample were doubled, how would the mean change? Would it increase or decrease?

b. Find and interpret the median of the data set. If the largest depth measurement in the sample were doubled, how would the median change? Would it increase or decrease?

c. Note that there is no single measurement that occurs more than once. How does this fact impact the mode?

2.68 **Mongolian desert ants.** The *Journal of Biogeography* (Dec. 2003) published an article on the first comprehensive study of ants in Mongolia (Central Asia). Botanists placed seed baits at 11 study sites and observed the ant species attracted to each site. Some of the data recorded at each study site are provided in the table on the next page and saved in the **GOBIANTS** file.

a. Find the mean, median, and mode for the number of ant species discovered at the 11 sites. Interpret each of these values.

b. Which measure of central tendency would you recommend to describe the center of the number-of-ant-species distribution? Explain.

c. Find the mean, median, and mode for the percentage of total plant cover at the five Dry Steppe sites only.

d. Find the mean, median, and mode for the percentage of total plant cover at the six Gobi Desert sites only.

Data for Exercise 2.68

Site	Region	Annual Rainfall (mm)	Max. Daily Temp. (°C)	Total Plant Cover (%)	Number of Ant Species	Species Diversity Index
1	Dry Steppe	196	5.7	40	3	.89
2	Dry Steppe	196	5.7	52	3	.83
3	Dry Steppe	179	7.0	40	52	1.31
4	Dry Steppe	197	8.0	43	7	1.48
5	Dry Steppe	149	8.5	27	5	.97
6	Gobi Desert	112	10.7	30	49	.46
7	Gobi Desert	125	11.4	16	5	1.23
8	Gobi Desert	99	10.9	30	4	
9	Gobi Desert	125	11.4	56	4	.76
10	Gobi Desert	84	11.4	22	5	1.26
11	Gobi Desert	115	11.4	14	4	.69

Based on Pfeiffer, M., et al. "Community organization and species richness of ants in Mongolia along an ecological gradient from steppe to Gobi desert." *Journal of Biogeography,* Vol. 30, No. 12, Dec. 2003 (Table 1 and 2).

e. On the basis of the results of parts **c** and **d,** does the center of the distribution for total plant cover percentage appear to be different at the two regions?

Applying the Concepts—Advanced

2.69 **Eye refractive study.** The conventional method of measuring the refractive status of an eye involves three quantities: (1) sphere power, (2) cylinder power, and (3) axis. Optometric researchers studied the variation in these three measures of refraction (*Optometry and Vision Science,* June 1995). Twenty-five successive refractive measurements were obtained on the eyes of over 100 university students. The cylinder power measurements for the left eye of one particular student (ID #11) are listed in the table and saved in the **LEFTEYE** file. [*Note:* All measurements are negative values.]

.08	.08	1.07	.09	.16	.04	.07	.17	.11
.06	.12	.17	.20	.12	.17	.09	.07	.16
.15	.16	.09	.06	.10	.21	.06		

Based on Rubin, A., and Harris, W. F. "Refractive variation during autorefraction: Multivariate distribution of refractive status." *Optometry and Vision Science: Official Publication of the American Academy of Optometry,* Vol. 72, No. 6, June 1995, p. 409 (Table 4).

a. Find measures of central tendency for the data and interpret their values.
b. Note that the data contain one unusually large (negative) cylinder power measurement relative to the other measurements in the data set. Find this measurement. (In Section 2.8, we call this value an **outlier**).
c. Delete the outlier of part **b** from the data set and recalculate the measures of central tendency. Which measure is most affected by the deletion of the outlier?

2.70 **Active nuclear power plants.** The U.S. Energy Information Administration monitors all nuclear power plants operating in the United States. The table below lists the number of active nuclear power plants operating in each of a sample of 20 states. All of the data are saved in the **NUCLEAR** file.

a. Find the mean, median, and mode of this data set.
b. Eliminate the largest value from the data set and repeat part **a.** What effect does dropping this measurement have on the measures of central tendency found in part **a**?
c. Arrange the 20 values in the table from lowest to highest. Next, eliminate the lowest two values and the highest two values from the data set, and find the mean of the remaining data values. The result is called a *10% trimmed mean,* since it is calculated after removing the highest 10% and the lowest 10% of the data values. What advantages does a trimmed mean have over the regular arithmetic mean?

Table for Exercise 2.70

State	Number of Power Plants	State	Number of Power Plants
Alabama	5	New Hampshire	1
Arizona	2	New York	6
California	4	North Carolina	5
Florida	5	Ohio	3
Georgia	4	Pennsylvania	9
Illinois	11	South Carolina	7
Kansas	1	Tennessee	3
Louisiana	2	Texas	4
Massachusetts	1	Vermont	1
Mississippi	1	Wisconsin	3

Based on *Statistical Abstract of the United States,* 2010 (Table 907). U.S. Energy Information Administration, *Electric Power Annual.*

2.5 Numerical Measures of Variability

a. Drug A

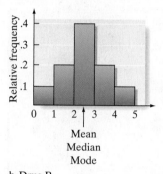

b. Drug B

Figure 2.19

Response time histograms for two drugs

Measures of central tendency provide only a partial description of a quantitative data set. The description is incomplete without a **measure of the variability,** or **spread,** of the data set. Knowledge of the data set's variability, along with knowledge of its center, can help us visualize the shape of the data set as well as its extreme values.

For example, suppose we want to compare response time to a stimulus for subjects treated with two different drugs, A and B. The histograms for the response times (in seconds) for each of the two drugs are shown in Figure 2.19. If you examine the two histograms, you'll notice that both data sets are symmetric, with equal modes, medians, and means. However, Drug A (Figure 2.19a) has response times spread with almost equal relative frequency over the measurement classes, while Drug B (Figure 2.19b) has most of its response times clustered about its center. Thus, the response times for Drug B are *less variable* than those for Drug A. Consequently, you can see that we need a measure of variability as well as a measure of central tendency to describe a data set.

Perhaps the simplest measure of the variability of a quantitative data set is its *range.*

> The **range** of a quantitative data set is equal to the largest measurement minus the smallest measurement.

The range is easy to compute and easy to understand, but it is a rather insensitive measure of data variation when the data sets are large. This is because two data sets can have the same range and be vastly different with respect to data variation. The phenomenon is demonstrated in Figure 2.19: Both distributions of data shown in the figure have the same range, but we already know that the response times for Drug B are much less variable than those for Drug A. Thus, you can see that the range does not always detect differences in data variation for large data sets.

Let's see if we can find a measure of data variation that is more sensitive than the range. Consider the two samples in Table 2.5: Each has five measurements. (We have ordered the numbers for convenience.) Note that both samples have a mean of 3 and

Table 2.5	Two Hypothetical Data Sets	
	Sample 1	Sample 2
Measurements	1, 2, 3, 4, 5	2, 3, 3, 3, 4
Mean	$\bar{x} = \dfrac{1 + 2 + 3 + 4 + 5}{5} = \dfrac{15}{5} = 3$	$\bar{x} = \dfrac{2 + 3 + 3 + 3 + 4}{5} = \dfrac{15}{5} = 3$
Deviations of measurement values from $\bar{x}$	$(1 - 3), (2 - 3), (3 - 3), (4 - 3),$ $(5 - 3)$ or $-2, -1, 0, 1, 2$	$(2 - 3), (3 - 3), (3 - 3), (3 - 3),$ $(4 - 3)$ or $-1, 0, 0, 0, 1$

a. Sample 1

b. Sample 2

Figure 2.20

Dot plots for deviations in Table 2.5

that we have calculated the distance and direction, or *deviation,* between each measurement and the mean. What information do these deviations contain? If they tend to be large in magnitude, as in sample 1, the data are spread out, or highly variable. If the deviations are mostly small, as in sample 2, the data are clustered around the mean, $\bar{x}$, and therefore do not exhibit much variability. You can see that these deviations, displayed graphically in Figure 2.20, provide information about the variability of the sample measurements.

The next step is to condense the information in these distances into a single numerical measure of variability. Averaging the deviations from $\bar{x}$ won't help, because the negative and positive deviations cancel; that is, the sum of the deviations (and thus the average deviation) is always equal to zero.

Two methods come to mind for dealing with the fact that positive and negative deviations from the mean cancel. The first is to treat all the deviations as though they were positive, ignoring the sign of the negative values. We won't pursue this line of thought because the resulting measure of variability (the mean of the absolute values of the deviations) presents analytical difficulties beyond the scope of this text. A second method of eliminating the minus signs associated with the deviations is to square them. The quantity we can calculate from the squared deviations provides a meaningful description of the variability of a data set and presents fewer analytical difficulties in making inferences.

To use the squared deviations calculated from a data set, we first calculate the *sample variance*.

The **sample variance** for a sample of n measurements is equal to the sum of the squared deviations from the mean, divided by $(n - 1)$. The symbol s^2 is used to represent the sample variance.

Formula for the Sample Variance:

$$s^2 = \frac{\sum_{i=1}^{n}(x_i - \bar{x})^2}{n - 1}$$

Note: A shortcut formula for calculating s^2 is

$$s^2 = \frac{\sum_{i=1}^{n}x_i^2 - \frac{\left(\sum_{i=1}^{n}x_i\right)^2}{n}}{n - 1}$$

Referring to the two samples in Table 2.5, you can calculate the variance for sample 1 as follows:

$$s^2 = \frac{(1 - 3)^2 + (2 - 3)^2 + (3 - 3)^2 + (4 - 3)^2 + (5 - 3)^2}{5 - 1}$$

$$= \frac{4 + 1 + 0 + 1 + 4}{4} = 2.5$$

The second step in finding a meaningful measure of data variability is to calculate the *standard deviation* of the data set.

The **sample standard deviation**, s, is defined as the positive square root of the sample variance, s^2, or, mathematically,

$$s = \sqrt{s^2}$$

The population variance, denoted by the symbol σ^2 (sigma squared), is the average of the squared deviations from the mean, μ, of the measurements on *all* units in the population, and σ (sigma) is the square root of this quantity.

Symbols for Variance and Standard Deviation

s^2 = Sample variance

s = Sample standard deviation

σ^2 = Population variance

σ = Population standard deviation

Notice that, unlike the variance, the standard deviation is expressed in the original units of measurement. For example, if the original measurements are in dollars, the variance is expressed in the peculiar units "dollars squared," but the standard deviation is expressed in dollars.

You may wonder why we use the divisor $(n - 1)$ instead of n when calculating the sample variance. Wouldn't using n seem more logical, so that the sample variance would be the average squared distance from the mean? The trouble is, using n tends to produce an underestimate of the population variance σ^2. So we use $(n - 1)$ in the denominator to provide the appropriate correction for this tendency.* Since sample statistics such as s^2 are used primarily to estimate population parameters such as σ^2, $(n - 1)$ is preferred to n in defining the sample variance.

Example 2.9
Computing Measures of Variation

Problem Calculate the variance and standard deviation of the following sample: 2, 3, 3, 3, 4.

Solution As the number of measurements increases, calculating s^2 and s becomes very tedious. Fortunately, as we show in Example 2.10, we can use a statistical software package (or a calculator) to find these values. If you must calculate these quantities by hand, it is advantageous to use the shortcut formula provided on page 62.

To do this, we need two summations: $\sum x$ and $\sum x^2$. These can easily be obtained from the following type of tabulation:

x	x^2
2	4
3	9
3	9
3	9
4	16
$\sum x = 15$	$\sum x^2 = 47$

Then we use[†]

$$s^2 = \frac{\sum_{i=1}^{n} x_i^2 - \frac{\left(\sum_{i=1}^{n} x_i\right)^2}{n}}{n - 1} = \frac{47 - \frac{(15)^2}{5}}{5 - 1} = \frac{2}{4} = .5$$

$$s = \sqrt{.5} = .71$$

Look Back As the sample size n increases, these calculations can become very tedious. As the next example shows, we can use the computer to find s^2 and s.

Now Work Exercise 2.78a

Example 2.10
Finding Measures of Variation on a Printout

Problem Use the computer to find the sample variance s^2 and the sample standard deviation s for the 100 gas mileage readings given in Table 2.2.

Solution The SAS printout describing the gas mileage data is reproduced in Figure 2.21. The variance and standard deviation, highlighted on the printout, are $s^2 = 5.85$ and $s = 2.42$ (rounded to two decimal places).

You now know that the standard deviation measures the variability of a set of data, and you know how to calculate the standard deviation. The larger the standard deviation,

---

*Appropriate here means that s^2, with a divisor of $(n - 1)$, is an *unbiased estimator* of σ^2. An unbiased estimator has a sampling distribution mean equal to the parameter being estimated. We define and discuss *sampling distributions* in Chapter 4.

[†]In calculating s^2, how many decimal places should you carry? Although there are no rules for the rounding procedure, it is reasonable to retain twice as many decimal places in s^2 as you ultimately wish to have in s. If, for example, you wish to calculate s to the nearest hundredth (two decimal places), you should calculate s^2 to the nearest ten-thousandth (four decimal places).

the more variable the data are. The smaller the standard deviation, the less variation there is in the data. But how can we practically interpret the standard deviation and use it to make inferences? This is the topic of Section 2.6.

Figure 2.21
Reproduction of SAS numerical descriptive measures for 100 EPA mileages

The MEANS Procedure

Analysis Variable : MPG

Mean	Std Dev	Variance	N	Minimum	Maximum	Median
36.9940000	2.4178971	5.8462263	100	30.0000000	44.9000000	37.0000000

Exercises 2.71–2.89

Understanding the Principles

2.71 What is the range of a data set?

2.72 What is the primary disadvantage of using the range to compare the variability of data sets?

2.73 Describe the sample variance in words rather than with a formula. Do the same with the population variance.

2.74 Can the variance of a data set ever be negative? Explain. Can the variance ever be smaller than the standard deviation? Explain.

2.75 If the standard deviation increases, does this imply that the data are more variable or less variable?

Learning the Mechanics

2.76 Calculate the variance and standard deviation for samples for which

   a. $n = 10, \sum x^2 = 84, \sum x = 20$

   b. $n = 40, \sum x^2 = 380, \sum x = 100$

   c. $n = 20, \sum x^2 = 18, \sum x = 17$

2.77 Calculate the range, variance, and standard deviation for the following samples:
   a. 39, 42, 40, 37, 41
   b. 100, 4, 7, 96, 80, 3, 1, 10, 2
   c. 100, 4, 7, 30, 80, 30, 42, 2

2.78 Calculate the range, variance, and standard deviation for
[NW] the following samples:
   a. 4, 2, 1, 0, 1
   b. 1, 6, 2, 2, 3, 0, 3
   c. 8, −2, 1, 3, 5, 4, 4, 1, 3
   d. 0, 2, 0, 0, −1, 1, −2, 1, 0, −1, 1, −1, 0, −3, −2, −1, 0, 1

2.79 Using only integers between 0 and 10, construct two data sets with at least 10 observations each such that the two sets have the same mean, but different variances. Construct dot plots for each of your data sets, and mark the mean of each data set on its dot plot.

2.80 Using only integers between 0 and 10, construct two data sets with at least 10 observations each such that the two sets have the same range, but different means. Construct a dot plot for each of your data sets, and mark the mean of each data set on its dot plot.

2.81 Consider the following sample of five measurements: 2, 1, 1, 0, 3.
   a. Calculate the range, s^2, and s.
   b. Add 3 to each measurement and repeat part **a**.

   c. Subtract 4 from each measurement and repeat part **a**.

   d. Considering your answers to parts **a**, **b**, and **c**, what seems to be the effect on the variability of a data set of adding the same number to or subtracting the same number from each measurement?

2.82 Compute s^2, and s for each of the data sets listed. Where appropriate, specify the units in which your answer is expressed.
   a. 3, 1, 10, 10, 4
   b. 8 feet, 10 feet, 32 feet, 5 feet
   c. −1, −4, −3, 1, −4, −6
   d. 1/5 ounce, 1/5 ounce, 1/5 ounce, 2/5 ounce, 1/5 ounce, 4/5 ounce

◉ Applet Exercise 2.4

Use the applet entitled *Standard Deviation* to find the standard deviation of each of the four data sets listed in Exercise 2.78. For each data set, set the lower limit to a number less than all of the data, set the upper limit to a number greater than all of the data, and then click on *Update*. Click on the approximate location of each data item on the number line. You can get rid of a point by dragging it to the trash can. To clear the graph between data sets, simply click on the trash can.
   a. Compare the standard deviations generated by the applet with those you calculated by hand in Exercise 2.78. If there are differences, explain why the applet might give values slightly different from the hand-calculated values.
   b. Despite the fact that it provides a slightly different value of the standard deviation of a data set, describe some advantages of using the applet.

◉ Applet Exercise 2.5

Use the applet *Standard Deviation* to study the effect that multiplying or dividing each number in a data set by the same number has on the standard deviation. Begin by setting appropriate limits and plotting the given data on the number line provided in the applet.

         0 1 1 2 2 3 4

   a. Record the standard deviation. Then multiply each data item by 2, plot the new data items, and record the standard deviation. Repeat the process, first multiplying each of the original data items by 3 and then by 4. Describe what happens to the standard deviation as the data items are multiplied by ever higher numbers. Divide each standard deviation by the standard deviation of the original data set. Do you see a pattern? Explain.

b. Divide each of the original data items by 2, plot the new data, and record the standard deviation. Repeat the process, first dividing each of the original data items by 3 and then by 4. Describe what happens to the standard deviation as the data items are divided by ever higher numbers. Divide each standard deviation by the standard deviation of the original data set. Do you see a pattern? Explain.

c. Using your results from parts **a** and **b,** describe what happens to the standard deviation of a data set when each of the data items in the set is multiplied or divided by a fixed number n. Experiment by repeating parts **a** and **b** for other data sets if you need to.

Applet Exercise 2.6

Use the applet *Standard Deviation* to study the effect that an extreme value has on the standard deviation. Begin by setting appropriate limits and plotting the following data on the number line provided in the applet:

0 6 7 7 8 8 8 9 9 9 10

a. Record the standard deviation. Replace the extreme value of 0 with 2, then 4, and then 6. Record the standard deviation each time. Describe what happens to the standard deviation as 0 is replaced by ever higher numbers.

b. How would the standard deviation of the data set compare with the original standard deviation if the 0 were replaced by 16? Explain.

Applying the Concepts—Basic

2.83 **Shell lengths of sea turtles.** *Aquatic Biology* (Vol. 9, 2010) reported on a study of green sea turtles inhabiting the Grand Cayman South Sound lagoon. The data on curved carapace (shell) length (in centimeters) for 76 captured turtles are saved in the **TURTLES** file. Descriptive statistics for the data are shown on the accompanying MINITAB printout.

Descriptive Statistics: Length

Variable	N	Mean	StDev	Variance	Minimum	Maximum	Range
Length	76	55.47	11.34	128.57	30.37	81.63	51.26

---

a. Locate the range of the shell lengths on the printout.

b. Locate the variance of the shell lengths on the printout.

c. Locate the standard deviation of the shell lengths on the printout.

d. If the target of your interest is these specific 76 captured turtles, what symbols would you use to represent the variance and standard deviation?

2.84 **Reading Japanese books.** Refer to the *Reading in a Foreign Language* (Apr. 2004) experiment to improve the Japanese reading comprehension levels of 14 University of Hawaii students, presented in Exercise 2.61 (p. 57). The data on number of books read and grade for each student are saved in the **JAPANESE** file.

a. Find the range, variance, and standard deviation of the number of books read by students who earned an A grade.

b. Find the range, variance, and standard deviation of the number of books read by students who earned either a B or C grade.

c. Refer to parts **a** and **b.** Which of the two groups of students has a more variable distribution for number of books read?

2.85 **Is honey a cough remedy?** Refer to the *Archives of Pediatrics and Adolescent Medicine* (Dec. 2007) study of honey as a remedy for coughing, Exercise 2.62 (p. 58). The coughing improvement scores (saved in the **HONEYCOUGH** file) for the patients in the over-the-counter cough medicine dosage (DM) group, honey dosage group, and control group are reproduced in the accompanying table.

Honey Dosage:	12	11	15	11	10	13	10	4	15	16	9
	14	10	6	10	8	11	12	12	8	12	9
	11	15	10	15	9	13	8	12	10	8	9
	5	12									
DM Dosage:	4	6	9	4	7	7	7	9	12	10	11
	6	3	4	9	12	7	6	8	12	12	4
	12	13	7	10	13	9	4	4	10	15	9
No Dosage (Control):	5	8	6	1	0	8	12	8	7	7	1
	6	7	7	12	7	9	7	9	5	11	9
	5	6	8	86	7	10	9	4	8	7	3
	1	4	3								

Based on Paul, I. M., et al. "Effect of honey, dextromethorphan, and no treatment on nocturnal cough and sleep quality for coughing children and their parents." *Archives of Pediatrics and Adolescent Medicine*, Vol. 161, No. 12, Dec. 2007 (data simulated).

a. Find the standard deviation of the improvement scores for the honey dosage group.

b. Find the standard deviation of the improvement scores for the DM dosage group.

c. Find the standard deviation of the improvement scores for the control group.

d. Based on the results, parts **a–c**, which group appears to have the most variability in coughing improvement scores? The least variability?

Applying the Concepts—Intermediate

2.86 **Characteristics of a rockfall.** Refer to the *Environmental Geology* (Vol. 58, 2009) study of how far a block from a collapsing rock wall will bounce, Exercise 2.59 (p. 57). The rebound lengths (meters) for a sample of 13 rock bounces are reproduced in the table below and saved in the **ROCKFALL** file.

10.94	13.71	11.38	7.26	17.83	11.92	11.87
5.44	13.35	4.90	5.85	5.10	6.77	

Based on Paronuzzi, P. "Rockfall-induced block propagation on a soil slope, northern Italy." *Environmental Geology*, Vol. 58, 2009 (Table 2).

a. Compute the range of the 13 rebound lengths. Give the units of measurement of the range.

b. Compute the variance of the 13 rebound lengths. Give the units of measurement of the variance.

c. Compute the standard deviation of the 13 rebound lengths. Give the units of measurement of the standard deviation.

2.87 **Cheek teeth of extinct primates.** Refer to the *American Journal of Physical Anthropology* (Vol. 142, 2010) study of the characteristics of cheek teeth (e.g., molars) in an extinct primate species, Exercise 2.67 (p. 59). The data on dentary depth of molars (in millimeters) for 18 cheek teeth extracted from skulls are reproduced in the next table (p. 66) and saved in the **CHEECKTEETH** file.

Data for Exercise 2.87

Data on Dentary Depth (mm) of Molars		
18.12	15.76	13.25
19.48	17.00	16.12
19.36	13.96	18.13
15.94	16.55	14.02
15.83	15.70	14.04
19.70	17.83	16.20

Based on Boyer, D. M., Evans, A. R., and Jernvall, J. "Evidence of dietary differentiation among Late Paleocene–Early Eocene Plesiadapids (Mammalia, Primates)." *American Journal of Physical Anthropology*, Vol. 142, 2010 (Table A3).

a. Find the range of the data set. If the largest depth measurement in the sample were doubled, how would the range change? Would it increase or decrease?

b. Find the variance of the data set. If the largest depth measurement in the sample were doubled, how would the variance change? Would it increase or decrease?

c. Find the standard deviation of the data set. If the largest depth measurement in the sample were doubled, how would the standard deviation change? Would it increase or decrease?

2.88 **Most powerful women in America.** Refer to Exercise 2.60 (p. 57) and *Fortune's* list of the 50 most powerful women in America. The data are stored in the **WPOWER50** file.

a. Find the range of the ages of these 50 women.

b. Find the variance of the ages of these 50 women.

c. Find the standard deviation of the ages for these 50 women.

d. Suppose the standard deviation of the ages of the most powerful women in Europe is 10 years. For which location, the United States or Europe, is the age data more variable?

e. If the largest age in the data set is omitted, would the standard deviation increase or decrease? Verify your answer.

2.89 **Active nuclear power plants.** Refer to Exercise 2.70 (p. 60) and the U.S. Energy Information Administration's data on the number of nuclear power plants operating in each of 20 states. The data are saved in the **NUCLEAR** file.

a. Find the range, variance, and standard deviation of this data set.

b. Eliminate the largest value from the data set and repeat part **a.** What effect does dropping this measurement have on the measures of variation found in part **a**?

c. Eliminate the smallest and largest value from the data set and repeat part **a.** What effect does dropping both of these measurements have on the measures of variation found in part **a**?

2.6 Interpreting the Standard Deviation

We've seen that if we are comparing the variability of two samples selected from a population, the sample with the larger standard deviation is the more variable of the two. Thus, we know how to interpret the standard deviation on a relative or comparative basis, but we haven't explained how it provides a measure of variability for a single sample.

To understand how the standard deviation provides a measure of variability of a data set, consider the following questions: How many measurements are within one standard deviation of the mean? How many measurements are within two standard deviations? For any particular data set, we can answer these questions by counting the number of measurement in each of the intervals. However, finding an answer that applies to *any* set of data, whether a population or a sample, is more problematic.

Tables 2.6 and 2.7 give two sets of answers to the questions of how many measurements fall within one, two, and three standard deviations of the mean. The first, which applies to *any* set of data, is derived from a theorem proved by the Russian mathematician P. L. Chebyshev (1821–1894). The second, which applies to **mound-shaped, symmetric distributions** of data (for which the mean, median, and mode are all about the same), is based upon empirical evidence that has accumulated over the years. However, the percentages given for the intervals in Table 2.7 provide remarkably good approximations even when the distribution of the data is slightly skewed or asymmetric. Note that the rules apply to either population or sample data sets.

BIOGRAPHY PAFNUTY L. CHEBYSHEV (1821–1894)

The Splendid Russian Mathematician

P. L. Chebyshev was educated in mathematical science at Moscow University, eventually earning his master's degree. Following his graduation, Chebyshev joined St. Petersburg (Russia) University as a professor, becoming part of the well-known "Petersburg mathematical school." It was here that he proved his famous theorem about the probability of a measurement being within *k* standard deviations of the mean (Table 2.6). His fluency in French allowed him to gain international recognition in probability theory. In fact, Chebyshev once objected to being described as a "splendid Russian mathematician," saying he surely was a "worldwide mathematician." One student remembered Chebyshev as "a wonderful lecturer" who "was always prompt for class. As soon as the bell sounded, he immediately dropped the chalk and, limping, left the auditorium." ∎

Table 2.6   Interpreting the Standard Deviation: Chebyshev's Rule

Chebyshev's rule applies to any data set, regardless of the shape of the frequency distribution of the data.

a. It is possible that very few of the measurements will fall within one standard deviation of the mean [i.e., within the interval $(\bar{x} - s, \bar{x} + s)$ for samples and $(\mu - \sigma, \mu + \sigma)$ for populations].

b. At least $\frac{3}{4}$ of the measurements will fall within two standard deviations of the mean [i.e., within the interval $(\bar{x} - 2s, \bar{x} + 2s)$ for samples and $(\mu - 2\sigma, \mu + 2\sigma)$ for populations].

c. At least $\frac{8}{9}$ of the measurements will fall within three standard deviations of the mean [i.e., within the interval $(\bar{x} - 3s, \bar{x} + 3s)$ for samples and $(\mu - 3\sigma, \mu + 3\sigma)$ for populations].

d. Generally, for any number k greater than 1, at least $(1 - 1/k^2)$ of the measurements will fall within k standard deviations of the mean [i.e., within the interval $(\bar{x} - ks, \bar{x} + ks)$ for samples and $(\mu - k\sigma, \mu + k\sigma)$ for populations].

Table 2.7   Interpreting the Standard Deviation: The Empirical Rule

The **empirical rule** is a rule of thumb that applies to data sets with frequency distributions that are mound shaped and symmetric, as follows:

a. Approximately 68% of the measurements will fall within one standard deviation of the mean [i.e., within the interval $(\bar{x} - s, \bar{x} + s)$ for samples and $(\mu - \sigma, \mu + \sigma)$ for populations].

b. Approximately 95% of the measurements will fall within two standard deviations of the mean [i.e., within the interval $(\bar{x} - 2s, \bar{x} + 2s)$ for samples and $(\mu - 2\sigma, \mu + 2\sigma)$ for populations].

c. Approximately 99.7% (essentially all) of the measurements will fall within three standard deviations of the mean [i.e., within the interval $(\bar{x} - 3s, \bar{x} + 3s)$ for samples and $(\mu - 3\sigma, \mu + 3\sigma)$ for populations].

Example 2.11

Interpreting the Standard Deviation— Rat-in-Maze Experiment

Problem Thirty students in an experimental psychology class use various techniques to train a rat to move through a maze. At the end of the course, each student's rat is timed as it negotiates the maze. The results (in minutes) are listed in Table 2.8. Determine the fraction of the 30 measurements in the intervals $\bar{x} \pm s, \bar{x} \pm 2s$, and $\bar{x} \pm 3s$, and compare the results with those predicted in Tables 2.6 and 2.7.

Table 2.8   Times (in Minutes) of 30 Rats Running through a Maze

1.97	.60	4.02	3.20	1.15	6.06	4.44	2.02	3.37	3.65
1.74	2.75	3.81	9.70	8.29	5.63	5.21	4.55	7.60	3.16
3.77	5.36	1.06	1.71	2.47	4.25	1.93	5.15	2.06	1.65

Data Set: RATMAZE

Solution First, we entered the data into the computer and used MINITAB to produce summary statistics. The mean and standard deviation of the sample data, highlighted on the printout shown in Figure 2.22, are

$$\bar{x} = 3.74 \text{ minutes} \qquad s = 2.20 \text{ minutes}$$

rounded to two decimal places.

Figure 2.22

MINITAB descriptive statistics for rat maze times

Descriptive Statistics: RUNTIME

Variable	N	Mean	StDev	Minimum	Median	Maximum
RUNTIME	30	3.744	2.198	0.600	3.510	9.700

Now we form the interval

$$(\bar{x} - s, \bar{x} + s) = (3.74 - 2.20, 3.74 + 2.20) = (1.54, 5.94)$$

A check of the measurements shows that 23 of the times are within this one standard-deviation interval around the mean. This number represents 23/30, or $\approx$ 77%, of the sample measurements.

The next interval of interest is

$$(\bar{x} - 2s, \bar{x} + 2s) = (3.74 - 4.40, 3.74 + 4.40) = (-.66, 8.14)$$

All but two of the times are within this interval, so 28/30, or approximately 93%, are within two standard deviations of $\bar{x}$.

Finally, the three-standard-deviation interval around $\bar{x}$ is

$$(\bar{x} - 3s, \bar{x} + 3s) = (3.74 - 6.60, 3.74 + 6.60) = (-2.86, 10.34)$$

All of the times fall within three standard deviations of the mean.

These one-, two-, and three-standard-deviation percentages (77%, 93%, and 100%) agree fairly well with the approximations of 68%, 95%, and 100% given by the empirical rule (Table 2.7).

Look Back If you look at the MINITAB frequency histogram for this data set (Figure 2.23), you'll note that the distribution is not really mound shaped, nor is it extremely skewed. Thus, we get reasonably good results from the mound-shaped approximations. Of course, we know from Chebyshev's rule (Table 2.6) that no matter what the shape of the distribution, we would expect at least 75% and at least 89% of the measurements to lie within two and three standard deviations of $\bar{x}$, respectively.

Figure 2.23

MINITAB histogram of rat maze times

Now Work Exercise 2.94

Example 2.12

Checking the Calculation of the Sample Standard Deviation

Problem Chebyshev's rule and the empirical rule are useful as a check on the calculation of the standard deviation. For example, suppose we calculated the standard deviation for the gas mileage data (Table 2.2) to be 5.85. Are there any "clues" in the data that enable us to judge whether this number is reasonable?

Solution The range of the mileage data in Table 2.2 is 44.9 − 30.0 = 14.9. From Chebyshev's rule and the empirical rule, we know that most of the measurements (approximately 95% if the distribution is mound shaped) will be within two standard deviations of the mean. And regardless of the shape of the distribution and the number of measurements, almost all of them will fall within three standard deviations of the mean. Consequently, we would expect the range of the measurements to be between 4 (i.e., $\pm 2s$) and 6 (i.e., $\pm 3s$)

Figure 2.24

The relation between the range and the standard deviation

standard deviations in length. (See Figure 2.24.) For the car mileage data, this means that s should fall between

$$\frac{\text{Range}}{6} = \frac{14.9}{6} = 2.48 \quad \text{and} \quad \frac{\text{Range}}{4} = \frac{14.9}{4} = 3.73$$

Hence, the standard deviation should not be much larger than 1/4 of the range, particularly for the data set with 100 measurements. Thus, we have reason to believe that the calculation of 5.85 is too large. A check of our work reveals that 5.85 is the variance s^2, not the standard deviation s. (See Example 2.10.) We "forgot" to take the square root (a common error); the correct value is $s = 2.42$. Note that this value is slightly smaller than the range divided by 6 (2.48). The larger the data set, the greater is the tendency for very large or very small measurements (extreme values) to appear, and when they do, the range may exceed six standard deviations.

Look Back In examples and exercises, we'll sometimes use $s \approx \text{range}/4$ to obtain a crude, and usually conservatively large, approximation for s. However, we stress that this is no substitute for calculating the exact value of s when possible.

Now Work Exercise 2.95

In the next example, we use the concepts in Chebyshev's rule and the empirical rule to build the foundation for making statistical inferences.

Example 2.13
Making a Statistical Inference—Car Battery Guarantee

Problem A manufacturer of automobile batteries claims that the average length of life for its grade A battery is 60 months. However, the guarantee on this brand is for just 36 months. Suppose the standard deviation of the life length is known to be 10 months and the frequency distribution of the life-length data is known to be mound shaped.

a. Approximately what percentage of the manufacturer's grade A batteries will last more than 50 months, assuming that the manufacturer's claim is true?

b. Approximately what percentage of the manufacturer's batteries will last less than 40 months, assuming that the manufacturer's claim is true?

c. Suppose your battery lasts 37 months. What could you infer about the manufacturer's claim?

Solution If the distribution of life length is assumed to be mound shaped with a mean of 60 months and a standard deviation of 10 months, it would appear as shown in Figure 2.25. Note that we can take advantage of the fact that mound-shaped distributions are (approximately) symmetric about the mean, so that the percentages given by the empirical rule can be split equally between the halves of the distribution on each side of the mean.

For example, since approximately 68% of the measurements will fall within one standard deviation of the mean, the distribution's symmetry implies that approximately

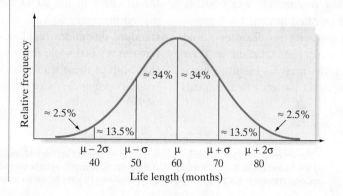

Figure 2.25

Battery life-length distribution: manufacturer's claim assumed true

(1/2)(68%) = 34% of the measurements will fall between the mean and one standard deviation on each side. This concept is illustrated in Figure 2.25. The figure also shows that 2.5% of the measurements lie beyond two standard deviations in each direction from the mean. This result follows from the fact that if approximately 95% of the measurements fall within two standard deviations of the mean, then about 5% fall outside two standard deviations; if the distribution is approximately symmetric, then about 2.5% of the measurements fall beyond two standard deviations on each side of the mean.

a. It is easy to see in Figure 2.25 that the percentage of batteries lasting more than 50 months is approximately 34% (between 50 and 60 months) plus 50% (greater than 60 months). Thus, approximately 84% of the batteries should have a life exceeding 50 months.

b. The percentage of batteries that last less than 40 months can also be easily determined from Figure 2.25: Approximately 2.5% of the batteries should fail prior to 40 months, assuming that the manufacturer's claim is true.

c. If you are so unfortunate that your grade A battery fails at 37 months, you can make one of two inferences: Either your battery was one of the approximately 2.5% that fail prior to 40 months, or something about the manufacturer's claim is not true. Because the chances are so small that a battery fails before 40 months, you would have good reason to have serious doubts about the manufacturer's claim. A mean smaller than 60 months or a standard deviation longer than 10 months would each increase the likelihood of failure prior to 40 months.*

Look Back The approximations given in Figure 2.25 are more dependent on the assumption of a **mound-shaped distribution** than are the assumptions stated in the empirical rule (Table 2.7), because the approximations in Figure 2.25 depend on the (approximate) symmetry of the mound-shaped distribution. We saw in Example 2.11 that the empirical rule can yield good approximations even for skewed distributions. This will *not* be true of the approximations in Figure 2.25; the distribution *must* be mound shaped and (approximately) symmetric.

Example 2.13 is our initial demonstration of the statistical inference-making process. At this point, you should realize that we'll use sample information (in Example 2.13, your battery's failure at 37 months) to make inferences about the population (in Example 2.13, the manufacturer's claim about the length of life for the population of all batteries). We'll build on this foundation as we proceed.

Statistics IN Action | **Revisited**   Interpreting Descriptive Statistics for the Body Image Data

We return to the analysis of the data from the *Body Image: An International Journal of Research* (Jan. 2010) study of 92 BDD patients. Recall that the quantitative variable of interest in the study is Total Appearance Evaluation score (ranging from 7 to 35 points). One of the questions of interest to the Brown University researchers was whether female BDD patients were less satisfied in their body image than normal females (i.e., females with no disorders), and whether male BDD patients were less satisfied in their body image than normal males. The analysis involved comparing the mean score for the BDD patients to the mean score for a normal group (called a "norm"). The appearance evaluation "norms" for females and males were 23.5 and 24.4, respectively.

MINITAB descriptive statistics for the data in the **BDD** file are displayed in Figure SIA2.5, with means and standard deviations highlighted. The sample mean score for females is 17.05 while the sample mean for males is 19.0. Note that both of these values fall well below the "norm" for each respective gender. Consequently, *for this sample of BDD*

---

*The assumption that the distribution is mound shaped and symmetric may also be incorrect. However, if the distribution were skewed to the right, as life-length distributions often tend to be, the percentage of measurements more than two standard deviations below the mean would be even less than 2.5%.

Statistics IN Action
(continued)

Figure SIA2.5
MINITAB descriptive statistics for appearance evaluation by gender

Descriptive Statistics: AEScore

Variable	Gender	N	Mean	StDev	Variance	Minimum	Maximum
AEScore	F	60	17.050	4.757	22.625	7.000	30.000
	M	32	19.000	5.424	29.419	7.000	32.000

patients, both females and males have a lower average opinion about their body image than do normal females and males.

To interpret the standard deviations, we substitute into the formula, $\bar{x} \pm 2s$, to obtain the intervals:

Females: $\bar{x} \pm 2s = 17.05 \pm 2(4.76) = 17.05 \pm 9.52$
　　　　　$= (7.53, 26.57)$

Males: $\bar{x} \pm 2s = 19.00 \pm 2(5.42) = 19.00 \pm 10.84$
　　　　$= (8.16, 29.84)$

From Chebyshev's rule (Table 2.6), we know that at least 75% of the females in the sample of BDD patients will have an

Appearance Evaluation score anywhere between 7.53 and 26.57 points. Similarly, we know that at least 75% of the males in the sample of BDD patients will have an Appearance Evaluation score anywhere between 8.16 and 29.84 points. You can see that the interval for the females lies slightly below the interval for males. How does this information help in determining whether the female BDD patients and male BDD patients have mean scores below the "norm" for their gender? In Chapters 5 and 6 we will learn about an inferential method that uses both the sample mean and sample standard deviation to determine whether the population mean of the BDD patients is really smaller than the "norm" value.

 Data Set: BBD

Exercises 2.90–2.107

Understanding the Principles

2.90 To what kind of data sets can Chebyshev's rule be applied? How about the empirical rule?

2.91 The output from a statistical computer program indicates that the mean and standard deviation of a data set consisting of 200 measurements are $1,500 and $300, respectively.
　a. What are the units of measurement of the variable of interest? On the basis of the units, what type of data is this, quantitative or qualitative?
　b. What can be said about the number of measurements between $900 and $2,100? between $600 and $2,400? between $1,200 and $1,800? between $1,500 and $2,100?

2.92 For any set of data, what can be said about the percentage of the measurements contained in each of the following intervals?
　a. $\bar{x} - s$ to $\bar{x} + s$
　b. $\bar{x} - 2s$ to $\bar{x} + 2s$
　c. $\bar{x} - 3s$ to $\bar{x} + 3s$

2.93 For a set of data with a mound-shaped relative frequency distribution, what can be said about the percentage of the measurements contained in each of the intervals specified in Exercise 2.92?

Learning the Mechanics

2.94 The following is a sample of 25 measurements. The data are saved in the **LM2_94** file.

7	6	6	11	8	9	11	9	10	8	7	7	5
9	10	7	7	7	7		9	12	10	10	8	6

NW　**a.** Compute $\bar{x}$, s^2, and s for this sample.
　b. Count the number of measurements in the intervals $\bar{x} \pm s$, $\bar{x} \pm 2s$, and $\bar{x} \pm 3s$. Express each count as a percentage of the total number of measurements.

　c. Compare the percentages found in part **b** with the percentages given by the empirical rule and Chebyshev's rule.
　d. Calculate the range and use it to obtain a rough approximation for s. Does the result compare favorably with the actual value for s found in part **a**?

2.95 Given a data set with a largest value of 760 and a smallest value of 135, what would you estimate the standard deviation to be? Explain the logic behind the procedure you used to estimate the standard deviation. Suppose the standard deviation is reported to be 25. Is this number reasonable? Explain.

Applying the Concepts—Basic

2.96 **Most powerful women in America.** Refer to the *Fortune* list of the 50 most powerful women in America, saved in the **WPOWER50** file. In Exercise 2.60 (p. 57) you found the mean age of the 50 women in the data set, and in Exercise 2.88 (p. 66) you found the standard deviation. Use these values to form an interval that will contain at least 75% of the ages in the data set.

2.97 **Sanitation inspection of cruise ships.** Refer to Exercise 2.35 (p. 47) and the Centers for Disease Control and Prevention listing of the sanitation scores for 186 cruise ships. The data are saved in the **SHIPSANIT** file.
　a. Find the mean and standard deviation of the sanitation scores.
　b. Calculate the intervals $\bar{x} \pm s$, $\bar{x} \pm 2s$, and $\bar{x} \pm 3s$.
　c. Find the percentage of measurements in the data set that fall within each of the intervals in part **b**. Do these percentages agree with either Chebyshev's rule or the empirical rule?

2.98 **Motivation of drug dealers.** Researchers at Georgia State University investigated the personality characteristics of drug dealers in order to shed light on their motivation for participating in the illegal drug market (*Applied Psychology in Criminal Justice*, Sept. 2009). The sample consisted of 100 convicted drug dealers who attended a court-mandated counseling program. Each dealer was scored on the Wanting Recognition (WR) Scale, which provides a quantitative measure of a person's level of need for approval and sensitivity to social situations. (Higher scores indicate a greater need for approval.) The sample of drug dealers had a mean WR score of 39, with a standard deviation of 6. Assume the distribution of WR scores for drug dealers is mound shaped and symmetric.

a. Give a range of WR scores that will contain about 95% of the scores in the drug dealer sample.

b. What proportion of the drug dealers will have WR scores above 51?

c. Give a range of WR sores that will contain nearly all the scores in the drug dealer sample.

2.99 Dentists' use of anesthetics. A study published in *Current Allergy & Clinical Immunology* (Mar. 2004) investigated allergic reactions of dental patients to local anesthetics. Based on a survey of dental practitioners, the study reported that the mean number of units (ampoules) of local anesthetics used per week by dentists was 79, with a standard deviation of 23. Suppose we want to determine the percentage of dentists who use less than 102 units of local anesthetics per week.

a. Assuming that nothing is known about the shape of the distribution for the data, what percentage of dentists use less than 102 units of local anesthetics per week?

b. Assuming that the data has a mound-shaped distribution, what percentage of dentists use less than 102 units of local anesthetics per week?

Applying the Concepts—Intermediate

2.100 Laptop use in middle school. Many middle schools have initiated a program that provides every student with a free laptop (notebook) computer. Student usage of laptops at a middle school that participates in the initiative was investigated in *American Secondary Education* (Fall 2009). In a sample of 106 students, the researchers reported the following statistics on how many minutes per day each student used his or her laptop for taking notes: $\bar{x} = 13.2, s = 19.5$

a. Compute the interval, $\bar{x} \pm 2s$.

b. Explain why the distribution of laptop usage for taking notes for these 106 students cannot be symmetric.

c. Given your answer to part **b**, what percentage of the 106 students have laptop usages that fall within the $\bar{x} \pm 2s$ interval?

2.101 Hand washing versus hand rubbing. In hospitals, washing the hands with soap is emphasized as the single most important measure in the prevention of infections. As an alternative to hand washing, some hospitals allow health workers to rub their hands with an alcohol-based antiseptic. The *British Medical Journal* (Aug. 17, 2002) reported on a study to compare the effectiveness of washing the hands with soap and rubbing the hands with alcohol. One group of health care workers used hand rubbing, while a second group used hand washing to clean their hands. The bacterial count (number of colony-forming units) on the hand of each worker was recorded. The table gives descriptive statistics on bacteria counts for the two groups of health care workers.

	Mean	Standard Deviation
Hand rubbing	35	59
Hand washing	69	106

a. For hand rubbers, form an interval that contains at least 75% of the bacterial counts. (*Note:* The bacterial count cannot be less than 0.)

b. Repeat part **a** for hand washers.

c. On the basis of your results in parts **a** and **b,** make an inference about the effectiveness of the two hand-cleaning methods.

2.102 Extinct New Zealand birds. Refer to the *Evolutionary Ecology Research* (July 2003) study of the patterns of extinction in the New Zealand bird population, presented in Exercise 2.20 (p. 37). Consider the data on the egg length (measured in millimeters) for the 132 bird species saved in the **NZBIRDS** file.

a. Find the mean and standard deviation of the egg lengths.

b. Form an interval that can be used to predict the egg length of a bird species found in New Zealand.

2.103 Velocity of Winchester bullets. The *American Rifleman* (June 1993) reported on the velocity of ammunition fired from the FEG P9R pistol, a 9-mm gun manufactured in Hungary. Field tests revealed that Winchester bullets fired from the pistol had a mean velocity (at 15 feet) of 936 feet per second and a standard deviation of 10 feet per second. Tests were also conducted with Uzi and Black Hills ammunition.

a. Describe the velocity distribution of Winchester bullets fired from the FEG P9R pistol.

b. A bullet whose brand is unknown is fired from the FEG P9R pistol. Suppose the velocity (at 15 feet) of the bullet is 1,000 feet per second. Is the bullet likely to be manufactured by Winchester? Explain.

2.104 Sentence complexity study. A study published in *Applied Psycholinguistics* (June 1998) compared the language skills of young children (16–30 months old) from low-income and middle-income families. A total of 260 children—65 in the low-income and 195 in the middle-income group—completed the Communicative Development Inventory (CDI) exam. One of the variables measured on each child was sentence complexity score. Summary statistics for the scores of the two groups are reproduced in the accompanying table. Use this information to sketch a graph of the sentence complexity score distribution for each income group. (Assume that the distributions are mound shaped and symmetric.) Compare the distributions. What can you infer?

	Low Income	Middle Income
Sample Size	65	195
Mean	7.62	15.55
Standard Deviation	8.91	12.24
Minimum	0	0
Maximum	36	37

Based on Arriaga, R. I., et al. "Scores on the MacArthur Communicative Development Inventory of children from low-income and middle-income families." *Applied Psycholinguistics,* Vol. 19, No. 2, June 1998, p. 217 (Table 7).

Applying the Concepts—Advanced

2.105 Improving SAT scores. The National Education Longitudinal Survey (NELS) tracks a nationally representative sample of U.S. students from eighth grade through high school and college. Research published in *Chance* (Winter 2001) examined the SAT scores of 265 NELS students who paid a private tutor to help them improve their scores. The table summarizes the changes in both the SAT-Mathematics and SAT-Verbal scores for these students.

	SAT-Math	SAT-Verbal
Mean change in score	19	7
Standard deviation of score changes	65	49

a. Suppose one of the 265 students who paid a private tutor is selected at random. Give an interval that is likely to contain the change in this student's SAT-Math score.

b. Repeat part **a** for the SAT-Verbal score.

c. Suppose the selected student's score increased on one of the SAT tests by 140 points. Which test, the SAT-Math or SAT-Verbal, is the one most likely to have had the 140-point increase? Explain.

2.106 Animal-assisted therapy for heart patients. Recall the *American Heart Association Conference* (Nov. 2005) study to gauge whether animal-assisted therapy can improve the physiological responses of heart failure patients. (See Chapter 1, Study 1.3, p. 4.) A team of nurses from the UCLA Medical Center randomly divided 76 heart patients into three groups. Each patient in group T was visited by a human volunteer accompanied by a trained dog, each patient in group V was visited by a volunteer only, and the patients in group C were not visited at all. The anxiety level of each patient was measured (in points) both before and after the visits. The accompanying table gives summary statistics for the drop in anxiety level for patients in the three groups. Suppose the anxiety level of a patient selected from the study had a drop of 22.5 points. From which group is the patient more likely to have come? Explain.

	Sample Size	Mean Drop	Std. Dev.
Group T: Volunteer + trained dog	26	10.5	7.6
Group V: Volunteer only	25	3.9	7.5
Group C: Control group (no visit)	25	1.4	7.5

Based on Cole, K., et al. "Animal assisted therapy decreases hemodynamics, plasma epinephrine and state anxiety in hospitalized heart failure patients." *American Journal of Critical Care*, 2007, 16: 575–585.

2.107 Land purchase decision. A buyer for a lumber company must decide whether to buy a piece of land containing 5,000 pine trees. If 1,000 of the trees are at least 40 feet tall, the buyer will purchase the land; otherwise, he won't. The owner of the land reports that the height of the trees has a mean of 30 feet and a standard deviation of 3 feet. On the basis of this information, what is the buyer's decision?

2.7 Numerical Measures of Relative Standing

We've seen that numerical measures of central tendency and variability describe the general nature of a quantitative data set (either a sample or a population). In addition, we may be interested in describing the *relative* quantitative location of a particular measurement within a data set. Descriptive measures of the relationship of a measurement to the rest of the data are called **measures of relative standing.**

One measure of the relative standing of a measurement is its **percentile ranking.** For example, suppose you scored an 80 on a test and you want to know how you fared in comparison with others in your class. If the instructor tells you that you scored at the 90th percentile, it means that 90% of the grades were lower than yours and 10% were higher. Thus, if the scores were described by the relative frequency histogram in Figure 2.26, the 90th percentile would be located at a point such that 90% of the total area under the relative frequency histogram lies below the 90th percentile and 10% lies above. If the instructor tells you that you scored in the 50th percentile (the median of the data set), 50% of the test grades would be lower than yours and 50% would be higher.

Percentile rankings are of practical value only with large data sets. Finding them involves a process similar to the one used in finding a median. The measurements are ranked in order, and a rule is selected to define the location of each percentile. Since we are interested primarily in interpreting the percentile rankings of measurements (rather than in finding particular percentiles for a data set), we define the *p*th *percentile* of a data set as follows.

Figure 2.26

Location of 90th percentile for test grades

For any set of *n* measurements (arranged in ascending or descending order), the ***p*th percentile** is a number such that *p*% of the measurements fall below that number and $(100 - p)$% fall above it.

Example 2.14

Finding and Interpreting Percentiles—The "Water Level Task"

Problem Refer to the water level task deviations for the 40 subjects in Table 2.4. An SPSS printout describing the data is shown in Figure 2.27. Locate the 25th percentile and 95th percentile on the printout, and interpret the associated values.

Solution Both the 25th percentile and 95th percentile are highlighted on the SPSS printout. The values in question are 2.0 and 24.8, respectively. Our interpretations are as follows: 25% of the 40 deviations fall below 2.0 and 95% of the deviations fall below 24.8.

Case Processing Summary

	Cases					
	Valid		Missing		Total	
	N	Percent	N	Percent	N	Percent
Deviation	40	100.0%	0	.0%	40	100.0%

Percentiles

		Percentiles						
		5	10	25	50	75	90	95
Weighted Average (Definition 1)	Deviation	-4.90	-1.00	2.00	7.00	10.00	15.80	24.80
Tukey's Hinges	Deviation			2.00	7.00	10.00		

Figure 2.27
SPSS percentile statistics for water-level task deviations

Look Back The method for computing percentiles with small data sets varies according to the software used. As the sample size increases, the percentiles from the different software packages will converge to a single number.

Now Work Exercise 2.109

Figure 2.28
The quartiles for a data set

Percentiles that partition a data set into four categories, each category containing exactly 25% of the measurements, are called **quartiles.** The *lower quartile (Q_L)* is the 25th percentile, the *middle quartile (M)* is the median or 50th percentile, and the *upper quartile (Q_U)* is the 75th percentile, as shown in Figure 2.28. Therefore, in Example 2.14, we have (from the SPSS printout, Figure 2.27), $Q_L = 2.0$, $M = 7.0$, and $Q_U = 10.0$. Quartiles will prove useful in finding unusual observations in a data set (Section 2.8).

> The **lower quartile (Q_L)** is the 25th percentile of a data set. The **middle quartile (M)** is the median or 50th percentile. The **upper quartile (Q_U)** is the 75th percentile.

Another measure of relative standing in popular use is the **z-score.** As you can see in the following definition, the z-score makes use of the mean and standard deviation of a data set in order to specify the relative location of the measurement.

> The **sample z-score** for a measurement x is
>
> $$z = \frac{x - \bar{x}}{s}$$
>
> The **population z-score** for a measurement x is
>
> $$z = \frac{x - \mu}{\sigma}$$

Note that the z-score is calculated by subtracting $\bar{x}$ (or μ) from the measurement x and then dividing the result by s (or σ). The final result, the **z-score**, represents the distance between a given measurement x and the mean, expressed in standard deviations.

Example 2.15 Finding a z-Score	**Problem** Suppose a sample of 2,000 high school seniors' verbal SAT scores is selected. The mean and standard deviation are $$\bar{x} = 550 \quad s = 75$$ Suppose Joe Smith's score is 475. What is his sample z-score? **Solution** Joe Smith's verbal SAT score lies below the mean score of the 2,000 seniors, as shown in Figure 2.29.

Figure 2.29

Verbal SAT scores of high school seniors

We compute

$$z = \frac{x - \bar{x}}{s} = \frac{475 - 550}{75} = -1.0$$

which tells us that Joe Smith's score is 1.0 standard deviation *below* the sample mean; in short, his sample z-score is -1.0.

Look Back The numerical value of the z-score reflects the relative standing of the measurement. A large positive z-score implies that the measurement is larger than almost all other measurements, whereas a large negative z-score indicates that the measurement is smaller than almost every other measurement. If a z-score is 0 or near 0, the measurement is located at or near the mean of the sample or population.

Now Work Exercise 2.111

We can be more specific if we know that the frequency distribution of the measurements is mound shaped. In this case, the following interpretation of the z-score can be given:

Interpretation of z-Scores for Mound-Shaped Distributions of Data

1. Approximately 68% of the measurements will have a z-score between −1 and 1.

2. Approximately 95% of the measurements will have a z-score between −2 and 2.

3. Approximately 99.7% (almost all) of the measurements will have a z-score between −3 and 3.

Note that this interpretation of z-scores is identical to that given by the empirical rule for mound-shaped distributions (Table 2.7). The statement that a measurement falls into the interval from $(\mu - \sigma)$ to $(\mu + \sigma)$ is equivalent to the statement that a measurement has a population z-score between -1 and 1, since all measurements between $(\mu - \sigma)$ and $(\mu + \sigma)$ are within one standard deviation of μ. These z-scores are displayed in Figure 2.30.

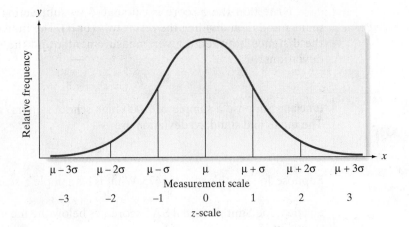

Figure 2.30
Population z-scores for a
mound-shaped distribution

Exercises 2.108–2.124

Understanding the Principles

2.108 For a quantitative data set
  a. What is the 50th percentile called?
  b. Define Q_L.
  c. Define Q_u.

2.109 Give the percentage of measurements in a data set that are
[NW] above and below each of the following percentiles:
  a. 75th percentile
  b. 50th percentile
  c. 20th percentile
  d. 84th percentile

2.110 For mound-shaped data, what percentage of measurements have a z-score between -2 and 2?

Learning the Mechanics

2.111 Compute the z-score corresponding to each of the follow-
[NW] ing values of x:
  a. $x = 40, s = 5, \bar{x} = 30$
  b. $x = 90, \mu = 89, \sigma = 2$
  c. $\mu = 50, \sigma = 5, x = 50$
  d. $s = 4, x = 20, \bar{x} = 30$
  e. In parts **a–d**, state whether the z-score locates x within a sample or within a population.
  f. In parts **a–d**, state whether each value of x lies above or below the mean and by how many standard deviations.

2.112 Compare the z-scores to decide which of the following x values lie the greatest distance above the mean and the greatest distance below the mean:
  a. $x = 100, \mu = 50, \sigma = 25$
  b. $x = 1, \mu = 4, \sigma = 1$
  c. $x = 0, \mu = 200, \sigma = 100$
  d. $x = 10, \mu = 5, \sigma = 3$

2.113 Suppose that 40 and 90 are two elements of a population data set and that their z-scores are -2 and 3, respectively. Using only this information, is it possible to determine the population's mean and standard deviation? If so, find them. If not, explain why it's not possible.

Applying the Concepts—Basic

2.114 **Math scores of eighth graders.** According to the National Center for Education Statistics (2009), scores on a mathematics assessment test for United States eighth graders have a mean of 283, a 25th percentile of 259, a 75th percentile of 308, and a 90th percentile of 329. Interpret each of these numerical descriptive measures.

2.115 **Drivers stopped by police.** According to the Bureau of Justice Statistics (June 2006), 74% of all licensed drivers stopped by police are 25 years or older. Give a percentile ranking for the age of 25 years in the distribution of all ages of licensed drivers stopped by police.

2.116 **Sanitation inspection of cruise ships.** Refer to the sanitation levels of cruise ships presented in Exercise 2.35 (p. 47) and saved in the **SHIPSANIT** file.
  a. Give a measure of relative standing for the *Nautilus Explorer's* score of 74. Interpret the result.
  b. Give a measure of relative standing for the *Rotterdam's* score of 92. Interpret the result.

2.117 **Motivation of drug dealers.** Refer to the *Applied Psychology in Criminal Justice* (Sept. 2009) study of convicted drug dealers' motivations, Exercise 2.98 (p. 71). Recall that the sample of drug dealers had a mean Wanting Recognition (WR) score of 39 points, with a standard deviation of 6 points.
  a. Find and interpret the z-score for a drug dealer with a WR score of 30 points.
  b. What proportion of the sampled drug dealers had WR scores below 39 points? (Assume the distribution of WR scores is mound shaped and symmetric.)

Applying the Concepts—Intermediate

2.118 **Reading Japanese books.** Refer to the *Reading in a Foreign Language* (Apr. 2004) experiment to improve the Japanese reading comprehension levels of 14 University of Hawaii students, presented in Exercise 2.61 (p. 57). The data on number of books read and grade for each student are saved in the **JAPANESE** file.

a. Find the mean and standard deviation of the number of books read by students who earned an A grade. Find the z-score for an A student who read 40 books. Interpret the result.

b. Find the mean and standard deviation of the number of books read by students who earned either a B or a C grade. Find the z-score for a B or C student who read 40 books. Interpret the result.

c. Refer to parts **a** and **b**. Which of the two groups of students is more likely to have read 40 books? Explain.

2.119 Extinct New Zealand birds. Refer to the *Evolutionary Ecology Research* (July 2003) study of the patterns of extinction in the New Zealand bird population, first presented in Exercise 2.20 (p. 37). Again, consider the data on the egg length (measured in millimeters) for the 132 bird species saved in the **NZBIRDS** file.

a. Find the 10th percentile for the egg length distribution and interpret its value.

b. The *moas (P. australis)* is a bird species with an egg length of 205 millimeters. Find the z-score for this species of bird and interpret its value.

2.120 Lead in drinking water. The U.S. Environmental Protection Agency (EPA) sets a limit on the amount of lead permitted in drinking water. The EPA *Action Level* for lead is .015 milligram per liter (mg/L) of water. Under EPA guidelines, if 90% of a water system's study samples have a lead concentration less than .015 mg/L, the water is considered safe for drinking. I (coauthor Sincich) received a report on a study of lead levels in the drinking water of homes in my subdivision. The 90th percentile of the study sample had a lead concentration of .00372 mg/L. Are water customers in my subdivision at risk of drinking water with unhealthy lead levels? Explain.

2.121 Ranking Ph.D. programs in economics. Thousands of students apply for admission to graduate schools in economics each year with the intention of obtaining a Ph.D. The *Southern Economic Journal* (Apr. 2008) published a guide to graduate study in economics by ranking the Ph.D. programs at 129 colleges and universities. Each program was evaluated according to the number of publications published by faculty teaching in the Ph.D. program and by the quality of the publications. Data obtained from the Social Science Citation Index (SSCI) were used to calculate an overall productivity score for each Ph.D. program. The mean and standard deviation of these 129 productivity scores were then used to compute a z-score for each economics program. Harvard University had the highest z-score ($z = 5.08$) and, hence, was the top-ranked school; Howard University was ranked last since it had the lowest z-score ($z = -0.81$). The data (z-scores) for all 129 economic programs are saved in the **ECOPHD** file.

a. Interpret the z-score for Harvard University.

b. Interpret the z-score for Howard University.

c. The authors of the *Southern Economic Journal* article note that "only 44 of the 129 schools have positive z-scores, indicating that the distribution of overall productivity is skewed to the right." Do you agree? (Check your answer by constructing a histogram for the z-scores in the **ECOPHD** file.)

2.122 Blue versus red exam study. In a study of how external clues influence performance, psychology professors at the University of Alberta and Pennsylvania State University gave two different forms of a midterm examination to a large group of introductory psychology students. The questions on the exam were identical and in the same order, but one exam was printed on blue paper and the other on red paper (*Teaching Psychology*, May 1998). Grading only the difficult questions on the exam, the researchers found that scores on the blue exam had a distribution with a mean of 53% and a standard deviation of 15%, while scores on the red exam had a distribution with a mean of 39% and a standard deviation of 12%. (Assume that both distributions are approximately mound shaped and symmetric.)

a. Give an interpretation of the standard deviation for the students who took the blue exam.

b. Give an interpretation of the standard deviation for the students who took the red exam.

c. Suppose a student is selected at random from the group of students who participated in the study and the student's score on the difficult questions is 20%. Which exam form is the student more likely to have taken, the blue or the red exam? Explain.

Applying the Concepts—Advanced

2.123 GPAs of students. At one university, the students are given z-scores at the end of each semester, rather than the traditional GPAs. The mean and standard deviation of all students' cumulative GPAs, on which the z-scores are based, are 2.7 and .5, respectively.

a. Translate each of the following z-scores to corresponding GPA scores: $z = 2.0, z = -1.0, z = .5, z = -2.5$.

b. Students with z-scores below -1.6 are put on probation. What is the corresponding probationary GPA?

c. The president of the university wishes to graduate the top 16% of the students with *cum laude* honors and the top 2.5% with *summa cum laude* honors. Where (approximately) should the limits be set in terms of z-scores? In terms of GPAs? What assumption, if any, did you make about the distribution of the GPAs at the university?

2.124 Ranking Ph.D. programs in economics (cont'd). Refer to the *Southern Economic Journal* (Apr. 2008) study of Ph.D. programs in economics, Exercise 2.121. The authors also made the following observation: "A noticeable feature of this skewness is that distinction between schools diminishes as the rank declines. For example, the top-ranked school, Harvard, has a z-score of 5.08, and the fifth-ranked school, Yale, has a z-score of 2.18, a substantial difference. However,…the 70th-ranked school, the University of Massachusetts, has a z-score of -0.43, and the 80th-ranked school, the University of Delaware, has a z-score of -0.50, a very small difference. [Consequently] the ordinal rankings presented in much of the literature that ranks economics departments miss the fact that below a relatively small group of top programs, the differences in [overall] productivity become fairly small." Do you agree?

2.8 Methods for Detecting Outliers: Box Plots and z-Scores

Sometimes it is important to identify inconsistent or unusual measurements in a data set. An observation that is unusually large or small relative to the data values we want to describe is called an **outlier**.

Outliers are often attributable to one of several causes. First, the measurement associated with the outlier may be invalid. For example, the experimental procedure used to generate the measurement may have malfunctioned, the experimenter may have misrecorded the measurement, or the data might have been coded incorrectly in the computer. Second, the outlier may be the result of a misclassified measurement. That is, the measurement belongs to a population different from that from which the rest of the sample was drawn. Finally, the measurement associated with the outlier may be recorded correctly and from the same population as the rest of the sample, but represent a rare (chance) event. Such outliers occur most often when the relative frequency distribution of the sample data is extremely skewed, because a skewed distribution has a tendency to include extremely large or small observations relative to the others in the data set.

An observation (or measurement) that is unusually large or small relative to the other values in a data set is called an **outlier**. Outliers typically are attributable to one of the following causes:

1. The measurement is observed, recorded, or entered into the computer incorrectly.
2. The measurement comes from a different population.
3. The measurement is correct, but represents a rare (chance) event.

Two useful methods for detecting outliers, one graphical and one numerical, are **box plots** and z-scores. The box plot is based on the quartiles (defined in Section 2.7) of a data set. Specifically, a box plot is based on the *interquartile range (IQR)*—the distance between the lower and upper quartiles:

$$IQR = Q_U - Q_L$$

The **interquartile range (IQR)** is the distance between the lower and upper quartiles:

$$IQR = Q_U - Q_L$$

An annotated MINITAB box plot for the gas mileage data (see Table 2.2) is shown in Figure 2.31.[*] Note that a rectangle (the *box*) is drawn, with the bottom and top of the rectangle (the **hinges**) drawn at the quartiles Q_L and Q_U, respectively. Recall that Q_L represents the 25th percentile and Q_u represents the 75th percentile. By definition, then, the "middle" 50% of the observations—those between Q_L and Q_U—fall inside the box. For the gas mileage data, these quartiles are at 35.625 and 38.375. Thus,

$$IQR = 38.375 - 35.625 = 2.75$$

The median is shown at 37 by a horizontal line within the box.

To guide the construction of the "tails" of the box plot, two sets of limits, called **inner fences** and **outer fences,** are used. Neither set of fences actually appears on the plot. Inner fences are located at a distance of 1.5(IQR) from the hinges. Emanating from the hinges of the box are vertical lines called the **whiskers.** The two whiskers extend to the most extreme observation inside the inner fences. For example, the inner fence on the lower side of the gas mileage box plot is

Lower inner fence = Lower hinge − 1.5(IQR)
= 35.625 − 1.5(2.75)
= 35.625 − 4.125 = 31.5

---

[*]Although box plots can be generated by hand, the amount of detail required makes them particularly well suited for computer generation. We use computer software to generate the box plots in this section.

Figure 2.31
Annotated MINITAB box plot
for EPA gas mileages

The smallest measurement *inside* this fence is the second-smallest measurement, 31.8. Thus, the lower whisker extends to 31.8. Similarly, the upper whisker extends to 42.1, the largest measurement inside the upper inner fence:

$$
\begin{aligned}
\text{Upper inner fence} &= \text{Upper hinge} + 1.5(\text{IQR}) \\
&= 38.375 + 1.5(2.75) \\
&= 38.375 + 4.125 = 42.5
\end{aligned}
$$

Values that are beyond the inner fences are deemed *potential outliers* because they are extreme values that represent relatively rare occurrences. In fact, for mound-shaped distributions, less than 1% of the observations are expected to fall outside the inner fences. Two of the 100 gas mileage measurements, 30.0 and 44.9, fall beyond the inner fences, one on each end of the distribution. Each of these potential outliers is represented by a common symbol (an asterisk in MINITAB).

The other two imaginary fences, the outer fences, are defined at a distance 3(IQR) from each end of the box. Measurements that fall beyond the outer fences (also represented by an asterisk in MINITAB) are very extreme measurements that require special analysis. Since less than one-hundredth of 1% (.01% or .0001) of the measurements from mound-shaped distributions are expected to fall beyond the outer fences, these measurements are considered to be *outliers*. Because there are no measurements of gas mileage beyond the outer fences, there are no outliers.

Recall that outliers may be incorrectly recorded observations, members of a population different from the rest of the sample, or, at the least, very unusual measurements from the same population. The box plot of Figure 2.31 detected two potential outliers: the two gas mileage measurements beyond the inner fences. When we analyze these measurements, we find that they are correctly recorded. Perhaps they represent mileages that correspond to exceptional models of the car being tested or to unusual gas mixtures. Outlier analysis often reveals useful information of this kind and therefore plays an important role in the statistical inference-making process.

In addition to detecting outliers, box plots provide useful information on the variation in a data set. The elements (and nomenclature) of box plots are summarized in the box. Some aids to the interpretation of box plots are also given.

Elements of a Box Plot

1. A rectangle (the **box**) is drawn with the ends (the **hinges**) drawn at the lower and upper quartiles (Q_L and Q_U). The median of the data is shown in the box, usually by a line.

2. The points at distances 1.5(IQR) from each hinge mark the **inner fences** of the data set. Lines (the **whiskers**) are drawn from each hinge to the most extreme measurement inside the inner fence. Thus,

$$\text{Lower inner fence} = Q_L - 1.5(\text{IQR})$$
$$\text{Upper inner fence} = Q_U + 1.5(\text{IQR})$$

3. A second pair of fences, the **outer fences**, appears at a distance of 3(IQR) from the hinges. One symbol (e.g., "*") is used to represent measurements falling between the inner and outer fences, and another (e.g., "0") is used to represent measurements that lie beyond the outer fences. Thus, outer fences are not shown unless one or more measurements lie beyond them. We have

$$\text{Lower outer fence} = Q_L - 3(\text{IQR})$$
$$\text{Upper outer fence} = Q_U + 3(\text{IQR})$$

4. The symbols used to represent the median and the extreme data points (those beyond the fences) will vary with the software you use to construct the box plot. (You may use your own symbols if you are constructing a box plot by hand.) You should consult the program's documentation to determine exactly which symbols are used.

Aids to the Interpretation of Box Plots

1. The line (median) inside the box represents the "center" of the distribution of data.

2. Examine the length of the box. The IQR is a measure of the sample's variability and is especially useful for the comparison of two samples. (See Example 2.16.)

3. Visually compare the lengths of the whiskers. If one is clearly longer, the distribution of the data is probably skewed in the direction of the longer whisker.

4. Analyze any measurements that lie beyond the fences. Less than 5% should fall beyond the inner fences, even for very skewed distributions. Measurements beyond the outer fences are probably outliers, with one of the following explanations:

   a. The measurement is incorrect. It may have been observed, recorded, or entered into the computer incorrectly.

   b. The measurement belongs to a population different from the population that the rest of the sample was drawn from. (See Example 2.17.)

   c. The measurement is correct *and* from the same population as the rest of the sample. Generally, we accept this explanation only after carefully ruling out all others.

Example 2.16
Computer-Generated Box Plot—The "Water-Level Task"

Problem Refer to the *Psychological Science* experiment, called the "water-level task," Examples 2.2 and 2.14. Use statistical software to produce a box plot for the 40 water-level task deviation angles shown in Table 2.4. Identify any outliers in the data set.

Solution The 40 deviations were entered into MINITAB, and a box plot produced in Figure 2.32. Recall (from Example 2.14) that $Q_L = 2$, $M = 7$, and $Q_U = 10$. The middle line in the box represents the median deviation of $M = 7$. The lower hinge of the box represents $Q_L = 2$, while the upper hinge represents $Q_U = 10$. Note that there are two extreme observations (represented by asterisks) that are beyond the upper inner fence, but inside the upper outer fence. These two values are outliers. Examination of the data reveals that these measurements correspond to deviation angles of 25° and 35°.

Look Ahead Before removing the outliers from the data set, a good analyst will make a concerted effort to find the cause of the outliers. For example, both the outliers are deviation measurements for waitresses. An investigation may discover that both waitresses were new on the job the day the experiment was conducted. If the study target population is experienced bartenders and waitresses, these two observations would be dropped from the analysis.

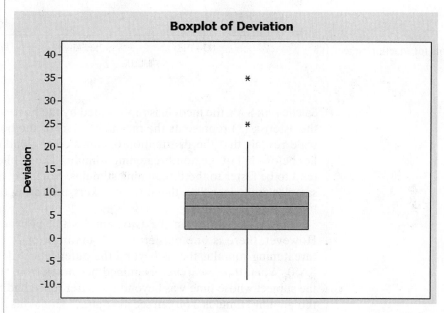

Figure 2.32
MINITAB box plot for water-level task deviations

Now Work Exercise 2.131

Example 2.17

Comparing Box Plots—Stimulus Reaction Study

Problem A Ph.D. student in psychology conducted a stimulus reaction experiment as a part of her dissertation research. She subjected 50 subjects to a threatening stimulus and 50 to a nonthreatening stimulus. The reaction times of all 100 students, recorded to the nearest tenth of a second, are listed in Table 2.9. Box plots of the two resulting samples of reaction times, generated with SAS, are shown in Figure 2.33. Interpret the box plots.

Table 2.9   Reaction Times of Students

Nonthreatening Stimulus									
2.0	1.8	2.3	2.1	2.0	2.2	2.1	2.2	2.1	2.1
2.0	2.0	1.8	1.9	2.2	2.0	2.2	2.4	2.1	2.0
2.2	2.1	2.2	1.9	1.7	2.0	2.0	2.3	2.1	1.9
2.0	2.2	1.6	2.1	2.3	2.0	2.0	2.0	2.2	2.6
2.0	2.0	1.9	1.9	2.2	2.3	1.8	1.7	1.7	1.8

Threatening Stimulus									
1.8	1.7	1.4	2.1	1.3	1.5	1.6	1.8	1.5	1.4
1.4	2.0	1.5	1.8	1.4	1.7	1.7	1.7	1.4	1.9
1.9	1.7	1.6	2.5	1.6	1.6	1.8	1.7	1.9	1.9
1.5	1.8	1.6	1.9	1.3	1.5	1.6	1.5	1.6	1.5
1.3	1.7	1.3	1.7	1.7	1.8	1.6	1.7	1.7	1.7

Data Set: REACTION

Box Plots of Time by Stimulus

Figure 2.33
SAS box plots for reaction-time data

Solution In SAS, the median is represented by the horizontal line through the box, while the asterisk (*) represents the mean. Analysis of the box plots on the same numerical scale reveals that the distribution of times corresponding to the threatening stimulus lies below that of the nonthreatening stimulus. The implication is that the reaction times tend to be faster to the threatening stimulus. Note, too, that the upper whiskers of both samples are longer than the lower whiskers, indicating that the reaction times are positively skewed.

No observations in the two samples fall between the inner and outer fences. However, there is one outlier: the observation of 2.5 seconds corresponding to the threatening stimulus that is beyond the outer fence (denoted by the square symbol in SAS). When the researcher examined her notes from the experiments, she found that the subject whose time was beyond the outer fence had mistakenly been given the nonthreatening stimulus. You can see in Figure 2.33 that his time would have been within the upper whisker if moved to the box plot corresponding to the nonthreatening stimulus. The box plots should be reconstructed, since they will both change slightly when this misclassified reaction time is moved from one sample to the other.

Look Ahead The researcher concluded that the reactions to the threatening stimulus were faster than those to the nonthreatening stimulus. However, she was asked by her Ph.D. committee whether the results were *statistically significant*. Their question addresses the issue of whether the observed difference between the samples might be attributable to chance or sampling variation rather than to real differences between the populations. To answer this question, the researcher must use inferential statistics rather than graphical descriptions. We discuss how to compare two samples using inferential statistics in Chapters 7 and 8.

The next example illustrates how z-scores can be used to detect outliers and make inferences.

Example 2.18
Inference Using
z-Scores

Problem Suppose a female bank employee believes that her salary is low as a result of sex discrimination. To substantiate her belief, she collects information on the salaries of her male counterparts in the banking business. She finds that their salaries have a mean of $64,000 and a standard deviation of $2,000. Her salary is $57,000. Does this information support her claim of sex discrimination?

Solution The analysis might proceed as follows: First, we calculate the z-score for the woman's salary with respect to those of her male counterparts. Thus,

$$z = \frac{\$57,000 - \$64,000}{\$2,000} = -3.5$$

The implication is that the woman's salary is 3.5 standard deviations *below* the mean of the male salary distribution. Furthermore, if a check of the male salary data shows that the frequency distribution is mound shaped, we can infer that very few salaries in this distribution should have a z-score less than -3, as shown in Figure 2.34. Clearly, a z-score of -3.5 represents an outlier. Either her salary is from a distribution different from the male salary distribution, or it is a very unusual (highly improbable) measurement from a salary distribution no different from the male distribution.

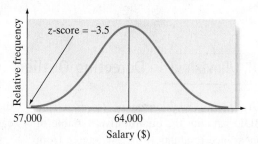

Figure 2.34
Male salary distribution

Look Back Which of the two situations do you think prevails? Statistical thinking would lead us to conclude that her salary does not come from the male salary distribution, lending support to the female bank employee's claim of sex discrimination. However, the careful investigator should require more information before inferring that sex discrimination is the cause. We would want to know more about the data collection technique the woman used and more about her competence at her job. Also, perhaps other factors, such as length of employment, should be considered in the analysis.

Now Work Exercise 2.129

Examples 2.17 and 2.18 exemplify an approach to statistical inference that might be called the **rare-event approach.** An experimenter hypothesizes a specific frequency distribution to describe a population of measurements. Then a sample of measurements is drawn from the population. If the experimenter finds it unlikely that the sample came from the hypothesized distribution, the hypothesis is judged to be false. Thus, in Example 2.18, the woman believes that her salary reflects discrimination. She hypothesizes that her salary should be just another measurement in the distribution of her male counterparts' salaries if no discrimination exists. However, it is so unlikely that the sample (in this case, her salary) came from the male frequency distribution that she rejects that hypothesis, concluding that the distribution from which her salary was drawn is different from the distribution for the men.

This rare-event approach to inference making is discussed further in later chapters. Proper application of the approach requires a knowledge of probability, the subject of our next chapter.

We conclude this section with some rules of thumb for detecting outliers.

Rules of Thumb for Detecting Outliers[*]

Box Plots: Observations falling between the inner and outer fences are deemed *suspect outliers*. Observations falling beyond the outer fence are deemed *highly suspect outliers*.

(continued)

---

[*]The z-score and box plot methods both establish rule-of-thumb limits outside of which a measurement is deemed to be an outlier. Usually, the two methods produce similar results. However, the presence of one or more outliers in a data set can inflate the computed value of s. Consequently, it will be less likely that an errant observation would have a z-score larger than $|3|$. In contrast, the values of the quartiles used to calculate the intervals for a box plot are not affected by the presence of outliers.

Suspect outliers

Between $Q_L - 1.5(\text{IQR})$

and $Q_L - 3(\text{IQR})$

or

Between $Q_U + 1.5(\text{IQR})$

and $Q_U + 3(\text{IQR})$

Highly suspect outliers

Below $Q_L - 3(\text{IQR})$

or

Above $Q_U + 3(\text{IQR})$

***z*-Scores:** Observations with z-scores greater than 3 in absolute value are considered *outliers*. For some highly skewed data sets, observations with z-scores greater than 2 in absolute value *may be outliers*.

Possible Outliers

$|z| > 2$

Outliers

$|z| > 3$

Statistics IN Action | Revisited   Detecting Outliers in the Body Image Data

In the *Body Image: An International Journal of Research* (Jan. 2010) study of 92 BDD patients, the quantitative variable of interest is Total Appearance Evaluation score (ranging from 7 to 35 points). Are there any unusual scores in the BDD data set? We will apply both the box plot and z-score method to aid in identifying any outliers in the data. Since from previous analyses, there appears to be a difference in the distribution of appearance evaluation scores for males and females, we will analyze the data by gender.

To employ the z-score method, we require the mean and standard deviation of the data for each gender. These values were already computed in the previous Statistics in Action Revisited section. For females, $\bar{x} = 17.05$ and $s = 4.76$; for males, $\bar{x} = 19.0$ and $s = 5.42$ (see Figure SIA2.5). Then, the 3-standard-deviation interval for each gender is:

Females: $\bar{x} \pm 3s = 17.05 \pm 3(4.76) = 17.05 \pm 14.28 = (2.77, 31.33)$

Males: $\bar{x} \pm 3s = 19.00 \pm 3(5.42) = 19.00 \pm 16.26 = (2.74, 35.26)$

If you examine the appearance evaluation scores in the **BDD** file, you will find that none of the scores fall beyond

the 3-standard-deviation interval for each group. Consequently, if we use the z-score approach, there are no highly suspect outliers in the data.

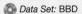

Box plots for the data are shown in Figure SIA2.6. Although several suspect outliers (asterisks) are shown on the box plot for each gender, there are no highly suspect outliers (zeros) shown. That is, no data points fall beyond the outer fences of the box plots. As with the z-score approach, the box plot method does not detect any highly suspect outliers.

[*Note*: If we were to detect one or more highly suspect outliers, we should investigate whether or not to include the observation in any analysis which leads to an inference about the population of BDD patients. Is the outlier a legitimate value (in which case it will remain in the data set for analysis) or is the outlier associated with a subject that is not a member of the population of interest—say, a person who is misdiagnosed with BDD (in which case it will be removed from the data set prior to analysis).]

Data Set: BBD

Figure SIA2.6

MINITAB box plots for Appearance Evaluation score by Gender

Exercises 2.125–2.142

Understanding the Principles

2.125 Define an outlier.

2.126 What is the interquartile range?

2.127 What are the hinges of a box plot?

2.128 With mound-shaped data, what proportion of the measurements have *z*-scores between −3 and 3?

Learning the Mechanics

2.129 A sample data set has a mean of 57 and a standard
NW deviation of 11. Determine whether each of the following sample measurements is an outlier.
  a. 65
  b. 21
  c. 72
  d. 98

2.130 Suppose a data set consisting of exam scores has a lower quartile $Q_L = 60$, a median $M = 75$, and an upper quartile $Q_U = 85$. The scores on the exam range from 18 to 100. Without having the actual scores available to you, construct as much of the box plot as possible.

2.131 Consider the following horizontal box plot:
NW

  a. What is the median of the data set (approximately)?
  b. What are the upper and lower quartiles of the data set (approximately)?
  c. What is the interquartile range of the data set (approximately)?
  d. Is the data set skewed to the left, skewed to the right, or symmetric?
  e. What percentage of the measurements in the data set lie to the right of the median? To the left of the upper quartile?
  f. Identify any outliers in the data.

2.132 Consider the following two sample data sets, saved in the **LM2_132** file:

Sample A			Sample B		
121	171	158	171	152	170
173	184	163	168	169	171
157	85	145	190	183	185
165	172	196	140	173	206
170	159	172	172	174	169
161	187	100	199	151	180
142	166	171	167	170	188

  a. Construct a box plot for each data set.
  b. Identify any outliers that may exist in the two data sets.

Applet Exercise 2.7

Use the applet *Standard Deviation* to determine whether an item in a data set may be an outlier. Begin by setting appropriate limits and plotting the given data on the number line provided in the applet. Here is the data set:

10 80 80 85 85 85 85 90 90 90 90 90 95 95 95 95 100 100

a. The green arrow shows the approximate location of the mean. Multiply the standard deviation given by the applet by 3. Is the data item 10 more than three standard deviations away from the green arrow (the mean)? Can you conclude that the 10 is an outlier?

b. Using the mean and standard deviation from part **a**, move the point at 10 on your plot to a point that appears to be about three standard deviations from the mean. Repeat the process in part **a** for the new plot and the new suspected outlier.

c. When you replaced the extreme value in part **a** with a number that appeared to be within three standard deviations of the mean, the standard deviation got smaller and the mean moved to the right, yielding a new data set whose extreme value was *not* within three standard deviations of the mean. Continue to replace the extreme value with higher numbers until the new value is within three standard deviations of the mean in the new data set. Use trial and error to estimate the smallest number that can replace the 10 in the original data set so that the replacement is not considered to be an outlier.

Applying the Concepts—Basic

2.133 Treating psoriasis with the "Doctorfish of Kangal." Psoriasis is a skin disorder with no known cure and no proven effective pharmacological treatment. An alternative treatment for psoriasis is ichthyotherapy also known as therapy with the "Doctorfish of Kangal." Fish from the hot pools of Kangal, Turkey, feed on the skin scales of bathers, reportedly reducing the symptoms of psoriasis. In one study, 67 patients diagnosed with psoriasis underwent three weeks of ichthyotherapy. (*Evidence-Based Research in Complementary and Alternative Medicine*, Dec. 2006). The Psoriasis Area Severity Index (PASI) of each patient was measured both before and after treatment. (The lower the PASI score, the better is the skin condition.) Box plots of the PASI scores, both before (baseline) and after three weeks of ichthyotherapy treatment, are shown in the accompanying diagram.

a. Find the approximate 25th percentile, the median, and the 75th percentile for the PASI scores before treatment.

b. Find the approximate 25th percentile, the median, and the 75th percentile for the PASI scores after treatment.

c. Comment on the effectiveness of ichthyotherapy in treating psoriasis.

Based on Grassberger, M., and Hoch, W. "Ichthyotherapy as alternative treatment for patients with psoriasis: A pilot study." *Evidence-Based Research in Complementary and Alternative Medicine*, Vol. 3, No. 4, Dec. 2004 (Figure 3). National Center for Complementary and Alternative Medicine.

2.134 Dentists' use of anesthetics. Refer to the *Current Allergy & Clinical Immunology* study of the use of local anesthetics in dentistry, presented in Exercise 2.99 (p. 72). Recall that the mean number of units (ampoules) of local anesthetics used per week by dentists was 79, with a standard deviation of 23. Consider a dentist who used 175 units of local anesthetics in a week.

a. Find the z-score for this measurement.

b. Would you consider the measurement to be an outlier? Explain.

c. Give several reasons the outlier may have occurred.

2.135 Research on brain specimens. Refer to the *Brain and Language* data on postmortem intervals (PMIs) of 22 human brain specimens, presented in Exercise 2.39 (p. 48). The data are saved in the **BRAINPMI** file. The mean and standard deviation of the PMI values are 7.3 and 3.18, respectively.

a. Find the z-score for the PMI value of 3.3.

b. Is the PMI value of 3:3 considered an outlier? Explain.

2.136 Estimating the age of glacial drifts. Tills are glacial drifts consisting of a mixture of clay, sand, gravel, and boulders. Engineers from the University of Washington's Department of Earth and Space Sciences studied the chemical makeup of buried tills in order to estimate the age of the glacial drifts in Wisconsin (*American Journal of Science*, Jan. 2005). The ratio of the elements aluminum (AI) and beryllium (Be) in sediment is related to the duration of burial. The Al–Be ratios for a sample of 26 buried till specimens are given in the accompanying table and saved in the **TILLRATIO** file. An SPSS printout with descriptive statistics for the Al–Be ratio is shown at the bottom of the page.

| 3.75 | 4.05 | 3.81 | 3.23 | 3.13 | 3.30 | 3.21 | 3.32 | 4.09 | 3.90 | 5.06 | 3.85 | 3.88 |
| 4.06 | 4.56 | 3.60 | 3.27 | 4.09 | 3.38 | 3.37 | 2.73 | 2.95 | 2.25 | 2.73 | 2.55 | 3.06 |

Based on *American Journal of Science*, Vol. 305, No. 1, Jan. 2005, p. 16 (Table 2).

a. Find and interpret the z-score associated with the largest ratio, the smallest ratio, and the mean ratio.

b. Would you consider the largest ratio to be unusually large? Why or why not?

c. Construct a box plot for the data and identify any outliers.

Applying the Concepts—Intermediate

2.137 Sanitation inspection of cruise ships. Refer to the data on sanitation levels of cruise ships, presented in Exercise 2.35 (p. 47). These data are saved in the **SHIPSANIT** file.

a. Use the box plot method to detect any outliers in the data.

b. Use the z-score method to detect any outliers in the data.

c. Do the two methods agree? If not, explain why.

2.138 Characteristics of a rockfall. Refer to the *Environmental Geology* (Vol. 58, 2009) study of how far a block from a collapsing rockwall will bounce, Exercise 2.59 (p. 57). The computer-simulated rebound lengths (in meters) for 13 block-soil impact marks left on a slope from an actual rock-fall (saved in the **ROCKWELL** file) are reproduced in the table. Do you detect any outliers in the data? Explain.

| 10.94 | 13.71 | 11.38 | 7.26 | 17.83 | 11.92 | 11.87 | 5.44 | 13.35 | 4.90 | 5.85 | 5.10 | 6.77 |

Based on Paronuzzi, P. "Rockfall-induced block propagation on a soil slope, northern Italy." *Environmental Geology*, Vol. 58, 2009 (Table 2).

2.139 Most powerful women in America. Refer to the *Fortune* ranking of the 50 most powerful women in America, presented in Exercise 2.60 (p. 57). (These data are saved in the **WPOWER50** file.) Use side-by-side box plots to compare the ages of the women in three groups, based on their position within the firm: Group 1 (CEO, CEO/ Chairman, or CEO/president); Group 2 (COO/President, EVP/Pres., chairman, COO, EVP/COO, or president); Group 3 (EVP, executive, SVP, SVP, VP, or direct). Do you detect outliers?

2.140 Is honey a cough remedy? Refer to the *Archives of Pediatrics and Adolescent Medicine* (Dec. 2007) study of honey as a remedy for coughing, Exercise 2.62 (p. 58). Recall that coughing improvement scores were recorded for children in three groups: the over-the-counter cough medicine dosage (DM) group, the honey dosage group, and the control (no medicine) group. The data are saved in the **HONEYCOUGH** file.

a. For each group, construct a box plot of the improvement scores.

b. How do the median improvement scores compare for the three groups?

c. How does the variability in improvement scores compare for the three groups?

d. Do you detect any outliers in any of the three coughing improvement score distributions?

2.141 Comparing SAT Scores. Refer to Exercise 2.42 (p. 48), in which we compared state average SAT scores in 2009 and 2005. The data are saved in the **SATSCORES** file.

a. Construct side-by-side box plots of the SAT scores for the two years.

b. Compare the variability of the SAT scores for the two years.

c. Are any state SAT scores outliers in either year? If so, identify them.

Applying the Concepts—Advanced

2.142 Library book checkouts. A city librarian claims that books have been checked out an average of seven (or more) times in the last year. You suspect he has exaggerated the checkout rate (book usage) and that the mean number of checkouts per book per year is, in fact, less than seven. Using the computerized card catalog, you randomly select one book and find that it has been checked out four times

Descriptive Statistics

	N	Minimum	Maximum	Mean	Std. Deviation
RATIO	26	2.25	5.06	3.5069	.63439
Valid N (listwise)	26				

SPSS Output for Exercise 2.136

in the last year. Assume that the standard deviation of the number of checkouts per book per year is approximately 1.

a. If the mean number of checkouts per book per year really is 7, what is the *z*-score corresponding to four?

b. Considering your answer to part **a**, do you have reason to believe that the librarian's claim is incorrect?

c. If you knew that the distribution of the number of checkouts was mound shaped, would your answer to part **b** change? Explain.

d. If the standard deviation of the number of checkouts per book per year were 2 (instead of 1), would your answers to parts **b** and **c** change? Explain.

2.9 Graphing Bivariate Relationships (Optional)

The claim is often made that the crime rate and the unemployment rate are "highly correlated." Another popular belief is that smoking and lung cancer are "related." Some people even believe that the Dow Jones Industrial Average and the lengths of fashionable skirts are "associated." The words *correlated*, *related*, and *associated* imply a relationship between two variables—in the examples just mentioned, two *quantitative* variables.

One way to describe the relationship between two quantitative variables—called a **bivariate relationship**—is to construct a **scatterplot**. A scatterplot is a two-dimensional plot, with one variable's values plotted along the vertical axis and the others along the horizontal axis. For example, Figure 2.35 is a scatterplot relating (1) the cost of mechanical work (heating, ventilating, and plumbing) to (2) the floor area of the building, for a sample of 26 factory and warehouse buildings. Note that the scatterplot suggests a general tendency for mechanical cost to increase as building floor area increases.

When an increase in one variable is generally associated with an increase in the second variable, we say that the two variables are "positively related" or "positively correlated."* Figure 2.35 implies that mechanical cost and floor area are positively correlated. Alternatively, if one variable has a tendency to decrease as the other increases, we say the variables are "negatively correlated." Figure 2.36 shows several hypothetical scatterplots that portray a positive bivariate relationship (Figure 2.36a), a negative bivariate relationship (Figure 2.36b), and a situation in which the two variables are unrelated (Figure 2.36c).

Figure 2.35

Scatterplot of cost vs. floor area

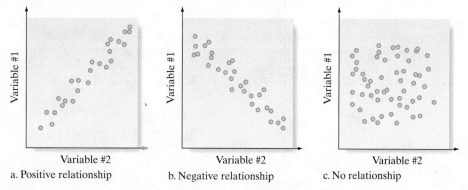

Figure 2.36

Hypothetical bivariate relationships

a. Positive relationship

b. Negative relationship

c. No relationship

*A formal definition of correlation is given in Chapter 9. There, we will learn that correlation measures the strength of the linear (or straight-line) relationship between two quantitative variables.

Example 2.19
Graphing Bivariate Data—Hospital Application

Problem A medical item used to administer to a hospital patient is called a *factor*. For example, factors can be intravenous (IV) tubing, IV fluid, needles, shave kits, bedpans, diapers, dressings, medications, and even code carts. The coronary care unit at Bayonet Point Hospital (St. Petersburg, Florida) recently investigated the relationship between the number of factors administered per patient and the patient's length of stay (in days). Data on these two variables for a sample of 50 coronary care patients are given in Table 2.10. Use a scatterplot to describe the relationship between the two variables of interest: number of factors and length of stay.

Table 2.10 Medfactors Data on Patients' Factors and Length of Stay

Number of Factors	Length of Stay (days)	Number of Factors	Length of Stay (days)
231	9	354	11
323	7	142	7
113	8	286	9
208	5	341	10
162	4	201	5
117	4	158	11
159	6	243	6
169	9	156	6
55	6	184	7
77	3	115	4
103	4	202	6
147	6	206	5
230	6	360	6
78	3	84	3
525	9	331	9
121	7	302	7
248	5	60	2
233	8	110	2
260	4	131	5
224	7	364	4
472	12	180	7
220	8	134	6
383	6	401	15
301	9	155	4
262	7	338	8

Based on Bayonet Point Hospital, Coronary Care Unit.

 Data Set: MEDFACTORS

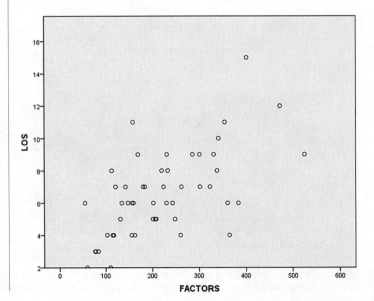

Figure 2.37

SPSS scatterplot for medical factors data in Table 2.10

Solution Rather than construct the plot by hand, we resort to a statistical software package. The SPSS plot of the data in Table 2.10, with length of stay (LOS) on the vertical axis and number of factors (FACTORS) on the horizontal axis, is shown in Figure 2.37.

Although the plotted points exhibit a fair amount of variation, the scatterplot clearly shows an increasing trend. It appears that a patient's length of stay is positively correlated with the number of factors administered to the patient.

Look Back If hospital administrators can be confident that the sample trend shown in Figure 2.37 accurately describes the trend in the population, then they may use this information to improve their forecasts of lengths of stay for future patients.

Now Work Exercise 2.147

The scatterplot is a simple, but powerful, tool for describing a bivariate relationship. However, keep in mind that it is only a graph. No measure of reliability can be attached to inferences made about bivariate populations based on scatterplots of sample data. The statistical tools that enable us to make inferences about bivariate relationships are presented in Chapter 9.

Exercises 2.143–2.158

Understanding the Principles

2.143 Define a bivariate relationship.

2.144 For what types of variables, quantitative or qualitative, are scatterplots useful?

2.145 What is the difference between positive association and negative association as it pertains to the relationship between two variables?

Learning the Mechanics

2.146 Construct a scatterplot for the data in the table that follows. Do you detect a trend?

Variable #1:	.5	1	1.5	2	2.5	3	3.5	4	4.5	5
Variable #2:	2	1	3	4	6	10	9	12	17	17

2.147 Construct a scatterplot for the data in the table that follows.
[NW] Do you detect a trend?

Variable #1:	5	3	−1	2	7	6	4	0	8
Variable #2:	14	3	10	1	8	5	3	2	12

Applying the Concepts—Basic

2.148 Does elevation impact hitting performance in baseball? The Colorado Rockies play their major league home baseball games in Coors Field, Denver. Each year, the Rockies are among the leaders in team batting statistics (e.g., home runs, batting average, and slugging percentage). Many baseball experts attribute this phenomenon to the "thin air" of Denver—called the "mile-high city" due to its elevation. *Chance* (Winter 2006) investigated the effects of elevation on slugging percentage in Major League Baseball. Data were compiled on players' composite slugging percentage at each of 29 cities for the 2003 season, as well as each city's elevation (feet above sea level). The data are saved in the **MLBPARKS** file. (Selected observations are shown in

the next table.) Construct a scatterplot for the data. Do you detect a trend?

City	Slug Pct	Elevation
Anaheim	.480	160
Arlington	.605	616
Atlanta	.530	1050
Baltimore	.505	130
Boston	.505	20
⋮	⋮	⋮
Denver	.625	5277
⋮	⋮	⋮
Seattle	.550	350
San Francisco	.510	63
St. Louis	.570	465
Tampa	.500	10
Toronto	.535	566

Based on Schaffer, J., and Heiny, E. L. "The effects of elevation on slugging percentage in Major League Baseball." *Chance*, Vol. 19, No. 1, Winter 2006 (Figure 2).

2.149 Comparing SAT scores. Refer to Exercise 2.42 (p. 48) and the data on state SAT scores saved in the **SATSCORES** file. Consider a scatterplot for the data, with 2009 SAT score on the vertical axis and 2005 SAT score on the horizontal axis. What type of trend would you expect to observe? Why? Create the scatterplot and check your answer.

2.150 Feeding behavior of fish. Zoologists at the University of Western Australia conducted a study of the feeding behavior of black bream (a type of fish) spawned in aquariums. (*Brain, Behavior and Evolution*, April 2000). In one experiment, the zoologists recorded the number of aggressive strikes of two black bream feeding at the bottom of the aquarium in the 10-minute period following the addition of food. The number of strikes and age of the fish (in days) were recorded approximately each week for nine

Data for Exercise 2.150

Week	Number of Strikes	Age of Fish (days)
1	85	120
2	63	136
3	34	150
4	39	155
5	58	162
6	35	169
7	57	178
8	12	184
9	15	190

Based on Shand, J., et al. "Variability in the location of the retinal ganglion cell area centralis is correlated with ontogenetic changes in feeding behavior in the Blackbream, *Acanthopagrus 'butcher.'*" *Brain and Behavior,* Vol. 55, No. 4, April 2000 (Figure 10H).

weeks, as shown in the table above. The data are saved in the **BLACKBREAM** file.

a. Construct a scatterplot of the data, with number of strikes on the *y*-axis and age of the fish on the *x*-axis.

b. Examine the scatterplot of part **a.** Do you detect a trend?

2.151 New method for blood typing. In *Analytical Chemistry* (May 2010), medical researchers tested a new method of typing blood using low cost paper. Blood drops were applied to the paper and the rate of absorption (called *blood wicking*) was measured. The table gives the wicking lengths (in millimeters) for six blood drops, each at a different antibody concentration. (The data are saved in the **BLOODTYPE** file.) Construct a plot to investigate the relationship between wicking length and antibody concentration. What do you observe?

Droplet	Length (mm)	Concentration
1	22.50	0.0
2	16.00	0.2
3	13.50	0.4
4	14.00	0.6
5	13.75	0.8
6	12.50	1.0

Based on Khan, M. S., "Paper diagnostic for instant blood typing." *Analytical Chemistry,* Vol. 82, No. 10, May 2010 (Figure 4b).

2.152 Sound waves from a basketball. Refer to the *American Journal of Physics* (June 2010) study of sound waves in a spherical cavity, Exercise 2.37 (p. 47). The frequencies of sound waves (estimated using a mathematical formula) resulting from the first 24 resonances (echoes) after striking a basketball with a metal rod are reproduced in the table in the next column. (The data are saved in the **BBALL** file.) Graph the data in a scatterplot, with frequency on the vertical axis and resonance number on the horizontal axis. Since a mathematical formula was used to estimate frequency, the researcher expects an increasing trend with very little variation. Does the graph support the researcher's theory?

Based on Russell, D. A. "Basketballs as spherical acoustic cavities." *American Journal of Physics,* Vol. 78, No. 6, June 2010 (Table I).

Applying the Concepts—Intermediate

2.153 College protests of labor exploitation. Refer to the *Journal of World-Systems Research* (Winter 2004) study of 14 student sit-ins for a "sweat-free campus," presented in

Resonance	Frequency
1	979
2	1572
3	2113
4	2122
5	2659
6	2795
7	3181
8	3431
9	3638
10	3694
11	4038
12	4203
13	4334
14	4631
15	4711
16	4993
17	5130
18	5210
19	5214
20	5633
21	5779
22	5836
23	6259
24	6339

Based on Russell, D. A. "Basketballs as spherical acoustic cavities." *American Journal of Physics,* Vol. 48, No. 6, June 2010 (Table I).

Exercise 2.41 (p. 48). The **SITIN** file contains data on the duration (in days) of each sit-in, as well as data on the number of student arrests.

a. Use a scatterplot to graph the relationship between duration and number of arrests. Do you detect a trend?

b. Repeat part **a,** but graph only the data for sit-ins in which there was at least one arrest. Do you detect a trend?

c. Comment on the reliability of the trend you detected in part **b.**

2.154 Are geography journals worth their cost? In *Geoforum* (Vol. 37, Nov. 2006), Simon Fraser University professor Nicholas Blomley assessed whether the price of a geography journal is correlated with quality. He collected pricing data (cost for a 1-year subscription, in U.S. dollars) for a sample of 28 geography journals. In addition to cost, three other variables were measured: journal impact factor (JIF), defined as the average number of times articles from the journal have been cited; number of citations for a journal over the past five years; and relative price index (RPI), a measure developed by economists. [*Note:* A journal with an RPI less than 1.25 is considered a "good value."] The data for the 28 geography journals are saved in the **GEOJRNL** file. (Selected observations are shown in the table on page 91.)

a. Construct a scatterplot for the variables JIF and cost. Do you detect a trend?

b. Construct a scatterplot for the variables number of cites and cost. Do you detect a trend?

c. Construct a scatterplot for the variables RPI and cost. Do you detect a trend?

Data for Exercise 2.154

Journal	Cost ($)	JIF	Cites	RPI
J.Econ.Geogr.	468	3.139	207	1.16
Prog. Hum. Geog.	624	2.943	544	0.77
T.I. Brit.Geogr.	499	2.388	249	1.11
Econ.Geogr.	90	2.325	173	0.30
A.A.A. Geogr.	698	2.115	377	0.93
⋮	⋮	⋮	⋮	⋮
Geogr.Anal.	213	0.902	106	0.88
Geogr.J.	223	0.857	81	0.94
Appl.Geogr.	646	0.853	74	3.38

Based on Blomley, N. "Is this journal worth US$118?" *Geoforum*, Vol. 37, No. 6, Nov. 2006 (Tables 1, 2, and 3).

2.155 Mongolian desert ants. Refer to the *Journal of Biogeography* (Dec. 2003) study of ants in Mongolia, presented in Exercise 2.68 (p. 59). Data on annual rainfall, maximum daily temperature, percentage of plant cover, number of ant species, and species diversity index recorded at each of 11 study sites are saved in the **GOBIANTS** file.

a. Construct a scatterplot to investigate the relationship between annual rainfall and maximum daily temperature. What type of trend (if any) do you detect?

b. Use scatterplots to investigate the relationship that annual rainfall has with each of the other four variables in the data set. Are the other variables positively or negatively related to rainfall?

2.156 Forest fragmentation study. Ecologists classify the cause of forest fragmentation as either anthropogenic (i.e., due to human development activities, such as road construction or logging) or natural in origin (e.g., due to wetlands or wildfire). *Conservation Ecology* (Dec. 2003) published an article on the causes of fragmentation for 54 South American forests. Using advanced high-resolution satellite imagery, the researchers developed two fragmentation indexes for each forest—one for anthropogenic fragmentation and one for fragmentation from natural causes. The values of these two indexes (where higher values indicate more fragmentation) for five of the forests in the sample are shown in the table below. The data for all 54 forests are saved in the **FORFRAG** file.

Ecoregion (forest)	Anthropogenic Index	Natural Origin Index
Araucaria moist forests	34.09	30.08
Atlantic Coast *restingas*	40.87	27.60
Bahia coastal forests	44.75	28.16
Bahia interior forests	37.58	27.44
Bolivian *Yungas*	12.40	16.75

Based on Wade, T. G., et al. "Distribution and causes of global forest fragmentation." *Conservation Ecology*, Vol. 72, No. 2, Dec. 2003 (Table 6).

a. Ecologists theorize that an approximately linear (straight-line) relationship exists between the two fragmentation indexes. Graph the data for all 54 forests. Does the graph support the theory?

b. Delete the data for the three forests with the largest anthropogenic indexes, and reconstruct the graph of part **a.** Comment on the ecologists' theory.

Applying the Concepts—Advanced

2.157 Ranking driving performance of professional golfers. Refer to *The Sport Journal* (Winter 2007) analysis of a new method for ranking the total driving performance of golfers on the PGA tour, Exercise 2.64 (p. 59). Recall that the method uses both the average driving distance (in yards) and the driving accuracy (percent of drives that land in the fairway). Data on these two variables for the top 40 PGA golfers are saved in the **PGADRIVER** file. A professional golfer is practicing a new swing to increase his average driving distance. However, he is concerned that his driving accuracy will be lower. Is his concern reasonable? Explain.

2.158 Spreading rate of spilled liquid. A contract engineer at DuPont Corp. studied the rate at which a spilled volatile liquid will spread across a surface (*Chemical Engineering Progress*, Jan. 2005). Suppose that 50 gallons of methanol spills onto a level surface outdoors. The engineer uses derived empirical formulas (assuming a state of turbulence-free convection) to calculate the mass (in pounds) of the spill after a period ranging from 0 to 60 minutes. The calculated mass values are given in the table below and saved in the **LIQUIDSPILL** file. Is there evidence to indicate that the mass of the spill tends to diminish as time increases?

Time (minutes)	Mass (pounds)
0	6.64
1	6.34
2	6.04
4	5.47
6	4.94
8	4.44
10	3.98
12	3.55
14	3.15
16	2.79
18	2.45
20	2.14
22	1.86
24	1.60
26	1.37
28	1.17
30	0.98
35	0.60
40	0.34
45	0.17
50	0.06
55	0.02
60	0.00

Based on Barry, J. "Estimating rates of spreading and evaporation of volatile liquids." *Chemical Engineering Progress*, Vol. 101, No. 1. Jan. 2005.

2.10 Distorting the Truth with Descriptive Statistics

A picture may be "worth a thousand words," but pictures can also color messages or distort them. In fact, the pictures displayed in statistics—histograms, bar charts, and other graphical images—are susceptible to distortion, so we have to examine each of them with care. Accordingly, we begin this section by mentioning a few of the pitfalls to watch for in interpreting a chart or a graph. Then we discuss how numerical descriptive statistics can be used to distort the truth.

Graphical Distortions

One common way to change the impression conveyed by a graph is to alter the scale on the vertical axis, the horizontal axis, or both. For example, consider the data on collisions of large marine vessels operating in European waters over a certain five-year period, summarized in Table 2.11. Figure 2.38 is a MINITAB bar graph showing the frequency of collisions for each of the three locations listed in the table. The graph shows that in-port collisions occur more often than collisions at sea or collisions in restricted waters.

Table 2.11	Collisions of Marine Vessels by Location
Location	Number of Ships
At Sea	376
In Restricted Waters	273
In Port	478
Total	1,127

Based on The Dock and Harbor Authority.   *Data Set:* COLLISION

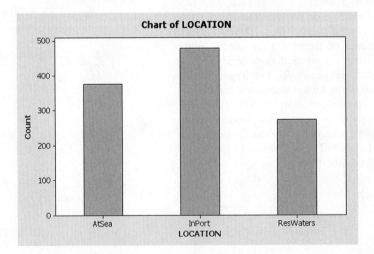

Figure 2.38
MINITAB bar graph of vessel collisions by location

Now, suppose you want to use the same data to exaggerate the difference between the number of in-port collisions and the number of collisions in restricted waters. One way to do this is to increase the distance between successive units on the vertical axis—that is, *stretch* the vertical axis by graphing only a few units per inch. A telltale sign of stretching is a long vertical axis, but this indication is often hidden by starting the vertical axis at some point above the origin, 0. Such a graph is shown in the SPSS printout in Figure 2.39. By starting the bar chart at 250 collisions (instead of 0), it appears that the frequency of in-port collisions is many times greater than the frequency of collisions in restricted waters.

Ethics IN Statistics

Intentionally distorting a graph to portray a particular viewpoint is considered unethical statistical practice.

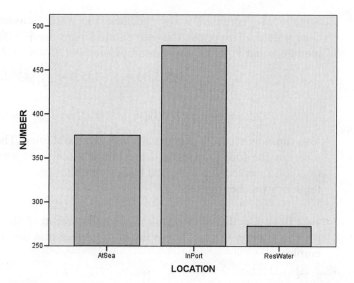

Figure 2.39
SPSS bar graph of vessel collisions by location, with adjusted vertical axis

Another method of achieving visual distortion with bar graphs is by making the width of the bars proportional to their height. For example, look at the bar chart in Figure 2.40a, which depicts the percentage of the total number of motor vehicle deaths in a year that occurred on each of four major highways. Now suppose we make both the width and the height grow as the percentage of fatal accidents grows. This change is shown in Figure 2.40b. The distortion is that the reader may tend to equate the *area* of the bars with the percentage of deaths occurring at each highway when, in fact, the true relative frequency of fatal accidents is proportional only to the *height* of the bars.

Figure 2.40
Relative frequency of fatal motor vehicle accidents on each of four major highways

Although we've discussed only a few of the ways that graphs can be used to convey misleading pictures of phenomena, the lesson is clear: Look at all graphical descriptions of data with a critical eye. In particular, check the axes and the size of the units on each axis. Ignore the visual changes, and concentrate on the actual numerical changes indicated by the graph or chart.

Misleading Numerical Descriptive Statistics

The information in a data set can also be distorted by using numerical descriptive measures, as Example 2.20 shows.

Example 2.20
Misleading Descriptive Statistics

Problem Suppose you're considering working for a small law firm—one that currently has a senior member and three junior members. You inquire about the salary you could expect to earn if you join the firm. Unfortunately, you receive two answers:

Answer A: The senior member tells you that an "average employee" earns $107,500.

Answer B: One of the junior members later tells you that an "average employee" earns $95,000.

Which answer can you believe?

Solution The confusion exists because the phrase "average employee" has not been clearly defined. Suppose the four salaries paid are $95,000 for each of the three junior members and $145,000 for the senior member. Then,

$$Mean = \frac{3(\$95,000) + \$145,000}{4} = \frac{\$430,000}{4} = \$107,500$$

Median = $95,000

You can now see how the two answers were obtained: The senior member reported the mean of the four salaries, and the junior member reported the median. The information you received was distorted because neither person stated which measure of central tendency was being used.

Look Back On the basis of our earlier discussion of the mean and median, we would probably prefer the median as the number that best describes the salary of the "average employee."

Another distortion of information in a sample occurs when *only* a measure of central tendency is reported. Both a measure of central tendency and a measure of variability are needed to obtain an accurate mental image of a data set.

Suppose, for instance, that you want to buy a new car and are trying to decide which of two models to purchase. Since energy and economy are both important issues, you decide to purchase model A because its EPA mileage rating is 32 miles per gallon in the city, whereas the mileage rating for model B is only 30 miles per gallon in the city.

However, you may have acted too quickly. How much variability is associated with the ratings? As an extreme example, suppose that further investigation reveals that the standard deviation for model A mileages is 5 miles per gallon, whereas that for model B is only 1 mile per gallon. If the mileages form a mound-shaped distribution, they might appear as shown in Figure 2.41. Note that the larger amount of variability associated with model A implies that more risk is involved in purchasing that model. That is, the particular car you purchase is more likely to have a mileage rating that will differ greatly from the EPA rating of 32 miles per gallon if you purchase model A, while a model B car is not likely to vary from the 30-miles-per-gallon rating by more than 2 miles per gallon.

We conclude this section with another example on distorting the truth with numerical descriptive measures.

Figure 2.41

Mileage distributions for two car models

Example 2.21

More Misleading Descriptive Statistics— Delinquent Children

Problem *Children Out of School in America* is a report on the delinquency of school-age children written by the Children's Defense Fund (CDF). Consider the following three reported results of the CDF survey.

- Reported result: Twenty-five percent of the 16- and 17-year-olds in the Portland, Maine, Bayside East Housing Project were out of school. Actual data: *Only eight children were surveyed; two were found to be out of school.*

- Reported result: Of all the secondary school students who had been suspended more than once in census tract 22 in Columbia, South Carolina, 33% had been suspended two times and 67% had been suspended three or more times. Actual data: *CDF found only three children in that entire census tract who had been suspended; one child was suspended twice and the other two children three or more times.*

- Reported result: In the Portland Bayside East Housing Project, 50% of all the secondary school children who had been suspended more than once had been suspended three or more times. Actual data: *The survey found just two secondary school children who had been suspended in that area; one of them had been suspended three or more times.*

Identify the potential distortions in the results reported by the CDF.

Solution In each of these examples, the reporting of percentages (i.e., relative frequencies) instead of the numbers themselves is misleading. No inference we might draw from the examples cited would be reliable. (We'll see how to measure the reliability of estimated percentages in Chapter 7.) In short, either the report should state the numbers alone instead of percentages, or, better yet, it should state that the numbers were too small to report by region.

Look Back If several regions were combined, the numbers (and percentages) would be more meaningful.

Ethics IN Statistics
Purposeful reporting of numerical descriptive statistics in order to mislead the target audience is considered *unethical statistical practice.*

CHAPTER NOTES

Key Terms

Note: Starred (*) terms are from the optional sections in this chapter.

Bar graph 29
Bivariate relationship* 87
Box plots* 80
Central tendency 50
Chebyshev's rule 67
Class 27
Class percentage 28
Class interval 40
Class frequency 27
Class relative frequency 27
Dot plot 43
Empirical rule 67
Hinge* 80
Histogram 44
Inner fence* 80
Interquartile range (IQR)* 78
Lower quartile (Q_L)* 74
Mean 50
Measures of central tendency 50
Measures of relative standing 73
Measures of variability
  (or spread) 61
Median 52
Middle quartile (M)* 74

Modal class 55
Mode 54
Mound-shaped, symmetrical
  distribution 66
Numerical descriptive measures 50
Outer fence* 80
Outlier* 78
Pareto diagram 29
Percentile ranking 73
Pie chart 29
population z-score 74
pth percentile 73
Quartiles* 74
Range 61
Rare-event approach* 83
sample standard deviation 62
sample variance 62
sample z-score 74
Scatterplot* 87
Skewness/skewed 53
Spread 61
Stem-and-leaf display 43
Summation notation 50
Upper quartile (Q_U)* 74
Variability 50
Whiskers* 80

Key Symbols

	Sample	Population
Mean:	$\bar{x}$	μ
Variance:	s^2	σ^2
Std. Dev.:	s	σ
Median:	M	
Lower Quartile:	Q_L	
Upper Quartile:	Q_U	
Interquartile Range:	IQR	

Key Ideas

Describing QUALITATIVE Data
1. Identify **category** classes
2. Determine **class frequencies**
3. **Class relative frequency** = (class frequency)/n
4. Graph relative frequencies

Pie Chart:

Bar Graph:

Pareto Diagram:

Graphing QUANTITATIVE Data

One Variable:
1. Identify class intervals
2. Determine class interval frequencies
3. Class interval relative frequency = (class interval frequency)/n
4. Graph class interval relative frequencies

Dot Plot:

Box plot:

Stem-and-Leaf Display:

1	3
2	2489
3	126678
4	37
5	2

Two Variables:

Scatterplot:

Histogram:

Numerical Description of QUANTITATIVE Data

Central Tendency

Mean: $\bar{x} = (\Sigma x_i)/n$

Median: Middle value (or mean of two middle values) when data ranked in order

Mode: Value that occurs most often

Variation

Range: Difference between largest and smallest value

Variance:

$$s^2 = \frac{\Sigma(x_i - \bar{x})^2}{n-1} = \frac{\Sigma x_i^2 - \frac{(\Sigma x_i)^2}{n}}{n-1}$$

Std Dev.: $s = \sqrt{s^2}$

Interquartile Range: IQR $= Q_U - Q_L$

Relative Standing

Percentile Score: Percentage of values that fall below x-score

z-score: $z = (x - \bar{x})/s$

Rules for Describing Quantitative Data

Interval	Chebyshev's Rule	Empirical Rule
$\bar{x} \pm s$	At least 0%	$\approx 68\%$
$\bar{x} \pm 2s$	At least 75%	$\approx 95\%$
$\bar{x} \pm 3s$	At least 89%	$\approx$ All

Rules for Detecting Quantitative Outliers

Method	Suspect	Highly Suspect
Box plot:	Values between inner and outer fences	Values beyond outer fences
z-score:	$2 < \|z\| < 3$	$\|z\| > 3$

Guide to Selecting the Data Description Method

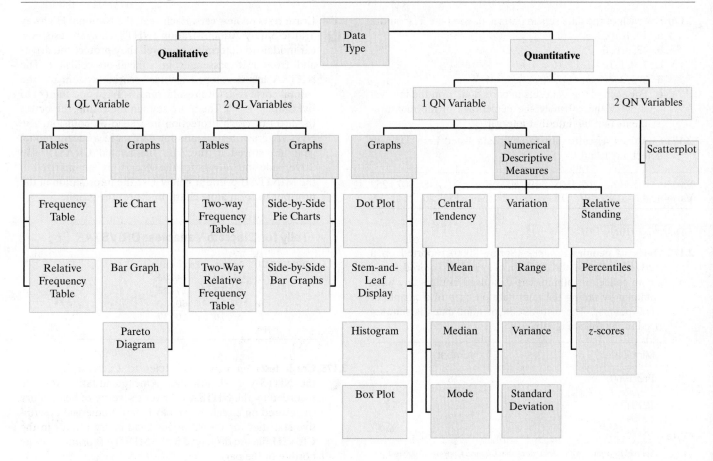

Supplementary Exercises 2.159–2.194

Note: Starred (*) exercises refer to the optional section in this chapter.

Understanding the Principles

2.159 Discuss conditions under which the median is preferred to the mean as a measure of central tendency.

2.160 Explain why we generally prefer the standard deviation to the range as a measure of variation.

2.161 Give a situation in which we would prefer using a stem-and-leaf display over a histogram in describing quantitative data graphically.

2.162 Give a situation in which we would prefer using a box plot over z-scores to detect an outlier.

2.163 Give a technique that is used to distort information shown on a graph.

Learning the Mechanics

2.164 Construct a relative frequency histogram for the data summarized in the following table:

Measurement Class	Relative Frequency	Measurement Class	Relative Frequency
.00–.75	.02	5.25–6.00	.15
.75–1.50	.01	6.00–6.75	.12
1.50–2.25	.03	6.75–7.50	.09
2.25–3.00	.05	7.50–8.25	.05
3.00–3.75	.10	8.25–9.00	.04
3.75–4.50	.14	9.00–9.75	.01
4.50–5.25	.19		

2.165 Consider the following three measurements: 50, 70, 80. Find the z-score for each measurement if they are from a population with a mean and standard deviation equal to
a. $\mu = 60, \sigma = 10$
b. $\mu = 60, \sigma = 5$
c. $\mu = 40, \sigma = 10$
d. $\mu = 40, \sigma = 100$

2.166 Refer to Exercise 2.165. For parts **a–d**, determine whether the three measurements 50, 70, and 80 are outliers.

2.167 Compute s^2 for data sets with the following characteristics:
a. $\sum_{i=1}^{n} x_i^2 = 246$, $\sum_{i=1}^{n} x_i = 63$, $n = 22$
b. $\sum_{i=1}^{n} x_i^2 = 666$, $\sum_{i=1}^{n} x_i = 106$, $n = 25$
c. $\sum_{i=1}^{n} x_i^2 = 76$, $\sum_{i=1}^{n} x_i = 11$, $n = 7$

2.168 If the range of a set of data is 20, find a rough approximation to the standard deviation of the data set.

2.169 For each of the data sets in parts **a–c**, compute $\bar{x}$, s^2, and s. If appropriate, specify the units in which your answers are expressed.
a. 4, 6, 6, 5, 6, 7
b. –$1, $4, –$3, $0, –$3, –$6
c. 3/5%, 4/5%, 2/5%, 1/5%, 1/16%
d. Calculate the range of each data set in parts **a–c**.

2.170 For each of the data sets in parts **a–d,** compute $\bar{x}$, s^2, and s.
  a. 13, 1, 10, 3, 3
  b. 13, 6, 6, 0
  c. 1, 0, 1, 10, 11, 11, 15
  d. 3, 3, 3, 3
  e. For each of the data sets in parts **a–d,** form the interval $\bar{x} \pm 2s$ and calculate the percentage of the measurements that fall into that interval.

*2.171 Construct a scatterplot for the data listed here. Do you detect any trends?

| Variable #1: | 174 | 268 | 345 | 119 | 400 | 520 | 190 | 448 | 307 | 252 |
| Variable #2: | 8 | 10 | 15 | 7 | 22 | 31 | 15 | 20 | 11 | 9 |

Applying the Concepts—Basic

2.172 Dates of pennies. *Chance* (Spring 2000) reported on a study to estimate the number of pennies required to fill a coin collector's album. The data used in the study were obtained by noting the mint date on each of a sample of 2,000 pennies. The distribution of mint dates is summarized in the following table:

Mint Date	Number
Pre-1960	18
1960s	125
1970s	330
1980s	727
1990s	800

Source: Reprinted with permission from *Chance.* Copyright 2000 by the American Statistical Association. All rights reserved.

  a. Identify the experimental unit for the study.
  b. Identify the variable measured.
  c. What proportion of pennies in the sample have mint dates in the 1960s?
  d. Construct a pie chart to describe the distribution of mint dates for the 2,000 sampled pennies.

2.173 Al Qaeda attacks on the United States. An article in *Studies in Conflict & Terrorism* (Vol. 29, 2006) conducted an empirical analysis of incidents involving suicide terrorist attacks. The data in the table that follows are the number of individual suicide bombings or attacks for each in a sample of 21 incidents involving an attack against the United States by the Al Qaeda terrorist group. (For example, the infamous September 11, 2001, Al Qaeda attack involved four separate attacks with hijacked airplanes: Two planes crashed into the World Trade Center towers, one battered into the Pentagon, and a fourth plane shattered in a field in Pennsylvania.) Summarize the data (saved in the **ALQAEDA** file) with a dot plot. Interpret the results.

1	4	2	1	2	1	1
2	1	1	3	1	5	1
1	4	2	1	2	1	2

Based on Moghadam, A. "Suicide terrorism, occupation, and the globalization of martyrdom: A critique of *Dying to Win,*" *Studies in Conflict & Terrorism,* Vol. 29, No. 8, 2006 (Table 3).

2.174 Crash tests on new cars. Each year, the National Highway Traffic Safety Administration (NHTSA) crash tests new car models to determine how well they protect the driver and front-seat passenger in a head-on collision. The NHTSA has developed a "star" scoring system for the frontal crash test, with results ranging from one star (*) to five stars (*****). The more stars in the rating, the better the level of crash protection in a head-on collision. The NHTSA crash test results for 98 cars in a recent model year are stored in the data file named **CRASH**. The driver-side star ratings for the 98 cars are summarized in the MINITAB printout below. Use the information in the printout to form a pie chart. Interpret the graph.

Tally for Discrete Variables: DRIVSTAR

DRIVSTAR	Count	Percent
2	4	4.08
3	17	17.35
4	59	60.20
5	18	18.37
N=	98	

2.175 Crash tests on new cars. Refer to Exercise 2.174 and the NHTSA crash test data. One quantitative variable recorded by the NHTSA is driver's severity of head injury (measured on a scale from 0 to 1,500). Numerical descriptive statistics for the 98 driver head-injury ratings in the **CRASH** file are displayed in the MINITAB printout at the bottom of the page.
  a. Interpret each of the statistics shown on the printout.
  b. Find the z-score for a driver head-injury rating of 408. Interpret the result.

2.176 Competition for bird nest holes. The *Condor* (May 1995) published a study of competition for nest holes among collared flycatchers, a bird species. The authors collected the data for the study by periodically inspecting nest boxes located on the island of Gotland in Sweden. The nest boxes were grouped into 14 discrete locations (called plots). The accompanying table gives the number of flycatchers killed and the number of flycatchers breeding at each plot. The data are saved in the **CONDOR** file.

	Number Killed					
	0	1	2	3	4	5
Number of Breeders	30	124	34	38	28	30
	46	68	26			
	132	86				
	100	32				
	6					

Based on Merila, J., and Wiggins, D. A. "Interspecific competition for nest holes causes adult mortality in the collared flycatcher." *The Condor,* Vol. 97, No. 2, May 1995, p. 447.

Descriptive Statistics: DRIVHEAD

Variable	N	Mean	StDev	Minimum	Q1	Median	Q3	Maximum
DRIVHEAD	98	603.7	185.4	216.0	475.0	605.0	724.3	1240.0

MINITAB Output for Exercise 2.175

a. Calculate the mean, median, and mode for the number of flycatchers killed at the 14 study plots.

b. Interpret the measures of central tendency you found in part **a.**

*c. Graphically examine the relationship between number killed and number of breeders. Do you detect a trend?

2.177 "Made in the USA" survey. "Made in the USA" is a claim stated in many product advertisements or on product labels. Advertisers want consumers to believe that their product is manufactured with 100% U.S. labor and materials—which is often not the case. What does "Made in the USA" mean to the typical consumer? To answer this question, a group of marketing professors conducted an experiment at a shopping mall in Muncie, Indiana. (*Journal of Global Business*, Spring 2002.) They asked every fourth adult entrant to the mall to participate in the study. A total of 106 shoppers agreed to answer the question, "'Made in the USA' means what percentage of U.S. labor and materials?" The responses of the 106 shoppers are summarized as follows: 64 shoppers responded 100%; 20 shoppers stated 75 to 99%; 18 shoppers stated 50 to 74%; and 4 shoppers said less than 50%.

a. What type of data collection method was used?

b. What type of variable, quantitative or qualitative, was measured?

c. Present the data in graphical form. Use the graph to make a statement about the percentage of consumers who believe that "Made in the USA" means 100% U.S. labor and materials.

Source: "'Made in the USA': Consumer perceptions, deception and policy alternatives." *Journal of Global Business,* Vol. 13, No. 24, Spring 2002 (Table 3). Reprinted with permission.

2.178 Groundwater contamination in wells. Refer to the *Environmental Science & Technology* (Jan. 2005) study of the factors related to MTBE contamination in 223 New Hampshire wells, presented in Exercise 2.21 (p. 37). The data are saved in the **MTBE** file. Two of the many quantitative variables measured for each well are the pH level (standard units) and the MTBE level (micrograms per liter).

a. Construct a histogram for the pH levels of the sampled wells. From the histogram, estimate the proportion of wells with pH values less than 7.0.

b. For those wells with detectible levels of MTBE, construct a histogram for the MTBE values. From the histogram, estimate the proportion of contaminated wells with MTBE values that exceed 5 micrograms per liter.

c. Find the mean and standard deviation for the pH levels of the sampled wells, and construct the interval $\bar{x} \pm 2s$. Estimate the percentage of wells with pH levels that fall within the interval. What rule did you apply to obtain the estimate? Explain.

d. Find the mean and standard deviation for the MTBE levels of the sampled wells, and construct the interval $\bar{x} \pm 2s$. Estimate the percentage of wells with MTBE levels that fall within the interval. What rule did you apply to obtain the estimate? Explain.

2.179 Achievement test scores. The distribution of scores on a nationally administered college achievement test has a median of 520 and a mean of 540.

a. Explain why it is possible for the mean to exceed the median for this distribution of measurements.

b. Suppose you are told that the 90th percentile is 660. What does this mean?

c. Suppose you are told that you scored at the 94th percentile. Interpret this statement.

Applying the Concepts—Intermediate

2.180 New-book reviews. *Choice* magazine provides new-book reviews in each issue. A random sample of 375 *Choice* book reviews in American history, geography, and area studies was selected, and the "overall opinion" of the book stated in each review was ascertained (*Library Acquisitions: Practice and Theory*, Vol. 19, 1995). Overall opinion was coded as follows: 1 = would not recommend, 2 = cautions or very little recommendation, 3 = little or no preference, 4 = favorable/recommended, 5 = outstanding/significant contribution. The data are summarized in the following table:

Opinion Code	Number of Reviews
1	19
2	37
3	35
4	238
5	46
Total	375

Based on Carlo, P. W., and Natowitz, A. "Choice book reviews in American history, geography, and area studies: An analysis for 1988–1993." *Library Acquisitions: Practice & Theory,* Vol. 19, No. 2, p. 159 (Figure 1).

a. Use a Pareto diagram to find the opinion that occurred most often. What proportion of the books reviewed received this opinion?

b. Do you agree with the following statement extracted from the study: "A majority (more than 75%) of books reviewed are evaluated favorably and recommended for purchase"?

2.181 Best-paid CEOs. Refer to the *Forbes* list of the 40 best-paid chief executive officers, Exercise 2.17 (p. 36). The salary and age of each CEO are saved in the **FORBES40** file.

*a. Use a graph to portray the relationship between a CEO's salary and age. Do you detect a trend?

b. According to Chebyshev's rule, what percentage of the age measurements would you expect to find in the intervals $\bar{x} \pm .75s$, $\bar{x} \pm 2.5s$, and $\bar{x} \pm 4s$?

c. What percentage of the age measurements actually fall into the intervals of part b? Compare your results with those of part **b.**

d. Repeat parts **b** and **c** for salary.

2.182 Ammonia in car exhaust. Three-way catalytic converters have been installed in new vehicles in order to reduce pollutants from motor vehicle exhaust emissions. However, these converters unintentionally increase the level of ammonia in the air. *Environmental Science & Technology* (Sept. 1, 2000) published a study on the ammonia levels near the exit ramp of a San Francisco highway tunnel. The data in the table (p. 100) represent daily ammonia concentrations (parts per million) on eight randomly selected days during the afternoon drive time in the summer of a recent year. The data are saved in the **AMMONIA** file. Find an interval that is likely to contain the ammonia level for a randomly selected day during afternoon drive time.

| 1.53 | 1.50 | 1.37 | 1.51 | 1.55 | 1.42 | 1.41 | 1.48 |

2.183 Whistling dolphins. Marine scientists who study dolphin communication have discovered that bottlenose dolphins exhibit an individualized whistle contour known as their *signature whistle*. A study was conducted to categorize the signature whistles of 10 captive adult bottlenose dolphins in socially interactive contexts (*Ethology*, July 1995). A total of 185 whistles was recorded during the study period; each whistle contour was analyzed and assigned to a category by a contour similarity (CS) technique. The results are reported in the table below and saved in the **DOLPHIN** file. Use a graphical method to summarize the results. Do you detect any patterns in the data that might be helpful to marine scientists?

Whistle Category	Number of Whistles
Type a	97
Type b	15
Type c	9
Type d	7
Type e	7
Type f	2
Type g	2
Type h	2
Type i	2
Type j	4
Type k	13
Other types	25

Based on McCowan, B., and Reiss, D. "Quantitative comparison of whistle repertoires from captive adult bottlenose dolphins (Delphiniae, *Tursiops truncates*): A re-evaluation of the signature whistle hypothesis." *Ethology*, Vol. 100, No. 3, July 1995, p. 200 (Table 2).

2.184 Speed of light from galaxies. Astronomers theorize that cold dark matter caused the formation of galaxies and clusters of galaxies in the universe. The theoretical model for cold dark matter requires an estimate of the velocities of light emitted from galaxy clusters. *The Astronomical Journal* (July 1995) published a study of observed velocities of galaxies in four different clusters. Galaxy velocity was measured in kilometers per second (km/s), using a spectrograph and high-power telescope.

Source: Oegerle, W. R., Hill, J. M., and Fitchett, M. J. "Observations of high dispersion clusters of galaxies: Constraints on cold dark matter." *The Astronomical Journal*, Vol. 110, No. 1, July 1995, p. 37. Reproduced by permission of the AAS.

a. The observed velocities of 103 galaxies located in the cluster named A2142 are summarized in the accompanying histogram. Comment on whether the empirical rule is applicable to describing the velocity distribution for this cluster.

b. The mean and standard deviation of the 103 velocities observed in galaxy cluster A2142 were reported as $\bar{x} = 27{,}117$ km/s and $s = 1{,}280$ km/s, respectively. Use this information to construct an interval that captures approximately 95% of the velocities of the galaxies in the cluster.

c. Recommend a single velocity value to be used in the CDM model for galaxy cluster A2142. Explain your reasoning.

*2.185 Freckling of superalloy ingots. Freckles are defects that sometimes form during the solidification of alloy ingots. A freckle index has been developed to measure the level of freckling on the ingot. A team of engineers conducted several experiments to measure the freckle index of a certain type of superalloy (*Journal of Metallurgy*, Sept. 2004). The data for $n = 18$ alloy tests are shown in the table and saved in the **FRECKLE** file.

12.6	22.0	4.1	16.4	1.4	2.4
16.8	10.0	3.2	30.1	6.8	14.6
2.5	12.0	33.4	22.2	8.1	15.1

Based on Yang, W. H. "A freckle criterion for the solidification of superalloys with a tilted solidification front." *JOM: Journal of the Minerals, Metals and Materials Society*, Vol. 56, No. 9, Sept. 2004.

a. Construct a box plot for the data and use it to find any outliers.

b. Find and interpret the z-scores associated with the alloys you identified in part **a**.

2.186 Benford's Law of Numbers. According to *Benford's law*, certain digits (1, 2, 3,…, 9) are more likely to occur as the first significant digit in a randomly selected number than are other digits. For example, the law predicts that the number "1" is the most likely to occur (30% of the time) as the first digit. In a study reported in the *American Scientist* (July–Aug. 1998) to test Benford's law, 743 first-year college students were asked to write down a six-digit number at random. The first significant digit of each number was recorded and its distribution summarized in the following table. These data are saved in the **DIGITS** file. Describe the first digit of the "random guess" data with an appropriate graph. Does the graph support Benford's law? Explain.

First Digit	Number of Occurrences
1	109
2	75
3	77
4	99
5	72
6	117
7	89
8	62
9	43
Total	743

Based on Hill, T. P. "The first digit phenomenon." *American Scientist*, Vol. 86, No. 4, July–Aug. 1998, p. 363 (Figure 5).

2.187 Oil spill impact on seabirds. The *Journal of Agricultural, Biological, and Environmental Statistics* (Sept. 2000) published a study on the impact of the *Exxon Valdez* tanker

MINITAB Output for Exercise 2.187

Descriptive Statistics: Density

Variable	Oil	N	Mean	StDev	Minimum	Q1	Median	Q3	Maximum
Density	no	36	3.27	6.70	0.000	0.000	0.890	3.87	36.23
	yes	60	3.495	5.968	0.0000	0.000	0.700	5.233	32.836

oil spill on the seabird population in Prince William Sound, Alaska. A subset of the data analyzed is stored in the **EVOS** file. Data were collected on 96 shoreline locations (called *transects*) of constant width, but variable length. For each transect, the number of seabirds found is recorded, as are the length (in kilometers) of the transect and whether or not the transect was in an oiled area. (The first five and last five observations in the **EVOS** file are listed in the accompanying table.)

Transect	Seabirds	Length	Oil
1	0	4.06	No
2	0	6.51	No
3	54	6.76	No
4	0	4.26	No
5	14	3.59	No
⋮	⋮	⋮	⋮
92	7	3.40	Yes
93	4	6.67	Yes
94	0	3.29	Yes
95	0	6.22	Yes
96	27	8.94	Yes

Based on McDonald, T. L., Erickson, W. P., and McDonald, L. L. "Analysis of count data from before–after control-impact studies." *Journal of Agricultural, Biological, and Environmental Statistics,* Vol. 5, No. 3, Sept. 2000, pp. 277–8 (Table A.1).

a. Identify the variables measured as quantitative or qualitative.

b. Identify the experimental unit.

c. Use a pie chart to describe the percentage of transects in oiled and unoiled areas.

*d. Use a graphical method to examine the relationship between observed number of seabirds and transect length.

e. Observed seabird density is defined as observed count divided by length of transect. MINITAB descriptive statistics for seabird densities in unoiled and oiled transects are displayed in the printout at the top of the page. Assess whether the distribution of seabird densities differs for transects in oiled and unoiled areas.

f. For unoiled transects, give an interval of values that is likely to contain at least 75% of the seabird densities.

g. For oiled transects, give an interval of values that is likely to contain at least 75% of the seabird densities.

h. Which type of transect, an oiled or unoiled one, is more likely to have a seabird density of 16? Explain.

Applying the Concepts—Advanced

2.188 Standardized test "average." *US News & World Report* reported on many factors contributing to the breakdown of public education. One study mentioned in the article found that over 90% of the nation's school districts reported that their students were scoring "above the national average"

on standardized tests. Using your knowledge of measures of central tendency, explain why the schools' reports are incorrect. Does your analysis change if the term "average" refers to the mean? To the median? Explain what effect this misinformation might have on the perception of the nation's schools.

2.189 Speed of light from galaxies. Refer to *The Astronomical Journal* study of galaxy velocities, presented in Exercise 2.184 (p. 100). A second cluster of galaxies, named A1775, is thought to be a *double cluster*—that is, two clusters of galaxies in close proximity. Fifty-one velocity observations (in kilometers per second, km/s) from cluster A1775 are listed in the table. The data are saved in the **GALAXY2** file.

22,922	20,210	21,911	19,225	18,792	21,993	23,059
20,785	22,781	23,303	22,192	19,462	19,057	23,017
20,186	23,292	19,408	24,909	19,866	22,891	23,121
19,673	23,261	22,796	22,355	19,807	23,432	22,625
22,744	22,426	19,111	18,933	22,417	19,595	23,408
22,809	19,619	22,738	18,499	19,130	23,220	22,647
22,718	22,779	19,026	22,513	19,740	22,682	19,179
19,404	22,193					

Based on Oegerle, W. R., Hill, J. M., and Fitchett, M. J. "Observations of high dispersion clusters of galaxies: Constraints on cold dark matter." *The Astronomical Journal,* Vol. 110, No. 1, July 1995 (Table 1 and Figure 1).

a. Use a graphical method to describe the velocity distribution of galaxy cluster A1775.

b. Examine the graph you created in part **a.** Is there evidence to support the double-cluster theory? Explain.

c. Calculate numerical descriptive measures (e.g., mean and standard deviation) for galaxy velocities in cluster A1775. Depending on your answer to part **b,** you may need to calculate two sets of numerical descriptive measures, one for each of the clusters (say, A1775A and A1775B) within the double cluster.

d. Suppose you observe a galaxy velocity of 20,000 km/s. Is this galaxy likely to belong to cluster A1775A or A1775B? Explain.

2.190 Risk of jail suicides. Suicide is the leading cause of death of Americans incarcerated in correctional facilities. To determine what factors increase the risk of suicide in urban jails, a group of researchers collected data on all 37 suicides that occurred over a 15-year period in the Wayne County Jail in Detroit, Michigan (*American Journal of Psychiatry*, July 1995). The data on each suicide victim are saved in the **SUICIDE** file. Selected observations are shown in the table at the top of page 102.

a. Identify the type (quantitative or qualitative) of each variable measured.

b. Are suicides at the jail more likely to be committed by inmates charged with murder/manslaughter or with lesser crimes? Illustrate with a graph.

Data for Exercise 2.190

Victim	Days in Jail before Suicide	Marital Status	Race	Murder/Manslaughter Charge	Time of Suicide	Year
1	3	Married	W	Yes	Night	1972
2	4	Single	W	Yes	Night	1987
3	5	Single	NW	Yes	Afternoon	1975
4	7	Widowed	NW	Yes	Night	1981
5	10	Single	NW	Yes	Afternoon	1982
⋮	⋮	⋮	⋮	⋮	⋮	⋮
36	41	Single	NW	No	Night	1985
37	86	Married	W	No	Night	1968

Based on DuRand, C. J., et al. "A quarter century of suicide in a major urban jail: Implications for community psychiatry." *American Journal of Psychiatry,* Vol. 152, No. 7, July 1995, p. 1078 (Table 1).

c. Are suicides at the jail more likely to be committed at night? Illustrate with a graph.

d. What is the mean length of time an inmate is in jail before committing suicide? What is the median? Interpret these two numbers.

e. Is it likely that a future suicide at the jail will occur after 200 days? Explain.

f. Have suicides at the jail declined over the years? Support your answer with a graph.

2.191 Zinc phosphide in pest control. A chemical company produces a substance composed of 98% cracked corn particles and 2% zinc phosphide for use in controlling rat populations in sugarcane fields. Production must be carefully controlled to maintain the 2% zinc phosphide, because too much zinc phosphide will cause damage to the sugarcane and too little will be ineffective in controlling the rat population. Records from past production indicate that the distribution of the actual percentage of zinc phosphide present in the substance is approximately mound shaped, with a mean of 2.0% and a standard deviation of .08%. Suppose one batch chosen randomly actually contains 1.80% zinc phosphide. Does this indicate that there is too little zinc phosphide in this production? Explain your reasoning.

2.192 Grades in statistics. The final grades given by two professors in introductory statistics courses have been carefully examined. The students in the first professor's class had a grade point average of 3.0 and a standard deviation of .2. Those in the second professor's class had grade points with an average of 3.0 and a standard deviation of 1.0. If you had a choice, which professor would you take for this course?

2.193 Salaries of professional athletes. The salaries of superstar professional athletes receive much attention in the media. The multimillion-dollar long-term contract is now commonplace among this elite group. Nevertheless, rarely does a season pass without negotiations between one or more of the players' associations and team owners for additional salary and fringe benefits for *all* players in their particular sports.

a. If a players' association wanted to support its argument for higher "average" salaries, which measure of central tendency do you think it should use? Why?

b. To refute the argument, which measure of central tendency should the owners apply to the players' salaries? Why?

Critical Thinking Challenge

2.194 The Hite Report. Researcher Shere Hite shocked conservative America with her famous Hite Report on the permissive sexual attitudes of American men and women. In her book *Women and Love: A Cultural Revolution in Progress* (Knopf Press, 1988), Hite reveals some startling statistics describing how women feel about contemporary relationships:

- Eighty-four percent are not emotionally satisfied with their relationship.
- Ninety-five percent report "emotional and psychological harassment" from their men.
- Seventy percent of those married five years or more are having extramarital affairs.
- Only 13% of those married more than two years are "in love."

Hite conducted the survey by mailing out 100,000 questionnaires to women across the country over a seven-year period. The questionnaires were mailed to a wide variety of organizations, including church groups, women's voting and political groups, women's rights organizations, and counseling and walk-in centers for women. Organizational leaders were asked to circulate the questionnaires to their members. Hite also relied on volunteer respondents who wrote in for copies of the questionnaire. Each questionnaire consisted of 127 open-ended questions, many with numerous subquestions and follow-ups. Hite's instructions read, "It is not necessary to answer every question! Feel free to skip around and answer those questions you choose." Approximately 4,500 completed questionnaires were returned, a response rate of 4.5%. These questionnaires form the data set from which the preceding percentages were determined. Hite claims that the 4,500 women respondents are a representative sample of all women in the United States and, therefore, that the survey results imply that vast numbers of women are "suffering a lot of pain in their love relationships with men." Critically assess the survey results. Do you believe they are reliable?

Activity   Describing Data from Popular Sources

We list here several sources of real-life data sets. (Many in the list have been obtained from Wasserman and Bernero's *Statistics Sources,* and many can be accessed via the Internet.) This index of data sources is far from complete, but is a useful reference for anyone interested in finding almost any type of data.

First we list some almanacs:

Baseball Almanac (www.baseball-almanac.com)

Information Please Almanac (www.infoplease.com)

World Almanac and Book of Facts (www.worldalmanac.com)

United States Government publications are also rich sources of data:
Agricultural Statistics (www.nass.usda.gov)
Digest of Educational Statistics (nces.ed.gov)
Handbook of Labor Statistics (www.bls.gov)
Housing and Urban Development Statistics (www.huduser.org)
Social Indicators (www.unstats.un.org)
Uniform Crime Reports (www.fbi.gov)
Vital Statistics of the United States (www.cdc.gov)
Economic Indicators (www.economicindicators.gov)
Monthly Labor Review (www.bls.gov)
Survey of Current Business (www.bea.gov)
Statistical Abstract of the United States (www.census.gov)

Many data sources are published on an annual basis:
Commodity Yearbook (www.crbtrader.com)
The World Bank Data Catalog (data.worldbank.org)
Yearbook of American and Canadian Churches (www.thearda.com)
Standard and Poor's Corporation, Trade and Securities: Statistics (published monthly)

Some sources contain data that are international in scope:
Compendium of Social Statistics (www.esewa.un.org)
Stokholm International Peace Research Institute Yearbook (www.sipri.org)

United Nations Statistical Yearbook (www.unstats.un.org)
World Handbook of Political and Social Indicators (www.lepsr.umich.edu)

Here are a few miscellaneous data sources:
Federal Drug Data Sources (www.whitehousedrugpolicy.gov)
Internet Data Sources for Social Scientists (www.ciser.cornell.edu)
Climate Data Sources (www.realclimate.org)
World Health Organization Statistics (www.who.int)

Utilizing the data sources listed, sources suggested by your instructor, or your own resourcefulness, find one real-life quantitative data set that stems from an area of particular interest to you.

a. Describe the data set by using a relative frequency histogram, stem-and-leaf display, or dot plot.

b. Find the mean, median, variance, standard deviation, and range of the data set.

c. Use Chebyshev's rule and the empirical rule to describe the distribution of this data set. Count the actual number of observations that fall within one, two, and three standard deviations of the mean of the data set, and compare these counts with the description of the data set you developed in part **b**.

References

Huff, D. *How to Lie with Statistics*. New York: Norton, 1954.

Mendenhall, W., Beaver, R. J., and Beaver, B. M. *Introduction to Probability and Statistics*, 12th ed. North Scituate, MA: Duxbury, 2006.

Tufte, E. R. *Beautiful Evidence.* Cheshire, CT: Graphics Press, 2006.

Tufte, E. R. *Envisioning Information.* Cheshire, CT: Graphics Press, 1990.

Tufte, E. R. *Visual Display of Quantitative Information*. Cheshire, CT: Graphics Press, 1983.

Tufte, E. R. *Visual Explanations*. Cheshire, CT: Graphics Press, 1997.

Tukey, J. *Exploratory Data Analysis*. Reading, MA: Addison-Wesley, 1977.

USING TECHNOLOGY

MINITAB: Describing Data

Graphing Data

Step 1 Click on the "Graph" button on the MINITAB menu bar.

Step 2 Click on the graph of your choice (bar, pie, scatterplot, histogram, dot plot, or stem-and-leaf) to view the appropriate dialog box. The dialog box for a histogram is shown in Figure 2.M.1.

Step 3 Make the appropriate variable selections and click "OK" to view the graph.

Figure 2.M.1 MINITAB histogram dialog box

Numerical Descriptive Statistics

Step 1 Click on the "Stat" button on the main menu bar, click on "Basic Statistics," and then click on "Display Descriptive Statistics." The resulting dialog box appears in Figure 2.M.2.

Step 2 Select the quantitative variables you want to analyze and place them in the "Variables" box. You can control which descriptive statistics appear by clicking the "Statistics" button on the dialog box and making your selections. (As an option, you can create histograms and dot plots for the data by clicking the "Graphs" button and making the appropriate selections.)

Figure 2.M.2 MINITAB descriptive statistics dialog box

Step 3 Click "OK" to view the descriptive statistics printout.

TI-83/TI-84 Plus Graphing Calculator: Describing Data

Histogram from Raw Data

Step 1 *Enter the data*

• Press **STAT** and select **1:Edit**

Note: If the list already contains data, clear the old data. Use the up arrow to highlight "**L1**." Press **CLEAR ENTER**. Use the arrow and **ENTER** keys to enter the data set into **L1**.

Step 2 *Set up the histogram plot*

• Press **2nd** and press **Y** = for **STAT PLOT**

• Press **1** for **Plot 1**

• Set the cursor so that **ON** is flashing

• For **Type,** use the arrow and Enter keys to highlight and select the histogram

• For **Xlist,** choose the column containing the data (in most cases, **L1**)

Note: Press **2nd 1** for **L1 Freq** should be set to 1.

Step 3 *Select your window settings*

• Press **WINDOW** and adjust the settings as follows:

$$\textbf{X min} = \text{lowest class boundary}$$
$$\textbf{X max} = \text{highest class boundary}$$
$$\textbf{X sel} = \text{class width}$$
$$\textbf{Y min} = 0$$
$$\textbf{Y max} \geq \text{greatest class frequency}$$
$$\textbf{Y scl} = 1$$
$$\textbf{X res} = 1$$

Step 4 *View the graph*

• Press **GRAPH**

Optional *Read class frequencies and class boundaries*

Step 5 You can press **TRACE** to read the class frequencies and class boundaries. Use the arrow keys to move between bars.

Example The following figures show TI-83/TI-84 window settings and histogram for the following sample data:

86, 70, 62, 98, 73, 56, 53, 92, 86, 37, 62, 83, 78, 49, 78, 37, 67, 79, 57

Histogram from a Frequency Table

Step 1 *Enter the data*

• Press **STAT** and select **1:Edit**

Note: If a list already contains data, clear the old data. Use the up arrow to highlight the list name, "**L1**" or "**L2**."

• Press **CLEAR ENTER**

• Enter the midpoint of each class into **L1**

• Enter the class frequencies or relative frequencies into **L2**

Step 2 *Set up the histogram plot*

• Press **2nd** and **Y** = for **STATPLOT**

• Press **1** for **Plot 1**

• Set the cursor so that **ON** is flashing

• For **Type,** use the arrow and Enter keys to highlight and select the histogram

• For **Xlist,** choose the column containing the midpoints

• For **Freq,** choose the column containing the frequencies or relative frequencies

Step 3–4 *Follow steps 3–4 given above.*

Note: To set up the Window for relative frequencies, be sure to set **Y max** to a value that is greater than or equal to the largest relative frequency.

One-Variable Descriptive Statistics

Step 1 *Enter the data*

• Press **STAT** and select **1:Edit**

Note: If the list already contains data, clear the old data. Use the up arrow to highlight "**L1.**" Press **CLEAR ENTER.**

- Use the arrow and ENTER keys to enter the data set into L1

Step 2 *Calculate descriptive statistics*

- Press **STAT**
- Press the right arrow key to highlight **CALC**
- Press **ENTER** for **1-Var Stats**
- Enter the name of the list containing your data
- Press **2nd 1** for **L1** (or **2nd 2** for **L2,** etc.)
- Press **ENTER**

You should see the statistics on your screen. Some of the statistics are off the bottom of the screen. Use the down arrow to scroll through to see the remaining statistics. Use the up arrow to scroll back up.

Example Consider finding descriptive statistics for the following sample data set:

86, 70, 62, 98, 73, 56, 53, 92, 86, 37, 62, 83, 78, 49, 78, 37, 67, 79, 57

The output screens for this example are shown below:

Sorting Data (to find the Mode)

The descriptive statistics do not include the mode. To find the mode, sort your data as follows:

- Press **STAT**
- Press **2** for **SORTA(**
- Enter the name of the list your data are in. If your data are in **L1,** press **2nd 1**
- Press **ENTER**
- The screen will say: **DONE**
- To see the sorted data, press **STAT** and select **1:Edit**
- Scroll down through the list and locate the data value that occurs most frequently

Box Plot

Step 1 *Enter the data*

- Press **STAT** and select **1:Edit**
  Note: If the list already contains data, clear the old data. Use the up arrow to highlight "**L1.**"
- Press **CLEAR ENTER**
- Use the arrow and **ENTER** keys to enter the data set into **L1**

Step 2 *Set up the box plot*

- Press **2nd Y** = for **STAT PLOT**
- Press **1** for **Plot 1**
- Set the cursor so that "**ON**" is flashing

- For **TYPE,** use the right arrow to scroll through the plot icons and select the box plot in the middle of the second row
- For **XLIST**, choose **L1**
- Set **FREQ** to **1**

Step 3 *View the graph*

- Press **ZOOM** and select **9:ZoomStat**

Optional *Read the five number summary*

- Press **TRACE**
- Use the left and right arrow keys to move between **minX, Q1, Med, Q3,** and **maxX**

Example Make a box plot for the given data:

86, 70, 62, 98, 73, 56, 53, 92, 86, 37, 62, 83, 78, 49, 78, 37, 67, 79, 57

The output screen for this example is shown below.

Scatterplots

Step 1 *Enter the data*

- Press **STAT** and select **1:Edit**

Note: If a list already contains data, clear the old data. Use the up arrow to highlight the list name, "**L1**" or "**L2.**"

- Press **CLEAR ENTER**
- Enter your *x*-data in **L1** and your *y*-data in **L2**

Step 2 *Set up the scatterplot*

- Press **2nd Y** = for **STAT PLOT**
- Press **1** for **Plot 1**
- Set the cursor so that **ON** is flashing
- For **Type,** use the arrow and Enter keys to highlight and select the scatterplot (first icon in the first row)
- For **Xlist,** choose the column containing the *x*-data
- For **Ylist,** choose the column containing the *y*-data

Step 3 *View the scatterplot*

- Press **ZOOM 9** for **ZoomStat**

Example The figures below show a table of data entered on the T1-84 and the scatterplot of the data obtained using the steps given above.

3 Probability

CONTENTS

Where We've Been

- Identified the objective of inferential statistics: to make inferences about a population on the basis of information in a sample

- Introduced graphical and numerical descriptive measures for both quantitative and qualitative data

Where We're Going

- Develop probability as a measure of uncertainty (3.1)
- Introduce basic rules for finding probabilities (3.2–3.6)
- Use a probability as a measure of reliability for an inference (3.2–3.6)
- Use of probability in random sampling (3.7)

Statistics IN Action Lotto Buster! Can You Improve Your Chance of Winning?

"Welcome to the Wonderful World of Lottery Busters." So began the premier issue of Lottery Buster, *a monthly publication for players of the state lottery games. Lottery Buster provides interesting facts and figures on the 42 state lotteries and 2 multistate lotteries currently operating in the United States and, more importantly, tips on how to increase a player's odds of winning the lottery.*

In 1963, New Hampshire became the first state in modern times to authorize a state lottery as an alternative to increasing taxes. (Prior to that time, beginning in 1895, lotteries were banned in America because they were thought to be corrupt.) Since then, lotteries have become immensely popular, for two reasons. First, they lure you with the opportunity to win millions of dollars with a $1 investment, and second, when you lose, at least you believe that your money is going to a good cause. Many state lotteries, like Florida's, earmark a high percentage of lottery revenues to fund state education.

The popularity of the state lottery has brought with it an avalanche of "experts" and "mathematical wizards" (such as the editors of *Lottery Buster*) who provide advice on how to win the lottery—for a fee, of course! Many offer guaranteed "systems" of winning through computer software products with catchy names such as Lotto Wizard, Lottorobics, Win4D, and Lotto-Luck.

For example, most knowledgeable lottery players would agree that the "golden rule" or "first rule" in winning lotteries is *game selection*. State lotteries generally offer three types of games: Instant (scratch-off tickets or online) games, Daily Numbers (Pick-3 or Pick-4), and the weekly Pick-6 Lotto game.

One version of the Instant game involves scratching off the thin opaque covering on a ticket with the edge of a coin to determine whether you have won or lost. The cost of the ticket ranges from 50¢ to $1, and the amount won ranges from $1 to $100,000 in most states to as much as $1 million in others. *Lottery Buster* advises against playing the Instant game because it is "a pure chance play, and you can win only by dumb luck. No skill can be applied to this game."

The Daily Numbers game permits you to choose either a three-digit (Pick-3) or four-digit (Pick-4) number at a cost of $1 per ticket. Each night, the winning number is drawn. If your number matches the winning number, you win a large sum of money, usually $100,000. You do have some control over the Daily Numbers game (since you pick the numbers that you play); consequently, there are strategies available to increase your chances of winning. However, the Daily Numbers game, like the Instant game, is not available for out-of-state play.

To play Pick-6 Lotto, you select six numbers of your choice from a field of numbers ranging from 1 to N, where N depends on which state's game you are playing. For example, Florida's current Lotto game involves picking six numbers ranging from 1 to 53. The cost of a ticket is $1, and the payoff, if your six numbers match the winning numbers drawn, is $7 million or more, depending on the number of tickets purchased. (To date, Florida has had the largest state weekly payoff, over $200 million.) In addition to capturing the grand prize, you can win second-, third-, and fourth-prize payoffs by matching five, four, and three of the six numbers drawn, respectively. And you don't have to be a resident of the state to play the state's Lotto game.

In this chapter, several Statistics in Action Revisited examples demonstrate how to use the basic concepts of probability to compute the odds of winning a state lottery game and to assess the validity of the strategies suggested by lottery "experts."

Statistics IN Action Revisited

- The Probability of Winning Lotto (p. 117)
- The Probability of Winning Lotto with a Wheeling System (p. 128)
- The Probability of Winning Cash 3 or Play 4 (p. 142)

Recall that one branch of statistics is concerned with decisions made about a population on the basis of information learned about a sample. You can see how this is accomplished more easily if you understand the relationship between population and sample—a relationship that becomes clearer if we reverse the statistical procedure of making inferences from sample to population. In this chapter, we assume that the population is *known* and calculate the chances of obtaining various samples from the population. Thus, we show that probability is the reverse of statistics: *In probability, we use information about the population to infer the probable nature of the sample.*

Probability plays an important role in making inferences. Suppose, for example, you have an opportunity to invest in an oil exploration company. Past records show that, out of 10 previous oil drillings (a sample of the company's experiences), all 10 came up dry. What do you conclude? Do you think the chances are better than 50:50 that the company will hit a gusher? Should you invest in this company? Chances are, your answer to these questions will be an emphatic "No!" However, if the company's exploratory prowess is sufficient to hit a producing well 50% of the time, a record of 10 dry wells out of 10 drilled is an event that is just too improbable.

Or suppose you're playing poker with what your opponents assure you is a well-shuffled deck of cards. In three consecutive five-card hands, the person on your right is dealt four aces. On the basis of this sample of three deals, do you think the cards are being adequately shuffled? Again, your answer is likely to be "No," because dealing three hands of four aces is just too improbable if the cards were properly shuffled.

Note that the decisions concerning the potential success of the oil-drilling company and the adequacy of card shuffling both involve knowing the chance—or probability—of a certain sample result. Both situations were contrived so that you could easily conclude that the probabilities of the sample results were small. Unfortunately, the probabilities of many observed sample results aren't so easy to evaluate intuitively. In these cases, we need the assistance of a theory of probability.

3.1 Events, Sample Spaces, and Probability

Let's begin our treatment of probability with straightforward examples that are easily described. With the aid of these simple examples, we can introduce important definitions that will help us develop the notion of probability more easily.

Suppose a coin is tossed once and the up face is recorded. The result we see is called an *observation*, or *measurement*, and the process of making an observation is called an *experiment*. Notice that our definition of experiment is broader than the one used in the physical sciences, which brings to mind test tubes, microscopes, and other laboratory equipment. Statistical experiments may include, in addition to these things, recording an Internet user's preference for a Web browser, recording a voter's opinion on an important political issue, measuring the amount of dissolved oxygen in a polluted river, observing the level of anxiety of a test taker, counting the number of errors in an inventory, and observing the fraction of insects killed by a new insecticide. The point is that a statistical experiment can be almost any act of observation, as long as the outcome is uncertain.

> An **experiment** is an act or process of observation that leads to a single outcome that cannot be predicted with certainty.

Consider another simple experiment consisting of tossing a die and observing the number on the up face. The six possible outcomes of this experiment are as follows:

1. Observe a 1.
2. Observe a 2.
3. Observe a 3.
4. Observe a 4.
5. Observe a 5.
6. Observe a 6.

Note that if this experiment is conducted once, *you can observe one and only one of these six basic outcomes, and the outcome cannot be predicted with certainty*. Also, these outcomes cannot be decomposed into more basic ones. Because observing the outcome of an experiment is similar to selecting a sample from a population, the basic possible outcomes of an experiment are called **sample points**.*

> A **sample point** is the most basic outcome of an experiment.

*Alternatively, the term *simple event* can be used.

Example 3.1

Listing Sample Points for a Coin-Tossing Experiment

Problem Two coins are tossed, and their up faces are recorded. List all the sample points for this experiment.

Solution Even for a seemingly trivial experiment, we must be careful when listing the sample points. At first glance, we might expect one of three basic outcomes: Observe two heads; Observe two tails; or Observe one head and one tail. However, further reflection reveals that the last of these, Observe one head and one tail, can be decomposed into two outcomes: Head on coin 1, Tail on coin 2; and Tail on coin 1, Head on coin 2.

A useful tool for illustrating this notion is a **tree diagram.** Figure 3.1 shows a tree diagram for this experiment. At the top of the "tree" there are two branches, representing

Figure 3.1
Tree diagram for the coin-tossing experiment

the two outcomes (H or T) for the first tossed coin. Each of these outcomes results in two more branches, representing the two outcomes (H or T) for the second tossed coin. Consequently, after tossing both coins, you can see that we have four sample points:

1. Observe HH
2. Observe HT
3. Observe TH
4. Observe TT

where H in the first position means "Head on coin 1," H in the second position means "Head on coin 2," and so on.

Look Back Even if the coins are identical in appearance, they are, in fact, two distinct coins. Thus, the sample points must account for this distinction.

Now Work Exercise 3.15a

We often wish to refer to the collection of all the sample points of an experiment. This collection is called the *sample space* of the experiment. For example, there are six sample points in the sample space associated with the die-toss experiment. The sample spaces for the experiments discussed thus far are shown in Table 3.1.

> The **sample space** of an experiment is the collection of all its sample points.

Just as graphs are useful in describing sets of data, a pictorial method for presenting the sample space will often be useful. Figure 3.2 shows such a representation for each of the experiments in Table 3.1. In each case, the sample space is shown as a closed figure, labeled S, containing all possible sample points. Each sample point is represented by a solid dot (i.e., a "point") and labeled accordingly. Such graphical representations are called **Venn diagrams.**

$\circ H \qquad \circ T$

S

a. Experiment: Observe the up face on a coin

$\circ 1 \quad \circ 2 \quad \circ 3$

$\circ 4 \quad \circ 5 \quad \circ 6$

S

b. Experiment: Observe the up face on a die

$\circ HH \qquad \circ HT$

$\circ TH \qquad \circ TT$

S

c. Experiment: Observe the up faces on two coins

Figure 3.2

Venn diagrams for the three experiments from Table 3.1

Table 3.1	**Experiments and Their Sample Spaces**

Experiment:	Observe the up face on a coin.
Sample Space:	1. Observe a head.
	2. Observe a tail.

This sample space can be represented in set notation as a set containing two sample points:

$$S: \{H, T\}$$

Here, H represents the sample point Observe a head and T represents the sample point Observe a tail.

Experiment:	Observe the up face on a die.
Sample Space:	1. Observe a 1.
	2. Observe a 2.
	3. Observe a 3.
	4. Observe a 4.
	5. Observe a 5.
	6. Observe a 6.

This sample space can be represented in set notation as a set of six sample points:

$$S: \{1, 2, 3, 4, 5, 6\}$$

Experiment:	Observe the up faces on two coins.
Sample Space:	1. Observe HH.
	2. Observe HT.
	3. Observe TH.
	4. Observe TT.

This sample space can be represented in set notation as a set of four sample points:

$$S: \{HH, HT, TH, TT\}$$

JOHN VENN (1834–1923)

The English Logician

Born in Hull, England, John Venn is probably best known for his pictorial representation of unions and intersections (the Venn diagram). While lecturing in moral science at Cambridge University, Venn wrote *The Logic of Chance* and two other treatises on logic. In these works, Venn probably had his greatest contribution to the field of probability and statistics: the notion that the probability of an event is simply the long-run proportion of times the event occurs. Besides being a well-known mathematician, Venn was a historian, philosopher, priest, and skilled machine builder. (His machine for bowling cricket balls once beat a top star on the Australian cricket team.) ■

Now that we know that an experiment will result in *only one* basic outcome—called a sample point—and that the sample space is the collection of all possible sample points, we're ready to discuss the probabilities of the sample points. You have undoubtedly used the term *probability* and have some intuitive idea about its meaning. Probability is generally used synonymously with "chance," "odds," and similar concepts. For example, if a fair coin is tossed, we might reason that both the sample points Observe a head and Observe a tail have the same *chance* of occurring. Thus, we might state, "The probability of observing a head is 50%" or "The odds of seeing a head are 50:50." Both of these statements are based on an informal knowledge of probability. We'll begin our treatment of probability by using such informal concepts and then solidify what we mean later.

The probability of a sample point is a number between 0 and 1 which measures the likelihood that the outcome will occur when the experiment is performed. This number is usually taken to be the relative frequency of the occurrence of a sample point in a very long series of repetitions of an experiment.* For example, if we are assigning probabilities to the two sample points in the coin-toss experiment (Observe a head and Observe a tail), we might reason that if we toss a balanced coin a very large number of times, the sample points Observe a head and Observe a tail will occur with the same relative frequency of .5.

Our reasoning is supported by Figure 3.3. The figure plots the relative frequency of the number of times that a head occurs in simulations (by computer) of the toss of a coin N times, where N ranges from as few as 25 tosses to as many as 1,500 tosses. You can see that when N is large (e.g., $N = 1,500$), the relative frequency is converging to .5. Thus, the probability of each sample point in the coin-tossing experiment is .5.

For some experiments, we may have little or no information on the relative frequency of occurrence of the sample points; consequently, we must assign probabilities to the sample points on the basis of general information about the experiment. For example,

---

*The result derives from an axiom in probability theory called the **law of large numbers.** Phrased informally, the law states that the relative frequency of the number of times that an outcome occurs when an experiment is replicated over and over again (i.e., a large number of times) approaches the true (or theoretical) probability of the outcome.

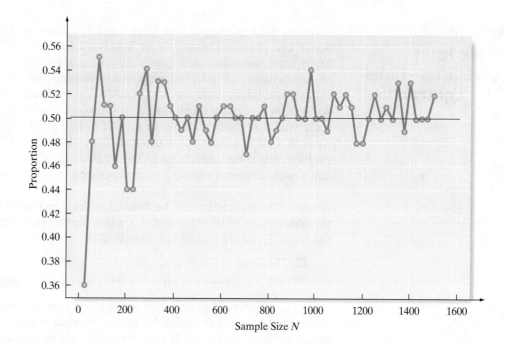

Figure 3.3

Proportion of heads in *N* tosses of a coin

Figure 3.4

Experiment: invest in a business venture and observe whether it succeeds (S) or fails (F)

if the experiment is about investing in a business venture and observing whether it succeeds or fails, the sample space would appear as in Figure 3.4.

We are unlikely to be able to assign probabilities to the sample points of this experiment based on a long series of repetitions, since unique factors govern each performance of this kind of experiment. Instead, we may consider factors such as the personnel managing the venture, the general state of the economy at the time, the rate of success of similar ventures, and any other information deemed pertinent. If we finally decide that the venture has an 80% chance of succeeding, we assign a probability of .8 to the sample point Success. This probability can be interpreted as a measure of our degree of belief in the outcome of the business venture; that is, it is a subjective probability. Notice, however, that such probabilities should be based on expert information that is carefully assessed. If not, we may be misled on any decisions based on these probabilities or based on any calculations in which they appear. [*Note:* For a text that deals in detail with the subjective evaluation of probabilities, see Winkler (1972) or Lindley (1985).]

No matter how you assign the probabilities to sample points, the probabilities assigned must obey two rules:

Probability Rules for Sample Points

Let p_i represent the probability of sample point i. Then

1. All sample point probabilities *must* lie between 0 and 1 (i.e., $0 \le p_i \le 1$).

2. The probabilities of all the sample points within a sample space *must* sum to 1 (i.e., $\Sigma p_i = 1$).

Assigning probabilities to sample points is easy for some experiments. For example, if the experiment is to toss a fair coin and observe the face, we would probably all agree to assign a probability of $\frac{1}{2}$ to the two sample points, Observe a head and Observe a tail. However, many experiments have sample points whose probabilities are more difficult to assign.

Example 3.2
Sample Point Probabilities—Hotel Water Conservation

Problem *Going Green* is the term used to describe water conservation programs at hotels and motels. Many hotels now offer their guests the option of participating in these Going Green programs by reusing towels and bed linens. Suppose you randomly select one hotel from a registry of all hotels in Orange County, California, and check whether or not the hotel participates in a water conservation program. Show how this problem might be formulated in the framework of an experiment with sample points and a sample space. Indicate how probabilities might be assigned to the sample points.

Solution The experiment can be defined as the selection of an Orange County hotel and the observation of whether or not a water conservation program is offered to guests at the hotel. There are two sample points in the sample space for this experiment:

  C: {The hotel offers a water conservation program.}

  N: {The hotel does not offer a water conservation program.}

The difference between this and the coin-toss experiment becomes apparent when we attempt to assign probabilities to the two sample points. What probability should we assign to sample point C? If you answer .5, you are assuming that the events C and N occur with equal likelihood, just like the sample points Observe a head and Observe a tail in the coin-toss experiment. But assigning sample point probabilities for the hotel water conservation experiment is not so easy. In fact, a recent report by the Orange County Water District stated that 70% of the hotels in the county now participate in a water conservation program of some type. In that case, it might be reasonable to approximate the probability of the sample point C as .7 and that of the sample point N as .3.

Look Back Here we see that the sample points are not always equally likely, so assigning probabilities to them can be complicated, particularly for experiments that represent real applications (as opposed to coin- and die-toss experiments).

Now Work Exercise 3.20

Although the probabilities of sample points are often of interest in their own right, it is usually probabilities of *collections* of sample points that are important. Example 3.3 demonstrates this point.

Example 3.3
Probability of a Collection of Sample Points—Die-Tossing Experiment

Problem A fair die is tossed, and the up face is observed. If the face is even, you win $1. Otherwise, you lose $1. What is the probability that you win?

Solution Recall that the sample space for this experiment contains six sample points:

$$S: \{1,2,3,4,5,6\}$$

Since the die is balanced, we assign a probability of $^1/_6$ to each of the sample points in this sample space. An even number will occur if one of the sample points Observe a 2, Observe a 4, or Observe a 6 occurs. A collection of sample points such as this is called an *event*, which we denote by the letter A. Since the event A contains three sample points—all with probability $^1/_6$—and since no sample points can occur simultaneously, we reason that the probability of A is the sum of the probabilities of the sample points in A. Thus, the probability of A is $^1/_6 + ^1/_6 + ^1/_6 = ^1/_2$.

Look Back On the basis of this notion of probability, *in the long run*, you will win $1 half the time and lose $1 half the time.

Now Work Exercise 3.14

Figure 3.5

Die-toss experiment with event A, observe an even number

Figure 3.5 is a Venn diagram depicting the sample space associated with a die-toss experiment and the event *A*, Observe an even number. The event *A* is represented by the closed figure inside the sample space **S.** This closed figure *A* comprises all of the sample points that make up event *A*.

To decide which sample points belong to the set associated with an event *A*, test each sample point in the sample space **S.** If event *A* occurs, then that sample point is in the event *A*. For example, the event *A*, Observe an even number, in the die-toss experiment will occur if the sample point Observe a 2 occurs. By the same reasoning, the sample points Observe a 4 and Observe a 6 are also in event *A*.

To summarize, we have demonstrated that an event can be defined in words, or it can be defined as a specific set of sample points. This leads us to the following general definition of an *event*:

An **event** is a specific collection of sample points.

Example 3.4
Probability of an Event—Coin-Tossing Experiment

Problem Consider the experiment of tossing two *unbalanced* coins. Because the coins are *not* balanced, their outcomes (*H* or *T*) are not equiprobable. Suppose the correct probabilities associated with the sample points are given in the accompanying table. [*Note:* The necessary properties for assigning probabilities to sample points are satisfied.] Consider the events

$$A: \{ \text{Observe exactly one head.} \}$$
$$B: \{ \text{Observe at least one head.} \}$$

Calculate the probability of *A* and the probability of *B*.

Sample Point	Probability
HH	$\frac{4}{9}$
HT	$\frac{2}{9}$
TH	$\frac{2}{9}$
TT	$\frac{1}{9}$

Solution Event *A* contains the sample points *HT* and *TH*. Since two or more sample points cannot occur at the same time, we can easily calculate the probability of event *A* by summing the probabilities of the two sample points. Thus, the probability of observing exactly one head (event *A*), denoted by the symbol *P(A)*, is

$$P(A) = P(\text{Observe } HT.) + P(\text{Observe } TH.) = \frac{2}{9} + \frac{2}{9} = \frac{4}{9}$$

Similarly, since *B* contains the sample points *HH, HT,* and *TH,* it follows that

$$P(B) = \frac{4}{9} + \frac{2}{9} + \frac{2}{9} = \frac{8}{9}$$

Look Back Again, these probabilities should be interpreted *in the long run*. For example, $P(B) = \frac{8}{9} \approx .89$ implies that if we were to toss two coins an infinite number of times, we would observe at least two heads on about 89% of the tosses.

Now Work Exercise 3.11

The preceding example leads us to a general procedure for finding the probability of an event *A*:

> **Probability of an Event**
>
> The probability of an event A is calculated by summing the probabilities of the sample points in the sample space for A.

Thus, we can summarize the steps for calculating the probability of any event, as indicated in the next box.

> **Steps for Calculating Probabilities of Events**
>
> 1. Define the experiment; that is, describe the process used to make an observation and the type of observation that will be recorded.
> 2. List the sample points.
> 3. Assign probabilities to the sample points.
> 4. Determine the collection of sample points contained in the event of interest.
> 5. Sum the sample point probabilities to get the probability of the event.

Example 3.5

Applying the Five Steps to Find a Probability—Study of Divorced Couples

Problem The American Association for Marriage and Family Therapy (AAMFT) is a group of professional therapists and family practitioners that treats many of the nation's couples and families. The AAMFT released the findings of a study that tracked the post-divorce history of 100 pairs of former spouses with children. Each divorced couple was classified into one of four groups, nicknamed "perfect pals (PP)," "cooperative colleagues (CC)," "angry associates (AA)," and "fiery foes (FF)." The proportions classified into each group are shown in Table 3.2.

Table 3.2 Results of AAMFT Study of Divorced Couples	
Group	Proportion
Perfect Pals (PP) (Joint-custody parents who get along well)	.12
Cooperative Colleagues (CC) (Occasional conflict, likely to be remarried)	.38
Angry Associates (AA) (Cooperate on children-related issues only, conflicting otherwise)	.25
Fiery Foes (FF) (Communicate only through children, hostile toward each other)	.25

Suppose one of the 100 couples is selected at random.

a. Define the experiment that generated the data in Table 3.2, and list the sample points.

b. Assign probabilities to the sample points.

c. What is the probability that the former spouses are "fiery foes?"

d. What is the probability that the former spouses have at least some conflict in their relationship?

Solution

a. The experiment is the act of classifying the randomly selected couple. The sample points—the simplest outcomes of the experiment—are the four groups (categories) listed in Table 3.2. They are shown in the Venn diagram in Figure 3.6.

b. If, as in Example 3.1, we were to assign equal probabilities in this case, each of the response categories would have a probability of one-fourth (1/4), or .25. But by examining Table 3.2, you can see that equal probabilities are not reasonable in this

Figure 3.6
Venn diagram for AAMFT survey

case, because the response percentages are not all the same in the four categories. It is more reasonable to assign a probability equal to the response proportion in each class, as shown in Table 3.3.*

Table 3.3	Sample Point Probabilities for AAMFT Survey
Sample Point	Probability
PP	.12
CC	.38
AA	.25
FF	.25

c. The event that the former spouses are "fiery foes" corresponds to the sample point FF. Consequently, the probability of the event is the probability of the sample point. From Table 3.3, we find that $P(FF) = .25$. Therefore, there is a .25 probability (or one-fourth chance) that the couple we select are "fiery foes."

d. The event that the former spouses have at least some conflict in their relationship, call it event C, is not a sample point, because it consists of more than one of the response classifications (the sample points). In fact, as shown in Figure 3.6, the event C consists of three sample points: CC, AA, and FF. The probability of C is defined to be the sum of the probabilities of the sample points in C:

$$P(C) = P(CC) + P(AA) + P(FF) = .38 + .25 + .25 = .88$$

Thus, the chance that we observe a divorced couple with some degree of conflict in their relationship is .88—a fairly high probability.

Look Back The key to solving this problem is to follow the steps outlined in the box. We defined the experiment (Step 1) and listed the sample points (Step 2) in part **a.** The assignment of probabilities to the sample points (Step 3) was done in part **b.** For each probability in parts **c** and **d,** we identified the collection of sample points in the event (Step 4) and summed their probabilities (Step 5).

Now Work Exercise 3.23

The preceding examples have one thing in common: The number of sample points in each of the sample spaces was small; hence, the sample points were easy to identify and list. How can we manage this when the sample points run into the thousands or millions? For example, suppose you wish to select 5 marines for a dangerous mission from a division of 1,000. Then each different group of 5 marines would represent a sample point. How can you determine the number of sample points associated with this experiment?

One method of determining the number of sample points for a complex experiment is to develop a counting system. Start by examining a simple version of the experiment. For example, see if you can develop a system for counting the number of ways to select 2 marines from a total of 4. If the marines are represented by the symbols $M_1, M_2, M_3,$ and M_4, the sample points could be listed in the following pattern:

$$(M_1, M_2) \qquad (M_2, M_3) \qquad (M_3, M_4)$$
$$(M_1, M_3) \qquad (M_2, M_4)$$
$$(M_1, M_4)$$

Note the pattern and now try a more complex situation—say, sampling 3 marines out of 5. List the sample points and observe the pattern. Finally, see if you can deduce the

---

*Since the response percentages were based on a sample of divorced couples, these assigned probabilities are estimates of the true population response percentages. You will learn how to measure the reliability of probability estimates in Chapter 5.

pattern for the general case. Perhaps you can program a computer to produce the matching and counting for the number of samples of 5 selected from a total of 1,000.

A second method of determining the number of sample points for an experiment is to use **combinatorial mathematics.** This branch of mathematics is concerned with developing counting rules for given situations. For example, there is a simple rule for finding the number of different samples of 5 marines selected from 1,000. This rule, called the **combinations rule,** is given in the box.

Combinations Rule

Suppose a sample of n elements is to be drawn without replacement from a set of N elements. Then the number of different samples possible is denoted by $\binom{N}{n}$ and is equal to

$$\binom{N}{n} = \frac{N!}{n!(N-n)!}$$

where

$$n! = n(n-1)(n-2)\cdots(3)(2)(1)$$

and similarly for $N!$ and $(N-n)!$ For example, $5! = 5 \cdot 4 \cdot 3 \cdot 2 \cdot 1$. [*Note:* The quantity $0!$ is defined to be equal to 1.]

Example 3.6

Using the Combinations Rule— Selecting 2 Marines from 4

Problem Consider the task of choosing 2 marines from a platoon of 4 to send on a dangerous mission. Use the combinations counting rule to determine how many different selections can be made.

Solution For this example, $N = 4, n = 2$, and

$$\binom{4}{2} = \frac{4!}{2!2!} = \frac{4 \cdot 3 \cdot 2 \cdot 1}{(2 \cdot 1)(2 \cdot 1)} = 6$$

Look Back You can see that this answer agrees with the number of sample points listed at the bottom of page 115.

Now Work Exercise 3.13

Example 3.7

Using The Combinations Rule— Selecting 5 movies from 20

Problem Suppose a movie reviewer for a newspaper reviews 5 movies each month. This month, the reviewer has 20 new movies from which to make the selection. How many different samples of 5 movies can be selected from the 20?

Solution For this example, $N = 20$ and $n = 5$. Then the number of different samples of 5 that can be selected from the 20 movies is

$$\binom{20}{5} = \frac{20!}{5!(20-5)!} = \frac{20!}{5!15!}$$

$$= \frac{20 \cdot 19 \cdot 18 \cdots \cdot 3 \cdot 2 \cdot 1}{(5 \cdot 4 \cdot 3 \cdot 2 \cdot 1)(15 \cdot 14 \cdot 13 \cdots \cdot 3 \cdot 2 \cdot 1)} = 15,504$$

Look Back You can see that attempting to list all the sample points for this experiment would be an extremely tedious and time-consuming, if not practically impossible, task.

The combinations rule is just one of a large number of counting rules that have been developed by combinatorial mathematicians. This counting rule applies to situations in which the experiment calls for selecting n elements from a total of N elements, without replacing each element before the next is selected. Several other basic counting rules are presented in optional Section 3.8.

Statistics IN Action | **Revisited**   ## The Probability of Winning Lotto

In Florida's state lottery game, called Pick-6 Lotto, you select six numbers of your choice from a set of numbers ranging from 1 to 53. We can apply the combinations rule to determine the total number of combinations of 6 numbers selected from 53 (i.e., the total number of sample points [or possible winning tickets]). Here, $N = 53$ and $n = 6$; therefore, we have

$$\binom{N}{n} = \frac{N!}{n!(N-n)!} = \frac{53!}{6!47!}$$
$$= \frac{(53)(52)(51)(50)(49)(48)(47!)}{(6)(5)(4)(3)(2)(1)(47!)}$$
$$= 22{,}957{,}480$$

Now, since the Lotto balls are selected at random, each of these 22,957,480 combinations is equally likely to occur. Therefore, the probability of winning Lotto is

$$P(\text{Win } 6/53 \text{ Lotto}) = 1/(22{,}957{,}480) = .00000004356$$

This probability is often stated as follows: The odds of winning the game with a single ticket are 1 in 22,957,480, or 1 in approximately 23 million. For all practical purposes, this probability is 0, implying that you have almost no chance of winning the lottery with a single ticket. Yet each week there is almost always a winner in the Florida Lotto. This apparent contradiction can be explained with the following analogy:

Suppose there is a line of minivans, front to back, from New York City to Los Angeles, California. Based on the distance between the two cities and the length of a standard minivan, there would be approximately 23 million minivans in line. Lottery officials will select, at random, one of the minivans and put a check for $10 million in the glove compartment. For a cost of $1, you may roam the country and select one (and only one) minivan and check the glove compartment. Do you think you will find $10 million in the minivan you choose? You can be almost certain that you won't. But now permit anyone to enter the lottery for $1 and suppose that 50 million people do so. With such a large number of participants, it is very likely that someone will find the minivan with the $10 million—but it almost certainly won't be you! (This example illustrates an axiom in statistics called the law of large numbers. See the footnote at the bottom of p. 110.)

Exercises 3.1–3.34

Understanding the Principles

3.1 What is an experiment?

3.2 What are the most basic outcomes of an experiment called?

3.3 Define the sample space.

3.4 What is a Venn diagram?

3.5 Give two probability rules for sample points.

3.6 What is an event?

3.7 How do you find the probability of an event made up of several sample points?

3.8 Give a scenario where the combinations rule is appropriate for counting the number of sample points.

Learning the Mechanics

3.9 An experiment results in one of the following sample points: E_1, E_2, E_3, E_4, and E_5.

a. Find $P(E_3)$ if $P(E_1) = .1, P(E_2) = .2, P(E_4) = .1$, and $P(E_5) = .1$

b. Find $P(E_3)$ if $P(E_1) = P(E_3), P(E_2) = .1$, $P(E_4) = .2$, and $P(E_5) = .1$

c. Find $P(E_3)$ if $P(E_1) = P(E_2) = P(E_4) = P(E_5) = .1$

3.10 The following Venn diagram describes the sample space of a particular experiment and events A and B:

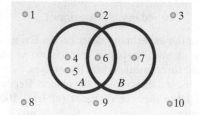

a. Suppose the sample points are equally likely. Find $P(A)$ and $P(B)$.

b. Suppose $P(1) = P(2) = P(3) = P(4) = P(5) = \frac{1}{20}$ and $P(6) = P(7) = P(8) = P(9) = P(10) = \frac{3}{20}$. Find $P(A)$ and $P(B)$.

3.11 The sample space for an experiment contains five sample points with probabilities as shown in the table. Find the probability of each of the following events:

A: {Either 1, 2, or 3 occurs.}

B: {Either 1, 3, or 5 occurs.}

C: {4 does not occur.}

Sample Points	Probabilities
1	.05
2	.20
3	.30
4	.30
5	.15

3.12 Compute each of the following:

a. $\binom{9}{4}$ **b.** $\binom{7}{2}$ **c.** $\binom{4}{4}$

d. $\binom{5}{0}$ **e.** $\binom{6}{5}$

3.13 Compute the number of ways you can select n elements from N elements for each of the following:

a. $n = 2, N = 5$

b. $n = 3, N = 6$

c. $n = 5, N = 20$

3.14 Two fair dice are tossed, and the up face on each die is recorded.

a. List the 36 sample points contained in the sample space.

b. Assign probabilities to the sample points.

c. Find the probability of observing each of the following events:

A: {A 3 appears on each of the two dice.}

B: {The sum of the numbers is even.}

C: {The sum of the numbers is equal to 7.}

D: {A 5 appears on at least one of the dice.}

E: {The sum of the numbers is 10 or more.}

3.15 Two marbles are drawn at random and without replacement from a box containing two blue marbles and three red marbles.

a. List the sample points.

b. Assign probabilities to the sample points.

c. Determine the probability of observing each of the following events:

A: {Two blue marbles are drawn.}

B: {A red and a blue marble are drawn.}

C: {Two red marbles are drawn.}

3.16 Simulate the experiment described in Exercise 3.15, using any five identically shaped objects, two of which are one color and three another. Mix the objects, draw two, record the results, and then replace the objects. Repeat the experiment a large number of times (at least 100). Calculate the proportion of times events A, B, and C occur. How do these proportions compare with the probabilities you calculated in Exercise 3.15? Should these proportions equal the probabilities? Explain.

Applet Exercise 3.1

Use the applet entitled *Simulating the Probability of Rolling a 6* to explore the relationship between the proportion of sixes rolled on several rolls of a die and the theoretical probability of rolling a 6 on a fair die.

a. To simulate rolling a die one time, click on the *Roll* button on the screen while $n = 1$. The outcome of the roll appears in the list at the right, and the cumulative proportion of sixes for one roll is shown above the graph and as a point on the graph corresponding to 1. Click *Reset* and repeat the process with $n = 1$ several times. What are the possible values of the cumulative proportion of sixes for one roll of a die? Can the cumulative proportion of sixes for one roll of a die equal the theoretical probability of rolling a 6 on a fair die? Explain.

b. Set $n = 10$ and click the *Roll* button. Repeat this several times, resetting after each time. Record the cumulative proportion of sixes for each roll. Compare the cumulative proportions for $n = 10$ with those for $n = 1$ in part **a**. Which tend to be closer to the theoretical probability of rolling a 6 on a fair die?

c. Repeat part **b** for $n = 1,000$, comparing the cumulative proportions for $n = 1,000$ with those for $n = 1$ in part **a** and for $n = 10$ in part **b**.

d. On the basis of your results for parts **a, b,** and **c**, do you believe that we could justifiably conclude that a die is unfair because we rolled it 10 times and didn't roll any sixes? Explain.

Applet Exercise 3.2

Use the applet entitled *Simulating the Probability of a Head with a Fair Coin* to explore the relationship between the proportion of heads on several flips of a coin and the theoretical probability of getting heads on one flip of a fair coin.

a. Repeat parts **a–c** of Applet Exercise 3.1 for the experiment of flipping a coin and the event of getting heads.

b. On the basis of your results for part **a**, do you believe that we could justifiably conclude that a coin is unfair because we flipped it 10 times and didn't roll any heads? Explain.

Applying the Concepts—Basic

3.17 **Colors of M&Ms candies.** In 1940, Forrest E. Mars, Sr., formed the Mars Corporation in order to produce chocolate candies with a sugar shell that could be sold throughout the year and wouldn't melt during the summer. Originally, M&Ms Plain Chocolate Candies came in only a brown color. Today, M&Ms in standard bags come in six colors: brown, yellow, red, blue, orange, and green. According to Mars Corporation, today 24% of all M&Ms produced are blue, 20% are orange, 16% are green, 14% are yellow, 13% are brown, and 13% are red. Suppose you purchase a randomly selected bag of M&Ms Plain Chocolate Candies and randomly select one of the M&Ms from the bag. The color of the selected M&M is of interest.

a. Identify the outcomes (sample points) of this experiment.

b. Assign reasonable probabilities to the outcomes, part **a**.

c. What is the probability that the selected M&M is brown (the original color)?

d. In 1960, the colors red, green, and yellow were added to brown M&Ms. What is the probability that the selected M&M is either red, green, or yellow?

e. In 1995, based on voting by American consumers, the color blue was added to the M&M mix. What is the probability that the selected M&M is not blue?

3.18 **Rare underwater sounds.** A study of underwater sounds in a specific region of the Pacific Ocean focused on scarce sounds, such as humpback whale screams, dolphin whistles, and sounds from passing ships (*Acoustical Physics,* Vol. 56, 2010). During the month of September (non–rainy season), research revealed the following probabilities of rare sounds: P(whale scream) = .03, P(ship sound) = .14, and P(rain) = 0. If a sound is picked up by the acoustical equipment placed in this region of the Pacific Ocean, is it more likely to be a whale scream or a sound from a passing ship? Explain.

3.19 **USDA chicken inspection.** The United States Department of Agriculture (USDA) reports that, under its standard inspection system, one in every 100 slaughtered chickens passes inspection for fecal contamination.

a. If a slaughtered chicken is selected at random, what is the probability that it passes inspection for fecal contamination?

b. The probability of part **a** was based on a USDA study which found that 306 of 32,075 chicken carcasses passed inspection for fecal contamination. Do you agree with the USDA's statement about the likelihood of a slaughtered chicken passing inspection for fecal contamination?

3.20 **African rhinos.** Two species of rhinoceros native to Africa are black rhinos and white rhinos. The International Rhino Federation estimates that the African rhinoceros population consists of 3,610 white rhinos and 11,330 black rhinos. Suppose one rhino is selected at random from the African rhino population and its species (black or white) is observed.

a. List the sample points for this experiment.

b. Assign probabilities to the sample points on the basis of the estimates made by the International Rhino Federation.

3.21 **Going online for health information.** A *cyberchondriac* is defined as a person who regularly searches the Web for health care information. A recent Harris Poll surveyed 1,010 U.S. adults by telephone and asked each respondent how often (in the past month) he or she looked for health care information online. The results are summarized in the following table. Consider the response category of a randomly selected person who participated in the Harris poll.

Response (# per month)	Percentage of Respondents
None	25
1 or 2	31
3–5	25
6–9	5
10 or more	14
Total	100%

a. List the sample points for the experiment.

b. Assign reasonable probabilities to the sample points.

c. Find the probability that the respondent looks for health care information online more than two times per month.

3.22 **Health risks to beachgoers.** According to a University of Florida veterinary researcher, the longer a beachgoer sits in wet sand or stays in the water, the higher the health risk (*University of Florida News*, Jan. 29, 2008). Using data collected at Fort Lauderdale, Hollywood, and Hobie Beach, the researcher discovered the following: (1) 6 out of 1,000 people exposed to wet sand for a 10-minute period will acquire gastroenteritis; (2) 12 out of 100 people exposed to wet sand for two consecutive hours will acquire gastroenteritis; (3) 7 out of 1,000 people exposed to ocean water for a 10-minute period will acquire gastroenteritis; and, (4) 7 out of 100 people exposed to ocean water for a 70-minute period will acquire gastroenteritis.

a. If a beachgoer spends 10 minutes in the wet sand, what is the probability that he or she will acquire gastroenteritis?

b. If a beachgoer spends two hours in the wet sand, what is the probability that he or she will acquire gastroenteritis?

c. If a beachgoer spends 10 minutes in the ocean water, what is the probability that he or she will acquire gastroenteritis?

d. If a beachgoer spends 70 minutes in the ocean water, what is the probability that he or she will acquire gastroenteritis?

Applying the Concepts—Intermediate

3.23 **Sociology fieldwork methods.** Refer to University of New Mexico professor Jane Hood's study of the fieldwork methods used by qualitative sociologists, presented in Exercise 2.16 (p. 36). Recall that she discovered that fieldwork methods could be classified into four distinct categories: Interview, Observation plus Participation, Observation Only, and Grounded Theory. The table that follows, reproduced from *Teaching Sociology* (July 2006), and saved in the **FIELDWORK** file, gives the number of sociology field research papers in each category. Suppose we randomly select one of these research papers and determine the method used. Find the probability that the method used is either Interview or Grounded Theory.

Fieldwork Method	Number of Papers
Interview	5,079
Observation plus Participation	1,042
Observation Only	848
Grounded Theory	537

Based on Hood, J. C. "Teaching against the text: The case of qualitative methods." *Teaching Sociology*, Vol. 34, Issue 3 (Exhibit 2).

3.24 **Chance of rain.** Answer the following question posed in the *Atlanta Journal-Constitution*: When a meteorologist says, "The probability of rain this afternoon is .4," does it mean that it will be raining 40% of the time during the afternoon?

3.25 **Cheek teeth of extinct primates.** Refer to the *American Journal of Physical Anthropology* (Vol. 142, 2010) study of the dietary habits of extinct mammals, Exercise 2.7

(p. 34). Recall that 18 cheek teeth extracted from skulls of an extinct primate species discovered in western Wyoming were analyzed. Each tooth was classified according to degree of wear (unworn, slight, light-moderate, moderate, moderate-heavy, or heavy). The 18 measurements are reproduced in the accompanying table and saved in the **CHEEKTEETH** file. One tooth is randomly selected from the 18 cheek teeth. What is the probability that the tooth shows a slight or moderate amount of wear?

Data on Degree of Wear	
Unknown	Slight
Unknown	Slight
Unknown	Heavy
Moderate	Unworn
Slight	Light-moderate
Unknown	Light-moderate
Moderate-heavy	Moderate
Moderate	Unworn
Slight	Unknown

3.26 **Museum management.** Refer to the *Museum Management and Curatorship* (June 2010) study of the criteria used to evaluate museum performance, Exercise 2.19 (p. 37). Recall that the managers of 30 leading museums of contemporary art were asked to provide the performance measure used most often. A summary of the results is reproduced in the table. One of the 30 museums is selected at random. Find the probability that the museum uses big shows most often as a performance measure.

Performance Measure	Number of Museums
Total visitors	8
Paying visitors	5
Big shows	6
Funds raised	7
Members	4

3.27 **Choosing portable grill displays.** University of Maryland marketing professor R. W. Hamilton studied how people attempt to influence the choices of others by offering undesirable alternatives (*Journal of Consumer Research*, Mar. 2003). Such a phenomenon typically occurs when family members propose a vacation spot, friends recommend a restaurant for dinner, and realtors show the buyer potential homesites. In one phase of the study, the researcher had each of 124 college students select showroom displays for portable grills. Five different displays (representing five different-sized grills) were available, but only three would be selected. The students were instructed to select the displays to maximize purchases of Grill #2 (a smaller grill).

a. In how many possible ways can the three-grill displays be selected from the 5 displays? List the possibilities.

b. The next table shows the grill display combinations and number of each selected by the 124 students. Use this information to assign reasonable probabilities to the different display combinations.

c. Find the probability that a student who participated in the study selected a display combination involving Grill #1.

Grill Display Combination	Number of Students
1–2–3	35
1–2–4	8
1–2–5	42
2–3–4	4
2–3–5	1
2–4–5	34

Based on Hamilton, R. W. "Why do people suggest what they do not want? Using context effects to influence others' choices." *Journal of Consumer Research*, Vol. 29, Mar. 2003, Table 1.

3.28 **ESL students and plagiarism.** The *Journal of Education and Human Development* (Vol. 3, 2009) investigated the causes of plagiarism among six English-as-a-second language (ESL) students taking a master's course in linguistics. All students in the class wrote two essays, one in the middle and one at the end of the semester. After the first essay, the students were instructed on how to avoid plagiarism in the second essay. Of the six ESL students, three admitted to plagiarizing on the first essay. Only one ESL student admitted to plagiarizing on the second essay. (This student, who also plagiarized on the first essay, claimed she misplaced her notes on plagiarism.)

a. If one of the six ESL students is randomly selected, what is the probability that he or she plagiarized on the first essay?

b. Consider the results (plagiarism or not) for the six ESL students on the second essay. List the possible outcomes (e.g., Students 1 and 3 plagiarize, the others do not).

c. Refer to part **b.** Assume that, despite the instruction on plagiarism, the ESL students are just as likely as not to plagiarize on the second essay. What is the probability that no more than one of the ESL students plagiarizes on the second essay?

3.29 **Jai alai Quinella bet.** The Quinella bet at the parimutuel game of jai alai consists of picking the jai alai players that will place first and second in a game, *irrespective of order*. In jai alai, eight players (numbered 1, 2, 3,..., 8) compete in every game.

a. How many different Quinella bets are possible?

b. Suppose you bet the Quinella combination 2–7. If the players are of equal ability, what is the probability that you win the bet?

3.30 **Groundwater contamination in wells.** Refer to the *Environmental Science & Technology* (Jan. 2005) study of methyl *tert*-butyl ether (MTBE) contamination in New Hampshire wells, presented in Exercise 2.21 (p. 37). Data collected for a sample of 223 wells are saved in the **MTBE** file. Recall that each well was classified according to its class (public or private), aquifer (bedrock or unconsolidated), and detectable level of MTBE (below limit or detectable).

a. Consider an experiment in which the class, aquifer, and detectable MTBE level of a well are observed. List the sample points for this experiment. [*Hint:* One sample point is Private/Bedrock/BelowLimit.]

b. Use statistical software to find the number of the 223 wells in each sample point outcome. Then use this information to compute probabilities for the sample points.

c. Find and interpret the probability that a well has a detectable level of MTBE.

Applying the Concepts—Advanced

3.31 **Lead bullets as forensic evidence.** *Chance* (Summer 2004) published an article on the use of lead bullets as forensic evidence in a federal criminal case. Typically, the Federal Bureau of Investigation (FBI) will use a laboratory method to match the lead in a bullet found at a crime scene with unexpended lead cartridges found in the possession of a suspect. The value of this evidence depends on the chance of a *false positive*—that is, the probability that the FBI finds a match, given that the lead at the crime scene and the lead in the possession of the suspect are actually from two different "melts," or sources. To estimate the false positive rate, the FBI collected 1,837 bullets that the agency was confident all came from different melts. Then, using its established criteria, the FBI examined every possible pair of bullets and counted the number of matches. According to *Chance*, the FBI found 693 matches. Use this information to compute the chance of a false positive. Is this probability small enough for you to have confidence in the FBI's forensic evidence?

3.32 **Matching socks.** Consider the following question posed to Marilyn vos Savant in her weekly newspaper column, "Ask Marilyn":

I have two pairs of argyle socks, and they look nearly identical—one navy blue and the other black. [When doing the laundry] my wife matches the socks incorrectly much more often than she does correctly.... If all four socks are in front of her, it seems to me that her chances are 50% for a wrong match and 50% for a right match. What do you think?

Source: Parade Magazine, Feb. 27, 1994.

Use your knowledge of probability to answer this question. [*Hint:* List the sample points in the experiment.]

3.33 **Post-op nausea study.** Nausea and vomiting after surgery are common side effects of anesthesia and painkillers. Six different drugs, varying in cost, were compared for their effectiveness in preventing nausea and vomiting (*New England Journal of Medicine,* June 10, 2004). The medical researchers looked at all possible combinations of the drugs as treatments, including a single drug, as well as two-drug, three-drug, four-drug, five-drug, and six-drug combinations.

a. How many two-drug combinations of the six drugs are possible?

b. How many three-drug combinations of the six drugs are possible?

c. How many four-drug combinations of the six drugs are possible?

d. How many five-drug combinations of the six drugs are possible?

e. The researchers stated that a total of 64 drug combinations were tested as treatments for nausea. Verify that there are 64 ways that the six drugs can be combined. (Remember to include the one-drug and six-drug combinations, as well as the control treatment of no drugs.)

3.34 **Dominant versus recessive traits.** An individual's genetic makeup is determined by the genes obtained from each parent. For every genetic trait, each parent possesses a gene pair, and each parent contributes one-half of this gene pair, with equal probability, to his or her offspring, forming a new gene pair. The offspring's traits (eye color, baldness, etc.) come from this new gene pair, each gene of which possesses some characteristic.

For the gene pair that determines eye color, each gene trait may be one of two types: dominant brown (*B*) or recessive blue (*b*). A person possessing the gene pair *BB* or *Bb* has brown eyes, whereas the gene pair *bb* produces blue eyes.

a. Suppose both parents of an individual are brown eyed, each with a gene pair of type *Bb*. What is the probability that a randomly selected child of this couple will have blue eyes?

b. If one parent has brown eyes, type *Bb*, and the other has blue eyes, what is the probability that a randomly selected child of this couple will have blue eyes?

c. Suppose one parent is brown eyed with a gene pair of type *BB*. What is the probability that a child has blue eyes?

3.2 Unions and Intersections

An event can often be viewed as a composition of two or more other events. Such events, which are called **compound events,** can be formed (composed) in two ways.

The **union** of two events *A* and *B* is the event that occurs if either *A* or *B* (or both) occurs on a single performance of the experiment. We denote the union of events *A* and *B* by the symbol $A \cup B$. $A \cup B$ consists of all the sample points that belong to *A or B or both*. (See Figure 3.7a.)

The **intersection** of two events *A* and *B* is the event that occurs if both *A* and *B* occur on a single performance of the experiment. We write $A \cap B$ for the intersection of *A* and *B*. $A \cap B$ consists of all the sample points belonging to *both A and B*. (See Figure 3.7b.)

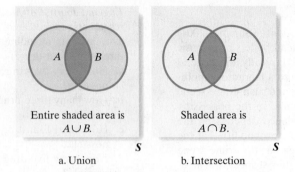

Figure 3.7

Venn diagrams for union and intersection

Entire shaded area is $A \cup B$.

Shaded area is $A \cap B$.

a. Union

b. Intersection

Example 3.8
Probabilities of Unions and Intersections— Die-Toss Experiment

Problem Consider a die-toss experiment in which the following events are defined:

A: {Toss an even number.}

B: {Toss a number less than or equal to 3.}

a. Describe $A \cup B$ for this experiment.

b. Describe $A \cap B$ for this experiment.

c. Calculate $P(A \cup B)$ and $P(A \cap B)$, assuming that the die is fair.

Solution Draw the Venn diagram as shown in Figure 3.8.

a. The union of A and B is the event that occurs if we observe either an even number, a number less than or equal to 3, or both on a single throw of the die. Consequently, the sample points in the event $A \cup B$ are those for which A occurs, B occurs, or both A and B occur. Checking the sample points in the entire sample space, we find that the collection of sample points in the union of A and B is

$$A \cup B = \{1, 2, 3, 4, 6\}$$

b. The intersection of A and B is the event that occurs if we observe *both* an even number and a number less than or equal to 3 on a single throw of the die. Checking the sample points to see which imply the occurrence of *both* events A and B, we see that the intersection contains only one sample point:

$$A \cap B = \{2\}$$

In other words, the intersection of A and B is the sample point Observe a 2.

c. Recalling that the probability of an event is the sum of the probabilities of the sample points of which the event is composed, we have

$$P(A \cup B) = P(1) + P(2) + P(3) + P(4) + P(6)$$

$$= \frac{1}{6} + \frac{1}{6} + \frac{1}{6} + \frac{1}{6} + \frac{1}{6} = \frac{5}{6}$$

and

$$P(A \cap B) = P(2) = \frac{1}{6}$$

Look Back Since the six sample points are equally likely, the probabilities in part **c** are simply the number of sample points in the event of interest, divided by 6.

Now Work Exercise 3.45a–d

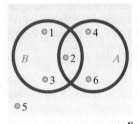

Figure 3.8

Venn diagram for die toss

Unions and intersections can be defined for more than two events. For example, the event $A \cup B \cup C$ represents the union of three events: A, B, and C. This event, which includes the set of sample points in A, B, or C, will occur if any one (or more) of the events A, B, and C occurs. Similarly, the intersection $A \cap B \cap C$ is the event that all three of the events A, B, and C occur. Therefore, $A \cap B \cap C$ is the set of sample points that are in all three of the events A, B, and C.

Example 3.9

Finding Probabilities From a Two-Way Table—Mother's Race versus Maternal Age

Problem *Family Planning Perspectives* reported on a study of over 200,000 births in New Jersey over a recent two-year period. The study investigated the link between the mother's race and the age at which she gave birth (called maternal age). The percentages of the total number of births in New Jersey, by the maternal age and race classifications, are given in Table 3.4.

Table 3.4	Percentage of New Jersey Birth Mothers, by Age and Race	
	Race	
Maternal Age (years)	White	Black
≤17	2%	2%
18–19	3%	2%
20–29	41%	12%
≥30	33%	5%

Based on Reichman, N. E., and Pagnini, D. L. "Maternal age and birth outcomes: Data from New Jersey." *Family Planning Perspectives*, Vol. 29, No. 6, Nov./Dec. 1997, p. 269 (adapted from Table 1).

This table is called a **two-way table,** since responses are classified according to two variables: maternal age (rows) and race (columns).

Define the following event:

A: {A New Jersey birth mother is white.}

B: {A New Jersey mother was a teenager when giving birth.}

a. Find $P(A)$ and $P(B)$.

b. Find $P(A \cup B)$.

c. Find $P(A \cap B)$.

Solution Following the steps for calculating probabilities of events, we first note that the objective is to characterize the race and maternal age distribution of New Jersey birth mothers. To accomplish this objective, we define the experiment to consist of selecting a birth mother from the collection of all New Jersey birth mothers during the two-year period of the study and observing her race and maternal age class. The sample points are the eight different age–race classifications:

E_1: {≤ 17 yrs., white} E_5: {≤ 17 yrs., black}

E_2: {18–19 yrs., white} E_6: {18–19 yrs., black}

E_3: {20–29 yrs., white} E_7: {20–29 yrs., black}

E_4: {≥ 30 yrs., white} E_8: {≥ 30 yrs., black}

Next, we assign probabilities to the sample points. If we blindly select one of the birth mothers, the probability that she will occupy a particular age–race classification is just the proportion, or relative frequency, of birth mothers in that classification. These proportions (as percentages) are given in Table 3.4. Thus,

$P(E_1)$ = Relative frequency of birth mothers in age–race class {≤ 17 yrs., white} = .02

$P(E_2)$ = .03

$P(E_3)$ = .41

$P(E_4)$ = .33

$P(E_5)$ = .02

$P(E_6)$ = .02

$P(E_7)$ = .12

$P(E_8)$ = .05

You may verify that the sample point probabilities sum to 1.

a. To find $P(A)$, we first determine the collection of sample points contained in event A. Since A is defined as {white}, we see from Table 3.4 that A contains the four sample points represented by the first column of the table. In other words, the event A consists of the race classification {white} in all four age classifications. The probability of A is the sum of the probabilities of the sample points in A:

$$P(A) = P(E_1) + P(E_2) + P(E_3) + P(E_4) = .02 + .03 + .41 + .33 = .79$$

Similarly, $B = $ {teenage mother, age $\leq$ 19 years} consists of the four sample points in the first and second rows of Table 3.4:

$$P(B) = P(E_1) + P(E_2) + P(E_5) + P(E_6) = .02 + .03 + .02 + .02 = .09$$

b. The union of events A and B, $A \cup B$, consists of all sample points in *either A or B (or both)*. That is, the union of A and B consists of all birth mothers who are white *or* who gave birth as a teenager. In Table 3.4, this is any sample point found in the first column *or* the first two rows. Thus,

$$P(A \cup B) = .02 + .03 + .41 + .33 + .02 + .02 = .83$$

c. The intersection of events A and B, $A \cap B$, consists of all sample points in *both A and B*. That is, the intersection of A and B consists of all birth mothers who are white *and* who gave birth as a teenager. In Table 3.4, this is any sample point found in the first column *and* the first two rows. Thus,

$$P(A \cap B) = .02 + .03 = .05.$$

Look Back As in previous problems, the key to finding the probabilities of parts **b** and **c** is to identify the sample points that make up the event of interest. In a two-way table such as Table 3.4, the total number of sample points will be equal to the number of rows times the number of columns.

Now Work Exercise 3.47f–g

3.3 Complementary Events

A very useful concept in the calculation of event probabilities is the notion of **complementary events:**

> The **complement** of an event A is the event that A does *not* occur—that is, the event consisting of all sample points that are not in event A. We denote the complement of A by A^c.

An event A is a collection of sample points, and the sample points included in A^c are those not in A. Figure 3.9 demonstrates this idea. Note from the figure that all sample points in S are included in *either* A or A^c and that *no* sample point is in both A and A^c. This leads us to conclude that the probabilities of an event and its complement *must sum to 1:*

Figure 3.9
Venn diagram of complementary events

> **Rule of Complements**
> The sum of the probabilities of complementary events equals 1; that is,
> $$P(A) + P(A^c) = 1.$$

In many probability problems, calculating the probability of the complement of the event of interest is easier than calculating the event itself. Then, because

$$P(A) + P(A^c) = 1$$

we can calculate $P(A)$ by using the relationship

$$P(A) = 1 - P(A^c).$$

Example 3.10

Probability of a Complementary Event—Coin-Toss Experiment

Problem Consider the experiment of tossing fair coins. Define the following event: A: {Observing at least one head}.

a. Find $P(A)$ if 2 coins are tossed.

b. Find $P(A)$ if 10 coins are tossed.

Solution

a. When 2 coins are tossed, we know that the event A: {Observe at least one head.} consists of the sample points

$$A: \{HH, HT, TH\}$$

The complement of A is defined as the event that occurs when A does not occur. Therefore,

$$A^c: \{\text{Observe no heads.}\} = \{TT\}$$

This complementary relationship is shown in Figure 3.10. Since the coins are balanced, we have

$$P(A^c) = P(TT) = \frac{1}{4}$$

and

$$P(A) = 1 - P(A^c) = 1 - \frac{1}{4} = \frac{3}{4}.$$

b. We solve this problem by following the five steps for calculating probabilities of events. (See Section 3.1.)

Step 1 Define the experiment. The experiment is to record the results of the 10 tosses of the coin.

Step 2 List the sample points. A sample point consists of a particular sequence of 10 heads and tails. Thus, one sample point is $HHTTTHTHTT$, which denotes head on first toss, head on second toss, tail on third toss, etc. Others are $HTHHHTTTTT$ and $THHTHTHTTH$. Obviously, the number of sample points is very large—too many to list. It can be shown (see Section 3.8) that there are $2^{10} = 1{,}024$ sample points for this experiment.

Step 3 Assign probabilities. Since the coin is fair, each sequence of heads and tails has the same chance of occurring; therefore, all the sample points are equally likely. Then

$$P(\text{Each sample point}) = \frac{1}{1{,}024}$$

Step 4 Determine the sample points in event A. A sample point is in A if at least one H appears in the sequence of 10 tosses. However, if we consider the complement of A, we find that

$$A^c = \{\text{No heads are observed in 10 tosses.}\}$$

Thus, A^c contains only one sample point:

$$A^c: \{TTTTTTTTTT\}$$

and $P(A^c) = \dfrac{1}{1{,}024}$

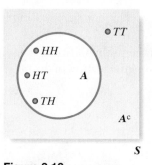

Figure 3.10
Complementary events in the toss of two coins

Step 5   Now we use the relationship of complementary events to find $P(A)$:

$$P(A) = 1 - P(A^c) = 1 - \frac{1}{1,024} = \frac{1,023}{1,024} = .999$$

Look Back   In part **a,** we can find $P(A)$ by summing the probabilities of the sample points HH, HT, and TH in A. Many times, however, it is easier to find $P(A^c)$ by using the rule of complements.

Look Forward   Since $P(A) = .999$ in part **b,** we are virtually certain of observing at least one head in 10 tosses of the coin.

Now Work Exercise 3.45e–f

3.4 The Additive Rule and Mutually Exclusive Events

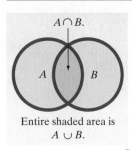

Figure 3.11
Venn diagram of union

In Section 3.2, we saw how to determine which sample points are contained in a union of two sets and how to calculate the probability of the union by summing the separate probabilities of the sample points in the union. It is also possible to obtain the probability of the union of two events by using the **additive rule of probability.**

By studying the Venn diagram in Figure 3.11, you can see that the probability of the union of two events A and B can be obtained by summing $P(A)$ and $P(B)$ and subtracting $P(A \cap B)$. We must subtract $P(A \cap B)$ because the sample point probabilities in $A \cap B$ have been included twice—once in $P(A)$ and once in $P(B)$.

The formula for calculating the probability of the union of two events is given in the next box.

Additive Rule of Probability

The probability of the union of events A and B is the sum of the probability of event A and the probability of event B, minus the probability of the intersection of events A and B; that is

$$P(A \cup B) = P(A) + P(B) - P(A \cap B)$$

Example 3.11

Applying the Additive Rule— Hospital Admissions Study

Problem   Hospital records show that 12% of all patients are admitted for surgical treatment, 16% are admitted for obstetrics, and 2% receive both obstetrics and surgical treatment. If a new patient is admitted to the hospital, what is the probability that the patient will be admitted for surgery, for obstetrics, or for both?

Solution   Consider the following events:

A: {A patient admitted to the hospital receives surgical treatment.}
B: {A patient admitted to the hospital receives obstetrics treatment.}

Then, from the given information,

$$P(A) = .12$$
$$P(B) = .16$$

and the probability of the event that a patient receives both obstetrics and surgical treatment is

$$P(A \cap B) = .02$$

The event that a patient admitted to the hospital receives either surgical treatment, obstetrics treatment, or both is the union $A \cup B$, the probability of which is given by the additive rule of probability:

$$P(A \cup B) = P(A) + P(B) - P(A \cap B)$$
$$= .12 + .16 - .02 = .26$$

Thus, 26% of all patients admitted to the hospital receive either surgical treatment, obstetrics treatment, or both.

Look Back From the information given, it is not possible to list and assign probabilities to all the sample points. Consequently, we cannot proceed through the five-step process (p. 114) for finding $P(A \cup B)$, and we must use the additive rule.

Now Work Exercise 3.43

A very special relationship exists between events A and B when $A \cap B$ contains no sample points. In this case, we call the events A and B *mutually exclusive events*.

Figure 3.12
Venn diagram of mutually exclusive events

Events A and B are **mutually exclusive** if $A \cap B$ contains no sample points—that is, if A and B have no sample points in common. For mutually exclusive events,

$$P(A \cap B) = 0$$

Figure 3.12 shows a Venn diagram of two mutually exclusive events. The events A and B have no sample points in common; that is, A and B cannot occur simultaneously, and $P(A \cap B) = 0$. Thus, we have the important relationship given in the next box.

Probability of Union of Two Mutually Exclusive Events

If two events A and B are *mutually exclusive*, the probability of the union of A and B equals the sum of the probability of A and the probability of B; that is, $P(A \cup B) = P(A) + P(B)$.

⚠ **CAUTION** The preceding formula is *false* if the events are *not* mutually exclusive. In that case (i.e., two non-mutually exclusive events), you must apply the general additive rule of probability. ▲

Example 3.12
Union of Two Mutually Exclusive Events—Coin-Tossing Experiment

Problem Consider the experiment of tossing two balanced coins. Find the probability of observing *at least* one head.

Solution Define the events

A: {Observe at least one head.}
B: {Observe exactly one head.}
C: {Observe exactly two heads.}

Note that

$$A = B \cup C$$

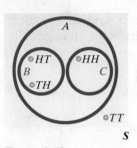

Figure 3.13
Venn diagram for coin-toss
experiment

and that $B \cap C$ contains no sample points. (See Figure 3.13.) Thus, B and C are mutually exclusive, and it follows that

$$P(A) = P(B \cup C) = P(B) + P(C) = \frac{1}{2} + \frac{1}{4} = \frac{3}{4}$$

Look Back Although this example is quite simple, it shows us that writing events with verbal descriptions that include the phrases "at least" or "at most" as unions of mutually exclusive events is very useful. This practice enables us to find the probability of the event by adding the probabilities of the mutually exclusive events.

Statistics IN Action   Revisited   The Probability of Winning Lotto with a Wheeling System

Refer to Florida's Pick-6 Lotto game, in which you select six numbers of your choice from a field of numbers ranging from 1 to 53. In Section 3.1, we learned that the probability of winning Lotto on a single ticket is only 1 in approximately 23 million. The "experts" at Lotto Buster recommend many strategies for increasing the odds of winning the lottery. One strategy is to employ a *wheeling system*. In a complete wheeling system, you select more than six numbers—say, seven—and play every combination of six of those seven numbers.

For example, suppose you choose to "wheel" the following seven numbers: 2, 7, 18, 23, 30, 32, and 51. Every combination of six of these seven numbers is listed in Table SIA3.1. You can see that there are seven different possibilities. (Use the combinations rule with $N = 7$ and $n = 6$ to verify this.) Thus, we would purchase seven tickets (at a cost of $7) corresponding to these different combinations in a complete wheeling system.

Table SIA3.1	Wheeling the Seven Numbers 2, 7, 18, 23, 30, 32, and 51					
Ticket #1	2	7	18	23	30	32
Ticket #2	2	7	18	23	30	51
Ticket #3	2	7	18	23	32	51
Ticket #4	2	7	18	30	32	51
Ticket #5	2	7	23	30	32	51
Ticket #6	2	18	23	30	32	51
Ticket #7	7	18	23	30	32	51

To determine whether this strategy does, in fact, increase our odds of winning, we need to find the probability that one of these seven combinations occurs during the 6/53 Lotto draw. That is, we need to find the probability that either Ticket #1 or Ticket #2 or Ticket #3 or Ticket #4 or Ticket #5 or Ticket #6 or Ticket #7 is the winning combination. Note that this probability is stated with the use of the word *or*, implying a *union* of seven events. Letting T_1 represent the event that Ticket #1 wins, and defining $T_2, T_3, \ldots, T_7$ in a similar fashion, we want to find

$$P(T_1 \text{ or } T_2 \text{ or } T_3 \text{ or } T_4 \text{ or } T_5 \text{ or } T_6 \text{ or } T_7)$$

Recall (Section 3.2) that the 22,957,480 possible combinations in Pick-6 Lotto are mutually exclusive and equally likely to occur. Consequently, the probability of the union of the seven events is simply the sum of the probabilities of the individual events, where each event has probability $1/(22,957,480)$:

$$P(\text{win Lotto with 7 Wheeled Numbers})$$
$$= P(T_1 \text{ or } T_2 \text{ or } T_3 \text{ or } T_4 \text{ or } T_5 \text{ or } T_6 \text{ or } T_7)$$
$$= 7/(22,957,480) = .0000003$$

In terms of odds, we now have 3 chances in 10 million of winning the Lotto with the complete wheeling system. Technically, the "experts" are correct: Our odds of winning Lotto have increased (from 1 in 23 million). However, the probability of winning is so close to 0 that we question whether the $7 spent on lottery tickets is worth the negligible increase in odds. In fact, it can be shown that to increase your chance of winning the 6/53 Lotto to 1 chance in 100 (i.e., .01) by means of a complete wheeling system, you would have to wheel 26 of your favorite numbers—a total of 230,230 combinations at a cost of $230,230!

Exercises 3.35–3.62

Understanding the Principles

3.35 Define in words the union of two events.

3.36 Define in words the intersection of two events.

3.37 Define in words the complement of an event.

3.38 State the rule of complements.

3.39 State the additive rule of probability for any two events.

3.40 Define in words mutually exclusive events.

3.41 State the additive rule of probability for mutually exclusive events.

Learning the Mechanics

3.42 Suppose $P(A) = .4, P(B) = .7$ and $P(A \cap B) = .3$. Find the following probabilities:
  a. $P(B^c)$
  b. $P(A^c)$
  c. $P(A \cup B)$

3.43 A fair coin is tossed three times, and the events A and B
  [NW] are defined as follows:

> A: {At least one head is observed.}
> B: {The number of heads observed is odd.}

  a. Identify the sample points in the events A, B, $A \cup B$, A^c, and $A \cap B$.
  b. Find $P(A), P(B), P(A \cup B), P(A^c)$, and $P(A \cap B)$ by summing the probabilities of the appropriate sample points.
  c. Use the additive rule to find $P(A \cup B)$. Compare your answer with the one you obtained in part **b.**
  d. Are the events A and B mutually exclusive? Why?

3.44 A pair of fair dice is tossed. Define the following events:

> A: {You will roll a 7 (i.e., the sum of the numbers of dots on the upper faces of the two dice is equal to 7).}
> B: {At least one of the two dice is showing a 4.}

  a. Identify the sample points in the events A, B, $A \cap B$, $A \cup B$, and A^c.
  b. Find $P(A), P(B), P(A \cap B), P(A \cup B)$, and $P(A^c)$ and by summing the probabilities of the appropriate sample points.
  c. Use the additive rule to find $P(A \cup B)$. Compare your answer with that for the same event in part **b.**
  d. Are A and B mutually exclusive? Why?

3.45 Consider the following Venn diagram, where
  [NW] $P(E_2) = P(E_3) = \frac{1}{5}, P(E_4) = P(E_5) = \frac{1}{20}$,
  $P(E_6) = \frac{1}{10}, P(E_7) = \frac{1}{5}$:

Find each of the following probabilities:
  a. $P(A)$ **e.** $P(A^c)$
  b. $P(B)$ **f.** $P(B^c)$
  c. $P(A \cup B)$ **g.** $P(A \cup A^c)$
  d. $P(A \cap B)$ **h.** $P(A^c \cap B)$

3.46 Consider the following Venn diagram, where
  $P(E_1) = .10, P(E_2) = .05, P(E_3) = P(E_4) = .2$,
  $P(E_5) = .06, P(E_6) = .3, P(E_7) = .06$, and
  $P(E_8) = .03$:

Find the following probabilities:
  a. $P(A^c)$
  b. $P(B^c)$
  c. $P(A^c \cap B)$
  d. $P(A \cup B)$
  e. $P(A \cap B)$
  f. $P(A^c \cup B^c)$
  g. Are events A and B mutually exclusive? Why?

3.47 The outcomes of two variables are (Low, Medium, High)
  [NW] and (On, Off), respectively. An experiment is conducted in which the outcomes of each of the two variables are observed. The probabilities associated with each of the six possible outcome pairs are given in the following table:

	Low	Medium	High
On	.50	.10	.05
Off	.25	.07	.03

Consider the following events:

> A: {On}
> B: {Medium or On}
> C: {Off and Low}
> D: {High}

  a. Find $P(A)$.
  b. Find $P(B)$.
  c. Find $P(C)$.
  d. Find $P(D)$.
  e. Find $P(A^c)$.
  f. Find $P(A \cup B)$.
  g. Find $P(A \cap B)$.
  h. Consider each possible pair of events taken from the events A, B, C, and D. List the pairs of events that are mutually exclusive. Justify your choices.

3.48 Three fair coins are tossed. We wish to find the probability of the event A: {Observe at least one head.}
  a. Express A as the union of three mutually exclusive events. Using the expression you wrote, find the probability of A.
  b. Express A as the complement of an event. Using the expression you wrote, find the probability of A.

Applet Exercise 3.3

Use the applets entitled *Simulating the Probability of Rolling a 6* and *Simulating the Probability of Rolling a 3 or 4* to explore the additive rule of probability.

  a. Explain why the applet *Simulating the Probability of Rolling a 6* can also be used to simulate the probability of rolling a 3. Then use the applet with $n = 1000$ to simulate the probability of rolling a 3. Record the cumulative proportion. Repeat the process to simulate the probability of rolling a 4.
  b. Use the applet *Simulating the Probability of Rolling a 3 or 4* with $n = 1000$ to simulate the probability of rolling a 3 or 4. Record the cumulative proportion.
  c. Add the two cumulative proportions from part **a.** How does this sum compare with the cumulative proportion in part **b**? How does your answer illustrate the additive rule for probability?

Applet Exercise 3.4

Use the applets entitled *Simulating the Probability of Rolling a 6* and *Simulating the Probability of Rolling a 3 or 4* to simulate the probability of the complement of an event.

a. Explain how the applet *Simulating the Probability of Rolling a 6* can also be used to simulate the probability of the event *rolling a 1, 2, 3, 4, or 5*. Then use the applet with $n = 1000$ to simulate this probability.

b. Explain how the applet *Simulating the Probability of Rolling a 3 or 4* can also be used to simulate the probability of the event *rolling a 1, 2, 5, or 6*. Then use the applet with $n = 1000$ to simulate this probability.

c. Which applet could be used to simulate the probability of the event *rolling a 1, 2, 3, or 4?* Explain.

Applying the Concepts—Basic

3.49 **Social networking Web sites in the United Kingdom.** In the United States, MySpace and FaceBook are considered the two most popular social networking Web sites. In the United Kingdom (UK), the competition for social networking is between MySpace and Bebo. According to Nielsen/NetRatings (April 2006), 4% of UK citizens visit MySpace, 3% visit Bebo, and 1% visit both MySpace and Bebo.

a. Draw a Venn diagram to illustrate the use of social networking sites in the United Kingdom.

b. Find the probability that a UK citizen visits either the MySpace or Bebo social networking sites.

c. Use your answer to part **b** to find the probability that a UK citizen does not visit either social networking site.

3.50 **Study of analysts' forecasts.** The *Journal of Accounting Research* (March 2008) published a study on relationship incentives and degree of optimism among analysts' forecasts. Participants were analysts at either a large or small brokerage firm who made their forecasts either early or late in the quarter. Also, some analysts were only concerned with making an accurate forecast, while others were also interested in their relationship with management. Suppose one of these analysts is randomly selected. Consider the following events:

$A = \{$The analyst is only concerned with making an accurate forecast.$\}$
$B = \{$The analyst makes the forecast early in the quarter.$\}$
$C = \{$the analyst is from a small brokerage firm.$\}$

Describe each of the following events in terms of unions, intersections, and complements (e.g., $A \cup B$, $A \cap B$, A^C, etc.).

a. The analyst makes an early forecast and is only concerned with accuracy.

b. The analyst is not only concerned with accuracy.

c. The analyst is from a small brokerage firm or makes an early forecast.

d. The analyst makes a late forecast and is not only concerned with accuracy.

3.51 **Toxic chemical incidents.** *Process Safety Progress* (Sept. 2004) reported on an emergency response system for incidents involving toxic chemicals in Taiwan. The system has logged over 250 incidents since being implemented. The next table gives a breakdown of the locations where these incidents occurred. Consider the location of a toxic chemical incident in Taiwan.

a. List the sample points for this experiment.

b. Assign reasonable probabilities to the sample points.

c. What is the probability that the incident occurs in a school laboratory?

d. What is the probability that the incident occurs in either a chemical or a nonchemical plant?

e. What is the probability that the incident does not occur in transit?

Location	Percent of Incidents
School laboratory	6%
In transit	26%
Chemical plant	21%
Nonchemical plant	35%
Other	12%
Total	100%

Based on Chen, J. R., et al. "Emergency response of toxic chemicals in Taiwan: The system and case studies." *Process Safety Progress*, Vol. 23, No. 3, Sept. 2004 (Figure 5a).

3.52 **Scanning errors at Wal-Mart.** The National Institute of Standards and Technology (NIST) mandates that for every 100 items scanned through the electronic checkout scanner at a retail store, no more than 2 should have an inaccurate price. A recent study of the accuracy of checkout scanners at Wal-Mart stores in California (*Tampa Tribune*, Nov. 22, 2005) showed that, of the 60 Wal-Mart stores investigated, 52 violated the NIST scanner accuracy standard. If 1 of the 60 Wal-Mart stores is randomly selected, what is the probability that that store does not violate the NIST scanner accuracy standard?

3.53 **Gene expression profiling.** Gene expression profiling is a state-of-the-art method for determining the biology of cells. In *Briefings in Functional Genomics and Proteomics* (Dec. 2006), biologists at Pacific Northwest National Laboratory reviewed several gene expression profiling methods. The biologists applied two of the methods (A and B) to data collected on proteins in human mammary cells. The probability that the protein is cross-referenced (i.e., identified) by method A is .41, the probability that the protein is cross-referenced by method B is .42, and the probability that the protein is cross-referenced by both methods is .40.

a. Draw a Venn diagram to illustrate the results of the gene-profiling analysis.

b. Find the probability that the protein is cross-referenced by either method A or method B.

c. On the basis of your answer to part **b**, find the probability that the protein is not cross-referenced by either method.

3.54 **Sleep apnea and sleep stage transitioning.** Sleep apnea is a common sleep disorder characterized by collapses of the upper airway during sleep. *Chance* (Winter 2009) investigated the role of sleep apnea in how people transition from one sleep stage to another. The various stages of sleep for a large group of sleep apnea patients were monitored in 30-second intervals, or "epochs." For each epoch, sleep stage was categorized as Wake, REM, or Non-REM. The table on the next page provides a summary of the results. Each cell of the table gives the number of epochs that occurred when transitioning from the previous sleep stage to the current sleep stage. Consider the previous and current sleep stage of a randomly selected epoch from the study.

	Previous Sleep Stage		
Current Stage	Non-REM	REM	Wake
Non-REM	31,880	160	1,733
REM	346	7,609	175
Wake	1,588	358	6,079
Totals	33,814	8,127	7,987

Based on Caffo, B. S., et al. "An overview of observational sleep research with application to sleep stage transitioning." *Chance*, Vol. 22, No. 1, Winter 2009 (Table 2).

a. List the sample points for this experiment.

b. Assign reasonable probabilities to the sample points.

c. What is the probability that the current sleep stage of the epoch is REM?

d. What is the probability that the previous sleep stage of the epoch is Wake?

e. What is the probability that the epoch transitioned from the REM sleep stage to the Non-REM sleep stage?

Applying the Concepts—Intermediate

3.55 Guilt in decision making. The effect of guilt emotion on how a decision maker focuses on the problem was investigated in the Jan. 2007 issue of the *Journal of Behavioral Decision Making* (see Exercise 1.27, p. 20). A total of 171 volunteer students participated in the experiment, where each was randomly assigned to one of three emotional states (guilt, anger, or neutral) through a reading/writing task. Immediately after the task, the students were presented with a decision problem where the stated option has predominantly negative features (e.g., spending money on repairing a very old car). The results (number responding in each category) are summarized in the accompanying table. Suppose one of the 171 participants is selected at random.

Emotional State	Choose Stated Option	Do Not Choose Stated Option	Totals
Guilt	45	12	57
Anger	8	50	58
Neutral	7	49	56
Totals	60	111	171

Based on Gangemi, A., and Mancini, F. "Guilt and focusing in decision-making." *Journal of Behavioral Decision Making*, Vol. 20, Jan. 2007 (Table 2).

a. Find the probability that the respondent is assigned to the guilty state.

b. Find the probability that the respondent chooses the stated option (repair the car).

c. Find the probability that the respondent is assigned to the guilty state and chooses the stated option.

d. Find the probability that the respondent is assigned to the guilty state or chooses the stated option.

3.56 Abortion provider survey. The Alan Guttmacher Institute Abortion Provider Survey is a survey of all 238 known nonhospital abortion providers in the United States (*Perspectives on Sexual and Reproductive Health*, Jan./Feb. 2003). For one part of the survey, the 358 providers were classified according to case load (number of abortions performed per year) and whether they permit their patients to take the abortion drug misoprostol at home or require the patients to return to the abortion facility to receive the drug. The responses are summarized in the accompanying table. Suppose we select, at random, one of the 358 providers and observe the provider's case load (fewer than 50, or 50 or more) and home use of the drug (yes or no).

	Number of Abortions		
Permit Drug at Home	Fewer than 50	50 or More	Totals
Yes	170	130	300
No	48	10	58
Totals	218	140	358

Based on Henshaw, S. K., and Finer, L. B. "The accessibility of abortion services in the United States, 2001." *Perspectives on Sexual and Reproductive Health*, Vol. 35, No. 1, Jan./Feb. 2003 (Table 4).

a. Find the probability that the provider permits home use of the abortion drug.

b. Find the probability that the provider permits home use of the drug or has a case load of fewer than 50 abortions.

c. Find the probability that the provider permits home use of the drug and has a case load of fewer than 50 abortions.

3.57 Fighting probability of fallow deer bucks. In *Aggressive Behavior* (Jan./Feb. 2007), zoologists investigated the likelihood of fallow deer bucks fighting during the mating season. During a 270-hour observation period, the researchers recorded 205 encounters between two bucks. Of these, 167 involved one buck clearly initiating the encounter with the other. In these 167 initiated encounters, the zoologists kept track of whether or not a physical contact fight occurred and whether the initiator ultimately won or lost the encounter. (The buck that is driven away by the other is considered the loser.) A summary of the 167 initiated encounters is provided in the accompanying table. Suppose we select one of these 167 encounters and note the outcome (fight status and winner).

a. What is the probability that a fight occurs and the initiator wins?

b. What is the probability that no fight occurs?

c. What is the probability that there is no clear winner?

d. What is the probability that a fight occurs or the initiator loses?

e. Are the events "No clear winner" and "initiator loses" mutually exclusive?

	Initiator Wins	No Clear Winner	Initiator Loses	Totals
Fight	26	23	15	64
No Fight	80	12	11	103
Totals	106	35	26	167

Based on Bartos, L. et al. "Estimation of the probability of fighting in fallow deer (*Dama dama*) during the rut." *Aggressive Behavior*, Vol. 33, Jan./Feb. 2007, pp. 7–13.

3.58 **Cell phone handoff behavior.** A "handoff" is a term used in wireless communications to describe the process of a cell phone moving from a coverage area of one base station to another. Each base station has multiple channels (called color codes) that allow it to communicate with the cell phone. The *Journal of Engineering, Computing and Architecture* (Vol. 3., 2009) published a study of cell phone handoff behavior. During a sample driving trip which involved crossing from one base station to another, the different color codes accessed by the cell phone were monitored and recorded. The table below shows the number of times each color code was accessed for two identical driving trips, each using a different cell phone model. (*Note:* The table is similar to the one published in the article.) Suppose you randomly select one point during the combined driving trips.

	Color Code				
	0	5	b	c	Total
Model 1	20	35	40	0	85
Model 2	15	50	6	4	75
Total	35	85	46	4	160

a. What is the probability that the cell phone is using color code 5?

b. What is the probability that the cell phone is using color code 5 or color code 0?

c. What is the probability that the cell phone used is Model 2 and the color code is 0?

3.59 **Federal civil trial appeals.** The *Journal of the American Law and Economics Association* (Vol. 3, 2001) published the results of a study of appeals of federal civil trials. The accompanying table, extracted from the article, gives a breakdown of 2,143 civil cases that were appealed by either the plaintiff or the defendant. The outcome of the appeal, as well as the type of trial (judge or jury), was determined for each civil case. Suppose one of the 2,143 cases is selected at random and both the outcome of the appeal and the type of trial are observed.

	Jury	Judge	Totals
Plaintiff trial win—reversed	194	71	265
Plaintiff trial win—affirmed/dismissed	429	240	669
Defendant trial win—reversed	111	68	179
Defendant trial win—affirmed/ dismissed	731	299	1,030
Totals	1,465	678	2,143

a. Find $P(A)$, where $A = \{$jury trial$\}$.
b. Find $P(B)$, where $B = \{$plaintiff trial win is reversed$\}$.
c. Are A and B mutually exclusive events?
d. Find $P(A^c)$.
e. Find $P(A \cup B)$.
f. Find $P(A \cap B)$.

3.60 **Employee behavior problems.** The *Organizational Development Journal* (Summer 2006) reported on the results of a survey of human resource officers (HROs) at major firms. The focus of the study was employee behavior, namely absenteeism, promptness to work, and turnover. The study found that 55% of the HROs had problems with absenteeism. Also, 41% of the HROs had problems with turnover. Suppose that 22% of the HROs had problems with both absenteeism and turnover.

a. Find the probability that a human resource officer selected from the group surveyed had problems with employee absenteeism or employee turnover.

b. Find the probability that a human resource officer selected from the group surveyed did not have problems with employee absenteeism.

c. Find the probability that a human resource officer selected from the group surveyed did not have problems with employee absenteeism nor with employee turnover.

3.61 **Chemical signals of mice.** The ability of a mouse to recognize the odor of a potential predator (e.g., a cat) is essential to the mouse's survival. The chemical makeup of these odors—called kairomones—was the subject of a study published in *Cell* (May 14, 2010). Typically, the source of these odors are major urinary proteins (Mups). Cells collected from lab mice were exposed to Mups from rodent species A, Mups from rodent species B, and kairomones (from a cat). The accompanying Venn diagram shows the proportion of cells that chemically responded to each of the three odors. (*Note*: A cell may respond to more than a single odor.)

a. What is the probability that a lab mouse responds to all three source odors?

b. What is the probability that a lab mouse responds to the kairomone?

c. What is the probability that a lab mouse responds to Mups A and Mups B, but not the kairomone?

Applying the Concepts—Advanced

3.62 **Galileo's passe-dix game.** Passe-dix is a game of chance played with three fair dice. Players bet whether the sum of the faces shown on the dice will be above or below 10. During the late 16th century, the astronomer and mathematician Galileo Galilei was asked by the Grand Duke of Tuscany to explain why "the chance of throwing a total of 9 with three fair dice was less than that of throwing a total of 10." (*Interstat*, Jan. 2004). The Grand Duke believed that the chance should be the same, since "there are an equal number of partitions of the numbers 9 and 10." Find the flaw in the Grand Duke's reasoning and answer the question posed to Galileo.

3.5 Conditional Probability

Figure 3.14
Reduced sample space for the die-toss experiment: given that event B has occurred

The event probabilities we've been discussing give the relative frequencies of the occurrences of the events when the experiment is repeated a very large number of times. Such probabilities are often called **unconditional probabilities**, because no special conditions are assumed other than those which define the experiment.

Often, however, we have additional knowledge that might affect the outcome of an experiment, so we may need to alter the probability of an event of interest. A probability that reflects such additional knowledge is called the **conditional probability** of the event. For example, we've seen that the probability of observing an even number (event A) on a toss of a fair die is $1/2$. But suppose we're given the information that on a particular throw of the die the result was a number less than or equal to 3 (event B). Would the probability of observing an even number on that throw of the die still be equal to $1/2$? It can't be, because making the assumption that B has occurred reduces the sample space from six sample points to three sample points (namely, those contained in event B). This reduced sample space is as shown in Figure 3.14.

Because the sample points for the die-toss experiment are equally likely, each of the three sample points in the reduced sample space is assigned an equal *conditional probability* of $1/3$. Since the only even number of the three in the reduced sample space B is the number 2 and the die is fair, we conclude that the probability that A occurs *given that B occurs* is $1/3$. We use the symbol $P(A \mid B)$ to represent the probability of event A given that event B occurs. For the die-toss example,

$$P(A \mid B) = \frac{1}{3}$$

To get the probability of event A given that event B occurs, we proceed as follows: We divide the probability of the part of A that falls within the reduced sample space B, namely, $P(A \cap B)$, by the total probability of the reduced sample space, namely, $P(B)$. Thus, for the die-toss example with event A: {Observe an even number.} and event B: {Observe a number less than or equal to 3.} we find that

$$P(A \mid B) = \frac{P(A \cap B)}{P(B)} = \frac{P(2)}{P(1) + P(2) + P(3)} = \frac{1/6}{3/6} = \frac{1}{3}$$

The formula for $P(A \mid B)$ is true in general:

Conditional Probability Formula

To find the *conditional probability that event A occurs given that event B occurs*, divide the probability that *both* A and B occur by the probability that B occurs; that is,

$$P(A \mid B) = \frac{P(A \cap B)}{P(B)} \quad \left[\text{We assume that } P(B) \neq 0.\right]$$

This formula adjusts the probability of $A \cap B$ from its original value in the complete sample space $\mathbf{S}$ to a conditional probability in the reduced sample space B. On the one hand, if the sample points in the complete sample space are equally likely, then the formula will assign equal probabilities to the sample points in the reduced sample space, as in the die-toss experiment. On the other hand, if the sample points have unequal probabilities, the formula will assign conditional probabilities proportional to the probabilities in the complete sample space. The latter situation is illustrated in the next three examples.

Example 3.13

The Conditional Probability Formula—Executives Who Cheat at Golf

Problem To develop programs for business travelers staying at convention hotels, Hyatt Hotels Corp. commissioned a study of executives who play golf. The study revealed that 55% of the executives admitted that they had cheated at golf. Also, 20% of the executives admitted that they had cheated at golf and had lied in business. Given that an executive had cheated at golf, what is the probability that the executive also had lied in business?

Solution Let's define events A and B as follows:

$$A = \{\text{Executive had cheated at golf.}\}$$
$$B = \{\text{Executive had lied in business.}\}$$

From the study, we know that 55% of executives had cheated at golf, so $P(A) = .55$. Now, executives who *both* cheat at golf (event A) *and* lie in business (event B) represent the compound event $A \cap B$. From the study, $P(A \cap B) = .20$. We want to know the probability that an executive lied in business (event B) given that he or she cheated at golf (event A); that is, we want to know the conditional probability $P(B|A)$. Applying the conditional probability formula, we have

$$P(B|A) = \frac{P(A \cap B)}{P(A)} = \frac{.20}{.55} = .364$$

Thus, given that a certain executive had cheated at golf, the probability that the executive also had lied in business is .364.

Look Back One of the keys to applying the formula correctly is to write the information in the study in the form of probability statements involving the events of interest. The word "and" in the clause "cheat at golf *and* lie in business" implies an intersection of the two events A and B. The word "given" in the phrase "*given* that an executive cheats at golf" implies that event A is the given event.

Now Work Exercise 3.79

Example 3.14

Applying the Conditional Probability Formula in a Two-Way Table—Smoking and Cancer

Problem Many medical researchers have conducted experiments to examine the relationship between cigarette smoking and cancer. Consider an individual randomly selected from the adult male population. Let A represent the event that the individual smokes, and let A^c denote the complement of A (the event that the individual does not smoke). Similarly, let B represent the event that the individual develops cancer, and let B^c be the complement of that event. Then the four sample points associated with the experiment are shown in Figure 3.15, and their probabilities for a certain section of the United States are given in Table 3.5. Use these sample point probabilities to examine the relationship between smoking and cancer.

$A \cap B$
(Smoker, cancer)

$A \cap B^c$
(Smoker, no cancer)

$A^c \cap B$
(Nonsmoker, cancer)

$A^c \cap B^c$
(Nonsmoker, no cancer)

S

Figure 3.15

Sample space for Example 3.15

Table 3.5	**Probabilities of Smoking and Developing Cancer**	
	Develops Cancer	
Smoker	Yes, B	No, B^c
Yes, A	.05	.20
No, A^c	.03	.72

Solution One method of determining whether the given probabilities indicate that smoking and cancer are related is to compare the *conditional probability* that an adult male acquires cancer given that he smokes with the conditional probability that an adult male acquires cancer given that he does not smoke [i.e., compare $P(B|A)$ with $P(B|A^c)$].

First, we consider the reduced sample space A corresponding to adult male smokers. This reduced sample space is highlighted in Figure 3.15. The two sample points $A \cap B$ and $A \cap B^c$ are contained in this reduced sample space, and the adjusted probabilities of these two sample points are the two conditional probabilities

$$P(B|A) = \frac{P(A \cap B)}{P(A)} \quad \text{and} \quad P(B^c|A) = \frac{P(A \cap B^c)}{P(A)}$$

The probability of event A is the sum of the probabilities of the sample points in A:

$$P(A) = P(A \cap B) + P(A \cap B^c) = .05 + .20 = .25$$

Then the values of the two conditional probabilities in the reduced sample space A are

$$P(B|A) = \frac{.05}{.25} = .20 \quad \text{and} \quad P(B^c|A) = \frac{.20}{.25} = .80$$

These two numbers represent the probabilities that an adult male smoker develops cancer and does not develop cancer, respectively.

In a like manner, the conditional probabilities of an adult male nonsmoker developing cancer and not developing cancer are

$$P(B|A^c) = \frac{P(A^c \cap B)}{P(A^c)} = \frac{.03}{.75} = .04$$

$$P(B^c|A^c) = \frac{P(A^c \cap B^c)}{P(A^c)} = \frac{.72}{.75} = .96$$

Two of the conditional probabilities give some insight into the relationship between cancer and smoking: the probability of developing cancer given that the adult male is a smoker, and the probability of developing cancer given that the adult male is not a smoker. The conditional probability that an adult male smoker develops cancer (.20) is five times the probability that a nonsmoker develops cancer (.04). This relationship does not imply that smoking *causes* cancer, but it does suggest a pronounced link between smoking and cancer.

Look Back Notice that the conditional probabilities $P(B^c|A) = .80$ and $P(B|A) = .20$ are in the same 4-to-1 ratio as the original unconditional probabilities, .20 and .05. The conditional probability formula simply adjusts the unconditional probabilities so that they add to 1 in the reduced sample space A of adult male smokers. Similarly, the conditional probabilities $P(B^c|A^c) = .96$ and $P(B|A^c) = .04$ are in the same 24-to-1 ratio as the unconditional probabilities .72 and .03.

Now Work Exercise 3.67a–b

Example 3.15
Conditional Probability in a Two-Way Table—Consumer Complaints

Problem The Federal Trade Commission's investigation of consumer product complaints has generated much interest on the part of manufacturers in the quality of their products. A manufacturer of an electromechanical kitchen utensil conducted an analysis of a large number of consumer complaints and found that they fell into the six categories shown in Table 3.6. If a consumer complaint is received, what is the probability that the cause of the complaint was the appearance of the product given that the complaint originated during the guarantee period?

Table 3.6   Distribution of Product Complaints

	Reason for Complaint			
	Electrical	Mechanical	Appearance	Totals
During Guarantee Period	18%	13%	32%	63%
After Guarantee Period	12%	22%	3%	37%
Totals	30%	35%	35%	100%

Solution Let A represent the event that the cause of a particular complaint is the appearance of the product, and let B represent the event that the complaint occurred during the guarantee period. Checking Table 3.6, you can see that $(18 + 13 + 32)\% = 63\%$ of the complaints occur during the guarantee period. Hence, $P(B) = .63$. The percentage of complaints that were caused by appearance and occurred during the guarantee period (the event $A \cap B$) is 32%. Therefore, $P(A \cap B) = .32$.

Using these probability values, we can calculate the conditional probability $P(A|B)$ that the cause of a complaint is appearance given that the complaint occurred during the guarantee time:

$$P(A|B) = \frac{P(A \cap B)}{P(B)} = \frac{.32}{.63} = .51$$

Thus, slightly more than half of the complaints that occurred during the guarantee period were due to scratches, dents, or other imperfections in the surface of the kitchen devices.

Look Back Note that the answer $\frac{.32}{.63}$ is the proportion for the event of interest A (.32), divided by the row total proportion for the given event B (.63). That is, it is the proportion of the time that A occurs together within the given event B.

Now Work Exercise 3.78a

3.6 The Multiplicative Rule and Independent Events

The probability of an intersection of two events can be calculated with the *multiplicative rule*, which employs the conditional probabilities we defined in the previous section. Actually, we've already developed the formula in another context. Recall that the conditional probability of B given A is

$$P(B|A) = \frac{P(A \cap B)}{P(A)}$$

Multiplying both sides of this equation by $P(A)$, we obtain a formula for the probability of the intersection of events A and B. This formula is often called the **multiplicative rule of probability.**

> **Multiplicative Rule of Probability**
> $P(A \cap B) = P(A)P(B|A)$ or, equivalently, $P(A \cap B) = P(B)P(A|B)$

Example 3.16
The Multiplicative Rule—a Famous Psychological Experiment

Problem In a classic psychology study conducted in the early 1960s, Stanley Milgram performed a series of experiments in which a teacher is asked to shock a learner who is attempting to memorize word pairs whenever the learner gives the wrong answer. The shock levels increase with each successive wrong answer. (Unknown to the teacher, the shocks are not real.) Two events of interest are

A: {The teacher " applies" a severe shock (450 volts).}
B: {The learner protests verbally prior to receiving the shock.}

A recent application of Milgram's shock study revealed that $P(B) = .5$ and $P(A|B) = .7$. On the basis of this information, what is the probability that a learner will protest verbally *and* a teacher will apply a severe shock? That is, find $P(A \cap B)$.

Solution We want to calculate $P(A \cap B)$. Using the formula for the multiplicative rule, we obtain

$$P(A \cap B) = P(B) \, P(A|B) = (.5)\,(.7) = .35$$

Thus, about 35% of the time the learner will give verbal protest and the teacher will apply a severe shock.

Look Back The multiplicative rule can be expressed in two ways: $P(A \cap B) = P(A) \, P(B|A)$, or $P(A \cap B) = P(B) \, P(A|B)$. Select the formula that involves a given event for which you know the probability (e.g., event B in the example).

Now Work Exercise 3.80

Intersections often contain only a few sample points. In this case, the probability of an intersection is easy to calculate by summing the appropriate sample point probabilities. However, the formula for calculating intersection probabilities is invaluable when the intersection contains numerous sample points, as the next example illustrates.

Example 3.17
Applying The Multiplicative Rule—Study of Welfare Workers

Problem A county welfare agency employs 10 welfare workers who interview prospective food stamp recipients. Periodically, the supervisor selects, at random, the forms completed by two workers and subsequently audits them for illegal deductions. Unknown to the supervisor, three of the workers have regularly been giving illegal deductions to applicants. What is the probability that both of the workers chosen have been giving illegal deductions?

Solution Define the following two events:

A: {First worker selected gives illegal deductions.}
B: {Second worker selected gives illegal deductions.}

We want to find the probability that both workers selected have been giving illegal deductions. This event can be restated as {First worker gives illegal deductions *and* second worker gives illegal deductions.}. Thus, we want to find the probability of the intersection $A \cap B$. Applying the multiplicative rule, we have

$$P(A \cap B) = P(A)P(B|A)$$

To find $P(A)$, it is helpful to consider the experiment as selecting 1 worker from the 10. Then the sample space for the experiment contains 10 sample points (representing the 10 welfare workers), in which the 3 workers giving illegal deductions are denoted by the symbol I (I_1, I_2, I_3) and the 7 workers not giving illegal deductions are denoted by the symbol N ($N_1, ..., N_7$). The resulting Venn diagram is illustrated in Figure 3.16.

Since the first worker is selected at random from the 10, it is reasonable to assign equal probabilities to the 10 sample points. Thus, each sample point has a probability of $^1/_{10}$. The sample points in event A are $\{I_1, I_2, I_3\}$—the three workers who are giving illegal deductions. Thus,

$$P(A) = P(I_1) + P(I_2) + P(I_3) = \frac{1}{10} + \frac{1}{10} + \frac{1}{10} = \frac{3}{10}$$

Figure 3.16
Venn diagram for finding $P(A)$

To find the conditional probability $P(B\,|\,A)$, we need to alter the sample space **S.** Since we know that A has occurred (i.e., the first worker selected is giving illegal deductions), only 2 of the 9 remaining workers in the sample space are giving illegal deductions. The Venn diagram for this new sample space is shown in Figure 3.17.

Each of these nine sample points is equally likely, so each is assigned a probability of $^1/_9$. Since the event $B\,|\,A$ contains the sample points $\{I_1, I_2\}$, we have

$$P(B\,|\,A) = P(I_1) + P(I_2) = \frac{1}{9} + \frac{1}{9} = \frac{2}{9}$$

Substituting $P(A) = ^3/_{10}$ and $P(B\,|\,A) = ^2/_9$ into the formula for the multiplicative rule, we find that

$$P(A \cap B) = P(A)P(B\,|\,A) = \left(\frac{3}{10}\right)\left(\frac{2}{9}\right) = \frac{6}{90} = \frac{1}{15}$$

Figure 3.17
Venn diagram for finding $P(B\,|\,A)$

Thus, there is a 1-in-15 chance that both workers chosen by the supervisor have been giving illegal deductions to food stamp recipients.

Look Back The key words *both* and *and* in the statement "both A and B occur" imply an intersection of two events. This, in turn, implies that we should *multiply* probabilities to obtain the answer.

Now Work Exercise 3.71

The sample-space approach is only one way to solve the problem posed in Example 3.17. An alternative method employs the **tree diagram** (introduced in Example 3.1). Tree diagrams are helpful for calculating the probability of an intersection.

To illustrate, a tree diagram for Example 3.17 is displayed in Figure 3.18. The tree begins at the far left with two branches. These branches represent the two possible outcomes N (no illegal deductions) and I (illegal deductions) for the first worker selected. The unconditional probability of each outcome is given (in parentheses) on the appropriate branch. That is, for the first worker selected, $P(N) = ^7/_{10}$ and $P(I) = ^3/_{10}$. (These unconditional probabilities can be obtained by summing sample point probabilities as in Example 3.17.)

The next level of the tree diagram (moving to the right) represents the outcomes for the second worker selected. The probabilities shown here are conditional probabilities, since the outcome for the first worker is assumed to be known. For example, if the first worker is giving illegal deductions (I), the probability that the second worker is also giving illegal deductions (I) is $^2/_9$, because, of the 9 workers left to be selected, only 2 remain who are giving illegal deductions. This conditional probability, $^2/_9$, is shown in parentheses on the bottom branch of Figure 3.18.

Finally, the four possible outcomes of the experiment are shown at the end of each of the four tree branches. These events are intersections of two events (outcome of first worker *and* outcome of second worker). Consequently, the multiplicative rule is applied to calculate each probability, as shown in Figure 3.18. You can see that the intersection $\{I \cap I\}$—the event that both workers selected are giving illegal deductions—has probability $^6/_{90} = ^1/_{15}$, which is the same as the value obtained in Example 3.17.

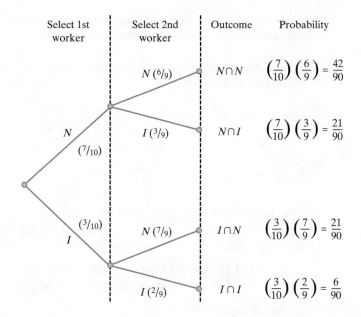

Figure 3.18

Tree diagram for Example 3.17

In Section 3.5, we showed that the probability of event A may be substantially altered by the knowledge that an event B has occurred. However, this will not always be the case; in some instances, the assumption that event B has occurred will *not* alter the probability of event A at all. When this occurs, we say that the two events A and B are *independent events*.

> Events A and B are **independent events** if the occurrence of B does not alter the probability that A has occurred; that is, events A and B are independent if
> $$P(A|B) = P(A)$$
> When events A and B are independent, it is also true that
> $$P(B|A) = P(B)$$
> Events that are not independent are said to be **dependent.**

Example 3.18

Checking for Independence— Die-Tossing Experiment

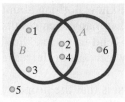

Figure 3.19

Venn diagram for die-toss experiment

Problem Consider the experiment of tossing a fair die, and let

$$A = \{\text{Observe an even number.}\}$$
$$B = \{\text{Observe a number less than or equal to 4.}\}$$

Are A and B independent events?

Solution The Venn diagram for this experiment is shown in Figure 3.19. We first calculate

$$P(A) = P(2) + P(4) + P(6) = \frac{1}{2}$$

$$P(B) = P(1) + P(2) + P(3) + P(4) = \frac{4}{6} = \frac{2}{3}$$

$$P(A \cap B) = P(2) + P(4) = \frac{2}{6} = \frac{1}{3}$$

Now, assuming that B has occurred, we see that the conditional probability of A given B is

$$P(A|B) = \frac{P(A \cap B)}{P(B)} = \frac{1/3}{2/3} = \frac{1}{2} = P(A)$$

Thus, assuming that the occurence of event B does not alter the probability of observing an even number, that probability remains $\frac{1}{2}$. Therefore, the events A and B are independent.

Look Back Note that if we calculate the conditional probability of B given A, our conclusion is the same:

$$P(B|A) = \frac{P(A \cap B)}{P(A)} = \frac{1/3}{1/2} = \frac{2}{3} = P(B)$$

Example 3.19
Checking for Independence— Consumer Complaint Study

Problem Refer to the consumer product complaint study in Example 3.15. The percentages of complaints of various types during and after the guarantee period are shown in Table 3.6 (p. 136). Define the following events:

> A: {The cause of the complaint is the appearance of the product.}
> B: {The complaint occurred during the guarantee term.}

Are A and B independent events?

Solution Events A and B are independent if $P(A|B) = P(A)$. In Example 3.15, we calculated $P(A|B)$ to be .51, and from Table 3.6, we see that

$$P(A) = .32 + .03 = .35$$

Therefore, $P(A|B)$ is not equal to $P(A)$, and A and B are dependent events.

Now Work Exercise 3.67c

BLAISE PASCAL (1623–1662)

Solver of the Chevalier's Dilemma

As a precocious child growing up in France, Blaise Pascal showed an early inclination toward mathematics. Although his father would not permit Pascal to study mathematics before the age of 15 (removing all math texts from his house), at age 12 Blaise discovered on his own that the sum of the angles of a triangle are two right angles.

Pascal went on to become a distinguished mathematician, as well as a physicist, a theologian, and the inventor of the first digital calculator. Most historians attribute the beginning of the study of probability to the correspondence between Pascal and Pierre de Fermat in 1654. The two solved the Chevalier's Dilemma, a gambling problem related to Pascal by his friend and Paris gambler the Chevalier de Mere. The problem involved determining the expected number of times one could roll two dice without throwing a double 6. (Pascal proved that the "break-even" point was 25 rolls.) ■

To gain an intuitive understanding of independence, think of situations in which the occurrence of one event does not alter the probability that a second event will occur. For example, new medical procedures are often tested on laboratory animals. The scientists conducting the tests generally try to perform the procedures on the animals so that the results for one animal do not affect the results for the others. That is, the event that the procedure is successful on one animal is *independent* of the result for another. In this way, the scientists can get a more accurate idea of the efficacy of the procedure than if the results were dependent, with the success or failure for one animal affecting the results for other animals.

As a second example, consider an election poll in which 1,000 registered voters are asked their preference between two candidates. Pollsters try to use procedures for selecting a sample of voters so that the responses are independent. That is, the objective of the pollster is to select the sample so that one polled voter's preference for candidate A does not alter the probability that a second polled voter prefers candidate A.

Now consider the world of sports. Do you think the results of a batter's successive trips to the plate in baseball, or of a basketball player's successive shots at the basket, are independent? For example, if a basketball player makes two successive shots, is the probability of making the next shot altered from its value if the result of the first shot is not known? If a player makes two shots in a row, the probability of a third successful shot is likely to be different from what we would assign if we knew nothing about the first two shots. Why should this be so? Research has shown that many such results in sports tend to be *dependent* because players (and even teams) tend to get on "hot" and "cold" streaks, during which their probabilities of success may increase or decrease significantly.

We will make three final points about independence. The first is that the property of independence, unlike the property of mutual exclusivity generally cannot be shown on, or gleaned from, a Venn diagram. This means that *you can't trust your intuition*. In general, the only way to check for independence is by performing the calculations of the probabilities in the definition.

Figure 3.20

Mutually exclusive events are dependent events

The second point concerns the relationship between the properties of mutual exclusivity and independence. Suppose that events A and B are mutually exclusive, as shown in Figure 3.20, and that both events have nonzero probabilities. Are these events independent or dependent? That is, does the assumption that B occurs alter the probability of the occurrence of A? It certainly does, because if we assume that B has occurred, it is impossible for A to have occurred simultaneously. That is, $P(A|B) = 0$. Thus, *mutually exclusive events are dependent events*, since $P(A) \neq P(A|B)$.

The third point is that the probability of the intersection of independent events is very easy to calculate. Referring to the formula for calculating the probability of an intersection, we find that

$$P(A \cap B) = P(A)P(B|A)$$

Thus, since $P(B|A) = P(B)$ when A and B are independent, we have the following useful rule:

Probability of Intersection of Two Independent Events

If events A and B are independent, then the probability of the intersection of A and B equals the product of the probabilities of A and B; that is,

$$P(A \cap B) = P(A)P(B)$$

The converse is also true: If $P(A \cap B) = P(A)P(B)$, then events A and B are independent.

In the die-toss experiment, we showed in Example 3.18 that the two events A: {Observe an even number.}   and B: {Observe a number less than or equal to 4.} are independent if the die is fair. Thus,

$$P(A \cap B) = P(A)P(B) = \left(\frac{1}{2}\right)\left(\frac{2}{3}\right) = \frac{1}{3}$$

This result agrees with the one we obtained in the example:

$$P(A \cap B) = P(2) + P(4) = \frac{2}{6} = \frac{1}{3}$$

Example 3.20

Probability of Independent Events Occurring Simultaneously— Divorced Couples Study

Problem Recall from Example 3.5 (p. 114) that the American Association for Marriage and Family Therapy (AAMFT) found that 25% of divorced couples are classified as "fiery foes" (i.e., they communicate through their children and are hostile toward each other).

a. What is the probability that in a sample of 2 divorced couples, both are classified as "fiery foes?"

b. What is the probability that in a sample of 10 divorced couples, all 10 are classified as "fiery foes?"

Solution

a. Let F_1 represent the event that divorced couple 1 is classified as a pair of "fiery foes" and F_2 represent the event that divorced couple 2 is also classified as a pair of "fiery foes." Then the event that *both* couples are "fiery foes" is the intersection of the two events, $F_1 \cap F_2$. On the basis of the AAMFT survey which found that 25% of divorced couples are "fiery foes," we could reasonably conclude that $P(F_1) = .25$ and $P(F_2) = .25$. However, in order to compute the probability of $F_1 \cap F_2$ from the multiplicative rule, we must make the assumption that the two events are independent. Since the classification of any divorced couple is not likely to affect the classification of another divorced couple, this assumption is reasonable. Assuming independence, we have

$$P(F_1 \cap F_2) = P(F_1)P(F_2) = (.25)(.25) = .0625$$

b. To see how to compute the probability that 10 of 10 divorced couples will all be classified as "fiery foes," first consider the event that 3 of 3 couples are "fiery foes." If F_3 represents the event that the third divorced couple are "fiery foes," then we want to compute the probability of the intersection $F_1 \cap F_2 \cap F_3$. Again assuming independence of the classifications, we have

$$P(F_1 \cap F_2 \cap F_3) = P(F_1)P(F_2)P(F_3) = (.25)(.25)(.25) = .015625$$

Similar reasoning leads us to the conclusion that the intersection of 10 such events can be calculated as follows:

$$P(F_1 \cap F_2 \cap F_3 \cap \ldots \cap F_{10}) = P(F_1)P(F_2)\ldots P(F_{10}) = (.25)^{10} = .000001$$

Thus, the probability that 10 of 10 divorced couples sampled are all classified as "fiery foes" is 1 in 1 million, assuming that the probability of each couple being classified as "fiery foes" is .25 and that the classification decisions are independent.

Look Back The very small probability in part **b** makes it extremely unlikely that 10 of 10 divorced couples are "fiery foes." If this event should actually occur, we should question the probability of .25 provided by the AAMFT and used in the calculation—it is likely to be much higher. (This conclusion is another application of the rare-event approach to statistical inference.)

Now Work Exercise 3.83

Statistics IN Action | Revisited | The Probability of Winning Cash 3 or Play 4

In addition to the biweekly Lotto 6/53, the Florida Lottery runs several other games. Two popular daily games are Cash 3 and Play 4. In Cash 3, players pay $1 to select three numbers in order, where each number ranges from 0 to 9. If the three numbers selected (e.g., 2–8–4) match exactly the order of the three numbers drawn, the player wins $500. Play 4 is similar to Cash 3, but players must match four numbers (each number ranging from 0 to 9). For a $1 Play 4 ticket (e.g., 3–8–3–0), the player will win $5,000 if the numbers match the order of the four numbers drawn.

During the official drawing for Cash 3, 10 table tennis balls numbered 0, 1, 2, 3, 4, 5, 6, 7, 8, and 9 are placed into each of three chambers. The balls in the first chamber are colored pink, the balls in the second chamber are blue, and the balls in the third chamber are yellow. One ball of each color is randomly drawn, with the official order as pink–blue–yellow. In Play 4, a fourth chamber with orange balls is added, and the official order is pink–blue–yellow–orange. Since the draws of the colored balls are random

and independent, we can apply an extension of the probability rule for the intersection of two independent events to find the odds of winning Cash 3 and Play 4. The probability of matching a numbered ball being drawn from a chamber is 1/10; therefore,

$$
\begin{aligned}
P(\text{Win Cash 3}) &= P(\text{match pink } and \text{ match blue} \\
&\quad and \text{ match yellow}) \\
&= P(\text{match pink}) \times P(\text{match blue}) \times \\
&\quad P(\text{match yellow}) \\
&= (1/10)(1/10)(1/10) = 1/1000 = .001
\end{aligned}
$$

$$
\begin{aligned}
P(\text{Win Play 4}) &= P(\text{match pink } and \text{ match blue} \\
&\quad \text{and match yellow } and \text{ match orange}) \\
&= P(\text{match pink}) \times P(\text{match blue}) \times \\
&\quad P(\text{match yellow}) \times P(\text{match orange}) \\
&= (1/10)(1/10)(1/10) \\
&= 1/10,000 = .0001
\end{aligned}
$$

Statistics IN Action
(continued)

Although the odds of winning one of these daily games is much better than the odds of winning Lotto 6/53, there is still only a 1 in 1,000 chance (for Cash 3) or 1 in 10,000 chance (for Play 4) of winning the daily game. And the payoffs ($500 or $5,000) are much smaller. In fact, it can be shown that you will lose an average of 50¢ every time you play either Cash 3 or Play 4!

Exercises 3.63–3.97

Understanding the Principles

3.63 Explain the difference between an unconditional probability and a conditional probability.

3.64 Give the multiplicative rule of probability for
  a. two independent events.
  b. any two events.

3.65 Give the formula for finding $P(B|A)$.

3.66 Defend or refute each of the following statements:
  a. Dependent events are always mutually exclusive.
  b. Mutually exclusive events are always dependent.
  c. Independent events are always mutually exclusive.

Learning the Mechanics

3.67 For two events A and B, $P(A) = .4, P(B) = .2$, and
[NW] $P(A \cap B) = .1$.
  a. Find $P(A|B)$.
  b. Find $P(B|A)$.
  c. Are A and B independent events?

3.68 For two events A and B, $P(A) = .4, P(B) = .2$, and $P(A|B) = .6$.
  a. Find $P(A \cap B)$.
  b. Find $P(B|A)$.

3.69 For two independent events A and B, $P(A) = .4$ and $P(B) = .2$.
  a. Find $P(A \cap B)$.
  b. Find $P(A|B)$.
  c. Find $P(A \cup B)$.

3.70 An experiment results in one of three mutually exclusive events A, B, and C. It is known that $P(A) = .30, P(B) = .55$, and $P(C) = .15$, Find each of the following probabilities:
  a. $P(A \cup B)$
  b. $P(A \cap B))$
  c. $P(A|B)$
  d. $P(B \cup C))$
  e. Are B and C independent events? Explain.

3.71 Consider the experiment defined by the accompanying
[NW] Venn diagram, with the sample space S containing five sample points. The sample points are assigned the following probabilities: $P(E_1) = .1, P(E_2) = .1, P(E_3) = .2, P(E_4) = .5, P(E_5) = .1$.

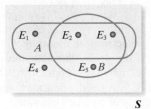

S

  a. Calculate $P(A), P(B)$, and $P(A \cap B)$.
  b. Suppose we know that event A has occurred, so the reduced sample space consists of the three sample points in A: E_1, E_2, and E_3. Use the formula for conditional probability to determine the probabilities of these three sample points given that A has occurred. Verify that the conditional probabilities are in the same ratio to one another as the original sample point probabilities and that they sum to 1.
  c. Calculate the conditional probability $P(B|A)$ in two ways: First, sum $P(E_2|A)$ and $P(E_3|A)$, since these sample points represent the event that B occurs given that A has occurred. Second, use the formula for conditional probability:

$$P(B|A) = \frac{P(A \cap B)}{P(A)}$$

  Verify that the two methods yield the same result.

3.72 An experiment results in one of five sample points with the following probabilities: $P(E_1) = .22, P(E_2) = .31, P(E_3) = .15, P(E_4) = .22$, and $P(E_5) = .1$. The following events have been defined:

$$A: \{E_1, E_3\}$$
$$B: \{E_2, E_3, E_4\}$$
$$C: \{E_1, E_5\}$$

  Find each of the following probabilities:
  a. $P(A)$
  b. $P(B)$
  c. $P(A \cap B)$
  d. $P(A|B)$
  e. $P(B \cap C)$
  f. $P(C|B)$
  g. Consider each pair of events A and B, A and C, and B and C. Are any of the pairs of events independent? Why?

3.73 Two fair dice are tossed, and the following events are defined:

  $A:$ {The sum of the numbers showing is odd.}
  $B:$ {The sum of the numbers showing is 9, 11, or 12.}

  Are events A and B independent? Why?

3.74 A sample space contains six sample points and events A, B, and C, as shown in the Venn diagram on page 144. The probabilities of the sample points are $P(1) = .20, P(2) = .05, P(3) = .30, P(4) = .10, P(5) = .10$, and $P(6) = .25$.

S

a. Which pairs of events, if any, are mutually exclusive? Why?

b. Which pairs of events, if any, are independen Why?

c. Find $P(A \cup B)$ by adding the probabilities of the sample points and then by using the additive rule. Verify that the answers agree. Repeat for $P(A \cup C)$.

3.75 A box contains two white, two red, and two blue poker chips. Two chips are randomly chosen without replacement, and their colors are noted. Define the following events:

> *A*: {Both chips are of the same color.}
> *B*: {Both chips are red.}
> *C*: {At least one chip is red or white.}

Find $P(B \mid A)$, $P(B \mid A^c)$, $P(B \mid C)$, $P(A \mid C)$, and $P(C \mid A^c)$.

Applet Exercise 3.5

Use the applet entitled *Simulating the Probability of Rolling a 6* to simulate conditional probabilities. Begin by running the applet twice with $n = 10$, without resetting between runs. The data on your screen represent 20 rolls of a die. The diagram above the *Roll* button shows the frequency of each of the six possible outcomes. Use this information to find each of the following probabilities:

a. The probability of 6 given that the outcome is 5 or 6

b. The probability of 6 given that the outcome is even

c. The probability of 4 or 6 given that the outcome is even

d. The probability of 4 or 6 given that the outcome is odd

Applying the Concepts—Basic

3.76 Do you have a library card? According to a September 2008 Harris poll, 68% of all American adults have a public library card. The percentages do differ by gender—62% of males have library cards compared to 73% of females.

a. Consider the three probabilities: $P(A) = .68$, $P(A \mid B) = .62$, and $P(A \mid C) = .73$. Define events *A*, *B*, and *C*.

b. Assuming that half of all American adults are males and half are females, what is the probability that an American adult is a female who owns a library card?

3.77 National firearms survey. The Harvard School of Public Health conducted a study of privately held firearm stock in the United States. In a representative household telephone survey of 2,770 adults, 26% reported that they own at least one gun (*Injury Prevention*, Jan. 2007). Of those that own a gun, 5% own a handgun. Suppose one of the 2,770 adults surveyed is randomly selected.

a. What is the probability that the adult owns at least one gun?

b. What is the probability that the adult owns at least one gun and the gun is a handgun?

3.78 Guilt in decision making. Refer to the *Journal of Behavioral Decision Making* (Jan. 2007) study of the effect of guilt emotion on how a decision maker focuses on the problem, Exercise 3.55 (p. 131). The results (number responding in each category) for the 171 study participants are reproduced in the accompanying table. Suppose one of the 171 participants is selected at random.

Emotional State	Choose Stated Option	Do Not Choose Stated Option	Totals
Guilt	45	12	57
Anger	8	50	58
Neutral	7	49	56
Totals	60	111	171

Based on Gangemi, A., and Mancini, F. "Guilt and focusing in decision-making." *Journal of Behavioral Decision Making,* Vol. 20, Jan. 2007 (Table 2).

a. Given the respondent is assigned to the guilty state, what is the probability that the respondent chooses the stated option?

b. If the respondent does not choose to repair the car, what is the probability of the respondent being in the anger state?

c. Are the events {repair the car} and {guilty state} independent?

3.79 Speeding linked to fatal car crashes. According to the National Highway Traffic and Safety Administration's National Center for Statistics and Analysis (NCSA), "Speeding is one of the most prevalent factors contributing to fatal traffic crashes" (*NHTSA Technical Report*, Aug. 2005). The probability that speeding is a cause of a fatal crash is .3. Furthermore, the probability that speeding and missing a curve are causes of a fatal crash is .12. Given that speeding is a cause of a fatal crash, what is the probability that the crash occurred on a curve?

3.80 Twitter and wireless Internet. According to the *Pew Internet & American Life Project* (Oct. 2009), 54% of Internet users have a wireless connection to the Internet via a laptop, cell phone, game console, or other mobile device. Of these wireless users, 25% use Twitter to share updates about themselves. What is the probability that an Internet user has a wireless connection and uses Twitter?

3.81 Appraisals and negative emotions. According to psychological theory, each negative emotion one has is linked to some thought (appraisal) about the environment. This theory was tested in the journal *Cognition and Emotion* (Vol. 24, 2010). Over a period of two days, undergraduate students were asked to respond to events that occurred within the past 15 minutes. Each student rated the negative emotions he or she felt (e.g., anger, sadness, fear, or guilt) and gave an appraisal of the event (e.g., unpleasant or unfair). A total of 10,797 responses was evaluated in the study, of which 594 indicated a perceived unfairness appraisal. Of these unfairness appraisals, the students reported feeling angry in 127 of them. Consider a response randomly selected from the study.

a. What is the probability that the response indicates a perceived unfairness appraisal?

b. Given a perceived unfairness appraisal, what is the probability of an angry emotion?

3.82 Study of ancient pottery. Refer to the *Chance* (Fall 2000) study of ancient pottery found at the Greek settlement of Phylakopi, presented in Exercise 2.14 (p. 36). Of the 837 pottery pieces uncovered at the excavation site, 183 were painted. These painted pieces included 14 painted in a curvilinear decoration, 165 painted in a geometric decoration, and 4 painted in a naturalistic decoration. Suppose 1 of the 837 pottery pieces is selected and examined.

a. What is the probability that the pottery piece is painted?

b. Given that the pottery piece is painted, what is the probability that it is painted in a curvilinear decoration?

Applying the Concepts—Intermediate

3.83 Health risks to beachgoers. Refer to the University of [NW] Florida study of health risks to beachgoers, Exercise 3.22 (p. 119). According to the study, 6 out of 1,000 people exposed to wet sand for a 10-minute period will acquire gastroenteritis and 7 out of 1,000 people exposed to ocean water for a 10-minute period will acquire gastroenteritis. Suppose you and a friend go to the beach. Over the next 10 minutes, you play in the wet sand while your friend swims in the ocean. What is the probability that at least one of you acquires gastroenteritis? (What assumption did you need to make in order to find the probability?)

3.84 Sleep apnea and sleep stage transitioning. Refer to the sleep apnea study published in *Chance* (Winter 2009), Exercise 3.54 (p. 130). Recall that the various stages of sleep for a large group of sleep apnea patients were monitored in 30-second "epochs," and the sleep stage for the previous and current epoch was determined. A summary table is reproduced below. One of the epochs in the study is selected.

Current Stage	Previous Sleep Stage		
	Non-REM	REM	Wake
Non-REM	31,880	160	1,733
REM	346	7,609	175
Wake	1,588	358	6,079
Totals	33,814	8,127	7,987

Based on Caffo, B. S., et al. "An overview of observational sleep research with application to sleep stage transitioning." *Chance*, Vol. 22, No. 1, Winter 2009 (Table 2).

a. Given that the previous sleep stage for the epoch was the Wake state, what is the probability that the current sleep stage is REM?

b. Given that the current sleep stage for the epoch is REM, what is the probability that the previous sleep stage was not the Wake state?

c. Are the events {previous stage is REM} and {current stage is REM} mutually exclusive?

d. Are the events {previous stage is REM} and {current stage is REM} independent?

e. Are the events {previous stage is Wake} and {current stage is Wake} independent?

3.85 Are you really being served red snapper? Red snapper is a rare and expensive reef fish served at upscale restaurants. Federal law prohibits restaurants from serving a cheaper, look-alike variety of fish (e.g., vermillion snapper or lane snapper) to customers who order red snapper. Researchers at the University of North Carolina used DNA analysis to examine fish specimens labeled "red snapper" that were purchased form vendors across the country (*Nature*, July 15, 2004). The DNA tests revealed that 77% of the specimens were not red snapper, but the cheaper, look-alike variety of fish.

a. Assuming that the results of the DNA analysis are valid, what is the probability that you are actually served red snapper the next time you order it at a restaurant?

b. If there are five customers at a restaurant, all who have ordered red snapper, what is the probability that at least one customer is actually served red snapper?

3.86 Fighting probability of fallow deer bucks. Refer to the *Aggressive Behavior* (Jan./Feb. 2007) study of fallow deer bucks fighting during the mating season, presented in Exercise 3.57 (p. 131). Recall that researchers recorded 167 encounters between two bucks, one of which clearly initiated the encounter with the other. A summary of the fight status of the initiated encounters is provided in the accompanying table. Suppose we select 1 of these 167 encounters and note the outcome (fight status and winner).

a. Given that a fight occurs, what is the probability that the initiator wins?

b. Given no fight, what is the probability that the initiator wins?

c. Are the events "no fight" and "initiator wins" independent?

	Initiator Wins	No Clear Winner	Initiator Loses	Totals
Fight	26	23	15	64
No Fight	80	12	11	103
Totals	106	35	26	167

Based on Bartos, L. et al. "Estimation of the probability of fighting in fallow deer (*Dama dama*) during the rut." *Aggressive Behavior*, Vol. 33, Jan./Feb., 2007, pp. 7–13.

3.87 Extinct New Zealand birds. Refer to the *Evolutionary Ecology Research* (July 2003) study of the patterns of extinction in the New Zealand bird population, presented in Exercise 2.20 (p. 37). Consider the data on extinction status (extinct, absent from island, present) for the 132 bird species. The data are saved in the **NZBIRDS** file and are summarized in the accompanying MINITAB printout. Suppose you randomly select 10 of the 132 bird species (without replacement) and record the extinction status of each.

a. What is the probability that the first species you select is extinct? (*Note*: Extinct = Yes on the MINITAB printout.)

b. Suppose the first 9 species you select are all extinct. What is the probability that the 10th species you select is extinct?

Tally for Discrete Variables: Extinct

Extinct	Count	Percent
Absent	16	12.12
No	78	59.09
Yes	38	28.79
N=	132	

3.88 Probability of winning a war. Before a country enters into a war, a prudent government will assess the cost, utility, and probability of a victory. University of Georgia professor P. L. Sullivan has developed a statistical model for determining the probability of winning a war based on a government's capabilities and resources (*Journal of Conflict Resolution,* Vol. 51, 2007). Now consider the current U.S.–Iraq conflict. One researcher used the model to estimate that the probability of a successful regime change in Iraq was .70 prior to the start of the war. Of course, we now know that the successful regime change was achieved. However, the model also estimates that given the mission is extended to support a weak Iraq government, the probability of ultimate success is only .26. Assume these probabilities are accurate.

a. Prior to the start of the U.S.–Iraq war, what is the probability that a successful regime change is not achieved?

b. Given that the mission is extended to support a weak Iraq government, what is the probability that a successful regime change is ultimately achieved?

c. Suppose the probability of the United States extending the mission to support a weak Iraq government was .55. Find the probability that the mission is extended and results in a successful regime change.

3.89 Ambulance response time. *Geographical Analysis* (Jan. 2010) presented a study of Emergency Medical Services (EMS) ability to meet the demand for an ambulance. In one example, the researchers presented the following scenario. An ambulance station has one vehicle and two demand locations, A and B. The probability that the ambulance can travel to a location in under eight minutes is .58 for location A and .42 for location B. The probability that the ambulance is busy at any point in time is .3.

a. Find the probability that EMS can meet demand for an ambulance at location A.

b. Find the probability that EMS can meet demand for an ambulance at location B.

3.90 Intrusion detection systems. A computer intrusion detection system (IDS) is designed to provide an alarm whenever someone intrudes (e.g., through unauthorized access) into a computer system. A probabilistic evaluation of a system with two independently operating intrusion detection systems (a double IDS) was published in the *Journal of Research of the National Institute of Standards and Technology* (Nov.–Dec. 2003). Consider a double IDS with system A and system B. If there is an intruder, system A sounds an alarm with probability .9 and system B sounds an alarm with probability .95. If there is no intruder, the probability that system A sounds an alarm (i.e., a false alarm) is .2 and the probability that system B sounds an alarm is .1.

a. Use symbols to express the four probabilities just given.

b. If there is an intruder, what is the probability that both systems sound an alarm?

c. If there is no intruder, what is the probability that both systems sound an alarm?

d. Given that there is an intruder, what is the probability that at least one of the systems sounds an alarm?

3.91 Detecting traces of TNT. University of Florida researchers in the Department of Materials Science and Engineering have invented a technique that rapidly detects traces of TNT (*Today,* Spring 2005). The method, which involves shining a laser on a potentially contaminated object, provides instantaneous results and gives no false positives. In this application, a false positive would occur if the laser detected traces of TNT when, in fact, no TNT were actually present on the object. Let A be the event that the laser light detects traces of TNT. Let B be the event that the object contains no traces of TNT. The probability of a false positive is 0. Write this probability in terms of A and B, using symbols such as "$\cup$", "$\cap$", and "$|$".

3.92 Forest fragmentation study. Refer to the *Conservation Ecology* (Dec. 2003) study of the causes of forest fragmentation, presented in Exercise 2.156 (p. 91). Recall that the researchers used advanced high-resolution satellite imagery to develop fragmentation indexes for each forest. A 3 × 3 grid was superimposed over an aerial photo of the forest, and each square (pixel) of the grid was classified as forest (F), as earmarked for anthropogenic land use (A), or as natural land cover (N). An example of one such grid is shown here. The edges of the grid (where an "edge" is an imaginary line that separates any two adjacent pixels) are classified as F–A, F–N, A–A, A–N, N–N, or F–F edges.

A	A	N
N	F	F
N	F	F

a. Note that there are 12 edges inside the grid. Classify each edge as F–A, F–N, A–A, A–N, N–N or F–F.

b. The researchers calculated the fragmentation index by considering only the F-edges in the grid. Count the number of F-edges. (These edges represent the sample space for the experiment.)

c. If an F-edge is selected at random, find the probability that it is an F–A edge. (This probability is proportional to the anthropogenic fragmentation index calculated by the researchers.)

d. If an F edge is selected at random, find the probability that it is an F–N edge. (This probability is proportional to the natural fragmentation index calculated by the researchers.)

3.93 Seeded players at a shuffleboard tournament. Shuffleboard is an outdoor game popular with senior citizens. In single-elimination shuffleboard tournaments, players are paired and play a match against each other, with the winners moving on to play another match. The tournament concludes in a final that matches two undefeated players. The winner of the final match is the overall tournament champion. *Chance* (Winter 2006) investigated the probability of winning a shuffleboard tournament with 64 players. As an example, the authors considered a small tournament with only 4 players, named A, B, C, and D. This tournament involves a total of three matches. In the first round, A plays B in one match and C plays D in the other match. Then the winners play in the final round.

a. List the different outcomes of the tournament. (For example, one outcome is "*A* and *C* win first-round matches and *A* wins the final.")

b. If the players are of equal ability, what is the probability that *A* wins the tournament?

c. Suppose the players are not of equal ability. The accompanying table gives the likelihood of one player defeating the other. (*Example:* The probability that *A* defeats *B* is .9.) What is the probability that *A* wins the tournament?

Outcome	Probability
A defeats *B*	.9
A defeats *C*	.7
A defeats *D*	.6
B defeats *C*	.1
B defeats *D*	.2
C defeats *D*	.4

Applying the Concepts—Advanced

3.94 **Random mutation of cells.** *Chance* (Spring 2010) presented an article on the random mutation hypothesis developed by microbiologists. Under this hypothesis, when a wild-type organic cell (e.g., a bacteria cell) divides, there is a chance that at least one of the two "daughter" cells is a mutant. When a mutant cell divides, both offspring will be mutant. The schematic below shows a possible pedigree from a single cell that has divided. Note that one "daughter" cell is mutant (•) and one is a normal cell (○).

a. Consider a single, normal cell that divides into two offspring. List the different possible pedigrees.

b. Assume that a "daughter" cell is equally likely to be mutant or normal. What is the probability that a single, normal cell that divides into two offspring will result in at least one mutant cell?

c. Now assume that the probability of a mutant "daughter" cell is .2. What is the probability that a single, normal cell that divides into two offspring will result in at least one mutant cell?

d. The schematic below shows a possible second-generation pedigree from a single cell that has divided. Note that the first generation mutant cell automatically produces two mutant cells in the second generation. List the different possible second generation pedigrees. (*Hint*: Use your answer to part **a.**)

e. Assume that a "daughter" cell is equally likely to be mutant or normal. What is the probability that a single, normal cell that divides into two offspring will result in at least one mutant cell after the second generation?

3.95 **Testing a psychic's ability.** Consider an experiment in which 10 identical small boxes are placed side by side on a table. A crystal is placed at random inside one of the boxes. A self-professed "psychic" is asked to pick the box that contains the crystal.

a. If the "psychic" simply guesses, what is the probability that she picks the box with the crystal?

b. If the experiment is repeated seven times, what is the probability that the "psychic" guesses correctly at least once?

c. A group called the Tampa Bay Skeptics recently tested a self-proclaimed "psychic" by administering the preceding experiment seven times. The "psychic" failed to pick the correct box all seven times (*Tampa Tribune*, Sept. 20, 1998). What would you infer about this person's psychic ability?

3.96 **Risk of a natural-gas pipeline accident.** *Process Safety Progress* (Dec. 2004) published a risk analysis for a natural-gas pipeline between Bolivia and Brazil. The most likely scenario for an accident would be natural-gas leakage from a hole in the pipeline. The probability that the leak ignites immediately (causing a jet fire) is .01. If the leak does not immediately ignite, it may result in the delayed ignition of a gas cloud. Given no immediate ignition, the probability of delayed ignition (causing a flash fire) is .01. If there is no delayed ignition, the gas cloud will disperse harmlessly. Suppose a leak occurs in the natural-gas pipeline. Find the probability that either a jet fire or a flash fire will occur. Illustrate with a tree diagram.

3.97 **NBA draft lottery.** The National Basketball Association (NBA) utilizes a lottery to determine the order in which teams draft amateur (high school and college) players. Teams with the poorest won–lost records have the best chance of obtaining the first selection (and, presumably, best player) in the draft. Under the current lottery system, 14 table tennis balls numbered 1 through 14 are placed in a drum, and 4 balls are drawn. There are 1,001 possible combinations when 4 balls are drawn out of 14 without regard to their order of selection. Prior to the drawing, 1,000 combinations are assigned to the 13 teams with the worst records—the lottery teams—based on their order of finish during the regular season. (The team with the worst record is assigned 250 of these combinations, the team with the second-worst record is assigned 200 of the combinations, the team with the third-worst record is assigned 157 of the combinations, etc. The table on page 148 gives the number of combinations assigned to each lottery team.) Once the 4 balls are drawn to the top of the drum to determine a four-digit combination, the team that has been assigned that combination will receive the first pick. The 4 balls are then placed back in the drum, and the process is repeated to determine the second, third, and subsequent draft picks. (*Note*: If the one unassigned combination is drawn, the balls are drawn to the top again.)

NBA Lottery Team	Number of Combinations
Worst record	250
2nd-worst record	200
3rd-worst record	157
4th-worst record	120
5th-worst record	89
6th-worst record	64
7th-worst record	44
8th-worst record	29
9th-worst record	18
10th-worst record	11
11th-worst record	7
12th-worst record	6
13th-worst record	5

Based on McCann, M. A. "Illegal defense: The irrational economics of banning high school players from the NBA draft." *Virginia Sports and Entertainment Law Journal,* Vol. 3, No. 2, Spring 2004 (Table 5).

a. Demonstrate that there are 1,001 four-number combinations of the numbers 1 through 14.

b. For each of the 13 Lottery teams, find the probability that the team obtains the first pick in the draft.

c. Given that the team with the second-worst record obtains the number-1 pick, find the probability that the team with the worst record obtains the second pick of the draft.

d. Given that the team with the third-worst record obtains the number-1 pick, find the probability that the team with the worst record obtains the second pick of the draft.

e. Find the probability that the team with the worst record obtains the second pick of the draft, assuming that it did not obtain the first pick.

3.7 Random Sampling

How a sample is selected from a population is of vital importance in statistical inference because the probability of an observed sample will be used to infer the characteristics of the sampled population. To illustrate, suppose you deal yourself four cards from a deck of 52 cards, and all four cards are aces. Do you conclude that your deck is an ordinary bridge deck, containing only four aces, or do you conclude that the deck is stacked with more than four aces? It depends on how the cards were drawn. If the four aces are always placed at the top of a standard bridge deck, drawing four aces is not unusual—it is certain. If, however, the cards are thoroughly mixed, drawing four aces in a sample of four cards is highly improbable. The point, of course, is that in order to use the observed sample of four cards to draw inferences about the population (the deck of 52 cards), you need to know how the sample was selected from the deck.

One of the simplest and most frequently employed sampling procedures is implied in the previous examples and exercises. It produces what is known as a *random sample*. We learned in Section 1.5 (p. 11) that a random sample is likely to be *representative* of the population that it is selected from.

> If *n* elements are selected from a population in such a way that every set of *n* elements in the population has an equal probability of being selected, then the *n* elements are said to be a **random sample.***

If a population is not too large and the elements can be numbered on slips of paper, poker chips, etc., you can physically mix the slips of paper or chips and remove *n* elements from the total. The numbers that appear on the chips selected would indicate the population elements to be included in the sample. Since it is often difficult to achieve a thorough mix, such a procedure only provides an approximation to random sampling. Most researchers rely on **random-number generators** to automatically generate a random sample. Random-number generators are available in tabular form, are built into most statistical software packages, and are available on the Internet (e.g., at www.random.org).

---

*Strictly speaking, this is a *simple random sample*. There are many different types of random samples. The simple random sample is the most common.

Example 3.21

Selecting a Random Sample—5 Households from 100,000

Problem Suppose you wish to randomly sample five households from a population of 100,000 households.

a. How many different samples can be selected?
b. Use a random-number generator to select a random sample.

Solution

a. Since we want to select $n = 5$ objects (households) from $N = 100,000$, we apply the combinations rule of Section 3.1:

$$\binom{N}{n} = \binom{100,000}{5} = \frac{100,000!}{5!\,99,995!}$$
$$= \frac{100,000 \cdot 99,999 \cdot 99,998 \cdot 99,997 \cdot 99,996}{5 \cdot 4 \cdot 3 \cdot 2 \cdot 1}$$
$$= 8.33 \times 10^{22}$$

Thus, there are 83.3 billion trillion different samples of five households that can be selected from 100,000.

b. To ensure that each of the possible samples has an equal chance of being selected, as is required for random sampling, we can employ a **random-number table**, such as Table I of Appendix A. Random-number tables are constructed in such a way that every number occurs with (approximately) equal probability. Furthermore, the occurrence of any one number in a position is independent of any of the other numbers that appear in the table. To use a table of random numbers, number the N elements in the population from 1 to N. Then turn to Table I and select a starting number in the table. Proceeding from this number either across the row or down the column, remove and record n numbers from the table.

To illustrate, first we number the households in the population from 1 to 100,000. Then we turn to a page of Table I—say, the first page. (A partial reproduction of the first page of Table I is shown in Table 3.7.) Now we arbitrarily select a starting number—say, the random number appearing in the third row, second column. This number is 48,360. Then we proceed down the second column to obtain the remaining four random numbers. In this case, we have selected five random numbers, which are highlighted in Table 3.7. Using the first five digits to represent households from 1 to 99,999 and the number 00000 to represent household 100,000, we can see that the households numbered

48,360   93,093   39,975   6,907   72,905

should be included in our sample.

Table 3.7	**Partial Reproduction of Table I in Appendix A**					
Row/Column	1	2	3	4	5	6
1	10480	15011	01536	02011	81647	91646
2	22368	46573	25595	85393	30995	89198
3	24130	48360	22527	97265	76393	64809
4	42167	93093	06243	61680	07856	16376
5	37670	39975	81837	16656	06121	91782
6	77921	06907	11008	42751	27756	53498
7	99562	72905	56420	69994	98872	31016
8	96301	91977	05463	07972	18876	20922
9	89579	14342	63661	10281	17453	18103
10	85475	36857	53342	53988	53060	59533
11	28918	69578	88231	33276	70997	79936
12	63553	40961	48235	03427	49626	69445
13	09429	93969	52636	92737	88974	33488

Note: Use only the necessary number of digits in each random number to identify the element to be included in the sample. If, in the course of recording the n numbers from the table, you select a number that has already been selected, simply discard the duplicate and select a replacement at the end of the sequence. Thus, you may have to record more than n numbers from the table to obtain a sample of n unique numbers.

Look Back Can we be sure that all 83.3 billion trillion samples have an equal chance of being selected? The fact is, we can't; but to the extent that the random number table contains truly random sequences of digits, the sample should be very close to random.

Now Work Exercise 3.102

Table I in Appendix A is just one example of a random-number generator. For most scientific studies that require a large random sample, computers are used to generate the random sample. The SAS, MINITAB, and SPSS statistical software packages all have easy-to-use random-number generators.

For example, suppose we required a random sample of $n = 50$ households from the population of 100,000 households in Example 3.21. Here, we might employ the MINITAB random-number generator. Figure 3.21 shows a MINITAB printout listing 50 random numbers (from a population of 100,000). The households with these identification numbers would be included in the random sample.

HOUSE50.MTW ***

	C1	C2	C3	C4	C5
	HouseID1	HouseID2	HouseID3	HouseID4	HouseID5
1	2036	22037	39884	62123	75750
2	3161	22055	44687	62421	80044
3	3930	22226	44689	63749	81857
4	9838	22773	45422	69079	81918
5	9961	22820	48104	69802	91680
6	11623	32989	50226	70895	94987
7	14114	33119	50443	71029	96110
8	17244	35045	50806	72879	97679
9	17598	35922	50813	73962	99232
10	18769	39272	61059	75517	99789
11					

Figure 3.21

MINITAB worksheet with random sample of 50 households

Recall our discussion of designed experiments in Section 1.5 (p. 11). The twin notions of random selection and randomization are key to conducting good research with a designed experiment. The next example illustrates a basic application.

Example 3.22

Randomization In a Designed Experiment— Clinical Trial

Problem A designed experiment in the medical field involving human subjects is referred to as a *clinical trial*. One recent clinical trial was designed to determine the potential of using aspirin in preventing heart attacks. Volunteer physicians were randomly divided into two groups: the *treatment* group and the *control* group. Each physician in the treatment group took one aspirin tablet a day for one year, while the physicians in the control group took an aspirin-free placebo made to look

ASPIRIN-STUDY.MTW ***		
↓	C1	C2	C3
	Physician	Treatment	
1	1	3	
2	2	11	
3	3	10	
4	4	14	
5	5	2	
6	6	7	
7	7	6	
8	8	16	
9	9	17	
10	10	9	
11	11		
12	12		
13	13		
14	14		
15	15		
16	16		
17	17		
18	18		
19	19		
20	20		
21			

Figure 3.22
MINITAB worksheet with random assignment of physicians

identical to an aspirin tablet. Since the physicians did not know to which group—treatment or control—they were assigned, the clinical trial is called a *blind study*. Assume that 20 physicians volunteered for the study. Use a random-number generator to randomly assign half of the physicians to the treatment group and half to the control group.

Solution Essentially, we want to select a random sample of 10 physicians from the 20. The first 10 selected will be assigned to the treatment group; the remaining 10 will be assigned to the control group. (Alternatively, we could randomly assign each physician, one by one, to either the treatment or control group. However, this would not guarantee exactly 10 physicians in each group.)

The MINITAB random-sample procedure was employed, producing the printout shown in Figure 3.22. Numbering the physicians from 1 to 20, we see that physicians 3, 11, 10, 14, 2, 7, 6, 16, 17, and 9 are assigned to receive the aspirin (the treatment). The remaining physicians are assigned the placebo (the control).

Now Work Exercise 3.107

Ethics IN Statistics Intentionally selecting a nonrandom sample in an effort to support a particular viewpoint is considered *unethical statistical practice*.

Exercises 3.98–3.111

Understanding the Principles

3.98 Define a random sample.

3.99 How are random samples related to representative samples?

3.100 What is a random-number generator?

3.101 Define a blind study.

Learning the Mechanics

3.102 Suppose you wish to sample $n = 2$ elements from a total of $N = 10$ elements.
[NW]
   a. Count the number of different samples that can be drawn, first by listing them and then by using combinatorial mathematics.
   b. If random sampling is to be employed, what is the probability that any particular sample will be selected?
   c. Show how to use the random-number table, Table I in Appendix A, to select a random sample of 2 elements from a population of 10 elements. Perform the sampling procedure 20 times. Do any two of the samples contain the same 2 elements? Given your answer to part **b**, did you expect repeated samples?

3.103 Suppose you wish to sample $n = 3$ elements from a total of $N = 600$ elements.
   a. Count the number of different samples by using combinatorial mathematics.
   b. If random sampling is to be employed, what is the probability that any particular sample will be selected?
   c. Show how to use the random-number table, Table I in Appendix A, to select a random sample of 3 elements from a population of 600 elements. Perform the sampling procedure 20 times. Do any two of the samples contain

the same 3 elements? Given your answer to part **b**, did you expect repeated samples?
   d. Use the computer to generate a random sample of 3 from the population of 600 elements.

3.104 Suppose that a population contains $N = 200{,}000$ elements. Use a computer or Table I of Appendix A to select a random sample of $n = 10$ elements from the population. Explain how you selected your sample.

Applying the Concepts—Basic

3.105 **Random-digit dialing.** To ascertain the effectiveness of their advertising campaigns, firms frequently conduct telephone interviews with consumers by using *random-digit dialing*. With this method, a random-number generator mechanically creates the sample of phone numbers to be called.
   a. Explain how the random-number table (Table I of Appendix A) or a computer could be used to generate a sample of seven-digit telephone numbers.
   b. Use the procedure you described in part **a** to generate a sample of 10 seven-digit telephone numbers.
   c. Use the procedure you described in part **a** to generate 5 seven-digit telephone numbers whose first three digits are 373.

3.106 **Census sampling.** In addition to its decennial enumeration of the population, the U.S. Census Bureau regularly samples the population for demographic information such as income, family size, employment, and marital status. Suppose the Bureau plans to sample 1,000 households in a city that has a total of 534,322 households. Show how the Bureau could use the random-number table in Appendix A or a computer to generate the sample. Select the first 10 households to be included in the sample.

Applying the Concepts—Intermediate

3.107 Kiwifruit as an iron supplement. Massey University (New
Zealand) researchers have designed a 16-week study to
determine the effectiveness of gold kiwifruit as an iron
supplement in women with an iron deficiency (*BMC Public
Health*, Vol. 10, 2010). Half the women in the study will
receive an iron-fortified breakfast cereal with a banana,
and half will receive an iron-fortified breakfast cereal with
gold kiwifruit. The researchers are planning for 100 women
to participate in the study. Number the women from 1 to
100. Then use a random-number generator to select 50 of
these women. These 50 will receive the kiwifruit breakfast;
the remaining 50 will receive the banana breakfast.

3.108 Stress and diet study. Researchers at the University of South
Florida investigated the effects of diet on the physiology and
behavior of rats (*USF News*, Oct. 17, 2009). One group of rats
was fed a typical American diet (40% fat, 40% protein, 20%
carbohydrates); a second group of rats was fed an Atkins-
type diet (70% fat, 20% protein, 10% carbohydrates); and, a
third (control) group was fed rat chow. Assume the experiment utilized 50 rats. Use a random-number generator to
randomly assign the rats to the three diet groups.

3.109 Study of the effect of songs with violent lyrics. A series of
designed experiments was conducted to examine the effects
of songs with violent lyrics on aggressive thoughts and hostile feelings (*Journal of Personality & Social Psychology*,
May 2003). In one of the experiments, college students were
randomly assigned to listen to one of four types of songs: (1)
a humorous song with violent lyrics, (2) a humorous song
with nonviolent lyrics, (3) a nonhumorous song with violent
lyrics, and (4) a nonhumorous song with nonviolent lyrics.
Assuming that 120 students participated in the study, use a
random-number generator to randomly assign 30 students
to each of the four experimental conditions.

3.110 Auditing an accounting system. In auditing a firm's financial
statements, an auditor is required to assess the operational
effectiveness of the accounting system. In performing the
assessment, the auditor frequently relies on a random
sample of actual transactions (Stickney and Weil, *Financial
Accounting: An Introduction to Concepts, Methods, and
Uses*, 2002). A particular firm has 5,382 customer accounts
that are numbered from 0001 to 5382.
  a. One account is to be selected at random for audit.
   What is the probability that account number 3,241 is
   selected?
  b. Draw a random sample of 10 accounts and explain in
   detail the procedure you used.
  c. Refer to part **b**. The following are two possible random
   samples of size 10:

Sample Number 1				
5011	0082	0963	0772	3415
2663	1126	0008	0026	4189

Sample Number 2				
0001	0003	0005	0007	0009
0002	0004	0006	0008	0010

Is one more likely to be selected than the other? Explain.

Applying the Concepts—Advanced

3.111 Selecting archaeological dig sites. Archaeologists plan to
perform test digs at a location they believe was inhab-
ited several thousand years ago. The site is approximately
10,000 meters long and 5,000 meters wide. They first draw
rectangular grids over the area, consisting of lines every
100 meters, creating a total of $100 \cdot 50 = 5,000$ intersections (not counting one of the outer boundaries). The plan
is to randomly sample 50 intersection points and dig at
the sampled intersections. Explain how you could use a
random-number generator to obtain a random sample of
50 intersections. Develop at least two plans: one that numbers the intersections from 1 to 5,000 prior to selection and
another that selects the row and column of each sampled
intersection (from the total of 100 rows and 50 columns).

CHAPTER NOTES

Key Terms

Additive rule of probability 126
Combinations rule 116
Combinatorial mathematics 116
Complement 124
Complementary events 124
Compound event 121
Conditional probability 133
Dependent events 139
Event 113
Experiment 108
Independent events 139
Intersection 121
Law of large numbers 110
Multiplicative rule of
   probability 136

Mutually exclusive 127
Odds 158
Probability rules for sample
   points 111
Random-number generator 148
Random-number table 149
Random sample 148
Sample point 108
Sample space 109
Tree diagram 109
Two-way table 123
Unconditional probabilities 133
Union 121
Venn diagram 109

Key Symbols

S	**Sample Space** (collection of all sample points)
$A: \{1,2\}$	Set of **sample points in event A**
$P(A)$	**Probability** of event A
$A \cup B$	**Union** of events A and B (either A or B occurs)
$A \cap B$	**Intersection** of events A and B (both A and B occur)
A^c	**Complement** of A (event A does not occur)
$A \mid B$	Event A occurs, **given that** event B occurs
$\binom{N}{n}$	Number of **combinations** of N elements taken n at a time
$N!$	**N factorial** $= N(N - 1)(N - 2) \ldots (2)(1)$

Key Ideas

Probability Rules for k Sample Points, $S_1, S_2, S_3, \ldots, S_k$

1. $0 \le P(E_i) \le 1$
2. $\Sigma P(E_i) = 1$

Random Sample

All possible such samples have equal probability of being selected

Combinations Rule

Counting number of samples of n elements selected from N elements

$$\binom{N}{n} = \frac{N!}{n!(N-n)!}$$

$$= \frac{N(N-1)(N-2)\ldots(N-n+1)}{n(n-1)(n-2)\ldots(2)(1)}$$

Guide to Selecting Probability Rules

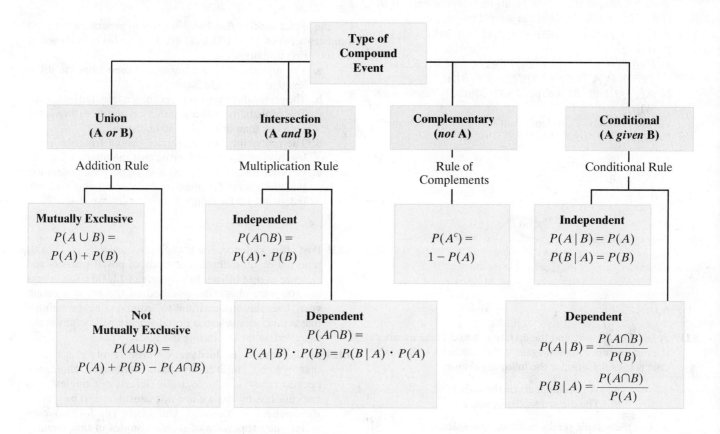

Supplementary Exercises 3.112–3.152

Understanding the Principles

3.112 Which of the following pairs of events are mutually exclusive?
  a. $A = \{$The Tampa Bay Rays win the World Series next year.$\}$
    $B = \{$Evan Longoria, Rays infielder, hits 75 home runs runs next year.$\}$
  b. $A = \{$Psychiatric patient Tony responds to a stimulus within 5 seconds.$\}$
    $B = \{$Psychiatric patient Tony has the fastest stimulus response time of 2.3 seconds.$\}$
  c. $A = \{$High school graduate Cindy enrolls at the University of South Florida next year.$\}$
    $B = \{$High school graduate Cindy does not enroll in college next year.$\}$

3.113 Use the symbols $\cap$, $\cup$, $|$, and ^{c}, to convert the following statements into compound events involving events A and B, where $A = \{$You purchase a notebook computer$\}$ and $B = \{$You vacation in Europe$\}$:
  a. You purchase a notebook computer or vacation in Europe.
  b. You will not vacation in Europe.
  c. You purchase a notebook computer and vacation in Europe.
  d. Given that you vacation in Europe, you will not purchase a notebook computer.

Learning the Mechanics

3.114 A sample space consists of four sample points S_1, S_2, S_3, and S_4, where $P(S_1) = .2$, $P(S_2) = .1$, $P(S_3) = .3$, and $P(S_4) = .4$.

a. Show that the sample points obey the two probability rules for a sample space.

b. If an event $A = \{S_1, S_4,\}$ find P(A).

3.115 A and B are mutually exclusive events, with $P(A) = .2$ and $P(B) = .3$.

a. Find $P(A|B)$.

b. Are A and B independent events?

3.116 For two events A and B, suppose $P(A) = .7, P(B) = .5$, and $P(A \cap B) = .4$. Find $P(A \cup B)$.

3.117 Given that $P(A \cap B) = .4$ and $P(A|B) = .8$, find $P(B)$.

3.118 The accompanying Venn diagram illustrates a sample space containing six sample points and three events, A, B, and C. The probabilities of the sample points are $P(1) = .3, P(2) = .2, P(3) = .1, P(4) = .1, P(5) = .1$, and $P(6) = .2$.

a. Find $P(A \cap B), P(B \cap C), P(A \cup C), P(A \cup B \cup C)$, $P(B^c), P(A^c \cap B), P(B|C)$, and $P(B|A)$.

b. Are A and B independent? Mutually exclusive? Why?

c. Are B and C independent? Mutually exclusive? Why?

S

3.119 A fair die is tossed, and the up face is noted. If the number is even, the die is tossed again; if the number is odd, a fair coin is tossed. Consider the following events:

A: {A head appears on the coin.}
B: {The die is tossed only one time.}

a. List the sample points in the sample space.

b. Give the probability for each of the sample points.

c. Find $P(A)$ and $P(B)$.

d. Identify the sample points in $A^c, B^c, A \cap B$, and $A \cup B$.

e. Find $P(A^c), P(B^c), P(A \cap B), \ P(A \cup B), P(A|B)$, and $P(B|A)$.

f. Are A and B mutually exclusive events? Independent events? Why?

3.120 A balanced die is thrown once. If a 4 appears, a ball is drawn from urn 1; otherwise, a ball is drawn from urn 2. Urn 1 contains four red, three white, and three black balls. Urn 2 contains six red and four white balls.

a. Find the probability that a red ball is drawn.

b. Find the probability that urn 1 was used given that a red ball was drawn.

3.121 Two events, A and B, are independent, with $P(A) = .3$ and $P(B) = .1$.

a. Are A and B mutually exclusive? Why?

b. Find $P(A|B)$ and $P(B|A)$.

c. Find $P(A \cup B)$.

3.122 Find the numerical value of

a. 6!

b. $\binom{10}{9}$

c. $\binom{10}{1}$

d. P_2^6

e. $\binom{6}{3}$

f. 0!

g. P_4^{10}

h. P_2^{50}

Applet Exercise 3.6

Use the applet entitled *Random Numbers* to generate a list of 50 numbers between 1 and 100, inclusive. Use the list to find each of the following probabilities:

a. The probability that a number chosen from the list is less than or equal to 50

b. The probability that a number chosen from the list is even

c. The probability that a number chosen from the list is both less than or equal to 50 and even

d. The probability that a number chosen from the list is less than or equal to 50, given that the number is even

e. Do your results from parts **a–d** support the conclusion that the events *less than or equal to 50* and *even* are independent? Explain.

Applying the Concepts—Basic

3.123 **Post office violence.** *The Wall Street Journal* (Sept. 1, 2000) reported on an independent study of postal workers and violence at post offices. In a sample of 12,000 postal workers, 600 were physically assaulted on the job in a recent year. Use this information to estimate the probability that a randomly selected postal worker will be physically assaulted on the job during the year.

3.124 **Sterile couples in Jordan.** A sterile family is a couple that has no children by their deliberate choice or because they are biologically infertile. Couples who are childless by chance are not considered to be sterile. Researchers at Yarmouk University (in Jordan) estimated the proportion of sterile couples in that country to be .06 (*Journal of Data Science*, July 2003). Also, 64% of the sterile couples in Jordan are infertile. Find the probability that a Jordanian couple is both sterile and infertile.

3.125 **NHTSA new car crash testing.** Refer to the National Highway Traffic Safety Administration (NHTSA) crash tests of new car models, presented in Exercise 2.174 (p. 98). Recall that the NHTSA has developed a "star" scoring system, with results ranging from one star (*) to five stars (*****). The more stars in the rating, the better the level of crash protection in a head-on collision. A summary of the driver-side star ratings for 98 cars (saved in the **CRASH** file) is reproduced in the MINITAB printout on p. 155. Assume that one of the 98 cars is selected at random. State whether each of the following is true or false:

a. The probability that the car has a rating of two stars is 4.

b. The probability that the car has a rating of four or five stars is .7857.

c. The probability that the car has a rating of one star is 0.

d. The car has a better chance of having a two-star rating than of having a five-star rating.

Tally for Discrete Variables: DRIVSTAR

DRIVSTAR	Count	Percent
2	4	4.08
3	17	17.35
4	59	60.20
5	18	18.37
N=	98	

3.126 Selecting a sample. A random sample of five students is to be selected from 50 sociology majors for participation in a special program.

a. In how many different ways can the sample be drawn?

b. Show how the random-number table, Table I of Appendix A, can be used to select the sample of students.

3.127 Mathematics achievement test. According to the National Center for Education Statistics, only 5% of U. S. eighth graders score above 655 on a mathematics assessment test. Find the probability that a randomly selected eighth grader has a score of 655 or below on the mathematics assessment test.

3.128 Oil spill impact on seabirds. Refer to the *Journal of Agricultural, Biological, and Environmental Statistics* (Sept. 2000) study of the impact of the *Exxon Valdez* tanker oil spill on the seabird population in Prince William Sound, Alaska, presented in Exercise 2.187 (p. 100). Recall that data were collected on 96 shoreline locations (called transects), and it was determined whether or not the transect was in an oiled area. The data are stored in the file called **EVOS.** From the data, estimate the probability that a randomly selected transect in Prince William Sound is contaminated with oil.

3.129 Fungi in beech forest trees. Beechwood forests in East Central Europe are being threatened by dynamic changes in land ownership and economic upheaval. The current status of the beech tree species in this area was evaluated by Hungarian university professors in *Applied Ecology and Environmental Research* (Vol. 1, 2003). Of 188 beech trees surveyed, 49 had been damaged by fungi. Depending on the species of fungus, damage will occur on either the trunk, branches, or leaves of the tree. In the damaged trees, the trunk was affected 85% of the time, the leaves 10% of the time, and the branches 5% of the time.

a. Give a reasonable estimate of the probability of a beech tree in East Central Europe being damaged by fungi.

b. A fungus-damaged beech tree is selected at random, and the area (trunk, leaf, or branch) affected is observed. List the sample points for this experiment, and assign a reasonable probability to each one.

3.130 Herbal medicines survey. Refer to *The American Association of Nurse Anesthetists Journal* (Feb. 2000) study on the use of herbal medicines before surgery, presented in Exercise 1.20 (p. 19). Recall that 51% of surgical patients use herbal medicines against their doctor's advice prior to surgery.

a. What is the probability that a randomly selected surgical patient will use herbal medicines against his or her doctor's advice?

b. What is the probability that in a sample of two independently selected surgical patients, both will use herbal medicines against their doctor's advice?

c. What is the probability that in a sample of five independently selected surgical patients, all five will use herbal medicines against their doctor's advice?

d. Would you expect the event in part **c** to occur? Explain.

3.131 Beach erosional hot spots. Beaches that exhibit high erosion rates relative to the surrounding beach are defined as *erosional hot spots*. The U.S. Army Corps of Engineers is conducting a study of beach hot spots. Through an online questionnaire, data are collected on six beach hot spots. The data are listed in the next table.

Beach Hot Spot	Beach Condition	Nearshore Bar Condition	Long-Term Erosion Rate (miles/year)
Miami Beach, FL	No dunes/flat	Single, shore parallel	4
Coney Island, NY	No dunes/flat	Other	13
Surfside, CA	Bluff/scarp	Single, shore parallel	35
Monmouth Beach, NJ	Single dune	Planar	Not estimated
Ocean City, NJ	Single dune	Other	Not estimated
Spring Lake, NJ	Not observed	Planar	14

Based on *Identification and Characterization of Erosional Hotspots*, William & Mary Virginia Institute of Marine Science, U.S. Army Corps of Engineers Project Report, Mar. 18, 2002.

a. Suppose you record the nearshore bar condition of each beach hot spot. Give the sample space for this experiment.

b. Find the probabilities of the sample points in the sample space you defined in part **a.**

c. What is the probability that a beach hot spot has either a planar or single, shore-parallel nearshore bar condition?

d. Now suppose you record the beach condition of each beach hot spot. Give the sample space for this experiment.

e. Find the probabilities of the sample points in the sample space you defined in part **d.**

f. What is the probability that the condition of the beach at a particular beach hot spot is not flat?

3.132 Chemical insect attractant. An entomologist is studying the effect of a chemical sex attractant (pheromone) on insects. Several insects are released at a site equidistant from the pheromone under study and a control substance. If the pheromone has an effect, more insects will travel toward it rather than toward the control. Otherwise, the insects are equally likely to travel in either direction. Suppose the pheromone under study has no effect, so that it is equally likely that an insect will move toward the pheromone or toward the control. Suppose five insects are released.

a. List or count the number of different ways the insects can travel.

b. What is the chance that all five travel toward the pheromone?

c. What is the chance that exactly four travel toward the pheromone?

d. What inference would you make if the event in part **c** actually occurs? Explain.

Applying the Concepts—Intermediate

3.133 Winning at roulette. *Roulette* is a very popular game in many American casinos. In Roulette, a ball spins on a circular wheel that is divided into 38 arcs of equal length, bearing the numbers 00, 0, 1, 2, ... , 35, 36. The number of the arc on which the ball stops is the outcome of one play of the game. The numbers are also colored in the manner shown in the following table.

Red: 1, 3, 5, 7, 9, 12, 14, 16, 18, 19, 21, 23, 25, 27, 30, 32, 34, 36
Black: 2, 4, 6, 8, 10, 11, 13, 15, 17, 20, 22, 24, 26, 28, 29, 31, 33, 35
Green: 00, 0

Players may place bets on the table in a variety of ways, including bets on odd, even, red, black, high, low, etc. Consider the following events:

A: {The outcome is an odd number (00 and 0 are considered neither odd nor even.)}
B: {The outcome is a black number.}
C: {The outcome is a low number (1–18).}

a. Define the event $A \cap B$ as a specific set of sample points.

b. Define the event $A \cup B$ as a specific set of sample points.

c. Find $P(A), P(B), P(A \cap B), P(A \cup B)$, and $P(C)$ by summing the probabilities of the appropriate sample points.

d. Define the event $A \cap B \cap C$ as a specific set of sample points.

e. Use the additive rule to find $P(A \cup B)$. Are events A and B mutually exclusive? Why?

f. Find $P(A \cap B \cap C)$ by summing the probabilities of the sample points given in part **d**.

g. Define the event $(A \cup B \cup C)$ as a specific set of sample points.

h. Find $P(A \cup B \cup C)$ by summing the probabilities of the sample points given in part **g**.

3.134 Stacking in the NBA. In professional sports, *stacking* is a term used to describe the practice of African-American players being excluded from certain positions because of race. To illustrate the stacking phenomenon, the *Sociology of Sport Journal* (Vol. 14, 1997) presented the table shown here. The table summarizes the race and positions of 368 National Basketball Association (NBA) players in 1993. Suppose an NBA player is selected at random from that year's player pool.

	Position			
	Guard	Forward	Center	Totals
White	26	30	28	84
Black	128	122	34	284
Totals	154	152	62	368

a. What is the probability that the player is white?

b. What is the probability that the player is a center?

c. What is the probability that the player is African-American and plays guard?

d. What is the probability that the player is not a guard?

e. What is the probability that the player is white or a center?

f. Given that the player is white, what is the probability that he is a center?

g. Given that the player is African-American, what is the probability that he is a center?

h. Are the events {White player} and {Center} independent?

i. Use your answers to parts **f–h** to make an inference about stacking in the NBA.

3.135 Shooting free throws. In college basketball games, a player may be afforded the opportunity to shoot two consecutive foul shots (free throws).

a. Suppose a player who makes (i.e., scores on) 80% of his foul shots has been awarded two free throws. If the two throws are considered independent, what is the probability that the player makes both shots? exactly one? neither shot?

b. Suppose a player who makes 80% of his first attempted foul shots has been awarded two free throws and the outcome on the second shot is dependent on the outcome of the first shot. In fact, if this player makes the first shot, he makes 90% of the second shots; and if he misses the first shot, he makes 70% of the second shots. In this case, what is the probability that the player makes both shots? exactly one? neither shot?

c. In parts **a** and **b**, we considered two ways of *modeling* the probability that a basketball player makes two consecutive foul shots. Which model do you think gives a more realistic explanation of the outcome of shooting foul shots; that is, do you think two consecutive foul shots are independent or dependent? Explain.

3.136 Elderly wheelchair user study. The *American Journal of Public Health* (Jan. 2002) reported on a study of elderly wheelchair users who live at home. A sample of 306 wheelchair users, age 65 or older, were surveyed about whether they had an injurious fall during the year and whether their home featured any one of five structural modifications: bathroom modifications, widened doorways/hallways, kitchen modifications, installed railings, and easy-open doors. The responses are summarized in the accompanying table. Suppose we select, at random, one of the 306 wheelchair users surveyed.

Home Features	Injurious Fall(s)	No Falls	Totals
All 5	2	7	9
At least 1, but not all	26	162	188
None	20	89	109
Totals	48	258	306

Based on Berg, K., Hines, M., and Allen, S. "Wheelchair users at home: Few home modifications and many injurious falls." *American Journal of Public Health*, Vol. 92, No. 1, Jan. 2002 (Table 1).

a. Find the probability that the wheelchair user had an injurious fall.

b. Find the probability that the wheelchair user had all five features installed in the home.

c. Find the probability that the wheelchair user had no falls and none of the features installed in the home.

d. Given no features installed in the home, find the probability of an injurious fall.

3.137 Maize seeds. The genetic origin and properties of maize (modern-day corn) were investigated in *Economic Botany*. Seeds from maize ears carry either single spikelets or paired spikelets, but not both. Progeny tests on approximately 600 maize ears revealed the following information: Forty percent of all seeds carry single spikelets, while 60% carry paired spikelets. A seed with single spikelets will produce maize ears with single spikelets 29% of the time and paired spikelets 71% of the time. A seed with paired spikelets will produce maize ears with single spikelets 26% of the time and paired spikelets 74% of the time.

a. Find the probability that a randomly selected maize ear seed carries a single spikelet and produces ears with single spikelets.

b. Find the probability that a randomly selected maize ear seed produces ears with paired spikelets.

3.138 Lie detector test. Consider a lie detector called the Computerized Voice Stress Analyzer (CVSA). The manufacturer claims that the CVSA is 98% accurate and, unlike a polygraph machine, will not be thrown off by drugs and medical factors. However, laboratory studies by the U.S. Defense Department found that the CVSA had an accuracy rate of 49.8%, slightly less than pure chance. Suppose the CVSA is used to test the veracity of four suspects. Assume that the suspects' responses are independent.

a. If the manufacturer's claim is true, what is the probability that the CVSA will correctly determine the veracity of all four suspects?

b. If the manufacturer's claim is true, what is the probability that the CVSA will yield an incorrect result for at least one of the four suspects?

c. Suppose that in a laboratory experiment conducted by the U.S. Defense Department on four suspects, the CVSA yielded incorrect results for two of the suspects. Make an inference about the true accuracy rate of the new lie detector.

3.139 Series and parallel systems. Consider the two systems shown in the schematic below. System A operates properly only if all three components operate properly. (The three components are said to operate *in series*.) The probability of failure for system A components 1, 2, and 3 is .12, .09, and .11, respectively. Assume that the components operate independently of each other. System B comprises two subsystems said to operate *in parallel*. Each subsystem has two components that operate in series. System B will operate properly as long as at least one of the subsystems functions

properly. The probability of failure for each component in the system is .1. Assume that the components operate independently of each other.

a. Find the probability that System A operates properly.

b. What is the probability that at least one of the components in System A will fail and therefore that the system will fail?

c. Find the probability that System B operates properly.

d. Find the probability that exactly one subsystem in System B fails.

e. Find the probability that System B fails to operate properly.

f. How many parallel subsystems like the two shown here would be required to guarantee that the system would operate properly at least 99% of the time?

3.140 Monitoring the quality of power equipment. *Mechanical Engineering* (Feb. 2005) reported on the need for wireless networks to monitor the quality of industrial equipment. For example, consider Eaton Corp., a company that develops distribution products. Eaton estimates that 90% of the electrical switching devices it sells can monitor the quality of the power running through the device. Eaton further estimates that, of the buyers of electrical switching devices capable of monitoring quality, 90% do not wire the equipment up for that purpose. Use this information to estimate the probability that an Eaton electrical switching device is capable of monitoring power quality and is wired up for that purpose.

a. It is known that the proportions of defective microchips produced by the seven suppliers are .001, .0003, .0007, .006, .0002, .0002, and .001, respectively. If a single PC microchip failure is observed, which supplier is most likely responsible?

b. Suppose the seven suppliers produce defective microchips at the same rate, .0005. If a single PC microchip failure is observed, which supplier is most likely responsible?

3.141 Cigar smoking and cancer. The *Journal of the National Cancer Institute* (Feb. 16, 2000) published the results of a study that investigated the association between cigar smoking and death from tobacco-related cancers. Data were obtained for a national sample of 137,243 American men. The results are summarized in the table on page 158. Each male in the study was classified according to his cigar-smoking status and whether or not he died from a tobacco-related cancer.

a. Find the probability that a randomly selected man never smoked cigars and died from cancer.

System A: Three Components in Series

Input — #1 — #2 — #3 — Output

System B: Two Parallel Subsystems

Subsystem C

#1 ——— #2

Input

#3 ——— #4

Subsystem D

Output

b. Find the probability that a randomly selected man was a former cigar smoker and died from cancer.

c. Find the probability that a randomly selected man was a current cigar smoker and died from cancer.

d. Given that a male was a current cigar smoker, find the probability that he died from cancer.

e. Given that a male never smoked cigars, find the probability that he died from cancer.

Cigars	Died from Cancer		
	Yes	No	Totals
Never Smoked	782	120,747	121,529
Former Smoker	91	7,757	7,848
Current Smoker	141	7,725	7,866
Totals	1,014	136,229	137,243

Based on Shapiro, J. A., Jacobs, E. J., and Thun, M. J. "Cigar smoking in men and risk of death from tobacco-related cancers." *Journal of the National Cancer Institute*, Vol. 92, No. 4, Feb. 16, 2000 (Table 2).

Applying the Concepts—Advanced

3.142 Sex composition patterns of children in families. In having children, is there a genetic factor that causes some families to favor one sex over the other? That is, does having boys or girls "run in the family"? This was the question of interest in *Chance* (Fall 2001). Using data collected on children's sex for over 4,000 American families that had at least two children, the researchers compiled the accompanying table. Make an inference about whether having boys or girls "runs in the family."

Sex Composition of First Two Children	Frequency
Boy–Boy	1,085
Boy–Girl	1,086
Girl–Boy	1,111
Girl–Girl	926
Total	4,208

Based on Rodgers, J. L., and Doughty, D. "Does having boys or girls run in the family?" *Chance*, Vol. 14, No. 4, Fall 2001 (Table 3).

3.143 Odds of winning a horse race. Handicappers for horse races express their beliefs about the probability of each horse winning a race in terms of **odds.** If the probability of event E is $P(E)$, then the *odds in favor of E* are $P(E)$ to $1 - P(E)$. Thus, if a handicapper assesses a probability of .25 that Smarty Jones will win the Belmont Stakes, the odds in favor of Smarty Jones are $^{25}/_{100}$ to $^{75}/_{100}$, or 1 to 3. It follows that the *odds against E* are $1 - P(E)$ to $P(E)$, or 3 to 1 against a win by Smarty Jones. In general, if the odds in favor of event E are a to b, then $P(E) = a/(a + b)$.

a. A second handicapper assesses the probability of a win by Smarty Jones to be $^1/_3$. According to the second handicapper, what are the odds in favor of a Smarty Jones win?

b. A third handicapper assesses the odds in favor of Smarty Jones to be 1 to 1. According to the third handicapper, what is the probability of a Smarty Jones win?

c. A fourth handicapper assesses the odds against Smarty Jones winning to be 3 to 2. Find this handicapper's assessment of the probability that Smarty Jones will win.

3.144 Chance of winning at blackjack. Blackjack, a favorite game of gamblers, is played by a dealer and at least one opponent. At the outset of the game, 2 cards of a 52-card bridge deck are dealt to the player and 2 cards to the dealer. Drawing an ace and a face card is called *blackjack*. If the dealer does not draw a blackjack and the player does, the player wins. If both the dealer and player draw blackjack, a "push" (i.e., a tie) occurs.

a. What is the probability that the dealer will draw a blackjack?

b. What is the probability that the player wins with a blackjack?

3.145 Finding an organ transplant match. One of the problems encountered with organ transplants is the body's rejection of the transplanted tissue. If the antigens attached to the tissue cells of the donor and receiver match, the body will accept the transplanted tissue. Although the antigens in identical twins always match, the probability of a match in other siblings is .25, and that of a match in two people from the population at large is .001. Suppose you need a kidney and you have two brothers and a sister.

a. If one of your three siblings offers a kidney, what is the probability that the antigens will match?

b. If all three siblings offer a kidney, what is the probability that all three antigens will match?

c. If all three siblings offer a kidney, what is the probability that none of the antigens will match?

d. Repeat parts **b** and **c**, this time assuming that the three donors were obtained from the population at large.

3.146 Language impairment in children. Children who develop unexpected difficulties with the spoken language are often diagnosed as *specifically language impaired* (SLI). A study published in the *Journal of Speech, Language, and Hearing Research* (Dec. 1997) investigated the incidence of SLI in kindergarten children. As an initial screen, each in a national sample of over 7,000 children was given a test for language performance. The percentages of children who passed and failed the screen were 73.8% and 26.2%, respectively. All children who failed the screen were tested clinically for SLI. About one-third of those who passed the screen were randomly selected and also tested for SLI. The percentage of children diagnosed with SLI in the "failed screen" group was 20.5%; the percentage diagnosed with SLI in the "pass screen" group was 2.8%.

a. For this problem, let "pass" represent a child who passed the language performance screen, "fail" represent a child who failed the screen, and "SLI" represent a child diagnosed with SLI. Now find each of the following probabilities: $P(\text{Pass})$, $P(\text{Fail})$, $P(\text{SLI}|\text{Pass})$, and $P(\text{SLI}|\text{Fail})$.

b. Use the probabilities from part **a** to find $P(\text{Pass} \cap \text{SLI})$ and $P(\text{Fail} \cap \text{SLI})$. What probability law did you use to calculate these probabilities?

c. Use the probabilities from part **b** to find $P(\text{SLI})$. What probability law did you use to calculate this probability?

3.147 Accuracy of pregnancy tests. Seventy-five percent of all women who submit to pregnancy tests are really pregnant. A certain pregnancy test gives a *false positive* result with probability .02 and a *valid positive result* with probability .99. If a particular woman's test is positive, what is the probability that she really is pregnant? [*Hint:* If A is the event that a woman is pregnant and B is the event that the pregnancy test is positive, then B is the union of the two mutually exclusive events $A \cap B$ and $A^c \cap B$. Also, the probability of a false positive result may be written as $P(B \mid A^c) = .02$.]

3.148 Chance of winning at "craps." A version of the dice game "craps" is played in the following manner. A player starts by rolling two balanced dice. If the roll (the sum of the two numbers showing on the dice) results in a 7 or 11, the player wins. If the roll results in a 2 or a 3 (called *craps*), the player loses. For any other roll outcome, the player continues to throw the dice until the original roll outcome recurs (in which case the player wins) or until a 7 occurs (in which case the player loses).

  a. What is the probability that a player wins the game on the first roll of the dice?

  b. What is the probability that a player loses the game on the first roll of the dice?

  c. If the player throws a total of 4 on the first roll, what is the probability that the game ends (win or lose) on the next roll?

3.149 Odd Man Out. Three people play a game called "Odd Man Out." In this game, each player flips a fair coin until the outcome (heads or tails) for one of the players is not the same as that for the other two players. This player is then "the odd man out" and loses the game. Find the probability that the game ends (i.e., either exactly one of the coins will fall heads or exactly one of the coins will fall tails) after only one toss by each player. Suppose one of the players, hoping to reduce his chances of being the odd man out, uses a two-headed coin. Will this ploy be successful? Solve by listing the sample points in the sample space.

Critical Thinking Challenges

3.150 Flawed Pentium computer chip. In October 1994, a flaw was discovered in the Pentium microchip installed in personal computers. The chip produced an incorrect result when dividing two numbers. Intel, the manufacturer of the Pentium chip, initially announced that such an error would occur once in 9 billion division operations, or "once every

27,000 years," for a typical user; consequently, Intel did not immediately offer to replace the chip.

  Depending on the procedure, statistical software packages (e.g., SAS) may perform an extremely large number of divisions to produce required output. For heavy users of the software, 1 billion divisions over a short time frame is not unusual. Will the flawed chip be a problem for a heavy SAS user? [*Note:* Two months after the flaw was discovered, Intel agreed to replace all Pentium chips free of charge.]

3.151 "Let's Make a Deal." Marilyn vos Savant, who is listed in the *Guinness Book of World Records Hall of Fame* as having the "Highest IQ," writes a weekly column in the Sunday newspaper supplement *Parade Magazine*. Her column, "Ask Marilyn," is devoted to games of skill, puzzles, and mind-bending riddles. In one issue (*Parade Magazine*, Feb. 24, 1991), vos Savant posed the following question:

  Suppose you're on a game show and you're given a choice of three doors. Behind one door is a car; behind the others, goats. You pick a door—say, #1—and the host, who knows what's behind the doors, opens another door—say #3—which has a goat. He then says to you, "Do you want to pick door #2?" Is it to your advantage to switch your choice?

  Marilyn's answer: "Yes, you should switch.The first door has a $\frac{1}{3}$ chance of winning [the car], but the second has a $\frac{2}{3}$ chance [of winning the car]." Predictably, vos Savant's surprising answer elicited thousands of critical letters, many of them from Ph.D. mathematicians, that disagreed with her. Who is correct, the Ph.D.s or Marilyn?

3.152 Most likely coin-toss sequence. In *Parade Magazine*'s (Nov. 26, 2000) column "Ask Marilyn," the following question was posed: "I have just tossed a [balanced] coin 10 times, and I ask you to guess which of the following three sequences was the result. One (and only one) of the sequences is genuine."

  1. H H H H H H H H H H

  2. H H T T H T T H H H

  3. T T T T T T T T T T

  Marilyn's answer to the question posed was "Though the chances of the three specific sequences occurring randomly are equal … it's reasonable for us to choose sequence (2) as the most likely genuine result." Do you agree?

| **Activity** | **Exit Polls** |

Exit polls are conducted in selected locations as voters leave their polling places after voting. In addition to being used to predict the outcome of elections before the votes are counted, these polls seek to gauge tendencies among voters. The results are usually stated in terms of conditional probabilities.

  The accompanying table shows the results of exit polling which suggest that men were almost evenly split on voting for John McCain or Barack Obama, while women were more likely to vote for Obama in the 2008 presidential election. In addition, the table suggests that more women than men voted in the election. The six percentages in the last three columns represent conditional probabilities, where the given event is gender.

2008 Presidential Election, Vote by Gender

	Obama	McCain	Other
Male (47%)	49%	48%	3%
Female (53%)	56%	43%	1%

Based on CNN.com

1. Find similar exit poll results in which voters are categorized by race, income, education, or some other criterion for a recent national, state, or local election. Choose two different examples, and interpret the percentages given as probabilities, or conditional probabilities where appropriate.

2. Use the multiplicative rule of probability to find the probabilities related to the percentages given. [For example, in the accompanying table, find P(Obama and Male).] Then interpret each of these probabilities and use them to determine the total percentage of the electorate that voted for each candidate.

3. Describe a situation in which a political group might use a form of exit polling to gauge voters' opinions on a "hot- button" topic (e.g., global warming). Identify the political group, the "hot-button" topic, the criterion used to categorize voters, and how the voters' opinions will be determined. Then describe how the results will be summarized as conditional probabilities. How might the results of the poll be used to support a particular agenda?

References

Bennett, D. J. *Randomness*. Cambridge, MA: Harvard University Press, 1998.

Epstein, R. A. *The Theory of Gambling and Statistical Logic*, rev. ed. New York: Academic Press, 1977.

Feller, W. *An Introduction to Probability Theory and Its Applications*, 3rd ed., Vol. 1. New York: Wiley, 1968.

Lindley, D. V. *Making Decisions*, 2nd ed. London: Wiley, 1985.

Parzen, E. *Modern Probability Theory and Its Applications*. New York: Wiley, 1960.

Wackerly, D., Mendenhall, W., and Scheaffer, R. L. *Mathematical Statistics with Applications*, 7th ed. Boston: Duxbury, 2008.

Williams, B. *A Sampler on Sampling*. New York: Wiley, 1978.

Winkler, R. L. *An Introduction to Bayesian Inference and Decision*. New York: Holt, Rinehart and Winston, 1972.

Wright, G., and Ayton, P., eds. *Subjective Probability*. New York: Wiley, 1994.

USING TECHNOLOGY

MINITAB: Generating a Random Sample

Step 1 Click on the "Calc" button on the MINITAB menu bar, then click on "Random Data," and finally, click on "Sample From Columns," as shown in Figure 3.M.1. The resulting dialog box appears as shown in Figure 3.M.2.

Figure 3.M.1
MINITAB menu options for sampling from a data set

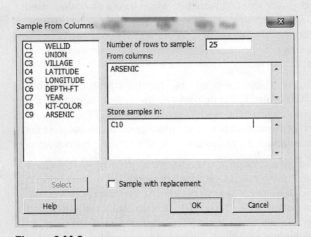

Figure 3.M.2
MINITAB options for selecting a random sample from worksheet columns

Step 2 Specify the sample size (i.e., number of rows), the variable(s) to be sampled, and the column(s) where you want to save the sample.

Step 3 Click "OK" and the MINITAB, worksheet will reappear with the values of the variable for the selected (sampled) cases in the column specified.

In MINITAB, you can also generate a sample of case numbers.

Step 1 From the MINITAB menu, click on the "Calc" button, and then click on "Random Data," and finally, click on the "Uniform" option (see Figure 3.M.1).

Step 2 In the resulting dialog box (shown in Figure 3.M.3), specify the number of cases (rows, i.e., the sample size), and the column where the case numbers selected will be stored.

Figure 3.M.3
MINITAB options for selecting a random sample of cases

Step 3 Click "OK" and the MINITAB worksheet will reappear with the case numbers for the selected (sampled) cases in the column specified.

[*Note:* If you want the option of generating the same (identical) sample multiple times from the data set, then first click on the "Set Base" option shown in Figure 3.M.1. Specify an integer in the resulting dialog box. If you always select the same integer,

MINITAB will select the same sample when you choose the random sampling options.]

TI-83/TI-84 Plus Graphing Calculator: Combinations and Permutations

Combinations—Choosing n Elements from N Elements

Step 1 Enter the value of N.

Step 2 Press **MATH**, then select **PRB**.

Step 3 Press **nCr** for combinations.

Step 4 Enter the value of n, then press **ENTER.**

Permutations—Choosing n Elements from N Elements

Step 1 Enter the value of N.

Step 2 Press **MATH.** Then select **PRB.**

Step 3 Press **nPr** for permutations.

Step 4 Enter the value of n, then press **ENTER.**

4 Random Variables and Probability Distributions

CONTENTS

Where We've Been

- Used probability to make an inference about a population from data in an observed sample
- Used probability to measure the reliability of the inference

Where We're Going

- Develop the notion of a random variable (4.1)
- Learn that many types of numerical data are observed values of discrete or continuous random variables (4.2, 4.4)
- Study two important types of random variables and their probability models: the binomial and the normal model (4.3, 4.5–4.7)
- Define a sampling distribution (4.8)
- Learn that the sampling distribution of $\bar{x}$ follows a normal model (4.9)

Statistics IN Action   Super Weapons Development–Is the Hit Ratio Optimized?

The U.S. Army is working with a major defense contractor to develop a "super" weapon. The weapon is designed to fire a large number of sharp tungsten bullets–called flechettes–with a single shot that will destroy a large number of enemy soldiers. Flechettes are about the size of an average nail, with small fins at one end to stabilize them in flight. Since World War I, when France dropped them in large quantities from aircraft on masses of ground troops, munitions experts have experimented with using flechettes in a variety of guns. The problem with using flechettes as ammunition is accuracy: Current weapons that fire large quantities of flechettes have unsatisfactory hit ratios when fired at long distances.

The defense contractor (not named here for both confidentiality and security reasons) has developed a prototype gun that fires 1,100 flechettes with a single round. In range tests, three 2-feet-wide targets were set up a distance of 500 meters (approximately 1,500 feet) from the weapon. With a number line used as a reference, the centers of the three targets were at 0, 5, and 10 feet, respectively, as shown in Figure SIA4.1. The prototype gun was aimed at the middle target (center at 5 feet) and fired once. The point where each of the 1,100 flechettes landed at the 500-meter distance was measured with the use of a horizontal and vertical grid. For the purposes of this application, only the horizontal measurements are considered. These 1,100 measurements are saved in the **MOAGUN** file. (The data are simulated for confidentiality reasons.) For example, a flechette with a value of $x = 5.5$ hit the middle target, but a flechette with a value of $x = 2.0$ did not hit any of the three targets. (See Figure SIA4.1.)

The defense contractor is interested in the likelihood of any one of the targets being hit by a flechette and, in particular, wants to set the gun specifications to maximize the number of hits. The weapon is designed to have a mean horizontal value equal to the aim point (e.g., $\mu = 5$ feet when aimed at the center target). By changing specifications, the contractor can vary the standard deviation σ. The **MOAGUN** file contains flechette measurements for three different range tests: one with a standard deviation of $\sigma = 1$ foot, one with $\sigma = 2$ feet, and one with $\sigma = 4$ feet.

In this chapter, two Statistics in Action Revisited examples demonstrate how we can use one of the probability models discussed in the chapter–the normal probability distribution–to aid the defense contractor in developing its "super" weapon.

Statistics IN Action Revisited

- Using the Normal Model to Maximize the Probability of a Hit with the Super Weapon (p. 198)
- Assessing whether the Normal Distribution Is Appropriate for Modeling the Super Weapon Hit Data (p. 205)

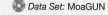 Data Set: MoaGUN

Figure SIA4.1

Target placement on gun range

You may have noticed that many of the examples of experiments in Chapter 3 generated quantitative (numerical) observations. The unemployment rate, the percentage of voters favoring a particular candidate, the cost of textbooks for a school term, and the amount of pesticide in the discharge waters of a chemical plant are all examples of numerical measurements of some phenomenon. Thus, most experiments have sample points that correspond to values of some numerical variable.

To illustrate, consider the coin-tossing experiment of Chapter 3. Figure 4.1 is a Venn diagram showing the sample points when two coins are tossed and the up faces (heads or tails) of the coins are observed. One possible numerical outcome is the total number of heads observed. This value (0, 1, or 2) is shown in parentheses on the Venn diagram, with one numerical value associated with each sample point. In the jargon of probability, the variable "total number of heads observed in two tosses of a coin" is called a *random variable*.

Figure 4.1

Venn diagram for coin-tossing experiment

> A **random variable** is a variable that assumes numerical values associated with the random outcomes of an experiment, where one (and only one) numerical value is assigned to each sample point.

The term *random variable* is more meaningful than the term *variable* alone, because the adjective *random* indicates that the coin-tossing experiment may result in one of the several possible values of the variable–0, 1, and 2–according to the *random* outcomes of the experiment: *HH, HT, TH,* and *TT.* Similarly, if the experiment is to count the number of customers who use the drive-up window of a bank each day, the random variable (the number of customers) will vary from day to day, partly because of the random phenomena that influence whether customers use the window. Thus, the possible values of this random variable range from 0 to the maximum number of customers the window can serve in a day.

We define two different types of random variables, *discrete* and *continuous*, in Section 4.1. Then we spend the remainder of the chapter discussing one specific type of discrete random variable and one specific type of continuous random variable.

4.1 Two Types of Random Variables

Recall that the sample-point probabilities corresponding to an experiment must sum to 1. Dividing one unit of probability among the sample points in a sample space and, consequently, assigning probabilities to the values of a random variable is not always as easy as the examples in Chapter 3 might lead you to believe. If the number of sample points can be completely listed, the job is straightforward. But if the experiment results in an infinite number of sample points that are impossible to list, the task of assigning probabilities to the sample points is impossible without the aid of a probability model. The next three examples demonstrate the need for different probability models, depending on the number of values that a random variable can assume.

Example 4.1
Values of a Discrete Random Variable–Wine Ratings

Problem A panel of 10 experts for the *Wine Spectator* (a national publication) is asked to taste a new white wine and assign it a rating of 0, 1, 2, or 3. A score is then obtained by adding together the ratings of the 10 experts. How many values can this random variable assume?

Solution A sample point is a sequence of 10 numbers associated with the rating of each expert. For example, one sample point is

$$\{1, 0, 0, 1, 2, 0, 0, 3, 1, 0\}$$

The random variable assigns a score to each one of these sample points by adding the 10 numbers together. Thus, the smallest score is 0 (if all 10 ratings are 0), and the largest score is 30 (if all 10 ratings are 3). Since every integer between 0 and 30 is a possible score, the random variable denoted by the symbol x can assume 31 values. Note that the value of the random variable for the sample point shown here is $x = 8$.[*]

Look Back This is an example of a *discrete random variable*, since there is a finite number of distinct possible values. Whenever all the possible values a random variable can assume can be listed (or *counted*), the random variable is *discrete*.

Example 4.2
Values of a Discrete Random Variable–EPA Application

Problem Suppose the Environmental Protection Agency (EPA) takes readings once a month on the amount of pesticide in the discharge water of a chemical company. If the amount of pesticide exceeds the maximum level set by the EPA, the company is forced to take corrective action and may be subject to penalty. Consider the random variable number x of months before the company's discharge exceeds the EPA's maximum level. What values can x assume?

---

[*]The standard mathematical convention is to use a capital letter (e.g., X) to denote the theoretical random variable. The possible values (or realizations) of the random variable are typically denoted with a lowercase letter (e.g., x). Thus, in Example 4.1, the random variable X can take on the values $x = 0, 1, 2, \ldots, 30$. Since this notation can be confusing for introductory statistics students, we simplify the notation by using the lowercase x to represent the random variable throughout.

Solution The company's discharge of pesticide may exceed the maximum allowable level on the first month of testing, the second month of testing, etc. It is possible that the company's discharge will *never* exceed the maximum level. Thus, the set of possible values for the number of months until the level is first exceeded is the set of all positive integers $1, 2, 3, 4, \ldots$.

Look Back If we can list the values of a random variable x, even though the list is never ending, we call the list **countable** and the corresponding random variable **discrete.** Thus, the number of months until the company's discharge first exceeds the limit is a **discrete random variable.**

Now Work Exercise 4.7

Example 4.3

Values of a Continuous Random Variable–Another EPA Application

Problem Refer to Example 4.2. A second random variable of interest is the amount x of pesticide (in milligrams per liter) found in the monthly sample of discharge waters from the same chemical company. What values can this random variable assume?

Solution Some possible values of x are 1.7, 28.42, and 100.987 milligrams per liter. Unlike the *number* of months before the company's discharge exceeds the EPA's maximum level, the set of all possible values for the *amount* of discharge *cannot* be listed (i.e., is not countable). The possible values for the amount x of pesticide would correspond to the points on the interval between 0 and the largest possible value the amount of the discharge could attain, the maximum number of milligrams that could occupy 1 liter of volume. (Practically, the interval would be much smaller, say, between 0 and 500 milligrams per liter.)

Look Ahead When the values of a random variable are not countable, but instead correspond to the points on some interval, we call the variable a **continuous random variable.** Thus, the *amount* of pesticide in the chemical plant's discharge waters is a *continuous random variable.*

Now Work Exercise 4.9

Random variables that can assume a *countable* number of values are called **discrete.**

Random variables that can assume values corresponding to any of the points contained in an interval are called **continuous.**

The following are examples of discrete random variables:

1. The number of seizures an epileptic patient has in a given week:
   $x = 0, 1, 2, \ldots$

2. The number of voters in a sample of 500 who favor impeachment of the president: $x = 0, 1, 2, \ldots, 500$

3. The shoe size of a tennis player: $x = \ldots \; 5, 5\frac{1}{2}, 6, 6\frac{1}{2}, 7, 7\frac{1}{2} \ldots$

4. The change received for paying a bill: $x = 1\text{¢}, 2\text{¢}, 3\text{¢}, \ldots, \$1, \$1.01, \$1.02, \ldots$

5. The number of customers waiting to be served in a restaurant at a particular time: $x = 0, 1, 2, \ldots$

Note that several of the examples of discrete random variables begin with the words *The number of. …* This wording is very common, since the discrete random variables most frequently observed are counts. The following are examples of continuous random variables:

1. The length of time (in seconds) between arrivals at a hospital clinic:
   $0 \leq x \leq \infty$ (infinity)

2. The length of time (in minutes) it takes a student to complete a one-hour exam:
   $0 \leq x \leq 60$

3. The amount (in ounces) of carbonated beverage loaded into a 12-ounce can in a can-filling operation: $0 \leq x \leq 12$

4. The depth (in feet) at which a successful oil-drilling venture first strikes oil: $0 \leq x \leq c$, where c is the maximum depth obtainable

5. The weight (in pounds) of a food item bought in a supermarket: $0 \leq x \leq 500$ [*Note:* Theoretically, there is no upper limit on x, but it is unlikely that it would exceed 500 pounds.]

Discrete random variables and their probability distributions are discussed in Section 4.2. Continuous random variables and their probability distributions are the topic of Section 4.4.

Exercises 4.1–4.14

Understanding the Principles

4.1 What is a random variable?

4.2 How do discrete and continuous random variables differ?

Applying the Concepts–Basic

4.3 **Type of Random Variable.** Classify the following random variables according to whether they are discrete or continuous:
a. The number of words spelled correctly by a student on a spelling test
b. The amount of water flowing through the Hoover Dam in a day
c. The length of time an employee is late for work
d. The number of bacteria in a particular cubic centimeter of drinking water
e. The amount of carbon monoxide produced per gallon of unleaded gas
f. Your weight

4.4 **Type of Random Variable.** Identify the following random variables as discrete or continuous:
a. The amount of flu vaccine in a syringe
b. The heart rate (number of beats per minute) of an American male
c. The time it takes a student to complete an examination
d. The barometric pressure at a given location
e. The number of registered voters who vote in a national election
f. Your score on the either the SAT or ACT

4.5 **Type of Random Variable.** Identify the following variables as discrete or continuous:
a. The difference in reaction time to the same stimulus before and after training
b. The number of violent crimes committed per month in your community
c. The number of commercial aircraft near-misses per month
d. The number of winners each week in a state lottery
e. The number of free throws made per game by a basketball team
f. The distance traveled by a school bus each day

4.6 **NHTSA crash tests.** The National Highway Traffic Safety Administration (NHTSA) has developed a driver-side "star" scoring system for crash-testing new cars. Each crash-tested car is given a rating ranging from one star (*) to five stars (*****): the more stars in the rating, the better is the level of crash protection in a head-on collision. Suppose that a car is selected and its driver-side star rating is determined. Let x equal the number of stars in the rating. Is x a discrete or continuous random variable?

4.7 **Customers in line at a Subway shop.** The number of customers, x, waiting in line to order sandwiches at a Subway shop at noon is of interest to the store manager. What values can x assume? Is x a discrete or continuous random variable?

4.8 **Sound waves from a basketball.** Refer to the *American Journal of Physics* (June 2010) experiment on sound waves produced from striking a basketball, Exercise 2.37 (p. 47). Recall that the frequencies of sound wave echoes resulting from striking a hanging basketball with a metal rod were recorded. Classify the random variable, frequency (measured in hertz) of an echo, as discrete or continuous.

4.9 **Mongolian desert ants.** Refer to the *Journal of Biogeography* (Dec. 2003) study of ants in Mongolia, presented in Exercise 2.68 (p. 59). Two of the several variables recorded at each of 11 study sites were annual rainfall (in millimeters) and number of ant species. Identify these variables as discrete or continuous.

4.10 **Motivation of drug dealers.** Refer to the *Applied Psychology in Criminal Justice* (Sept. 2009) study of the personality characteristics of drug dealers, Exercise 2.98 (p. 71). For each of 100 convicted drug dealers, the researchers measured several variables, including the number of prior felony arrests x. Is x a discrete or continuous random variable? Explain.

Applying the Concepts–Intermediate

4.11 **Psychology.** Give an example of a discrete random variable of interest to a psychologist.

4.12 **Sociology.** Give an example of a discrete random variable of interest to a sociologist.

4.13 **Nursing.** Give an example of a discrete random variable of interest to a hospital nurse.

4.14 **Art history.** Give an example of a discrete random variable of interest to an art historian.

4.2 Probability Distributions for Discrete Random Variables

A complete description of a discrete random variable requires that we *specify all the values the random variable can assume* and *the probability associated with each value.* To illustrate, consider Example 4.4.

Example 4.4

Finding a Probability Distribution— Coin-Tossing Experiment

Figure 4.2
Venn diagram for the two-coin-toss experiment

Problem Recall the experiment of tossing two coins (Example 3.1, p. 109), and let *x* be the number of heads observed. Find the probability associated with each value of the random variable *x*, assuming that the two coins are fair.

Solution The sample space and sample points for this experiment are reproduced in Figure 4.2. Note that the random variable *x* can assume values 0, 1, 2. Recall (from Chapter 3) that the probability associated with each of the four sample points is $\frac{1}{4}$. Then, identifying the probabilities of the sample points associated with each of these values of *x*, we have

$$P(x = 0) = P(TT) = \frac{1}{4}$$

$$P(x = 1) = P(TH) + P(HT) = \frac{1}{4} + \frac{1}{4} = \frac{1}{2}$$

$$P(x = 2) = P(HH) = \frac{1}{4}$$

Thus, we now know the values the random variable can assume (0, 1, 2) and how the probability is *distributed over* those values ($\frac{1}{4}$, $\frac{1}{2}$, $\frac{1}{4}$). This dual specification completely describes the random variable and is referred to as the *probability distribution*, denoted by the symbol $p(x)$.* The probability distribution for the coin-toss example is shown in tabular form in Table 4.1 and in graphic form in Figure 4.3 Since the probability distribution for a discrete random variable is concentrated at specific points (values of *x*), the graph in Figure 4.3a represents the probabilities as the heights of vertical lines over the corresponding values of *x*. Although the representation of the probability distribution as a histogram, as in Figure 4.3b, is less precise (since the probability is spread over a unit interval), the histogram representation will prove useful when we approximate probabilities of certain discrete random variables in Section 4.4.

a. Point representation of $p(x)$　　b. Histogram representation of $p(x)$

Figure 4.3
Probability distribution for coin-toss experiment: graphical form

Table 4.1	
Probability Distribution for Coin-Toss Experiment: Tabular Form	
x	p(x)
0	$\frac{1}{4}$
1	$\frac{1}{2}$
2	$\frac{1}{4}$

Look Ahead We could also present the probability distribution for *x* as a formula, but this would unnecessarily complicate a very simple example. We give the formulas for the probability distributions of some common discrete random variables later in the chapter.

Now Work Exercise 4.22

*In standard mathematical notation, the probability that a random variable *X* takes on a value *x* is denoted $P(X = x) = p(x)$ Thus, $P(X = 0) = p(0)$ $P(X = 1) = p(1)$ etc. In this text, we adopt the simpler $p(x)$ notation.

> The **probability distribution of a discrete random variable** is a graph, table, or formula that specifies the probability associated with each possible value that the random variable can assume.

Two requirements must be satisfied by all probability distributions for discrete random variables:

> **Requirements for the Probability Distribution of a Discrete Random Variable x**
> 1. $p(x) \geq 0$ for all values of x.
> 2. $\sum p(x) = 1$
>
> where the summation of $p(x)$ is over all possible values of x.[*]

Example 4.5
Probability Distribution from a Graph–Playing Craps

Problem Craps is a popular casino game in which a player throws two dice and bets on the outcome (the sum total of the dots showing on the upper faces of the two dice). Consider a $5 wager. On the first toss (called the *come-out* roll), if the total is 7 or 11 the roller wins $5. If the outcome is a 2, 3, or 12, the roller loses $5 (i.e., the roller wins − $5). For any other outcome (4, 5, 6, 8, 9, or 10) a *point* is established and no money is lost or won on that roll (i.e., the roller wins $0). In a computer simulation of repeated tosses of two dice, the outcome x of the come-out roll wager (−$5, $0, or +$5) was recorded. A relative frequency histogram summarizing the results is shown in Figure 4.4. Use the histogram to find the approximate probability distribution of x.

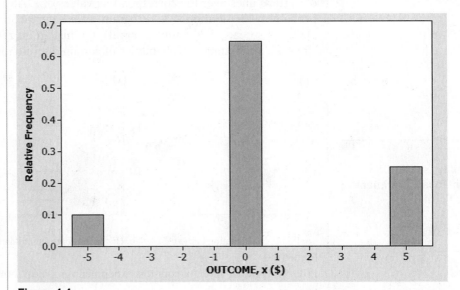

Figure 4.4
MINITAB Histogram for $5 Wager on Come-Out Roll in Craps

Solution The histogram shows that the relative frequencies of the outcomes $x = -\$5$, $x = \$0$, and $x = \$5$ are .1, .65, and .25, respectively. For example, in repeated tosses of two dice, 25% of the outcomes resulted in a sum of 7 or 11 (a $5 win for

[*]Unless otherwise indicated, summations will always be over all possible values of x.

the roller). Based on our long-run definition of probability given in Chapter 3, these relative frequencies estimate the probabilities of the three outcomes. Consequently, the approximate probability distribution of x, the outcome of the come-out wager in craps, is $p(-\$5) = .1, p(\$0) = .65$, and $p(\$5) = .25$. Note that these probabilities sum to 1.

Look Back When two dice are tossed, there is a total of 36 possible outcomes. (Can you list these outcomes, or sample points?) Of these, 4 result in a sum of 2, 3, or 12; 24 result in a sum of 4, 5, 6, 8, 9, or 10; and 8 result in a sum of 7 or 11. Using the rules of probability established in Chapter 3, you can show that the actual probability distribution for x is $p(-\$5) = 4/36 = .1111, p(\$0) = 24/36 = .6667$, and $p(\$5) = 8/36 = .2222$.

Now Work Exercise 4.19

    Examples 4.4 and 4.5 illustrate how the probability distribution for a discrete random variable can be derived, but for many practical situations the task is much more difficult. Fortunately, numerous experiments and associated discrete random variables observed in nature possess identical characteristics. Thus, you might observe a random variable in a psychology experiment that would possess the same probability distribution as a random variable observed in an engineering experiment or a social sample survey. We classify random variables according to type of experiment, derive the probability distribution for each of the different types, and then use the appropriate probability distribution when a particular type of random variable is observed in a practical situation. The probability distributions for most commonly occurring discrete random variables have already been derived. This fact simplifies the problem of finding the probability distributions for random variables, as the next example illustrates.

Example 4.6
Probability Distribution Using a Formula–Texas Droughts

Problem A drought is a period of abnormal dry weather that causes serious problems in the farming industry of the region. University of Arizona researchers used historical annual data to study the severity of droughts in Texas (*Journal of Hydrologic Engineering*, Sept./Oct. 2003). The researchers showed that the distribution of x, the number of consecutive years that must be sampled until a dry (drought) year is observed, can be modeled using the formula

$$p(x) = (.3)(.7)^{x-1}, x = 1, 2, 3, \ldots$$

Find the probability that exactly 3 years must be sampled before a drought year occurs.

Solution We want to find the probability that $x = 3$. Using the formula, we have

$$p(3) = (.3)(.7)^{3-1} = (.3)(.7)^2 = (.3)(.49) = .147$$

Thus, there is about a 15% chance that exactly 3 years must be sampled before a drought year occurs in Texas.

Look Back The probability of interest can also be derived using the principles of probability developed in Chapter 3. The event of interest is $N_1 N_2 D_3$, where N^1 represents no drought occurs in the first sampled year, N_2 represents no drought occurs in the second sampled year, and D_3 represents a drought occurs in the third sampled year. The researchers discovered that the probability of a drought occurring in any sampled year is .3 (and, consequently, the probability of no drought occurring in any sampled year is .7). Using the multiplicative rule of probability for independent events, the probability of interest is $(.7)(.7)(.3) = .147$.

Now Work Exercise 4.30

    Since probability distributions are analogous to the relative frequency distributions of Chapter 2, it should be no surprise that the mean and standard deviation are useful descriptive measures.

    If a discrete random variable x were observed a very large number of times and the data generated were arranged in a relative frequency distribution, the relative frequency distribution would be indistinguishable from the probability distribution for the random variable. Thus, the probability distribution for a random variable is a theoretical model for the relative frequency distribution of a population. To the extent that the two distributions are equivalent (and we will assume that they are), the probability distribution for x possesses a mean μ and a variance σ^2 that are identical to the corresponding descriptive measures for the population. The procedures for finding μ and σ^2 of a random variable follow.

Examine the probability distribution for x (the number of heads observed in the toss of two fair coins) in Figure 4.5. Try to locate the mean of the distribution intuitively. We may reason that the mean μ of this distribution is equal to 1 as follows: In a large number of experiments–say, 100,000–$\frac{1}{4}$ (or 25,000) should result in $x = 0$ heads, $\frac{1}{2}$ (or 50,000) in $x = 1$ head, and $\frac{1}{4}$ (or 25,000) in $x = 2$ heads. Therefore, the average number of heads is

$$\mu = \frac{0(25{,}000) + 1(50{,}000) + 2(25{,}000)}{100{,}000} = 0\left(\frac{1}{4}\right) + 1\left(\frac{1}{2}\right) + 2\left(\frac{1}{4}\right)$$

$$= 0 + \frac{1}{2} + \frac{1}{2} = 1$$

Note that to get the population mean of the random variable x, we multiply each possible value of x by its probability $p(x)$, and then we sum this product over all possible values of x. The *mean of x* is also referred to as the *expected value of x*, denoted $E(x)$.

Figure 4.5
Probability distribution for a two-coin toss

> The **mean**, or **expected value**, of a discrete random variable x is
>
> $$\mu = E(x) = \Sigma x p(x)$$

Expected is a mathematical term and should not be interpreted as it is typically used. Specifically, *a random variable might never be equal to its "expected value."* Rather, the expected value is the mean of the probability distribution, or a measure of its central tendency. You can think of μ as the mean value of x in a *very large* (actually, *infinite*) number of repetitions of the experiment in which the values of x occur in proportions equivalent to the probabilities of x.

Example 4.7

Finding an Expected Value–An Insurance Application

Problem Suppose you work for an insurance company and you sell a $10,000 one-year term insurance policy at an annual premium of $290. Actuarial tables show that the probability of death during the next year for a person of your customer's age, sex, health, etc., is .001. What is the expected gain (amount of money made by the company) for a policy of this type?

Solution The experiment is to observe whether the customer survives the upcoming year. The probabilities associated with the two sample points, Live and Die, are .999 and .001, respectively. The random variable you are interested in is the gain x, which can assume the values shown in the following table:

Gain x	Sample Point	Probability
$290	Customer lives	.999
− $9,710	Customer dies	.001

If the customer lives, the company gains the $290 premium as profit. If the customer dies, the gain is negative because the company must pay $10,000, for a net "gain" of $(290 − 10{,}000) = −\$9{,}710$. The expected gain is therefore

$$\mu = E(x) = \Sigma x p(x) = (290)(.999) + (-9{,}710)(.001) = \$280$$

In other words, if the company were to sell a very large number of $10,000 one-year policies to customers possessing the characteristics described, it would (on the average) net $280 per sale in the next year.

> **Look Back** Note that $E(x)$ need not equal a possible value of x. That is, the expected value is $280, but x will equal either $290 or $-9,710$ each time the experiment is performed (a policy is sold and a year elapses). The expected value is a measure of central tendency–and in this case represents the average over a very large number of one-year policies–but is not a possible value of x.
>
> Now Work Exercise 4.26d

We learned in Chapter 2 that the mean and other measures of central tendency tell only part of the story about a set of data. The same is true about probability distributions: We need to measure variability as well. Since a probability distribution can be viewed as a representation of a population, we will use the population variance to measure its variability.

The *population variance* σ^2 is defined as the average of the squared distance of x from the population mean μ. Since x is a random variable, the squared distance, $(x - \mu)^2$, is also a random variable. Using the same logic we employed to find the mean value of x, we calculate the mean value of $(x - \mu)^2$ by multiplying all possible values of $(x - \mu)^2$ by $p(x)$ and then summing over all possible x values.* This quantity,

$$E\big[(x - \mu)^2 \big] = \sum_{\text{all } x} (x - \mu)^2 p(x)$$

is also called the *expected value of the squared distance from the mean;* that is, $\sigma^2 = E\big[(x - \mu)^2 \big]$. The standard deviation of x is defined as the square root of the variance σ^2.

The **variance** of a random variable x is

$$\sigma^2 = E\big[(x - \mu)^2 \big] = \Sigma (x - \mu)^2 p(x) = \Sigma x^2 p(x) - \mu^2$$

The **standard deviation** of a discrete random variable is equal to the square root of the variance, or $\sigma = \sqrt{\sigma^2}$.

Knowing the mean μ and standard deviation σ of the probability distribution of x, in conjunction with Chebyshev's rule (Table 2.6) and the Empirical rule (Table 2.7), we can make statements about the likelihood of values of x falling within the intervals $\mu \pm \sigma, \mu \pm 2\sigma$, and $\mu \pm 3\sigma$. These probabilities are given in the following box:

Chebyshev's Rule and Empirical Rule for a Discrete Random Variable

Let x be a discrete random variable with probability distribution $p(x)$, mean μ, and standard deviation σ. Then, depending on the shape of $p(x)$, the following probability statements can be made:

	Chebyshev's Rule	Empirical Rule
	Applies to any probability distribution (see Figure 4.6a)	Applies to probability distributions that are mound shaped and symmetric (see Figure 4.6b)
$P(\mu - \sigma < x < \mu + \sigma)$	≥ 0	$\approx .68$
$P(\mu - 2\sigma < x < \mu + 2\sigma)$	$\geq {}^3\!/_4$	$\approx .95$
$P(\mu - 3\sigma < x < \mu + 3\sigma)$	$\geq {}^8\!/_9$	≈ 1.00

---

*It can be shown that $E[(x - \mu)^2] = E(x^2) - \mu^2$ where $E(x^2) = \Sigma x^2 p(x)$ Note the similarity between this expression and the shortcut formula $\Sigma (x - \bar{x})^2 = \Sigma x^2 - (\Sigma x)^2/n$ given in Chapter 2.

Figure 4.6
Shapes of two probability distributions for a discrete random variable x

---

Example 4.8

Finding μ and σ–Skin Cancer Treatment

Problem Medical research has shown that a certain type of chemotherapy is successful 70% of the time when used to treat skin cancer. Suppose five skin cancer patients are treated with this type of chemotherapy, and let x equal the number of successful cures out of the five. The probability distribution for the number x of successful cures out of five is given in the following table:

x	0	1	2	3	4	5
$p(x)$	.002	.029	.132	.309	.360	.168

a. Find $\mu = E(x)$. Interpret the result.

b. Find $\sigma = \sqrt{E[(x - \mu)^2]}$. Interpret the result.

c. Graph $p(x)$. Locate μ and the interval $\mu \pm 2\sigma$ on the graph. Use either Chebyshev's rule or the empirical rule to approximate the probability that x falls into this interval. Compare your result with the actual probability.

d. Would you expect to observe fewer than two successful cures out of five?

Solution

a. Applying the formula for μ, we obtain

$$\mu = E(x) = \Sigma x p(x)$$
$$= 0(.002) + 1(.029) + 2(.132) + 3(.309) + 4(.360) + 5(.168) = 3.50$$

On average, the number of successful cures out of five skin cancer patients treated with chemotherapy will equal 3.5. Remember that this expected value has meaning only when the experiment–treating five skin cancer patients with chemotherapy–is repeated a large number of times.

b. Now we calculate the variance of x:

$$\sigma^2 = E[(x - \mu)^2] = \Sigma(x - \mu)^2 p(x)$$
$$= (0 - 3.5)^2(.002) + (1 - 3.5)^2(.029) + (2 - 3.5)^2(.132)$$
$$+ (3 - 3.5)^2(.309) + (4 - 3.5)^2(.360) + (5 - 3.5)^2(.168)$$
$$= 1.05$$

Thus, the standard deviation is

$$\sigma = \sqrt{\sigma^2} = \sqrt{1.05} = 1.02$$

This value measures the spread of the probability distribution of x, the number of successful cures out of five. A more useful interpretation is obtained by answering parts **c** and **d**.

c. The graph of $p(x)$ is shown in Figure 4.7, with the mean μ and the interval $\mu \pm 2\sigma = 3.50 \pm 2(1.02) = 3.50 \pm 2.04 = (1.46, 5.54)$ also indicated. Note particularly that $\mu = 3.5$ locates the center of the probability distribution. Since this distribution is a theoretical relative frequency distribution that is moderately mound shaped (see Figure 4.7), we expect (from Chebyshev's rule) at least 75% and, more likely (from the empirical rule), approximately 95%, of observed x values to fall between 1.46 and 5.54. You can see from the figure that the actual probability that x falls in the interval $\mu \pm 2\sigma$ includes the sum of $p(x)$ for the values $x = 2, x = 3, x = 4,$ and $x = 5$. This probability is $p(2) + p(3) + p(4) + p(5) = .132 + .309 + .360 + .168 = .969$. Therefore, 96.9% of the probability distribution lies within two standard deviations of the mean. This percentage is consistent with both Chebyshev's rule and the empirical rule.

d. Fewer than two successful cures out of five implies that $x = 0$ or $x = 1$. Both of these values of x lie outside the interval $\mu \pm 2\sigma$, and the empirical rule tells us that such a result is unlikely (approximate probability of .05). The exact probability, $P(x \le 1)$, is $p(0) + p(1) = .002 + .029 = .031$. Consequently, in a single experiment in which five skin cancer patients are treated with chemotherapy, we would not expect to observe fewer than two successful cures.

Now Work Exercise 4.24

Figure 4.7
Graph of $p(x)$ for Example 4.8

Exercises 4.15–4.41

Understanding the Principles

4.15 Give three different ways of representing the probability distribution of a discrete random variable.

4.16 What does the expected value of a random variable represent?

4.17 Will $E(x)$ always be equal to a specific value of the random variable x?

4.18 For a mound-shaped, symmetric distribution, what is the probability that x falls into the interval $\mu \pm 2\sigma$?

Learning the Mechanics

4.19 A discrete random variable x can assume five possible val-
[NW] ues: 20, 21, 22, 23, and 24. The MINITAB histogram below shows the likelihood of each value.

a. What is $p(22)$?
b. What is the probability that x equals 20 or 24?
c. What is $P(x \le 23)$?

4.20 Explain why each of the following is or is not a valid probability distribution for a discrete random variable x:

a.

x	0	1	2	3
$p(x)$	.2	.3	.3	.2

b.

x	-2	-1	0
$p(x)$	.25	.50	.20

Histogram for Exercise 4.19

c.

x	4	9	20
$p(x)$	$-.3$	1.0	.3

d.

x	2	3	5	6
$p(x)$	.15	.20	.40	.35

4.21 The random variable x has the following discrete probability distribution:

x	10	11	12	13	14
$p(x)$	.2	.3	.2	.1	.2

Since the values that x can assume are mutually exclusive events, the event $\{x \leq 12\}$ is the union of three mutually exclusive events:

$$\{x = 10\} \cup \{x = 11\} \cup \{x = 12\}$$

a. Find $P(x \leq 12)$.
b. Find $P(x > 12)$.
c. Find $P(x \leq 14)$.
d. Find $P(x = 14)$.
e. Find $P(x \leq 11 \text{ or } x > 12)$.

4.22 Toss three fair coins, and let x equal the number of heads
NW observed.
a. Identify the sample points associated with this experiment, and assign a value of x to each sample point.
b. Calculate $p(x)$ for each value of x.
c. Construct a probability histogram for $p(x)$.
d. What is $P(x = 2 \text{ or } x = 3)$?

4.23 Consider the probability distribution shown for the random variable x here:

x	1	2	4	10
$p(x)$	.2	.4	.2	.2

a. Find $\mu = E(x)$.
b. Find $\sigma^2 = E[(x - \mu)^2]$.
c. Find σ.
d. Interpret the value you obtained for μ.
e. In this case, can the random variable x ever assume the value μ? Explain.
f. In general, can a random variable ever assume a value equal to its expected value? Explain.

4.24 Consider the probability distribution shown here:
NW

x	-4	-3	-2	-1	0	1	2	3	4
$p(x)$	.02	.07	.10	.15	.30	.18	.10	.06	.02

a. Calculate μ, σ^2, and σ.
b. Graph $p(x)$. Locate $\mu, \mu - 2\sigma$, and $\mu + 2\sigma$ on the graph.
c. What is the probability that x will fall into the interval $\mu \pm 2\sigma$?

4.25 Consider the probability distributions shown here:

x	0	1	2
$p(x)$	.3	.4	.3

y	0	1	2
$p(y)$	.1	.8	.1

a. Use your intuition to find the mean for each distribution. How did you arrive at your choice?
b. Which distribution appears to be more variable? Why?
c. Calculate μ and σ^2 for each distribution. Compare these answers with your answers in parts **a** and **b**.

● Applet Exercise 4.1

Use the applet entitled *Random Numbers* to generate a list of 25 numbers between 1 and 3, inclusive. Let x represent a number chosen from this list.
a. What are the possible values of x?
b. Give the probability distribution for x in table form.
c. Let y be a number randomly chosen from the set $\{1, 2, 3\}$. Give the probability distribution for y in table form.
d. Compare the probability distributions of x and y in parts **b** and **c**. Why should these distributions be approximately the same?

● Applet Exercise 4.2

Run the applet entitled *Simulating the Probability of a Head with a Fair Coin* 10 times with $n = 2$, resetting between runs, to simulate flipping two coins 10 times. Count and record the number of heads each time. Let x represent the number of heads on a single flip of the coins.
a. What are the possible values of x?
b. Use the results of the simulation to find the probability distribution for x in table form.
c. Explain why the probability of exactly two heads should be close to .25.

4.26 **NHTSA crash tests.** Refer to the National Highway Traffic
● Safety Administration (NHTSA) crash tests of new car models, presented in Exercise 4.6 (p. 166). A summary of the driver-side star ratings for the 98 cars in the **CRASH** file is reproduced in the accompanying MINITAB printout. Assume that 1 of the 98 cars is selected at random, and let x equal the number of stars in the car's driver-side star rating.
a. Use the information in the printout to find the probability distribution for x.
b. Find $P(x = 5)$.
c. Find $P(x \leq 2)$.
NW **d.** Find $\mu = E(x)$ and interpret the result.

Tally for Discrete Variables: DRIVSTAR

DRIVSTAR	Count	Percent
2	4	4.08
3	17	17.35
4	59	60.20
5	18	18.37
N=	98	

4.27 **Dust mite allergies.** A dust mite allergen level that exceeds 2 micrograms per gram ($\mu g/g$) of dust has been associated with the development of allergies. Consider a random sample of four homes, and let x be the number of homes with a dust mite level that exceeds 2 $\mu g/g$. The probability distribution for x, based on a study by the National Institute of Environmental Health Sciences, is shown in the following table:

x	0	1	2	3	4
$p(x)$	.09	.30	.37	.20	.04

a. Verify that the probabilities for x in the table sum to 1.
b. Find the probability that three or four of the homes in the sample have a dust mite level that exceeds 2 $\mu g/g$.
c. Find the probability that fewer than two homes in the sample have a dust mite level that exceeds 2 $\mu g/g$.

d. Find $E(x)$. Give a meaningful interpretation of the result.

e. Find σ.

f. Find the exact probability that x is in the interval $\mu \pm 2\sigma$. Compare your answer with Chebyshev's rule and the empirical rule.

4.28 **Controlling the water hyacinth.** Entomologists are continually searching for new biological agents to control the water hyacinth, one of the world's worst aquatic weeds. An insect that naturally feeds on the water hyacinth is the delphacid. Female delphacids lay anywhere from one to four eggs onto a water hyacinth blade. The *Annals of the Entomological Society of America* (Jan. 2005) published a study of the life cycle of a South American delphacid species. The following table gives the percentages of water hyacinth blades that have one, two, three, and four delphacid eggs:

	One Egg	Two Eggs	Three Eggs	Four Eggs
Percentage of Blades	40	54	2	4

Based on Sosa, A. J., et al. "Life history of *Megamelus scutellaris* with description of immature stages," *Annals of the Entomological Society of America*, Vol. 98, No. 1, Jan. 2005 (adapted from Table 1).

a. One of the water hyacinth blades in the study is randomly selected, and x, the number of delphacid eggs on the blade, is observed. Give the probability distribution of x.

b. What is the probability that the blade has at least three delphacid eggs?

c. Find $E(x)$ and interpret the result.

4.29 **Gender in two-child families.** *Human Biology* (Feb. 2009) published a study on the gender of children in two-child families. In populations where it is just as likely to have a boy as a girl, the probabilities of having two girls, two boys, or a boy and a girl are well known. Let x represent the number of boys in a two-child family.

a. List the possible ways (sample points) in which a two-child family can be gender-configured. (For example, BG represents the event that the first child is a boy and the second is a girl.)

b. Assuming boys are just as likely as girls to be born, assign probabilities to the sample points in part **a.**

c. Use the probabilities, part **a,** to find the probability distribution for x.

d. The article reported on the results of the National Health Interview Survey (NHIS) of almost 43,000 two-child families. The table gives the proportion of families with each gender configuration. Use this information to revise the probability distribution for x.

Gander Configuration	Proportion
Girl-girl (GG)	.222
Boy-girl (BG)	.259
Girl-boy (GB)	.254
Boy-boy (BB)	.265

e. Refer to part **d.** Find $E(x)$ and give a practical interpretation of its value.

Applying the Concepts–Intermediate

4.30 **Solar energy cells.** According to *Wired* (June 2008), 35% of [NW] the world's solar energy cells are manufactured in China.

Consider a random sample of 5 solar energy cells, and let x represent the number in the sample that are manufactured in China. In the next section, we show that the probability distribution for x is given by the formula,

$$p(x) = \frac{(5!)(.35)^x(.65)^{5-x}}{(x!)(5-x)!},$$

where $n! = (n)(n-1)(n-2)\ldots(2)(1)$

a. Explain why x is a discrete random variable.

b. Find p(x) for $x = 0, 1, 2, 3, 4$, and 5.

c. Show that the properties for a discrete probability distribution are satisfied.

d. Find the probability that at least 4 of the 5 solar energy cells in the sample are manufactured in China.

4.31 **Contaminated gun cartridges.** A weapons manufacturer [NW] uses a liquid propellant to produce gun cartridges. During the manufacturing process, the propellant can get mixed with another liquid to produce a contaminated cartridge. A University of South Florida statistician hired by the company to investigate the level of contamination in the stored cartridges found that 23% of the cartridges in a particular lot were contaminated. Suppose you randomly sample (without replacement) gun cartridges from this lot until you find a contaminated one. Let x be the number of cartridges sampled until a contaminated one is found. It is known that the probability distribution for x is given by the formula

$$p(x) = (.23)(.77)^{x-1}, x = 1, 2, 3, \ldots$$

a. Find $p(1)$. Interpret this result.

b. Find $p(5)$. Interpret this result.

c. Find $P(x \geq 2)$. Interpret this result.

4.32 **Ryder Cup miracle in golf.** The Ryder Cup is a three-day golf tournament played between a team of golf professionals from the United States and a team from Europe. A total of 28 matches are played between the teams; one point is awarded to the team winning a match and half a point is awarded to each team if the match ends in a tie (draw). The team with the most points wins the tournament. In 1999, the United States was losing 10 points to 6 when it miraculously won 8.5 of a possible 12 points on

Wins	Ties	Points	Probability
5	7	8.5	.000123
6	5	8.5	.008823
6	6	9.0	.000456
7	3	8.5	.128030
7	4	9.0	.020086
7	5	9.5	.001257
8	1	8.5	.325213
8	2	9.0	.153044
8	3	9.5	.032014
8	4	10.0	.002514
9	0	9.0	.115178
9	1	9.5	.108400
9	2	10.0	.032901
9	3	10.5	.003561
10	0	10.0	.034552
10	1	10.5	.021675
10	2	11.0	.003401
11	0	11.0	.006284
11	1	11.0	.001972
12	0	12.0	.000518

the last day of the tournament to seal the win. On the last day, 12 single matches are played. A total of 8.5 points can be won in a variety of ways, as shown in the table (p. 175). Given one team scores at least 8.5 points on the last day of the tournament, *Chance* (Fall 2009) determined the probabilities of each of these outcomes assuming each team is equally likely to win a match. Let x be the points scored by the winning team on the last day of the tournament when the team scores at least 8.5 points. Find the probability distribution of x.

4.33   **Beach erosional hot spots.** Refer to the U.S. Army Corps of Engineers' study of beach erosional hot spots, presented in Exercise 3.131 (p. 155). The data on the nearshore bar condition for six beach hot spots are reproduced in the accompanying table. Suppose you randomly select two of these six beaches and count x, the total number in the sample with a planar nearshore bar condition.

Beach Hot Spot	Nearshore Bar Condition
Miami Beach, FL	Single, shore parallel
Coney Island, NY	Other
Surfside, CA	Single, shore parallel
Monmouth Beach, NJ	Planar
Ocean City, NJ	Other
Spring Lake, NJ	Planar

Based on *Identification and Characterization of Erosional Hotspots*, William & Mary Virginia Institute of Marine Science, U.S. Army Corps of Engineers Project Report, March 18, 2002.

a. List all possible pairs of beach hot spots that can be selected from the six.

b. Assign probabilities to the outcomes in part **a.**

c. For each outcome in part **a,** determine the value of x.

d. Form a probability distribution table for x.

e. Find the expected value of x. Interpret the result.

4.34   **Expected Lotto winnings.** The chance of winning Florida's Pick-6 Lotto game is 1 in approximately 23 million. Suppose you buy a \$1 Lotto ticket in anticipation of winning the \$7 million grand prize. Calculate your expected net winnings for this single ticket. Interpret the result.

4.35   **Expected winnings in roulette.** In the popular casino game of roulette, you can bet on whether the ball will fall in an arc on the wheel colored red, black, or green. You showed (Exercise 3.133, p. 156) that the probability of a red outcome is 18/38, that of a black outcome is 18/38, and that of a green outcome is 2/38. Suppose you make a \$5 bet on red. Find your expected net winnings for this single bet. Interpret the result.

4.36   **Federal civil trial appeals.** Refer to the *Journal of the American Law and Economics Association* (Vol. 3, 2001) study of appeals of federal civil trials, presented in Exercise 3.59 (p. 132). A breakdown of the 678 civil cases that were originally tried in front of a judge (rather than a jury) and appealed by either the plaintiff or defendant is reproduced in the next table. Suppose each civil case is awarded points (positive or negative) on the basis of the outcome of the appeal for the purpose of evaluating federal judges. If the appeal is affirmed or dismissed, +5 points are awarded. If the appeal of a plaintiff trial win is reversed, −1 point is awarded. If the appeal of a defendant trial win is reversed, −3 points are awarded. Suppose 1 of the 678 cases is selected at random, and the number x of points awarded is determined. Find and graph the probability distribution for x.

Outcome of Appeal	Number of Cases
Plaintiff trial win–reversed	71
Plaintiff trial win–affirmed/ dismissed	240
Defendant trial win–reversed	68
Defendant trial win–affirmed/ dismissed	299
Total	678

4.37   **Choosing portable grill displays.** Refer to the *Journal of Consumer Research* (Mar. 2003) marketing study of influencing consumer choices by offering undesirable alternatives, presented in Exercise 3.27 (p. 120). Recall that each of 124 college students selected showroom displays for portable grills. Five different displays (representing five different-sized grills) were available, but the students were instructed to select only three displays in order to maximize purchases of Grill #2 (a smaller grill). The table that follows shows the grill display combinations and number of each selected by the 124 students. Suppose 1 of the 124 students is selected at random. Let x represent the sum of the grill numbers selected by that student. (This sum is an indicator of the size of the grills selected.)

a. Find the probability distribution for x.

b. What is the probability that x exceeds 10?

Grill Display Combination	Number of Students
1–2–3	35
1–2–4	8
1–2–5	42
2–3–4	4
2–3–5	1
2–4–5	34

Based on Hamilton, R. W. "Why do people suggest what they do not want? Using context effects to influence others' choices." *Journal of Consumer Research*, Vol. 29, Mar. 2003 (Table 1).

4.38   **The Showcase Showdown.** On the popular television game show *The Price is Right*, contestants can play "The Showcase Showdown." The game involves a large wheel with 20 nickel values, 5, 10, 15, 20,..., 95, 100, marked on it. Contestants spin the wheel once or twice, with the objective of obtaining the highest total score *without going over a dollar (100)*. [According to the *American Statistician* (Aug. 1995), the optimal strategy for the first spinner in a three-player game is to spin a second time only if the value of the initial spin is 65 or less.] Let x represent the score of a single contestant playing "The Showcase Showdown." Assume a "fair" wheel (i.e., a wheel with equally likely outcomes). If the total of the player's spins exceeds 100, the total score is set to 0.

a. If the player is permitted only one spin of the wheel, find the probability distrubtion for x.

b. Refer to part **a.** Find $E(x)$ and interpret this value.

c. Refer to part **a.** Give a range of values within which x is likely to fall.

d. Suppose the player will spin the wheel twice, no matter what the outcome of the first spin. Find the probability distribution for x.

e. What assumption did you make to obtain the probability distribution in part **d**? Is it a reasonable assumption?

f. Find μ and σ for the probability distribution of part **d,** and interpret the results.

g. Refer to part **d.** What is the probability that in two spins the player's total score exceeds a dollar (i.e., is set to 0)?

h. Suppose the player obtains a 20 on the first spin and decides to spin again. Find the probability distribution for x.

i. Refer to part **h.** What is the probability that the player's total score exceeds a dollar?

j. Given that the player obtains a 65 on the first spin and decides to spin again, find the probability that the player's total score exceeds a dollar.

k. Repeat part **j** for different first-spin outcomes. Use this information to suggest a strategy for the one-player game.

Applying the Concepts–Advanced

4.39 **Parlay card betting.** Odds makers try to predict which professional and college football teams will win and by how much (the *spread*). If the odds makers do this accurately, adding the spread to the underdog's score should make the final score a tie. Suppose a bookie will give you $6 for every $1 you risk if you pick the winners in three different football games (adjusted by the spread) on a "parlay" card. What is the bookie's expected earnings per dollar wagered? Interpret this value.

4.40 **Punnett square for earlobes.** Geneticists use a grid–called a *Punnett square*–to display all possible gene combinations in genetic crosses. (The grid is named for Reginald Punnett, a British geneticist who developed the method in the early 1900s.) The accompanying figure is a Punnett square for a cross involving human earlobes. In humans, free earlobes (E) are dominant over attached earlobes (e). Consequently, the gene pairs EE and Ee will result in free earlobes, while the gene pair ee results in attached earlobes. Consider a couple with genes as shown in the accompanying Punnett square. Suppose the couple has seven children. Let x represent the number of children with attached earlobes (i.e., with the gene pair ee). Find the probability distribution of x.

4.41 **Robot-sensor system configuration.** Engineers at Broadcom Corp. and Simon Fraser University collaborated on research involving a robot-sensor system in an unknown environment (*The International Journal of Robotics Research*, Dec. 2004). As an example, the engineers presented the three-point, single-link robotic system shown in the accompanying figure. Each point $(A, B,$ or $C)$ in the physical space of the system has either an "obstacle" status or a "free" status. There are two single links in the system: $A \leftrightarrow B$ and $B \leftrightarrow C$. A link has a "free" status if and only if both points in the link are "free"; otherwise the link has an "obstacle" status. Of interest is the random variable x: the total number of links in the system that are "free."

a. List the possible values of x for the system.

b. The researchers stated that the probability of any point in the system having a "free" status is .5. Assuming that the three points in the system operate independently, find the probability distribution for x.

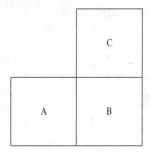

4.3 The Binomial Random Variable

Many experiments result in *dichotomous* responses (i.e., responses for which there exist two possible alternatives, such as Yes–No, Pass–Fail, Defective–Nondefective, or Male–Female). A simple example of such an experiment is the coin-toss experiment. A coin is tossed a number of times, say, 10. Each toss results in one of two outcomes, Head or Tail, and the probability of observing each of these two outcomes remains the same for each of the 10 tosses. Ultimately, we are interested in the probability distribution of x, the number of heads observed. Many other experiments are equivalent to tossing a coin (either balanced or unbalanced) a fixed number n of times and observing the number x of times that one of the two possible outcomes occurs. Random variables that possess these characteristics are called **binomial random variables.**

Public opinion and consumer preference polls (e.g., the CNN, Gallup, and Harris polls) frequently yield observations on binomial random variables. For example, suppose a sample of 100 students is selected from a large student body and each person is asked whether he or she favors (a Head) or opposes (a Tail) a certain campus issue. Suppose we are interested in x, the number of students in the sample who favor the issue. Sampling 100 students is analogous to tossing the coin 100 times. Thus, you can see that opinion polls which record the number of people who favor a certain issue are real-life equivalents of coin-toss experiments. We have been describing a **binomial experiment,** identified by the following characteristics:

Characteristics of a Binomial Random Variable

1. The experiment consists of n identical trials.
2. There are only two possible outcomes on each trial. We will denote one outcome by S (for Success) and the other by F (for Failure).
3. The probability of S remains the same from trial to trial. This probability is denoted by p, and the probability of F is denoted by $q = 1 - p$.
4. The trials are independent.
5. The binomial random variable x is the number of S's in n trials.

BIOGRAPHY JACOB BERNOULLI (1654–1705)

The Bernoulli Distribution

Son of a magistrate and spice maker in Basel, Switzerland, Jacob Bernoulli completed a degree in theology at the University of Basel. While at the university, however, he studied mathematics secretly and against the will of his father. Jacob taught mathematics to his younger brother Johan, and they both went on to become distinguished European mathematicians. At first the brothers collaborated on the problems of the time (e.g., calculus); unfortunately, they later became bitter mathematical rivals. Jacob applied his philosophical training and mathematical intuition to probability and the theory of games of chance, where he developed the law of large numbers. In his book *Ars Conjectandi*, published in 1713 (eight years after his death), the binomial distribution was first proposed. Jacob showed that the binomial distribution was a sum of independent 0–1 variables, now known as Bernoulli random variables. ∎

Example 4.9

Assessing Whether x Is a Binomial

Problem For the following examples, decide whether x is a binomial random variable:

a. A university scholarship committee must select two students to receive a scholarship for the next academic year. The committee receives 10 applications for the scholarships–6 from male students and 4 from female students. Suppose the applicants are all equally qualified, so that the selections are randomly made. Let x be the number of female students who receive a scholarship.

b. Before marketing a new product on a large scale, many companies conduct a consumer-preference survey to determine whether the product is likely to be successful. Suppose a company develops a new diet soda and then conducts a taste-preference survey in which 100 randomly chosen consumers state their preferences from among the new soda and the two leading sellers. Let x be the number of the 100 who choose the new brand over the two others.

c. Some surveys are conducted by using a method of sampling other than simple random sampling (defined in Chapter 3). For example, suppose a television cable company plans to conduct a survey to determine the fraction of households in a certain city that would use its new fiber-optic service (FiOS). The sampling method is to choose a city block at random and then survey every household on that block. This sampling technique is called *cluster sampling.* Suppose 10 blocks are so sampled, producing a total of 124 household responses. Let x be the number of the 124 households that would use FiOS.

Solution

a. In checking the binomial characteristics, a problem arises with both characteristic 3 (probabilities the same across trials) and characteristic 4 (independence). On the one hand, given that the first student selected is female, the probability that the second chosen is female is $^3/_9$. On the other hand, given that the first selection is a male student, the probability that the second is female is $^4/_9$. Thus, the conditional probability of a Success (choosing a female student to receive a scholarship) on the second trial (selection) depends on the outcome of the first trial, and the trials are therefore dependent. Since the trials are *not independent*, this variable is not a binomial random variable. (This variable is actually a *hypergeometric* random variable, the topic of optional Section 4.6.)

b. Surveys that produce dichotomous responses and use random-sampling techniques are classic examples of binomial experiments. In this example, each randomly selected consumer either states a preference for the new diet soda or does not. The sample of 100 consumers is a very small proportion of the totality of potential consumers, so the response of one would be, for all practical purposes, independent of another.[*] Thus, x is a binomial random variable.

c. This example is a survey with dichotomous responses (Yes or No to using FiOS), but the sampling method is not simple random sampling. Again, the binomial characteristic of independent trials would probably not be satisfied. The responses of households within a particular block would be dependent, since households within a block tend to be similar with respect to income, level of education, and general interests. Thus, the binomial model would not be satisfactory for x if the cluster sampling technique were employed.

Look Back Nonbinomial variables with two outcomes on every trial typically occur because they do not satisfy characteristic 3 or characteristic 4 of a binomial distribution listed in the previous box.

Now Work Exercise 4.64a

Example 4.10

Deriving the Binomial Probability Distribution–Passing a Physical Fitness Exam

Problem The Heart Association claims that only 10% of U.S. adults over 30 years of age meet the minimum requirements established by the President's Council on Fitness, Sports, and Nutrition. Suppose four adults are randomly selected and each is given the fitness test.

a. Use the steps given in Chapter 3 (box on p. 114) to find the probability that none of the four adults passes the test.

b. Find the probability that three of the four adults pass the test.

c. Let x represent the number of the four adults who pass the fitness test. Explain why x is a binomial random variable.

d. Use the answers to parts **a** and **b** to derive a formula for $p(x)$, the probability distribution of the binomial random variable x.

Solution

a. 1. The first step is to define the experiment. Here we are interested in observing the fitness test results of each of the four adults: pass (S) or fail (F).

2. Next, we list the sample points associated with the experiment. Each sample point consists of the test results of the four adults. For example, $SSSS$ represents the sample point denoting that all four adults pass, while $FSSS$ represents the sample point denoting that adult 1 fails, while adults 2, 3, and 4 pass the test. The 16 sample points are listed in Table 4.2.

---

[*]In most real-life applications of the binomial distribution, the population of interest has a finite number of elements (trials), denoted N. When N is large and the sample size n is small relative to N, say, $n/N \leq .05$, the sampling procedure, for all practical purposes, satisfies the conditions of a binomial experiment.

Table 4.2	Sample Points for Fitness Test of Example 4.10			
SSSS	FSSS	FFSS	SFFF	FFFF
	SFSS	FSFS	FSFF	
	SSFS	FSSF	FFSF	
	SSSF	SFFS	FFFS	
		SFSF		
		SSFF		

3. We now assign probabilities to the sample points. Note that each sample point can be viewed as the intersection of four adults' test results, and assuming that the results are independent, the probability of each sample point can be obtained by the multiplicative rule as follows:

$$P(SSSS) = P[\,(\text{adult 1 passes}) \cap (\text{adult 2 passes})$$
$$\cap \,(\text{adult 3 passes}) \cap (\text{adult 4 passes})\,]$$
$$= P(\text{adult 1 passes}) \times P(\text{adult 2 passes})$$
$$\times P(\text{adult 3 passes}) \times P(\text{adult 4 passes})$$
$$= (.1)(.1)(.1)(.1) = (.1)^4 = .0001$$

All other sample-point probabilities are calculated by similar reasoning. For example,

$$P(FSSS) = (.9)(.1)(.1)(.1) = .0009$$

You can check that this reasoning results in sample-point probabilities that add to 1 over the 16 points in the sample space.

4. Finally, we add the appropriate sample-point probabilities to obtain the desired event probability. The event of interest is that all four adults fail the fitness test. In Table 4.2, we find only one sample point, $FFFF$, contained in this event. All other sample points imply that at least one adult passes. Thus,

$$P(\text{All four adults fail}) = P(FFFF) = (.9)^4 = .6561$$

b. The event that three of the four adults pass the fitness test consists of the four sample points in the second column of Table 4.2: $FSSS, SFSS, SSFS,$ and $SSSF$. To obtain the event probability, we add the sample-point probabilities:

$$P(3 \text{ of } 4 \text{ adults pass}) = P(FSSS) + P(SFSS) + P(SSFS) + P(SSSF)$$
$$= (.1)^3(.9) + (.1)^3(.9) + (.1)^3(.9) + (.1)^3(.9)$$
$$= 4(.1)^3(.9) = .0036$$

Note that each of the four sample-point probabilities is the same, because each sample point consists of three S's and one F; the order does not affect the probability because the adults' test results are (assumed) independent.

c. We can characterize this experiment as consisting of four identical trials: the four test results. There are two possible outcomes to each trial, S or F, and the probability of passing, $p = .1$, is the same for each trial. Finally, we are assuming that each adult's test result is independent of all others, so that the four trials are independent. Then it follows that x, the number of the four adults who pass the fitness test, is a binomial random variable.

d. The event probabilities in parts **a** and **b** provide insight into the formula for the probability distribution $p(x)$. First, consider the event that three adults pass (part **b**). We found that

$$P(x = 3) =$$

$$(\text{Number of sample points for which } x = 3) \times (.1)^{\text{Number of successes}} \times (.9)^{\text{Number of failures}}$$
$$= 4(.1)^3(.9)^1$$

In general, we can use combinatorial mathematics to count the number of sample points. For example,

Number of sample points for which $x = 3$

$= $ Number of different ways of selecting 3 successes in the 4 trials

$$= \binom{4}{3} = \frac{4!}{3!(4-3)!} = \frac{4 \cdot 3 \cdot 2 \cdot 1}{(3 \cdot 2 \cdot 1) \cdot 1} = 4$$

The formula that works for any value of x can be deduced as follows:

$$P(x = 3) = \binom{4}{3}(.1)^3(.9)^1 = \binom{4}{x}(.1)^x(.9)^{4-x}$$

The component $\binom{4}{x}$ counts the number of sample points with x successes, and the component $(.1)^x(.9)^{4-x}$ is the probability associated with each sample point having x successes.

For the general binomial experiment, with n trials and probability p of Success on each trial, the probability of x successes is

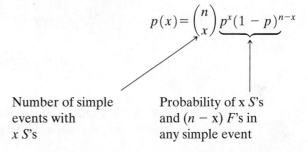

$$p(x) = \binom{n}{x} p^x(1-p)^{n-x}$$

Number of simple events with x S's

Probability of x S's and $(n - x)$ F's in any simple event

Look Ahead In theory, you could always resort to the principles developed in this example to calculate binomial probabilities; just list the sample points and sum their probabilities. However, as the number of trials (n) increases, the number of sample points grows very rapidly. (The number of sample points is 2^n.) Thus, we prefer the formula for calculating binomial probabilities, since its use avoids listing sample points.

The **binomial distribution*** is summarized in the following box:

The Binomial Probability Distribution

$$p(x) = \binom{n}{x} p^x q^{n-x} \quad (x = 0, 1, 2, \dots, n)$$

where

$p = $ Probability of a success on a single trial

$q = 1 - p$

$n = $ Number of trials

$x = $ Number of successes in n trials

$$\binom{n}{x} = \frac{n!}{x!(n-x)!}$$

As noted in Chapter 3, the symbol 5! Means $5 \cdot 4 \cdot 3 \cdot 2 \cdot 1 = 120$. Similarly, $n! = n(n-1)(n-2) \dots 3 \cdot 2 \cdot 1$. (Remember, $0! = 1$.)

---

*The binomial distribution is so named because the probabilities, $p(x), x = 0, 1, \dots, n$, are terms of the binomial expansion, $(q + p)^n$.

Example 4.11
Applying the Binomial Distribution– Physical Fitness Problem

Problem Refer to Example 4.10. Use the formula for a binomial random variable to find the probability distribution of x, where x is the number of adults who pass the fitness test. Graph the distribution.

Solution For this application, we have $n = 4$ trials. Since a success S is defined as an adult who passes the test, $p = P(S) = .1$ and $q = 1 - p = .9$. Substituting $n = 4, p = .1$, and $q = .9$ into the formula for $p(x)$, we obtain

$$p(0) = \frac{4!}{0!(4-0)!}(.1)^0(.9)^{4-0} = \frac{4 \cdot 3 \cdot 2 \cdot 1}{(1)(4 \cdot 3 \cdot 2 \cdot 1)}(.1)^0(.9)^4 = 1(.1)^0(.9)^4 = .6561$$

$$p(1) = \frac{4!}{1!(4-1)!}(.1)^1(.9)^{4-1} = \frac{4 \cdot 3 \cdot 2 \cdot 1}{(1)(3 \cdot 2 \cdot 1)}(.1)^1(.9)^3 = 4(.1)(.9)^3 = .2916$$

$$p(2) = \frac{4!}{2!(4-2)!}(.1)^2(.9)^{4-2} = \frac{4 \cdot 3 \cdot 2 \cdot 1}{(2 \cdot 1)(2 \cdot 1)}(.1)^2(.9)^2 = 6(.1)^2(.9)^2 = .0486$$

$$p(3) = \frac{4!}{3!(4-3)!}(.1)^3(.9)^{4-3} = \frac{4 \cdot 3 \cdot 2 \cdot 1}{(3 \cdot 2 \cdot 1)(1)}(.1)^3(.9)^1 = 4(.1)^3(.9) = .0036$$

$$p(4) = \frac{4!}{4!(4-4)!}(.1)^4(.9)^{4-4} = \frac{4 \cdot 3 \cdot 2 \cdot 1}{(4 \cdot 3 \cdot 2 \cdot 1)(1)}(.1)^4(.9)^0 = 1(.1)^4(.9) = .0001$$

Look Back Note that these probabilities, listed in Table 4.3, sum to 1. A graph of this probability distribution is shown in Figure 4.8.

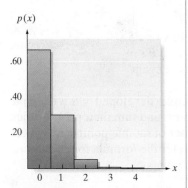

Figure 4.8
Probability distribution for physical fitness example: graphical form

Table 4.3	Probability Distribution for Physical Fitness Example: Tabular Form
x	p(x)
0	.6561
1	.2916
2	.0486
3	.0036
4	.0001

Now Work Exercise 4.46

Example 4.12
Finding μ and σ–Physical Fitness Problem

Problem Refer to Examples 4.10 and 4.11. Calculate μ and σ, the mean and standard deviation, respectively, of the number of the four adults who pass the test. Interpret the results.

Solution From Section 4.3, we know that the mean of a discrete probability distribution is

$$\mu = \sum xp(x)$$

Referring to Table 4.3, the probability distribution for the number x who pass the fitness test, we find that

$$\mu = 0(.6561) + 1(.2916) + 2(.0486) + 3(.0036) + 4(.0001) = .4$$
$$= 4(.1) = np$$

Thus, in the long run, the average number of adults (out of four) who pass the test is only .4. [*Note: The relationship $\mu = np$ holds in general for a binomial random variable.*]

The variance is

$$\sigma^2 = \Sigma(x - \mu)^2 p(x) = \Sigma(x - .4)^2 p(x)$$
$$= (0 - .4)^2(.6561) + (1 - .4)^2(.2916) + (2 - .4)^2(.0486)$$
$$+ (3 - .4)^2(.0036) + (4 - .4)^2(.0001)$$
$$= .104976 + .104976 + .124416 + .024336 + .001296$$
$$= .36 = 4(.1)(.9) = npq$$

[*Note: The relationship $\sigma^2 = npq$ holds in general for a binomial random variable.*] Finally, the standard deviation of the number who pass the fitness test is

$$\sigma = \sqrt{\sigma^2} = \sqrt{.36} = .6$$

Since the distribution shown in Figure 4.8 is skewed right, we should apply Chebyshev's rule to describe where most of the x-values fall. According to the rule, at least 75% of the x values will fall into the interval $\mu \pm 2\sigma = .4 \pm 2(.6) = (-.8, 1.6)$. Since x cannot be negative, we expect (i.e., in the long run) the number of adults out of four who pass the fitness test to be less than 1.6.

Look Back Examining Figure 4.8, you can see that all observations equal to 0 or 1 will fall within the interval $(-.8, 1.6)$. The probabilities corresponding to these values (from Table 4.3) are .6561 and .2916, respectively. Summing them, we obtain $.6561 + .2916 = .9477 \approx .95$. This result is closer to the value stated by the empirical rule. In practice, researchers have found that the proportion of observations that fall within two standard deviations of the mean for many skewed distributions will be close to .95.

We emphasize that you need not use the expectation summation rules to calculate μ and σ^2 for a binomial random variable. You can find them easily from the formulas $\mu = np$ and $\sigma^2 = npq$.

> **Mean, Variance, and Standard Deviation for a Binomial Random Variable**
>
> Mean: $\mu = np$
> Variance: $\sigma^2 = npq$
> Standard deviation: $\sigma = \sqrt{npq}$

Using Binomial Tables

Calculating binomial probabilities becomes tedious when n is large. For some values of n and p, the binomial probabilities have been tabulated in Table II of Appendix A. Part of that table is shown in Table 4.4; a graph of the binomial probability distribution for $n = 10$ and $p = .10$ is shown in Figure 4.9.

Table 4.4   Reproduction of Part of Table II of Appendix A: Binomial Probabilities for $n = 10$

k \ p	.01	.05	.10	.20	.30	.40	.50	.60	.70	.80	.90	.95	.99
0	.904	.599	.349	.107	.028	.006	.001	.000	.000	.000	.000	.000	.000
1	.996	.914	.736	.376	.149	.046	.011	.002	.000	.000	.000	.000	.000
2	1.000	.988	.930	.678	.383	.167	.055	.012	.002	.000	.000	.000	.000
3	1.000	.999	.987	.879	.650	.382	.172	.055	.011	.001	.000	.000	.000
4	1.000	1.000	.998	.967	.850	.633	.377	.166	.047	.006	.000	.000	.000
5	1.000	1.000	1.000	.994	.953	.834	.623	.367	.150	.033	.002	.000	.000
6	1.000	1.000	1.000	.999	.989	.945	.828	.618	.350	.121	.013	.001	.000
7	1.000	1.000	1.000	1.000	.998	.988	.945	.833	.617	.322	.070	.012	.000
8	1.000	1.000	1.000	1.000	1.000	.988	.989	.954	.851	.624	.264	.086	.004
9	1.000	1.000	1.000	1.000	1.000	1.000	.999	.994	.972	.893	.651	.401	.096

Table **II** actually contains a total of nine tables, labeled **(a)** through **(i)**, one each corresponding to $n = 5, 6, 7, 8, 9, 10, 15, 20$, and 25, respectively. In each of these tables, the columns correspond to values of p and the rows correspond to values of the random variable x. The entries in the table represent **cumulative binomial probabilities.** For example, the entry in the column corresponding to $p = .10$ and the row corresponding to $x = 2$ is .930 (highlighted), and its interpretation is

$$P(x \le 2) = P(x = 0) + P(x = 1) + P(x = 2) = .930$$

This probability is also highlighted in the graphical representation of the binomial distribution with $n = 10$ and $p = .10$ in Figure 4.9.

You can also use Table II to find the probability that x equals a specific value. For example, suppose you want to find the probability that $x = 2$ in the binomial distribution with $n = 10$ and $p = .10$ This probability is found by subtraction as follows:

$$P(x = 2) = \left[P(x = 0) + P(x = 1) + P(x = 2) \right] - \left[P(x = 0) + P(x = 1) \right]$$
$$= P(x \le 2) - P(x \le 1) = .930 - .736 = .194$$

The probability that a binomial random variable exceeds a specified value can be found from Table II together with the notion of complementary events. For example, to find the probability that x exceeds 2 when $n = 10$ and $p = .10$, we use

$$P(x > 2) = 1 - P(x \le 2) = 1 - .930 = .070$$

Note that this probability is represented by the *un*highlighted portion of the graph in Figure 4.9

All probabilities in Table II are rounded to three decimal places. Thus, although none of the binomial probabilities in the table is exactly zero, some are small enough (less than .0005) to round to .000. For example, using the formula to find $P(x = 0)$ when $n = 10$ and $p = .6$, we obtain

$$P(x = 0) = \binom{10}{0}(.6)^0(.4)^{10-0} = .4^{10} = .00010486$$

but this is rounded to .000 in Table II of Appendix A. (See Table 4.4.)

Similarly, none of the table entries is exactly 1.0, but when the cumulative probabilities exceed .9995, they are rounded to 1.000. The row corresponding to the largest possible value for $x, x = n$, is omitted, because all the cumulative probabilities in that row are equal to 1.0 (exactly). For example, in Table 4.4 with $n = 10$, $P(x \le 10) = 1.0$, no matter what the value of p.

The next example further illustrates the use of Table II.

Figure 4.9
Binomial probability distribution for $n = 10$ and $p = .10$, with $P(x \le 2)$ highlighted

---

Example 4.13
Using the Binomial Table–Voting for Mayor

Problem Suppose a poll of 20 voters is taken in a large city. The purpose is to determine x, the number who favor a certain candidate for mayor. Suppose that 60% of all the city's voters favor the candidate.

a. Find the mean and standard deviation of x.

b. Use Table II of Appendix A to find the probability that $x \le 10$.

c. Use Table II to find the probability that $x > 12$.

d. Use Table II to find the probability that $x = 11$.

e. Graph the probability distribution of x, and locate the interval $\mu \pm 2\sigma$ on the graph.

Solution

a. The number of voters polled is presumably small compared with the total number of eligible voters in the city. Thus, we may treat x, the number of the 20 who favor the mayoral candidate, as a binomial random variable. The value of p is the fraction of the total number of voters who favor the candidate (i.e., $p = .6$). Therefore, we calculate the mean and variance:

$$\mu = np = 20(.6) = 12$$
$$\sigma^2 = npq = 20(.6)(.4) = 4.8$$
$$\sigma = \sqrt{4.8} = 2.19$$

b. Looking in the row for $k = 10$ and the column for $p = .6$ of Table II (Appendix A) for $n = 20$, we find the value .245. Thus,

$$P(x \le 10) = .245$$

c. To find the probability

$$P(x > 12) = \sum_{x=13}^{20} p(x)$$

we use the fact that for all probability distributions,

$$\sum_{\text{all } x} p(x) = 1.$$

Therefore,

$$P(x > 12) = 1 - P(x \le 12) = 1 - \sum_{x=0}^{12} p(x)$$

Consulting Table II of Appendix A, we find the entry in row $k = 12$, column $p = .6$ to be .584. Thus,

$$P(x > 12) = 1 - .584 = .416$$

d. To find the probability that exactly 11 voters favor the candidate, recall that the entries in Table II are cumulative probabilities and use the relationship

$$P(x = 11) = [p(0) + p(1) + \cdots + p(11)] - [p(0) + p(1) + \cdots + p(10)]$$
$$= P(x \le 11) - P(x \le 10)$$

Then

$$P(x = 11) = .404 - .245 = .159$$

e. The probability distribution for x is shown in Figure 4.10. Note that

$$\mu - 2\sigma = 12 - 2(2.2) = 7.6 \quad \mu + 2\sigma = 12 + 2(2.2) = 16.4$$

Figure 4.10

The binomial probability distribution for x in Example 4.13; $n = 20$ and $p = .6$

The interval $\mu - 2\sigma$ to $\mu + 2\sigma$ also is shown in Figure 4.10. The probability that x falls into the interval $\mu + 2\sigma$ is $P(x = 8, 9, 10, \ldots, 16) = P(x \le 16) - P(x \le 7)$ .984 - .021 = .963. Note that this probability is very close to the .95 given by the empirical rule. Thus, we expect the number of voters in the sample of 20 who favor the mayoral candidate to be between 8 and 16.

Now Work Exercise 4.48

Exercises 4.42–4.66

Understanding the Principles

4.42 Give the five characteristics of a binomial random variable.

4.43 Give the formula for $p(x)$ for a binomial random variable with $n = 7$ and $p = .2$.

4.44 Consider the following binomial probability distribution:

$$p(x) = \binom{5}{x}(.7)^x(.3)^{5-x} \quad (x = 0, 1, 2, \ldots, 5)$$

   a. How many trials (n) are in the experiment?

   b. What is the value of p, the probability of success?

Learning the Mechanics

4.45 Compute the following:

   a. $\dfrac{6!}{2!(6-2)!}$ **b.** $\binom{5}{2}$ **c.** $\binom{7}{0}$

   d. $\binom{6}{6}$ **e.** $\binom{4}{3}$

4.46 If x is a binomial random variable, compute $p(x)$ for each
[NW] of the following cases:

   a. $n = 5, x = 1, p = .2$

   b. $n = 4, x = 2, q = .4$

   c. $n = 3, x = 0, p = .7$

   d. $n = 5, x = 3, p = .1$

   e. $n = 4, x = 2, q = .6$

   f. $n = 3, x = 1, p = .9$

4.47 If x is a binomial random variable, calculate μ, σ^2, and σ
for each of the following:

   a. $n = 25, p = .5$

   b. $n = 80, p = .2$

   c. $n = 100, p = .6$

   d. $n = 70, p = .9$

   e. $n = 60, p = .8$

   f. $n = 1,000, p = .04$

4.48 If x is a binomial random variable, use Table II in Appendix
[NW] A to find the following probabilities:

   a. $P(x = 2)$ for $n = 10, p = .4$

   b. $P(x \le 5)$ for $n = 15, p = .6$

   c. $P(x > 1)$ for $n = 5, p = .1$

4.49 If x is a binomial random variable, use Table II in Appendix
A to find the following probabilities:

   a. $P(x < 10)$ for $n = 25, p = .7$

   b. $P(x \ge 10)$ for $n = 15, p = .9$

   c. $P(x = 2)$ for $n = 20, p = .2$

4.50 Suppose x is a binomial random variable with $n = 5$ and
$p = .5$. Compute $p(x)$ for $x = 0, 1, 2, 3, 4$, and 5, using
the following two methods:

   a. List the sample points (take S for Success and F for
Failure on each trial) corresponding to each value of
x, assign probabilities to each sample point, and obtain
$p(x)$ by adding sample–point probabilities.

   b. Use the formula for the binomial probability distribution to obtain $p(x)$.

4.51 The binomial probability distribution is a family of probability distributions with each single distribution depending
on the values of n and p. Assume that x is a binomial
random variable with $n = 4$.

   a. Determine a value of p such that the probability distribution of x is symmetric.

   b. Determine a value of p such that the probability
distribution of x is skewed to the right.

   c. Determine a value of p such that the probability
distribution of x is skewed to the left.

   d. Graph each of the binomial distributions you obtained
in parts **a**, **b**, and **c**. Locate the mean for each distribution on its graph.

   e. In general, for what values of p will a binomial distribution be symmetric? skewed to the right? skewed to
the left?

Applet Exercise 4.3

Use the applets entitled *Simulating the Probability of a Head with
an Unfair Coin* ($P(H) = 0.2$) and *Simulating the Probability of
a Head with an Unfair Coin* ($P(H) = 0.8$) to study the mean μ
of a binomial distribution.

   a. Run each applet once with $n = 1,000$ and record
the cumulative proportions. How does the cumulative
proportion for each applet compare with the value of
$P(H)$ given for the applet?

   b. Using the cumulative proportion from each applet as
p, compute $\mu = np$ for each applet, where $n = 1,000$.
What does the value of μ represent in terms of the
results obtained from running each applet in part **a**?

   c. In your own words, describe what the mean μ of a binomial distribution represents.

Applet Exercise 4.4

Open the applet entitled *Sample from a Population*. On the
pull-down menu to the right of the top graph, select *Binary*. Set
$n = 10$ as the sample size and repeatedly choose samples from
the population. For each sample, record the number of 1's in the
sample. Let x be the number of 1's in a sample of size 10. Explain
why x is a binomial random variable.

Applet Exercise 4.5

Use the applet entitled *Simulating the Stock Market* to estimate
the probability that the stock market will go up each of the next
two days. Repeatedly run the applet for $n = 2$, recording the
number of ups each time. Use the proportion of 2's among your
results as the estimate of the probability. Compare your answer
with the binomial probability where $x = 2, n = 2$, and $p = 0.5$.

Applying the Concepts–Basic

4.52 **Where will you get your next pet?** According to an
Associated Press/Petside.com poll, half of all pet owners
would get their next dog or cat from a shelter (*USA Today*,
May 12, 2010). Consider a random sample of 10 pet owners and define x as the number of pet owners who would
acquire their next dog or cat from a shelter. Assume that x
is a binomial random variable.

   a. For this binomial experiment, define a success.

   b. For this binomial experiment, what is n?

   c. For this binomial experiment, what is p?

   d. Find $P(x = 7)$.

   e. Find $P(x \ge 3)$.

   f. Find $P(x > 8)$.

4.53 **Chemical signals of mice.** Refer to the *Cell* (May 14, 2010)
study of the ability of a mouse to recognize the odor of
a potential predator, Exercise 3.61 (p. 132). Recall that

the sources of these odors are typically major urinary proteins (Mups). In an experiment, 40% of lab mice cells exposed to chemically produced cat Mups responded positively (i.e., recognized the danger of the lurking predator). Consider a sample of 100 lab mice cells, each exposed to chemically produced cat Mups. Let x represent the number of cells that respond positively.

   a. Explain why the probability distribution of x can be approximated by the binomial distribution.

   b. Find $E(x)$ and interpret its value, practically.

   c. Find the variance of x.

   d. Give an interval that is likely to contain the value of x.

4.54 **Analysis of bottled water.** Is the bottled water you're drinking really purified water? A study of various brands of bottled water conducted by the Natural Resources Defense Council found that 25% of bottled water is just tap water packaged in a bottle (*Scientific American*, July 2003). Consider a sample of five bottled-water brands, and let x equal the number of these brands that use tap water.

   NW **a.** Explain why x is (approximately) a binomial random variable.

   b. Give the probability distribution for x as a formula.

   c. Find $P(x = 2)$.

   d. Find $P(x \le 1)$.

4.55 **Tracking missiles with satellite imagery.** The U.S. government has devoted considerable funding to missile defense research over the past 20 years. The latest development is the Space-Based Infrared System (SBIRS), which uses satellite imagery to detect and track missiles (*Chance*, Summer 2005). The probability that an intruding object (e.g., a missile) will be detected on a flight track by SBIRS is .8. Consider a sample of 20 simulated tracks, each with an intruding object. Let x equal the number of these tracks on which SBIRS detects the object.

   a. Demonstrate that x is (approximately) a binomial random variable.

   b. Give the values of p and n for the binomial distribution.

   c. Find $P(x = 15)$, the probability that SBIRS will detect the object on exactly 15 tracks.

   d. Find $P(x \ge 15)$, the probability that SBIRS will detect the object on at least 15 tracks.

   e. Find $E(x)$ and interpret the result.

4.56 **Caesarian births.** The American College of Obstetricians and Gynecologists reports that 32% of all births in the United States take place by Caesarian section each year. (*National Vital Statistics Reports*, Mar. 2010).

   a. In a random sample of 1,000 births, how many, on average, will take place by Caesarian section?

   b. What is the standard deviation of the number of Caesarian section births in a sample of 1,000 births?

   c. Use your answers to parts **a** and **b** to form an interval that is likely to contain the number of Caesarian section births in a sample of 1,000 births.

Apply the Concepts–Intermediate

4.57 **Hotel guest satisfaction.** Each year, J. D. Power and Associates publishes the results of its North American Hotel Guest Satisfaction Index Study. For 2009, the study revealed that 66% of hotel guests were aware of the hotel's "green" conservation program. Among these guests, 72%

actually participate in the program by reusing towels and bed linens. In a random sample of 15 hotel guests, consider the number (x) of guests who are aware and participate in the hotel's conservation efforts.

   a. Explain why x is (approximately) a binomial random variable.

   b. Use the rules of probability to determine the value of p for this binomial experiment.

   c. Assume $p = .45$. Find the probability that at least 10 of the 15 hotel guests are aware of and participate in the hotel's conservation efforts.

4.58 **Immediate feedback to incorrect exam answers.** Researchers from the Educational Testing Service (ETS) found that providing immediate feedback to students answering open-ended questions can dramatically improve students' future performance on exams (*Educational and Psychological Measurement*, Feb. 2010). The ETS researchers used questions from the Graduate Record Examination (GRE) in the experiment. After obtaining feedback, students could revise their answers. Consider one of these questions. Initially, 50% of the students answered the question correctly. After providing immediate feedback to students who answered incorrectly, 70% answered correctly. Consider a bank of 100 open-ended questions similar to those on the GRE.

   a. In a random sample of 20 students, what is the probability that more than half initially answer the question correctly?

   b. Refer to part **a**. After providing immediate feedback, what is the probability that more than half of the students answer the question correctly?

4.59 **Making your vote count.** The recent Democratic and Republican presidential state primary elections were highlighted by the difference in the way winning candidates were awarded delegates. In Republican states, the winner is awarded all the state's delegates; conversely, the Democratic state winner is awarded delegates in proportion to the percentage of votes. This led to a *Chance* (Fall 2007) article on making your vote count. Consider a scenario where you are one of five county commissioners voting on an issue, where each commissioner is equally likely to vote for or against.

   a. Your vote counts (i.e., is the decisive vote) only if the other four voters split, 2 in favor and 2 against. Use the binomial distribution to find the probability that your vote counts.

   b. If you convince two other commissioners to "vote in bloc" (i.e., you all agree to vote among yourselves first, and whatever the majority decides is the way all three will vote, guaranteeing that the issue is decided by the bloc), your vote counts only if these 2 commissioners split their bloc votes, 1 in favor and 1 against. Again, use the binomial distribution to find the probability that your vote counts.

4.60 **Student gambling on sports.** A study of gambling activity at the University of West Georgia (UWG) discovered that 60% of the male students wagered on sports the previous year (*The Sport Journal*, Fall 2006). Consider a random sample of 50 UWG male students. How many of these students would you expect to have gambled on sports the previous year? Give a range that is likely to include the number of male students who have gambled on sports.

4.61 **Victims of domestic abuse.** According to researchers at Dan Jones & Associates, 1 in every 3 women has been a victim of domestic abuse (Domestic Violence: Incidence and Prevalence Study, Sept.–Dec., 2005). This probability was obtained from a survey of 1,000 adult women residing in Utah. Suppose we randomly sample 15 women and find that 4 have been abused.

a. What is the probability of observing 4 or more abused women in a sample of 15 if the proportion p of women who are victims of domestic abuse is really $p = \frac{1}{3}$?

b. Many experts on domestic violence believe that the proportion of women who are domestically abused is closer to $p = .10$. Calculate the probability of observing 4 or more abused women in a sample of 15 if $p = .10$.

c. Why might your answers to parts **a** and **b** lead you to believe that $p = \frac{1}{3}$?

4.62 **Chickens with fecal contamination.** The United States Department of Agriculture (USDA) reports that, under its standard inspection system, one in every 100 slaughtered chickens passes inspection with fecal contamination (*Tampa Tribune*, Mar. 31, 2000). In Exercise 3.19 (p. 119), you found the probability that a randomly selected slaughtered chicken passes inspection with fecal contamination. Now find the probability that, in a random sample of 5 slaughtered chickens, at least one passes inspection with fecal contamination.

4.63 **Testing a psychic's ESP.** Refer to Exercise 3.95 (p. 147) and the experiment conducted by the Tampa Bay Skeptics to see whether an acclaimed psychic has extrasensory perception (ESP). Recall that a crystal was placed, at random, inside 1 of 10 identical boxes lying side by side on a table. The experiment was repeated seven times, and x, the number of correct decisions, was recorded. (Assume that the seven trials are independent.)

a. If the psychic is guessing (i.e., if the psychic does *not* possess ESP), what is the value of p, the probability of a correct decision on each trial?

b. If the psychic is guessing, what is the expected number of correct decisions in seven trials?

c. If the psychic is guessing, what is the probability of no correct decisions in seven trials?

d. Now suppose the psychic has ESP and $p = .5$. What is the probability that the psychic guesses incorrectly in all seven trials?

e. Refer to part **d**. Recall that the psychic failed to select the box with the crystal on all seven trials. Is this evidence against the psychic having ESP? Explain.

4.64 **Assigning a passing grade.** A literature professor decides to give a 20-question true–false quiz to determine who has read an assigned novel. She wants to choose the passing grade such that the probability of passing a student who guesses on every question is less than .05. What score should she set as the lowest passing grade?

4.65 **USGA golf ball specifications.** According to the U.S. Golf Association (USGA), "The weight of the [golf] ball shall not be greater than 1.620 ounces avoirdupois (45.93 grams).... The diameter of the ball shall not be less than 1.680 inches.... The velocity of the ball shall not be greater than 250 feet per second" (USGA, 2006). The USGA periodically checks the specifications of golf balls sold in the United States by randomly sampling balls from pro shops around the country. Two dozen of each kind are sampled, and if more than three do not meet size or velocity requirements, that kind of ball is removed from the USGA's approved-ball list.

a. What assumptions must be made and what information must be known in order to use the binomial probability distribution to calculate the probability that the USGA will remove a particular kind of golf ball from its approved-ball list?

b. Suppose 10% of all balls produced by a particular manufacturer are less than 1.680 inches in diameter, and assume that the number of such balls, x, in a sample of two dozen balls can be adequately characterized by a binomial probability distribution. Find the mean and standard deviation of the binomial distribution.

c. Refer to part **b**. If x has a binomial distribution, then so does the number, y, of balls in the sample that meet the USGA's minimum diameter. [*Note:* $x + y = 24$.] Describe the distribution of y. In particular, what are p, q, and n? Also, find $E(y)$ and the standard deviation of y.

4.66 **Does having boys run in the family?** *Chance* (Fall 2001) reported that the eight men in the Rodgers family produced 24 biological children over four generations. Of these 24 children, 21 were boys and 3 were girls. How likely is it for a family of 24 children to have 21 boys? Use the binomial distribution and the fact that 50% of the babies born in the United States are male to answer the question. Do you agree with the statement, "Rodgers men produce boys"?

4.4 Probability Distributions for Continuous Random Variables

The graphical form of the probability distribution for a continuous random variable x is a smooth curve that might appear as shown in Figure 4.11. This curve, a function of x, is denoted by the symbol $f(x)$ and is variously called a **probability density function,** a **frequency function,** or a **probability distribution.**

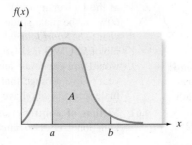

Figure 4.11

A probability distribution $f(x)$ for a continuous random variable x

The areas under a probability distribution correspond to probabilities for x. For example, the area A beneath the curve between the two points a and b, as shown in Figure 4.11, is the probability that x assumes a value between a and b ($a < x < b$). Because there is no area over a single point, say, $x = a$, it follows that (according to our model) the probability associated with a particular value of x is equal to 0; that is, $P(x = a) = 0$, and hence $P(a < x < b) = P(a \leq x \leq b)$. In other words, the probability is the same regardless of whether or not you include the endpoints of the interval. Also, because areas over intervals represent probabilities, it follows that the total area under a probability distribution–the total probability assigned to the set of all values of x–should equal 1. Note that probability distributions for continuous random variables possess different shapes, depending on the relative frequency distributions of real data that the probability distributions are supposed to model.

> The **probability distribution for a continuous random variable,** x, can be represented by a smooth curve–a function of x, denoted $f(x)$. The curve is called a **density function** or **frequency function.** The probability that x falls between two values, a and b, i.e., $P(a < x < b)$, is the area under the curve between a and b.

Example 4.14

Finding a Continuous Probability—Paper Friction in a Photocopier

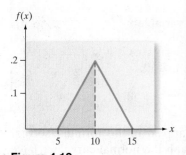

Figure 4.12

Density Function for Friction Coefficient, Example 4.14

Problem Researchers at the University of Rochester studied the friction that occurs in the paper-feeding process of a photocopier and published their results in the *Journal of Engineering for Industry*. The friction coefficient, x, is a continuous random variable that measures the degree of friction between two adjacent sheets of paper in the feeder stack. The random variable can be modeled using the smooth curve (called a *triangular density function*) shown in Figure 4.12. Note that the friction coefficient, x, ranges between the values 5 and 15. Find the probability that the friction coefficient is less than 10.

Solution To find the probability that the friction coefficient is less than 10, we need to compute $P(5 < x < 10)$. This is the area between $x = 5$ and $x = 10$ shaded on the graph in Figure 4.12. Note that this area is represented by a right triangle with a base of length 5 and a height of .2. Since the area of a triangle is $(1/2)$ (base) $\times$ (height), then

$$P(5 < x < 10) = (1/2)(\text{base}) \times (\text{height}) = (.5)(5)(.2) = .5$$

Look Ahead The areas under most probability distributions are obtained by means of calculus or numerical methods.[*] Because these methods often involve difficult procedures, we give the areas for some of the most common probability distributions in tabular form in Appendix A. Then, to find the area between two values of x, say, $x = a$ and $x = b$, you simply have to consult the appropriate table.

For the continuous random variable presented in this chapter, we give the formula for the probability distribution, along with its mean μ and standard deviation σ. These two numbers will enable you to make some approximate probability statements about a random variable even when you do not have access to a table of areas under the probability distribution.

4.5 The Normal Distribution

One of the most commonly observed continuous random variables has a **bell-shaped probability distribution** (or **bell curve**), as shown in Figure 4.13. It is known as a **normal random variable** and its probability distribution is called a **normal distribution**.

[*]Students with a knowledge of calculus should note that the probability that x assumes a value in the interval $a < x < b$ is $P(a < x < b) = \int_a^b f(x) \, dx$, assuming that the integral exists. As with the requirements for a discrete probability distribution, we require that $f(x) \geq 0$ and $\int_{-\infty}^{\infty} f(x) \, dx = 1$.

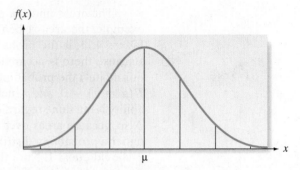

Figure 4.13
A normal probability distribution

f(x)

Figure 4.14
Several normal distributions with different means and standard deviations

CARL F. GAUSS (1777–1855)

The Gaussian Distribution

The normal distribution began in the 18th century as a theoretical distribution for errors in disciplines in which fluctuations in nature were believed to behave randomly. Although he may not have been the first to discover the formula, the normal distribution was named the Gaussian distribution after Carl Friedrich Gauss. A well-known and respected German mathematician, physicist, and astronomer, Gauss applied the normal distribution while studying the motion of planets and stars. Gauss's prowess as a mathematician was exemplified by one of his most important discoveries: At the young age of 22, Gauss constructed a regular 17-gon by ruler and compasses—a feat that was the most major advance in mathematics since the time of the ancient Greeks. In addition to publishing close to 200 scientific papers, Gauss invented the heliograph as well as a primitive telegraph. ∎

The normal distribution plays a very important role in the science of statistical inference. Moreover, many phenomena generate random variables with probability distributions that are very well approximated by a normal distribution. For example, the error made in measuring a person's blood pressure may be a normal random variable, and the probability distribution for the yearly rainfall in a certain region might be approximated by a normal probability distribution. You can determine the adequacy of the normal approximation to an existing population of data by comparing the relative frequency distribution of a large sample of the data with the normal probability distribution. Methods for detecting disagreement between a set of data and the assumption of normality are presented in Section 4.6

The normal distribution is perfectly symmetric about its mean μ, as can be seen in the examples in Figure 4.14. Its spread is determined by the value of its standard deviation σ. The formula for the normal probability distribution is shown in the next box. When plotted, this formula yields a curve like that shown in Figure 4.13.

Probability Distribution for a Normal Random Variable x

Probability density function: $f(x) = \dfrac{1}{\sigma\sqrt{2\pi}} e^{-(1/2)[(x-\mu)/\sigma]^2}$

where

μ = Mean of the normal random variable x
σ = Standard deviation
π = 3.1416...
e = 2.71828...

$P(x < a)$ is obtained from a table of normal probabilities.

Note that the mean μ and standard deviation σ appear in this formula, so that no separate formulas for μ and σ are necessary. To graph the normal curve, we have to know the numerical values of μ and σ. Computing the area over intervals under the normal probability distribution is a difficult task.* Consequently, we will use the computed areas listed in Table III of Appendix A. Although there are an infinitely large number of normal curves—one for each pair of values of μ and σ—we have formed a single table that will apply to any normal curve.

Table III is based on a normal distribution with mean $\mu = 0$ and standard deviation $\sigma = 1$, called a *standard normal distribution*. A random variable with a standard normal distribution is typically denoted by the symbol z. The formula for the probability distribution of z is

$$f(z) = \frac{1}{\sqrt{2\pi}} e^{-(1/2)z^2}$$

Figure 4.15 shows the graph of a standard normal distribution.

---

*The student with a knowledge of calculus should note that there is no closed-form expression for $P(a < x < b) = \int_a^b f(x)\,dx$ for the normal probability distribution. The value of this definite integral can be obtained to any desired degree of accuracy by numerical approximation. For this reason, it is tabulated for the user.

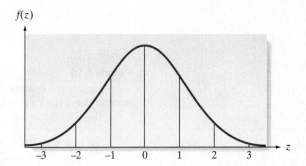

Figure 4.15
Standard normal distribution:
$\mu = 0, \sigma = 1$

> The **standard normal distribution** is a normal distribution with $\mu = 0$ and $\sigma = 1$.
> A random variable with a standard normal distribution, denoted by the symbol z, is
> called a **standard normal random variable**.

Since we will ultimately convert all normal random variables to standard normal variables in order to use Table III to find probabilities, it is important that you learn to use Table III well. A partial reproduction of that table is shown in Table 4.5. Note that the values of the standard normal random variable z are listed in the left-hand column. The entries in the body of the table give the area (probability) between 0 and z. Examples 4.15–4.18 illustrate the use of the table.

Table 4.5   Reproduction of Part of Table III in Appendix A

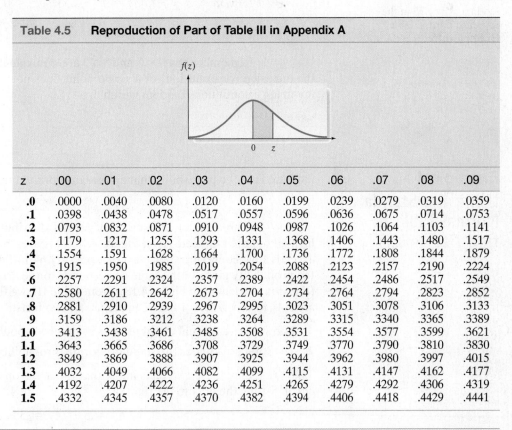

z	.00	.01	.02	.03	.04	.05	.06	.07	.08	.09
.0	.0000	.0040	.0080	.0120	.0160	.0199	.0239	.0279	.0319	.0359
.1	.0398	.0438	.0478	.0517	.0557	.0596	.0636	.0675	.0714	.0753
.2	.0793	.0832	.0871	.0910	.0948	.0987	.1026	.1064	.1103	.1141
.3	.1179	.1217	.1255	.1293	.1331	.1368	.1406	.1443	.1480	.1517
.4	.1554	.1591	.1628	.1664	.1700	.1736	.1772	.1808	.1844	.1879
.5	.1915	.1950	.1985	.2019	.2054	.2088	.2123	.2157	.2190	.2224
.6	.2257	.2291	.2324	.2357	.2389	.2422	.2454	.2486	.2517	.2549
.7	.2580	.2611	.2642	.2673	.2704	.2734	.2764	.2794	.2823	.2852
.8	.2881	.2910	.2939	.2967	.2995	.3023	.3051	.3078	.3106	.3133
.9	.3159	.3186	.3212	.3238	.3264	.3289	.3315	.3340	.3365	.3389
1.0	.3413	.3438	.3461	.3485	.3508	.3531	.3554	.3577	.3599	.3621
1.1	.3643	.3665	.3686	.3708	.3729	.3749	.3770	.3790	.3810	.3830
1.2	.3849	.3869	.3888	.3907	.3925	.3944	.3962	.3980	.3997	.4015
1.3	.4032	.4049	.4066	.4082	.4099	.4115	.4131	.4147	.4162	.4177
1.4	.4192	.4207	.4222	.4236	.4251	.4265	.4279	.4292	.4306	.4319
1.5	.4332	.4345	.4357	.4370	.4382	.4394	.4406	.4418	.4429	.4441

Example 4.15

Using the Standard Normal Table to find
$P(-z_0 < z < z_0)$

Problem Find the probability that the standard normal random variable z falls between -1.33 and $+1.33$.

Solution The standard normal distribution is shown again in Figure 4.16. Since all probabilities associated with standard normal random variables can be depicted as areas under the standard normal curve, you should always draw the curve and then equate the desired probability to an area.

Figure 4.16

Areas under the standard normal curve for Example 4.15

In this example, we want to find the probability that z falls between -1.33 and $+1.33$, which is equivalent to the area between -1.33 and $+1.33$, shown highlighted in Figure 4.16. Table III gives the area between $z = 0$ and any positive value of z, so that if we look up $z = 1.33$ (the value in the 1.3 row and .03 column, as shown in Figure 4.17), we find that the area between $z = 0$ and $z = 1.33$ is .4082. This is the area labeled A_1 in Figure 4.16. To find the area A_2 located between $z = 0$ and $z = -1.33$, we note that the symmetry of the normal distribution implies that the area between $z = 0$ and any point to the left is equal to the area between $z = 0$ and the point equidistant to the right. Thus, in this example the area between $z = 0$ and $z = -1.33$ is equal to the area between $z = 0$ and $z = +1.33$. That is,

$$A_1 = A_2 = .4082$$

The probability that z falls between -1.33 and $+1.33$ is the sum of the areas of A_1 and A_2. We summarize in probabilistic notation:

$$P(-1.33 < z < +1.33) = P(-1.33 < z < 0) + P(0 < z \leq 1.33)$$
$$= A_1 + A_2 = .4082 + .4082 = .8164$$

Figure 4.17

Finding $z = 1.33$ in the standard normal table, Example 4.15

Look Back Remember that "$<$" and "$\leq$" are equivalent in events involving z, because the inclusion (or exclusion) of a single point does not alter the probability of an event involving a continuous random variable.

Now Work Exercise 4.73 e–f

Example 4.16

Using the Standard Normal Table to find $P(z > z_0)$

Problem Find the probability that a standard normal random variable exceeds 1.64; that is, find $P(z > 1.64)$.

Solution The area under the standard normal distribution to the right of 1.64 is the highlighted area labeled A_1 in Figure 4.18. This area represents the probability that z exceeds 1.64. However, when we look up $z = 1.64$ in Table III, we must remember that the probability given in the table corresponds to the area between $z = 0$ and $z = 1.64$ (the area labeled A_2 in Figure 4.18). From Table III, we find that $A_2 = .4495$. To find the area A_1 to the right of 1.64, we make use of two facts:

1. The standard normal distribution is symmetric about its mean, $z = 0$.

2. The total area under the standard normal probability distribution equals 1.

Taken together, these two facts imply that the areas on either side of the mean, $z = 0$, equal .5; thus, the area to the right of $z = 0$ in Figure 4.18 is $A_1 + A_2 = .5$. Then

$$P(z > 1.64) = A_1 = .5 - A_2 = .5 - .4495 = .0505$$

Figure 4.18

Areas under the standard normal curve for Example 4.16

Look Back To attach some practical significance to this probability, note that the implication is that the chance of a standard normal random variable exceeding 1.64 is only about .05.

Now Work Exercise 4.72a

Example 4.17
Using the Standard Normal Table to find $P(z < z_0)$

Problem Find the probability that a standard normal random variable lies to the left of .67.

Solution The event sought is shown as the highlighted area in Figure 4.19. We want to find $P(z < .67)$. We divide the highlighted area into two parts: the area A_1 between $z = 0$ and $z = .67$, and the area A_2 to the left of $z = 0$. We must always make such a division when the desired area lies on both sides of the mean $(z = 0)$ because Table III contains areas between $z = 0$ and the point you look up. We look up $z = .67$ in Table III to find that $A_1 = .2486$. The symmetry of the standard normal distribution also implies that half the distribution lies on each side of the mean, so the area A_2 to the left of $z = 0$ is .5. Then

$$P(z < .67) = A_1 + A_2 = .2486 + .5 = .7486$$

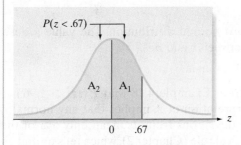

Figure 4.19

Areas under the standard normal curve for Example 4.17

Look Back Note that this probability is approximately .75. Thus, about 75% of the time, the standard normal random variable z will fall below .67. This statement implies that $z = .67$ represents the approximate 75th percentile (or upper quartile) of the standard normal distribution.

Now Work Exercise 4.72h

Example 4.18
Using the Standard Normal Table to find $P(|z| > z_0)$

Problem Find the probability that a standard normal random variable exceeds 1.96 in absolute value.

Solution The event sought is shown highlighted in Figure 4.20. We want to find

$$P(|z| > 1.96) = P(z < -1.96 \text{ or } z > 1.96)$$

Figure 4.20

Areas under the standard normal curve for Example 4.18

Note that the total highlighted area is the sum of the two areas A_1 and A_2—areas that are equal because of the symmetry of the normal distribution.

We look up $z = 1.96$ and find the area between $z = 0$ and $z = 1.96$ to be .4750. Then A_2, the area to the right of 1.96, is $.5 - .4750 = .0250$, so that

$$P(|z| > 1.96) = A_1 + A_2 = .0250 + .0250 = .05$$

Look Back We emphasize, again, the importance of sketching the standard normal curve in finding normal probabilities.

To apply Table III to a normal random variable x with any mean μ and any standard deviation σ, we must first convert the value of x to a z-score. The population z-score for a measurement was defined in Section 2.7 as the *distance* between the measurement and the population mean, divided by the population standard deviation. Thus, the z-score gives the distance between a measurement and the mean in units equal to the standard deviation. In symbolic form, the z-score for the measurement x is

$$z = \frac{x - \mu}{\sigma}$$

Note that when $x = \mu$, we obtain $z = 0$.

An important property of the normal distribution is that if x is normally distributed with any mean and any standard deviation, z is *always* normally distributed with mean 0 and standard deviation 1. That is, z is a standard normal random variable.

Property of Normal Distributions

If x is a normal random variable with mean μ and standard deviation σ, then the random variable z defined by the formula

$$z = \frac{x - \mu}{\sigma}$$

has a standard normal distribution. The value z describes the number of standard deviations between x and μ.

Recall from Example 4.18 that $P(|z| > 1.96) = .05$. This probability, coupled with our interpretation of z, implies that any normal random variable lies more than 1.96 standard deviations from its mean only 5% of the time. Compare this statement with the empirical rule (Chapter 2), which tells us that about 5% of the measurements in mound-shaped distributions will lie beyond two standard deviations from the mean. The normal distribution actually provides the model on which the empirical rule is based, along with much "empirical" experience with real data that often approximately obey the rule, whether drawn from a normal distribution or not.

Example 4.19
Finding a Normal Probability—Cell Phone Application

Problem Assume that the length of time, x, between charges of a cellular phone is normally distributed with a mean of 10 hours and a standard deviation of 1.5 hours. Find the probability that the cell phone will last between 8 and 12 hours between charges.

Solution The normal distribution with mean $\mu = 10$ and $\sigma = 1.5$ is shown in Figure 4.21. The desired probability that the cell phone lasts between 8 and 12 hours is highlighted. In order to find that probability, we must first convert the distribution to a standard normal distribution, which we do by calculating the z-score:

$$z = \frac{x - \mu}{\sigma}$$

The z-scores corresponding to the important values of x are shown beneath the x values on the horizontal axis in Figure 4.21. Note that $z = 0$ corresponds to the mean of $\mu = 10$ hours, whereas the x values 8 and 12 yield z-scores of -1.33 and $+1.33$, respectively. Thus, the event that the cell phone lasts between 8 and 12 hours is equivalent to the event that a standard normal random variable lies between -1.33 and $+1.33$. We found this probability in Example 4.15 (see Figure 4.16) by doubling the area corresponding to $z = 1.33$ in Table III. That is,

$$P(8 \leq x \leq 12) = P(-1.33 \leq z \leq 1.33) = 2(.4082) = .8164$$

Now Work Exercise 4.78a

Figure 4.21
Areas under the normal curve for Example 4.19

The steps to follow in calculating a probability corresponding to a normal random variable are shown in the following box:

Steps for Finding a Probability Corresponding to a Normal Random Variable

1. Sketch the normal distribution and indicate the mean of the random variable x. Then shade the area corresponding to the probability you want to find.

2. Convert the boundaries of the shaded area from x values to standard normal random variable z values by using the formula

$$z = \frac{x - \mu}{\sigma}$$

   Show the z values under the corresponding x values on your sketch.

3. Use Table III in Appendix A to find the areas corresponding to the z values. If necessary, use the symmetry of the normal distribution to find areas corresponding to negative z values and the fact that the total area on each side of the mean equals .5 to convert the areas from Table III to the probabilities of the event you have shaded.

Example 4.20

Using Normal Probabilities to Make an Inference— Advertised Gas Mileage

Problem Suppose an automobile manufacturer introduces a new model that has an advertised mean in-city mileage of 27 miles per gallon. Although such advertisements seldom report any measure of variability, suppose you write the manufacturer for the details of the tests and you find that the standard deviation is 3 miles per gallon. This information leads you to formulate a probability model for the random variable x, the in-city mileage for this car model. You believe that the probability distribution of x can be approximated by a normal distribution with a mean of 27 and a standard deviation of 3.

a. If you were to buy this model of automobile, what is the probability that you would purchase one that averages less than 20 miles per gallon for in-city driving? In other words, find $P(x < 20)$.

b. Suppose you purchase one of these new models and it does get less than 20 miles per gallon for in-city driving. Should you conclude that your probability model is incorrect?

Solution

a. The probability model proposed for x, the in-city mileage, is shown in Figure 4.22. We are interested in finding the area A to the left of 20, since that area corresponds to the probability that a measurement chosen from this distribution falls below 20. In other words, if this model is correct, the area A represents the fraction of cars that can be expected to get less than 20 miles per gallon for in-city driving. To find A, we first calculate the z value corresponding to $x = 20$. That is,

$$z = \frac{x - \mu}{\sigma} = \frac{20 - 27}{3} = -\frac{7}{3} = -2.33$$

Then

$$P(x < 20) = P(z < -2.33)$$

as indicated by the highlighted area in Figure 4.22. Since Table III gives only areas to the right of the mean (and because the normal distribution is symmetric about its mean), we look up 2.33 in Table III and find that the corresponding area is .4901. This is equal to the area between $z = 0$ and $z = -2.33$, so we find that

$$P(x < 20) = A = .5 - .4901 = .0099 \approx .01$$

Figure 4.22
Area under the normal curve for Example 4.20

According to this probability model, you should have only about a 1% chance of purchasing a car of this make with an in-city mileage under 20 miles per gallon.

b. Now you are asked to make an inference based on a sample: the car you purchased. You are getting less than 20 miles per gallon for in-city driving. What do you infer? We think you will agree that one of two possibilities exists:

1. The probability model is correct. You simply were unfortunate to have purchased one of the cars in the 1% that get less than 20 miles per gallon in the city.

2. The probability model is incorrect. Perhaps the assumption of a normal distribution is unwarranted, or the mean of 27 is an overestimate, or the standard deviation of 3 is an underestimate, or some combination of these errors occurred. At any rate, the form of the actual probability model certainly merits further investigation.

You have no way of knowing with certainty which possibility is correct, but the evidence points to the second one. We are again relying on the rare-event approach to statistical inference that we introduced earlier. The sample (one measurement in this case) was so unlikely to have been drawn from the proposed probability model that it casts serious doubt on the model. We would be inclined to believe that the model is somehow in error.

Look Back In applying the rare-event approach, the calculated probability must be small (say, less than or equal to .05) in order to infer that the observed event is, indeed, unlikely.

Now Work Exercise 4.84

Occasionally you will be given a probability and will want to find the values of the normal random variable that correspond to that probability. For example, suppose the scores on a college entrance examination are known to be normally distributed and a certain prestigious university will consider for admission only those applicants whose scores exceed the 90th percentile of the test score distribution. To determine the minimum score for consideration for admission, you will need to be able to use Table III in reverse, as demonstrated in the next example.

Example 4.21

Using the Normal Table in Reverse

Problem Find the value of z—call it z_0—in the standard normal distribution that will be exceeded only 10% of the time. That is, find z_0 such that $P(z \geq z_0) = .10$.

Solution In this case, we are given a probability, or an area, and are asked to find the value of the standard normal random variable that corresponds to the area. Specifically, we want to find the value z_0 such that only 10% of the standard normal distribution exceeds z_0. (See Figure 4.23.)

Figure 4.23
Areas under the standard normal curve for Example 4.21

We know that the total area to the right of the mean $z = 0$, is .5, which implies that z_0 must lie to the right of 0 ($z_0 > 0$). To pinpoint the value, we use the fact that the area to the right of z_0 is .10, which implies that the area between $z = 0$ and z_0 is $.5 - .1 = .4$. But areas between $z = 0$ and some other z value are exactly the types given in Table III. Therefore, we look up the area .4000 in the body of Table III and find that the corresponding z value is (to the closest approximation) $z_0 = 1.28$. The implication is that the point 1.28 standard deviations above the mean is the 90th percentile of a normal distribution.

Look Back As with earlier problems, it is critical to correctly draw the normal probability sought on the normal curve. The placement of z_0 to the left or right of 0 is the key. Be sure to shade the probability (area) involving z_0. If it does not agree with the probability sought (i.e., if the shaded area is greater than .5 and the probability sought is smaller than .5), then you need to place z_0 on the opposite side of 0.

Example 4.22
Using the Normal Table in Reverse

Problem Find the value of z_0 such that 95% of the standard normal z values lie between $-z_0$ and $+z_0$; that is, find $P(-z_0 \le z \le z_0) = .95$.

Solution Here we wish to move an equal distance z_0 in the positive and negative directions from the mean $z = 0$ until 95% of the standard normal distribution is enclosed. This means that the area on each side of the mean will be equal to $\frac{1}{2}(.95) = .475$, as shown in Figure 4.24. Since the area between $z = 0$ and z_0 is .475, we look up .475 in the body of Table III to find the value $z_0 = 1.96$. Thus, as we found in reverse order in Example 4.18, 95% of a normal distribution lies between $+1.96$ and -1.96 standard deviations of the mean.

Figure 4.24
Areas under the standard normal curve for Example 4.22

Now Work Exercise 4.76

Now that you have learned to use Table III to find a standard normal z value that corresponds to a specified probability, we demonstrate a practical application in Example 4.23.

Example 4.23
The Normal Table in Reverse—College Entrance Exam Application

Problem Suppose the scores x on a college entrance examination are normally distributed with a mean of 550 and a standard deviation of 100. A certain prestigious university will consider for admission only those applicants whose scores exceed the 90th percentile of the distribution. Find the minimum score an applicant must achieve in order to receive consideration for admission to the university.

Solution In this example, we want to find a score x_0 such that 90% of the scores (x values) in the distribution fall below x_0 and only 10% fall above x_0. That is,

$$P(x \le x_0) = .90$$

Converting x to a standard normal random variable where $\mu = 550$ and $\sigma = 100$, we have

$$P(x \le x_0) = P\left(z \le \frac{x_0 - \mu}{\sigma}\right)$$

$$= P\left(z \le \frac{x_0 - 550}{100}\right) = .90$$

In Example 4.21 (see Figure 4.23), we found the 90th percentile of the standard normal distribution to be $z_0 = 1.28$. That is, we found that $P(z \le 1.28) = .90$. Consequently, we know that the minimum test score x_0 corresponds to a z-score of 1.28; in other words,

$$\frac{x_0 - 550}{100} = 1.28$$

If we solve this equation for x_0, we find that

$$x_0 = 550 + 1.28(100) = 550 + 128 = 678$$

Figure 4.25

Area under the normal curve for Example 4.23

This x value is shown in Figure 4.25. Thus, the 90th percentile of the test score distribution is 678. That is to say, an applicant must score at least 678 on the entrance exam to receive consideration for admission by the university.

Look Back As the example shows, in practical applications of the normal table in reverse, first find the value of z_0 and then use the z-score formula in reverse to convert the value to the units of x.

Now Work Exercise 4.87

Statistics IN Action | Revisited   **Using the Normal Model to Maximize the Probability of a Hit with the Super Weapon**

Recall that a defense contractor has developed a prototype gun for the U.S. Army that fires 1,100 flechettes with a single round. The specifications of the weapon are set so that when the gun is aimed at a target 500 meters away, the mean horizontal grid value of the flechettes is equal to the aim point. In the range test, the weapon was aimed at the center target in Figure SIA4.1; thus, $\mu = 5$ feet. For three different tests, the standard deviation was set at $\sigma = 1$ foot, $\sigma = 2$ feet, and $\sigma = 4$ feet. From past experience, the defense contractor has found that the distribution of the horizontal flechette measurements is closely approximated by a normal distribution. Therefore, we can use the normal distribution to find the probability that a single flechette shot from the weapon will hit any one of the three targets. Recall from Figure SIA4.1 that the three targets range from -1 to 1, 4 to 6, and 9 to 11 feet on the horizontal grid.

Consider first the middle target. Letting x represent the horizontal measurement for a flechette shot from the gun, we see that the flechette will hit the target if $4 \le x \le 6$. Using the normal probability table (Table III, Appendix A), we then find that the probability that this flechette will hit the target when $\mu = 5$ and $\sigma = 1$ is

Middle: $P(4 \le x \le 6) = P\left(\dfrac{4-5}{1} < z < \dfrac{6-5}{1}\right)$

$\sigma = 1$

$$= P(-1 < z < 1)$$
$$= 2(.3413) = .6826$$

Similarly, we find the probabilities that the flechette hits the left and right targets shown in Figure SIA4.1:

Left: $P(-1 \le x \le 1) = P\left(\dfrac{-1-5}{1} < z < \dfrac{1-5}{1}\right)$

$\sigma = 1$

$$= P(-6 < z < -4) \approx 0$$

Right: $P(9 \le x \le 11) = P\left(\dfrac{9-5}{1} < z < \dfrac{11-5}{1}\right)$

$\sigma = 1$

$$= P(4 < z < 6) \approx 0$$

You can see that there is about a 68% chance that a flechette will hit the middle target, but virtually no chance that one will hit the left or right target when the standard deviation is set at 1 foot.

To find these three probabilities for $\sigma = 2$ and $\sigma = 4$, we use the normal probability function in MINITAB. Figure SIA4.2 is a MINITAB worksheet giving the cumulative probabilities of a normal random variable falling below the x values in the first column. The cumulative probabilities for $\sigma = 2$ and $\sigma = 4$ are given in the columns named "sigma2" and "sigma4," respectively.

Using the cumulative probabilities in the figure to find the three probabilities when $\sigma = 2$, we have

Middle: $P(4 \le x \le 6) = P(x \le 6) - P(x \le 4)$

$\sigma = 2$

$$= .6915 - .3085 = .3830$$

Left: $P(-1 \le x \le 1) = P(x \le 1) - P(x \le -1)$

$\sigma = 2$

$$= .0227 - .0013 = .0214$$

Right: $P(9 \le x \le 11) = P(x \le 11) - P(x \le 9)$

$\sigma = 2$

$$= .9987 - .9773 = .0214$$

↓	C1	C2	C3	C4	C5
	x	sigma1	sigma2	sigma4	
1	-1	0.00000	0.001350	0.066807	
2	1	0.00003	0.022750	0.158655	
3	4	0.15866	0.308538	0.401294	
4	6	0.84134	0.691462	0.598706	
5	9	0.99997	0.977250	0.841345	
6	11	1.00000	0.998650	0.933193	
7					

Figure SIA4.2

MINITAB worksheet with cumulative normal probabilities

Statistics IN Action
(continued)

Thus, when $\sigma = 2$, there is about a 38% chance that a flechette will hit the middle target, a 2% chance that one will hit the left target, and a 2% chance that one will hit the right target. The probability that a flechette will hit either the middle or the left or the right target is simply the sum of these three probabilities (an application of the additive rule of probability). This sum is .3830 + .0214 + .0214 = .4258; consequently, there is about a 42% chance of hitting any one of the three targets when specifications are set so that $\sigma = 2$.

Now we use the cumulative probabilities in Figure SIA4.2 to find the three hit probabilities when $\sigma = 4$:

Middle: $P(4 \leq x \leq 6) = P(x \leq 6) - P(x \leq 4)$
$\sigma = 4$
$$= .5987 - .4013 = .1974$$

Left: $P(-1 \leq x \leq 1) = P(x \leq 1) - P(x \leq -1)$
$\sigma = 4$
$$= .1587 - .0668 = .0919$$

Right: $P(9 \leq x \leq 11) = P(x \leq 11) - P(x \leq 9)$
$\sigma = 4$
$$= .9332 - .8413 = .0919$$

Thus, when $\sigma = 4$, there is about a 20% chance that a flechette will hit the middle target, a 9% chance that one will hit the left target, and a 9% chance that one will hit the right target. The probability that a flechette will hit any one of the three targets is .1974 + .0919 + .0919 = .3812.

These probability calculations reveal a few patterns. First, the probability of hitting the middle target (the target at which the gun is aimed) is reduced as the standard deviation is increased. Obviously, then, if the U.S. Army wants to maximize the chance of hitting the target that the prototype gun is aimed at, it will want specifications set with a small value of σ. But if the Army wants to hit multiple targets with a single shot of the weapon, σ should be increased. With a larger σ, not as many of the flechettes will hit the target aimed at, but more will hit peripheral targets. Whether σ should be set at 4 or 6 (or some other value) depends on how high of a hit rate is required for the peripheral targets.

Exercises 4.67–4.96

Understanding the Principles

4.67 Describe the shape of a normal probability distribution.

4.68 If x has a normal distribution with mean μ and standard deviation σ, describe the distribution of $z = (x - \mu)/\sigma$.

4.69 What is the name given to a normal distribution when $\mu = 0$ and $\sigma = 1$?

Learning the Mechanics

4.70 Find the area under the standard normal distribution between the following pairs of z-scores:
   a. $z = -2.00$ and $z = 0$
   b. $z = -1.00$ and $z = 0$
   c. $z = -1.69$ and $z = 0$
   d. $z = -.58$ and $z = 0$

4.71 Find the area under the standard normal probability distribution between the following pairs of z-scores:
   a. $z = 0$ and $z = 2.00$
   b. $z = 0$ and $z = 1.00$
   c. $z = 0$ and $z = 3$
   d. $z = 0$ and $z = .58$

4.72 Find the following probabilities for the standard normal random variable z:
   NW **a.** $P(z > 1.46)$
   b. $P(z < -1.56)$
   c. $P(.67 \leq z \leq 2.41)$
   d. $P(-1.96 \leq z < -.33)$
   e. $P(z \geq 0)$
   f. $P(-2.33 < z < 1.50)$
   g. $P(z \geq -2.33)$
   NW **h.** $P(z < 2.33)$

4.73 Find each of the following probabilities for a standard normal random variable z:
   a. $P(z = 1)$
   b. $P(z \leq 1)$
   c. $P(z < 1)$
   d. $P(z > 1)$
   NW **e.** $P(-1 \leq z \leq 1)$
   NW **f.** $P(-2 \leq z \leq 2)$
   g. $P(-2.16 \leq z \leq .55)$
   NW **h.** $P(-.42 < z < 1.96)$

4.74 Suppose the random variable x is best described by a normal distribution with $\mu = 25$ and $\sigma = 5$. Find the z-score that corresponds to each of the following x values:
   a. $x = 25$
   b. $x = 30$
   c. $x = 37.5$
   d. $x = 10$
   e. $x = 50$
   f. $x = 32$

4.75 Give the z-score for a measurement from a normal distribution for the following:
   a. 1 standard deviation above the mean
   b. 1 standard deviation below the mean
   c. Equal to the mean
   d. 2.5 standard deviations below the mean
   e. 3 standard deviations above the mean

4.76 Find a value z_0 of the standard normal random variable z such that
   NW **a.** $P(z \leq z_0) = .0401$
   b. $P(-z_0 \leq z \leq z_0) = .95$
   c. $P(-z_0 \leq z \leq z_0) = .90$

d. $P(-z_0 \leq z \leq z_0) = .8740$
e. $P(-z_0 \leq z \leq 0) = .2967$
f. $P(-2 < z < z_0) = .9710$
g. $P(z \geq z_0) = .5$
h. $P(z \geq z_0) = .0057$

4.77 Find a value z_0 of the standard normal random variable z such that
a. $P(z \geq z_0) = .05$
b. $P(z \geq z_0) = .025$
c. $P(z \leq z_0) = .025$
d. $P(z \geq z_0) = .10$
e. $P(z > z_0) = .10$

4.78 Suppose x is a normally distributed random variable with $\mu = 11$ and $\sigma = 2$. Find each of the following:
[NW] **a.** $P(10 \leq x \leq 12)$
b. $P(6 \leq x \leq 10)$
c. $P(13 \leq x \leq 16)$
d. $P(7.8 \leq x \leq 12.6)$
e. $P(x \geq 13.24)$
f. $P(x \geq 7.62)$

4.79 Suppose x is a normally distributed random variable with $\mu = 30$ and $\sigma = 8$. Find a value x_0 of the random variable x such that
a. $P(x \geq x_0) = .5$
b. $P(x < x_0) = .025$
c. $P(x > x_0) = .10$
d. $P(x > x_0) = .95$
e. 10% of the values of x are less than x_0.
f. 80% of the values of x are less than x_0.
g. 1% of the values of x are greater than x_0.

4.80 The random variable x has a normal distribution with standard deviation 25. It is known that the probability that x exceeds 150 is .90. Find the mean μ of the probability distribution.

⦿ Applet Exercise 4.6

Open the applet entitled *Sample from a Population*. On the pull-down menu to the right of the top graph, select *Bell shaped*. The box to the left of the top graph displays the population mean, median, and standard deviation.
a. Run the applet for each available value of n on the pull-down menu for the sample size. Go from the smallest to the largest value of n. For each value of n, observe the shape of the graph of the sample data and record the mean, median, and standard deviation of the sample.
b. Describe what happens to the shape of the graph and the mean, median, and standard deviation of the sample as the sample size increases.

Applying the Concepts—Basic

4.81 **Dental anxiety study.** To gauge their fear of going to a dentist, a random sample of adults completed the Modified Dental Anxiety Scale questionnaire (*BMC Oral Health*, Vol. 9, 2009). Scores on the scale range from zero (no anxiety) to 25 (extreme anxiety). The mean score was 11 and the standard deviation was 4. Assume that the distribution of all scores on the Modified Dental Anxiety Scale is approximately normal with $\mu = 11$ and $\sigma = 4$.
a. Suppose you score a 10 on the Modified Dental Anxiety Scale. Find the z-value for your score.

b. Find the probability that someone scores between 10 and 15 on the Modified Dental Anxiety Scale.
c. Find the probability that someone scores above 20 on the Modified Dental Anxiety Scale.

4.82 **Tomato as a taste modifier.** Miraculin—a protein naturally produced in a rare tropical fruit—can convert a sour taste into a sweet taste. Consequently, miraculin has the potential to be an alternative low-calorie sweetener. In *Plant Science* (May 2010), a group of Japanese environmental scientists investigated the ability of a hybrid tomato plant to produce miraculin. For a particular generation of the tomato plant, the amount x of miraculin produced (measured in micrograms per gram of fresh weight) had a mean of 105.3 and a standard deviation of 8.0. Assume that x is normally distributed.
a. Find $P(x > 120)$.
b. Find $P(100 < x < 110)$.
c. Find the value a for which $P(x < a) = .25$.

4.83 **Most powerful American women.** Refer to the *Fortune* (Oct. 16, 2008) list of the 50 most powerful women in America, presented in Exercise 2.60 (p. 57). Recall that the data on age (in years) of each woman is stored in the **WPOWER50** file. The ages in the data set can be shown to be approximately normally distributed with a mean of 50 years and a standard deviation of 6.4 years. A powerful woman is randomly selected from the data, and her age is observed.
a. Find the probability that her age will fall between 55 and 60 years.
b. Find the probability that her age will fall between 48 and 52 years.
c. Find the probability that her age will be less than 35 years.
d. Find the probability that her age will exceed 40 years.
e. A 25-year-old woman claims she is one of the 50 most powerful women. Do you believe her? Explain.

4.84 **Casino gaming.** Casino gaming yields over $35 billion in revenue each year in the United States. In *Chance* (Spring 2005), University of Denver statistician R. C. Hannum discussed the business of casino gaming and its reliance on the laws of probability. Casino games of pure chance (e.g., craps, roulette, baccarat, and keno) always yield a "house advantage." For example, in the game of double-zero roulette, the expected casino win percentage is 5.26% on bets made on whether the outcome will be either black or red. (This percentage implies that for every $5 bet on black or red, the casino will earn a net of about 25 cents.) It can be shown that in 100 roulette plays on black/red, the average casino win percentage is normally distributed with mean 5.26% and standard deviation 10%. Let x represent the average casino win percentage after 100 bets on black/red in double-zero roulette.
a. Find $P(x > 0)$. (This is the probability that the casino wins money.)
b. Find $P(5 < x < 15)$.
c. Find $P(x < 1)$.
d. If you observed an average casino win percentage of -25% after 100 roulette bets on black/red, what would you conclude?

4.85 **Transmission delays in wireless technology.** Resource Reservation Protocol (RSVP) was originally designed to establish signaling links for stationary networks. In *Mobile Networks and Applications* (Dec. 2003), RSVP was applied to mobile wireless technology (e.g., a PC notebook with wireless LAN card for Internet access). A simulation study revealed that the transmission delay (measured in milliseconds) of an RSVP–linked wireless device has an approximate normal distribution with mean $\mu = 48.5$ milliseconds and $\sigma = 8.5$ milliseconds.

a. What is the probability that the transmission delay is less than 57 milliseconds?

b. What is the probability that the transmission delay is between 40 and 60 milliseconds?

4.86 **Shell lengths of sea turtles.** Refer to the *Aquatic Biology* (Vol. 9, 2010) study of green sea turtles inhabiting the Grand Cayman South Sound lagoon, Exercise 2.83 (p. 65). Researchers discovered that the curved carapace (shell) length of these turtles is approximately normally distributed with mean 55.7 centimeters and standard deviation 11.5 centimeters.

a. The minimum and maximum size limits for captured turtles in the legal marine turtle fishery are 40 cm and 60 cm, respectively. How likely are you to capture a green sea turtle that is considered illegal?

b. What maximum limit, L, should be set so that only 10% of the turtles captured have shell lengths greater than L?

4.87 **NHTSA crash safety tests.** Refer to the National Highway Traffic Safety Administration (NHTSA) crash test data for new cars, presented in Exercise 2.174 (p. 98) and saved in the **CRASH** file. One of the variables measured is the severity of a driver's head injury when the car is in a head-on collision with a fixed barrier while traveling at 35 miles per hour. The more points assigned to the head injury rating, the more severe is the injury. The head injury ratings can be shown to be approximately normally distributed with a mean of 605 points and a standard deviation of 185 points. One of the crash-tested cars is randomly selected from the data, and the driver's head injury rating is observed.

a. Find the probability that the rating will fall between 500 and 700 points.

b. Find the probability that the rating will fall between 400 and 500 points.

c. Find the probability that the rating will be less than 850 points.

d. Find the probability that the rating will exceed 1,000 points.

NW **e.** What rating will only 10% of the crash-tested cars exceed?

Applying the Concepts—Intermediate

4.88 **Optimal goal target in soccer.** When attempting to score a goal in soccer, where should you aim your shot? Should you aim for a goalpost (as some soccer coaches teach), the middle of the goal, or some other target? To answer these questions, *Chance* (Fall 2009) utilized the normal probability distribution. Suppose the accuracy x of a professional soccer player's shots follows a normal distribution with a mean of zero feet and a standard deviation of 3 feet. (For example, if the player hits his target, $x = 0$; if he misses his target by 2 feet to the right, $x = 2$; and if he misses 1 foot to the left, $x = -1$.) Now, a regulation soccer goal is 24 feet wide. Assume that a goalkeeper will stop (save) all shots within 9 feet of where he is standing; all other shots

on goal will score. Consider a goalkeeper who stands in the middle of the goal.

a. If the player aims for the right goalpost, what is the probability that he will score?

b. If the player aims for the center of the goal, what is the probability that he will score?

c. If the player aims for halfway between the right goalpost and the outer limit of the goalkeeper's reach, what is the probability that he will score?

4.89 **Ambulance response time.** Ambulance response time is measured as the time (in minutes) between the initial call to emergency medical services (EMS) and when the patient is reached by ambulance. *Geographical Analysis* (Vol. 41, 2009) investigated the characteristics of ambulance response time for EMS calls in Edmonton, Alberta. For a particular EMS station (call it Station A), ambulance response time is known to be normally distributed with $\mu = 7.5$ minutes and $\sigma = 2.5$ minutes.

a. Regulations require that 90% of all emergency calls should be reached in 9 minutes or less. Are the regulations met at EMS Station A? Explain.

b. A randomly selected EMS call in Edmonton has an ambulance response time of 2 minutes. Is it likely that this call was serviced by Station A? Explain.

4.90 **Rating employee performance.** Almost all companies utilize some type of year-end performance review for their employees. Human Resources (HR) at the University of Texas Health Science Center provides guidelines for supervisors rating their subordinates. For example, raters are advised to examine their ratings for a tendency to be either too lenient or too harsh. According to HR, "if you have this tendency, consider using a normal distribution—10% of employees (rated) exemplary, 20% distinguished, 40% competent, 20% marginal, and 10% unacceptable." Suppose you are rating an employee's performance on a scale of 1 (lowest) to 100 (highest). Also, assume the ratings follow a normal distribution with a mean of 50 and a standard deviation of 15.

a. What is the lowest rating you should give to an "exemplary" employee if you follow the University of Texas HR guidelines?

b. What is the lowest rating you should give to a "competent" employee if you follow the University of Texas HR guidelines?

4.91 **Range of women's heights.** In *Chance* (Winter 2007), Yale Law School professor Ian Ayres published the results of a study he conducted with his son and daughter on whether college students could estimate a range for women's heights. The students were shown a graph of a normal distribution of heights and were asked, "The average height of women over 20 years old in the United States is 64 inches. Using your intuition, please give your best estimate of the range of heights that would include $C\%$ of women over 20 years old. Please make sure that the center of the range is the average height of 64 inches." The value of C was randomly selected as 50%, 75%, 90%, 95%, or 99% for each student surveyed.

a. Give your estimate of the range for $C = 50\%$ of women's heights.

b. Give your estimate of the range for $C = 75\%$ of women's heights.

c. Give your estimate of the range for $C = 90\%$ of women's heights.

d. Give your estimate of the range for $C = 95\%$ of women's heights.

e. Give your estimate of the range for $C = 99\%$ of women's heights.

f. The standard deviation of heights for women over 20 years old is known to be 2.6 inches. Use this information to revise your answers to parts **a–e**.

g. Which value of C has the most accurate estimated range? (*Note*: The researchers found that college students were most accurate for $C = 90\%$ and $C = 95\%$.)

4.92 **Alcohol, threats, and electric shocks.** A group of Florida State University psychologists examined the effects of alcohol on the reactions of people to a threat (*Journal of Abnormal Psychology*, Vol. 107, 1998). After obtaining a specified blood alcohol level, the psychologists placed experimental subjects in a room and threatened them with electric shocks. Using sophisticated equipment to monitor the subjects' eye movements, the psychologists recorded the startle response (measured in milliseconds) of each subject. The mean and standard deviation of the startle responses were 37.9 and 12.4, respectively. Assume that the startle response x for a person with the specified blood alcohol level is approximately normally distributed.

a. Find the probability that x is between 40 and 50 milliseconds.

b. Find the probability that x is less than 30 milliseconds.

c. Give an interval for x centered around 37.9 milliseconds so that the probability that x falls in the interval is .95.

d. Ten percent of the experimental subjects have startle responses above what value?

4.93 **California's electoral college votes**. During a presidential election, each state is allotted a different number of votes to the electoral college depending on population. For example, California is allotted 55 votes (the most) while several states (including the District of Columbia) are allotted 3 votes each (the least). When a presidential candidate wins the popular vote in a state, the candidate wins all the electoral college votes in that state. To become president, a candidate must win 270 of the total of 538 votes in the electoral college. *Chance* (Winter 2010) demonstrated the impact of winning California on the presidential election. Assuming a candidate wins California's 55 votes, the number of additional electoral college votes the candidate will win can be approximated by a normal distribution with $\mu = 241.5$ votes and $\sigma = 49.8$ votes. If a presidential candidate wins the popular vote in California, what are the chances that he or she becomes the next U.S. president?

Applying the Concepts—Advanced

4.94 **Industrial filling process.** The characteristics of an industrial filling process in which an expensive liquid is injected into a container were investigated in the *Journal of Quality Technology* (July 1999). The quantity injected per container is approximately normally distributed with mean 10 units and standard deviation .2 unit. Each unit of fill costs $20. If a container contains less than 10 units (i.e., is underfilled), it must be reprocessed at a cost of $10. A properly filled container sells for $230.

a. Find the probability that a container is underfilled.

b. A container is initially underfilled and must be reprocessed. Upon refilling, it contains 10.6 units. How much profit will the company make on this container?

c. The operations manager adjusts the mean of the filling process upward to 10.5 units in order to make the probability of underfilling approximately zero. Under these conditions, what is the expected profit per container?

4.95 **Box plots and the standard normal distribution.** What relationship exists between the standard normal distribution and the box-plot methodology (optional Section 2.8) for describing distributions of data by means of quartiles? The answer depends on the true underlying probability distribution of the data. Assume for the remainder of this exercise that the distribution is normal.

a. Calculate the values z_L and z_U of the standard normal random variable z that correspond, respectively, to the hinges of the box plot (i.e., the lower and upper quartiles Q_L and Q_U) of the probability distribution.

b. Calculate the z values that correspond to the inner fences of the box plot for a normal probability distribution.

c. Calculate the z values that correspond to the outer fences of the box plot for a normal probability distribution.

d. What is the probability that an observation lies beyond the inner fences of a normal probability distribution? The outer fences?

e. Can you now better understand why the inner and outer fences of a box plot are used to detect outliers in a distribution? Explain.

4.96 **Load on frame structures.** In the *Journal of the International Association for Shell and Spatial Structures* (April 2004), Japanese environmental researchers studied the performance of truss-and-frame structures subjected to uncertain loads. The load was assumed to have a normal distribution with a mean of 20 thousand pounds. Also, the probability that the load is between 10 and 30 thousand pounds is .95. On the basis of this information, find the standard deviation of the load distribution.

4.6 Descriptive Methods for Assessing Normality

In the chapters that follow, we learn how to make inferences about the population on the basis of information contained in the sample. Several of these techniques are based on the assumption that the population is approximately normally distributed. Consequently, it will be important to determine whether the sample data come from a normal population before we can apply these techniques properly.

A number of descriptive methods can be used to check for normality. In this section, we consider the four methods summarized in the following box:

Determining whether the Data are from an Approximately Normal Distribution

1. Construct either a histogram or stem-and-leaf display for the data, and note the shape of the graph. If the data are approximately normal, the shape of the histogram or stem-and-leaf display will be similar to the normal curve shown in Figure 4.13 (i.e., the display will be mound shaped and symmetric about the mean).

2. Compute the intervals $\bar{x} \pm s$, $\bar{x} \pm 2s$, and $\bar{x} \pm 3s$, and determine the percentage of measurements falling into each. If the data are approximately normal, the percentages will be approximately equal to 68%, 95%, and 100%, respectively.

3. Find the interquartile range IQR and standard deviation s for the sample, and then calculate the ratio IQR/s. If the data are approximately normal, then IQR/$s \approx 1.3$.

4. Construct a *normal probability plot* for the data. If the data are approximately normal, the points will fall (approximately) on a straight line.

The first two methods come directly from the properties of a normal distribution established in Section 4.5. Method 3 is based on the fact that for normal distributions, the z values corresponding to the 25th and 75th percentiles are $-.67$ and $.67$, respectively. (See Example 4.17.) Since $\sigma = 1$ for a standard normal distribution,

$$\frac{\text{IQR}}{\sigma} = \frac{Q_U - Q_L}{\sigma} = \frac{.67 - (-.67)}{1} = 1.34$$

The final descriptive method for checking normality is based on a *normal probability plot*. In such a plot, the observations in a data set are ordered from smallest to largest and are then plotted against the expected z-scores of observations calculated under the assumption that the data come from a normal distribution. When the data are, in fact, normally distributed, a linear (straight-line) trend will result. A nonlinear trend in the plot suggests that the data are nonnormal.

A **normal probability plot** for a data set is a scatterplot with the ranked data values on one axis and their corresponding expected z-scores from a standard normal distribution on the other axis. [*Note:* Computation of the expected standard normal z-scores is beyond the scope of this text. Therefore, we will rely on available statistical software packages to generate a normal probability plot.]

Example 4.24

Checking for Normal Data—EPA Estimated Gas Mileages

Problem The EPA mileage ratings on 100 cars, first presented in Chapter 2 (p. 38), are reproduced in Table 4.6. Recall that these data are saved in the **EPAGAS** file. Numerical and graphical descriptive measures for the data are shown on the MINITAB and SPSS printouts presented in Figure 4.26. Determine whether the EPA mileage ratings are from an approximate normal distribution.

Solution As a first check, we examine the MINITAB histogram of the data shown in Figure 4.26. Clearly, the mileages fall into an approximately mound shaped, symmetric distribution centered around the mean of about 37 mpg. Note that a normal curve is superimposed on the figure. Therefore, check #1 in the box indicates that the data are approximately normal.

Table 4.6		EPA Gas Mileage Ratings for 100 Cars (miles per gallon)							
36.3	41.0	36.9	37.1	44.9	36.8	30.0	37.2	42.1	36.7
32.7	37.3	41.2	36.6	32.9	36.5	33.2	37.4	37.5	33.6
40.5	36.5	37.6	33.9	40.2	36.4	37.7	37.7	40.0	34.2
36.2	37.9	36.0	37.9	35.9	38.2	38.3	35.7	35.6	35.1
38.5	39.0	35.5	34.8	38.6	39.4	35.3	34.4	38.8	39.7
36.3	36.8	32.5	36.4	40.5	36.6	36.1	38.2	38.4	39.3
41.0	31.8	37.3	33.1	37.0	37.6	37.0	38.7	39.0	35.8
37.0	37.2	40.7	37.4	37.1	37.8	35.9	35.6	36.7	34.5
37.1	40.3	36.7	37.0	33.9	40.1	38.0	35.2	34.8	39.5
39.9	36.9	32.9	33.8	39.8	34.0	36.8	35.0	38.1	36.9

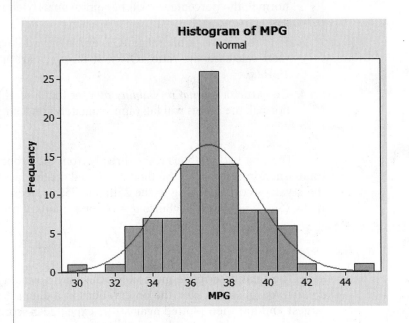

Figure 4.26a

MINITAB histogram for gas mileage data

Figure 4.26b

MINITAB descriptive statistics for gas mileage data

Descriptive Statistics: MPG

Variable	N	Mean	StDev	Minimum	Q1	Median	Q3	Maximum
MPG	100	36.994	2.418	30.000	35.625	37.000	38.375	44.900

To apply check #2, we obtain $\bar{x} = 37$ and $s = 2.4$ from the MINITAB printout of Figure 4.26b. The intervals $\bar{x} \pm s$, $\bar{x} \pm 2s$, and $\bar{x} \pm 3s$ are shown in Table 4.7, as is the percentage of mileage ratings that fall into each interval. (We obtained these results in Section 2.6, p. 66.) These percentages agree almost exactly with those from a normal distribution.

Check #3 in the box requires that we find the ratio IQR/s. From Figure 4.26b, the 25th percentile (labeled Q_1 by MINITAB) is $Q_L = 35.625$ and the 75th percentile (labeled Q_3 by MINITAB) is $Q_U = 38.375$. Then IQR = $Q_U - Q_L = 2.75$, and the ratio is

$$\frac{\text{IQR}}{s} = \frac{2.75}{2.4} = 1.15$$

Since this value is approximately equal to 1.3, we have further confirmation that the data are approximately normal.

A fourth descriptive method is to interpret a normal probability plot. An SPSS normal probability plot of the mileage data is shown in Figure 4.26c. Notice that the ordered mileage values (shown on the horizontal axis) fall reasonably close to a straight line when plotted against the expected values from a normal distribution. Thus, check #4 also suggests that the EPA mileage data are approximately normally distributed.

Figure 4.26c

SPSS normal probability plot for gas mileage data

Table 4.7	Describing the 100 EPA Mileage Ratings
Interval	Percentage in Interval
$\bar{x} \pm s = (34.6, 39.4)$	68
$\bar{x} \pm 2s = (32.2, 41.8)$	96
$\bar{x} \pm 3s = (29.8, 44.2)$	99

Look Back The checks for normality given in the box are simple, yet powerful, techniques to apply, but they are only descriptive in nature. It is possible (although unlikely) that the data are nonnormal even when the checks are reasonably satisfied. Thus, we should be careful not to claim that the 100 EPA mileage ratings are, in fact, normally distributed. We can only state that it is reasonable to believe that the data are from a normal distribution.*

Now Work Exercise 4.104

As we will learn in the next chapter, several inferential methods of analysis require the data to be approximately normal. If the data are clearly nonnormal, inferences derived from the method may be invalid. Therefore, it is advisable to check the normality of the data prior to conducting any analysis.

Statistics IN Action | **Revisited**

Assessing whether the Normal Distribution Is Appropriate for Modeling the Super Weapon Hit Data

In *Statistics in Action Revisited* in Section 4.5, we used the normal distribution to find the probability that a single flechette from a super weapon that shoots 1,100 flechettes at once hits one of three targets at 500 meters. Recall that for three range tests, the weapon was always aimed at the center target (i.e., the specification mean was set at $\mu = 5$ feet), but the specification standard deviation was varied at $\sigma = 1$ foot, $\sigma = 2$ feet, and $\sigma = 4$ feet. Table SIA4.1 shows the calculated normal probabilities of hitting the three targets for the different values of σ, as well as the actual results of the three range tests. (Recall that the actual data are saved in the **MOAGUN** file.) You can see that the proportion of the

(continued)

---

*Statistical tests of normality that provide a measure of reliability for the inference are available. However, these tests tend to be very sensitive to slight departures from normality (i.e., they tend to reject the hypothesis of normality for any distribution that is not perfectly symmetrical and mound shaped). Consult the references (see especially Ramsey & Ramsey, 1990) if you want to learn more about these tests.

Statistics IN Action
(continued)

1,100 flechettes that actually hit each target—called the hit ratio—agrees very well with the estimated probability of a hit derived from the normal distribution.

Table SIA4.1	Summary of Normal Probability Calculations and Actual Range Test Results			
Target	Specification	Normal Probability	Actual Number of Hits	Hit Ratio (Hits/1,100)
LEFT (− 1 to 1)	$\sigma = 1$	.0000	0	.000
	$\sigma = 2$	.0214	30	.027
	$\sigma = 4$	.0919	73	.066
MIDDLE (4 to 6)	$\sigma = 1$	.6826	764	.695
	$\sigma = 2$	.3820	409	.372
	$\sigma = 4$	.1974	242	.220
RIGHT (9 to 11)	$\sigma = 1$	.0000	0	.000
	$\sigma = 2$	.0214	23	.021
	$\sigma = 4$	.0919	93	.085

Consequently, it appears that our assumption that the horizontal hit measurements are approximately normally distributed is reasonably satisfied. Further evidence that this assumption is satisfied is provided by the MINITAB histograms of the horizontal hit measurements shown in Figures SIA4.3a–c. The normal curves superimposed on the histograms fit the data very well.

Data Set: MOAGUN

Figure SIA4.3a
MINITAB histogram for the horizontal hit measurements when $\sigma = 1$

Figure SIA4.3b
MINITAB histogram for the horizontal hit measurements when $\sigma = 2$

Figure SIA4.3c
MINITAB histogram for the horizontal hit measurements when $\sigma = 4$

Exercises 4.97–4.116

Understanding the Principles

4.97 Why is it important to check whether the sample data come from a normal population?

4.98 Give four methods for determining whether the sample data come from a normal population.

4.99 If a population data set is normally distributed, what is the proportion of measurements you would expect to fall within the following intervals?
a. $\mu \pm \sigma$ **b.** $\mu \pm 2\sigma$ **c.** $\mu \pm 3\sigma$

4.100 What is a normal probability plot and how is it used?

Learning the Mechanics

4.101 Normal probability plots for three data sets are shown below. Which plot indicates that the data are approximately normally distributed?

4.102 Consider a sample data set with the following summary statistics: $s = 95$, $Q_L = 72$, and $Q_U = 195$.
a. Calculate IQR.
b. Calculate IQR/s.
c. Is the value of IQR/s approximately equal to 1.3? What does this imply?

4.103 Examine the following sample data saved in the **LM4_103** file.

32	48	25	135	53	37	5	39	213	165
109	40	1	146	39	25	21	66	64	57
197	60	112	10	155	134	301	304	107	82
35	81	60	95	401	308	180	3	200	59

a. Construct a stem-and-leaf plot to assess whether the data are from an approximately normal distribution.
b. Find the values of Q_L, Q_U, and s for the sample data.
c. Use the results from part **b** to assess the normality of the data.
d. Generate a normal probability plot for the data, and use it to assess whether the data are approximately normal.

4.104 Examine the following sample data saved in the **LM4_104** file.

5.9	5.3	1.6	7.4	8.6	3.2	2.1
4.0	7.3	8.4	5.9	6.7	4.5	6.3
6.0	9.7	3.5	3.1	4.3	3.3	8.4
4.6	8.2	6.5	1.1	5.0	9.4	6.4

a. Construct a stem-and-leaf plot to assess whether the data are from an approximately normal distribution.
b. Compute s for the sample data.
c. Find the values of Q_L and Q_U and the value of s from part **b** to assess whether the data come from an approximately normal distribution.
d. Generate a normal probability plot for the data, and use it to assess whether the data are approximately normal.

Applying the Concepts—Basic

4.105 Most powerful American women. Refer to the *Fortune* (Oct. 16, 2008) list of the 50 most powerful women in America. In Exercise 4.83 (p. 200), you assumed that the ages (in years) of these women are approximately normally distributed. A MINITAB printout with summary statistics for the age variable saved in the **WPOWER50** file is reproduced below.

Descriptive Statistics: AGE

Variable	N	Mean	StDev	Minimum	Q1	Median	Q3	Maximum
AGE	50	50.020	6.444	28.000	47.000	51.000	54.000	64.000

a. Use the relevant statistics on the printout to find the interquartile range IQR.
b. Locate the value of the standard deviation s on the printout.
c. Use the results from parts **a** and **b** to demonstrate that the age distribution is approximately normal.
d. Construct a relative frequency histogram for the age data. Use this graph to support your assumption of normality.

4.106 Shear strength of rock fractures. Understanding the characteristics of rock masses, especially the nature of the fractures, are essential when building dams and power plants. The shear strength of rock fractures was investigated in *Engineering Geology* (May 12, 2010). The Joint Roughness Coefficient (JRC) was used to measure shear strength. Civil engineers collected JRC data for over 750 rock fractures. The results (simulated from information provided in the article) are summarized in the SPSS histogram shown below. Should the engineers use the normal probability distribution to model the behavior of shear strength for rock fractures? Explain.

Plots for Exercise 4.101

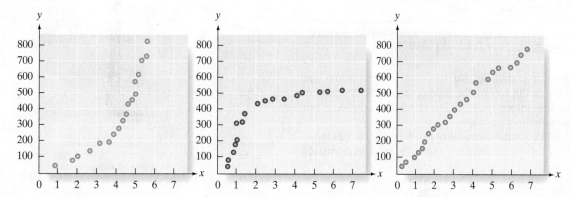

SPSS Output for Exercise 4.106

Based on Pooyan Asadollahi and Fulvio Tonon, "Constitutive model for rock fractures: Revisiting Bartoris empirical model." *Engineering Geology*, vol. 113, no. 1, pp. 11–32.

4.107 Software file updates. Software configuration management was used to monitor a software engineering team's performance at Motorola, Inc. (*Software Quality Professional*, Nov. 2004). One of the variables of interest was the number of updates to a file changed because of a problem report. Summary statistics for $n = 421$ files yielded the following results: $\bar{x} = 4.71$, $s = 6.09$, $Q_L = 1$, and $Q_U = 6$. Are these data approximately normally distributed? Explain.

4.108 Drug content assessment. Scientists at GlaxoSmithKline Medicines Research Center used high-performance liquid chromatography (HPLC) to determine the amount of drug in a tablet produced by the company (*Analytical Chemistry*, Dec. 15, 2009). Drug concentrations (measured as a percentage) for 50 randomly selected tablets are listed in the accompanying table and saved in the **DRUGCON** file.

91.28	92.83	89.35	91.90	82.85	94.83	89.83	89.00	84.62
86.96	88.32	91.17	83.86	89.74	92.24	92.59	84.21	89.36
90.96	92.85	89.39	89.82	89.91	92.16	88.67	89.35	86.51
89.04	91.82	93.02	88.32	88.76	89.26	90.36	87.16	91.74
86.12	92.10	83.33	87.61	88.20	92.78	86.35	93.84	91.20
93.44	86.77	83.77	93.19	81.79				

Based on Borman, P. J., et al., "Design and analysis of method equivalence studies." *Analytical Chemistry*, Vol. 81, No. 24, December 15, 2009 (Table 3).

a. Descriptive statistics for the drug concentrations are shown at the top of the accompanying SPSS printout. Use this information to assess whether the data are approximately normal.

b. An SPSS normal probability plot follows. Use this information to assess whether the data are approximately normal.

SPSS Output for Exercise 4.108

Descriptives

			Statistic	Std. Error
Content	Mean		89.2906	.45021
	95% Confidence Interval for Mean	Lower Bound	88.3859	
		Upper Bound	90.1953	
	5% Trimmed Mean		89.3963	
	Median		89.3750	
	Variance		10.134	
	Std. Deviation		3.18344	
	Minimum		81.79	
	Maximum		94.83	
	Range		13.04	
	Interquartile Range		4.84	
	Skewness		-.544	.337
	Kurtosis		-.389	.662

4.109 Estimating glacier elevations. Digital elevation models (DEMs) are now used to estimate elevations and slopes of remote regions. In *Arctic, Antarctic, and Alpine Research* (May 2004), geographers analyzed reading errors from maps produced by DEMs. Two readers of a DEM map of White Glacier (in Canada) estimated elevations at 400 points in the area. The difference between the elevation

"Uncertainty in digital elevation models of Axel Heiberg Island. Arctic Canada," *Arctic, Antarctic, and Alpine Research*, Vol. 36, No. 2, May 2004 (Figure 3). Regents of the University of Colorado. Reprinted with permission.

estimates of the two readers had a mean of $\mu = .28$ meter and a standard deviation of $\sigma = 1.6$ meters. A histogram of the difference (with a normal histogram superimposed on the graph) is shown on p. 208.

a. On the basis of the histogram, the researchers concluded that the difference between elevation estimates is not normally distributed. Why?

b. Will the interval $\mu \pm 2\sigma$ contain more than 95%, exactly 95%, or less than 95% of the 400 elevation differences? Explain.

Applying the Concepts—Intermediate

4.110 Habitats of endangered species. An evaluation of the habitats of endangered salmon species was performed in *Conservation Ecology* (December 2003). The researchers identified 734 sites (habitats) for Chinook, coho, or steelhead salmon species in Oregon and assigned a habitat quality score to each site. (Scores range from 0 to 36 points, with lower scores indicating poorly maintained or degraded habitats.) The data are saved in the **HABITAT** file. Give your opinion on whether the habitat quality score is normally distributed.

4.111 Baseball batting averages. Major League Baseball (MLB) has two leagues: the American League (AL), which utilizes the designated hitter (DH) to bat for the pitcher, and the National League (NL), which does not allow the DH. A player's batting average is computed by dividing the player's total number of hits by his official number of at bats. The batting averages for all AL and NL players with at least 100 official at bats during the 2006 season are stored in the **MLB2009-AL** and **MLB2009-NL** files, respectively. Determine whether each batting average distribution is approximately normal.

4.112 Ranking the driving performance of professional golfers. Refer to *The Sport Journal* (Winter 2007) article on a new method for ranking the driving performance of PGA golfers, presented in Exercise 2.64 (p. 59). Recall that the method incorporates a golfer's average driving distance (yards) and driving accuracy (percentage of drives that land in the fairway) into a driving performance index. Data on these three variables for the top 40 PGA golfers are saved in the **PGADRIVER** file. Determine which of the variables—driving distance, driving accuracy, and driving performance index—are approximately normally distributed.

4.113 NHTSA crash tests. Refer to the National Highway Traffic Safety Administration (NHTSA) crash test data for new cars. In Exercise 4.87 (p. 201), you assumed that the driver's head injury rating is approximately normally distributed. Apply the methods of this chapter to the data saved in the **CRASH** file to support that assumption.

4.114 Cruise ship sanitation scores. Refer to the data on the Jan. 2010 sanitation scores for 186 cruise ships, presented in Exercise 2.35 (p. 47). The data are saved in the **SHIPSANIT** file. Assess whether the sanitation scores are approximately normally distributed.

Applying the Concepts—Advanced

4.115 Laptop use in middle school. Refer to the *American Secondary Education* (Fall 2009) study of student usage of laptops at a middle school, Exercise 2.100 (p. 72). Recall that x, the number of minutes per day each student used his or her laptop for taking notes, has a mean of 13.2 with a standard deviation of 19.5. Demonstrate why the probability distribution of x is unlikely to be normal.

4.116 Ranking Ph.D. programs in economics. Refer to the *Southern Economic Journal* (Apr. 2008) rankings of Ph.D. programs in economics at 129 colleges and universities, Exercise 2.121 (p. 77). Recall that the number of publications published by faculty teaching in the Ph.D. program and the quality of the publications were used to calculate an overall productivity score for each program. The mean and standard deviation of these 129 productivity scores were then used to compute a z-score for each economics program. The data (z-scores) for all 129 economic programs are saved in the **ECOPHD** file. A MINITAB normal probability plot for the z-scores is shown in the accompanying printout.

a. Use the graph to assess whether the data are approximately normal.

b. Based on the graph, determine the nature of the skewness of the data.

4.7 Approximating a Binomial Distribution with a Normal Distribution (Optional)

When the discrete binomial random variable (Section 4.3) can assume a large number of values, the calculation of its probabilities may become tedious. To contend with this problem, we provide tables in Appendix A that give the probabilities for some values of n and p, but these tables are by necessity incomplete. Recall that the binomial probability table (Table II) can be used only for $n = 5, 6, 7, 8, 9, 10, 15, 20,$ or 25. To deal with this limitation, we seek approximation procedures for calculating the probabilities associated with a binomial probability distribution.

When n is large, a normal probability distribution may be used to provide a good approximation to the probability distribution of a binomial random variable. To show how this approximation works, we refer to Example 4.13, in which we used the binomial distribution to model the number x of 20 eligible voters who favor a candidate. We assumed that 60% of all the eligible voters favored the candidate. The mean and standard deviation of x were found to be $\mu = 12$ and $\sigma = 2.2$, respectively. The binomial distribution for $n = 20$ and $p = .6$ is shown in Figure 4.27, and the approximating normal distribution with mean $\mu = 12$ and standard deviation $\sigma = 2.2$ is superimposed.

Figure 4.27

Binomial distribution for $n = 20, p = .6$ and normal distribution with $\mu = 12, \sigma = 2.2$

As part of Example 4.13, we used Table II to find the probability that $x \leq 10$. This probability, which is equal to the sum of the areas contained in the rectangles (shown in Figure 4.27) that correspond to $p(0), p(1), p(2), \ldots, p(10)$, was found to equal .245. The portion of the normal curve that would be used to approximate the area $p(0) + p(1) + p(2) + \cdots + p(10)$ is highlighted in Figure 4.27. *Note that this highlighted area lies to the left of 10.5 (not 10), so we may include all of the probability in the rectangle corresponding to* $p(10)$. Because we are approximating a discrete distribution (the binomial) with a continuous distribution (the normal), we call the use of 10.5 (instead of 10 or 11) a **correction for continuity.** That is, we are correcting the discrete distribution so that it can be approximated by the continuous one. The use of the correction for continuity leads to the calculation of the following standard normal z-value:

$$z = \frac{x - \mu}{\sigma} = \frac{10.5 - 12}{2.2} = -.68$$

Using Table III, we find the area between $z = 0$ and $z = .68$ to be .2517. Then the probability that x is less than or equal to 10 is approximated by the area under the normal distribution to the left of 10.5, shown highlighted in Figure 4.27. That is,

$$P(x \leq 10) \approx P(z \leq -.68) = .5 - P(-.68 < z \leq 0) = .5 - .2517 = .2483$$

The approximation differs only slightly from the exact binomial probability, .245. Of course, when tables of exact binomial probabilities are available, we will use the exact value rather than a normal approximation.

The normal distribution will not always provide a good approximation to binomial probabilities. The following is a useful rule of thumb to determine when n is large enough for the approximation to be effective: *The interval $\mu \pm 3\sigma$ should lie within the range of the binomial random variable x (i.e., from 0 to n) in order for the normal approximation to be adequate.* The rule works well because almost all of the normal distribution falls within three standard deviations of the mean, so if this interval is contained within the range of x values, there is "room" for the normal approximation to work.

As shown in Figure 4.28a for the preceding example with $n = 20$ and $p = .6$, the interval $\mu \pm 3\sigma = 12 + 3(2.2) = (5.4, 18.6)$ lies within the range from 0 to 20. However, if we were to try to use the normal approximation with $n = 10$ and $p = .1$, the interval $\mu \pm 3\sigma$ becomes $1 \pm 3(.95)$, or $(-1.85, 3.85)$. As shown in Figure 4.28b, this interval is not contained within the range of x, since $x = 0$ is the lower bound for a binomial random variable. Note in Figure 4.28b that the normal distribution will not "fit" in the range of x; therefore, it will not provide a good approximation to the binomial probabilities.

a. $n = 20, p = .6$: Normal approximation is good

b. $n = 10, p = .1$: Normal approximation is poor

Figure 4.28

Rule of thumb for normal approximation to binomial probabilities

---

BIOGRAPHY **ABRAHAM DE MOIVRE (1667–1754)**

Advisor to Gamblers

French-born mathematician Abraham de Moivre moved to London when he was 21 years old to escape religious persecution. In England, he earned a living first as a traveling teacher of mathematics and then as an advisor to gamblers, underwriters, and annuity brokers. De Moivre's major contributions to probability theory are contained in two of his books: *The Doctrine of Chances* (1718) and *Miscellanea Analytica* (1730). In these works, he defines statistical independence, develops the formula for the normal probability distribution, and derives the normal curve as an approximation to the binomial distribution. Despite his eminence as a mathematician, de Moivre died in poverty. He is famous for using an arithmetic progression to predict the day of his death. ∎

Example 4.25

Approximating a Binomial Probability with the Normal Distribution— Lot Acceptance Sampling

Problem One problem with any product that is mass produced (e.g., a graphing calculator) is quality control. The process must be monitored or audited to be sure that the output of the process conforms to require- ments. One monitoring method is *lot acceptance sampling,* in which items being produced are sampled at various stages of the production process and are carefully inspected. The lot of items from which the sample is drawn is then accepted or rejected on the basis of the number of defectives in the sample. Lots that are accepted may be sent forward for further processing or may be shipped to customers; lots that are rejected may be reworked or scrapped. For example, suppose a manufacturer of calculators chooses 200 stamped circuits from the day's production and determines x, the number of defective circuits in the sample. Suppose that up to a 6% rate of defectives is considered acceptable for the process.

a. Find the mean and standard deviation of x, assuming that the rate of defectives is 6%.

b. Use the normal approximation to determine the probability that 20 or more defec- tives are observed in the sample of 200 circuits (i.e., find the approximate probability that $x \geq 20$).

Solution

a. The random variable x is binomial with $n = 200$ and the fraction defective $p = .06$. Thus,

$$\mu = np = 200(.06) = 12$$

$$\sigma = \sqrt{npq} = \sqrt{200(.06)(.94)} = \sqrt{11.28} = 3.36$$

We first note that

$$\mu \pm 3\sigma = 12 \pm 3(3.36) = 12 \pm 10.08 = (1.92, 22.08)$$

lies completely within the range from 0 to 200. Therefore, a normal probability distribution should provide an adequate approximation to this binomial distribution.

b. By the rule of complements, $P(x \geq 20) = 1 - P(x \leq 19)$. To find the approximat- ing area corresponding to $x \leq 19$, refer to Figure 4.29. Note that we want to include all the binomial probability histograms from 0 to 19, inclusive. Since the event is of the form $x \leq a$, the proper correction for continuity is $a + .5 = 19 + .5 = 19.5$. Thus, the z value of interest is

$$z = \frac{(a + .5) - \mu}{\sigma} = \frac{19.5 - 12}{3.36} = 2.23$$

$f(x)$

$P(x \leq 19)$

3 4 5 6 7 8 9 10 11 12 13 14 15 16 17 18 19 20

x

Figure 4.29

Normal approximation to the binomial distribution with $n = 200, p = .06$

Referring to Table III in Appendix A, we find that the area to the right of the mean, 0, corresponding to $z = 2.23$ (see Figure 4.30) is .4871. So the area $A = P(z \leq 2.23)$ is

$$A = .5 + .4871 = .9871$$

Figure 4.30
Standard normal distribution

Thus, the normal approximation to the binomial probability we seek is

$$P(x \geq 20) = 1 - P(x \leq 19) \approx 1 - .9871 = .0129$$

In other words, *if, in fact, the true fraction of defectives is .06*, then the probability that 20 or more defectives will be observed in a sample of 200 circuits is extremely small.

Look Back If the manufacturer observes $x \geq 20$, the likely reason is that the process is producing more than the acceptable 6% defectives. The lot acceptance sampling procedure is another example of using the rare-event approach to make inferences.

Now Work Exercise 4.119

The steps for approximating a binomial probability by a normal probability are given in the following box:

Using a Normal Distribution to Approximate Binomial Probabilities

1. After you have determined n and p for the binomial distribution, calculate the interval

$$\mu \pm 3\sigma = np \pm 3\sqrt{npq}$$

If the interval lies in the range from 0 to n, the normal distribution will provide a reasonable approximation to the probabilities of most binomial events.

2. Express the binomial probability to be approximated in the form $P(x \leq a)$ or $P(x \leq b) - P(x \leq a)$. For example,

$$P(x < 3) = P(x \leq 2)$$
$$P(x \geq 5) = 1 - P(x \leq 4)$$
$$P(7 \leq x \leq 10) = P(x \leq 10) - P(x \leq 6)$$

3. For each value of interest, a, the correction for continuity is $(a + .5)$ and the corresponding standard normal z value is

$$z = \frac{(a + .5) - \mu}{\sigma} \text{ (see Figure 4.31)}$$

4. Sketch the approximating normal distribution and shade the area corresponding to the probability of the event of interest, as in Figure 4.31. Verify that the rectangles you have included in the shaded area correspond to the probability you wish to approximate. Use Table III and the z value(s) you calculated in step 3 to find the shaded area. This is the approximate probability of the binomial event.

Figure 4.31
Approximating binomial probabilities by normal probabilities

Exercises 4.117–4.136

Understanding the Principles

4.117 For large n (say, $n = 100$), why is it advantageous to use the normal distribution to approximate a binomial probability?

4.118 Why do we need a correction for continuity when approximating a binomial probability with the normal distribution?

Learning the Mechanics

4.119 Suppose x is a binomial random variable with $p = .4$ and
NW $n = 25$.
  a. Would it be appropriate to approximate the probability distribution of x with a normal distribution? Explain.
  b. Assuming that a normal distribution provides an adequate approximation to the distribution of x, what are the mean and variance of the approximating normal distribution?
  c. Use Table II of Appendix A to find the exact value of $P(x \geq 9)$.
  d. Use the normal approximation to find $P(x \geq 9)$.

4.120 Assume that x is a binomial random variable with n and p as specified in parts **a–f** that follow. For which cases would it be appropriate to use a normal distribution to approximate the binomial distribution?
  a. $n = 100, p = .01$ **b.** $n = 20, p = .6$
  c. $n = 10, p = .4$ **d.** $n = 1,000, p = .05$
  e. $n = 100, p = .8$ **f.** $n = 35, p = .7$

4.121 Assume that x is a binomial random variable with $n = 25$ and $p = .5$. Use Table II of Appendix A and the normal approximation to find the exact and approximate values, respectively, of the following probabilities:
  a. $P(x \leq 11)$ **b.** $P(x \geq 16)$
  c. $P(8 \leq x \leq 16)$

4.122 Assume that x is a binomial random variable with $n = 1,000$ and $p = .50$. Find each of the following probabilities:
  a. $P(x > 500)$ **b.** $P(490 \leq x < 500)$
  c. $P(x > 550)$

4.123 Assume that x is a binomial random variable with $n = 100$ and $p = .40$. Use a normal approximation to find the following:
  a. $P(x \leq 35)$ **b.** $P(40 \leq x \leq 50)$
  c. $P(x \geq 38)$

Applying the Concepts—Basic

4.124 **Where will you get your next pet?** Refer to Exercise 4.52 (p. 186) and the Associated Press/Petside.com poll that revealed that half of all pet owners would get their next dog or cat from a shelter (*USAToday*, May 12, 2010). Consider a random sample of 500 pet owners and define x as the number of pet owners who would acquire their next dog or cat from a shelter. In Exercise 4.52, you determined that x is a binomial random variable.
  a. Compute μ and σ for the probability distribution of x.
  b. Compute the z-score for the value $x = 240$.
  c. Compute the z-score for the value $x = 270$.
  d. Use the technique of this section to approximate $P(240 < x < 270)$.

4.125 **Analysis of bottled water.** Refer to the *Scientific American* (July 2003) report on whether bottled water is really purified water, presented in Exercise 4.54 (p. 187). Recall that the Natural Resources Defense Council found that 25% of bottled-water brands fill their bottles with just tap water. In a random sample of 65 bottled-water brands, let x be the number that contain tap water.
  a. Find the mean of x.
  b. Find the standard deviation of x.
  c. Find the z-score for the value $x = 20$.
  d. Find the approximate probability that 20 or more of the 65 sampled bottled-water brands will contain tap water.

4.126 **LASIK surgery complications.** According to recent studies, 1% of all patients who undergo laser surgery (i.e., LASIK) to correct their vision have serious post-laser vision problems (*All About Vision*, 2006). In a random sample of 100,000 LASIK patients, let x be the number who experience serious post-laser vision problems.
  a. Find $E(x)$.
  b. Find $\text{Var}(x)$.
  c. Find the z-score for $x = 950$.
  d. Find the approximate probability that fewer than 950 patients in a sample of 100,000 will experience serious post-laser vision problems.

4.127 **Melanoma deaths.** According to the *American Cancer Society*, melanoma, a form of skin cancer, kills 15% of Americans who suffer from the disease each year. Consider a sample of 10,000 melanoma patients.
  a. What are the expected value and variance of x, the number of the 10,000 melanoma patients who die of the affliction this year?
  b. Find the probability that x will exceed 1,600 patients per year.
  c. Would you expect x, the number of patients dying of melanoma, to exceed 6,500 in any single year? Explain.

4.128 **Caesarian birth study.** In Exercise 4.56 (p. 187), you learned that 32% of all births in the United States occur by Caesarian section each year (*National Vital Statistics Reports*, Mar. 2010). In a random sample of 1,000 births this year, let x be the number that occur by Caesarian section.
  a. Find the mean of x. (This value should agree with your answer to Exercise 4.56 **a.**)
  b. Find the standard deviation of x. (This value should agree with your answer to Exercise 4.56 **b.**)
  c. Find the z-score for the value $x = 200.5$.
  d. Find the approximate probability that the number of Caesarian sections in a sample of 1,000 births is less than or equal to 200.

Applying the Concepts—Intermediate

4.129 **Ecotoxicological survival study.** The *Journal of Agricultural, Biological and Environmental Statistics* (Sept. 2000) gave an evaluation of the risk posed by hazardous pollutants. In the experiment, guppies (all the same age and size) were released into a tank of natural seawater polluted with the pesticide dieldrin and the number of guppies surviving after five days was determined. The researchers estimated that the probability of any single guppy surviving is .60. If 300 guppies are released into the polluted tank, estimate the probability that fewer than 100 guppies survive after five days.

4.130 Chemical signals of mice. Refer to the *Cell* (May 14, 2010) study of the ability of a mouse to recognize the odor of a potential predator, Exercise 4.53 (p. 186). You learned that 40% of lab mice cells exposed to chemically produced major urinary proteins (Mups) from a cat responded positively (i.e., recognized the danger of the lurking predator). Again, consider a sample of 100 lab mice cells, each exposed to chemically produced cat Mups, and let x represent the number of cells that respond positively. How likely is it that less than half of the cells respond positively to cat Mups?

4.131 Defects in semiconductor wafers. The computer chips in notebook and laptop computers are produced from semiconductor wafers. Certain semiconductor wafers are exposed to an environment that generates up to 100 possible defects per wafer. The number of defects per wafer, x, was found to follow a binomial distribution if the manufacturing process is stable and generates defects that are randomly distributed on the wafers (*IEEE Transactions on Semiconductor Manufacturing*, May 1995). Let p represent the probability that a defect occurs at any one of the 100 points of the wafer. For each of the following cases, determine whether the normal approximation can be used to characterize x:

a. $p = .01$ **b.** $p = .50$ **c.** $p = .90$

4.132 Victims of domestic abuse. In Exercise 4.61 (p. 187), you learned that some researchers believe that one in every three women has been a victim of domestic abuse (Domestic Vidence: Incidence and Prevalance study, Sep–Dec., 2005).

a. For a random sample of 150 women, what is the approximate probability that more than half are victims of domestic abuse?

b. For a random sample of 150 women, what is the approximate probability that fewer than 50 are victims of domestic abuse?

c. Would you expect to observe fewer than 30 domestically abused women in a sample of 150? Explain.

4.133 Hotel guest satisfaction. Refer to the 2009 North American Hotel Guest Satisfaction Index Study, Exercise 4.57 (p. 187). You determined that the probability of a hotel guest participating in the hotel's "green" conservation program by reusing towels and bed linens is .45. Suppose a large hotel chain randomly samples 200 of its guests. The chain's national director claims that more than 110 of these

guests participated in the conservation program. Do you believe this claim? Explain.

Applying the Concepts—Advanced

4.134 Body fat in men. The percentage of fat in the bodies of American men is an approximately normal random variable with mean equal to 15% and standard deviation equal to 2%.

a. If these values were used to describe the body fat of men in the U.S. Army, and if a measure of 20% or more body fat characterizes the person as obese, what is the approximate probability that a random sample of 10,000 soldiers will contain fewer than 50 who would actually be characterized as obese?

b. If the army actually were to check the percentage of body fat for a random sample of 10,000 men, and if only 30 contained 20% (or higher) body fat, would you conclude that the army was successful in reducing the percentage of obese men below the percentage in the general population? Explain your reasoning.

4.135 Luggage inspection at Newark airport. *New Jersey Business* reports that Newark International Airport's terminal handles an average of 3,000 international passengers an hour, but is capable of handling twice that number. Also, after scanning all luggage, 20% of arriving international passengers are detained for intrusive luggage inspection. The inspection facility can handle 600 passengers an hour without unreasonable delays for the travelers.

a. When international passengers arrive at the rate of 1,500 per hour, what is the expected number of passengers who will be detained for luggage inspection?

b. In the future, it is expected that as many as 4,000 international passengers will arrive per hour. When that occurs, what is the expected number of passengers who will be detained for luggage inspection?

c. Refer to part **b**. Find the approximate probability that more than 600 international passengers will be detained for luggage inspection. (This is also the probability that travelers will experience unreasonable luggage inspection delays.)

4.136 Waiting time at an emergency room. According to *Health Affairs* (Oct. 28, 2004), the median time a patient waits to see a doctor in a typical U.S. emergency room is 30 minutes. On a day when 150 patients visit the emergency room, what is the approximate probability that

a. More than half will wait more than 30 minutes?

b. More than 85 will wait more than 30 minutes?

c. More than 60, but fewer than 90, will wait more than 30 minutes?

4.8 Sampling Distributions

In previous sections, we assumed that we knew the probability distribution of a random variable, and using this knowledge, we were able to compute the mean, variance, and probabilities associated with the random variable. However, in most practical applications, the true mean and standard deviation are unknown quantities that have to be estimated. Numerical quantities that describe probability distributions are called *parameters*. Thus, p, the probability of a success in a binomial experiment, and μ and σ, the mean and standard deviation, respectively, of a normal distribution, are examples of parameters.

A **parameter** is a numerical descriptive measure of a population. Because it is based on the observations in the population, its value is almost always unknown.

We have also discussed the sample mean $\bar{x}$, sample variance s^2, sample standard deviation s, and the like, which are numerical descriptive measures calculated from the sample. (See Table 4.8 for a list of the statistics covered so far in this text.) We will often use the information contained in these *sample statistics* to make inferences about the parameters of a population.

Table 4.8   List of Population Parameters and Corresponding Sample Statistics

	Population Parameter	Sample Statistic
Mean:	μ	$\bar{x}$
Variance:	σ^2	s^2
Standard deviation:	σ	s
Binomial proportion:	p	$\hat{p}$

A **sample statistic** is a numerical descriptive measure of a sample. It is calculated from the observations in the sample.

Note that the term *statistic* refers to a *sample* quantity and the term *parameter* refers to a *population* quantity.

Before we can show you how to use sample statistics to make inferences about population parameters, we need to be able to evaluate their properties. Does one sample statistic contain more information than another about a population parameter? On what basis should we choose the "best" statistic for making inferences about a parameter? For example, if we want to estimate a parameter of a population—say, the population mean μ—we can use a number of sample statistics for our estimate. Two possibilities are the sample mean $\bar{x}$ and the sample median M. Which of these do you think will provide a better estimate of μ?

Before answering this question, consider the following example: Toss a fair die, and let x equal the number of dots showing on the up face. Suppose the die is tossed three times, producing the sample measurements 2, 2, 6. The sample mean is then $\bar{x} = 3.33$, and the sample median is $M = 2$. Since the population mean of x is $\mu = 3.5$, you can see that, for this sample of three measurements, the sample mean $\bar{x}$ provides an estimate that falls closer to μ than does the sample median (see Figure 4.32a). Now suppose we toss the die three more times and obtain the sample measurements 3, 4, 6. Then the mean and median of this sample are $\bar{x} = 4.33$ and $M = 4$, respectively. This time M is closer to μ. (See Figure 4.32b.)

Figure 4.32

Comparing the sample mean ($\bar{x}$) and sample median (M) as estimators of the population mean (μ)

a. Sample 1: $\bar{x}$ is closer than M to μ b. Sample 2: M is closer than $\bar{x}$ to μ

This simple example illustrates an important point: Neither the sample mean nor the sample median will *always* fall closer to the population mean. Consequently, we cannot compare these two sample statistics or, in general, any two sample statistics on the basis of their performance with a single sample. Instead, we need to recognize that sample statistics are themselves random variables, because different samples can lead to different values for the sample statistics. As random variables, sample statistics must be judged and compared on the basis of their probability distributions (i.e., the *collection* of values and associated probabilities of each statistic that would be obtained if the sampling experiment were repeated a *very large number of times*). We will illustrate this concept with another example.

Suppose it is known that in a certain part of Canada the daily high temperature recorded for all past months of January has a mean $\mu = 10°F$ and a standard deviation $\sigma = 5°F$. Consider an experiment consisting of randomly selecting 25 daily high temperatures from the records of past months of January and calculating the sample mean $\bar{x}$. If this experiment were repeated a very large number of times, the value of $\bar{x}$ would vary from sample to sample. For example, the first sample of 25 temperature measurements might have a mean $\bar{x} = 9.8$, the second sample a mean $\bar{x} = 11.4$, the third sample a mean $\bar{x} = 10.5$, etc. If the sampling experiment were repeated a very large number of times, the resulting histogram of sample means would be approximately the probability distribution of $\bar{x}$. If $\bar{x}$ is a good estimator of μ, we would expect the values of $\bar{x}$ to cluster around μ as shown in Figure 4.33. This probability distribution is called a *sampling distribution*, because it is generated by repeating a sampling experiment a very large number of times.

> The **sampling distribution** of a sample statistic calculated from a sample of n temperature measurements is the probability distribution of the statistic.

In actual practice, the sampling distribution of a statistic is obtained mathematically or (at least approximately) by simulating the sample on a computer, using a procedure similar to that just described.

If $\bar{x}$ has been calculated from a sample of $n = 25$ measurements selected from a population with mean $\mu = 10$ and standard deviation $\sigma = 5$, the sampling distribution (Figure 4.33) provides information about the behavior of $\bar{x}$ in repeated sampling.

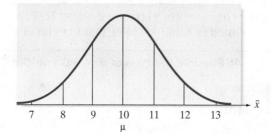

Figure 4.33

Sampling distribution for $\bar{x}$ based on a sample of $n = 25$ temperature measurements

For example, the probability that you will draw a sample of 25 measurements and obtain a value of $\bar{x}$ in the interval $9 \leq \bar{x} \leq 10$ will be the area under the sampling distribution over that interval.

Since the properties of a statistic are typified by its sampling distribution, it follows that, to compare two sample statistics, you compare their sampling distributions. For example, if you have two statistics, A and B, for estimating the same parameter (for purposes of illustration, suppose the parameter is the population variance σ^2), and if their sampling distributions are as shown in Figure 4.34, you would prefer statistic A over statistic B. You would do so because the sampling distribution for statistic A centers over σ^2 and has less spread (variation) than the sampling distribution for statistic B. Then, when you draw a single sample in a practical sampling situation, the probability is higher that statistic A will fall nearer σ^2.

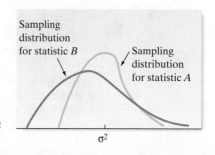

Figure 4.34

Two sampling distributions for estimating the population variance σ^2

Remember that, in practice, we will not know the numerical value of the unknown parameter σ^2, so we will not know whether statistic A or statistic B is closer to σ^2 for a particular sample. We have to rely on our knowledge of the theoretical sampling distributions to choose the best sample statistic and then use it sample after sample. The procedure for finding the sampling distribution for a statistic is demonstrated in Example 4.26.

Example 4.26

Finding a Sampling Distribution— Come-Out Roll in Craps

Problem Consider the popular casino game of craps, in which a player throws two dice and bets on the outcome (the sum total of the dots showing on the upper faces of the two dice). In Example 4.5 (p. 168), we looked at the possible outcomes of a $5 wager on the first toss (called the *come-out* roll). Recall that if the sum total of the dice is 7 or 11, the roller wins $5; if the total is a 2, 3, or 12, the roller loses $5 (i.e., the roller "wins" −$5); and, for any other total (4, 5, 6, 8, 9, or 10) no money is lost or won on that roll (i.e., the roller wins $0). Let x represent the result of the come-out roll wager (−$5, $0, or + $5). We showed in Example 4.5 that the actual probability distribution of x is:

Outcome of wager, x	−5	0	5
$p(x)$	1/9	6/9	2/9

Now, consider a random sample of $n = 3$ come-out rolls.

a. Find the sampling distribution of the sample mean, $\bar{x}$.

b. Find the sampling distribution of the sample median, M.

Solution The outcomes for every possible sample of $n = 3$ come-out rolls are listed in Table 4.9, along with the sample mean and median. The probability of each sample is obtained using the Multiplicative Rule. For example, the probability of the sample $(0, 0, 5)$ is $p(0) \cdot p(0) \cdot p(5) = (6/9)(6/9)(2/9) = 72/729 = .099$. The probability for each sample is also listed in Table 4.9. Note that the sum of these probabilities is equal to 1.

Table 4.9	All Possible Samples of $n = 3$ Come-Out Rolls in Craps		
Possible Samples	$\bar{x}$	M	**Probability**
−5, −5, −5	−5	−5	$(1/9)(1/9)(1/9) = 1/729$
−5, −5, 0	−3.33	−5	$(1/9)(1/9)(6/9) = 6/729$
−5, −5, 5	−1.67	−5	$(1/9)(1/9)(2/9) = 2/729$
−5, 0, −5	−3.33	−5	$(1/9)(6/9)(1/9) = 6/729$
−5, 0, 0	−1.67	0	$(1/9)(6/9)(6/9) = 36/729$
−5, 0, 5	0	0	$(1/9)(6/9)(2/9) = 12/729$
−5, 5, −5	−1.67	−5	$(1/9)(2/9)(1/9) = 2/729$
−5, 5, 0	0	0	$(1/9)(2/9)(6/9) = 12/729$
−5, 5, 5	1.67	5	$(1/9)(2/9)(2/9) = 4/729$
0, −5, −5	−3.33	−5	$(6/9)(1/9)(1/9) = 6/729$
0, −5, 0	−1.67	0	$(6/9)(1/9)(6/9) = 36/729$
0, −5, 5	0	0	$(6/9)(1/9)(2/9) = 12/729$
0, 0, −5	−1.67	0	$(6/9)(6/9)(1/9) = 36/729$
0, 0, 0	0	0	$(6/9)(6/9)(6/9) = 216/729$
0, 0, 5	1.67	0	$(6/9)(6/9)(2/9) = 72/729$
0, 5, −5	0	0	$(6/9)(2/9)(1/9) = 12/729$
0, 5, 0	1.67	0	$(6/9)(2/9)(6/9) = 72/729$
0, 5, 5	3.33	5	$(6/9)(2/9)(2/9) = 24/729$
5, −5, −5	−1.67	−5	$(2/9)(1/9)(1/9) = 2/729$
5, −5, 0	0	0	$(2/9)(1/9)(6/9) = 12/729$
5, −5, 5	1.67	5	$(2/9)(1/9)(2/9) = 4/729$
5, 0, −5	0	0	$(2/9)(6/9)(1/9) = 12/729$
5, 0, 0	1.67	0	$(2/9)(6/9)(6/9) = 72/729$
5, 0, 5	3.33	5	$(2/9)(6/9)(2/9) = 24/729$
5, 5, −5	1.67	5	$(2/9)(2/9)(1/9) = 4/729$
5, 5, 0	3.33	5	$(2/9)(2/9)(6/9) = 24/729$
5, 5, 5	5	5	$(2/9)(2/9)(2/9) = 8/729$

$Sum = 729/729 = 1$

a. From Table 4.9, you can see that $\bar{x}$ can assume the values $-5, -3.33, -1.67, 0, 1.67,$ 3.33, and 5. Because $\bar{x} = -5$ occurs only in one sample, $P(\bar{x} = -5) = 1/729$. Similarly, $\bar{x} = -3.33$ occurs in three samples, $(-5, -5, 0), (-5, 0, -5)$ and, $(0, -5, -5)$. Therefore, $P(\bar{x} = -3.33) = 6/729 + 6/729 + 6/729 = 18/729$. Calculating the probabilities of the remaining values of $\bar{x}$ and arranging them in a table, we obtain the following probability distribution:

$\bar{x}$	-5	-3.33	-1.67	0	1.67	3.33	5
$p(\bar{x})$	1/729 = .0014	18/729 = .0247	114/729 = .1564	288/729 = .3951	228/729 = .3127	72/729 = .0988	8/729 = .0110

This is the sampling distribution for $\bar{x}$ because it specifies the probability associated with each possible value of $\bar{x}$. You can see that the most likely mean outcome after 3 randomly selected come-out rolls is $\bar{x} = \$0$; this result occurs with probability $288/729 = .3951$.

b. In Table 4.9, you can see that the median M can assume one of three values: $-5, 0,$ and 5. The value $M = -5$ occurs in 7 different samples. Therefore, $P(M = -5)$ is the sum of the probabilities associated with these 7 samples; that is, $P(M = -5) = 1/729 + 6/729 + 2/729 + 6/729 + 2/729 + 6/729 + 2/729 = 25/729$. Similarly, $M = 0$ occurs in 13 samples and $M = 5$ occurs in 7 samples. These probabilities are obtained by summing the probabilities of their respective sample points. After performing these calculations, we obtain the following probability distribution for the median M:

M	-5	0	5
$p(M)$	25/729 = .0343	612/729 = .8395	92/729 = .1262

Once again, the most likely median outcome after 3 randomly selected come-out rolls is $\bar{x} = \$0$—a result that occurs with probability $612/729 = .8395$.

Look Back The sampling distributions of parts **a** and **b** are found by first listing all possible distinct values of the statistic and then calculating the probability of each value. Note that if the values of x were equally likely, the 27 sample points in Table 4.9 would all have the same probability of occurring, namely, 1/27.

Now Work Exercise 4.139

Example 4.26 demonstrates the procedure for finding the exact sampling distribution of a statistic when the number of different samples that could be selected from the population is relatively small. In the real world, populations often consist of a large number of different values, making samples difficult (or impossible) to enumerate. When this situation occurs, we may choose to obtain the approximate sampling distribution for a statistic by simulating the sampling over and over again and recording the proportion of times different values of the statistic occur. Example 4.27 illustrates this procedure.

Example 4.27

Simulating a Sampling Distribution— Thickness of Steel Sheets

Problem The rolling machine of a steel manufacturer produces sheets of steel of varying thickness. The thickness of a steel sheet ranges between 150 and 200 millimeters, with distribution shown in Figure 4.35. (This distribution is known as the *uniform distribution*.) Suppose we perform the following experiment over and over again: Randomly sample 11 steel sheets from the production line and record the thickness x of each. Calculate the two sample statistics

$$\bar{x} = \text{Sample mean} = \frac{\Sigma x}{11}$$

$M = \text{Median} = \text{Sixth sample measurement when the 11 thicknesses}$ are arranged in ascending order

Obtain approximations to the sampling distributions of $\bar{x}$ and M.

Figure 4.35
Uniform distribution for
thickness of steel sheets

Solution We used MINITAB to generate 1,000 samples from this population, each with $n = 11$ observations. Then we computed $\bar{x}$ and M for each sample. Our goal is to obtain approximations to the sampling distributions of $\bar{x}$ and M in order to find out which sample statistic ($\bar{x}$ or M) contains more information about μ. [*Note:* In this particular example, it is known that the population mean is $\mu = 175$ mm.] The first 10 of the 1,000 samples generated are presented in Table 4.10. For instance, the first computer-generated sample from the uniform distribution contained the following measurements (arranged in ascending order): 151, 157, 162, 169, 171, 173, 181, 182, 187, 188, and 193 millimeters. The sample mean $\bar{x}$ and median M computed for this sample are

$$\bar{x} = \frac{151 + 157 + \cdots + 193}{11} = 174.0$$

$$M = \text{Sixth ordered measurement} = 173$$

Table 4.10   First 10 Samples of $n = 11$ Thickness Measurements from Uniform Distribution

Sample	Thickness Measurements											Mean	Median
1	173	171	187	151	188	181	182	157	162	169	193	174.00	173
2	181	190	182	171	187	177	162	172	188	200	193	182.09	182
3	192	195	187	187	172	164	164	189	179	182	173	180.36	182
4	173	157	150	154	168	174	171	182	200	181	187	172.45	173
5	169	160	167	170	197	159	174	174	161	173	160	169.46	169
6	179	170	167	174	173	178	173	170	173	198	187	176.55	173
7	166	177	162	171	154	177	154	179	175	185	193	172.09	175
8	164	199	152	153	163	156	184	151	198	167	180	169.73	164
9	181	193	151	166	180	199	180	184	182	181	175	179.27	181
10	155	199	199	171	172	157	173	187	190	185	150	176.18	173

Data Set: SIMUNI

The MINITAB relative frequency histograms for $\bar{x}$ and M for the 1,000 samples of size $n = 11$ are shown in Figure 4.36. These histograms represent approximations to the true sampling distributions of $\bar{x}$ and M.

Figure 4.36
MINITAB histograms for
sample mean and sample
median, Example 4.27

Look Back You can see that the values of $\bar{x}$ tend to cluster around μ to a greater extent than do the values of M. Thus, on the basis of the observed sampling distributions, we conclude that $\bar{x}$ contains more information about μ than M does—at least for samples of $n = 11$ measurements from the uniform distribution.

Now Work Exercise 4.144

As noted earlier, many sampling distributions can be derived mathematically, but the theory necessary to do so is beyond the scope of this text. Consequently, when we need to know the properties of a statistic, we will present its sampling distribution and simply describe its properties. Several of the important properties we look for in sampling distributions are discussed in the next section.

Exercises 4.137–4.145

Understanding the Principles

4.137 What is the difference between a population parameter and a sample statistic?

4.138 What is a sampling distribution of a sample statistic?

Learning the Mechanics

4.139 The probability distribution shown here describes a popula-
NW tion of measurements that can assume values of 0, 2, 4, and 6, each of which occurs with the same relative frequency:

x	0	2	4	6
$p(x)$	$1/4$	$1/4$	$1/4$	$1/4$

a. List all the different samples of $n = 2$ measurements that can be selected from this population.
b. Calculate the mean of each different sample listed in part **a**.
c. If a sample of $n = 2$ measurements is randomly selected from the population, what is the probability that a specific sample will be selected?
d. Assume that a random sample of $n = 2$ measurements is selected from the population. List the different values of $\bar{x}$ found in part **b**, and find the probability of each. Then give the sampling distribution of the sample mean $\bar{x}$ in tabular form.
e. Construct a probability histogram for the sampling distribution of $\bar{x}$.

4.140 Simulate sampling from the population described in Exercise 4.139 by marking the values of x, one on each of four identical coins (or poker chips, etc.). Place the coins (marked 0, 2, 4, and 6) into a bag, randomly select one, and observe its value. Replace this coin, draw a second coin, and observe its value. Finally, calculate the mean $\bar{x}$ for this sample of $n = 2$ observations randomly selected from the population (Exercise 4.139, part **b**). Replace the coins, mix them, and, using the same procedure, select a sample of $n = 2$ observations from the population. Record the numbers and calculate $\bar{x}$ for this sample. Repeat this sampling process until you acquire 100 values of $\bar{x}$. Construct a relative frequency distribution for these 100 sample means. Compare this distribution with the exact sampling distribution of $\bar{x}$ found in part **e** of Exercise 4.139. [*Note:* The distribution obtained in this exercise is an approximation to the exact sampling distribution. However, if you were to repeat the sampling procedure, drawing two coins not 100 times, but 10,000 times, then the relative frequency distribution for the 10,000 sample means would be almost identical to the sampling distribution of $\bar{x}$ found in Exercise 4.139, part **e**.]

4.141 Consider the population described by the probability distribution shown here:

x	1	2	3	4	5
$p(x)$	.2	.3	.2	.2	.1

The random variable x is observed twice. If these observations are independent, verify that the different samples of size 2 and their probabilities are as follows:

Sample	Probability	Sample	Probability
1, 1	.04	3, 4	.04
1, 2	.06	3, 5	.02
1, 3	.04	4, 1	.04
1, 4	.04	4, 2	.06
1, 5	.02	4, 3	.04
2, 1	.06	4, 4	.04
2, 2	.09	4, 5	.02
2, 3	.06	5, 1	.02
2, 4	.06	5, 2	.03
2, 5	.03	5, 3	.02
3, 1	.04	5, 4	.02
3, 2	.06	5, 5	.01
3, 3	.04		

a. Find the sampling distribution of the sample mean $\bar{x}$.
b. Construct a probability histogram for the sampling distribution of $\bar{x}$.
c. What is the probability that $\bar{x}$ is 4.5 or larger?
d. Would you expect to observe a value of $\bar{x}$ equal to 4.5 or larger? Explain.

4.142 Refer to Exercise 4.141 and find $E(x) = \mu$. Then use the sampling distribution of $\bar{x}$ found in Exercise 4.141 to find the expected value of $\bar{x}$. Note that $E(\bar{x}) = \mu$.

4.143 Refer to Exercise 4.141. Assume that a random sample of $n = 2$ measurements is randomly selected from the population.
a. List the different values that the sample median M may assume, and find the probability of each. Then give the sampling distribution of the sample median.
b. Construct a probability histogram for the sampling distribution of the sample median, and compare it with the probability histogram for the sample mean (Exercise 4.141, part **b**).

4.144 In Example 4.27, we use the computer to generate 1,000
NW samples, each containing $n = 11$ observations, from a uniform distribution over the interval from 150 to 200. Now use the computer to generate 500 samples, each containing $n = 15$ observations, from that same population.
a. Calculate the sample mean for each sample. To approximate the sampling distribution of $\bar{x}$, construct a relative frequency histogram for the 500 values of $\bar{x}$.
b. Repeat part **a** for the sample median. Compare this approximate sampling distribution with the approximate sampling distribution of $\bar{x}$ found in part **a**.

4.145 Consider a population that contains values of x equal to 00, 01, 02, 03, ... , 96, 97, 98, 99. Assume that these values occur with equal probability. Use the computer to generate 500 samples, each containing $n = 25$ measurements, from this population. Calculate the sample mean $\bar{x}$ and sample variance s^2 for each of the 500 samples.
a. To approximate the sampling distribution of $\bar{x}$, construct a relative frequency histogram for the 500 values of $\bar{x}$.
b. Repeat part **a** for the 500 values of s^2.

4.9 The Sampling Distribution of $\bar{x}$ and the Central Limit Theorem

Estimating the mean useful life of automobiles, the mean number of crimes per month in a large city, and the mean yield per acre of a new soybean hybrid are practical problems with something in common. In each case, we are interested in making an inference about the mean μ of some population. As we mentioned in Chapter 2, the sample mean $\bar{x}$ is, in general, a good estimator of μ. We now develop pertinent information about the sampling distribution for this useful statistic.

Example 4.28

Describing the Sampling Distribution of $\bar{x}$

Problem Suppose a population has the uniform probability distribution given in Figure 4.37. It can be shown (proof omitted) that the mean and standard deviation of this probability distribution are, respectively, $\mu = 175$ and $\sigma = 14.43$. Now suppose a sample of 11 measurements is selected from this population. Describe the sampling distribution of the sample mean $\bar{x}$ based on the 1,000 sampling experiments discussed in Example 4.27.

Figure 4.37
Sampled uniform population

Solution Recall that in Example 4.27 we generated 1,000 samples of $n = 11$ measurements each. The MINITAB histogram for the 1,000 sample means is shown in Figure 4.38, with a normal probability distribution superimposed. You can see that this normal probability distribution approximates the computer-generated sampling distribution very well.

To fully describe a normal probability distribution, it is necessary to know its mean and standard deviation. MINITAB gives these statistics for the 1,000 $\bar{x}$'s in the upper right corner of the histogram of Figure 4.38. You can see that the mean is 175.2 and the standard deviation is 4.383.

To summarize our findings based on 1,000 samples, each consisting of 11 measurements from a uniform population, the sampling distribution of $\bar{x}$ appears to be approximately normal with a mean of about 175 and a standard deviation of about 4.38.

Figure 4.38
MINITAB histogram for sample mean in 1,000 samples

Look Back Note that the simulated value $\mu_{\bar{x}} = 175.2$ is very close to $\mu = 175$ for the uniform distribution; that is, the simulated sampling distribution of $\bar{x}$ appears to provide an unbiased estimate of μ.

The true sampling distribution of $\bar{x}$ has the properties given in the next box, assuming only that a random sample of n observations has been selected from *any* population.

Properties of the Sampling Distribution of $\bar{x}$

1. The mean of the sampling distribution of $\bar{x}$ equals the mean of the sampled population. That is, $\mu_{\bar{x}} = E(\bar{x}) = \mu$.*

2. The standard deviation of the sampling distribution of $\bar{x}$ equals

$$\frac{\text{Standard deviation of sampled population}}{\text{Square root of sample size}}$$

That is, $\sigma_{\bar{x}} = \sigma/\sqrt{n}$**

The standard deviation $\sigma_{\bar{x}}$ is often referred to as the **standard error of the mean.**

You can see that our approximation to $\mu_{\bar{x}}$ in Example 4.28 was precise, since property 1 assures us that the mean is the same as that of the sampled population: 175. Property 2 tells us how to calculate the standard deviation of the sampling distribution of $\bar{x}$. Substituting $\sigma = 14.43$ (the standard deviation of the sampled uniform distribution) and the sample size $n = 11$ into the formula for $\sigma_{\bar{x}}$, we find that

$$\sigma_{\bar{x}} = \frac{\sigma}{\sqrt{n}} = \frac{14.43}{\sqrt{11}} = 4.35$$

Thus, the approximation we obtained in Example 6.6, $\sigma_{\bar{x}} \approx 4.38$, is very close to the exact value, $\sigma_{\bar{x}} = 4.35$.***

What about the shape of the sampling distribution? Two theorems provide this information. One is applicable whenever the original population data are normally distributed. The other, applicable when the sample size n is large, represents one of the most important theoretical results in statistics: the *Central Limit Theorem*.

Theorem 4.1

If a random sample of n observations is selected from a population with a normal distribution, the sampling distribution of $\bar{x}$ will be a normal distribution.

BIOGRAPHY **PIERRE-SIMON LAPLACE (1749–1827)**

The Originator of the Central Limit Theorem

As a boy growing up in Normandy, France, Pierre-Simon Laplace attended a Benedictine priory school. Upon graduation, he entered Caen University to study theology. During his two years there, he discovered his mathematical talents and began his career as an eminent mathematician. In fact, he considered himself the best mathematician in France. Laplace's contributions to mathematics ranged from introducing new methods of solving differential equations to presenting complex analyses of motions of astronomical bodies. While studying the angles of inclination of comet orbits in 1778, Laplace showed that the sum of the angles was normally distributed. Consequently, he is considered to be the originator of the Central Limit Theorem. (A rigorous proof of the theorem, however, was not provided until the early 1930s by another French mathematician, Paul Levy.) Laplace also discovered Bayes's theorem and established Bayesian statistical analysis as a valid approach to many practical problems of his time. ∎

---

*When this property helds, we say that $\bar{x}$ is an *unbiased* estimator of μ.

**If the sample size n is large relative to the number N of elements in the population (e.g., 5% or more), $\sigma/\sqrt{n}$ must be multiplied by the finite population correction factor $\sqrt{(N-n)/(N-1)}$. In most sampling situations, this correction factor will be close to 1 and can be ignored.

***It can be shown (proof emitted) that the value of $\sigma_{\bar{x}}^2$ is the smallest variance of all unbiased estimators of μ; thus, $\bar{x}$ is the minimum-variance unbiased estimator (MVUE) of μ.

> **Theorem 4.2: Central Limit Theorem**
>
> Consider a random sample of n observations selected from a population (*any* population) with mean μ and standard deviation σ. Then, when n is sufficiently large, the sampling distribution of $\bar{x}$ will be approximately a normal distribution with mean $\mu_{\bar{x}} = \mu$ and standard deviation $\sigma_{\bar{x}} = \sigma/\sqrt{n}$. The larger the sample size, the better will be the normal approximation to the sampling distribution of $\bar{x}$.*

Thus, for sufficiently large samples, the sampling distribution of $\bar{x}$ is approximately normal. How large must the sample size n be so that the normal distribution provides a good approximation to the sampling distribution of $\bar{x}$? The answer depends on the shape of the distribution of the sampled population, as shown by Figure 4.39. Generally speaking, the greater the skewness of the sampled population distribution, the larger the sample size must be before the normal distribution is an adequate approximation to the sampling distribution of $\bar{x}$. For most sampled populations, sample sizes of $n \geq 30$ will uffice for the normal approximation to be reasonable.

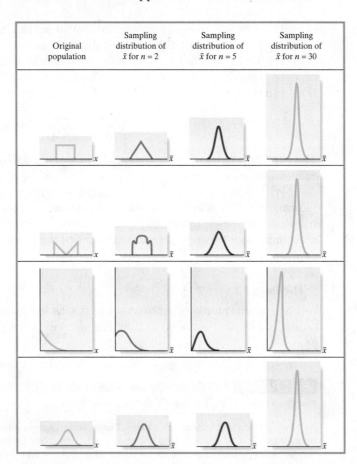

Figure 4.39
Sampling distributions of $\bar{x}$ for different populations and different sample sizes

Example 4.29

Using the Central Limit Theorem to Find a Probability

Problem Suppose we have selected a random sample of $n = 36$ observations from a population with mean equal to 80 and standard deviation equal to 6. It is known that the population is not extremely skewed.

a. Sketch the relative frequency distributions for the population and for the sampling distribution of the sample mean $\bar{x}$.

b. Find the probability that $\bar{x}$ will be larger than 82.

*Moreover, because of the Central Limit Theorem, the sum of a random sample of n observations, Σx, will possess a sampling distribution that is approximately normal for large samples. This distribution will have a mean equal to $n\mu$ and a variance equal to $n\sigma^2$. Proof of the Central Limit Theorem is beyond the scope of this book, but it can be found in many mathematical statistics texts.

Solution

a. We do not know the exact shape of the population relative frequency distribution, but we do know that it should be centered about $\mu = 80$, its spread should be measured by $\sigma = 6$, and it is not highly skewed. One possibility is shown in Figure 4.40a. From the Central Limit Theorem, we know that the sampling distribution of $\bar{x}$ will be approximately normal, since the sampled population distribution is not extremely skewed. We also know that the sampling distribution will have mean

and standard deviation

$$\mu_{\bar{x}} = \mu = 80$$

$$\sigma_{\bar{x}} = \frac{\sigma}{\sqrt{n}} = \frac{6}{\sqrt{36}} = 1$$

The sampling distribution of $\bar{x}$ is shown in Figure 4.40b.

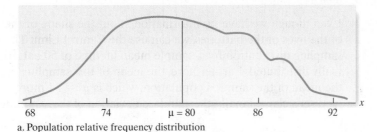

a. Population relative frequency distribution

b. Sampling distribution of $\bar{x}$

Figure 4.40

A population relative frequency distribution and the sampling distribution for $\bar{x}$

b. The probability that $\bar{x}$ will exceed 82 is equal to the highlighted area in Figure 4.41. To find this area, we need to find the z value corresponding to $\bar{x} = 82$. Recall that the standard normal random variable z is the difference of any normally distributed random variable and its mean, expressed in units of its standard deviation. Since $\bar{x}$

Figure 4.41

The sampling distribution of $\bar{x}$

is a normally distributed random variable with mean $\mu_{\bar{x}} = \mu$ and standard deviation $\sigma_{\bar{x}} = \sigma/\sqrt{n}$, it follows that the standard normal z value corresponding to the sample mean $\bar{x}$ is

$$z = \frac{(\text{Normal random variable}) - (\text{Mean})}{\text{Standard deviation}} = \frac{\bar{x} - \mu_{\bar{x}}}{\sigma_{\bar{x}}}$$

Therefore, for $\bar{x} = 82$, we have

$$z = \frac{\bar{x} - \mu_{\bar{x}}}{\sigma_{\bar{x}}} = \frac{82 - 80}{1} = 2$$

The area A in Figure 4.41 corresponding to $z = 2$ is given in the table of areas under the normal curve (see Table III of Appendix A) as .4772. Therefore, the tail area corresponding to the probability that $\bar{x}$ exceeds 82 is

$$P(\bar{x} > 82) = P(z > 2) = .5 - .4772 = .0228$$

Look Back The key to finding the probability in part **b** is to recognize that the distribution of $\bar{x}$ is normal with $\mu_{\bar{x}} = \mu$ and $\sigma_{\bar{x}} = \sigma/\sqrt{n}$.

Now Work Exercise 4.154

Example 4.30

Application of the Central Limit Theorem—Testing a Manufacturer's Claim

Problem A manufacturer of automobile batteries claims that the distribution of the lengths of life of its best battery has a mean of 54 months and a standard deviation of 6 months. Suppose a consumer group decides to check the claim by purchasing a sample of 50 of the batteries and subjecting them to tests that estimate the battery's life.

a. Assuming that the manufacturer's claim is true, describe the sampling distribution of the mean lifetime of a sample of 50 batteries.

b. Assuming that the manufacturer's claim is true, what is the probability that the consumer group's sample has a mean life of 52 or fewer months?

Solution

a. Even though we have no information about the shape of the probability distribution of the lives of the batteries, we can use the Central Limit Theorem to deduce that the sampling distribution for a sample mean lifetime of 50 batteries is approximately normally distributed. Furthermore, the mean of this sampling distribution is the same as the mean of the sampled population, which is $\mu = 54$ months according to the manufacturer's claim. Finally, the standard deviation of the sampling distribution is given by

$$\sigma_{\bar{x}} = \frac{\sigma}{\sqrt{n}} = \frac{6}{\sqrt{50}} = .85 \text{ month}$$

Note that we used the claimed standard deviation of the sampled population, $\sigma = 6$ months. Thus, if we assume that the claim is true, then the sampling distribution for the mean life of the 50 batteries sampled must be as shown in Figure 4.42.

Figure 4.42
Sampling distribution of $\bar{x}$ in Example 6.8 for $n = 50$

b. If the manufacturer's claim is true, the probability that the consumer group observes a mean battery life of 52 or fewer months for its sample of 50 batteries, $P(\bar{x} \leq 52)$, is equivalent to the highlighted area in Figure 4.42. Since the sampling distribution is approximately normal, we can find this area by computing the standard normal z value:

$$z = \frac{\bar{x} - \mu_{\bar{x}}}{\sigma_{\bar{x}}} = \frac{\bar{x} - \mu}{\sigma_{\bar{x}}} = \frac{52 - 54}{.85} = -2.35$$

Here, $\mu_{\bar{x}}$, the mean of the sampling distribution of $\bar{x}$, is equal to μ, the mean of the lifetimes of the sampled population, and $\sigma_{\bar{x}}$ is the standard deviation of the sampling distribution of $\bar{x}$. Note that z is the familiar standardized distance (z-score) of Section 2.7, and since $\bar{x}$ is approximately normally distributed, it will possess the standard normal distribution of Section 4.5.

The area A shown in Figure 4.42 between $\bar{x} = 52$ and $\bar{x} = 54$ (corresponding to $z = -2.35$) is found in Table III of Appendix A to be .4906. Therefore, the area to the left of $\bar{x} = 52$ is

$$P(\bar{x} \leq 52) = .5 - A = .5 - .4906 = .0094$$

Thus, the probability that the consumer group will observe a sample mean of 52 or less is only .0094 if the manufacturer's claim is true.

Look Back If the 50 batteries tested do exhibit a mean of 52 or fewer months, the consumer group will have strong evidence that the manufacturer's claim is untrue, because such an event is very unlikely to occur if the claim is true. (This is still another application of the *rare-event approach* to statistical inference.)

Now Work Exercise 4.159

We conclude this section with two final comments on the sampling distribution of $\bar{x}$. First, from the formula $\sigma_{\bar{x}} = \sigma/\sqrt{n}$, we see that the standard deviation of the sampling distribution of $\bar{x}$ gets smaller as the sample size n gets larger. For example, we computed $\sigma_{\bar{x}} = .85$ when $n = 50$ in Example 4.30. However, for $n = 100$, we obtain $\sigma_{\bar{x}} = \sigma/\sqrt{n} = 6/\sqrt{100} = .60$. This relationship will hold true for most of the sample statistics encountered in this text. That is, *the standard deviation of the sampling distribution decreases as the sample size increases.* Consequently, the larger the sample size, the more accurate the sample statistic (e.g., $\bar{x}$) is in estimating a population parameter (e.g., μ). We will use this result in Chapter 5 to help us determine the sample size needed to obtain a specified accuracy of estimation.

Our second comment concerns the Central Limit Theorem. In addition to providing a very useful approximation for the sampling distribution of a sample mean, the Central Limit Theorem offers an explanation for the fact that many relative frequency distributions of data possess mound-shaped distributions. Many of the measurements we take in various areas of research are really means or sums of a large number of small phenomena. For example, a year's growth of a pine seedling is the total of the numerous individual components that affect the plant's growth. Similarly, we can view the length of time a construction company takes to build a house as the total of the times taken to complete a multitude of distinct jobs, and we can regard the monthly demand for blood at a hospital as the total of the many individual patients' needs. Whether or not the observations entering into these sums satisfy the assumptions basic to the Central Limit Theorem is open to question; however, it is a fact that many distributions of data in nature are mound shaped and possess the appearance of normal distributions.

Exercises 4.146–4.169

Understanding the Principles

4.146 What do the symbols $\mu_{\bar{x}}$ and $\sigma_{\bar{x}}$ represent?

4.147 How does the mean of the sampling distribution of $\bar{x}$ relate to the mean of the population from which the sample is selected?

4.148 How does the standard deviation of the sampling distribution of $\bar{x}$ relate to the standard deviation of the population from which the sample is selected?

4.149 State the Central Limit Theorem.

4.150 Will the sampling distribution of $\bar{x}$ always be approximately normally distributed? Explain.

Learning the Mechanics

4.151 Suppose a random sample of n measurements is selected from a population with mean $\mu = 100$ and variance $\sigma^2 = 100$. For each of the following values of n, give the mean and standard deviation of the sampling distribution of the sample mean $\bar{x}$.
  a. $n = 4$ **b.** $n = 25$
  c. $n = 100$ **d.** $n = 50$
  e. $n = 500$ **f.** $n = 1,000$

4.152 Suppose a random sample of $n = 25$ measurements is selected from a population with mean μ and standard deviation σ. For each of the following values of μ and σ, give the values of $\mu_{\bar{x}}$ and $\sigma_{\bar{x}}$.
  a. $\mu = 10, \sigma = 3$
  b. $\mu = 100, \sigma = 25$
  c. $\mu = 20, \sigma = 40$
  d. $\mu = 10, \sigma = 100$

4.153 Consider the following probability distribution:

x	1	2	3	8
$p(x)$	.1	.4	.4	.1

  a. Find μ, σ^2, and σ.
  b. Find the sampling distribution of $\bar{x}$ for random samples of $n = 2$ measurements from this distribution by listing all possible values of $\bar{x}$, and find the probability associated with each.
  c. Use the results of part **b** to calculate $\mu_{\bar{x}}$ and $\sigma_{\bar{x}}$. Confirm that $\mu_{\bar{x}} = \mu$ and that $\sigma_{\bar{x}} = \sigma/\sqrt{n} = \sigma/\sqrt{2}$.

4.154 A random sample of $n = 64$ observations is drawn from a population with a mean equal to 20 and standard deviation equal to 16.
  a. Give the mean and standard deviation of the (repeated) sampling distribution of $\bar{x}$.
  b. Describe the shape of the sampling distribution of $\bar{x}$. Does your answer depend on the sample size?
  c. Calculate the standard normal z-score corresponding to a value of $\bar{x} = 16$.
  d. Calculate the standard normal z-score corresponding to $\bar{x} = 23$.
  e. Find $P(\bar{x} < 16)$.
  f. Find $P(\bar{x} > 23)$.
  g. Find $P(16 < \bar{x} < 23)$.

4.155 A random sample of $n = 100$ observations is selected from a population with $\mu = 30$ and $\sigma = 16$.
  a. Find $\mu_{\bar{x}}$ and $\sigma_{\bar{x}}$.
  b. Describe the shape of the sampling distribution of $\bar{x}$.
  c. Find $P(\bar{x} \geq 28)$.
  d. Find $P(22.1 \leq \bar{x} \leq 26.8)$.
  e. Find $P(\bar{x} \leq 28.2)$.
  f. Find $P(\bar{x} \geq 27.0)$.

Applet Exercise 4.7

Open the applet entitled *Sampling Distribution*. On the pull-down menu to the right of the top graph, select *Binary*.
  a. Run the applet for the sample size $n = 10$ and the number of samples $N = 1000$. Observe the shape of the graph

of the sample proportions, and record the mean, median, and standard deviation of the sample proportions.

b. How does the mean of the sample proportions compare with the mean $\mu = 0.5$ of the original distribution?

c. Use the formula $\sigma = \sqrt{np(1-p)}$, where $n = 1$ and $p = 0.5$, to compute the standard deviation of the original distribution. Divide the result by $\sqrt{10}$, the square root of the sample size used in the sampling distribution. How does this result compare with the standard deviation of the sample proportions?

d. Explain how the graph of the distribution of sample proportions suggests that the distribution may be approximately normal.

e. Explain how the results of parts **b–d** illustrate the Central Limit Theorem.

⊙ Applet Exercise 4.8

Open the applet entitled *Sampling Distributions*. On the pull-down menu to the right of the top graph, select *Uniform*. The box to the left of the top graph displays the population mean, median, and standard deviation of the original distribution.

a. Run the applet for the sample size $n = 30$ and the number of samples $N = 1000$. Observe the shape of the graph of the sample means, and record the mean, median, and standard deviation of the sample means.

b. How does the mean of the sample means compare with the mean of the original distribution?

c. Divide the standard deviation of the original distribution by $\sqrt{30}$, the square root of the sample size used in the sampling distribution. How does this result compare with the standard deviation of the sample proportions?

d. Explain how the graph of the distribution of sample means suggests that the distribution may be approximately normal.

e. Explain how the results of parts **b–d** illustrate the Central Limit Theorem.

Applying the Concepts—Basic

4.156 **Phishing attacks to e-mail accounts.** In Exercise 2.43 (p. 49), you learned that *phishing* describes an attempt to extract personal/financial information from unsuspecting people through fraudulent e-mail. Data from an actual phishing attack against an organization were presented in *Chance* (Summer 2007). The interarrival times, i.e., the time differences (in seconds), for 267 fraud box e-mail notifications were recorded and are saved in the **PHISHING** file. For this exercise, consider these interarrival times to represent the population of interest.

a. In Exercise 2.43 you constructed a histogram for the interarrival times. Describe the shape of the population of interarrival times.

b. Find the mean and standard deviation of the population of interarrival times.

c. Now consider a random sample of $n = 40$ interarrival times selected from the population. Describe the shape of the sampling distribution of $\bar{x}$, the sample mean. Theoretically, what are $\mu_{\bar{x}}$ and $\sigma_{\bar{x}}$?

d. Find $P(\bar{x} < 90)$.

e. Use a random number generator to select a random sample of $n = 40$ interarrival times from the population, and calculate the value of $\bar{x}$. (Every student in the class should do this.)

f. Refer to part **e**. Obtain the values of $\bar{x}$ computed by the students and combine them into a single data set. Form a histogram for these values of $\bar{x}$. Is the shape approximately normal?

g. Refer to part **f**. Find the mean and standard deviation of the $\bar{x}$-values. Do these values approximate $\mu_{\bar{x}}$ and $\sigma_{\bar{x}}$, respectively?

4.157 **Physical activity of obese young adults.** In a study on the physical activity of young adults, pediatric researchers measured overall physical activity as the total number of registered movements (counts) over a period of time and then computed the number of counts per minute (cpm) for each subject (*International Journal of Obesity*, January 2007). The study revealed that the overall physical activity of obese young adults has a mean of $\mu = 320$ cpm and a standard deviation of $\sigma = 100$ cpm. (In comparison, the mean for young adults of normal weight is 540 cpm.) In a random sample of $n = 100$ obese young adults, consider the sample mean counts per minute, $\bar{x}$.

a. Describe the sampling distribution of $\bar{x}$.

b. What is the probability that the mean overall physical activity level of the sample is between 300 and 310 cpm?

c. What is the probability that the mean overall physical activity level of the sample is greater than 360 cpm?

4.158 **Research on eating disorders.** Refer to *The American Statistician* (May 2001) study of female students who suffer from bulimia, presented in Exercise 2.40 (p. 48). Recall that each student completed a questionnaire from which a "fear of negative evaluation" (FNE) score was produced. (The higher the score, the greater is the fear of negative evaluation.) Suppose the FNE scores of bulimic students have a distribution with mean $\mu = 18$ and standard deviation $\sigma = 5$. Now, consider a random sample of 45 female students with bulimia.

a. What is the probability that the sample mean FNE score is greater than 17.5?

b. What is the probability that the sample mean FNE score is between 18 and 18.5?

c. What is the probability that the sample mean FNE score is less than 18.5?

4.159 **Dentists' use of anesthetics.** Refer to the *Current Allergy & Clinical Immunology* (March 2004) study of allergic reactions of dental patients to local anesthetics, presented in Exercise 2.99 (p. 72). The study reported that the mean number of units (ampoules) of local anesthetics used per week by dentists was 79, with a standard deviation of 23. Consider a random sample of 100 dentists, and let $\bar{x}$ represent the mean number of units of local anesthetic used per week for the sample.

a. What is $\mu_{\bar{x}}$?

b. What is $\sigma_{\bar{x}}$?

c. Describe the shape of the sampling distribution of $\bar{x}$.

d. Find the z-score for the value $\bar{x} = 80$ ampoules.

e. Find $P(\bar{x} > 80)$.

4.160 **Shell lengths of sea turtles.** Refer to the *Aquatic Biology* (Vol. 9, 2010) study of green sea turtles inhabiting the Grand Cayman South Sound lagoon, Exercise 2.83 (p. 65). Research shows that the curved carapace (shell) lengths of these turtles has a distribution with mean $\mu = 50$ cm and standard deviation $\sigma = 10$ cm. In the study, $n = 76$ green sea turtles were captured from the lagoon; the mean shell length for the sample was $\bar{x} = 55.5$ cm. How likely is it to observe a sample mean of 55.5 cm or larger?

4.161 Tomato as a taste modifier. Miraculin is a protein naturally produced in a rare tropical fruit that can convert a sour taste into a sweet taste. Refer to the *Plant Science* (May 2010) investigation of the ability of a hybrid tomato plant to produce miraculin, Exercise 4.82 (p. 200). Recall that the amount x of miraculin produced in the plant had a mean of 105.3 micro-grams per gram of fresh weight with a standard deviation of 8.0. Consider a random sample of $n = 64$ hybrid tomato plants and let $\bar{x}$ represent the sample mean amount of miraculin produced. Would you expect to observe a value of $\bar{x}$ less than 103 micrograms per gram of fresh weight? Explain.

Applying the Concepts—Intermediate

4.162 Uranium in the Earth's crust. In an *American Mineralogist* (October 2009) study of the evolution of uranium minerals in the Earth's crust, researchers estimated that the trace amount of uranium x in reservoirs follows a uniform distribution ranging between 1 and 3 parts per million. This distribution has mean $\mu = 2$ ppm and variance $\sigma^2 = \frac{1}{3}$ ppm. In a random sample of $n = 60$ reservoirs, let $\bar{x}$ represent the sample mean amount of uranium.
a. Find $E(\bar{x})$ and interpret its value.

b. Find $\text{Var}(\bar{x})$.

c. Describe the shape of the sampling distribution of $\bar{x}$.
d. Find the probability that $\bar{x}$ is between 1.5 ppm and 2.5 ppm.
e. Find the probability that $\bar{x}$ exceeds 2.2 ppm.

4.163 Critical part failures in NASCAR vehicles. *The Sport Journal* (Winter, 2007) published a study of critical part failures at NASCAR races. The researchers found that the time x (in hours) until the first critical part failure has a skewed distribution with $\mu = .10$ and $\sigma = .10$. Now, consider a random sample of $n = 50$ NASCAR races and let $\bar{x}$ represent the sample mean time until the first critical part failure.
a. Find $E(\bar{x})$ and $\text{Var}(\bar{x})$.
b. Although x has a skewed distribution, the sampling distribution of $\bar{x}$ is approximately normal. Why?
c. Find the probability that the sample mean time until the first critical part failure exceeds .13 hour.

4.164 Cost of unleaded fuel. According to the American Automobile Association (AAA), the average cost of a gallon of regular unleaded fuel at gas stations in August 2010 was $2.78 (*AAA Fuel Gauge Report*). Assume that the standard deviation of such costs is $.15. Suppose that a random sample of $n = 100$ gas stations is selected from the population and the cost per gallon of regular unleaded fuel is determined for each. Consider $\bar{x}$, the sample mean cost per gallon.
a. Calculate $\mu_{\bar{x}}$ and $\sigma_{\bar{x}}$.
b. What is the approximate probability that the sample has a mean fuel cost between $2.78 and $2.80?
c. What is the approximate probability that the sample has a mean fuel cost that exceeds $2.80?
d. How would the sampling distribution of $\bar{x}$ change if the sample size n were doubled from 100 to 200? How do your answers to parts b and c change when the sample size is doubled?

4.165 Levelness of concrete slabs. Geotechnical engineers use water-level "manometer" surveys to assess the levelness of newly constructed concrete slabs. Elevations are typically measured at eight points on the slab; of interest is the maximum differential between elevations. The *Journal of Performance of Constructed Facilities* (Feb. 2005) published an article on the levelness of slabs in California residential developments. Elevation data collected on over 1,300 concrete slabs *before tensioning* revealed that the maximum differential x has a mean of $\mu = .53$ inch and a standard deviation of $\sigma = .193$ inch. Consider a sample of $n = 50$ slabs selected from those surveyed, and let $\bar{x}$ represent the mean of the sample.
a. Describe fully the sampling distribution of $\bar{x}$.
b. Find $P(\bar{x} > .58)$.
c. The study also revealed that the mean maximum differential of concrete slabs measured *after tensioning and loading* is $\mu = .58$ inch. Suppose the sample data yield $\bar{x} = .59$ inch. Comment on whether the sample measurements were obtained before tensioning or after tensioning and loading.

4.166 Motivation of drug dealers. Refer to the *Applied Psychology in Criminal Justice* (Sept. 2009) investigation of the personality characteristics of drug dealers, Exercise 2.98 (p. 71). Convicted drug dealers were scored on the Wanting Recognition (WR) Scale—a scale which provides a quantitative measure of a person's level of need for approval and sensitivity to social situations. (Higher scores indicate a greater need for approval.) Based on the study results, we can assume that the WR scores for the population of convicted drug dealers has a mean of 40 and a standard deviation of 5. Suppose that in a sample of $n = 100$ people, the mean WR scale score is $\bar{x} = 42$. Is this sample likely to have been selected from the population of convicted drug dealers? Explain.

4.167 Is exposure to a chemical in Teflon-coated cookware hazardous? Perfluorooctanoic acid (PFOA) is a chemical used in Teflon-coated cookware to prevent food from sticking. The Environmental Protection Agency is investigating the potential risk of PFOA as a cancer-causing agent (*Science News Online*, August 27, 2005). It is known that the blood concentration of PFOA in the general population has a mean of $\mu = 6$ parts per billion (ppb) and a standard deviation of $\sigma = 10$ ppb. *Science News Online* reported on tests for PFOA exposure conducted on a sample of 326 people who live near DuPont's Teflon-making Washington (West Virginia) Works facility.
a. What is the probability that the average blood concentration of PFOA in the sample is greater than 7.5 ppb?
b. The actual study resulted in $\bar{x} = 300$ ppb. Use this information to make an inference about the true mean (μ) PFOA concentration for the population that lives near DuPont's Teflon facility.

Applying the Concepts—Advanced

4.168 Test of Knowledge about Epilepsy. The Test of Knowledge about Epilepsy (KAE), which is designed to measure attitudes toward persons with epilepsy, uses 20 multiple-choice items, all of which are incorrect. For each person, two scores (ranging from 0 to 20) are obtained: an attitude score (KAE-A) and a general-knowledge score (KAE-GK). On the basis of a large-scale study of college students, the distribution of KAE-A scores has a mean of $\mu = 11.92$

and a standard deviation of $\sigma = 2.95$ while the distribution of KAE-GK scores has a mean of $\mu = 6.35$ and a standard deviation of $\sigma = 2.12$ (*Rehabilitative Psychology*, Spring 1995). Consider a random sample of 100 college students, and suppose you observe a sample mean KAE score of 6.5. Is this result more likely to be the mean of the attitude scores (KAE-A) or the general-knowledge scores (KAE-GK)? Explain.

4.169 Hand washing versus hand rubbing. Refer to the *British Medical Journal* (August 17, 2002) study comparing the effectiveness of hand washing with soap and hand

rubbing with alcohol, presented in Exercise 2.101 (p. 72). Health care workers who used hand rubbing had a mean bacterial count of 35 per hand with a standard deviation of 59. Health care workers who used hand washing had a mean bacterial count of 69 per hand with a standard deviation of 106. In a random sample of 50 health care workers, all using the same method of cleaning their hands, the mean bacterial count per hand ($\bar{x}$) is less than 30. Give your opinion on whether this sample of workers used hand rubbing with alcohol or hand washing with soap.

CHAPTER NOTES

Key Terms

Note: Starred (*) terms are from the optional sections in this chapter.

Bell curve 189
Bell-shaped probability distribution 189
Binomial distribution 181
Binomial experiment 178
Binomial random variable 177
Central Limit Theorem 223
Continuous random variable 165
*Correction for continuity 210
Countable 165
Cumulative binomial probabilities 184
Discrete random variable 165
Expected value 170
Frequency function 188
Mean value of a discrete random variable 170
Normal distribution 189

Normal probability plot 203
Normal random variable 189
Parameter 216
Probability density function 188
Probability distribution 188
probability distribution for a continuous random variable 189
Probability distribution of a discrete random variable 168
Random variable 163
Sample statistic 216
Sampling distribution 217
Standard deviation of a discrete random variable 171
Standard error of the mean 223
Standard normal distribution 191
Standard normal random variable 191
Variance of a random variable 171

Key Symbols

$P(x)$	Probability distribution for discrete random variable x
S	Outcome of binomial trial denoted "success"
F	Outcome of binomial trial denoted "failure"
P	$P(S)$ in binomial trial
q	$P(F)$ in binomial trial $= 1 - p$
$f(x)$	Probability distribution (density function) for continuous random variable x
e	Constant used in normal probability distributions, $e = 2.71828\ldots$
π	Constant used in normal probability distributions, $\pi = 3.1416\ldots$
$\mu_{\bar{x}}$	True mean of sampling distribution of $\bar{x}$
$\sigma_{\bar{x}}$	True standard deviation of sampling distribution of $\bar{x}$

Key Ideas

Properties of Discrete Probability Distributions

1. $p(x) \geq 0$

2. $\sum_{\text{all } x} p(x) = 1$

Properties of Continuous Probability Distributions

1. $P(x = a) = 0$
2. $P(a < x < b)$ is area under curve between a and b

Methods for Assessing Normality

(1) *Histogram*

(2) *Stem-and-leaf display*

1	7
2	3389
3	245677
4	19
5	2

(3) $(IQR)/ \approx 1.3$

(4) *Normal probability plot*

Normal Approximation to Binomial

x is binomial (n, p)

$$P(\leq a) \approx P\{z < (a + .5) - \mu\}$$

Sampling distribution of a statistic—the theoretical probability distribution of the statistic in repeated sampling

Central Limit Theorem—the sampling distribution of the sample mean, $\bar{x}$, is approximately normal for large n (e.g., $n \geq 30$).

Key Formulas

Random Variable	Prob. Dist'n	Mean	Variance
General Discrete:	Table, formula, or graph for $p(x)$	$\sum_{\text{all } x} x \cdot p(x)$	$\sum_{\text{all } x} (x - \mu)^2 \cdot p(x)$
Binomial:	$p(x) = \binom{n}{x} p^x q^{n-x}$	np	npq
Normal:	$f(x) = \dfrac{1}{\sigma\sqrt{2\pi}} e^{-1/2[(x-\mu)/\sigma]^2}$	μ	σ^2
Standard Normal:	$f(z) = \dfrac{1}{\sqrt{2\pi}} e^{-1/2(z)^2}$ $z = (x - \mu)/\sigma$	$\mu = 0$	$\sigma^2 = 1$
Sampling distribution of $\bar{x}$	(See Normal)	$\mu_{\bar{x}} = \mu$	$\sigma_{\bar{x}}^2 = \sigma^2/n$

Guide to Selecting a Discrete Probability Distribution

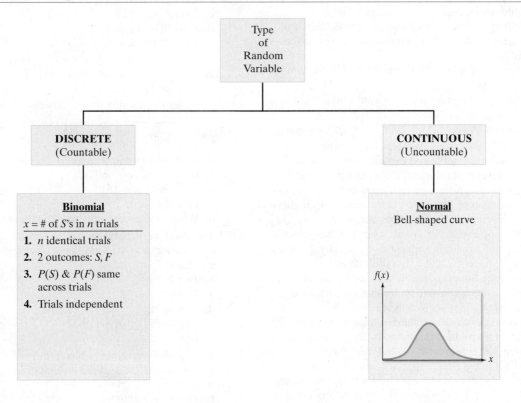

Generating the Sampling Distribution of $\bar{x}$

Generating the Sampling Distribution of $\bar{x}$

Select sample size n (large) from target population

Calculate $\bar{x}$

$\bar{x}_1$
$\bar{x}_2$
$\bar{x}_3$
$\vdots$

Population:
Mean $= \mu$
Std. Dev. $= \sigma$
Unknown shape

Repeat this process an infinite number of times

Sampling distribution of $\bar{x}$
(i.e., theoretical population of x's)
Mean $= \mu_{\bar{x}} = \mu$
Std. Dev. $= \sigma_{\bar{x}} = \sigma/\sqrt{n}$
Normal distribution (Central Limit Theorem)

Supplementary Exercises 4.170–4.204

Note: Starred (*) exercises refer to the optional section in this chapter.

Understanding the Principles

4.170 Which of the following describe discrete random variables, and which describe continuous random variables?
  a. The length of time that an exercise physiologist's program takes to elevate her client's heart rate to 140 beats per minute
  b. The number of crimes committed on a college campus per year
  c. The number of square feet of vacant office space in a large city
  d. The number of voters who favor a new tax proposal

4.171 For each of the following examples, decide whether x is a binomial random variable and explain your decision:
  a. A manufacturer of computer chips randomly selects 100 chips from each hour's production in order to estimate the proportion of defectives. Let x represent the number of defectives in the 100 chips sampled.
  b. Of five applicants for a job, two will be selected. Although all applicants appear to be equally qualified, only three have the ability to fulfill the expectations of the company. Suppose that the two selections are made at random from the five applicants, and let x be the number of qualified applicants selected.
  c. A software developer establishes a support hot line for customers to call in with questions regarding use of the software. Let x represent the number of calls received on the hot line during a specified workday.
  d. Florida is one of a minority of states with no state income tax. A poll of 1,000 registered voters is conducted to determine how many would favor a state income tax in light of the state's current fiscal condition. Let x be the number in the sample who would favor the tax.

4.172 Describe how you could obtain the simulated sampling distribution of a sample statistic.

4.173 True or False. The sample mean, $\bar{x}$, will always be equal to $\mu_{\bar{x}}$.

4.174 True or False. The sampling distribution of $\bar{x}$ is normally distributed, regardless of the size of the sample n.

Learning the Mechanics

4.175 Suppose x is a binomial random variable with $n = 20$ and $p = .7$.
  a. Find $P(x = 14)$.
  b. Find $P(x \le 12)$.
  c. Find $P(x > 12)$.
  d. Find $P(9 \le x \le 18)$.
  e. Find $P(8 < x < 18)$.
  f. Find μ, σ^2, and σ.
  g. What is the probability that x is in the interval $\mu \pm 2\sigma$?

4.176 Consider the discrete probability distribution shown here:

x	10	12	18	20
$p(x)$	.2	.3	.1	.4

  a. Calculate μ, σ^2, and σ.
  b. What is $P(x < 15)$?
  c. Calculate $\mu \pm 2\sigma$.
  d. What is the probability that x is in the interval $\mu \pm 2\sigma$?

4.177 The random variable x has a normal distribution with $\mu = 70$ and $\sigma = 10$. Find the following probabilities:
  a. $P(x \le 75)$
  b. $P(x \ge 90)$
  c. $P(60 \le x \le 75)$
  d. $P(x > 75)$
  e. $P(x = 75)$
  f. $P(x \le 95)$

4.178 The random variable x has a normal distribution with $\mu = 40$ and $\sigma^2 = 36$. Find a value of x, say, x_0, such that
  a. $P(x \ge x_0) = .5$
  b. $P(x \le x_0) = .9911$
  c. $P(x \le x_0) = .0028$
  d. $P(x \ge x_0) = .0228$
  e. $P(x \le x_0) = .1003$
  f. $P(x \ge x_0) = .7995$

*4.179 Assume that x is a binomial random variable with $n = 100$ and $p = .5$. Use the normal probability distribution to approximate the following probabilities:
  a. $P(x \le 48)$
  b. $P(50 \le x \le 65)$

c. $P(x \geq 70)$
d. $P(55 \leq x \leq 58)$
e. $P(x = 62)$
f. $P(x \leq 49 \text{ or } x \geq 72)$

4.180 A random sample of $n = 68$ observations is selected from a population with $\mu = 19.6$ and $\sigma = 3.2$. Approximate each of the following probabilities:
a. $P(\bar{x} \leq 19.6)$
b. $P(\bar{x} \leq 19)$
c. $P(\bar{x} \geq 20.1)$
d. $P(19.2 \leq \bar{x} \leq 20.6)$

Applying the Concepts—Basic

4.181 New book reviews. In Exercise 2.181 (p. 99), you read about a study of book reviews in American history, geography, and area studies published in *Choice* magazine (*Library Acquisitions: Practice and Theory*, Vol. 19, 1995). The overall rating stated in each review was ascertained and recorded as follows: 1 = would not recommend, 2 = cautious or very little recommendation, 3 = little or no preference, 4 = favorable/recommended, 5 = outstanding/significant contribution. Based on a sample of 375 book reviews, the probability distribution of rating, x, follows.

Booking Rating x	$p(x)$
1	.051
2	.099
3	.093
4	.635
5	.122

a. Is this a valid probability distribution?
b. What is the probability that a book reviewed in *Choice* has a rating of 1?
c. What is the probability that a book reviewed in *Choice* has a rating of at least 4?
d. What is the probability that a book reviewed in *Choice* has a rating of 2 or 3?
e. Find $E(x)$.
f. Interpret the result, part **e**.

4.182 Parents who condone spanking. According to a nationwide survey, 60% of parents with young children condone spanking their child as a regular form of punishment (*Tampa Tribune*, Oct. 5, 2000). Consider a random sample of three people, each of whom is a parent with young children. Assume that x, the number in the sample who condone spanking, is a binomial random variable.
a. What is the probability that none of the three parents condones spanking as a regular form of punishment for their children?
b. What is the probability that at least one condones spanking as a regular form of punishment?
c. Give the mean and standard deviation of x. Interpret the results.

4.183 Quit-smoking program. According to the University of South Florida's Tobacco Research and Intervention Program, only 5% of the nation's cigarette smokers ever enter into a treatment program to help them quit smoking (*USF Magazine*, Spring 2000). In a random sample of 200 smokers, let x be the number who enter into a treatment program.
a. Explain why x is a binomial random variable (to a reasonable degree of approximation).

b. What is the value of p? Interpret this value.
c. What is the expected value of x? Interpret this value.

4.184 Married-women study. According to *Women's Day* magazine (Jan. 2007), only 50% of married women would marry their current husbands again if given the chance. Consider the number x in a random sample of 10 married women who would marry their husbands again. Identify the discrete probability distribution that best models the distribution of x. Explain.

4.185 Passing the FCAT math test. All Florida high schools require their students to demonstrate competence in mathematics by scoring 70% or above on the FCAT mathematics achievement test. The FCAT math scores of those students taking the test for the first time are normally distributed with a mean of 77% and a standard deviation of 7.3%. What percentage of students who take the test for the first time will pass it?

*4.186 Where will you get your next pet? In Exercise 4.52 (p. 186), you learned that 50% of pet owners will get their next dog or cat from a Shelter (*USA Today*, May 12, 2010). In a random sample of 200 pet owners, let x be the number who will get their next dog or cat from a shelter.
a. Find the mean of x.
b. Find the standard deviation of x.
c. Find the z-score for the value $x = 110$.
d. Find the approximate probability that x is less than or equal to 110.

4.187 Alkalinity of river water. The alkalinity level of water specimens collected from the Han River in Seoul, Korea, has a mean of 50 milligrams per liter and a standard deviation of 3.2 milligrams per liter (*Environmental Science & Engineering*, Sept., 1, 2000). Assume that the distribution of alkalinity levels is approximately normal, and find the probability that a water specimen collected from the river has an alkalinity level
a. exceeding 45 milligrams per liter.
b. below 55 milligrams per liter.
c. between 51 and 52 milligrams per liter.

4.188 Salaries of travel professionals. According to a Natimal Business Travel Association (NBTA) 2008 survey, the average salary of a travel management professional is $97,300. Assume that the standard deviation of such salaries is $30,000. Consider a random sample of 50 travel management professionals, and let $\bar{x}$ represent the mean salary for the sample.
a. What is $\mu_{\bar{x}}$?
b. What is $\sigma_{\bar{x}}$?
c. Describe the shape of the sampling distribution of $\bar{x}$.
d. Find the z-score for the value $\bar{x} = \$89,500$.
e. Find $P(\bar{x} > 89,500)$.

4.189 Children's attitude toward reading. In the journal *Knowledge Quest* (January/February 2002), education professors at the University of Southern California investigated children's attitudes toward reading. One study measured third through sixth graders' attitudes toward recreational reading on a 140-point scale (where higher scores indicate a more positive attitude). The mean score for this population of children was 106 points, with a standard deviation of 16.4 points. Consider a random sample of 36 children from this population, and let $\bar{x}$ represent the mean recreational reading attitude score for the sample.
a. What is $\mu_{\bar{x}}$?
b. What is $\sigma_{\bar{x}}$?

c. Describe the shape of the sampling distribution of $\bar{x}$.

d. Find the z-score for the value $\bar{x} = 100$ points.

e. Find $P(\bar{x} < 100)$.

Applying the Concepts—Intermediate

4.190 Fungi in beech forest trees. Refer to the *Applied Ecology and Environmental Research* (Vol. 1, 2003) study of beech trees damaged by fungi, presented in Exercise 3.129 (p. 155). The researchers found that 25% of the beech trees in east central Europe had been damaged by fungi. Consider a sample of 20 beech trees from this area.

a. What is the probability that fewer than half are damaged by fungi?

b. What is the probability that more than 15 are damaged by fungi?

c. How many of the sampled trees would you expect to be damaged by fungi?

4.191 Parents' behavior at a gym meet. *Pediatric Exercise Science* (Feb. 1995) published an article on the behavior of parents at competitive youth gymnastic meets. On the basis of a survey of the gymnasts, the researchers estimated the probability of a parent "yelling" at his or her child before, during, or after the meet as .05. In a random sample of 20 parents attending a gymnastic meet, find the probability that at least 1 parent yells at his or her child before, during, or after the meet.

*4.192 Countries that allow a free press. The degree to which democratic and nondemocratic countries attempt to control the news media was examined in the *Journal of Peace Research* (Nov. 1997). The article reported that 80% of all democratic regimes allow a free press. In contrast, 10% of all nondemocratic regimes allow a free press.

a. In a random sample of 50 democratic regimes, how many would you expect to allow a free press? Give a range that is highly likely to include the number of democratic regimes with a free press.

b. In a random sample of 50 nondemocratic regimes, how many would you expect to allow a free press? Give a range that is highly likely to include the number of nondemocratic regimes with a free press.

4.193 Visually impaired students. The *Journal of Visual Impairment & Blindness* (May–June 1997) published a study of the lifestyles of visually impaired students. Using diaries, the students kept track of several variables, including number of hours of sleep obtained in a typical day. These visually impaired students had a mean of 9.06 hours and a standard deviation of 2.11 hours. Assume that the distribution of the number of hours of sleep for this group of students is approximately normal.

a. Find the probability that a visually impaired student obtains less than 6 hours of sleep on a typical day.

b. Find the probability that a visually impaired student gets between 8 and 10 hours of sleep on a typical day.

c. Twenty percent of all visually impaired students obtain less than how many hours of sleep on a typical day?

4.194 Comparison of exam scores: red versus blue exam. Refer to the *Teaching Psychology* (May 1998) study of how external clues influence performance, presented in Exercise 2.122 (p. 77). Recall that two different forms of a midterm psychology examination were given, one printed on blue paper and the other on red paper. Grading only the difficult questions, the researchers found that scores on the blue exam had a distribution with a mean of 53% and a standard deviation of 15%, while scores on the red exam had a distribution with a mean of 39% and a standard deviation of 12%. Assuming that both distributions are approximately normal, on which exam is a student more likely to score below 20% on the difficult questions, the blue one or the red one? (Compare your answer with that of Exercise 2.122**c.**)

*4.195 Fungi in beech forest trees. Refer to the *Applied Ecology and Environmental Research* (Vol. 1, 2003) study of beech trees damaged by fungi, presented in Exercise 4.190 (p. 234). Recall that the researchers found that 25% of the beech trees in east central Europe have been damaged by fungi. In an east central European forest with 200 beech trees, how likely is it that more than half of the trees have been damaged by fungi?

4.196 Wear-out of used display panels. Wear-out failure time of electronic components is often assumed to have a normal distribution. Can the normal distribution be applied to the wear-out of used manufactured products, such as colored display panels? A lot of 50 used display panels was purchased by an outlet store. Each panel displays 12 to 18 color characters. Prior to acquisition, the panels had been used for about one-third of their expected lifetimes. The data in the table below (saved in the **PANELFAIL** file) give the failure times (in years) of the 50 used panels. Use the techniques of this section to determine whether the used panel wear-out times are approximately normally distributed.

4.197 Susceptibility to hypnosis. The Computer-Assisted Hypnosis Scale (CAHS) is designed to measure a person's susceptibility to hypnosis. In computer-assisted hypnosis, the computer serves as a facilitator of hypnosis by using digitized speech processing coupled with interactive involvement with the hypnotic subject. CAHS scores range from 0 (no susceptibility) to 12 (extremely high susceptibility). A study in *Psychological Assessment* (March 1995) reported a mean CAHS score of 4.59 and a standard deviation of 2.95 for University of Tennessee undergraduates.

Data for Exercise 4.196

0.01	1.21	1.71	2.30	2.96	0.19	1.22	1.75	2.30	2.98	0.51
1.24	1.77	2.41	3.19	0.57	1.48	1.79	2.44	3.25	0.70	1.54
1.88	2.57	3.31	0.73	1.59	1.90	2.61	1.19	0.75	1.61	1.93
2.62	3.50	0.75	1.61	2.01	2.72	3.50	1.11	1.62	2.16	2.76
3.50	1.16	1.62	2.18	2.84	3.50					

Based on Irony, T. Z., Lauretto, M., Pereira, C., and Stern, J. M. "A Weibull Wearout Test: Full Bayesian Approach," paper presented at *Mathematical Sciences Colloquium*, Binghamton University, Binghamton, UK, December 2001.

Assume that $\mu = 4.59$ and $\sigma = 2.95$ for this population. Suppose a psychologist uses the CAHS to test a random sample of 50 subjects.

a. Would you expect to observe a sample mean CAHS score of $\bar{x} = 6$ or higher? Explain.

b. Suppose the psychologist actually observes $\bar{x} = 6.2$. On the basis of your answer to part **a,** make an inference about the population from which the sample was selected.

4.198 Supercooling temperature of frogs. Many species of terrestrial tree frogs that hibernate at or near the ground surface can survive prolonged exposure to low winter temperatures. In freezing conditions, the frog's body temperature, called its *supercooling temperature*, remains relatively higher because of an accumulation of glycerol in its body fluids. A study in *Science* revealed that the supercooling temperature of terrestrial frogs frozen at $-6°C$ has a relative frequency distribution with a mean of $-2°C$ and a standard deviation of $.3°C$. Consider the mean supercooling temperature $\bar{x}$ of a random sample of $n = 42$ terrestrial frogs frozen at $-6°C$.

a. Find the probability that $\bar{x}$ exceeds $-2.05°C$.

b. Find the probability that $\bar{x}$ falls between $-2.20°C$ and $-2.10°C$.

4.199 Surface roughness of pipe. The journal *Anti-Corrosion Methods and Materials* (Vol. 50, 2003) published a study of the surface roughness of oil field pipes. A scanning probe instrument was used to measure the surface roughness x (in micrometers) of 20 sampled sections of coated interior pipe. Consider the sample mean $\bar{x}$.

a. Assume that the surface roughness distribution has a mean of $\mu = 1.8$ micrometers and a standard deviation of $\sigma = .5$ micrometer. Use this information to find the probability that $\bar{x}$ exceeds 1.85 micrometers.

b. The sample data from the study are listed in the table and saved in the **ROUGHPIPE** file. Compute $\bar{x}$.

c. On the basis of the result from part **b**, comment on the validity of the assumptions made in part **a**.

1.72	2.50	2.16	2.13	1.06	2.24	2.31	2.03	1.09	1.40
2.57	2.64	1.26	2.05	1.19	2.13	1.27	1.51	2.41	1.95

Based on Farshad, F., and Pesacreta, T. "Coated pipe interior surface roughness as measured by three scanning probe instruments." *Anti-Corrosion Methods and Materials*, Vol. 50, No. 1, 2003 (Table II).

Applying the Concepts—Advanced

4.200 Dye discharged in paint. A machine used to regulate the amount of dye dispensed for mixing shades of paint can be set so that it discharges an average of μ milliliters (mL) of dye per can of paint. The amount of dye discharged is known to have a normal distribution with a standard deviation of .4 mL. If more than 6 mL of dye are discharged when making a certain shade of blue paint, the shade is unacceptable. Determine the setting for μ so that only 1% of the cans of paint will be unacceptable.

4.201 How many questionnaires to mail? The probability that a person responds to a mailed questionnaire is .4. How many questionnaires should be mailed if you want to be reasonably certain that at least 100 will be returned?

Critical Thinking Challenges

*4.202 Space shuttle disaster. On January 28, 1986, the space shuttle *Challenger* exploded, killing all seven astronauts aboard. An investigation concluded that the explosion was caused by the failure of the O ring seal in the joint between the two lower segments of the right solid rocket booster. In a report made one year prior to the catastrophe, the National Aeronautics and Space Administration (NASA) claimed that the probability of such a failure was about $^1/_{60,000}$, or about once in every 60,000 flights. But a risk-assessment study conducted for the Air Force at about the same time assessed the probability to be $^1/_{35}$, or about once in every 35 missions. (*Note*: The shuttle had flown 24 successful missions prior to the disaster.) Given the events of January 28, 1986, which risk assessment–NASA's or the Air Force's–appears to be more appropriate?

4.203 IQs and *The Bell Curve*. In their controversial book *The Bell Curve* (Free Press, 1994), Professors Richard J. Herrnstein (a Harvard psychologist who died while the book was in production) and Charles Murray (a political scientist at MIT) explored, as the subtitle states, "intelligence and class structure in American life." *The Bell Curve* employs statistical analyses heavily in an attempt to support the authors' positions. Since the book's publication, many expert statisticians have raised doubts about the authors' statistical methods and the inferences drawn from them. (See, for example, "Wringing *The Bell Curve*: A cautionary tale about the relationships among race, genes, and IQ," *Chance*, Summer 1995.) One of the many controversies sparked by the book is the authors' tenet that level of intelligence (or lack thereof) is a cause of a wide range of intractable social problems, including constrained economic mobility. The measure of intelligence chosen by the authors is the well-known intelligent quotient (IQ). Numerous tests have been developed to measure IQ; Herrnstein and Murray use the Armed Forces Qualification Test (AFQT), originally designed to measure the cognitive ability of military recruits. Psychologists traditionally treat IQ as a random variable having a normal distribution with mean $\mu = 100$ and standard deviation $\sigma = 15$.

In their book, Herrnstein and Murray refer to five cognitive classes of people defined by percentiles of the normal distribution. Class I ("very bright") consists of those with IQs above the 95th percentile; Class II ("bright") are those with IQs between the 75th and 95th percentiles; Class III ("normal") includes IQs between the 25th and 75th percentiles; Class IV ("dull") are those with IQs between the 5th and 25th percentiles; and Class V ("very dull") are IQs below the 5th percentile.

a. Assuming that the distribution of IQ is accurately represented by the normal curve, determine the proportion of people with IQs in each of the five cognitive classes defined by Herrnstein and Murray.

b. Although Herrnstein and Murray define the cognitive classes in terms of percentiles, they stress that IQ scores should be compared with z-scores, not percentiles. In other words, it is more informative to give the difference in z-scores for two IQ scores than it is to give the difference in percentiles. Do you agree?

c. Researchers have found that scores on many intelligence tests are decidedly nonnormal. Some distributions are skewed toward higher scores, others toward lower scores. How would the proportions in the five cognitive classes defined in part **a** differ for an IQ distribution that is skewed right? Skewed left?

4.204 **Fecal pollution at Huntington Beach.** The state of California mandates fecal indicator bacteria monitoring at all public beaches. When the concentration of fecal bacteria in the water exceeds a certain limit (400 colony-forming units of fecal coliform per 100 milliliters), local health officials must post a sign (called surf zone posting) warning beachgoers of potential health risks upon entering the water. For fecal bacteria, the state uses a single-sample standard; that is, if the fecal limit is exceeded in a single sample of water, surf zone posting is mandatory. This single-sample standard policy has led to a recent rash of beach closures in California.

Joon Ha Kim and Stanley B. Grant, engineers at the University of California at Irvine, conducted a study of the surf water quality at Huntington Beach in California and reported the results in *Environmental Science & Technology* (September 2004). The researchers found that beach closings were occurring despite low pollution levels in some instances, while in others signs were not posted when the fecal limit was exceeded. They attributed these "surf zone posting errors" to the variable nature of water quality in the surf zone (for example, fecal bacteria concentration tends to be higher during ebb tide and at night) and the inherent time delay between when a water sample is collected and when a sign is posted or removed. In order to prevent posting errors, the researchers recommend using an averaging method, rather than a single sample, to determine unsafe water quality. (For example, one simple averaging method is to take a random sample of multiple water specimens and compare the average fecal bacteria level of the sample with the limit of 400 cfu/100 mL in order to determine whether the water is safe.)

Discuss the pros and cons of using the single-sample standard versus the averaging method. Part of your discussion should address the probability of posting a sign when in fact the water is safe and the probability of posting a sign when in fact the water is unsafe. (Assume that the fecal bacteria concentrations of water specimens at Huntington Beach follow an approximately normal distribution.)

Activity   Simulating a Sampling Distribution—Cell Phone Usage

[*Note:* This activity is designed for small groups or the entire class.] Consider the length of time a student spends making a cell phone call, sending/retrieving a text message, or accessing e-mail on his/her cell phone. Let *x* represent the length of time, in seconds, for a single cell phone activity (call, text, or e-mail). Here, we are interested in the sampling distribution of $\bar{x}$, the mean length of time for a sample of size *n* cell phone activities.

1. Keep track of the time lengths for all cell phone activities you engage in over the next week.

2. Pool your time length data with data from other class members or the entire class so that the pooled data set has at least 100 observations. Designate someone in the group to calculate the mean and standard deviation of the pooled data set.

3. Devise a convenient way to choose random samples from the pooled data set. (For example, you could assign each observation a number beginning with "1" and use a random number generator to select a sample.)

4. Choose a random sample of size *n* = 30 from the pooled data, and find the mean of the sample. Group members should repeat the process of choosing a sample of size *n* = 30 from the pooled data and finding the sample mean until the group has accumulated at least 25 sample means. (Call this data set *Sample Means*.)

5. Find the mean and standard deviation of the *Sample Means* data set. Also, form a histogram for the *Sample Means* data set. Explain how the Central Limit Theorem is illustrated in this activity.

References

Hogg, R. V., McKean, J. W., and Craig, A. T. *Introduction to Mathematical Statistics*, 6th ed. Upper Saddle River, NJ: Prentice Hall, 2005.

Larsen, R. J., and Marx, M. L. An *Introduction to Mathematical Statistics and Its Applications,* 4th ed. Upper Saddle River, N.J.: Prentice Hall, 2005.

Lindgren, B. W. *Statistical Theory*, 4d ed. New York: Chapman & Hall, 1993.

Mood, A. M., Graybill, F. A., and Boes, D. C. *Introduction to the Theory of Statistics*, 3rd ed. New York: McGraw-Hill, 1974.

Parzen, E. *Modern Probability Theory and Its Applications*. New York: Wiley, 1960.

Ramsey, P. P., and Ramsey, P. H. "Simple tests of normality in small samples." *Journal of Quality Technology*, Vol. 22, 1990.

Ross, S. M. *Stochastic Processes*, 2nd ed. New York: Wiley, 1996.

Wackerly, D., Mendenhall, W., and Scheaffer, R. L. *Mathematical Statistics with Applications*, 7th ed. North Scituate, MA: Duxbury, 2008.

Winkler, R. L., and Hays, W. *Statistics: Probability, Inference, and Decision*, 2nd ed. New York: Holt, Rinehart and Winston, 1975.

USING TECHNOLOGY

MINITAB: Binomial Probabilities, Normal Probabilities, and Simulated Sampling Distributions

Binomial Probabilities

Step 1 Select the "Calc" button on the MINITAB menu bar, click on "Probability Distributions," and then finally select the discrete distribution of your choice (e.g., "Binomial"), as shown in Figure 4.M.1.

Figure 4.M.1
MINITAB menu options for discrete probabilities

Step 2 Select either "Probability" or "Cumulative probability" on the resulting dialog box.

Step 3 Specify the parameters of the distribution (e.g., sample size n, probability of success p).

Step 4 Specify the value x of interest in the "Input constant" box.

Figure 4.M.2
MINITAB binomial distribution dialog box

Step 5 Click "OK." The probability for the value of x will appear on the MINITAB session window.

[*Note*: Figure 4.M.2 gives the specifications for finding $p(x = 2)$ in a binomial distribution with $n = 5$ and $p = .3$.]

Normal Probabilities

Step 1 Select the "Calc" button on the MINITAB menu bar, click on "Probability Distributions," and then finally select the Continuous distribution of your choice (e.g., "Normal"), as shown in Figure 4 M.3.

Figure 4.M.3
MINITAB menu options for continuous probabilities

Step 2 Select either "Probability" or "Cumulative probability" on the resulting dialog box.

Step 3 Specify the parameters of the distribution (e.g., μ and σ for a normal distributions).

Step 4 Specify the value x of interest in the "Input constant" box.

Step 5 Click "OK". The probability for the value of x will appear on the MINITAB session window. [*Note*: Figure 4.M.4 gives the specifications for finding $P(x \leq 23)$ in a normal distribution with $\mu = 24.5$ and $\sigma = 1$.]

Figure 4.M.4
MINITAB normal distribution dialog box

Normal Probability Plot

Step 1 Select "Graph" on the MINITAB menu bar, and then click on "Probability Plot," as shown in Figure 4.M.5.

Figure 4.M.5
MINITAB options for a normal probability plot

Step 2 Select "Single" (for one variable) on the next box, and the dialog box will appear as shown in Figure 4.M.6.

Figure 4.M.6
MINITAB normal probability plot dialog box

Step 3 Specify the variable of interest in the "Graph variables" box, and then click the "Distribution" button and select the "Normal" option. Click "OK" to return to the Probability Plot dialog box.

Step 4 Click "OK" to generate the normal probability plot.

Simulating a Sampling Distribution

Step 1 Select "Calc" on the MINITAB menu bar, and then click on "Random Data" (see Figure 4.M.7)

Step 2 On the resulting menu list, click on the distribution of your choice (e.g., "Uniform"). A dialog box similar to the one (the Uniform Distribution) shown in Figure 4.M.8 will appear.

Step 3 Specify the number of samples (e.g., 1,000) to generate in the "Number of rows to generate" box and the columns where the data will be stored in the "Store in columns" box. (The number of columns will be equal to the sample size, e.g., 40.) Finally, specify the parameters of the distribution (e.g., the lower and upper range of the uniform distribution). When you click "OK," the simulated data will appear on the MINITAB worksheet.

Figure 4.M.7
MINITAB options for generating random data

Figure 4.M.8
MINITAB dialog box for simulating the uniform distribution

Step 4 Calculate the value of the sample statistic of interest for each sample. To do this, click on the "Calc" button on the MINITAB menu bar, and then click on "Row Statistics," as shown in Figure 4.M.9. The resulting dialog box appears in Figure 4.M.10.

Figure 4.M.9
MINITAB selections for generating sample statistics for the simulated data

Figure 4.M.10
MINITAB row statistics dialog box

Step 5 Check the sample statistic (e.g., the mean) you want to calculate, specify the "Input variables" (or columns), and the column where you want the value of the sample statistic to be saved. Click "OK" and the value of the statistic for each sample will appear on the MINITAB worksheet.

[*Note:* Use the MINITAB menu choices provided in the Using Technology section in Chapter 2, p. 103 to generate a histogram of the sampling distribution of the statistic or to find the mean and variance of the sampling distribution.]

TI–83/TI–84 Plus Graphing Calculator: Discrete Random Variables, Binomial Probabilities, and Normal Probabilities

Calculating the Mean and Standard Deviation of a Discrete Random Variable

Step 1 *Enter the data*
- Press **STAT** and select **1:Edit**
  (*Note:* If the lists already contain data, clear the old data. Use the up **ARROW** to highlight 'L1.')
- Press **CLEAR ENTER**.
- Use the up **ARROW** to highlight 'L2'
- Press **CLEAR ENTER**
- Use the **ARROW** and **ENTER** keys to enter the x-values of the variable into **L1**
- Use the **ARROW** and **ENTER** keys to enter the probabilities, $P(x)$, into **L2**

Step 2 *Access the Calc Menu*
- Press **STAT**
- Arrow right to **CALC**
- Select 1-**Var Stats**
- Press **ENTER**
- Press **2nd 1** for **L1**
- Press **COMMA**
- Select **2nd 2** for **L2**
- Press **ENTER**

The mean and standard deviation will be displayed on the screen, as will the quartiles, min, and max.

Calculating Binomial Probabilities

I. $P(x = k)$

To compute the probability of k successes in n trials where p is the probability of success for each trial, use the **binompdf(** command. *Binompdf* stands for "binomial probability density function." This command is under the **DISTR**ibution menu and has the format **binompdf(*n*, *p*, *k*)**.

Example Compute the probability of 5 successes in 8 trials where the probability of success for a single trial is 40%. In this example, $n = 8$, $p = .4$, and $k = 5$.

Step 1 *Enter the binomial parameter*
- Press **2nd VARS** for **DISTR**
- Press the down **ARROW** key until **binompdf** is highlighted
- Press **ENTER**
- After **binompdf(**, type **8, .4, 5)** (*Note:* Be sure to use the **COMMA** key between each parameter)
- Press **ENTER**
- You should see

```
binompdf(8,.4,5)
          .12386304
```

Thus, $P(x = k)$ is about 12.4%.

II. $P(x \le k)$

To compute the probability of k or fewer successes in n trials where the p is probability of success for each trial, use the **binomcdf(** command. *Binomcdf* stands for "binomial cumulative probability density function." This command is under the **DISTR**ibution menu and has the format **binomcdf(*n*, *p*, *k*)**.

Example Compute the probability of 5 or fewer successes in 8 trials where the probability of success for a single trial is 40%. In this example, $n = 8$, $p = .4$, and $k = 5$.

Step 2 *Enter the binomial parameters*
- Press **2nd VARS** for **DISTR**
- Press down the **ARROW** key until **binomcdf** is highlighted
- Press **ENTER**
- After **binomcdf(**, type **8, .4, 5)**
- Press **ENTER**
- You should see

```
binomcdf(8,.4,5)
          .95019264
```

Thus, $P(x \le 5)$ is about 95%.

III. $P(x < k)$, $P(x > k)$, $P(x \ge k)$

To find the probability of less than k successes $P(x < k)$, more than k successes $P(x > k)$, or at least k successes $P(x \ge k)$, variations of the **binomcdf(** command must be used as shown below.

- $P(x < k)$ use **binomcdf(*n*, *p*, *k* – 1)**
- $P(x > k)$ use **1 – binomcdf(*n*, *p*, *k*)**
- $P(x \ge k)$ use **1 – binomcdf(*n*, *p*, *k* – 1)**

Graphing the Area under the Standard Normal Curve

Step 1 *Turn off all plots*
- Press **2nd PRGM** and select **1:ClrDraw**

- Press **ENTER,** and 'Done' will appear on the screen
- Press **2nd Y=** and select **4:PlotsOff**
- Press **ENTER,** and 'Done' will appear on the screen

Step 2 *Set the viewing window (Recall that almost all of the area under the standard normal curve falls between −5 and 5. A height of 0.5 is a good choice for Ymax.)*

Note: When entering a negative number, be sure to use the negative sign (−), and not the minus sign.
- Set **Xmin =** − 5
- **Xmax =** 5
- **Xscl =** 1
- **Ymin =** 0
- **Ymax =** .5
- **Yscl =** 0
- **Xres =** 1

Step 3 *View graph*
- Press **2nd VARS**
- Arrow right to **DRAW**
- Press **ENTER** to select **1:ShadeNorm(**
- Enter your lower limit (e.g., −5)
- Press **COMMA**
- Enter your upper limit (e.g., 1.5)
- Press)
- Press **ENTER** The graph will be displayed along with the area, lower limit, and upper limit. Thus, $P(Z < 1.5) = .9332$

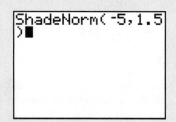

Finding Normal Probabilities without a Graph

To compute probabilities for a normal distribution, use the **normalcdf(** command. *Normalcdf* stands for "normal cumulative density function." This command is under the **DISTR**ibution menu and has the format **normalcdf(*lower limit, upper limit, mean, standard deviation*)**.

Step 1 *Find the probability*
- Press **2nd VARS** for **DISTR** and select **Normalcdf(**
- After **Normalcdf(,** type in the lower limit
- Press **COMMA**
- Enter the upper limit
- Press **COMMA**
- Enter the mean
- Press **COMMA**
- Enter the standard deviation
- Press)
- Press **ENTER**

The probability will be displayed on the screen.

Example What is $P(x < 115)$ for a normal distribution with $\mu = 100$ and $\sigma = 10$? In this example, the lower limit is − ∞, the upper limit is 115, the mean is 100, and the standard deviation is 10. To represent − ∞ on

the calculator, enter (−) **1**, press **2nd** and press the **COMMA** key for **EE,** and then press **99.** The screen appears as follows: Thus, $P(x < 115)$ *is* = .9332

Finding Normal Probabilities with a Graph

Step 1 *Turn off all plots*
- Press **Y =** and **CLEAR** all functions from the Y registers
- Press **2nd Y =** and select **4:PlotsOff**
- Press **ENTER ENTER,** and 'Done' will appear on the screen

Step 2 *Set the viewing window (These values depend on the mean and standard deviation of the data.) Note*: When entering a negative number, be sure to use the negative sign (−), not the minus sign.
- Press **WINDOW**
- Set **Xmin =** $\mu - 5\sigma$
- **Xmax =** $\mu + 5\sigma$
- **Xscl =** σ
- **Ymin =** − $.125/\sigma$
- **Ymax =** $.5/\sigma$
- **Yscl =** 1
- **Xres =** 1

Step 3 *View graph*
- Press **2nd VARS**
- **ARROW** right to **DRAW**
- Press **ENTER** to select **1:ShadeNorm(**
- Enter the lower limit
- Press **COMMA**
- Enter the upper limit
- Press **COMMA**
- Enter the mean
- Press **COMMA**
- Enter the standard deviation
- Press)
- Press **ENTER**

The graph will be displayed along with the area, lower limit, and upper limit

Example What is $P(x < 115)$ for a normal distribution with $\mu = 100$ and $\sigma = 10$? In this example, the lower limit is − ∞, the upper limit is 115, the mean is 100, and the standard deviation is 10. To represent − ∞ on the calculator, enter (−) **1**, press **2nd** and press the **comma** key for **EE,** and then press **99.** The screens appear as follows:

Graphing a Normal Probability Plot

Step 1 *Enter the data*
- Press **STAT** and select **1:Edit**

  Note: If the list already contains data, clear the old data. Use the up arrow to highlight **L1**
- Press **CLEAR ENTER**
- Use the **ARROW** and **ENTER** keys to enter the data set into **L1**

Step 2 *Set up the normal probability plot*
- Press **Y =** and **CLEAR** all functions from the Y registers
- Press **2nd** and press **Y =** for **STAT PLOT**
- Press **1** for **Plot 1**
- Set the cursor so that **ON** is flashing
- For **Type**, use the **ARROW** and **ENTER** keys to highlight and select the last graph in the bottom row

- For **Data List**, choose the column containing the data (in most cases, L1)
  (*Note*: Press **2nd 1** for **L1**)
- For **Data Axis**, choose **X** and press **ENTER**

Step 3 *View plot*
- Press **ZOOM 9**

Your data will be displayed against the expected *z*-scores from a normal distribution. If you see a "generally" linear relationship, your data set is approximately normal.

5 Inferences Based on a Single Sample
Estimation with Confidence Intervals

CONTENTS

Where We've Been

- Learned that populations are characterized by numerical descriptive measures—called *parameters*
- Found that decisions about population parameters are based on *statistics* computed from the sample
- Discovered that *inferences* about parameters are subject to uncertainty and that this uncertainty is reflected in the *sampling distribution* of a statistic

Where We're Going

- Estimate a population parameter (mean, proportion, or variance) on the basis of a sample selected from the population (5.1)
- Use the sampling distribution of a statistic to form a confidence interval for the population parameter (5.2–5.4, 5.6)
- Show how to select the proper sample size for estimating a population parameter (5.5)

Statistics IN Action Medicare Fraud Investigations

United States Department of Justice (USDOJ) press release (May 8, 2008): *Eleven people have been indicted in . . . a targeted criminal, civil and administrative effort against individuals and health care companies that fraudulently bill the Medicare program. The indictments in the Central District of California resulted from the creation of a multi-agency team of federal, state and local investigators designed specifically to combat Medicare fraud through the use of real-time analysis of Medicare billing data.*

USDOJ press release (Jun 27, 2008): *The owners of four Miami-based health care corporations were sentenced and remanded to prison yesterday for their roles in schemes to defraud the Medicare program. Collectively, the three defendants through their companies collected more than $14 million from the Medicare program for unnecessary medicine, durable medical equipment (DME) and home health care services.*

USDOJ press release (Oct. 7, 2008): *Eight Miami–Dade County residents have been charged in a 16-count indictment for their alleged roles in a Medicare fraud scheme involving fake HIV infusion treatments.*

As the above press releases imply, the U.S. Department of Justice (USDOJ) and its Medicare Fraud Strike Force conduct investigations into suspected fraud and abuse of the Medicare system by health care providers. According to published reports, the Strike Force was responsible for almost 25% of the Medicare fraud charges brought nationwide in 2007.

One way in which Medicare fraud occurs is through the use of "upcoding," which refers to the practice of providers coding Medicare claims at a higher level of care than was actually provided to the patient. For example, suppose a particular kind of claim can be coded at three levels, where Level 1 is a routine office visit, Level 2 is a thorough examination involving advanced diagnostic tests, and Level 3 involves performing minor surgery. The amount of Medicare payment is higher for each increased level of claim. Thus, upcoding would occur if Level 1 services were billed at Level 2 or Level 3 payments, or if Level 2 services were billed at Level 3 payment.

The USDOJ relies on sound statistical methods to help identify Medicare fraud. Once the USDOJ has determined that possible upcoding has occurred, it next seeks to further investigate whether it is the result of legitimate practice (perhaps the provider is a specialist giving higher levels of care) or the result of fraudulent action on the part of the provider. To further its investigation, the USDOJ will next ask a statistician to select a sample of the provider's claims. For example, the statistician might determine that a random sample of 52 claims from the 1,000 claims in question will provide a sufficient sample to estimate the overcharge reliably. The USDOJ then asks a health care expert to audit each of the medical files corresponding to the sampled claims and determine whether the level of care matches the level billed by the provider, and, if not, to determine what level should have been billed. Once the audit has been completed, the USDOJ will calculate the overcharge.

In this chapter, we present a recent Medicare fraud case investigated by the USDOJ. Results for the audit of 52 sampled claims, with the amount paid for each claim, the amount disallowed by the auditor, and the amount that should have been paid for each claim, are saved in the **MCFRAUD** file.* Knowing that a total of $103,500 was paid for the 1,000 claims, the USDOJ wants to use the sample results to extrapolate the overpayment amount to the entire population of 1,000 claims.

Statistics IN Action Revisited

- Estimating the Mean Overpayment (p. 259)
- Estimating the Coding Error Rate (p. 268)
- Determining Sample Size (p. 275)

Data Set: MCFRAUD

5.1 Identifying and Estimating the Target Parameter

In this chapter, our goal is to estimate the value of an unknown population parameter, such as a population mean or a proportion from a binomial population. For example, we might want to know the mean gas mileage for a new car model, the average expected life of a flat-screen computer monitor, or the proportion of Iraq War veterans with post-traumatic stress syndrome.

You'll see that different techniques are used for estimating a mean or proportion, depending on whether a sample contains a large or small number of measurements. Nevertheless, our objectives remain the same: We want to use the sample information to

*Data provided (with permission) from Info Tech, Inc., Gainesville, Florida.

estimate the population parameter of interest (called the *target parameter*) and to assess the reliability of the estimate.

> The unknown population parameter (e.g., mean or proportion) that we are interested in estimating is called the **target parameter.**

Often, there are one or more key words in the statement of the problem that indicate the appropriate target parameter. Some key words associated with the parameters covered in this chapter are listed in the following box:

> ### Determining the Target Parameter
>
> | Parameter | Key Words or Phrases | Type of Data |
> |-----------|---------------------|--------------|
> | μ | Mean; average | Quantitative |
> | p | Proportion; percentage; fraction; rate | Qualitative |
> | σ^2 (optional) | Variance; variability; spread | Quantitative |

For the examples given in the first paragraph of this section, the words *mean* in "mean gas mileage" and *average* in "average life expectancy" imply that the target parameter is the population mean μ. The word *proportion* in "proportion of Iraq War veterans with post-traumatic stress syndrome" indicates that the target parameter is the binomial proportion p.

In addition to key words and phrases, the type of data (quantitative or qualitative) collected is indicative of the target parameter. With quantitative data, you are likely to be estimating the mean or variance of the data. With qualitative data with two outcomes (success or failure), the binomial proportion of successes is likely to be the parameter of interest.

A single number calculated from the sample that estimates a target population parameter is called a **point estimator.** For example, we'll use the sample mean, $\bar{x}$, to estimate the population mean μ. Consequently, $\bar{x}$ is a point estimator. Similarly, we'll learn that the sample proportion of successes, denoted $\hat{p}$, is a point estimator for the binomial proportion p and that the sample variance s^2 is a point estimator for the population variance σ^2. Also, we will attach a measure of reliability to our estimate by obtaining an **interval estimator**—a range of numbers that contain the target parameter with a high degree of confidence. For this reason the interval estimate is also called a **confidence interval.**

> A **point estimator** of a population parameter is a rule or formula that tells us how to use the sample data to calculate a *single* number that can be used as an *estimate* of the target parameter.

> An **interval estimator (or confidence interval)** is a formula that tells us how to use the sample data to calculate an *interval* that *estimates* the target parameter.

We consider methods for estimating a population mean in Sections 5.2 and 5.3. Estimating a population proportion is presented in Section 5.4. We show how to determine the sample sizes necessary for reliable estimates of the target parameters in Section 5.5. Finally, in optional Section 5.6 we discuss estimation of a population variance.

5.2 Confidence Interval for a Population Mean: Normal (z) Statistic

Suppose a large hospital wants to estimate the average length of time patients remain in the hospital. Hence, the hospital's target parameter is the population mean μ. To accomplish this objective, the hospital administrators plan to randomly sample 100 of all previous patients' records and to use the sample mean $\bar{x}$ of the lengths of stay to estimate μ, the mean of *all* patients' visits. The sample mean $\bar{x}$ represents a *point estimator* of the population mean μ. How can we assess the accuracy of this large-sample point estimator?

According to the Central Limit Theorem, the sampling distribution of the sample mean is approximately normal for large samples, as shown in Figure 5.1. Let us calculate the interval estimator:

$$\bar{x} \pm 1.96\sigma_{\bar{x}} = \bar{x} \pm \frac{1.96\sigma}{\sqrt{n}}$$

Figure 5.1
Sampling distribution of $\bar{x}$

That is, we form an interval from 1.96 standard deviations below the sample mean to 1.96 standard deviations above the mean. *Prior to drawing the sample*, what are the chances that this interval will enclose μ, the population mean?

To answer this question, refer to Figure 5.1. If the 100 measurements yield a value of $\bar{x}$ that falls between the two lines on either side of μ (i.e., within 1.96 standard deviations of μ), then the interval $\bar{x} \pm 1.96\sigma_{\bar{x}}$ will contain μ; if $\bar{x}$ falls outside these boundaries, the interval $\bar{x} \pm 1.96\sigma_{\bar{x}}$ will not contain μ. From Section 4.9, we know the area under the normal curve (the sampling distribution of $\bar{x}$) between these boundaries is exactly .95. Thus, the interval $\bar{x} \pm 1.96\sigma_{\bar{x}}$ will contain μ with a probability equal to .95.

Example 5.1
Estimating the Mean Hospital Length of Stay, σ Known

Problem Consider the large hospital that wants to estimate the average length of stay of its patients, μ. The hospital randomly samples $n = 100$ of its patients and finds that the sample mean length of stay is $\bar{x} = 4.5$ days. Also, suppose it is known that the standard deviation of the length of stay for all hospital patients is $\sigma = 4$ days. Use the interval estimator $\bar{x} \pm 1.96\sigma_{\bar{x}}$ to calculate a confidence interval for the target parameter, μ.

Solution Substituting $\bar{x} = 4.5$ and $\sigma = 4$ into the interval estimator formula, we obtain:

$$\bar{x} \pm 1.96\sigma_{\bar{x}} = \bar{x} \pm (1.96)\sigma/\sqrt{n} = 4.5 \pm (1.96)(4/\sqrt{100}) = 4.5 \pm .78$$

Or, (3.72, 5.28).

Look Back Since we know the probability that the interval, $\bar{x} \pm 1.96\sigma_{\bar{x}}$, will contain μ is .95, we call the interval estimator a *95% confidence interval* for μ.

Now Work Exercise 5.9a

The interval $\bar{x} \pm 1.96\sigma_{\bar{x}}$ in Example 5.1 is called a *large-sample 95% confidence interval* for the population mean μ. The term *large-sample* refers to the sample being of sufficiently large size that we can apply the Central Limit Theorem and the **normal (z) statistic** to determine the form of the sampling distribution of $\bar{x}$. Empirical research suggests that a sample size n exceeding a value between 20 and 30 will usually yield a sampling distribution of $\bar{x}$ that is approximately normal. This result led many practitioners to adopt the rule of thumb that a sample size of $n \geq 30$ is required to use large-sample confidence interval procedures. Keep in mind, though, that 30 is not a magical number and, in fact, is quite arbitrary.

Also, note that the large-sample interval estimator requires knowing the value of the population standard deviation, σ. In most (if not nearly all) practical applications, however, the value of σ will be unknown. For large samples, the fact that σ is unknown poses only a minor problem since the sample standard deviation s provides a very good approximation to σ. The next example illustrates the more realistic large-sample confidence interval procedure.

Example 5.2

Estimating the Mean Hospital Length of Stay, σ Unknown

Problem Refer to Example 5.1 and the problem of estimating μ, the average length of stay of a hospital's patients. The lengths of stay (in days) for the $n = 100$ sampled patients are shown in Table 5.1. Use the data to find a 95% confidence interval for μ and interpret the result.

Table 5.1	Lengths of Stay (in Days) for 100 Patients								
2	3	8	6	4	4	6	4	2	5
8	10	4	4	4	2	1	3	2	10
1	3	2	3	4	3	5	2	4	1
2	9	1	7	17	9	9	9	4	4
1	1	1	3	1	6	3	3	2	5
1	3	3	14	2	3	9	6	6	3
5	1	4	6	11	22	1	9	6	5
2	2	5	4	3	6	1	5	1	6
17	1	2	4	5	4	4	3	2	3
3	5	2	3	3	2	10	2	4	2

Data Set: HOSPLOS

Solution The hospital almost surely does not know the true standard deviation, σ, of the population of lengths of stay. However, since the sample size is large, we will use the sample standard deviation, s, as an estimate for σ in the confidence interval formula. An SAS printout of summary statistics for the sample of 100 lengths of stay is shown at the top of Figure 5.2. From the shaded portion of the printout, we find $\bar{x} = 4.53$ days and $s = 3.68$ days. Substituting these values into the interval estimator formula, we obtain:

$$\bar{x} \pm (1.96)\sigma/\sqrt{n} \approx \bar{x} \pm (1.96)s/\sqrt{n} = 4.53 \pm (1.96)(3.68)/\sqrt{100} = 4.53 \pm .72$$

Or, (3.81, 5.25). That is, we estimate the mean length of stay for all hospital patients to fall within the interval 3.81 to 5.25 days.

Look Back The confidence interval is also shown at the bottom of the SAS printout, Figure 5.2. Note that the endpoints of the interval vary slightly from those computed in

Figure 5.2

SAS printout with summary statistics and 95% confidence interval for data on 100 hospital stays

```
Sample Statistics for LOS

      N           Mean        Std. Dev.        Std. Error
    -------------------------------------------------------
     100          4.53          3.68             0.37

Hypothesis Test

     Null hypothesis:    Mean of LOS  =  0
     Alternative:        Mean of LOS ^= 0

           t Statistic        Df        Prob > t
         ------------------------------------------
             12.318           99         <.0001


95 % Confidence Interval for the Mean

        Lower Limit:              3.80
        Upper Limit:              5.26
```

the example. This is due to the fact that when σ is unknown and n is large, the sampling distribution of $\bar{x}$ will deviate slightly from the normal (z) distribution. In practice, these differences can be ignored.

Now Work Exercise 5.13

Can we be sure that μ, the true mean, is in the interval from 3.81 to 5.25 in Example 5.2? We cannot be certain, but we can be reasonably confident that it is. This confidence is derived from the knowledge that if we were to draw repeated random samples of 100 measurements from this population and form the interval $\bar{x} \pm 1.96\sigma_{\bar{x}}$ each time, 95% of the intervals would contain μ. We have no way of knowing (without looking at all the patients' records) whether our sample interval is one of the 95% that contain μ or one of the 5% that do not, but the odds certainly favor its containing μ. The probability, .95, that measures the confidence we can place in the interval estimate is called a *confidence coefficient*. The percentage, 95%, is called the *confidence level* for the interval estimate.

> The **confidence coefficient** is the probability that an interval estimator encloses the population parameter—that is, the relative frequency with which the interval estimator encloses the population parameter when the estimator is used repeatedly a very large number of times. The **confidence level** is the confidence coefficient expressed as a percentage.

Figure 5.3

Confidence intervals for μ: 10 samples

Now we have seen how an interval can be used to estimate a population mean. When we use an interval estimator, we can usually calculate the probability that the estimation *process* will result in an interval that contains the true value of the population mean. That is, the probability that the interval contains the parameter in repeated usage is usually known. Figure 5.3 shows what happens when 10 different samples are drawn from a population and a confidence interval for μ is calculated from each. The location of μ is indicated by the vertical line in the figure. Ten confidence intervals, each based on one of 10 samples, are shown as horizontal line segments. Note that the confidence intervals move from sample to sample, sometimes containing μ and other times missing μ. *If our confidence level is 95%, then in the long run, 95% of our sample confidence intervals will contain* μ.

Suppose you wish to choose a confidence coefficient other than .95. Notice in Figure 5.1 that the confidence coefficient .95 is equal to the total area under the sampling distribution, less .05 of the area, which is divided equally between the two tails. Using this idea, we can construct a confidence interval with any desired confidence coefficient by increasing or decreasing the area (call it α) assigned to the tails of the sampling distribution. (See Figure 5.4.) For example, if we place the area $\alpha/2$ in each tail and if $z_{\alpha/2}$ is the z value such that $\alpha/2$ will lie to its right, then the confidence interval with confidence coefficient $(1 - \alpha)$ is

$$\bar{x} \pm z_{\alpha/2}\sigma_{\bar{x}}$$

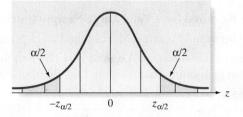

Figure 5.4

Locating $z_{\alpha/2}$ on the standard normal curve

> The value z_α is defined as the value of the standard normal random variable z such
> that the area α will lie to its right. In other words, $P(z > z_\alpha) = \alpha$.

To illustrate, for a confidence coefficient of .90, we have $(1 - \alpha) = .90, \alpha = .10$,
and $\alpha/2 = .05$; $z_{.05}$ is the z value that locates area .05 in the upper tail of the sampling
distribution. Recall that Table III in Appendix A gives the areas between the mean and
a specified z-value. Since the total area to the right of the mean is .5, we find that $z_{.05}$ will
be the z value corresponding to an area of $.5 - .05 = .45$ to the right of the mean. (See
Figure 5.5.) This z value is $z_{.05} = 1.645$.

Confidence coefficients used in practice usually range from .90 to .99. The most
commonly used confidence coefficients with corresponding values of α and $z_{\alpha/2}$ are
shown in Table 5.2.

Figure 5.5

The z value ($z_{.05}$) corresponding
to an area equal to .05 in the
upper tail of the z-distribution

Table 5.2	**Commonly Used Values of $z_{\alpha/2}$**		
Confidence Level			
$100(1 - \alpha)\%$	α	$\alpha/2$	$z_{\alpha/2}$
90%	.10	.05	1.645
95%	.05	.025	1.96
99%	.01	.005	2.575

Now Work Exercise 5.7

> **Large-Sample $100(1 - \alpha)\%$ Confidence Interval for μ, Based on a Normal (z) Statistic**
>
> σ *known:* $\bar{x} \pm (z_{\alpha/2})\sigma_{\bar{x}} = \bar{x} \pm (z_{\alpha/2})(\sigma/\sqrt{n})$
>
> σ *unknown:* $\bar{x} \pm (z_{\alpha/2})\sigma_{\bar{x}} \approx \bar{x} \pm (z_{\alpha/2})(s/\sqrt{n})$
>
> where $z_{\alpha/2}$ is the z-value corresponding to an area $\alpha/2$ in the tail of a standard
> normal distribution (see Figure 5.4), $\sigma_{\bar{x}}$ is the standard deviation of the sampling
> distribution of $\bar{x}$, σ is the standard deviation of the population, and s is the standard
> deviation of the sample.

> **Conditions Required for a Valid Large-Sample Confidence Interval for μ**
>
> **1.** A random sample is selected from the target population.
> **2.** The sample size n is large (i.e., $n \geq 30$). (Due to the Central Limit Theorem, this
>   condition guarantees that the sampling distribution of $\bar{x}$ is approximately normal.
>   Also, for large n, s will be a good estimator of σ.)

Example 5.3
A Large-Sample Confidence Interval for μ— Mean Number of Unoccupied Seats per Flight

Problem Unoccupied seats on flights cause airlines to lose revenue. Suppose a large airline wants to estimate its average number of unoccupied seats per flight over the past year. To accomplish this, the records of 225 flights are randomly selected, and the number of unoccupied seats is noted for each of the sampled flights. (The data are saved in the **AIRNOSHOWS** file.) Descriptive statistics for the data are displayed in the MINITAB printout of Figure 5.6.

Estimate μ, the mean number of unoccupied seats per flight during the past year, using a 90% confidence interval.

Solution The form of a large-sample 90% confidence interval for a population mean (based on the z-statistic) is:

$$\bar{x} \pm z_{\alpha/2}\sigma_{\bar{x}} = \bar{x} \pm z_{.05}\sigma_{\bar{x}} = \bar{x} \pm 1.645\left(\frac{\sigma}{\sqrt{n}}\right)$$

From Figure 5.6, we find (after rounding) that $\bar{x} = 11.6$. Since we do not know the value of σ (the standard deviation of the number of unoccupied seats per flight for all flights of the year), we use our best approximation—the sample standard deviation, $s = 4.1$, shown on the MINITAB printout. Then the 90% confidence interval is approximately

$$11.6 \pm 1.645\left(\frac{4.1}{\sqrt{225}}\right) = 11.6 \pm .45$$

or from 11.15 to 12.05. That is, at the 90% confidence level, we estimate the mean number of unoccupied seats per flight to be between 11.15 and 12.05 during the sampled year. This result is verified (except for rounding) on the right side of the MINITAB printout in Figure 5.6.

Figure 5.6

MINITAB printout with descriptive statistics and 90% confidence interval for Example 5.3

Variable	N	Mean	StDev	SE Mean	90% CI
NOSHOWS	225	11.596	4.103	0.274	(11.144, 12.047)

Look Back We stress that the confidence level for this example, 90%, refers to the procedure used. If we were to apply that procedure repeatedly to different samples, approximately 90% of the intervals would contain μ. Although we do not know for sure whether this particular interval (11.15, 12.05) is one of the 90% that contain μ or one of the 10% that do not, our knowledge of probability gives us "confidence" that the interval contains μ.

Now Work Exercise 5.14a

The interpretation of confidence intervals for a population mean is summarized in the next box.

Interpretation of a Confidence Interval for a Population Mean

When we form a $100(1 - \alpha)\%$ confidence interval for μ, we usually express our confidence in the interval with a statement such as "We can be $100(1 - \alpha)\%$ confident that μ lies between the lower and upper bounds of the confidence interval," where, for a particular application, we substitute the appropriate numerical values for the level of confidence and for the lower and upper bounds. *The statement reflects our confidence in the estimation process, rather than in the particular interval that is calculated from the sample data.* We know that repeated application of the same procedure will result in different lower and upper bounds on the interval. Furthermore, we know that $100(1 - \alpha)\%$ of the resulting intervals will contain μ. There is (usually) no way to determine whether any particular interval is one of those which contain μ or one of those which do not. However, unlike point estimators, confidence intervals have some measure of reliability—the confidence coefficient—associated with them. For that reason, they are generally preferred to point estimators.

Sometimes, the estimation procedure yields a confidence interval that is too wide for our purposes. In this case, we will want to reduce the width of the interval to obtain a more precise estimate of μ. One way to accomplish that is to decrease the confidence coefficient, $1 - \alpha$. For example, consider the problem of estimating the mean length of stay, μ for hospital patients. Recall that for a sample of 100 patients, $\bar{x} = 4.53$ days and $s = 3.68$ days. A 90% confidence interval for μ is

$$\bar{x} \pm 1.645(\sigma/\sqrt{n}) \approx 4.53 \pm (1.645)(3.68)/\sqrt{100} = 4.53 \pm .61$$

or (3.92, 5.14). You can see that this interval is narrower than the previously calculated 95% confidence interval, (3.81, 5.25). Unfortunately, we also have "less confidence" in the 90% confidence interval. An alternative method used to decrease the width of an interval without sacrificing "confidence" is to increase the sample size n. We demonstrate this method in Section 5.5.

Exercises 5.1–5.26

Understanding the Principles

5.1 Define the target parameter.

5.2 What is the confidence coefficient in a 90% confidence interval for μ?

5.3 Explain the difference between an interval estimator and a point estimator for μ.

5.4 Explain what is meant by the statement "We are 95% confident that an interval estimate contains μ."

5.5 Will a large-sample confidence interval be valid if the population from which the sample is taken is not normally distributed? Explain.

5.6 What conditions are required to form a valid large-sample confidence interval for μ?

Learning the Mechanics

5.7 Find $z_{\alpha/2}$ for each of the following:
NW
a. $\alpha = .10$ **b.** $\alpha = .01$
c. $\alpha = .05$ **d.** $\alpha = .20$

5.8 What is the confidence level of each of the following confidence intervals for μ?

a. $\bar{x} \pm 1.96\left(\dfrac{\sigma}{\sqrt{n}}\right)$

b. $\bar{x} \pm 1.645\left(\dfrac{\sigma}{\sqrt{n}}\right)$

c. $\bar{x} \pm 2.575\left(\dfrac{\sigma}{\sqrt{n}}\right)$

d. $\bar{x} \pm 1.28\left(\dfrac{\sigma}{\sqrt{n}}\right)$

e. $\bar{x} \pm .99\left(\dfrac{\sigma}{\sqrt{n}}\right)$

5.9 A random sample of n measurements was selected from a
NW population with unknown mean μ and standard deviation $\sigma = 20$. Calculate a 95% confidence interval for μ for each of the following situations:
a. $n = 75, \bar{x} = 28$
b. $n = 200, \bar{x} = 102$
c. $n = 100, \bar{x} = 15$
d. $n = 100, \bar{x} = 4.05$
e. Is the assumption that the underlying population of measurements is normally distributed necessary to ensure the validity of the confidence intervals in parts **a–d**? Explain.

5.10 A random sample of 90 observations produced a mean $\bar{x} = 25.9$ and a standard deviation $s = 2.7$.
a. Find a 95% confidence interval for the population mean μ.
b. Find a 90% confidence interval for μ.
c. Find a 99% confidence interval for μ.

5.11 A random sample of 100 observations from a normally distributed population possesses a mean equal to 83.2 and a standard deviation equal to 6.4.
a. Find a 95% confidence interval for μ.
b. What do you mean when you say that a confidence coefficient is .95?
c. Find a 99% confidence interval for μ.
d. What happens to the width of a confidence interval as the value of the confidence coefficient is increased while the sample size is held fixed?
e. Would your confidence intervals of parts **a** and **c** be valid if the distribution of the original population were not normal? Explain.

5.12 The mean and standard deviation of a random sample of n measurements are equal to 33.9 and 3.3, respectively.
a. Find a 95% confidence interval for μ if $n = 100$.
b. Find a 95% confidence interval for μ if $n = 400$.
c. Find the widths of the confidence intervals you calculated in parts **a** and **b**. What is the effect on the width of a confidence interval of quadrupling the sample size while holding the confidence coefficient fixed?

● Applet Exercise 5.1

Use the applet entitled *Confidence Intervals for a Mean (the impact of confidence level)* to investigate the situation in Exercise 5.11 further. For this exercise, assume that $\mu = 83.2$ is the population mean and $\sigma = 6.4$ is the population standard deviation.
a. Using $n = 100$ and the normal distribution with mean and standard deviation as just given, run the applet one time. How many of the 95% confidence intervals contain the mean? How many would you expect to contain the mean? How many of the 99% confidence intervals contain the mean? How many would you expect to contain the mean?
b. Which confidence level has a greater frequency of intervals that contain the mean? Is this result what you would expect? Explain.
c. Without clearing, run the applet several more times. What happens to the proportion of 95% confidence

intervals that contain the mean as you run the applet more and more? What happens to the proportion of 99% confidence intervals that contain the mean as you run the applet more and more? Interpret these results in terms of the meanings of the 95% confidence interval and the 99% confidence interval.

d. Change the distribution to *right skewed*, clear, and run the applet several more times. Do you get the same results as in part c? Would you change your answer to part e of Exercise 5.11? Explain.

Applet Exercise 5.2

Use the applet entitled *Confidence Intervals for a Mean (the impact of confidence level)* to investigate the effect of the sample size on the proportion of confidence intervals that contain the mean when the underlying distribution is skewed. Set the distribution to *right skewed*, the mean to 10, and the standard deviation to 1.

a. Using $n = 30$, run the applet several times without clearing. What happens to the proportion of 95% confidence intervals that contain the mean as you run the applet more and more? What happens to the proportion of 99% confidence intervals that contain the mean as you run the applet more and more? Do the proportions seem to be approaching the values that you would expect?

b. Clear and run the applet several times, using $n = 100$. What happens to the proportions of 95% confidence intervals and 99% confidence intervals that contain the mean this time? How do these results compare with your results in part **a**?

c. Clear and run the applet several times, using $n = 1000$. How do the results compare with your results in parts **a** and **b**?

d. Describe the effect of sample size on the likelihood that a confidence interval contains the mean for a skewed distribution.

Applying the Concepts—Basic

5.13 Latex allergy in health care workers. Health care workers who use latex gloves with glove powder on a daily basis are particularly susceptible to developing a latex allergy. Symptoms of a latex allergy include conjunctivitis, hand eczema, nasal congestion, a skin rash, and shortness of breath. Each in a sample of 46 hospital employees who were diagnosed with latex allergy based on a skin-prick test reported on their exposure to latex gloves (*Current Allergy & Clinical Immunology,* March 2004). Summary statistics for the number of latex gloves used per week are $\bar{x} = 19.3$ and $s = 11.9$.

a. Give a point estimate for the average number of latex gloves used per week by all health care workers with a latex allergy.

b. Form a 95% confidence interval for the average number of latex gloves used per week by all health care workers with a latex allergy.

c. Give a practical interpretation of the interval you found in part **b**.

d. Give the conditions required for the interval in part **b** to be valid.

5.14 Lipid profiles of hypertensive patients. Hypertension is diagnosed if a patient's systolic blood pressure exceeds 140 mmHg and diastolic blood pressure exceeds 90 mmHg.

A study of the lipid profiles of hypertensive patients was carried out and the results published in *Biology and Medicine* (Vol. 2, 2010). Data on fasting blood sugar (milligrams/deciliter) and magnesium (milligrams/deciliter) in blood specimens collected from 50 patients diagnosed with hypertension were collected and are stored in the **HYPER** file. Biochemists used these data to establish a benchmark for fasting blood sugar (FBS) and magnesium (MAG) levels in hypertensive patients. The accompanying MINITAB printout gives 90% confidence intervals for the mean fasting blood sugar and mean magnesium level.

a. Locate and interpret the 90% confidence interval for mean fasting blood sugar on the printout.

b. Locate and interpret the 90% confidence interval for mean magnesium level on the printout.

c. If the biochemists increase the confidence level to 95%, what will happen to the width of the intervals?

d. If the biochemists increase the sample of hypertensive patients from 50 to 100, what will likely happen to the width of the intervals?

```
Variable   N    Mean    StDev   SE Mean      90% CI
FBS        50  101.62   34.13    4.83    ( 93.53,  109.71)
MAG        50  1.94100 0.05607  0.00793  (1.92771, 1.95429)
```

5.15 Motivation of drug dealers. Refer to the *Applied Psychology in Criminal Justice* (Sept. 2009) study of the personality characteristics of drug dealers, Exercise 2.98 (p. 71). Recall that each in a sample of 100 convicted drug dealers was scored on the Wanting Recognition (WR) Scale, which provides a quantitative measure of a person's level of need for approval and sensitivity to social situations. (Higher scores indicate a greater need for approval.) The sample of drug dealers had a mean WR score of 39, with a standard deviation of 6. Use this information to find an interval estimate of the mean WR score for all convicted drug dealers. Use a confidence coefficient of 99%. Interpret the result.

5.16 Laptop use in middle school. Refer to the *American Secondary Education* (Fall 2009) study of middle school students' usage of laptop computers, Exercise 2.100 (p. 72). Recall that for a sample of 106 students, the researchers reported the following statistics on how many minutes per day each student used a laptop for taking notes: $\bar{x} = 13.2, s = 19.5$. Now the researchers want to estimate the average amount of time per day laptops are used for taking notes for all middle school students across the country.

a. Define the target parameter for the researchers in words and symbols.

b. In Exercise 2.100**a**, you computed the interval, $\bar{x} \pm 2s$. Explain why this formula should not be used as an interval estimate for the target parameter.

c. Calculate a 90% confidence interval for the target parameter. Interpret the result.

d. Explain what the phrase "90% confidence" implies in your answer to part **c**.

e. In Exercise 2.100**b** you concluded that the distribution of laptop usage for taking notes for the sample of 106 students cannot be symmetric. Consequently, the distribution cannot be normal. Does this compromise the validity of the interval estimate, part **c**? Explain.

5.17 Albedo of ice melt ponds. Refer to the National Snow and Ice Data Center (NSIDC) collection of data on the albedo, depth, and physical characteristics of ice-melt ponds in the Canadian Arctic, presented in Exercise 2.13 (p. 35). Albedo is the ratio of the light reflected by the ice to that received by it. (High albedo values give a white appearance to the ice.) Visible albedo values were recorded for a sample of 504 ice-melt ponds located in the Barrow Strait in the Canadian Arctic; these data are saved in the **PONDICE** file.

a. Find a 90% confidence interval for the true mean visible albedo value of all Canadian Arctic ice ponds.

b. Give both a practical and a theoretical interpretation of the interval.

c. Recall from Exercise 2.13 that the type of ice for each pond was classified as first-year ice, multiyear ice, or landfast ice. Find 90% confidence intervals for the mean visible albedo for each of the three types of ice. Interpret the intervals.

5.18 Extinct New Zealand birds. Refer to the *Evolutionary Ecology Research* (July 2003) study of the patterns of extinction in the New Zealand bird population, presented in Exercise 2.102 (p. 72). Suppose you are interested in estimating the mean egg length (in millimeters) for the New Zealand bird population.

a. What is the target parameter?

b. Recall that the egg lengths for 132 bird species are saved in the **NZBIRDS** file. Obtain a random sample of 50 egg lengths from the data set.

c. Find the mean and standard deviation of the 50 egg lengths you obtained in part **b.**

d. Use the information from part **c** to form a 99% confidence interval for the true mean egg length of a bird species found in New Zealand.

e. Give a practical interpretation of the interval you found in part **d.**

Applying the Concepts—Intermediate

5.19 Personality and aggressive behavior. How does personality impact aggressive behavior? A team of university psychologists conducted a review of studies that examined the relationship between personality and aggressive behavior (*Psychological Bulletin*, Vol. 132, 2006). One variable of interest to the researchers was the difference between the aggressive behavior level of individuals in the study who scored high on a personality test and those who scored low on the test. This variable, standardized to be between −7 and 7, was called "effect size". (A large positive effect size indicates that those who score high

on the personality test are more aggressive than those who score low.) The researchers collected the effect sizes for a sample of $n = 109$ studies published in psychology journals. This data is saved in the **PERAGGR** file. A dot plot and summary statistics for effect size are shown in the MINITAB printouts at the bottom of the page. Of interest to the researchers is the true mean effect size μ for all psychological studies of personality and aggressive behavior.

a. Identify the parameter of interest to the researchers.

b. Examine the dot plot. Does effect size have a normal distribution? Explain why your answer is irrelevant to the subsequent analysis.

c. Locate a 95% confidence interval for μ on the accompanying printout. Interpret the result.

d. If the true mean effect size exceeds 0, then the researchers will conclude that in the population, those who score high on a personality test are more aggressive than those who score low. Can the researchers draw this conclusion? Explain.

5.20 Shell lengths of sea turtles. Refer to the *Aquatic Biology* (Vol. 9, 2010) study of green sea turtles inhabiting the Grand Cayman South Sound lagoon, Exercise 2.83 (p. 65). The data on curved carapace (shell) length, measured in centimeters, for 76 captured turtles are displayed in the table and saved in the **TURTLES** file. Environmentalists want to estimate the true mean shell length of all green sea turtles in the lagoon.

33.96	30.37	32.57	31.50	36.46	35.54	36.16	35.32	35.99
39.55	44.33	42.73	42.15	42.43	49.96	46.04	48.76	47.78
45.81	49.05	49.65	49.71	54.29	52.01	51.15	54.42	52.62
53.27	54.07	50.40	53.69	51.30	54.29	54.58	55.11	57.65
56.35	55.68	58.40	58.06	57.79	56.54	57.03	57.64	59.27
64.79	61.96	60.08	62.34	63.84	60.61	64.91	60.35	62.63
63.33	63.00	64.55	60.03	64.75	60.24	69.01	65.07	65.77
65.30	68.24	65.28	67.54	68.49	66.98	65.67	70.26	70.94
70.52	72.01	74.34	81.63					

a. Define the parameter of interest to the environmentalists.

b. Use the data in the **TURTLES** file to find a point estimate of the target parameter.

c. Compute a 95% confidence interval for the target parameter. Interpret the result.

d. Suppose a biologist claims that the mean shell length of all green sea turtles in the lagoon is 60 cm. Make an inference about the validity of this claim.

5.21 Colored string preferred by chickens. Animal behaviorists have discovered that the more domestic chickens

MINITAB Output for Exercise 5.19

```
Variable    N    Mean    StDev   SE Mean      95% CI
EffSize    109   0.6477  0.8906   0.0853   (0.4786, 0.8167)
```

peck at objects placed in their environment, the healthier the chickens seem to be. White string has been found to be a particularly attractive pecking stimulus. In one experiment, 72 chickens were exposed to a string stimulus. Instead of white string, blue string was used. The number of pecks each chicken took at the blue string over a specified interval of time was recorded. Summary statistics for the 72 chickens were $\bar{x} = 1.13$ pecks and $s = 2.21$ pecks (*Applied Animal Behaviour Science*, October 2000).

a. Use a 99% confidence interval to estimate the population mean number of pecks made by chickens pecking at blue string. Interpret the result.

b. Previous research has shown that $\mu = 7.5$ pecks if chickens are exposed to white string. Based on the results you found in part **a,** is there evidence that chickens are more apt to peck at white string than blue string? Explain.

5.22 **Speed training in football.** A key statistic used by football coaches to evaluate players is a player's 40-yard sprint time. Can a drill be developed for improving a player's speed in the sprint? Researchers at Northern Kentucky University designed and tested a speed-training program for junior varsity and varsity high school football players (*The Sport Journal*, Winter 2004). The training program included 50-yard sprints run at varying speeds, high knee running sprints, butt kick sprints, "crazy legs" straddle runs, quick feet drills, jumping, power skipping, and all-out sprinting. Each in a sample of 38 high school athletes was timed in a 40-yard sprint prior to the start of the training program and timed again after completing the program. The decreases in times (measured in seconds) are listed in the table and saved in the **SPRINT** file. [*Note:* A negative decrease implies that the athlete's time after completion of the program was higher than his time prior to training.] The goal of the research is to demonstrate that the training program is effective in improving 40-yard sprint times.

−.01	.1	.1	.24	.25	.05	.28	.25	.2	.14
.32	.34	.3	.09	.05	0	.04	.17	0	.21
.15	.3	.02	.12	.14	.1	.08	.5	.36	.1
.01	.9	.34	.38	.44	.08	0	0		

Based on Gray, M., & Sauerbeck, J. A. "Speed training program for high school football players." *The Sport Journal*, Vol. 7, No. 1, Winter 2004 (Table 2).

a. Find a 95% confidence interval for the true mean decrease in sprint times for the population of all football players who participate in the speed-training program.

b. Based on the confidence interval, is the training program really effective in improving the mean 40-yard sprint time of high school football players? Explain.

5.23 **Improving SAT scores.** Refer to the *Chance* (Winter 2001) and National Education Longitudinal Survey (NELS) study of 265 students who paid a private tutor to help them improve their SAT scores, presented in Exercise 2.105 (p. 73). The changes in both the SAT-Mathematics and SAT-Verbal scores for these students are reproduced in the following table:

	SAT-Math	SAT-Verbal
Mean change in score	19	7
Standard deviation of score changes	65	49

a. Construct and interpret a 95% confidence interval for the population mean change in SAT-Mathematics score for students who pay a private tutor.

b. Repeat part **a** for the population mean change in SAT-Verbal score.

c. Suppose the true population mean change in score on one of the SAT tests for all students who paid a private tutor is 15. Which of the two tests, SAT-Mathematics or SAT-Verbal, is most likely to have this mean change? Explain.

5.24 **Attention time given to twins.** Psychologists have found that twins, in their early years, tend to have lower IQs and pick up language more slowly than nontwins. (*Wisconsin Twin Research Newsletter*, Winter 2004). The slower intellectual growth of most twins may be caused by benign parental neglect. Suppose it is desired to estimate the mean attention time given to twins per week by their parents. A sample of 50 sets of $2\frac{1}{2}$-year-old twin boys is taken, and at the end of 1 week, the attention time given to each pair is recorded. The data (in hours) are listed in the following table and saved in the **ATTIMES** file. Find a 90% confidence interval for the mean attention time given to all twin boys by their parents. Interpret the confidence interval.

20.7	16.7	22.5	12.1	2.9
23.5	6.4	1.3	39.6	35.6
10.9	7.1	46.0	23.4	29.4
44.1	13.8	24.3	9.3	3.4
15.7	46.6	10.6	6.7	5.4
14.0	20.7	48.2	7.7	22.2
20.3	34.0	44.5	23.8	20.0
43.1	14.3	21.9	17.5	9.6
36.4	0.8	1.1	19.3	14.6
32.5	19.1	36.9	27.9	14.0

Applying the Concepts—Advanced

5.25 **Study of undergraduate problem drinking.** In *Alcohol & Alcoholism* (Jan/Feb. 2007), psychologists at the University of Pennsylvania compared the levels of alcohol consumption of male and female freshman students. Each student was asked to estimate the amount of alcohol (beer, wine, or liquor) they consume in a typical week. Summary statistics for 128 males and 184 females are provided in the accompanying table.

a. For each gender, find a 95% confidence interval for mean weekly alcohol consumption.

b. Prior to sampling, what is the probability that at least one of the two confidence intervals will not contain the population mean it estimates? Assume that the two intervals are independent.

c. Based on the two confidence intervals, what inference can you make about which gender consumes the most alcohol, on average, per week? (*Caution:* In Chapter 7, we will learn about a more valid method of comparing population means.)

	Males	Females
Sample size, n	128	184
Mean (ounces), $\bar{x}$	16.79	10.79
Standard deviation, s	13.57	11.53

Based on Leeman, R. F., Fenton, M., & Volpicelli, J. R. "Impaired control and undergraduate problem drinking." *Alcohol & Alcoholism*, Vol. 42, No. 1, Jan./Feb. 2007 (Table 1).

5.26 **The Raid test kitchen.** According to scientists, the cockroach has had 300 million years to develop a resistance to destruction. In a study conducted by researchers for S. C. Johnson & Son, Inc. (manufacturers of Raid® and Off®), 5,000 roaches (the expected number in a roach-infested house) were released in the Raid test kitchen. One week later, the kitchen was fumigated and 16,298 dead roaches were counted, a gain of 11,298 roaches for the 1-week period. Assume that none of the original roaches died during the 1-week period and that the standard deviation of x, the number of roaches produced per roach in a 1-week period, is 1.5. Use the number of roaches produced by the sample of 5,000 roaches to find a 95% confidence interval for the mean number of roaches produced per week for each roach in a typical roach-infested house.

5.3 Confidence Interval for a Population Mean: Student's *t*-Statistic

Federal legislation requires pharmaceutical companies to perform extensive tests on new drugs before they can be marketed. Initially, a new drug is tested on animals. If the drug is deemed safe after this first phase of testing, the pharmaceutical company is then permitted to begin human testing on a limited basis. During this second phase, inferences must be made about the safety of the drug on the basis of information obtained from very small samples.

Suppose a pharmaceutical company must estimate the average increase in blood pressure of patients who take a certain new drug. Assume that only six patients (randomly selected from the population of all patients) can be used in the initial phase of human testing. The use of a *small sample* in making an inference about μ presents two immediate problems when we attempt to use the standard normal z as a test statistic.

BIOGRAPHY

WILLIAM S. GOSSET (1876–1937)

Student's t-Distribution

At the age of 23, William Gosset earned a degree in chemistry and mathematics at prestigious Oxford University. He was immediately hired by the Guinness Brewing Company in Dublin, Ireland, for his expertise in chemistry. However, Gosset's mathematical skills allowed him to solve numerous practical problems associated with brewing beer. For example, Gosset applied the Poisson distribution to model the number of yeast cells per unit volume in the fermentation process. His most important discovery was the *t*-distribution in 1908. Since most applied researchers worked with small samples, Gosset was interested in the behavior of the mean in the case of small samples. He tediously took numerous small sets of numbers, calculated the mean and standard deviation of each, obtained their *t*-ratio, and plotted the results on graph paper. The shape of the distribution was always the same: the *t*-distribution. Under company policy, employees were forbidden to publish their research results, so Gosset used the pen name *Student* to publish a paper on the subject. Hence, the distribution has been called Student's *t*-distribution. ■

Problem 1   The shape of the sampling distribution of the sample mean $\bar{x}$ (and the z-statistic) now depends on the shape of the population that is sampled. We can no longer assume that the sampling distribution of $\bar{x}$ is approximately normal, because the Central Limit Theorem ensures normality only for samples that are sufficiently large.

Solution to Problem 1   The sampling distribution of $\bar{x}$ (and z) is exactly normal even for relatively small samples *if the sampled population is normal*. It is approximately normal if the sampled population is approximately normal.

Problem 2   The population standard deviation σ is almost always unknown. Although it is still true that $\sigma_{\bar{x}} = \sigma/\sqrt{n}$, the sample standard deviation s may provide a poor approximation for σ when the sample size is small.

Solution to Problem 2   Instead of using the standard normal statistic

$$z = \frac{\bar{x} - \mu}{\sigma_{\bar{x}}} = \frac{\bar{x} - \mu}{\sigma/\sqrt{n}}$$

which requires knowledge of, or a good approximation to, σ, we define and use the statistic

$$t = \frac{\bar{x} - \mu}{s/\sqrt{n}}$$

in which the sample standard deviation s replaces the population standard deviation σ.

If we are sampling from a normal distribution, the **_t_-statistic** has a sampling distribution very much like that of the z-statistic: mound shaped, symmetric, and with mean 0. The primary difference between the sampling distributions of t and z is that the t-statistic is more variable than the z, a property that follows intuitively when you realize that t contains two random quantities ($\bar{x}$ and s), whereas z contains only one ($\bar{x}$).

The actual amount of variability in the sampling distribution of t depends on the sample size n. A convenient way of expressing this dependence is to say that the t statistic has $(n - 1)$ **degrees of freedom (df).*** Recall that the quantity $(n - 1)$ is the divisor that appears in the formula for s^2. This number plays a key role in the sampling distribution of

*Since degrees of freedom are related to the sample size n, it is helpful to think of the number of degrees of freedom as the amount of information in the sample available for estimating the target parameter.

s^2 and appears in discussions of other statistics in later chapters. In particular, the smaller the number of degrees of freedom associated with the *t*-statistic, the more variable will be its sampling distribution.

In Figure 5.7, we show both the sampling distribution of *z* and the sampling distribution of a *t*-statistic with both 4 and 20 df. You can see that the *t*-distribution is more variable than the *z*-distribution and that this variability increases as the degrees of freedom decrease. You can also see that the increased variability of the *t*-statistic means that the *t*-value, t_α, that locates an area α in the upper tail of the *t*-distribution is larger than the corresponding value, z_α. For any given value of α the *t*-value, t_α, increases as the number of degrees of freedom (df) decreases. Values of *t* that will be used in forming small-sample confidence intervals of μ are given in Table IV of Appendix A. A partial reproduction of this table is shown in Table 5.3. Note that t_α values are listed for various degrees of freedom, where α refers to the tail area under the *t*-distribution to the right of t_α.

For example, if we want the *t*-value with an area of .025 to its right and 4 df, we look in the table under the column $t_{.025}$ for the entry in the row corresponding to 4 df. This entry is $t_{.025} = 2.776$, shaded in Table 5.3 and shown in Figure 5.8. The corresponding standard normal *z*-score is $z_{.025} = 1.96$. Note that the last row of Table IV, where df $= \infty$ (infinity), contains the standard normal *z*-values. This follows from the fact that as the sample size *n* grows very large, *s* becomes closer to σ and thus *t* becomes closer in distribution to *z*. In fact,

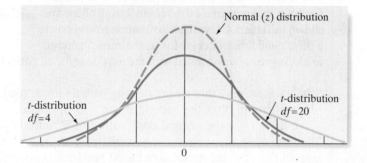

Figure 5.7

Standard normal (*z*) distribution and *t*-distributions

Table 5.3   Reproduction of Part of Table IV in Appendix A

Degrees of Freedom	$t_{.100}$	$t_{.050}$	$t_{.025}$	$t_{.010}$	$t_{.005}$	$t_{.001}$	$t_{.0005}$
1	3.078	6.314	12.706	31.821	63.657	318.31	636.62
2	1.886	2.920	4.303	6.965	9.925	22.326	31.598
3	1.638	2.353	3.182	4.541	5.841	10.213	12.924
4	1.533	2.132	2.776	3.747	4.604	7.173	8.610
5	1.476	2.015	2.571	3.365	4.032	5.893	6.869
6	1.440	1.943	2.447	3.143	3.707	5.208	5.959
7	1.415	1.895	2.365	2.998	3.499	4.785	5.408
8	1.397	1.860	2.306	2.896	3.355	4.501	5.041
9	1.383	1.833	2.262	2.821	3.250	4.297	4.781
10	1.372	1.812	2.228	2.764	3.169	4.144	4.587
11	1.363	1.796	2.201	2.718	3.106	4.025	4.437
12	1.356	1.782	2.179	2.681	3.055	3.930	4.318
13	1.350	1.771	2.160	2.650	3.012	3.852	4.221
14	1.345	1.761	2.145	2.624	2.977	3.787	4.140
15	1.341	1.753	2.131	2.602	2.947	3.733	4.073
$\vdots$	$\vdots$	$\vdots$	$\vdots$	$\vdots$	$\vdots$	$\vdots$	$\vdots$
∞	1.282	1.645	1.960	2.326	2.576	3.090	3.291

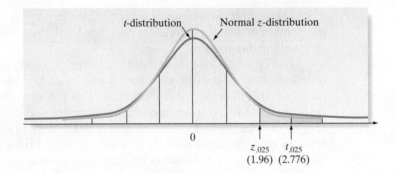

Figure 5.8

The $t_{.025}$ value in a t-distribution with 4 df, and the corresponding $z_{.025}$ value

when df = 29, there is little difference between corresponding tabulated values of z and t. Thus, we choose the arbitrary cutoff of $n = 30$ (df = 29) to distinguish between large-sample and small-sample inferential techniques when σ is unknown.

Example 5.4

Confidence Interval for Mean Blood Pressure Increase, t–Statistic

Problem Consider the pharmaceutical company that desires an estimate of the mean increase in blood pressure of patients who take a new drug. The blood pressure increases (measured in points) for the $n = 6$ patients in the human testing phase are shown in Table 5.4. Use this information to construct a 95% confidence interval for μ, the mean increase in blood pressure associated with the new drug for all patients in the population.

Solution First, note that we are dealing with a sample too small to assume, by the Central Limit Theorem, that the sample mean $\bar{x}$ is approximately normally distributed. That is, we do not get the normal distribution of $\bar{x}$ "automatically" from the Central Limit Theorem when the sample size is small. Instead, we must *assume* that the measured variable, in this case the increase in blood pressure, is normally distributed in order for the distribution of $\bar{x}$ to be normal.

Second, unless we are fortunate enough to know the population standard deviation σ, which in this case represents the standard deviation of *all* the patients' increases in blood pressure when they take the new drug, we cannot use the standard normal z-statistic to form our confidence interval for μ. Instead, we must use the t-distribution, with $(n - 1)$ degrees of freedom.

In this case, $n - 1 = 5$ df, and the t-value is found in Table 5.3 to be

$$t_{.025} = 2.571 \text{ with 5 df}$$

Recall that the large-sample confidence interval would have been of the form

$$\bar{x} \pm z_{\alpha/2}\sigma_{\bar{x}} = \bar{x} \pm z_{\alpha/2}\frac{\sigma}{\sqrt{n}} = \bar{x} \pm z_{.025}\frac{\sigma}{\sqrt{n}}$$

where 95% is the desired confidence level. To form the interval for a small sample from *a normal distribution, we simply substitute t for z and s for σ in the preceding formula, yielding*

$$\bar{x} \pm t_{\alpha/2}\frac{s}{\sqrt{n}}$$

An SPSS printout showing descriptive statistics for the six blood pressure increases is displayed in Figure 5.9. Note that $\bar{x} = 2.283$ and $s = .950$. Substituting these numerical values into the confidence interval formula, we get

$$2.283 \pm (2.571)\left(\frac{.950}{\sqrt{6}}\right) = 2.283 \pm .997$$

or 1.286 to 3.280 points. Note that this interval agrees (except for rounding) with the confidence interval highlighted on the SPSS printout in Figure 5.9.

Table 5.4   Blood Pressure Increases (Points) for Six Patients

1.7	3.0	.8	3.4	2.7	2.1

Data Set: BPINCR

Descriptives

			Statistic	Std. Error
BPINCR	Mean		2.283	.3877
	95% Confidence Interval for Mean	Lower Bound	1.287	
		Upper Bound	3.280	
	5% Trimmed Mean		2.304	
	Median		2.400	
	Variance		.902	
	Std. Deviation		.9496	
	Minimum		.8	
	Maximum		3.4	
	Range		2.6	
	Interquartile Range		1.625	
	Skewness		-.573	.845
	Kurtosis		-.389	1.741

Figure 5.9

SPSS confidence interval for mean blood pressure increase

We interpret the interval as follows: We can be 95% confident that the mean increase in blood pressure associated with taking this new drug is between 1.286 and 3.28 points. As with our large-sample interval estimates, our confidence is in the *process*, not in this particular interval. We know that if we were to repeatedly use this estimation procedure, 95% of the confidence intervals produced would contain the true mean μ, *assuming that the probability distribution of changes in blood pressure from which our sample was selected is normal.* The latter assumption is necessary for the small-sample interval to be valid.

Look Back What price did we pay for having to utilize a small sample to make the inference? First, we had to assume that the underlying population is normally distributed, and if the assumption is invalid, our interval might also be invalid.* Second, we had to form the interval by using a *t* value of 2.571 rather than a *z* value of 1.96, resulting in a wider interval to achieve the same 95% level of confidence. If the interval from 1.286 to 3.28 is too wide to be of much use, we know how to remedy the situation: Increase the number of patients sampled in order to decrease the width of the interval (on average).

Now Work Exercise 5.29

The procedure for forming a small-sample confidence interval is summarized in the accompanying boxes.

Small-Sample $100(1-\alpha)$ Confidence Interval for μ, *t*-Statistic

σ unknown: $\bar{x} \pm (t_{\alpha/2})(s/\sqrt{n})$

where $t_{\alpha/2}$ is the *t*-value corresponding to an area $\alpha/2$ in the upper tail of the Student's *t*-distribution based on $(n-1)$ degrees of freedom.

Note: If σ *is known* and condition #2 (below) is met, then use the *z*-statistic to find the confidence interval: $\bar{x} \pm (z_{\alpha/2})(\sigma/\sqrt{n})$

Conditions Required for a Valid Small-Sample Confidence Interval for μ

1. A random sample is selected from the target population.
2. The population has a relative frequency distribution that is approximately normal.

---

*By *invalid*, we mean that the probability that the procedure will yield an interval that contains μ is not equal to $(1 - \alpha)$. Generally, if the underlying population is approximately normal, the confidence coefficient will approximate the probability that the interval contains μ.

Example 5.5
A Small–Sample Confidence Interval for μ—Destructive Sampling

Problem Some quality control experiments require *destructive sampling* (i.e., the test to determine whether the item is defective destroys the item) in order to measure a particular characteristic of the product. The cost of destructive sampling often dictates small samples. Suppose a manufacturer of printers for personal computers wishes to estimate the mean number of characters printed before the printhead fails. The printer manufacturer tests $n = 15$ printheads and records the number of characters printed until failure for each. These 15 measurements (in millions of characters) are listed in Table 5.5, followed by a MINITAB summary statistics printout in Figure 5.10.

a. Form a 99% confidence interval for the mean number of characters printed before the printhead fails. Interpret the result.

b. What assumption is required for the interval you found in part **a** to be valid? Is that assumption reasonably satisfied?

Table 5.5 **Number of Characters (in Millions) for $n = 15$ Printhead Tests**

1.13	1.55	1.43	.92	1.25	1.36	1.32	.85	1.07	1.48	1.20	1.33	1.18	1.22	1.29

Data Set: PRINTHEAD

Figure 5.10
MINITAB printout with descriptive statistics and 99% confidence interval for Example 5.5

```
Variable   N     Mean    StDev  SE Mean       99% CI
NUMCHAR    15  1.23867  0.19316  0.04987  (1.09020, 1.38714)
```

Solution

a. For this small sample ($n = 15$), we use the *t*-statistic to form the confidence interval. We use a confidence coefficient of .99 and $n - 1 = 14$ degrees of freedom to find $t_{\alpha/2}$ in Table IV:

$$t_{\alpha/2} = t_{.005} = 2.977$$

[*Note:* The small sample forces us to extend the interval almost three standard deviations (of $\bar{x}$) on each side of the sample mean in order to form the 99% confidence interval.] From the MINITAB printout shown in Figure 5.10, we find that $\bar{x} = 1.24$ and $s = .19$. Substituting these (rounded) values into the confidence interval formula, we obtain

$$\bar{x} \pm t_{.005}\left(\frac{s}{\sqrt{n}}\right) = 1.24 \pm 2.977\left(\frac{.19}{\sqrt{15}}\right)$$

$$= 1.24 \pm .15 \text{ or } (1.09, 1.39)$$

This interval is highlighted in Figure 5.10.

Our interpretation is as follows: The manufacturer can be 99% confident that the printhead has a mean life of between 1.09 and 1.39 million characters. If the manufacturer were to advertise that the mean life of its printheads is (at least) 1 million characters, the interval would support such a claim. Our confidence is derived from the fact that 99% of the intervals formed in repeated applications of this procedure will contain μ.

b. Since n is small, we must assume that the number of characters printed before the printhead fails is a random variable from a normal distribution. That is, we assume that the population from which the sample of 15 measurements is selected is distributed normally. One way to check this assumption is to graph the distribution of data in Table 5.5. If the sample data are approximately normally distributed, then the population from which the sample is selected is very likely to be normal. A MINITAB stem-and-leaf plot for the sample data is displayed in Figure 5.11. The

Stem-and-Leaf Display: NUMBER

```
Stem-and-leaf of NUMBER   N  = 15
Leaf Unit = 0.010

    1     8    5
    2     9    2
    3    10    7
    5    11    38
   (4)   12    0259
    6    13    236
    3    14    38
    1    15    5
```

Figure 5.11
MINITAB stem-and-leaf
display of data in Table 5.5

distribution is mound shaped and nearly symmetric. Therefore, the assumption of normality appears to be reasonably satisfied.

Look Back Other checks for normality, such as a normal probability plot and the ratio IQR/*s*, may also be used to verify the normality condition.

Now Work Exercise 5.39

We have emphasized throughout this section that an assumption that the population is approximately normally distributed is necessary for making small-sample inferences about μ when the *t*-statistic is used. Although many phenomena do have approximately normal distributions, it is also true that many random phenomena have distributions that are not normal or even mound shaped. Empirical evidence acquired over the years has shown that the *t*-distribution is rather insensitive to moderate departures from normality. That is, the use of the *t*-statistic when sampling from slightly skewed mound-shaped populations generally produces credible results; however, for cases in which the distribution is distinctly nonnormal, we must either take a large sample or use a *nonparametric method*.

> **What Do You Do When the Population Relative Frequency Distribution Departs Greatly from Normality?**
>
> *Answer:* Use the nonparametric statistical method of Optional Section 6.8.

Statistics IN Action | **Revisited** | ## Estimating the Mean Overpayment

Refer to the Medicare fraud investigation described in the Statistics in Action (p. 243). Recall that the United States Department of Justice (USDOJ) obtained a random sample of 52 claims from a population of 1,000 Medicare claims. For each claim, the amount paid, the amount disallowed (denied) by the auditor, and the amount that should have been paid (allowed) were recorded and saved in the **MCFRAUD** file. The USDOJ wants to use these data to calculate an estimate of the overpayment for all 1,000 claims in the population.

One way to do this is to first use the sample data to estimate the mean overpayment per claim for the population, then use the estimated mean to extrapolate the overpayment amount to the population of all 1,000 claims. The difference

between the amount paid and the amount allowed by the auditor represents the overpayment for each claim. This value is recorded as the amount denied in the **MCFRAUD** file. These overpayment amounts are listed in the accompanying table.

MINITAB software is used to find a 95% confidence interval for μ, the mean overpayment amount. The MINITAB printout is displayed in Figure SIA5.1. The 95% confidence interval for μ, (highlighted on the printout) is (16.51, 27.13). Thus, the USDOJ can be 95% confident that the mean overpayment

(continued)

Statistics IN Action
(continued)

amount for the population of 1,000 claims is between $16.51 and $27.13.

Now, let x_i represent the overpayment amount for the i the claim. If the true mean μ were known, then the total overpayment amount for all 1,000 claims would be equal to

$$\sum_{i=1}^{1,000} x_i = (1,000)\left[\sum_{i=1}^{1,000} x_i/(1,000)\right] = (1,000)\mu$$

Consequently, to estimate the total overpayment amount for the 1,000 claims, the USDOJ will simply multiply the end points of the interval by 1,000. This yields the 95% confidence interval ($16,510, $27,130).* Typically, the USDOJ is willing to give the Medicare provider in question the benefit of the doubt by demanding a repayment equal to the lower 95% confidence bound—in this case, $16,510.

Table SIA5.1	Overpayment Amounts for Sample of 52 Claims						
$0.00	$31.00	$0.00	$37.20	$37.20	$0.00	$43.40	$0.00
$37.20	$43.40	$0.00	$37.20	$0.00	$24.80	$0.00	$0.00
$37.20	$0.00	$37.20	$0.00	$37.20	$37.20	$37.20	$0.00
$37.20	$37.20	$0.00	$0.00	$37.20	$0.00	$43.40	$37.20
$0.00	$37.20	$0.00	$37.20	$37.20	$0.00	$37.20	$37.20
$0.00	$37.20	$43.40	$0.00	$37.20	$37.20	$37.20	$0.00
$43.40	$0.00	$43.40	$0.00				

Data Set: MCFRAUD

Figure SIA5.1
MINITAB confidence interval for mean overpayment

```
Variable        N   Mean   StDev  SE Mean     95% CI
Denied_Amount  52  21.82  19.08     2.65   (16.51, 27.13)
```

Exercises 5.27–5.47

Understanding the Principles

5.27 State the two problems (and corresponding solutions) that arise with using a small sample to estimate μ.

5.28 Compare the shapes of the z- and t-distributions.

5.29 Explain the differences in the sampling distributions [NW] of $\bar{x}$ for large and small samples under the following assumptions:
a. The variable of interest, x, is normally distributed.
b. Nothing is known about the distribution of the variable x.

Applet Exercise 5.3

Use the applet entitled *Confidence Intervals for a Mean (the impact of not knowing the standard deviation)* to compare proportions of z-intervals and t-intervals that contain the mean for a population that is normally distributed.
a. Using $n = 5$ and the normal distribution with mean 50 and standard deviation 10, run the applet several times. How do the proportions of z-intervals and t-intervals that contain the mean compare?
b. Repeart part **a** first for $n = 10$ and then for $n = 20$. Compare your results with those you obtained in part **a**.
c. Describe any patterns you observe between the proportion of z-intervals that contain the mean and the

proportion of t-intervals that contain the mean as the sample size increases.

Applet Exercise 5.4

Use the applet entitled *Confidence Intervals for a Mean (the impact of not knowing the standard deviation)* to compare proportions of z-intervals and t-intervals that contain the mean for a population with a skewed distribution.
a. Using $n = 5$ and the right-skewed distribution with mean 50 and standard deviation 10, run the applet several times. How do the proportions of z-intervals and t-intervals that contain the mean compare?
b. Repeat part **a** first for $n = 10$ and then for $n = 20$. Compare your results with those you obtained in part **a**.
c. Describe any patterns you observe between the proportion of z-intervals that contain the mean and the proportion of t-intervals that contain the mean as the sample size increases.
d. How does the skewness of the underlying distribution affect the proportions of z-intervals and t-intervals that contain the mean?

Learning the Mechanics

5.30 Suppose you have selected a random sample of $n = 7$ measurements from a normal distribution. Compare the

*This interval represents an approximation to the true 95% confidence interval for the total amount of overpayment. The precise interval involves use of a continuity correction for population size. Consult the references for the use of a finite population correction factor in interval estimation.

standard normal *z*-values with the corresponding *t*-values if you were forming the following confidence intervals:
  a. 80% confidence interval
  b. 90% confidence interval
  c. 95% confidence interval
  d. 98% confidence interval
  e. 99% confidence interval
  f. Use the table values you obtained in parts **a–e** to sketch the *z*- and *t*-distributions. What are the similarities and differences?

5.31 Let t_0 be a specific value of *t*. Use Table IV in Appendix A to find t_0 values such that the following statements are true:
  a. $P(t \geq t_0) = .025$, where df $= 10$
  b. $P(t \geq t_0) = .01$, where df $= 17$
  c. $P(t \leq t_0) = .005$, where df $= 6$
  d. $P(t \leq t_0) = .05$, where df $= 13$

5.32 Let t_0 be a particular value of *t*. Use Table IV of Appendix A to find t_0 values such that the following statements are true:
  a. $P(-t_0 < t < t_0) = .95$, where df $= 16$
  b. $P(t \leq -t_0 \text{ or } t \geq t_0) = .05$, where df $= 16$
  c. $P(t \leq t_0) = .05$, where df $= 16$
  d. $P(t \leq -t_0 \text{ or } t \geq t_0) = .10$, where df $= 12$
  e. $P(t \leq -t_0 \text{ or } t \geq t_0) = .01$, where df $= 8$

5.33 The following random sample was selected from a normal distribution: 4, 6, 3, 5, 9, 3.
  a. Construct a 90% confidence interval for the population mean μ.
  b. Construct a 95% confidence interval for the population mean μ.
  c. Construct a 99% confidence interval for the population mean μ.
  d. Assume that the sample mean $\bar{x}$ and sample standard deviation *s* remain exactly the same as those you just calculated, but that they are based on a sample of $n = 25$ observations rather than $n = 6$ observations. Repeat parts **a–c**. What is the effect of increasing the sample size on the width of the confidence intervals?

5.34 The following sample of 16 measurements, saved in the **LM5_34** file, was selected from a population that is approximately normally distributed:

| 91 | 80 | 99 | 110 | 95 | 106 | 78 | 121 | 106 | 100 |
| 97 | 82 | 100 | 83 | 115 | 104 | | | | |

  a. Construct an 80% confidence interval for the population mean.
  b. Construct a 95% confidence interval for the population mean, and compare the width of this interval with that of part **a.**
  c. Carefully interpret each of the confidence intervals, and explain why the 80% confidence interval is narrower.

Applying the Concepts—Basic

5.35 **Lobster trap placement.** Strategic placement of lobster traps is one of the keys for a successful lobster fisherman. An observational study of teams fishing for the red spiny lobster in Baja California Sur, Mexico, was conducted and the results published in *Bulletin of Marine Science* (April, 2010). One of the variables of interest was the average distance separating traps—called *trap spacing*—deployed by the same team of fishermen. Trap spacing measurements (in meters) for a sample of seven teams of red spiny lobster fishermen are shown in the accompanying table (and saved in the **TRAPSPACE** file). Of interest is the mean trap spacing for the population of red spiny lobster fishermen fishing in Baja California Sur, Mexico.

| 93 | 99 | 105 | 94 | 82 | 70 | 86 |

From Shester, G. G. "Explaining catch variation among Baja California lobster fishers through spatial analysis of trap-placement decisions." *Bulletin of Marine Science*, Vol. 86, No. 2, April 2010 (Table 1), pp. 479–498. Reprinted with permission from the University of Miami – *Bulletin of Marine Science*.

  a. Identify the target parameter for this study.
  b. Compute a point estimate of the target parameter.
  c. What is the problem with using the normal (*z*) statistic to find a confidence interval for the target parameter?
  d. Find a 95% confidence interval for the target parameter.
  e. Give a practical interpretation of the interval, part **d.**
  f. What conditions must be satisfied for the interval, part **d**, to be valid?

5.36 **Shell lengths of sea turtles.** Refer to the *Aquatic Biology* (Vol. 9, 2010) study of green sea turtles inhabiting the Grand Cayman South Sound lagoon, Exercise 5.20 (p. 252). Time-depth recorders were deployed on 6 of the 76 captured turtles. The time-depth recorders allowed the environmentalists to track the movement of the sea turtles in the lagoon. These 6 turtles had a mean shell length of 52.9 cm with a standard deviation of 6.8 cm.
  a. Use the information on the 6 tracked turtles to estimate, with 99% confidence, the true mean shell length of all green sea turtles in the lagoon. Interpret the result.
  b. What assumption about the distribution of shell lengths must be true in order for the confidence interval, part a, to be valid? Is this assumption reasonably satisfied? (Use the data saved in the **TURTLES** file to help you answer this question.)

5.37 **Duration of daylight in western Pennsylvania.** What area of the United States has the least amount of daylight, on average? Having grown up in western Pennsylvania, co-author Sincich wonders if it is his hometown of Sharon, PA. Data on the number of minutes of daylight per day in Sharon, PA, for 12 randomly selected days (one each month) in 2009 were obtained from the *Naval Oceanography Portal* Web site (*aa.usno.navy.mil/USNO/astronomical-applications/data-services*). The data are listed in the table and saved in the **DAYLIGHT** file. Descriptive statistics and a 95% confidence interval for the mean are produced in the SPSS printout on page 262.
  a. Locate the confidence interval on the printout and give the value of the confidence coefficient.
  b. Use the descriptive statistics on the printout to calculate the 95% confidence interval. Be sure your answer agrees with the interval shown on the printout.
  c. Practically interpret the confidence interval.
  d. Comment on the method of sampling. Do you think the sample is representative of the target population?

| 591 | 618 | 704 | 831 | 896 | 909 | 865 | 839 | 748 | 672 | 583 | 565 |

Based on Astronomical Applications Dept., U.S. Naval Observatory, Washington, DC.

SPSS Output for Exercise 5.37

Descriptives

			Statistic	Std. Error
DAYLIGHT	Mean		735.08	37.402
	95% Confidence Interval for Mean	Lower Bound	652.76	
		Upper Bound	817.40	
	5% Trimmed Mean		734.87	
	Median		726.00	
	Variance		16786.992	
	Std. Deviation		129.565	
	Minimum		565	
	Maximum		909	
	Range		344	
	Interquartile Range		261	
	Skewness		.016	.637
	Kurtosis		-1.752	1.232

5.38 **Assessing the bending strength of a wooden roof.** The white wood material used for the roof of an ancient Japanese temple is imported from Northern Europe. The wooden roof must withstand as much as 100 centimeters of snow in the winter. Architects at Tohoku University (in Japan) conducted a study to estimate the mean bending strength of the white wood roof (*Journal of the International Association for Shell and Spatial Structures,* Aug. 2004). A sample of 25 pieces of the imported wood was tested and yielded the following statistics on breaking strength (in MPa): $\bar{x} = 75.4, s = 10.9$. Estimate the true mean breaking strength of the white wood with a 90% confidence interval. Interpret the result.

5.39 **Radioactive lichen.** Refer to the Lichen Radionuclide Baseline Research project at the University of Alaska, presented in Exercise 2.36 (p. 47). Recall that the researchers collected 9 lichen specimens and measured the amount (in microcuries per milliliter) of the radioactive element cesium-137 for each. (The natural logarithms of the data values are saved in the **LICHEN** file.) A MINITAB printout with summary statistics for the actual data is shown below.

a. Give a point estimate for the mean amount of cesium in lichen specimens collected in Alaska.

b. Give the *t*-value used in a small-sample 95% confidence interval for the true mean amount of cesium in Alaskan lichen specimens.

c. Use the result you obtained in part **b** and the values of $\bar{x}$ and *s* shown on the MINITAB printout to form a 95% confidence interval for the true mean amount of cesium in Alaskan lichen specimens.

d. Check the interval you found in part **c** with the 95% confidence interval shown on the MINITAB printout.

e. Give a practical interpretation for the interval you obtained in part **c.**

Variable	N	Mean	StDev	SE Mean	95% CI
CESIUM	9	0.009027	0.004854	0.001618	(0.005296, 0.012759)

5.40 **Rainfall and desert ants.** Refer to the *Journal of Biogeography* (December 2003) study of ants and their habitat in the desert of Central Asia, presented in Exercise 2.68 (p. 59). Recall that botanists randomly selected five sites in the Dry Steppe region and six sites in the Gobi Desert where ants were observed. One of the variables of interest is the annual rainfall (in millimeters) at each site. (The data are saved in the **GOBIANTS** file.) Summary statistics for the annual rainfall at each site are provided in the SAS printout below.

a. Give a point estimate for the average annual rainfall amount at ant sites in the Dry Steppe region of Central Asia.

b. Give the *t*-value used in a small-sample 90% confidence interval for the true average annual rainfall amount at ant sites in the Dry Steppe region.

c. Use the result you obtained in part **b** and the values of $\bar{x}$ and *s* shown on the SAS printout to form a 90% confidence interval for the target parameter.

d. Give a practical interpretation for the interval you found in part **c.**

e. Use the data in the **GOBIANTS** file to check the validity of the confidence interval you found in part **c.**

f. Repeat parts **a–e** for the Gobi Desert region of Central Asia.

The MEANS Procedure

Analysis Variable : RAIN

REGION	N Obs	Mean	Std Dev
DryStepp	5	183.4000000	20.6470337
Gobi	6	110.0000000	15.9749804

Applying the Concepts—Intermediate

5.41 **Pitch memory of amusiacs.** Congenital amusia is a disorder that impacts one's perception of music. A team of psychologists and neuroscientists tested the pitch memory of individuals diagnosed with amusia and reported their results in *Advances in Cognitive Psychology* (Vol. 6, 2010). Each in a sample of 17 amusiacs listened to a series of tone pairs, where each tone pair was a standard tone followed by a comparison tone. For each tone pair, the subjects were asked to determine if the tones were the same or different. In one trial, the tones were separated by 1 second. In a second trial, the tones were separated by 5 seconds. (In theory, the longer the delay between tones, the less likely one is to detect a difference between the tones.) Scores in the two trials were compared for each amusiac. The mean score difference was .11 with a standard deviation of .19. Use this information to form a 90% confidence interval for the true mean score difference for all amusiacs. Interpret the result. What assumption about the population of score differences must hold true for the interval to be valid?

5.42 **Studies on treating Alzheimer's disease.** Alzheimer's disease is a progressive disease of the brain. Much research has been conducted on how to treat Alzheimer's. The journal *eCAM* (November 2006) published an article that critiqued the quality of the methodology used in studies on Alzheimer treatment. For each in a sample of 13 studies, the quality of the methodology was measured on the Wong scale, with scores ranging from 9 (low quality) to 27 (high quality). The data are shown in the table (p. 263) and saved in the **TREATAD** file. Estimate, with a 99% confidence interval, the mean quality μ of all studies on the treatment of Alzheimer's disease. Interpret the result.

Data for Exercise 5.42

22	21	18	19	20	15	19	20	15	20	17	20	21		

Based on Chiappelli, F., et al. "Evidence-based research in complementary and alternative medicine III: Treatment of patients with Alzheimer's disease." *eCAM*, Vol. 3, No. 4, Nov. 2006 (Table 1).

5.43 **Reproduction of bacteria-infected spider mites.** Zoologists in Japan investigated the reproductive traits of spider mites with a bacterial infection (*Heredity*, Jan. 2007). Male and female pairs of infected spider mites were mated in a laboratory and the number of eggs produced by each female recorded. Summary statistics for several samples are provided in the accompanying table. Note that, in some samples, one or both infected spider mites were treated with antibiotic prior to mating.

   a. For each type of female–male pair, construct and interpret a 90% confidence interval for the population mean number of eggs produced by the female spider mite.

   b. Identify the type of female–male pair that appears to produce the highest mean number of eggs.

Female–Male Pairs	Sample Size	Mean # of Eggs	Standard Deviation
Both untreated	29	20.9	3.34
Male treated	23	20.3	3.50
Female treated	18	22.9	4.37
Both treated	21	18.6	2.11

Based on Gotoh, T., Noda, H., and Ito, S. "Cardinium symbionts cause cytoplasmic incompatibility in spider mites." *Heredity*, Vol. 98, No. 1, Jan. 2007 (Table 2).

5.44 **Minimizing tractor skidding distance.** In planning for a new forest road to be used for tree harvesting, planners must select the location that will minimize tractor skidding distance. In the *Journal of Forest Engineering* (July 1999), researchers wanted to estimate the true mean skidding distance along a new road in a European forest. The skidding distances (in meters) were measured at 20 randomly selected road sites. These values are given in the accompanying table and saved in the **SKIDDING** file.

488	350	457	199	285	409	435	574	439	546
385	295	184	261	273	400	311	312	141	425

Based on Tujek, J., and Pacola, E. "Algorithms for skidding distance modeling on a raster Digital Terrain Model." *Journal of Forest Engineering*, Vol. 10, No. 1, July 1999 (Table 1).

   a. Estimate, with a 95% confidence interval, the true mean skidding distance of the road.

   b. Give a practical interpretation of the interval you found in part **a.**

   c. What conditions are required for the inference you made in part b to be valid? Are these conditions reasonably satisfied?

   d. A logger working on the road claims that the mean skidding distance is at least 425 meters. Do you agree?

5.45 **Antigens for a parasitic roundworm in birds.** *Ascaridia galli* is a parasitic roundworm that attacks the intestines of birds, especially chickens and turkeys. Scientists are working on a synthetic vaccine (antigen) for the parasite. The success of the vaccine hinges on the characteristics of DNA in peptide (protein) produced by the antigen. In the journal *Gene Therapy and Molecular Biology* (June 2009), scientists tested alleles of antigen-produced protein for level of peptide. For a sample of 4 alleles, the mean peptide score was 1.43 and the standard deviation was .13.

   a. Use this information to construct a 90% confidence interval for the true mean peptide score in alleles of the antigen-produced protein.

   b. Interpret the interval for the scientists.

   c. What is meant by the phrase "90% confidence"?

5.46 **Eating disorders in females.** The "fear of negative evaluation" (FNE) scores for 11 bulimic female students and 14 normal female students, first presented in Exercise 2.40 and saved in the **BULIMIA** file (p. 48), are reproduced at the bottom of the page. (Recall that the higher the score, the greater is the FNE.)

   a. Construct a 95% confidence interval for the mean FNE score of the population of bulimic female students. Interpret the result.

   b. Construct a 95% confidence interval for the mean FNE score of the population of normal female students. Interpret the result.

   c. What assumptions are required for the intervals of parts **a** and **b** to be statistically valid? Are these assumptions reasonably satisfied? Explain.

Applying the Concepts—Advanced

5.47 **Study on waking sleepers early.** Scientists have discovered increased levels of the hormone adrenocorticotropin in people just before they awake from sleeping (*Nature*, January 7, 1999). In the study described, 15 subjects were monitored during their sleep after being told that they would be woken at a particular time. One hour prior to the designated wake-up time, the adrenocorticotropin level (pg/mL) was measured in each, with the following results:

$$\bar{x} = 37.3 \qquad s = 13.9$$

   a. Use a 95% confidence interval to estimate the true mean adrenocorticotropin level of sleepers one hour prior to waking.

   b. Interpret the interval you found in part **a** in the words of the problem.

   c. The researchers also found that if the subjects were woken three hours earlier than they anticipated, the average adrenocorticotropin level was 25.5 pg/mL. Assume that $\mu = 25.5$ for all sleepers who are woken three hours earlier than expected. Use the interval from part **a** to make an inference about the mean adrenocorticotropin level of sleepers under two conditions: one hour before the anticipated wake-up time and three hours before the anticipated wake-up time.

Table for Exercise 5.46

Bulimic students:	21	13	10	20	25	19	16	21	24	13	14	
Normal students:	13	6	16	13	8	19	23	18	11	19	7	10	15	20

Based on Randles, R. H. "On neutral responses (zeros) in the sign test and ties in the Wilcoxon–Mann–Whitney Test." *The American Statistician*, Vol. 55, No. 2, May 2001 (Figure 3).

5.4 Large-Sample Confidence Interval for a Population Proportion

The number of public-opinion polls has grown at an astounding rate in recent years. Almost daily, the news media report the results of some poll. Pollsters regularly determine the percentage of people in favor of the president's welfare-reform program, the fraction of voters in favor of a certain candidate, the fraction of customers who favor a particular brand of wine, and the proportion of people who smoke cigarettes. In each case, we are interested in estimating the percentage (or proportion) of some group with a certain characteristic. In this section, we consider methods for making inferences about population proportions when the sample is large.

Example 5.6

Estimating a Population Proportion—Fraction Who Trust the President

Problem Public-opinion polls are conducted regularly to estimate the fraction of U.S. citizens who trust the president. Suppose 1,000 people are randomly chosen and 637 answer that they trust the president. How would you estimate the true fraction of *all* U.S. citizens who trust the president?

Solution What we have really asked is how you would estimate the probability p of success in a binomial experiment in which p is the probability that a person chosen trusts the president. One logical method of estimating p for the population is to use the proportion of successes in the sample. That is, we can estimate p by calculating

$$\hat{p} = \frac{\text{Number of people sampled who trust the president}}{\text{Number of people sampled}}$$

where $\hat{p}$ is read "p hat." Thus, in this case,

$$\hat{p} = \frac{637}{1,000} = .637$$

Look Back To determine the reliability of the estimator $\hat{p}$, we need to know its sampling distribution. That is, if we were to draw samples of 1,000 people over and over again, each time calculating a new estimate $\hat{p}$, what would be the frequency distribution of all the $\hat{p}$-values? The answer lies in viewing $\hat{p}$ as the average, or mean, number of successes per trial over the n trials. If each success is assigned a value equal to 1 and each failure is assigned a value of 0, then the sum of all n sample observations is x, the total number of successes, and $\hat{p} = x/n$ is the average, or mean, number of successes per trial in the n trials. The Central Limit Theorem tells us that the *relative frequency distribution of the sample mean for any population is approximately normal for sufficiently large samples.*

Now Work Exercise 5.56a

The repeated sampling distribution of $\hat{p}$ has the characteristics listed in the next box and shown in Figure 5.12.

Figure 5.12
Sampling distribution of $\hat{p}$

> **Sampling Distribution of $\hat{p}$**
>
> 1. The mean of the sampling distribution of $\hat{p}$ is p; that is, $\hat{p}$ is an unbiased estimator of p.
> 2. The standard deviation of the sampling distribution of $\hat{p}$ is $\sqrt{pq/n}$; that is, $\sigma_p = \sqrt{pq/n}$, where $q = 1 - p$.
> 3. For large samples, the sampling distribution of $\hat{p}$ is approximately normal. A sample size is considered large if both $n\hat{p} \geq 15$ and $n\hat{q} \geq 15$.

The fact that $\hat{p}$ is a "sample mean fraction of successes" allows us to form confidence intervals about p in a manner that is completely analogous to that used for large-sample estimation of μ.

> **Large-Sample Confidence Interval for p**
>
> $$\hat{p} \pm z_{\alpha/2}\sigma_{\hat{p}} = \hat{p} \pm z_{\alpha/2}\sqrt{\frac{pq}{n}} \approx \hat{p} \pm z_{\alpha/2}\sqrt{\frac{\hat{p}\hat{q}}{n}}$$
>
> where $\hat{p} = \dfrac{x}{n}$ and $\hat{q} = 1 - \hat{p}$
>
> *Note:* When n is large, $\hat{p}$ can approximate the value of p in the formula for $\sigma_{\hat{p}}$.

> **Conditions Required for a Valid Large-Sample Confidence Interval for p**
>
> 1. A random sample is selected from the target population.
> 2. The sample size n is large. (This condition will be satisfied if both $n\hat{p} \geq 15$ and $n\hat{q} \geq 15$. Note that $n\hat{p}$ and $n\hat{q}$ are simply the number of successes and number of failures, respectively, in the sample.

Thus, if 637 of 1,000 U.S. citizens say they trust the president, a 95% confidence interval for the proportion of *all* U.S. citizens who trust the president is

$$\hat{p} \pm z_{\alpha/2}\sigma_{\hat{p}} = .637 \pm 1.96\sqrt{\frac{pq}{1,000}}$$

where $q = 1 - p$. Just as we needed an approximation for σ in calculating a large-sample confidence interval for μ, we now need an approximation for p. As Table 5.6 shows, the approximation for p does not have to be especially accurate, because the value of $\sqrt{pq}$ needed for the confidence interval is relatively insensitive to changes in p. Therefore, we can use $\hat{p}$ to approximate p. Keeping in mind that $\hat{q} = 1 - \hat{p}$, we substitute these values into the formula for the confidence interval:

$$\hat{p} \pm 1.96\sqrt{pq/1,000} \approx \hat{p} \pm 1.96\sqrt{\hat{p}\hat{q}/1,000}$$
$$= .637 \pm 1.96\sqrt{(.637)(.363)/1,000} = .637 \pm .030$$
$$= (.607, .667)$$

Then we can be 95% confident that the interval from 60.7% to 66.7% contains the true percentage of *all* U.S. citizens who trust the president. That is, in repeated constructions of confidence intervals, approximately 95% of all samples would produce confidence

Table 5.6 Values of pq for Several Different Values of p

P	pq	$\sqrt{pq}$
.5	.25	.50
.6 or .4	.24	.49
.7 or .3	.21	.46
.8 or .2	.16	.40
.9 or .1	.09	.30

intervals that enclose p. Note that the guidelines for interpreting a confidence interval about μ also apply to interpreting a confidence interval for p, because p is the "population mean fraction of successes" in a binomial experiment.

Example 5.7

A Large-Sample Confidence Interval for p—Proportion Optimistic about the Economy

Problem Many public polling agencies conduct surveys to determine the current consumer sentiment concerning the state of the economy. For example, the Bureau of Economic and Business Research (BEBR) at the University of Florida conducts quarterly surveys to gauge consumer sentiment in the Sunshine State. Suppose that BEBR randomly samples 484 consumers and finds that only 157 are optimistic about the state of the economy. Use a 90% confidence interval to estimate the proportion of all consumers in Florida who are optimistic about the state of the economy. Based on the confidence interval, can BEBR infer that a minority of Florida consumers is optimistic about the economy?

Solution The number x of the 484 sampled consumers who are optimistic about the Florida economy is a binomial random variable if we can assume that the sample was randomly selected from the population of Florida consumers and that the poll was conducted identically for each consumer sampled.

The point estimate of the proportion of Florida consumers who are optimistic about the economy is

$$\hat{p} = \frac{x}{n} = \frac{157}{484} = .324$$

We first check to be sure that the sample size is sufficiently large that the normal distribution provides a reasonable approximation to the sampling distribution of $\hat{p}$. We require the number of successes in the sample, $n\hat{p}$, and the number of failures, $n\hat{q}$, both to be at least 15. Since the number of successes is $n\hat{p} = 157$ and the number of failures is $n\hat{q} = 327$, we may conclude that the normal approximation is reasonable.

We now proceed to form the 90% confidence interval for p, the true proportion of Florida consumers who are optimistic about the state of the economy:

$$\hat{p} \pm z_{\alpha/2}\sigma_{\hat{p}} = \hat{p} \pm z_{\alpha/2}\sqrt{\frac{pq}{n}} \approx \hat{p} \pm z_{\alpha/2}\sqrt{\frac{\hat{p}\hat{q}}{n}}$$

$$= .324 \pm 1.645\sqrt{\frac{(.324)(.676)}{484}} = .324 \pm .035 = (.289, .359)$$

(This interval is also shown on the MINITAB printout of Figure 5.13.) Thus, we can be 90% confident that the proportion of all Florida consumers who are confident about the economy is between .289 and .359. As always, our confidence stems from the fact that 90% of all similarly formed intervals will contain the true proportion p and not from any knowledge about whether this particular interval does.

Figure 5.13

MINITAB printout with 90% confidence interval for p

Test and CI for One Proportion

Sample	X	N	Sample p	90% CI
1	157	484	0.324380	(0.289379, 0.359381)

Using the normal approximation.

Can we conclude on the basis of this interval that a minority of Florida consumers is optimistic about the economy? If we wished to use this interval to infer that a minority is optimistic, the interval would have to support the inference that p is less than .5—that is, that less than 50% of Florida consumers are optimistic about the economy. Note that the interval contains only values below .5. Therefore, we can conclude on the basis of this 90% confidence interval that the true value of p is less than .5.

Look Ahead If the confidence interval includes .5 (e.g., an interval from .42 to .54), then we could not conclude that the true proportion of consumers who are optimistic is less than .5. (This is because it is possible that p is as high as .54.)

Now Work Exercise 5.59

We conclude this section with a warning and an illustrative example.

⚠ **CAUTION** Unless n is extremely large, the large-sample procedure presented in this section performs poorly when p is near 0 or near 1. ▲

The problem stated in the above warning can be illustrated as follows. Suppose you want to estimate the proportion of people who die from a bee sting using a sample size of $n = 100$. This proportion is likely to be near 0, say $p \approx .001$. If so, then $np \approx 100(.001) = .1$ is less than the recommended value of 15 (see conditions in the box on p. 265). Consequently, a confidence interval for p based on a sample of $p = 100$ will probably be misleading.

To overcome this potential problem, an *extremely* large sample size is required. Since the value of n required to satisfy "extremely large" is difficult to determine, statisticians (see Agresti & Coull, 1998) have proposed an alternative method, based on the Wilson (1927) point estimator of p. **Wilson's adjustment for estimating p** is outlined in the next box. Researchers have shown that this confidence interval works well for any p, even when the sample size n is very small.

Adjusted $(1 - \alpha)$ 100% Confidence Interval for a Population Proportion p

$$\tilde{p} \pm z_{\alpha/2}\sqrt{\frac{\tilde{p}(1 - \tilde{p})}{n + 4}}$$

where $\tilde{p} = \dfrac{x + 2}{n + 4}$ is the adjusted sample proportion of observations with the characteristic of interest, x is the number of successes in the sample, and n is the sample size.

Example 5.8

Using the Adjusted Confidence Interval Procedure— Proportion Who are Victims of a Violent Crime

Problem According to *True Odds: How Risk Affects Your Everyday Life* (Walsh, 1997), the probability of being the victim of a violent crime is less than .01. Suppose that, in a random sample of 200 Americans, 3 were victims of a violent crime. Use a 95% confidence interval to estimate the true proportion of Americans who were victims of a violent crime.

Solution Let p represent the true proportion of Americans who were victims of a violent crime. Since p is near 0, an "extremely large" sample is required to estimate its value by the usual large-sample method. Here, the number of "successes," 3, is less than 15. Thus, we doubt whether the sample size of 200 is large enough, to apply the large-sample method. Alternatively, we will apply Wilson's adjustment outlined in the box.

Because the number of "successes" (i.e., number of victims of a violent crime) in the sample is $x = 3$, the adjusted sample proportion is

$$\tilde{p} = \frac{x + 2}{n + 4} = \frac{3 + 2}{200 + 4} = \frac{5}{204} = .025$$

Note that this adjusted sample proportion is obtained by adding a total of four observations—two "successes" and two "failures"—to the sample data. Substituting $\widetilde{p} = .025$ into the equation for a 95% confidence interval, we obtain

$$\widetilde{p} \pm 1.96\sqrt{\frac{\widetilde{p}(1 - \widetilde{p})}{n + 4}} = .025 \pm 1.96\sqrt{\frac{(.025)(.975)}{204}}$$

$$= .025 \pm .021$$

or (.004, .046). Consequently, we are 95% confident that the true prfoportion of Americans who are victims of a violent crime falls between .004 and .046.

Look Back: If we apply the standard large-sample confidence interval formula, where $\hat{p} = 3/200 = .015$, we obtain

$$\hat{p} \pm 1.96\sqrt{\frac{\hat{p}\hat{q}}{200}} = .015 \pm 1.96\sqrt{\frac{(.015)(.985)}{200}}$$

$$= .015 \pm .017$$

or, $(-.002, .032)$. Note that the interval estimate contains negative (nonsensical) values for the true proportion. Such a result is typical when the large-sample method is misapplied.

Now Work Exercise 5.65b

Statistics IN Action | **Revisited** | **Estimating the Coding Error Rate**

In the previous Statistics in Action Revisited (p. 259), we showed how to estimate the mean overpayment amount for claims in a Medicare fraud study. In addition to estimating overcharges, the USDOJ also is interested in estimating the *coding error rate* of a Medicare provider. The coding error rate is defined as the proportion of Medicare claims that are coded incorrectly. Thus, for this inference, the USDOJ is interested in estimating a population proportion, p. Typically, the USDOJ finds that about 50% of the claims in a Medicare fraud case are incorrectly coded.

If you examine the sample data in Table SIA5.1, you can verify that of the 52 audited claims, 31 were determined to be coded incorrectly, resulting in an overcharge. These are the claims with a disallowed amount greater than $0. Therefore, an estimate of the coding error rate, p, for this Medicare provider is

$$\hat{p} = 31/52 = .596$$

A 95% confidence interval for p can be obtained with the use of a confidence interval formula or with statistical software.

The **MCFRAUD** file includes a qualitative variable (called "Coding Error") at two levels, where "Yes" represents that the claim is coded incorrectly and "No" represents that the claim is coded correctly. Thus, the USDOJ desires an estimate of the proportion of "Yes" values for the "Coding Error" variable. A confidence interval for the proportion of incorrectly coded claims is highlighted on the accompanying MINITAB printout, Figure SIA5.2. The interval, (.45, .73), implies that the true coding error rate for the 1,000 claims in the population falls between .45 and .73, with 95% confidence. Note that the value .5—the proportion of incorrectly coded claims expected by the USDOJ—falls within the 95% confidence interval.

Data Set: MCFRAUD

Figure SIA5.2

MINITAB confidence interval for coding error rate

```
Event = Yes

Variable       X   N  Sample p        95% CI
Coding_Error  31  52  0.596154  (0.451016, 0.729940)
```

Exercises 5.48–5.67

Understanding the Principles

5.48 Describe the sampling distribution of $\hat{p}$ on the basis of large samples of size n. That is, give the mean, the standard deviation, and the (approximate) shape of the distribution of $\hat{p}$ when large samples of size n are (repeatedly) selected from the binomial distribution with probability p of success.

5.49 Explain the meaning of the phrase "$\hat{p}$ is an unbiased estimator of p."

5.50 If p is near 0 or 1, how large a sample is needed to employ the large-sample confidence interval procedure?

⊚ Applet Exercise 5.5

Use the applet entitled *Confidence Intervals for a Proportion* to investigate the effect of the value of p on the number of confidence intervals that contain the population proportion p for a fixed sample size. For this exercise, use sample size $n = 10$.

a. Run the applet several times without clearing for $p = .1$. What proportion of the 95% confidence intervals contain p? What proportion of the 99% confidence intervals contain p? Do the results surprise you? Explain.

b. Repeat part **a** for each value of p: $p = .2, p = .3, p = .4. p = .5, p = .6, p = .7, p = .8$, and $p = .9$.

c. Which value of p yields the greatest proportion of each type of interval that contain p?

d. Based on your results, what values of p will yield more reliable confidence intervals for a fixed sample size n? Explain.

⊚ Applet Exercise 5.6

Use the applet entitled *Confidence Intervals for a Proportion* to investigate the effect of the sample size on the number of confidence intervals that contain the population proportion p for a value of p close to 0 or 1.

a. Run the applet several times without clearing for $p = .5$ and $n = 50$. Record the proportion of the 99% confidence intervals containing p.

b. Now set $p = .1$ and run the applet several times without clearing for $n = 50$. How does the proportion of the 99% confidence intervals containing p compare with that in part **a**?

c. Repeat part **b**, keeping $p = .1$ and increasing the sample size by 50 until you find a sample size that yields a similar proportion of the 99% confidence intervals containing p as that in part **a**.

d. Based on your results, describe how the value of p affects the sample size needed to guarantee a certain level of confidence.

Learning the Mechanics

5.51 A random sample of size $n = 196$ yielded $\hat{p} = .64$.

a. Is the sample size large enough to use the methods of this section to construct a confidence interval for p? Explain.

b. Construct a 95% confidence interval for p.

c. Interpret the 95% confidence interval.

d. Explain what is meant by the phrase "95% confidence interval."

5.52 A random sample of size $n = 144$ yielded $\hat{p} = .76$.

a. Is the sample size large enough to use the methods of this section to construct a confidence interval for p? Explain.

b. Construct a 90% confidence interval for p.

c. What assumption is necessary to ensure the validity of this confidence interval?

5.53 For the binomial sample information summarized in each part, indicate whether the sample size is large enough to use the methods of this chapter to construct a confidence interval for p.

a. $n = 500, \hat{p} = .05$
b. $n = 100, \hat{p} = .05$
c. $n = 10, \hat{p} = .5$
d. $n = 10, \hat{p} = .3$

5.54 A random sample of 50 consumers taste-tested a new snack food. Their responses were coded (0: do not like; 1: like; 2: indifferent), recorded and saved in the **SNACK** file as follows:

1	0	0	1	2	0	1	1	0	0
0	1	0	2	0	2	2	0	0	1
1	0	0	0	0	1	0	2	0	0
0	1	0	0	1	0	0	1	0	1
0	2	0	0	1	1	0	0	0	1

a. Use an 80% confidence interval to estimate the proportion of consumers who like the snack food.

b. Provide a statistical interpretation for the confidence interval you constructed in part **a**.

Applying the Concepts—Basic

5.55 **Is Starbucks coffee overpriced?** The *Minneapolis Star Tribune* (August 12, 2008) reported that 73% of Americans say that Starbucks coffee is overpriced. The source of this information was a national telephone survey of 1,000 American adults conducted by Rasmussen Reports.

a. Identify the population of interest in this study.

b. Identify the sample for the study.

c. Identify the parameter of interest in the study.

d. Find and interpret a 95% confidence interval for the parameter of interest.

5.56 **Satellite radio in cars.** Refer to the June 2007 survey on satellite radio subscriber service and usage conducted for the National Association of Broadcasters, Exercise 1.24 (p. 20). Recall that a random sample of 501 satellite radio subscribers were asked, "Do you have a satellite radio receiver in your car?" The survey found that 396 subscribers did, in fact, have a satellite receiver in their car.

[NW] a. From the sample, calculate an estimate of the true proportion of satellite radio subscribers who have a satellite radio receiver in their car.

b. Form a 90% confidence interval for the estimate, part **a**.

c. Give a practical interpretation of the interval, part **b**.

5.57 **National Firearms Survey.** Refer to the Harvard School of Public Health survey to determine the size and composition of privately held firearm stock in the United States, presented in Exercise 3.77 (p. 144). Recall that, in a representative household telephone survey of 2,770 adults, 26% reported that they own at least one gun (*Injury Prevention*, Jan. 2007). The researchers want to estimate the true percentage of adults in the United States that own at least one gun.

a. Identify the population of interest to the researchers.

b. Identify the parameter of interest to the researchers.

c. Compute an estimate of the population parameter.

d. Form a 99% confidence interval around the estimate.

e. Interpret the confidence interval practically.

f. Explain the meaning of the phrase "99% confident."

5.58 **Are you really being served red snapper?** Refer to the *Nature* (July 15, 2004) study of fish specimens labeled "red snapper," presented in Exercise 3.85 (p. 145). Recall that federal law prohibits restaurants from serving a cheaper look-alike variety of fish (e.g., vermillion snapper or lane snapper) to customers who order red snapper. In an effort to estimate the true proportion of fillets that are really red snapper, a team of University of North Carolina (UNC) researchers analyzed the meat from each in a sample of 22 "red snapper" fish fillets purchased from vendors across the United States. DNA tests revealed that 17 of the 22 fillets (or 77%) were not red snapper, but the cheaper look-alike variety of fish.

a. Identify the parameter of interest to the UNC researchers.

b. Explain why a large-sample confidence interval is inappropriate to apply in this study.

c. Use Wilson's adjustment to construct a 95% confidence interval for the parameter of interest.

d. Give a practical interpretation of the confidence interval.

5.59 **What we do when we are sick at home.** *USA Today* (Feb. 15, 2007) reported on the results of an opinion poll in which adults were asked what one thing they are most likely to do when they are home sick with a cold or the flu. In the survey, 63% said that they are most likely to sleep and 18% said that they would watch television. Although the sample size was not reported, typically opinion polls include approximately 1,000 randomly selected respondents.

a. Assuming a sample size of 1,000 for this poll, construct a 95% confidence interval for the true percentage of all adults who would choose to sleep when they are at home sick.

b. If the true percentage of adults who would choose to sleep when they are at home sick is 70%, would you be surprised? Explain.

5.60 **Speed training in football.** Refer to *The Sport Journal* (Winter 2004) study on the effectiveness of a speed-training program for football players, Exercise 5.22 (p. 253). The decreases in 40-yard sprint times (time after training minus time before training) for 38 players are saved in the **SPRINT** file. An athlete's sprint performance will be classified as "Improved" if the "after" time is less than the "before" time, and classified as "Not Improved" if otherwise.

a. Find an estimate for the true proportion of all high school athletes who attain improved sprint times after participating in the speed-training program.

b. Convert the estimate, part **a**, into a 95% confidence interval. Give a practical interpretation of the result.

Applying the Concepts—Intermediate

5.61 **Teenagers' use of emoticons in school writing.** Due to the popularity of instant messaging and social networking sites among American teenagers, informal elements such as emoticons (e.g., the symbol ":)" to represent a smile) and abbreviations (e.g., "LOL" for "laughing out loud") have worked their way into teenagers' school writing assignments. In a survey cosponsored by the National Commission on Writing at the College Board, the *Pew Internet and American Life Project* (April 2008) interviewed 700 randomly selected U.S. teenagers by telephone on their writing habits. Overall, 448 of the teenagers admitted using at least one informal element in school writing assignments. Based on the survey results, construct a 99% confidence interval for the proportion of all U.S. teenagers who have used at least one informal element in school writing assignments. Give a practical interpretation of the interval, one that is relevant to the National Commission on Writing at the College Board.

5.62 **Characteristics of ice-melt ponds.** Refer to the University of Colorado study of ice-melt ponds in the Canadian Arctic, presented in Exercise 2.13 (p. 35). Environmental engineers are using data collected by the National Snow and Ice Data Center to learn how climate affects the sea ice. Data on 504 ice melt ponds are saved in the **PONDICE** file. Of these 504 melt ponds, 88 were classified as having "first-year ice." Recall that the researchers estimated that about 17% of melt ponds in the Canadian Arctic have first-year ice. Use the methodology of this chapter to estimate, with 90% confidence, the percentage of all ice-melt ponds in the Canadian Arctic that have first-year ice. Give a practical interpretation of the results.

5.63 **Do you believe in the Bible?** Refer to the National Opinion Research Center's General Social Survey (GSS), presented in Exercise 2.18 (p. 36). Data on the approximately 2,800 Americans who participated in the 2004 GSS are saved in the **BIBLE** file. Recall that one question in the survey asked about a person's belief in the Bible. Suppose you want to estimate the proportion of all Americans who believe that the Bible is the actual word of God and is to be taken literally. (*Note:* The variable "Bible1" contains the responses to this question.)

a. Use a 95% confidence interval to estimate the proportion of interest.

b. Give a practical interpretation of the interval you used in part **a.**

c. Discuss how the survey methodology can affect the validity of the results.

5.64 **Do you think you smell?** If you falsely believe you emit an unpleasant, foul, or offensive body odor, you may suffer from olfactory reference syndrome (ORS). The disorder disables patients, who often isolate themselves and consider suicide. Psychiatrists disagree over how prevalent ORS is in the human population. *Depression and Anxiety* (June 2010) discussed one self-reported survey of 2,481 university students in Japan. The study reported that 52 of the students were "concerned with emitting a strange bodily odor."

a. Use the survey results to estimate the true proportion of all people in the world who suffer from ORS. Place 95% confidence bounds around the estimate and interpret the result.

b. Give several reasons why the inference, part **a**, may be invalid. Explain.

5.65 **Splinting in mountain-climbing accidents.** The most common injury that occurs among mountain climbers is trauma to the lower extremity (leg). Consequently, rescuers must be proficient in immobilizing and splinting of fractures. In *High Altitude Medicine & Biology* (Vol. 10, 2009), researchers provided official recommendations for mountain emergency medicine. As part of the document, the researchers examined the likelihood of needing certain types of splints. A Scottish Mountain Rescue study reported that there was 1 femoral shaft splint needed among 333 live casualties. The researchers will use this study to estimate the proportion of all mountain casualties that require a femoral shaft splint.

a. Is the sample large enough to apply the large-sample estimation method of this section? Show why or why not.

[NW] **b.** Use Wilson's adjustment to find a 95% confidence interval for the true proportion of all mountain casualties that require a femoral shaft splint. Interpret the result.

5.66 **Studies on treating Alzheimer's disease.** Refer to the journal *eCAM* (November 2006) assessment of the quality of the methodology used in studies of the treatment for Alzheimer's disease, presented in Exercise 5.42 (p. 262). Data on the quality of the methodology (using the Wong scale) for each in a sample of 13 studies are reproduced in the next table and saved in the **TREATAD** file. According to the researchers, a study with a Wong score below 18 used

a methodology that "fails to support the author's conclusions" about the treatment of Alzheimer's. Use Wilson's adjustment to estimate the proportion of all studies on the treatment of Alzheimer's disease with a Wong score below 18. Construct a 99% confidence interval around the estimate and interpret the result.

22	21	18	19	20	15	19	20	15	20	17	20	21

Based on Chiappelli, F., et al. "Evidence-based research in complementary and alternative medicine III: Treatment of patients with Alzheimer's disease." *eCAM*, Vol. 3, No. 4, Nov. 2006 (Table 1).

Applying the Concepts—Advanced

5.67 **Latex allergy in health care workers.** Refer to the *Current Allergy & Clinical Immunology* (March 2004) study of health care workers who use latex gloves, presented in Exercise 5.13 (p. 251). In addition to the 46 hospital employees who were diagnosed with a latex allergy on the basis of a skin-prick test, another 37 health care workers were diagnosed with the allergy by means of a latex-specific serum test. Of these 83 workers with a confirmed latex allergy, only 36 suspected that they had the allergy when they were asked about it on a questionnaire. Make a statement about the likelihood that a health care worker with a latex allergy suspects that he or she actually has the allergy. Attach a measure of reliability to your inference.

5.5 Determining the Sample Size

Recall from Section 1.5 that one way to collect the relevant data for a study used to make inferences about a population is to implement a designed (planned) experiment. Perhaps the most important design decision faced by the analyst is to determine the size of the sample. We show in this section that the appropriate sample size for making an inference about a population mean or proportion depends on the desired reliability.

Estimating a Population Mean

Consider Example 5.1 (p. 245), in which we estimated the mean length of stay for patients in a large hospital. A sample of 100 patients' records produced the 95% confidence interval $\bar{x} \pm 1.96\sigma_{\bar{x}} = 4.5 \pm .78$. Consequently, our estimate $\bar{x}$ was within .78 day of the true mean length of stay, μ, for all the hospital's patients at the 95% confidence level. That is, the 95% confidence interval for μ was $2(.78) = 1.56$ days wide when 100 accounts were sampled. This is illustrated in Figure 5.14a.

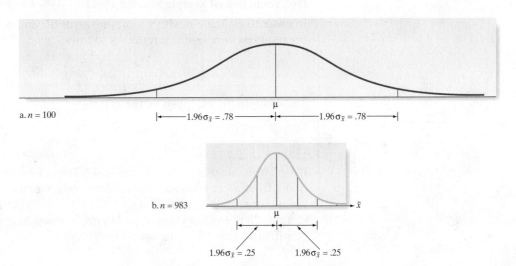

Figure 5.14

Relationship between sample size and width of confidence interval: hospital-stay example

Now suppose we want to estimate μ to within .25 day with 95% confidence. That is, we want to narrow the width of the confidence interval from 1.56 days to .50 day, as shown in Figure 5.14b. How much will the sample size have to be increased to accomplish this? If we want the estimator $\bar{x}$ to be within .25 day of μ, then we must have

$$1.96\sigma_{\bar{x}} = .25 \text{ or, equivalently, } 1.96\left(\frac{\sigma}{\sqrt{n}}\right) = .25$$

Note that we are using $\sigma_{\bar{x}} = \dfrac{\sigma}{\sqrt{n}}$ in the formula, since we are dealing with the sampling distribution of $\bar{x}$ (the estimator of μ).

The necessary sample size is obtained by solving this equation for n. First, we need the value of σ. In Example 5.1, we assumed that the standard deviation of length of stay was $\sigma = 4$ days. Thus,

$$1.96\left(\frac{\sigma}{\sqrt{n}}\right) = 1.96\left(\frac{4}{\sqrt{n}}\right) = .25$$

$$\sqrt{n} = \frac{1.96(4)}{.25} = 31.36$$

$$n = (31.36)^2 = 983.45$$

Consequently, over 983 patients' records will have to be sampled to estimate the mean length of stay, μ, to within .25 day with (approximately) 95% confidence. The confidence interval resulting from a sample of this size will be approximately .50 day wide. (See Figure 5.14b.)

In general, we express the reliability associated with a confidence interval for the population mean μ by specifying the **sampling error** within which we want to estimate μ with $100(1 - \alpha)\%$ confidence. The sampling error (denoted SE) is then equal to the half-width of the confidence interval, as shown in Figure 5.15.

The procedure for finding the sample size necessary to estimate μ with a specific sampling error is given in the following box. Note that if σ is unknown (as is usually the case, in practice) you will need to estimate the value of σ.

Figure 5.15

Specifying the sampling error SE as the half-width of a confidence interval

$\mu - z_{\alpha/2}\sigma_{\bar{x}}$ μ $\mu + z_{\alpha/2}\sigma_{\bar{x}}$

|← SE ──→|←── SE →|

Determination of Sample Size for $100(1 - \alpha)\%$ Confidence Intervals for μ

In order to estimate μ with a sampling error SE and with $100(1 - \alpha)\%$ confidence, the required sample size is found as follows:

$$z_{\alpha/2}\left(\frac{\sigma}{\sqrt{n}}\right) = \text{SE}$$

The solution for n is given by the equation

$$n = \frac{(z_{\alpha/2})^2\sigma^2}{(\text{SE})^2}$$

Note: The value of σ is usually unknown. It can be estimated by the standard deviation s from a previous sample. Alternatively, we may approximate the range R of observations in the population and (conservatively) estimate $\sigma \approx R/4$. In any case, you should round the value of n obtained *upward* to ensure that the sample size will be sufficient to achieve the specified reliability.

Example 5.9
Sample Size for Estimating μ— Mean Inflation Pressure of Footballs

Problem Suppose the manufacturer of official NFL footballs uses a machine to inflate the new balls to a pressure of 13.5 pounds. When the machine is properly calibrated, the mean inflation pressure is 13.5 pounds, but uncontrollable factors cause the pressures of individual footballs to vary randomly from about 13.3 to 13.7 pounds. For quality control purposes, the manufacturer wishes to estimate the mean inflation pressure to within .025 pound of its true value with a 99% confidence interval. What sample size should be specified for the experiment?

Solution We desire a 99% confidence interval that estimates μ with a sampling error of SE = .025 pound. For a 99% confidence interval, we have $z_{\alpha/2} = z_{.005} = 2.575$. To estimate σ, we note that the range of observations is $R = 13.7 - 13.3 = .4$ and we use $\sigma \approx R/4 = .1$. Next, we employ the formula derived in the box to find the sample size n:

$$n = \frac{(z_{\alpha/2})^2 \sigma^2}{(SE)^2} \approx \frac{(2.575)^2 (.1)^2}{(.025)^2} = 106.09$$

We round this up to $n = 107$. Realizing that σ was approximated by $R/4$, we might even advise that the sample size be specified as $n = 110$ to be more certain of attaining the objective of a 99% confidence interval with a sampling error of .025 pound or less.

Look Back To determine the value of the sampling error SE, look for the value that follows the key words "estimate μ to within . . . ."

Now Work Exercise 5.83

Sometimes the formula will lead to a solution that indicates a small sample size is sufficient to achieve the confidence interval goal. Unfortunately, the procedures and assumptions for small samples differ from those for large samples, as we discovered in Section 5.3. Therefore, if the formulas yield a small sample size, one simple strategy is to select a sample size $n = 30$.

Estimating a Population Proportion

The method just outlined is easily applied to a population proportion p. To illustrate, in Example 5.6 (p. 264) a pollster used a sample of 1,000 U.S. citizens to calculate a 95% confidence interval for the proportion who trust the president, obtaining the interval $.637 \pm .03$. Suppose the pollster wishes to estimate more precisely the proportion who trust the president, say, to within .015 with a 95% confidence interval.

The pollster wants a confidence interval for p with a sampling error SE = .015. The sample size required to generate such an interval is found by solving the following equation for n:

$$z_{\alpha/2}\sigma_{\hat{p}} = SE \qquad \text{or} \qquad z_{\alpha/2}\sqrt{\frac{pq}{n}} = .015 \qquad \text{(see Figure 5.16)}$$

Since a 95% confidence interval is desired, the appropriate z value is $z_{\alpha/2} = z_{.025} = 1.96$. We must approximate the value of the product pq before we can solve the equation

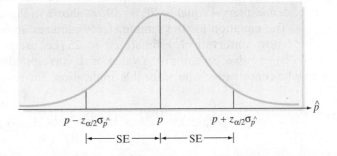

Figure 5.16

Specifying the sampling error SE of a confidence interval for a population proportion p

for n. As shown in Table 5.6 (p. 265), the closer the values of p and q to .5, the larger is the product pq. Thus, to find a conservatively large sample size that will generate a confidence interval with the specified reliability, we generally choose an approximation of p close to .5. In the case of the proportion of U.S. citizens who trust the president, how-ever, we have an initial sample estimate of $\hat{p} = .637$. A conservatively large estimate of pq can therefore be obtained by using, say, $p = .60$. We now substitute into the equation and solve for n:

$$1.96 \sqrt{\frac{(.60)(.40)}{n}} = .015$$

$$n = \frac{(1.96)^2(.60)(.40)}{(.015)^2} = 4{,}097.7 \approx 4{,}098$$

The pollster must sample about 4,098 U.S. citizens to estimate the percentage who trust the president with a confidence interval of width .03.

The procedure for finding the sample size necessary to estimate a population proportion p with a specified sampling error SE is given in the following box:

Determination of Sample Size for $100(1 - \alpha)\%$ Confidence Interval for p

In order to estimate a binomial probability p with sampling error SE and with $100(1 - \alpha)\%$ confidence, the required sample size is found by solving the following equation for n:

$$z_{\alpha/2} \sqrt{\frac{pq}{n}} = \text{SE}$$

The solution for n can be written as follows:

$$n = \frac{(z_{\alpha/2})^2(pq)}{(\text{SE})^2}$$

Note: Because the value of the product pq is unknown, it can be estimated by the sample fraction of successes, $\hat{p}$, from a previous sample. Remember (Table 5.6) that the value of pq is at its maximum when p equals .5, so you can obtain conservatively large values of n by approximating p by .5 or values close to .5. In any case, you should round the value of n obtained *upward* to ensure that the sample size will be sufficient to achieve the specified reliability.

Example 5.10

Sample Size For Estimating p — Fraction of Defective Cell Phones

Problem A cellular telephone manufacturer that entered the post-regulation market quickly has an initial problem with excessive customer complaints and consequent returns of cell phones for repair or replacement. The manufacturer wants to estimate the magnitude of the problem in order to design a quality control program. How many cellular telephones should be sampled and checked in order to estimate the fraction defective, p, to within .01 with 90% confidence?

Solution In order to estimate p to within .01 of its true value, we set the half-width of the confidence interval equal to $\text{SE} = .01$, as shown in Figure 5.17.

The equation for the sample size n requires an estimate of the product pq. We could most conservatively estimate $pq = .25$ (i.e., use $p = .5$), but this estimate may be too conservative. By contrast, a value of .1, corresponding to 10% defective, will probably be conservatively large for this application. The solution is therefore

$$n = \frac{(z_{\alpha/2})^2(pq)}{(\text{SE})^2} = \frac{(1.645)^2(.1)(.9)}{(.01)^2} = 2{,}435.4 \approx 2{,}436$$

Figure 5.17
Specified reliability for estimate of fraction defective in Example 5.10

Thus, the manufacturer should sample 2,436 telephones in order to estimate the fraction defective, p, to within .01 with 90% confidence.

Look Back Remember that this answer depends on our approximation of pq, for which we used .09. If the fraction defective is closer to .05 than .10, we can use a sample of 1,286 telephones (check this) to estimate p to within .01 with 90% confidence.

Now Work Exercise 5.82

Ethics IN Statistics

In sampling, intentional omission of experimental units (e.g., respondents in a survey) in order to bias the results towards a particular view or outcome is considered *unethical statistical practice*.

The cost of sampling will also play an important role in the final determination of the sample size to be selected to estimate either μ or p. Although more complex formulas can be derived to balance the reliability and cost considerations, we will solve for the necessary sample size and note that the sampling budget may be a limiting factor. (Consult the references for a more complete treatment of this problem.) Once the sample size n is determined, be sure to devise a sampling plan that will ensure that a representative sample is selected from the target population.

Statistics IN Action | Revisited — Determining Sample Size

In the previous Statistics in Action applications in this chapter, we used confidence intervals (1) to estimate μ, the mean overpayment amount for claims in a Medicare fraud study, and (2) to estimate p, the coding error rate (i.e., proportion of claims that are incorrectly coded) of a Medicare provider. Both of these confidence intervals were based on selecting a random sample of 52 claims from the population of claims handled by the Medicare provider. How does the USDOJ determine how many claims to sample for auditing?

Consider the problem of estimating the coding error rate, p. As stated in a previous Statistics in Action Revisited, the USDOJ typically finds that about 50% of the claims in a Medicare fraud case are incorrectly coded. Suppose the USDOJ wants to estimate the true coding error rate of a Medicare provider to within .1 with 95% confidence. How many claims should be randomly sampled for audit in order to attain the desired estimate?

Here, the USDOJ desires a sampling error of SE = .1, a confidence level of $1-\alpha = .95$ (for which $z_{\alpha/2} = 1.96$), and

uses an estimate $p \approx .50$. Substituting these values into the sample size formula (p. 272), we obtain.

$$n = (z_{\alpha/2})^2(pq)/(\text{SE})^2$$
$$= (1.96)^2(.5)(.5)/(.1)^2$$
$$= 96.04$$

Consequently, the USDOJ should audit about 97 randomly selected claims to attain a 95% confidence interval for p with a sampling error of .10.

[*Note:* You may wonder why the sample actually used in the fraud analysis included only 52 claims. The sampling strategy employed involved more than selecting a simple random sample; rather, it used a more sophisticated sampling scheme, called stratified random sampling. The 52 claims represented the sample for just one of the strata.]

Exercises 5.68–5.92

Understanding the Principles

5.68 How does the sampling error SE compare with the width of a confidence interval?

5.69 **True or false.** For a specified sampling error SE, increasing the confidence level $(1 - \alpha)$ will lead to a larger n in determining the sample size.

5.70 **True or false.** For a fixed confidence level $(1 - \alpha)$, increasing the sampling error SE will lead to a smaller n in determining the sample size.

Learning the Mechanics

5.71 If you wish to estimate a population mean to within .2 with a 95% confidence interval and you know from previous sampling that σ^2 is approximately equal to 5.4, how many observations would you have to include in your sample?

5.72 If nothing is known about p, .5 can be substituted for p in the sample-size formula for a population proportion. But when this is done, the resulting sample size may be larger than needed. Under what circumstances will using $p = .5$ in the sample-size formula yield a sample size larger than is needed to construct a confidence interval for p with a specified bound and a specified confidence level?

5.73 Suppose you wish to estimate a population mean correct to within .15 with a confidence level of .90. You do not know σ^2, but you know that the observations will range in value between 31 and 39.
   a. Find the approximate sample size that will produce the desired accuracy of the estimate. You wish to be conservative to ensure that the sample size will be ample for achieving the desired accuracy of the estimate. [*Hint:* Using your knowledge of data variation from Section 2.5, assume that the range of the observations will equal 4σ.]
   b. Calculate the approximate sample size, making the less conservative assumption that the range of the observations is equal to 6σ.

5.74 In each case, find the approximate sample size required to construct a 95% confidence interval for p that has sampling error SE = .06.
   a. Assume that p is near .3.
   b. Assume that you have no prior knowledge about p, but you wish to be certain that your sample is large enough to achieve the specified accuracy for the estimate.

5.75 The following is a 90% confidence interval for p: (.26, .54). How large was the sample used to construct this interval?

5.76 It costs you $10 to draw a sample of size $n = 1$ and measure the attribute of interest. You have a budget of $1,200.
   a. Do you have sufficient funds to estimate the population mean for the attribute of interest with a 95% confidence interval 4 units in width? Assume that $\sigma = 12$.
   b. If a 90% confidence level were used, would your answer to part **a** change? Explain.

5.77 Suppose you wish to estimate the mean of a normal population with a 95% confidence interval and you know from prior information that $\sigma^2 \approx 1$.
   a. To see the effect of the sample size on the width of the confidence interval, calculate the width of the confidence interval for $n = 16, 25, 49, 100,$ and 400.

   b. Plot the width as a function of sample size n on graph paper. Connect the points by a smooth curve, and note how the width decreases as n increases.

Applying the Concepts—Basic

5.78 **Risk of home burglary in cul-de-sacs.** Research published in the *Journal of Quantitative Criminology* (March 2010) revealed that the risk of burglaries in homes located on cul-de-sacs is lower than for homes on major roads. Suppose you want to estimate the true percentage of cul-de-sac homes in your home city that were burglarized in the past year. Devise a sampling plan so that your estimate will be accurate to within 2% of the true value using a confidence coefficient of 95%. How many cul-de-sac homes need to be sampled and what information do you need to collect for each sampled home?

5.79 **Lobster trap placement.** Refer to the *Bulletin of Marine Science* (April 2010) study of lobster trap placement, Exercise 5.35 (p. 261). Recall that you used a 95% confidence interval to estimate the mean trap spacing (in meters) for the population of red spiny lobster fishermen fishing in Baja California Sur, Mexico. How many teams of fishermen would need to be sampled in order to reduce the width of the confidence interval to 5 meters? Use the sample standard deviation from Exercise 5.35 in your calculation.

5.80 **Radioactive lichen.** Refer to the Alaskan Lichen Radionuclide Baseline Research study, presented in Exercise 5.39 (p. 262). In a sample of $n = 9$ lichen specimens, the researchers found the mean and standard deviation of the amount of the radioactive element, cesium-137, that was present to be .009 and .005 microcurie per milliliter, respectively. Suppose the researchers want to increase the sample size in order to estimate the mean μ to within .001 microcurie per milliliter of its true value, using a 95% confidence interval.
   a. What is the confidence level desired by the researchers?
   b. What is the sampling error desired by the researchers?
   c. Compute the sample size necessary to obtain the desired estimate.

5.81 **Scanning errors at Wal-Mart.** Refer to the National Institute for Standards and Technology (NIST) study of the accuracy of checkout scanners at Wal-Mart stores in California, presented in Exercise 3.52 (p. 130). NIST sets standards so that no more than 2 of every 100 items scanned through an electronic checkout scanner can have an inaccurate price. Recall that in a sample of 60 Wal-Mart stores, 52 violated the NIST scanner accuracy standard (*Tampa Tribune*, Nov. 22, 2005). Suppose you want to estimate the true proportion of Wal-Mart stores in California that violate the NIST standard.
   a. Explain why the large-sample methodology of Section 7.4 is inappropriate for this study.
   b. Determine the number of Wal-Mart stores that must be sampled in order to estimate the true proportion to within .05 with 90% confidence, using the large-sample method.

5.82 **Aluminum cans contaminated by fire.** A gigantic warehouse located in Tampa, Florida, stores approximately

60 million empty aluminum beer and soda cans. Recently, a fire occurred at the warehouse. The smoke from the fire contaminated many of the cans with blackspot, rendering them unusable. A University of South Florida statistician was hired by the insurance company to estimate p, the true proportion of cans in the warehouse that were contaminated by the fire. How many aluminum cans should be randomly sampled to estimate the true proportion to within .02 with 90% confidence?

5.83 **Assessing the bending strength of a wooden roof.** Refer
NW to the *Journal of the International Association for Shell and Spatial Structures* (Aug. 2004) study to estimate the mean bending strength of imported white wood used on the roof of an ancient Japanese temple, presented in Exercise 5.38 (p. 262). Suppose you want to estimate the true mean breaking strength of the white wood to within 4 MPa, using a 90% confidence interval. How many pieces of the imported wood need to be tested? Recall that the sample standard deviation of the breaking strengths found in the study was 10.9 Mpa.

Applying the Concepts—Intermediate

5.84 **Duration of daylight in western Pennsylvania.** Refer to the *Naval Oceanography Portal* data on number of minutes of daylight per day in Sharon, PA, Exercise 5.37 (p. 261). An estimate of the mean number of minutes of daylight per day was obtained using data collected for 12 randomly selected days (one each month) in 2009.
   a. Determine the number of days in 2009 that need to be sampled in order to estimate the desired mean to within 45 minutes of its true value with 95% confidence.
   b. Based on your answer, part **a**, develop a sampling plan that will likely result in a random sample that is representative of the population.
   c. Go to the Web site, http://aa.usno.navy.mil/USNO/astronomical-applications/data-services, and collect the data for Sharon, PA in year 2009 using your sampling plan.
   d. Use the data, part **c**, to construct a 95% confidence interval for the desired mean. Does your interval have the desired width?

5.85 **Pitch memory of amusiacs.** Refer to the *Advances in Cognitive Psychology* (Vol. 6, 2010) study of pitch memory of amusiacs, Exercise 5.41 (p. 262). Recall that diagnosed amusiacs listened to a series of tone pairs and were asked to determine if the tones were the same or different. In the first trial, the tones were separated by 1 second; in the second trial, the tones were separated by 5 seconds. The variable of interest was the difference between scores on the two trials. How many amusiacs would need to participate in the study in order to estimate the true mean score difference for all amusiacs to within .05 with 90% confidence?

5.86 **Speed training in football.** Refer to *The Sport Journal* (Winter 2004) study on the effectiveness of a speed training program for football players, Exercise 5.60 (p. 270). The confidence interval for the population proportion of athletes who improve their sprint times after participating in the speed-training program was fairly wide due to the relatively small number of athletes ($n = 38$) who participated in the program. In order to find a valid, narrower confidence interval for the true proportion, the researchers would need to include more athletes in the study. How many high school athletes should be sampled

to estimate the true proportion to within .03 with 95% confidence?

5.87 **Bacteria in bottled water.** Is the bottled water you drink safe? The Natural Resources Defense Council warns that the bottled water you are drinking may contain more bacteria and other potentially carcinogenic chemicals than are allowed by state and federal regulations. Of the more than 1,000 bottles studied, nearly one-third exceeded government levels. Suppose that the Natural Resources Defense Council wants an updated estimate of the population proportion of bottled water that violates at least one government standard. Determine the sample size (number of bottles) needed to estimate this proportion to within ± 0.01 with 99% confidence.

5.88 **Asthma drug study.** The chemical benzalkonium chloride (BAC) is an antibacterial agent that is added to some asthma medications to prevent contamination. Researchers at the University of Florida College of Pharmacy have discovered that adding BAC to asthma drugs can cause airway constriction in patients. In a sample of 18 asthmatic patients, each of whom received a heavy dose of BAC, 10 experienced a significant drop in breathing capacity (*Journal of Allergy and Clinical Immunology*, January 2001). Based on this information, a 95% confidence interval for the true percentage of asthmatic patients who experience breathing difficulties after taking BAC is (.326, .785).
   a. Why might the confidence interval lead to an erroneous inference?
   b. How many asthma patients must be included in the study in order to estimate the true percentage who experience a significant drop in breathing capacity to within 4% with a 95% confidence interval?

5.89 **Do you think you smell?** Refer to the *Depression and Anxiety* (June 2010) study of patients who suffer from olfactory reference syndrome (ORS), Exercise 5.64 (p. 270). Recall that psychiatrists disagree over how prevalent ORS is in the human population. Suppose you want to estimate the true proportion of U.S. adults who suffer from ORS using a 99% confidence interval. Determine the size of the sample necessary to attain a sampling error no larger than .04.

5.90 **Caffeine content of coffee.** According to a Food and Drug Administration (FDA) study, a cup of coffee contains an average of 115 milligrams (mg) of caffeine, with the amount per cup ranging from 60 to 180 mg. Suppose you want to repeat the FDA experiment in order to obtain an estimate of the mean caffeine content in a cup of coffee correct to within 5 mg with 95% confidence. How many cups of coffee would have to be included in your sample?

5.91 **USGA golf ball tests.** The United States Golf Association (USGA) tests all new brands of golf balls to ensure that they meet USGA specifications. One test conducted is intended to measure the average distance traveled when the ball is hit by a machine called "Iron Byron." Suppose the USGA wishes to estimate the mean distance for a new brand to within 1 yard with 90% confidence. Assume that past tests have indicated that the standard deviation of the distances Iron Byron hits golf balls is approximately 10 yards. How many golf balls should be hit by Iron Byron to achieve the desired accuracy in estimating the mean?

5.92 **Preventing production of defective items.** It costs more to produce defective items—since they must be scrapped or reworked—than it does to produce nondefective items. This simple fact suggests that manufacturers should ensure the quality of their products by perfecting their production processes instead of depending on inspection of finished products (Deming, 1986). In order to better understand a particular metal stamping process, a manufacturer wishes to estimate the mean length of items produced by the process during the past 24 hours.

    a. How many parts should be sampled in order to estimate the population mean to within .1 millimeter (mm) with 90% confidence? Previous studies of this machine have indicated that the standard deviation of lengths produced by the stamping operation is about 2 mm.

    b. Time permits the use of a sample size no larger than 100. If a 90% confidence interval for is constructed with $n = 100$, will it be wider or narrower than would have been obtained using the sample size determined in part **a**? Explain.

    c. If management requires that μ be estimated to within .1 mm and that a sample size of no more than 100 be used, what is (approximately) the maximum confidence level that could be attained for a confidence interval that meets management's specifications?

5.6 Confidence Interval for a Population Variance (Optional)

In the previous sections, we considered interval estimation for population means or proportions. In this optional section, we discuss a confidence interval for a population variance, σ^2.

    Recall Example 1.4 (p. 10) and the U.S. Army Corps of Engineers study of contaminated fish in the Tennessee River, Alabama. It is important for the Corps of Engineers to know how stable the weights of the contaminated fish are. That is, how large is the variation in the fish weights? The keyword "variation" indicates that the target population parameter is σ^2, the variance of the weights of all contaminated fish inhabiting the Tennessee River. Of course, the exact value of σ^2 will be unknown. Consequently, the Corps of Engineers wants to estimate its value with a high level of confidence.

    Intuitively, it seems reasonable to use the sample variance, s^2, to estimate σ^2. However, unlike with sample means and proportions, the sampling distribution of s^2 does not follow a normal (z) distribution or a Student's t distribution. Rather, when certain assumptions are satisfied (we discuss these later), the sampling distribution of s^2 possesses approximately a **chi-square (χ^2) distribution**. The chi-square probability distribution, like the t distribution, is characterized by a quantity called the *degrees of freedom* (df) associated with the distribution. Several chi-square distributions with different df values are shown in Figure 5.18. You can see that unlike z and t distributions, the chi-square distribution is not symmetric about 0.

    The upper-tail areas for this distribution have been tabulated and are given in Table V in Appendix A, a portion of which is reproduced in Table 5.7. The table gives the values of χ^2, denoted as χ^2_α, that locate an area of α in the upper tail of the chi-square distribution; that is, $P(\chi^2 > \chi^2_\alpha) = \alpha$. As with the t-statistic, the degrees of freedom associated with s^2 are $(n-1)$. Thus, for $n = 10$ and an upper-tail value $\alpha = .05$, you will have $n-1 = 9$ df and $\chi^2_{.05} = 16.9190$ (highlighted in Table 5.7).

    The chi-square distribution is used to find a confidence interval for σ^2, as shown in the box. An illustrative example follows.

$f(\chi^2)$

df = 1

df = 4

df = 6

0 2 4 6 8 10 12 14 16 χ^2

Figure 5.18

Several χ^2 Probability Distributions

A $100(1-\alpha)$ Confidence Interval for σ^2

$$\frac{(n-1)s^2}{\chi^2_{\alpha/2}} \le \sigma^2 \le \frac{(n-1)s^2}{\chi^2_{(1-\alpha/2)}}$$

where $\chi^2_{\alpha/2}$ and $\chi^2_{(1-\alpha/2)}$ are values corresponding to an area of $\alpha/2$ in the right (upper) and left (lower) tails, respectively, of the chi-square distribution based on $(n-1)$ degrees of freedom.

Table 5.7	Reproduction of Part of Table V in Appendix A: Critical Values of Chi Square

Degrees of Freedom	$\chi^2_{.100}$	$\chi^2_{.050}$	$\chi^2_{.025}$	$\chi^2_{.010}$	$\chi^2_{.005}$
1	2.70554	3.84146	5.02389	6.63490	7.87944
2	4.60517	5.99147	7.37776	9.21034	10.5966
3	6.25139	7.81473	9.34840	11.3449	12.8381
4	7.77944	9.48773	11.1433	13.2767	14.8602
5	9.23635	11.0705	12.8325	15.0863	16.7496
6	10.6446	12.5916	14.4494	16.8119	18.5476
7	12.0170	14.0671	16.0128	18.4753	20.2777
8	13.3616	15.5073	17.5346	20.0902	21.9550
9	14.6837	16.9190	19.0228	21.6660	23.5893
10	15.9871	18.3070	20.4831	23.2093	25.1882
11	17.2750	19.6751	21.9200	24.7250	25.7569
12	18.5494	21.0261	23.3367	26.2170	28.2995
13	19.8119	22.3621	24.7356	27.6883	29.8194
14	21.0642	23.6848	26.1190	29.1413	31.3193
15	22.3072	24.9958	27.4884	30.5779	32.8013
16	23.5418	26.2862	28.8454	31.9999	34.2672
17	24.7690	27.5871	30.1910	33.4087	35.7185
18	25.9894	28.8693	31.5264	34.8053	37.1564
19	27.2036	30.1435	32.8523	36.1908	38.5822

Conditions Required for a Valid Confidence Interval for σ^2

1. A random sample is selected from the target population.
2. The population of interest has a relative frequency distribution that is approximately normal.

Example 5.11
Estimating the Weight Variance, σ^2, of Contaminated Fish

Problem Refer to the U.S. Army Corps of Engineers study of contaminated fish in the Tennessee River. The Corps of Engineers has collected data for a random sample of 144 fish contaminated with DDT. (The engineers made sure to capture contaminated fish in several different randomly selected streams and tributaries of the river.) The fish weights (in grams) are saved in the **FISHDDT** file. The Army Corps of Engineers wants to estimate the true variation in fish weights in order to determine whether the fish are stable enough to allow further testing for DDT contamination.

a. Use the sample data to find a 95% confidence interval for the parameter of interest.

b. Determine whether the confidence interval, part **a,** is valid.

Solution

a. Here, the target parameter is σ^2, the variance of the population of weights of contaminated fish. First, we need to find the sample variance, s^2, to compute the interval estimate. The MINITAB printout, Figure 5.19, gives descriptive statistics for

Figure 5.19

MINITAB descriptive statistics for fish weights, Example 5.11

Descriptive Statistics: WEIGHT

Variable	N	Mean	StDev	Minimum	Median	Maximum
WEIGHT	144	1049.7	376.5	173.0	1000.0	2302.0

the sample weights saved in the **FISHDDT** file. You can see that $s = 376.5$ grams. Consequently, $s^2 = (376.5)^2 = 141,752.25$.

Next, we require the critical values $\chi^2_{\alpha/2}$ and $\chi^2_{(1-\alpha/2)}$ for a chi-square distribution. For a 95% confidence interval, $\alpha = .05$, $\alpha/2 = .025$, and $(1 - \alpha/2) = .975$. Therefore, we need $\chi^2_{.025}$ and $\chi^2_{.975}$. Now, for a sample size $n = 144$, the degrees of freedom associated with the distribution is df $= (n - 1) = 143$. Looking in the df $= 150$ row of Table V, Appendix A (the row with the df value closest to 143), we find $\chi^2_{.025} = 185.800$ and $\chi^2_{.975} = 117.985$.

Substituting the appropriate values into the formula given in the box, we obtain

$$\frac{(144 - 1)(376.5)^2}{185.500} \leq \sigma^2 \leq \frac{(144 - 1)(376.5)^2}{117.985}$$

Or,

$$109,275 \leq \sigma^2 \leq 171,806$$

Thus, the Army Corps of Engineers can be 95% confident that the variance in weights of the population of contaminated fish ranges between 109,275 and 171,806.

b. According to the box, two conditions are required for the confidence interval to be valid. First, the sample must be randomly selected from the population. The Army Corps of Engineers did, indeed, collect a random sample of contaminated fish, making sure to sample fish from different locations in the Tennessee River. Second, the population data (the fish weights) must be approximately normally distributed. A MINITAB histogram for the sampled fish weights (with a normal curve superimposed) is displayed in Figure 5.20. Clearly, the data appear to be approximately normally distributed. Thus, the confidence interval is valid.

Figure 5.20

MINITAB histogram of fish weights, Example 5.11

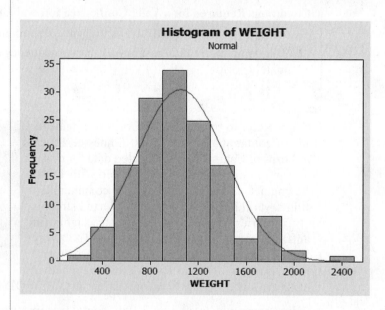

Look Ahead Will this confidence interval be practically useful in helping the Corps of Engineers decide whether the weights of the fish are stable? Only if it is clear what a weight variance of, say, 150,000 grams^2 implies. Most likely, the Corps of Engineers will want the interval in the same units as the weight measurement—grams. Consequently, a confidence interval for σ, the standard deviation of the population of fish weights, is desired. We demonstrate how to obtain this interval estimate in the next example.

Now Work Exercise 5.103a

Example 5.12

Estimating the Weight Standard Deviation, σ, of Contaminated Fish

Problem Refer to Example 5.11. Find a 95% confidence interval for σ, the true standard deviation of the contaminated fish weights.

Solution A confidence interval for σ is obtained by taking the square roots of the lower and upper endpoints of a confidence interval for σ^2. Consequently, the 95% confidence interval for σ is:

$$\sqrt{109{,}275} \le \sigma \le \sqrt{171{,}806}$$

Or,

$$330.5 \le \sigma \le 414.5$$

Thus, the engineers can be 95% confident that the true standard deviation of fish weights is between 330.5 grams and 414.5 grams.

Look Back Suppose the Corps of Engineers' threshold is $\sigma = 500$ grams. That is, if the standard deviation in fish weights is 500 grams or higher, further DDT contamination tests will be suspended due to the unstableness of the fish weights. Since the 95% confidence interval for σ lies below 500 grams, the engineers will continue the DDT contamination tests on the fish.

Now Work Exercise 5.103b

⚠ **CAUTION** The procedure for estimating either σ^2 or σ requires an assumption regardless of whether the sample size n is large or small (see the conditions in the box). The sampled data must come from a population that has an approximate normal distribution. Unlike small sample confidence intervals for μ based on the t-distribution, *slight to moderate departures from normality will render the chi-square confidence interval for σ^2 invalid.* ▲

Exercises 5.93–5.108

Understanding the Principles

5.93 What sampling distribution is used to find an interval estimate for σ^2?

5.94 What conditions are required for a valid confidence interval for σ^2?

5.95 How many degrees of freedom are associated with a chi-square sampling distribution for a sample of size n?

Learning the Mechanics

5.96 For each of the following combinations of α and degrees of freedom (df), use Table V in Appendix B to find the values of $\chi^2_{\alpha/2}$ and $\chi^2_{(1-\alpha/2)}$ that would be used to form a confidence interval for σ^2.
a. $\alpha = .05, df = 7$
b. $\alpha = .10, df = 16$
c. $\alpha = .01, df = 20$
d. $\alpha = .05, df = 20$

5.97 Given the following values of $\bar{x}$, s, and n, form a 90% confidence interval for σ^2.
a. $\bar{x} = 21, s = 2.5, n = 50$
b. $\bar{x} = 1.3, s = .02, n = 15$
c. $\bar{x} = 167, s = 31.6, n = 22$
d. $\bar{x} = 9.4, s = 1.5, n = 5$

5.98 Refer to Exercise 5.97. For each part, **a–d,** form a 90% confidence interval for σ.

5.99 A random sample of $n = 6$ observations from a normal distribution resulted in the data shown in the table and saved in the **LM5_99** file. Compute a 95% confidence interval for σ^2.

8	2	3	7	11	6

Applying the Concepts—Basics

5.100 Motivation of drug dealers. Refer to the *Applied Psychology in Criminal Justice* (Sept. 2009) study of the personality characteristics of convicted drug dealers, Exercise 5.15 (p. 251). A random sample of 100 drug dealers had a mean Wanting Recognition (WR) score of 39 points, with a standard deviation of 6 points. The researchers also are interested in σ^2, the variation in WR scores for all convicted drug dealers.
a. Identify the target parameter, in symbols and words.
b. Compute a 99% confidence interval for σ^2.
c. What does it mean to say that the target parameter lies within the interval with "99% confidence"?
d. What assumption about the data must be satisfied in order for the confidence interval to be valid?
e. To obtain a practical interpretation of the interval, part **b,** explain why a confidence interval for the standard deviation, σ, is desired.

f. Use the results, part **b,** to compute a 99% confidence interval for σ. Give a practical interpretation of the interval.

5.101 Laptop use in middle school. Refer to the *American Secondary Education* (Fall 2009) study of 106 middle school students' usage of laptop computers, Exercise 5.16 (p. 251). The number of minutes per day each student used a laptop for taking notes was recorded, with the following results: $\bar{x} = 13.2, s = 19.5$.

a. Find and interpret a 95% confidence interval for the variance in the amount of time per day laptops are used for taking notes for all middle school students across the country.

b. Recall that in Exercise 2.100b you concluded that the distribution of laptop usage for taking notes cannot be normally distributed. Does this compromise the validity of the interval estimate, part **a**? Explain.

5.102 Antigens for a parasitic roundworm in birds. Refer to the *Gene Therapy and Molecular Biology* (June 2009) study of DNA in peptide (protein) produced by antigens for a parasitic roundworm in birds, Exercise 5.45 (p. 263). Recall that scientists tested each in a sample of 4 alleles of antigen-produced protein for level of peptide. The results were: $\bar{x} = 1.43$ and $s = .13$. Use this information to construct a 90% confidence interval for the true variation in peptide scores for alleles of the antigen-produced protein. Interpret the interval for the scientists.

5.103 Characteristics of a rockfall. Refer to the *Environmental Geology* (Vol. 58, 2009) simulation study of how far a block from a collapsing rock wall will bounce down a soil slope, Exercise 2.59 (p. 57). Rebound lengths (in meters) were estimated for 13 rock bounces. The data are repeated in the table and saved in the **ROCKFALL** file. A MINITAB analysis of the data is shown in the printout below.

a. Locate a 95% confidence interval for σ^2 on the printout. Interpret the result.

b. Locate a 95% confidence interval for σ on the printout. Interpret the result.

c. What conditions are required for the intervals, parts **a** and **b,** to be valid?

10.94	13.71	11.38	7.26	17.83	11.92	11.87
5.44	13.35	4.90	5.85	5.10	6.77	

Based on Paronuzzi, P. "Rockfall-induced block propagation on a soil slope, northern Italy." *Environmental Geology*, Vol. 58, 2009 (Table 2).

Test and CI for One Variance: REB-LENGTH

```
Method

The chi-square method is only for the normal distribution.
The Bonett method is for any continuous distribution.


Statistics

Variable    N  StDev  Variance
REB-LENGTH  13  4.09    16.8


95% Confidence Intervals

                          CI for       CI for
Variable   Method         StDev        Variance
REB-LENGTH Chi-Square  (2.94, 6.76)  (8.6, 45.7)
           Bonett      (2.97, 6.64)  (8.8, 44.1)
```

Applying the Concepts—Intermediate

5.104 Jitter in a water power system. *Jitter* is a term used to describe the variation in conduction time of a water power system. Low throughput jitter is critical to successful waterline technology. An investigation of throughput jitter in the opening switch of a prototype system (*Journal of Applied Physics*) yielded the following descriptive statistics on conduction time for $n = 18$ trials: $\bar{x} = 334.8$ nanoseconds, $s = 6.3$ nanoseconds. (Conduction time is defined as the length of time required for the downstream current to equal 10% of the upstream current.)

a. Construct a 95% confidence interval for the true standard deviation of conduction times of the prototype system.

b. Practically interpret the confidence interval, part **a.**

c. A system is considered to have low throughput jitter if the true conduction time standard deviation is less than 7 nanoseconds. Does the prototype system satisfy this requirement? Explain.

5.105 Cheek teeth of extinct primates. Refer to the *American Journal of Physical Anthropology* (Vol. 142, 2010) study of the characteristics of cheek teeth (e.g., molars) in an extinct primate species, Exercise 2.34 (p. 46). Recall that the researchers recorded the dentary depth of molars (in millimeters) for a sample of 18 cheek teeth extracted from skulls. The data are repeated in the table and saved in the **CHEEKTEETH** file. Estimate the true standard deviation in molar depths for the population of cheek teeth in extinct primates using a 95% confidence interval. Give a practical interpretation of the result. Are the conditions required for a valid confidence interval reasonably satisfied?

18.12	16.55
19.48	15.70
19.36	17.83
15.94	13.25
15.83	16.12
19.70	18.13
15.76	14.02
17.00	14.04
13.96	16.20

Based on Boyer, D. M., Evans, A. R., and Jernvall, J. "Evidence of dietary differentiation among Late Paleocene–Early Eocene Plesiadapids (Mammalia, Primates)." *American Journal of Physical Anthropology*, Vol. 142, 2010 (Table A3).

5.106 Shell lengths of sea turtles. Refer to the *Aquatic Biology* (Vol. 9, 2010) study of green sea turtles inhabiting the Grand Cayman South Sound lagoon, Exercise 5.20 (p. 252). Recall that the data on shell length, measured in centimeters, for 76 captured turtles are saved in the **TURTLES** file. Use the sample data to estimate the true variance in shell lengths of all green sea turtles in the lagoon with 90% confidence. Interpret the result.

5.107 Lobster trap placement. Refer to the *Bulletin of Marine Science* (April 2010) observational study of teams fishing for the red spiny lobster in Baja California Sur, Mexico, Exercise 5.35 (p. 261). Trap spacing measurements (in meters) for a sample of seven teams of red spiny lobster fishermen are repeated in the table (p. 283) and saved in the **TRAPSPACE** file. The researchers want to know how variable the trap spacing measurements are for the population of red spiny lobster fishermen fishing in Baja California Sur, Mexico. Provide the researchers with an estimate of the target parameter using a 99% confidence interval.

93	99	105	94	82	70	86

Based on Shester, G. G. "Explaining catch variation among Baja California lobster fishers through spatial analysis of trap-placement decisions." *Bulletin of Marine Science,* Vol. 86, No. 2, April 2010 (Table 1).

5.108 Is honey a cough remedy? Refer to the *Archives of Pediatrics and Adolescent Medicine* (Dec. 2007) study of honey as a remedy for coughing, Exercise 2.32 (p. 45). Recall that the 105 ill children in the sample were randomly divided into groups. One group received a dosage of an over-the-counter cough medicine (DM); another group received a dosage of honey (H). The coughing improvement scores (as determined by the children's parents) for the patients in the two groups are reproduced in the accompanying table and saved in the **HONEYCOUGH** file. The pediatric researchers desire information on the variation in coughing improvement scores for each of the two groups.

a. Find a 90% confidence interval for the standard deviation in improvement scores for the honey dosage group.
b. Repeat part **a** for the DM dosage group.
c. Based on the results, parts **a** and **b**, what conclusions can the pediatric researchers draw about which group has the smaller variation in improvement scores?

Honey Dosage:	12 11 15 11 10 13 10 4 15 16 9 14 10 6
	10 8 11 12 12 8 12 9 11 15 10 15 9 13
	8 12 10 8 9 5 12

DM Dosage:	4 6 9 4 7 7 7 9 12 10 11 6 3 4
	9 12 7 6 8 12 12 4 12 13 7 10 13 9
	4 4 10 15 9

Based on Paul, I. M., et al. "Effect of honey, dextromethorphan, and no treatment on nocturnal cough and sleep quality for coughing children and their parents." *Archives of Pediatrics and Adolescent Medicine,* Vol. 161, No. 12, Dec. 2007 (data simulated).

CHAPTER NOTES

Key Terms

Note: Asterisks (*) denote terms from the optional section in this chapter.

*Chi-square (χ^2) distribution 278
Confidence coefficient 247
Confidence interval 244
Confidence level 247
Degrees of freedom (df) 254

Interval estimator 244
normal (z) statistic 245
Point estimator 244
Sampling error 272
Target parameter 244
t-statistic 254
Wilson adjustment for estimating p 267

Key Symbols

θ	General population parameter (theta)
μ	Population mean
σ^2	Population variance
σ	Population standard deviation
p	Population proportion; P (Success) in binomial trial
q	$1 - p$
$\bar{x}$	Sample mean (estimator of μ)
s^2	Sample variance (estimator of σ^2)
$\hat{p}$	Sample proportion (estimator of p)
$\mu_{\bar{x}}$	Mean of the population sampling distribution of $\bar{x}$
$\sigma_{\bar{x}}$	Standard deviation of the sampling distribution of $\bar{x}$
$\sigma_{\hat{p}}$	Standard deviation of the sampling distribution of $\hat{p}$
SE	Sampling error in estimation
α	$(1 - \alpha)$ represents the confidence coefficient
$z_{\alpha/2}$	z-value used in a $100(1 - \alpha)\%$ large-sample confidence interval for μ or p
$t_{\alpha/2}$	Student's t-value used in a $100(1 - \alpha)\%$ small-sample confidence interval for μ
$\chi^2_{\alpha/2}$	*Chi-Square value used in a $100(1 - \alpha)\%$ confidence interval for σ^2

Key Ideas

Confidence Interval: An interval that encloses an unknown population parameter with a certain level of confidence, $(1 - \alpha)$
Confidence Coefficient: The probability $(1 - \alpha)$ that a randomly selected confidence interval encloses the true value of the population parameter.

Population Parameters, Estimators, & Standard Errors

Parameter (θ)	Estimator ($\hat{\theta}$)	Standard Error of Estimator ($\hat{\sigma}_{\hat{\theta}}$)	Estimated Std. Error ($\hat{\sigma}_{\hat{\theta}}$)
Mean, μ	$\bar{x}$	$\sigma/\sqrt{n}$	$s/\sqrt{n}$
Proportion, p	$\hat{p}$	$\sqrt{pq/n}$	$\sqrt{\hat{p}\hat{q}/n}$
*Variance, σ^2	s^2	—	—

Determining the Sample Size *n*:

Estimating μ: $\quad n = (z_{\alpha/2})^2(\sigma^2)/(\text{SE})^2$

Estimating p: $\quad n = (z_{\alpha/2})^2(pq)/(\text{SE})^2$

Key Words for Identifying the Target Parameter:

μ—Mean, Average
p—Proportion, Fraction, Percentage, Rate, Probability
*σ^2—Variance, Variation, Spread

Commonly Used *z*-values for a Large-Sample Confidence Interval for μ or *p*:

90% CI:	$(1 - \alpha) = .10$	$z_{.05} = 1.645$
95% CI:	$(1 - \alpha) = .05$	$z_{.025} = 1.96$
99% CI:	$(1 - \alpha) = .01$	$z_{.005} = 2.575$

Illustrating the Notion of "95% Confidence"

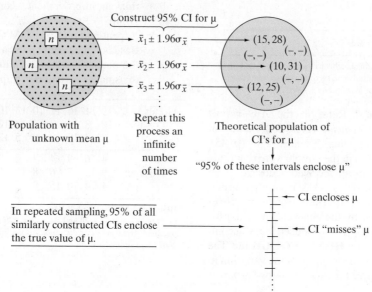

In repeated sampling, 95% of all similarly constructed CIs enclose the true value of μ.

Guide to Forming a Confidence Interval

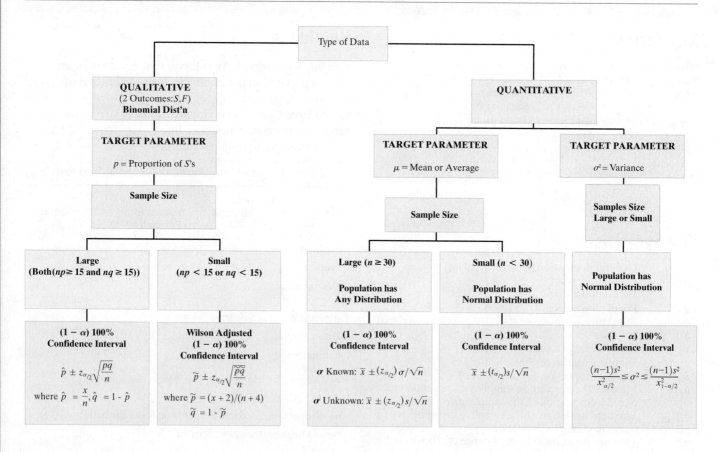

Supplementary Exercises 5.109–5.141

Note: List the assumptions necessary for the valid implementation of the statistical procedures you use in solving all these exercises. Starred (*) exercises are from the optimal section in this chapter.

Understanding the Principles

5.109 For each of the following, identify the target parameter as μ, p, or σ^2.

a. Average score on the SAT
b. Mean time waiting at a supermarket checkout lane
c. Proportion of voters in favor of legalizing marijuana
d. Percentage of NFL players who have ever made the Pro Bowl
e. Dropout rate of American college students
*f. Variation in IQ scores of sociopaths

5.110 Interpret the phrase "95% confident" in the following statement:"We are 95% confident that the proportion of all PCs with a computer virus falls between .12 and .18."

5.111 In each of the following instances, determine whether you would use a z- or t-statistic (or neither) to form a 95% confidence interval for μ; then look up the appropriate z or t value.
   a. Random sample of size $n = 21$ from a normal distribution with unknown mean μ and standard deviation σ
   b. Random sample of size $n = 175$ from a normal distribution with unknown mean μ and standard deviation σ
   c. Random sample of size $n = 12$ from a normal distribution with unknown mean and standard deviation $\sigma = 5$
   d. Random sample of size $n = 65$ from a distribution about which nothing is known
   e. Random sample of size $n = 8$ from a distribution about which nothing is known

Learning the Mechanics

5.112 Let t_0 represent a particular value of t from Table IV of Appendix A. Find the table values such that the following statements are true:
   a. $P(t \le t_0) = .05$, where df $= 17$
   b. $P(t \ge t_0) = .005$, where df $= 14$
   c. $P(t \le -t_0 \text{ or } t \ge t_0) = .10$, where df $= 6$
   d. $P(t \le -t_0 \text{ or } t \ge t_0) = .01$, where df $= 22$

5.113 In a random sample of 400 measurements, 227 possess the characteristic of interest, A.
   a. Use a 95% confidence interval to estimate the true proportion p of measurements in the population with characteristic A.
   b. How large a sample would be needed to estimate p to within .02 with 95% confidence?

5.114 A random sample of 225 measurements is selected from a population, and the sample mean and standard deviation are $\bar{x} = 32.5$ and $s = 30.0$, respectively.
   a. Use a 99% confidence interval to estimate the mean of the population, μ.
   b. How large a sample would be needed to estimate μ to within .5 with 99% confidence?
   c. What is meant by the phrase "99% confidence" as it is used in this exercise?
   *d.** Find a 99% confidence interval for σ^2.

Applying the Concepts—Basic

5.115 CDC health survey. The Centers for Disease Control and Prevention (CDCP) in Atlanta, Georgia, conduct an annual survey of the general health of the U.S. population as part of their Behavioral Risk Factor Surveillance System. Using random-digit dialing, the CDCP telephones U.S. citizens over 18 years of age and asks them the following four questions:
   1. Is your health generally excellent, very good, good, fair, or poor?
   2. How many days during the previous 30 days was your physical health not good because of injury or illness?
   3. How many days during the previous 30 days was your mental health not good because of stress, depression, or emotional problems?
   4. How many days dur- ing the previous 30 days did your physical or mental health prevent you from performing your usual activities?

Identify the parameter of interest for each question.

5.116 Personal networks of older adults. In sociology, a personal network is defined as the people with whom you make frequent contact. The Living Arrangements and Social Networks of Older Adults (LSN) research program used a stratified random sample of men and women born between 1908 and 1937 to gauge the size of the personal network of older adults. Each adult in the sample was asked to "please name the people (e.g., in your neighborhood) you have frequent contact with and who are also important to you." Based on the number of people named, the personal network size for each adult was determined. The responses of 2,819 adults in the LSN sample yielded the following statistics on network size: $\bar{x} = 14.6; s = 9.8$ (*Sociological Methods & Research*, August 2001).
   a. Give a point estimate for the mean personal network size of all older adults.
   b. Form a 95% confidence interval for the mean personal network size of all older adults.
   c. Give a practical interpretation of the interval you found in part **b.**
   d. Give the conditions required for the interval in part **b** to be valid.

5.117 Ancient Greek pottery. Refer to the *Chance* (Fall 2000) study of 837 pieces of pottery found at the ancient Greek settlement at Phylakopi, presented in Exercise 2.14 (p. 36). Of the 837 pieces, 183 were painted with either a curvilinear, geometric, or naturalistic decoration. Find a 90% confidence interval for the population proportion of all pottery artifacts at Phylakopi that are painted. Interpret the resulting interval.

5.118 Tax-exempt charities. Donations to tax-exempt organizations such as the Red Cross, the Salvation Army, the

Organization	Charitable Commitment (%)
American Cancer Society	71
American National Red Cross	89
Big Brothers Big Sisters of America	78
Boy Scouts of America	87
Boys & Girls Clubs of America	80
CARE USA	90
Covenant House	67
Disabled American Veterans	69
Ducks Unlimited	86
Feed The Children	84
Girl Scouts of the USA	82
Goodwill Industries International	88
Habitat for Humanity International	83
Mayo Clinic	97
March of Dimes Foundation	77
Multiple Sclerosis Society	78
Museum of Modern Art	80
Nature Conservancy	80
Paralyzed Veterans of America	62
Planned Parenthood Federation	82
Salvation Army	82
Shriners Hospital for Children	92
Smithsonian Institution	82
Special Olympics	74
Trust for Public Land	89
United States Olympic Committee	71
United Way	86
WGBH Educational Foundation	79
World Trade Center Memorial Foundation	79
YMCA of the USA	81

Based on "The 200 largest U.S. charities." Forbes.com, November 19, 2008.

YMCA, and the American Cancer Society not only go to the stated charitable purpose, but are used to cover fund-raising expenses and overhead. The **CHARITY** file (see table on p. 285) contains the charitable commitments (i.e., the percentage of expenses that goes toward the stated charitable purpose) for a sample of 30 charities.

a. Give a point estimate for the mean charitable commitment of tax-exempt organizations.
b. Construct a 95% confidence interval for the mean charitable commitment.
c. What assumption(s) must hold for the method of estimation used in part **b** to be appropriate?
d. Why is the confidence interval of part **b** a better estimator of the mean charitable commitment than the point estimator of part **a**?
*e. Constract a 95% confidence interval for the variance in charitable commitments for all tax-exempt organizations.

5.119 Cell phone use by drivers. The U.S. Department of Transportation reported the results of the *National Occupant Protection Use Survey*. One focus of the survey was to determine the level of cell phone use by drivers while they are in the act of driving a motor passenger vehicle. Data collected by observers at randomly selected intersections across the country revealed that in a sample of 1,165 drivers, 35 were using their cell phone.

a. Give a point estimate of p, the true driver cell phone use rate (i.e., the true proportion of drivers who are using a cell phone while driving).
b. Compute a 95% confidence interval for p.
c. Give a practical interpretation of the interval you found in part **b.**
d. How many drivers need to be sampled to estimate the true cell phone use rate to within .005 with 95% confidence?

5.120 "Made in the USA" survey. Refer to Exercise 2.177 (p. 99) and the *Journal of Global Business* (Spring 2002) survey to determine what "Made in the USA" means to consumers. Recall that 106 shoppers at a shopping mall in Muncie, Indiana, responded to the question "'Made in the USA' means what percentage of U.S. labor and materials?" Sixty-four shoppers answered "100%."

a. Define the population of interest in the survey.
b. What is the characteristic of interest in the population?
c. Use a 90% confidence interval to estimate the true proportion of consumers who believe that "Made in the USA" means 100% U.S. labor and materials.
d. Give a practical interpretation of the interval you used in part **c.**
e. Explain what the phrase "90% confidence" means for this interval.

5.121 "Made in the USA" survey (cont'd). Refer to Exercise 5.120. Suppose the researchers want to increase the sample size in order to estimate the true proportion p to within .05 of its true value with a 90% confidence interval.

a. What is the confidence level desired by the researchers?
b. What is the sampling error desired by the researchers?
c. Compute the sample size necessary to obtain the desired estimate.

5.122 Water pollution testing. The EPA wants to test a randomly selected sample of n water specimens and estimate μ, the mean daily rate of pollution produced by a mining operation. If the EPA wants a 95% confidence interval estimate with a sampling error of 1 milligram per liter (mg/L), how many water specimens are required in the sample? Assume that prior knowledge indicates that pollution readings in water samples taken during a day are approximately normally distributed with a standard deviation equal to 5 (mg/L).

5.123 Al Qaeda attacks on the United States. Refer to the *Studies in Conflict & Terrorism* (Vol. 29, 2006) analysis of incidents involving suicide terrorist attacks, presented in Exercise 2.173 (p. 98). Data on the number of individual suicide bombings or attacks for each in a sample of 21 incidents involving an attack against the United States by the Al Qaeda terrorist group (see table below) are saved in the **ALQAEDA** file.

a. Find the mean and standard deviation of the sample data.
b. Describe the population from which the sample is selected.
c. Use the information from part **a** to find a 90% confidence interval for the mean μ of the population.
d. Give a practical interpretation of the result you obtained in part **c.**
e. In repeatedly sampling, where intervals similar to the one computed in part **c** are generated, what proportion of the intervals will enclose the true value of μ?

1	1	2	1	2	4	1	1	1	1	2	3	4	5	1	1	1	2	2	2	1

Based on Moghadam, A. "Suicide terrorism, occupation, and the globalization of martyrdom: A critique of *Dying to Win*." *Studies in Conflict & Terrorism*, Vol. 29, No. 8, 2006 (Table 3).

5.124 Ammonia in car exhaust. Refer to the *Environmental Science & Technology* (September 1, 2000) study on the ammonia levels near the exit ramp of a San Francisco highway tunnel, presented in Exercise 2.182 (p. 99). The ammonia concentration (parts per million) data for eight randomly selected days during the afternoon drive time are reproduced in the accompanying table and saved in the **AMMONIA** file. Find a 99% confidence interval for the population mean daily ammonia level in air in the tunnel. Interpret your result.

1.53	1.50	1.37	1.51	1.55	1.42	1.41	1.48

5.125 Oven cooking study. A group of Harvard University School of Public Health researchers studied the impact of cooking on the size of indoor air particles (*Environmental Science & Technology*, September 1, 2000). The decay rate (measured in $\mu m/hour$) for fine particles produced from oven cooking or toasting was recorded on six randomly selected days. The six measurements (shown below) are saved in the **DECAY** file.

.95	.83	1.20	.89	1.45	1.12

Based on Abt, E., et al. "Relative contribution of outdoor and indoor particle sources to indoor concentrations." *Environmental Science & Technology*, Vol. 34, No. 17, Sept. 1, 2000 (Table 3).

a. Find and interpret a 95% confidence interval for the true average decay rate of fine particles produced from oven cooking or toasting.

b. Explain what the phrase "95% confident" implies in the interpretation of part **a**.

*__c.__ Estimate the true standard deviation of decay rate with a 95% confidence interval. Interpret the result.

d. What must be true about the distribution of the population of decay rates for the inferences you made in parts **a** and **c** to be valid?

e. Suppose that we want to estimate the average decay rate of fine particles produced from oven cooking or toasting to within .04 with 95% confidence. How large a sample should be selected?

5.126 Scary-movie study. According to a University of Michigan study, many adults have experienced lingering "fright" effects from a scary movie or TV show they saw as a teenager (*Tampa Tribune*, March 10, 1999). In a survey of 150 college students, 39 said they still experience "residual anxiety" from a scary TV show or movie.

a. Give a point estimate $\hat{p}$ for the true proportion of college students who experience "residual anxiety" from a scary TV show or movie.

b. Find a 95% confidence interval for p.

c. Interpret the interval you found in part **b**.

Applying the Concepts—Intermediate

5.127 Brown-bag lunches at work. In a study reported in *The Wall Street Journal*, the Tupperware Corporation surveyed 1,007 U.S. workers. Of the people surveyed, 665 indicated that they take their lunch to work with them. Of these 665 taking their lunch, 200 reported that they take it in brown bags.

a. Find a 95% confidence interval estimate of the population proportion of U.S. workers who take their lunch to work with them. Interpret the interval.

b. Consider the population of U.S. workers who take their lunch to work with them. Find a 95% confidence interval estimate of the population proportion who take brown-bag lunches. Interpret the interval.

5.128 Research on brain specimens. In Exercise 2.39 (p. 48), you learned that the postmortem interval (PMI) is the elapsed time between death and the performance of an autopsy on the cadaver. *Brain and Language* (June 1995) reported on the PMIs of 22 randomly selected human brain specimens obtained at autopsy. The data, reproduced in the following table, are saved in the **BRAINPMI** file. A coroner claims that the true mean PMI of human brain specimens obtained at autopsy is 10 days. Do you agree? Use a 95% confidence interval to make an inference.

5.5	14.5	6.0	5.5	5.3	5.8	11.0	6.4
7.0	14.5	10.4	4.6	4.3	7.2	10.5	6.5
3.3	7.0	4.1	6.2	10.4	4.9		

Based on Hayes, T. L., and Lewis, D. A. "Anatomical specialization of the anterior motor speech area: Hemispheric differences in magnopyramidal neurons," *Brain and Language*, Vol. 49, No. 3, p. 292 (Table 1).

5.129 Inbreeding of tropical wasps. Tropical swarm-founding wasps rely on female workers to raise their offspring. One possible explanation for this strange behavior is inbreeding, which increases relatedness among the wasps, presumably making it easier for the workers to pick out their closest relatives as propagators of their own genetic material. To test this theory, 197 swarm-founding wasps were captured in Venezuela, frozen at $-70°C$, and then subjected to a series of genetic tests. The data were used to generate an inbreeding coefficient x for each wasp specimen, with the following results: $\bar{x} = .044$ and $s = .884$.

a. Construct a 99% confidence interval for the mean inbreeding coefficient of this species of wasp.

b. A coefficient of 0 implies that the wasp has no tendency to inbreed. Use the confidence interval you constructed in part **a** to make an inference about the tendency for this species of wasp to inbreed.

5.130 Exercise workout dropouts. Researchers at the University of Florida's Department of Exercise and Sport Sciences conducted a study of variety in exercise workouts (*Journal of Sport Behavior*, 2001). A sample of 120 men and women were randomly divided into three groups, with 40 people per group. Group 1 members varied their exercise routine in workouts, group 2 members performed the same exercise at each workout, and group 3 members had no set schedule or regulations for their workouts.

a. By the end of the study, 15 people had dropped out of the first exercise group. Estimate the dropout rate for exercisers who vary their routine in workouts. Use a 90% confidence interval and interpret the result.

b. By the end of the study, 23 people had dropped out of the third exercise group. Estimate the dropout rate for exercisers who have no set schedule for their workouts. Use a 90% confidence interval and interpret the result.

5.131 Time to solve a math programming problem. *IEEE Transactions* presented a hybrid algorithm for solving a polynomial zero–one mathematical programming problem. The algorithm incorporates a mixture of pseudo-Boolean concepts and time-proven implicit enumeration procedures. Fifty-two random problems were solved by the hybrid algorithm; the times to solution (CPU time in seconds) are listed in the accompanying table and saved in the **MATHCPU** file.

a. Estimate, with 95% confidence, the mean solution time for the hybrid algorithm. Interpret the result.

b. How many problems must be solved to estimate the mean μ to within .25 second with 95% confidence?

*__c.__ Form a 95% confidence interval for the true standard deviation of the solution times for the hybrid algorithm. Interpret the result.

.045	1.055	.136	1.894	.379	.136	.336	.258	1.070
.506	.088	.242	1.639	.912	.412	.361	8.788	.579
1.267	.567	.182	.036	.394	.209	.445	.179	.118
.333	.554	.258	.182	.070	3.985	.670	3.888	.136
.091	.600	.291	.327	.130	.145	4.170	.227	.064
.194	.209	.258	3.046	.045	.049	.079		

Source: Snyder, W. S., and Chrissis, J. W. "A hybrid algorithm for solving zero–one mathematical programming problems." *IEEE Transactions*, Vol. 22, No. 2, June 1990. Copyright © 1990. IEEE Reprinted with permission.

5.132 Sentence complexity study. Refer to the *Applied Psycholinguistics* (June 1998) study of language skills of low-income children, presented in Exercise 2.104 (p. 72). Each in a sample of 65 low-income children was administered the Communicative Development Inventory (CDI) exam. The sentence complexity scores had a mean of 7.62 and a standard deviation of 8.91.

a. Construct a 90% confidence interval for the mean sentence complexity score of all low-income children.

b. Interpret the interval you found in part **a** in the words of the problem.

c. Suppose we know that the true mean sentence complexity score of middle-income children is 15.55. Is there evidence that the true mean for low-income children differs from 15.55? Explain.

5.133 Jail suicide study. In Exercise 2.190 (p. 101) you considered data on suicides in correctional facilities. The data were published in the *American Journal of Psychiatry* (July 1995). The researchers wanted to know what factors increase the risk of suicide by those incarcerated in urban jails. The data on 37 suicides that occurred in Wayne County Jail, Detroit, Michigan, are saved in the **SUICIDE** file. Answer each of the questions that follow, using a 95% confidence interval for the target parameter implied in the question. Interpret each confidence interval fully.

a. What proportion of suicides at the jail are committed by inmates charged with murder or manslaughter?

b. What proportion of suicides at the jail are committed at night?

c. What is the average length of time an inmate is in jail before committing suicide?

d. What percentage of suicides are committed by white inmates?

5.134 Training zoo animals. Refer to Exercise 2.65 (p. 59) and the *Teaching of Psychology* (May 1998) study in which students assisted in the training of zoo animals. A sample of 15 psychology students rated "The Training Game" as a "great" method of understanding the animal's perspective during training on a 7–1 point scale (where 1 = strongly disagree and 7 = strongly agree). The mean response was 5.87, with a standard deviation of 1.51.

a. Construct a 95% confidence interval for the true mean response of the students.

b. Suppose you want to reduce the width of the 95% confidence interval to half the size obtained in part **a.** How many students are required in the sample in order to obtain the desired confidence interval width?

5.135 Salmonella in ice cream bars. Recently, a case of salmonella (bacterial) poisoning was traced to a particular brand of ice cream bar, and the manufacturer removed the bars from the market. Despite this response, many consumers refused to purchase *any* brand of ice cream bars for some time after the event (McClave, personal correspondence). One manufacturer conducted a survey of consumers 6 months after the poisoning. A sample of 244 ice cream bar consumers was contacted, and 23 indicated that they would not purchase ice cream bars because of the potential for food poisoning.

a. What is the point estimate of the true fraction of the entire market who refuse to purchase bars 6 months after the poisoning?

b. Is the sample size large enough to use the normal approximation for the sampling distribution of the estimator of the binomial probability? Justify your response.

c. Construct a 95% confidence interval for the true proportion of the market who still refuse to purchase ice cream bars 6 months after the event.

d. Interpret both the point estimate and confidence interval in terms of this application.

5.136 Salmonella in ice cream bars (cont'd). Refer to Exercise 5.135. Suppose it is now 1 year after the poisoning was traced to ice cream bars. The manufacturer wishes to estimate the proportion who still will not purchase bars to within .02, using a 95% confidence interval. How many consumers should be sampled?

5.137 Removing a soil contaminant. A common hazardous compound found in contaminated soil is benzo(a)pyrene [B(a)p]. An experiment was conducted to determine the effectiveness of a method designed to remove B(a)p from soil (*Journal of Hazardous Materials*, June 1995). Three soil specimens contaminated with a known amount of B(a)p were treated with a toxin that inhibits microbial growth. After 95 days of incubation, the percentage of B(a)p removed from each soil specimen was measured. The experiment produced the following summary statistics: $\bar{x} = 49.3$ and $s = 1.5$.

a. Use a 99% confidence interval to estimate the mean percentage of B(a)p removed from a soil specimen in which the toxin was used.

b. Interpret the interval in terms of this application.

c. What assumption is necessary to ensure the validity of this confidence interval?

d. How many soil specimens must be sampled to estimate the mean percentage removed to within .5, using a 99% confidence interval?

*e. Find and interpret a 90% confidence interval for the true variance in the percentages of B(a)p removed.

5.138 Material safety data sheets. For over 20 years, the Occupational Safety & Health Administration has required companies that handle hazardous chemicals to complete material safety data sheets (MSDSs). These sheets have been criticized for being too hard to understand and complete by workers. Although improvements were implemented in 1990, a recent study of 150 MSDSs revealed that only 11% were completed satisfactorily (*Chemical & Engineering News*, Feb. 7, 2005). Find an interval that contains the true proportion of all MSDSs that are completed satisfactorily, with 95% confidence. Would it surprise you if the true proportion was as high as 20%? Explain.

Applying the Concepts—Advanced

5.139 IMA salary survey. Each year, *Management Accounting* reports the results of a salary survey of the members of the Institute of Management Accountants (IMA). One year, the 2,112 members responding had a salary distribution with a 20th percentile of $35,100, a median of $50,000, and an 80th percentile of $73,000.

a. Use this information to determine the minimum sample size that could be used in next year's survey to estimate the mean salary of IMA members to within $2,000 with 98% confidence. [*Hint:* To estimate s, first apply Chebyshev's theorem to find k such that at least 60% of the data fall within k standard deviations of μ. Then find $s \approx$ (80th percentile –20th percentile)/$2k$.]

b. Explain how you estimated the standard deviation required for the calculation of the sample size.

c. List any assumptions you make.

5.140 Air bags pose danger for children. By law, all new cars must be equipped with both driver-side and passenger-side safety air bags. There is concern, however, over whether air bags pose a danger for children sitting on the passenger side. In a National Highway Traffic Safety Administration (NHTSA) study of 55 people killed by the explosive force of air bags, 35 were children seated on the front-passenger side. This study led some car owners with the information about children to disconnect the passenger-side air bag.

a. Use the study to estimate the risk of an air bag fatality on a child seated on the front passenger seat.

b. NHTSA investigators determined that 24 of 35 children killed by the air bags were not wearing seat belts or were improperly restrained. How does this information affect your assessment of the risk of an air bag fatality?

Critical Thinking Challenge

5.141 Scallops, sampling, and the law. *Interfaces* (March–April 1995), discussed the case of a ship that fishes for scallops off the coast of New England. In order to protect baby scallops from being harvested, the U.S. Fisheries and Wildlife Service requires that "the average meat per scallop weigh at least $\frac{1}{36}$ of a pound." The ship was accused of violating this weight standard. Author Arnold Barnett lays out the scenario:

The vessel arrived at a Massachusetts port with 11,000 bags of scallops, from which the harbormaster randomly selected 18 bags for weighing. From each such bag, his agents took a large scoopful of scallops; then, to estimate the bag's average meat per scallop, they divided the total weight of meat in the scoopful by the number of scallops it contained. Based on the 18 [numbers] thus generated, the harbormaster estimated that each of the ship's scallops possessed an average $\frac{1}{39}$ of a pound of meat (that is, they were about seven percent lighter than the minimum requirement). Viewing this outcome as conclusive evidence that the weight standard had been

violated, federal authorities at once confiscated 95 percent of the catch (which they then sold at auction). The fishing voyage was thus transformed into a financial catastrophe for its participants.

The actual scallop weight measurements for each of the 18 sampled bags are listed in the accompanying table and saved in the **SCALLOPS** file. For ease of exposition, Bennett expressed each number as a multiple of $\frac{1}{36}$ of a pound, the minimum permissible average weight per scallop. Consequently, numbers below 1 indicate individual bags that do not meet the standard.

The ship's owner filed a lawsuit against the federal government, declaring that his vessel had fully complied with the weight standard. A Boston law firm was hired to represent the owner in legal proceedings and Bennett was retained by the firm to provide statistical litigation support and, if necessary, expert witness testimony.

a. Recall that the harbormaster sampled only 18 of the ship's 11,000 bags of scallops. One of the questions the lawyers asked Bennett was "Can a reliable estimate of the mean weight of all the scallops be obtained from a sample of size 18?" Give your opinion on this issue.

b. As stated in the article, the government's decision rule is to confiscate a catch if the sample mean weight of the scallops is less than $\frac{1}{36}$ of a pound. Do you see any flaws in this rule?

c. Develop your own procedure for determining whether a ship is in violation of the minimum-weight restriction. Apply your rule to the data. Draw a conclusion about the ship in question.

.93	.88	.85	.91	.91	.84	.90	.98	.88
.89	.98	.87	.91	.92	.99	1.14	1.06	.93

Based on Barnett, A. "Misapplications review: Jail terms." *Interfaces*, Vol. 25, No. 2, Mar.–Apr. 1995, p. 20.

Activity Conducting a Pilot Study

Choose a population pertinent to your major area of interest—a population that has an unknown mean or, if the population is binomial, that has an unknown probability of success. For example, a marketing major may be interested in the proportion of consumers who prefer a certain product. A sociology major may be interested in estimating the proportion of people in a certain socioeconomic group or the mean income of people living in a particular part of a city. A political science major may wish to estimate the proportion of an electorate in favor of a certain candidate, a certain amendment, or a certain presidential policy. A pre-med student might want to find the average length of time patients stay in the hospital or the average number of people treated daily in the emergency room. We could continue with

examples, but the point should be clear: Choose something of interest to you.

Define the parameter you want to estimate and conduct a *pilot study* to obtain an initial estimate of the parameter of interest and, more importantly, an estimate of the variability associated with the estimator. A pilot study is a small experiment (perhaps 20 to 30 observations) used to gain some information about the population of interest. The purpose of the study is to help plan more elaborate future experiments. Using the results of your pilot study, determine the sample size necessary to estimate the parameter to within a reasonable bound (of your choice) with a 95% confidence interval.

References

Agresti, A., and Coull, B. A. "Approximate is better than 'exact' for interval estimation of binomial proportions." *The American Statistician*, Vol. 52, No. 2, May 1998, pp. 119–126.

Cochran, W. G. *Sampling Techniques*, 3rd ed. New York: Wiley, 1977.

Freedman, D., Pisani, R., and Purves, R. *Statistics*. New York: Norton, 1978.

Kish, L. *Survey Sampling*. New York: Wiley, 1965.

Mendenhall, W., Beaver, R. J., and Beaver, B. *Introduction to Probability and Statistics*, 13th ed. Belmont, CA: Brooks/Cole, 2009.

Wilson, E. G. "Probable inference, the law of succession, and statistical inference." *Journal of the American Statistical Association*, Vol. 22, 1927, pp. 209–212.

USING TECHNOLOGY

MINITAB: Confidence Intervals

MINITAB can be used to obtain one-sample confidence intervals for a population mean, a population proportion, and a population variance.

Confidence Interval for a Mean

Step 1 Access the MINITAB data worksheet that contains the quantitative variable of interest.

Step 2 Click on the "Stat" button on the MINITAB menu bar and then click on "Basic Statistics" and "1-Sample t," as shown in Figure 5.M.1.

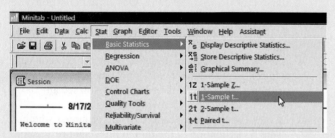

Figure 5.M.1 MINITAB menu options—confidence interval for the mean

Step 3 On the resulting dialog box (shown in Figure 5.M.2), click on "Samples in Columns," and then specify the quantitative variable of interest in the open box.

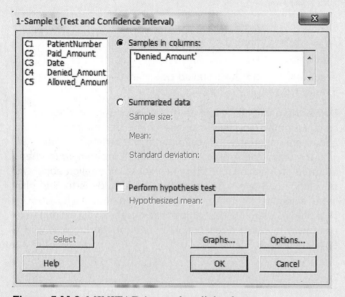

Figure 5.M.2 MINITAB 1-sample t dialog box

Step 4 Click on the "Options" button at the bottom of the dialog box and specify the confidence level in the resulting dialog box, as shown in Figure 5.M.3.

Step 5 Click "OK" to return to the "1-Sample t" dialog box, and then click "OK" again to produce the confidence interval.

Figure 5.M.3 MINITAB 1-sample t options

Note: If you want to produce a confidence interval for the mean from summary information (e.g., the sample mean, sample standard deviation, and sample size), click on "Summarized data" in the "1-Sample t" dialog box, as shown in Figure 5.M.4. Enter the values of the summary statistics and then click "OK."

Figure 5.M.4 MINITAB 1-sample t using summary statistics

Important: The MINITAB 1-sample t procedure uses the *t*-statistic to generate the confidence interval. When the sample size *n* is small, this is the appropriate method. When the sample size *n* is large, the *t*-value will be approximately equal to the large-sample *z*-value and the resulting interval will still be valid. If you have a large sample and you know the value of the population standard deviation σ (which is rarely the case), select "1-sample Z" from the "Basic Statistics" menu options (see Figure 5.M.1) and make the appropriate selections.

Confidence Interval for a Proportion

Step 1 Access the MINITAB data worksheet that contains the qualitative variable of interest.

Step 2 Click on the "Stat" button on the MINITAB menu bar and then click on "Basic Statistics" and "1-Proportion."

Step 3 On the resulting dialog box (shown in Figure 5.M.5), click on "Samples in Columns" and then specify the qualitative variable of interest in the open box.

Figure 5.M.5 MINITAB 1 proportion dialog box

Step 4 Click on the "Options" button at the bottom of the dialog box and specify the confidence level in the resulting dialog box, as shown in Figure 5.M.6. Also, check the "Use test and interval based on normal distribution" box at the bottom.

Figure 5.M.6 MINITAB 1 proportion options

Step 5 Click "OK" to return to the "1-Proportion" dialog box and then click "OK" again to produce the confidence interval.

Note: If you want to produce a confidence interval for a proportion from summary information (e.g., the number of successes and the sample size), click on "Summarized data" in the "1-Proportion" dialog box (see Figure 5.M.5). Enter the value for the number of trials (i.e., the sample size) and the number of events (i.e., the number of successes) and then click "OK."

Confidence Interval for a Variance

Step 1 Access the MINITAB data worksheet that contains the quantitative variable of interest.

Step 2 Click on the "Stat" button on the main MINITAB menu bar, then click on "Basic Statistics" and "1 Variance."

Step 3 On the resulting dialog box (shown in Figure 5.M.7), select "Samples in Columns" in the "Data" box, then specify the quantitative variable of interest in the "Columns" box.

Step 4 Click the "Options" button at the bottom of the dialog box and specify the confidence level in the resulting menu, as shown in Figure 5.M.8.

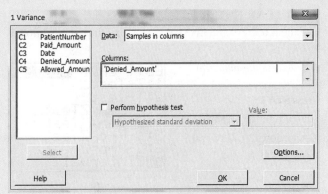

Figure 5.M.7 MINITAB 1 variance dialog box

Figure 5.M.8 MINITAB 1 variance options

Step 5 Click "OK" to return to the "1 Variance" dialog box and then click "OK" again to produce the confidence interval.

Note: If you want to produce a confidence interval for a variance from summary information (e.g., the sample variance), select "Sample variance" in the "Data" box of the "1 Variance" dialog box (see Figure 5.M.9). Enter the values of the sample size and sample variance in the appropriate boxes, then click "OK."

Figure 5.M.9 MINITAB 1 variance dialog box with summary statistics option

TI-83/TI-84 Plus Graphing Calculator: Confidence Intervals

Creating a Confidence Interval for a Population Mean (Known σ or n $\geq$ 30)

Step 1 *Enter the data (Skip to Step 2 if you have summary statistics, not raw data.)*

- Press **STAT** and select **1:Edit**

Note: If the list already contains data, clear the old data. Use the up **ARROW** to highlight **"L1."**

- Press **CLEAR ENTER**
- Use the **ARROW** and **ENTER** kys to enter the data set into **L1**

Step 2 *Access the statistical tests menu*

- Press **STAT**
- Arrow right to **TESTS**
- Arrow down to **Zinterval**
- Press **ENTER**

Step 3 *Choose "Data," or "Stats" ("Data" is selected when you have entered the raw data into a List. "Stats" is selected when you are given only the mean, standard deviation, and sample size.)*

- Press **ENTER**

  If you selected "Data," enter a value for σ. (The best approximation is s, the sample standard deviation.)

- Set **List** to **L1**
- Set **Freq** to **1**
- Set **C-Level** to the confidence level
- Arrow down to "**Calculate**"
- Press **ENTER**

If you selected "Stats," enter a value for σ. (The best approximation is s, the sample standard deviation.)

- Enter the sample mean and sample size
- Set **C-Level** to the confidence level
- Arrow down to "**Calculate**"
- Press **ENTER**

(The screen at the right is set up for an example with a standard deviation of 20, a mean of 200, and a sample size of 40.) The confidence interval will be displayed along with the sample mean and the sample size.

Creating a Confidence Interval for a Population Mean (n < 30)

Step 1 *Enter the data (Skip to Step 2 if you have summary statistics, not raw data.)*

- Press **STAT** and select **1:Edit**

Note: If the list already contains data, clear the old data. Use the up **ARROW** to highlight "**L1**."

- Press **CLEAR ENTER**
- Use the **ARROW** and **ENTER** keys to enter the data set into **L1**

Step 2 *Access the statistical tests menu*

- Press **STAT**

- Arrow right to **TESTS**
- Arrow down to **TInterval**
- Press **ENTER**

Step 3 *Choose "**Data**" or "**Stats**" ("Data" is selected when you have entered the raw data into a List. "Stats" is selected when you are given only the mean, standard deviation, and sample size.)*

- Press **ENTER**
- If you selected "Data," set **List** to **L1**
- Set **Freq** to **1**
- Set **C-Level** to the confidence level
- Arrow down to "**Calculate**"
- Press **ENTER**

  If you selected "Stats" enter the mean, standard deviation, and sample size.

- Set **C-Level** to the confidence level
- Arrow down to "**Calculate**"
- Press **ENTER** (The screen here is set up for an example with a mean of 100 and a standard deviation of 10.)

  The confidence interval will be displayed with the mean, standard deviation, and sample size.

Creating a Confidence Interval for a Population Proportion (Large Samples)

Step 1 *Access the statistical tests menu*

- Press **STAT**
- Arrow right to **TESTS**
- Arrow down to **1-PropZInt**
- Press **ENTER**

Step 2 *Enter the values for* **x, n**, *and* **C-Level**

- where **x** = number of successes
- **n** = sample size
- **C-Level** = level of confidence
- Arrow down to "**Calculate**"
- Press **ENTER**

(The screens at the right are set up for an example with $x = 532$, $n = 1,100$, and confidence level of .95.)

Note: A confidence interval for a population variance is not available in the TI-83/T1–84 plus graphing calculator.

6 Inferences Based on a Single Sample *Tests of Hypothesis*

CONTENTS

Where We've Been

- Used sample information to provide a *point estimate* of a population parameter
- Used the sampling distribution of a statistic to assess the reliability of an estimate through a *confidence interval*

Where We're Going

- Introduce the concepts of a *test of hypothesis* (6.1–6.2)
- Provide a measure of reliability for the hypothesis test—called the *significance level* of the test (6.2, 6.4)
- Test a specific value of a population parameter (mean, proportion or variance) (6.3, 6.5–6.6, 6.7–6.8)

In 1924, Kimberly-Clark Corporation invented a facial tissue for removing cold cream and began marketing it as KLEENEX® brand tissues. Today, KLEENEX® is recognized as the top-selling brand of tissue in the world. A wide variety of KLEENEX® products is available, ranging from extra-large tissues to tissues with lotion. Over the past 80 years, Kimberly-Clark Corporation has packaged the tissues in boxes of different sizes and shapes and varied the number of tissues packaged in each box. For example, currently a family-size box contains 144 two-ply tissues, a cold-care box contains 70 tissues (coated with lotion), and a convenience pocket pack contains 15 miniature tissues.

How does Kimberly-Clark Corp. decide how many tissues to put in each box? According to the *Wall Street Journal*, marketing experts at the company use the results of a survey of KLEENEX® customers to help determine how many tissues are packed in a box. In the mid-1980s, when Kimberly-Clark Corp. developed the cold-care box, designed especially for people who have a cold, the company conducted its initial survey of customers for this purpose. Hundreds of customers were asked to keep count of their KLEENEX® use in diaries. According to the *Wall Street Journal* report, the survey results left "little doubt that the company should put 60 tissues in each box." The number 60 was "the average number of times people blow their nose during a cold." In 2000, the company increased the number of tissues packaged in a cold-care box to 70 based on the results of a more recent survey.

From summary information provided in the *Wall Street Journal* article, we constructed a data set that represents the results of a survey similar to the one just described. In the data file named **TISSUES**, we recorded the number of tissues used by each of 250 consumers during a period when they had a cold. We apply the hypothesis-testing methodology presented in this chapter to that data set in several Statistics in Action Revisited examples.

Statistics IN Action Revisited

- Identifying the Key Elements of a Hypothesis Test Relevant to the KLEENEX® Survey (p. 303)

- Testing a Population Mean in the KLEENEX® Survey (p. 315)

- Testing a Population Proportion in the KLEENEX® Survey (p. 328)

Data Set: TISSUES

Suppose you wanted to determine whether the mean level of a driver's blood alcohol exceeds the legal limit after two drinks, or whether the majority of registered voters approve of the president's performance. In both cases, you are interested in making an inference about how the value of a parameter relates to a specific numerical value. Is it less than, equal to, or greater than the specified number? This type of inference, called a **test of hypothesis,** is the subject of this chapter.

We introduce the elements of a test of hypothesis in Sections 6.1 and 6.2. We then show how to conduct tests of hypothesis about a population mean in Sections 6.3–6.5. Large-sample tests about binomial probabilities are the subject of Section 6.6, and we show how to conduct a test about a population variance in optional Section 6.7. Finally, a nonparametric test for a population median is covered in optional Section 6.8.

6.1 The Elements of a Test of Hypothesis

Suppose building specifications in a certain city require that the average breaking strength of residential sewer pipe be more than 2,400 pounds per foot of length (i.e., per linear foot). Each manufacturer who wants to sell pipe in that city must demonstrate that its product meets the specification. Note that we are interested in making an inference about the mean μ of a population. However, in this example we are less interested in estimating the value of μ than we are in testing a **hypothesis** about its value—that is, *we want to decide whether the mean breaking strength of the pipe exceeds 2,400 pounds per linear foot.*

> A statistical **hypothesis** is a statement about the numerical value of a population parameter.

The method used to reach a decision is based on the rare-event concept explained in earlier chapters. We define two hypotheses: (1) The **null hypothesis** represents the status quo to the party performing the sampling experiment—the hypothesis that will be accepted unless the data provide convincing evidence that it is false. (2) The **alternative, or research, hypothesis** is that which will be accepted only if the data provide convincing evidence of its truth. From the point of view of the city conducting the tests, the null hypothesis is that the manufacturer's pipe does *not* meet specifications unless the tests provide convincing evidence otherwise. The null and alternative hypotheses are therefore

Null hypothesis(H_0): $\mu \leq 2{,}400$

(i.e., the manufacturer's pipe does not meet specifications)

Alternative (research) hypothesis(H_a): $\mu > 2{,}400$

(i.e., the manufacturer's pipe meets specifications)

> The **null hypothesis,** denoted H_0, represents the hypothesis that will be accepted unless the data provide convincing evidence that it is false. This usually represents the "status quo" or some claim about the population parameter that the researcher wants to test.

> The **alternative (research) hypothesis,** denoted H_a, represents the hypothesis that will be accepted only if the data provide convincing evidence of its truth. This usually represents the values of a population parameter for which the researcher wants to gather evidence to support.

How can the city decide when enough evidence exists to conclude that the manufacturer's pipe meets specifications? Because the hypotheses concern the value of the population mean μ, it is reasonable to use the sample mean $\bar{x}$ to make the inference, just as we did when we formed confidence intervals for μ in Sections 5.2 and 5.3. The city will conclude that the pipe meets specifications only when the sample mean $\bar{x}$ convincingly indicates that the population mean exceeds 2,400 pounds per linear foot.

"Convincing" evidence in favor of the alternative hypothesis will exist when the value of $\bar{x}$ exceeds 2,400 by an amount that cannot be readily attributed to sampling variability. To decide, we compute a **test statistic,** i.e., a numerical value computed from the sample. Here, the test statistic is the z-value that measures the distance between the value of $\bar{x}$ and the value of μ specified in the alternative hypothesis. When the null hypothesis contains more than one values of μ, as in this case $(H_0: \mu \leq 2{,}400)$, we use the value of μ closest to the values specified in the alternative hypothesis. The idea is that if the hypothesis that μ *equals* 2,400 can be rejected in favor of $\mu > 2{,}400$, then μ *less than or equal to* 2,400 can certainly be rejected. Thus, the test statistic is

$$z = \frac{\bar{x} - 2{,}400}{\sigma_{\bar{x}}} = \frac{\bar{x} - 2{,}400}{\sigma/\sqrt{n}}$$

Note that a value of $z = 1$ means that $\bar{x}$ is 1 standard deviation above $\mu = 2{,}400$, a value of $z = 1.5$ means that $\bar{x}$ is 1.5 standard deviations above $\mu = 2{,}400$, and so on. How large must z be before the city can be convinced that the null hypothesis can be rejected in favor of the alternative and conclude that the pipe meets specifications?

> The **test statistic** is a sample statistic, computed from information provided in the sample, that the researcher uses to decide between the null and alternative hypotheses.

If you examine Figure 6.1, you will note that the chance of observing $\bar{x}$ more than 1.645 standard deviations above 2,400 is only .05—*if in fact the true mean μ is 2,400.* Thus, if the sample mean is more than 1.645 standard deviations above 2,400, either H_0

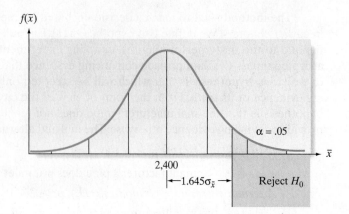

Figure 6.1
The sampling distribution of $\bar{x}$, assuming $\mu = 2,400$

is true and a relatively rare event has occurred (.05 probability), or H_a is true and the population mean exceeds 2,400. Because we would most likely reject the notion that a rare event has occurred, we would reject the null hypothesis ($\mu \leq 2,400$) and conclude that the alternative hypothesis ($\mu > 2,400$) is true. What is the probability that this procedure will lead us to an incorrect decision?

Such an incorrect decision—deciding that the null hypothesis is false when in fact it is true—is called a **Type I error.** As indicated in Figure 6.1, the risk of making a Type I error is denoted by the symbol α—that is,

$$\alpha = P \text{ (Type I error)}$$
$$= P \text{ (Rejecting the null hypothesis when in fact the null hypothesis is true)}$$

> A **Type I error** occurs if the researcher rejects the null hypothesis in favor of the alternative hypothesis when, in fact, H_0 is true. The probability of committing a Type I error is denoted by α.

In our example,

$$\alpha = P \text{ (}z > 1.645 \text{ when in fact } \mu = 2,400) = .05$$

We now summarize the elements of the test:

$H_0: \mu \leq 2,400$ (Pipe does not meet specifications.)
$H_a: \mu > 2,400$ (Pipe meets specifications.)

Test statistic: $z = \dfrac{\bar{x} - 2,400}{\sigma_{\bar{x}}}$

Rejection region: $z > 1.645$, which corresponds to $\alpha = .05$

Note that the **rejection region** refers to the values of the test statistic for which we will *reject the null hypothesis.*

> The **rejection region** of a statistical test is the set of possible values of the test statistic for which the researcher will reject H_0 in favor of H_a.

To illustrate the use of the test, suppose we test 50 sections of sewer pipe and find the mean and standard deviation for these 50 measurements to be

$$\bar{x} = 2,460 \text{ pounds per linear foot}$$
$$s = 200 \text{ pounds per linear foot}$$

As in the case of estimation, we can use s to approximate σ when s is calculated from a large set of sample measurements.

The test statistic is

$$z = \frac{\bar{x} - 2{,}400}{\sigma_{\bar{x}}} = \frac{\bar{x} - 2{,}400}{\sigma/\sqrt{n}} \approx \frac{\bar{x} - 2{,}400}{s/\sqrt{n}}$$

Substituting $\bar{x} = 2{,}460$, $n = 50$, and $s = 200$, we have

$$z \approx \frac{2{,}460 - 2{,}400}{200/\sqrt{50}} = \frac{60}{28.28} = 2.12$$

Therefore, the sample mean lies $2.12\sigma_{\bar{x}}$ above the hypothesized value of μ, 2,400, as shown in Figure 6.2. Because this value of z exceeds 1.645, it falls into the rejection region. That is, we reject the null hypothesis that $\mu = 2{,}400$ and conclude that $\mu > 2{,}400$. Thus, it appears that the company's pipe has a mean strength that exceeds 2,400 pounds per linear foot.

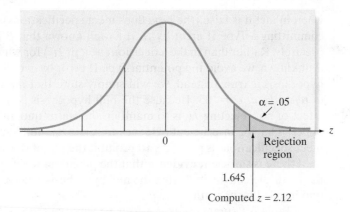

Figure 6.2

Location of the test statistic for a test of the hypothesis H_0: $\mu = 2{,}400$

EGON S. PEARSON (1895–1980)

The Neyman-Pearson Lemma

Egon Pearson was the only son of noteworthy British statistician Karl Pearson (see Biography, p. 441). As you might expect, Egon developed an interest in the statistical methods developed by his father and, upon completing graduate school, accepted a position to work for Karl in the Department of Applied Statistics at University College, London. Egon is best known for his collaboration with Jerzey Neyman (see Biography, p. 248) on the development of the theory of hypothesis testing. One of the basic concepts in the Neyman-Pearson approach was that of the "null" and "alternative" hypotheses. Their famous Neyman-Pearson lemma was published in *Biometrika* in 1928. Egon Pearson had numerous other contributions to statistics and was known as an excellent teacher and lecturer. In his last major work, Egon fulfilled a promise made to his father by publishing an annotated version of Karl Pearson's lectures on the early history of statistics. ∎

How much faith can be placed in this conclusion? What is the probability that our statistical test could lead us to reject the null hypothesis (and conclude that the company's pipe meets the city's specifications) when in fact the null hypothesis is true? The answer is $\alpha = .05$—that is, we selected the level of risk, α, of making a Type I error when we constructed the test. Thus, the chance is only 1 in 20 that our test would lead us to conclude the manufacturer's pipe satisfies the city's specifications when in fact the pipe does *not* meet specifications.

Now, suppose the sample mean breaking strength for the 50 sections of sewer pipe turned out to be $\bar{x} = 2{,}430$ pounds per linear foot. Assuming that the sample standard deviation is still $s = 200$, the test statistic is

$$z = \frac{2{,}430 - 2{,}400}{200/\sqrt{50}} = \frac{30}{28.28} = 1.06$$

Therefore, the sample mean $\bar{x} = 2{,}430$ is only 1.06 standard deviations above the null hypothesized value of $\mu = 2{,}400$. As shown in Figure 6.3, this value does not fall into the rejection region ($z > 1.645$). Therefore, we know that we cannot reject H_0 using $\alpha = .05$. Even though the sample mean exceeds the city's specification of 2,400 by 30 pounds per linear foot, it does not exceed the specification by enough to provide *convincing* evidence that the *population mean* exceeds 2,400.

A **Type II error** occurs if the researcher accepts the null hypothesis when, in fact, H_0 is false. The probability of committing a Type II error is denoted by β.

Should we accept the null hypothesis H_0: $\mu \leq 2{,}400$ and conclude that the manufacturer's pipe does not meet specifications? To do so would be to risk a **Type II error**— that of concluding that the null hypothesis is true (the pipe does not meet specifications)

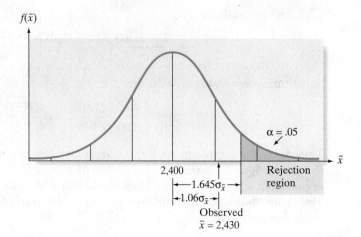

Figure 6.3
Location of test statistic
when $\bar{x} = 2{,}430$

when in fact it is false (the pipe does meet specifications). We denote the probability of committing a Type II error by β. It is well known that β is often difficult to determine precisely. Rather than make a decision (accept H_0) for which the probability of error (β) is unknown, we avoid the potential Type II error by avoiding the conclusion that the null hypothesis is true. Instead, we will simply state that *the sample evidence is insufficient to reject H_0 at $\alpha = .05$*. Because the null hypothesis is the "status-quo" hypothesis, the effect of not rejecting H_0 is to maintain the status quo. In our pipe-testing example, the effect of having insufficient evidence to reject the null hypothesis that the pipe does not meet specifications is probably to prohibit the use of the manufacturer's pipe unless and until there is sufficient evidence that the pipe does meet specifications—that is, until the data indicate convincingly that the null hypothesis is false, we usually maintain the status quo implied by its truth.

Table 6.1 summarizes the four possible outcomes (i.e., **conclusions**) of a test of hypothesis. The "true state of nature" columns in Table 6.1 refer to the fact that either the null hypothesis H_0 is true or the alternative hypothesis H_a is true. Note that the true state of nature is unknown to the researcher conducting the test. The "decision" rows in Table 6.1 refer to the action of the researcher, assuming that he or she will either conclude that H_0 is true or that H_a is true, based on the results of the sampling experiment. Note that a Type I error can be made *only* when the null hypothesis is rejected in favor of the alternative hypothesis, and a Type II error can be made *only* when the null hypothesis is accepted. Our policy will be to make a decision only when we know the probability of making the error that corresponds to that decision. Because α is usually specified by the analyst, we will generally be able to reject H_0 (accept H_a) when the sample evidence supports that decision. However, because β is usually not specified, *we will generally avoid the decision to accept H_0, preferring instead to state that the sample evidence is insufficient to reject H_0 when the test statistic is not in the rejection region.*

Table 6.1	Conclusions and Consequences for a Test of Hypothesis	
	True State of Nature	
Conclusion	H_0 True	H_a True
Accept H_0 (Assume H_0 True)	Correct decision	Type II error (probability β)
Reject H_0 (Assume H_a True)	Type I error (probability α)	Correct decision

⚠ **CAUTION** Be careful not to "accept H_0" when conducting a test of hypothesis because the measure of reliability, $\beta = P$ (Type II error), is almost always unknown. If the test statistic does not fall into the rejection region, it is better to state the conclusion as "insufficient evidence to reject H_0."* ▲

*In many practical business applications of hypothesis testing, nonrejection leads management to behave as if the null hypothesis were accepted. Accordingly, the distinction between acceptance and nonrejection is frequently blurred in practice.

The elements of a test of hypothesis are summarized in the following box. Note that the first four elements are all specified *before* the sampling experiment is performed. In no case will the results of the sample be used to determine the hypotheses; the data are collected to test the predetermined hypotheses, not to formulate them.

Elements of a Test of Hypothesis

1. *Null hypothesis* (H_0): A theory about the specific values of one or more population parameters. The theory generally represents the status quo, which we adopt until it is proven false. The theory is always stated as H_0: parameter $=$ value.

2. *Alternative (research) hypothesis* (H_a): A theory that contradicts the null hypothesis. The theory generally represents that which we will adopt only when sufficient evidence exists to establish its truth.

3. *Test statistic:* A sample statistic used to decide whether to reject the null hypothesis.

4. *Rejection region:* The numerical values of the test statistic for which the null hypothesis will be rejected. The rejection region is chosen so that the probability is α that it will contain the test statistic when the null hypothesis is true, thereby leading to a Type I error. The value of α is usually chosen to be small (e.g., .01, .05, or .10) and is referred to as the **level of significance** of the test.

5. *Assumptions:* Clear statement(s) of any assumptions made about the population(s) being sampled.

6. *Experiment and calculation of test statistic:* Performance of the sampling experiment and determination of the numerical value of the test statistic.

7. *Conclusion:*

   a. If the numerical value of the test statistic falls in the rejection region, we reject the null hypothesis and conclude that the alternative hypothesis is true. We know that the hypothesis-testing process will lead to this conclusion incorrectly (Type I error) only $100\alpha\%$ of the time when H_0 is true.

   b. If the test statistic does not fall in the rejection region, we do not reject H_0. Thus, we reserve judgment about which hypothesis is true. We do not conclude that the null hypothesis is true because we do not (in general) know the probability β that our test procedure will lead to an incorrect acceptance of H_0 (Type II error).

As with confidence intervals, the methodology for testing hypotheses varies depending on the target population parameter. In this chapter, we develop methods for testing a population mean, a population proportion, and (optionally) a population variance. As a reminder, the key words and the type of data associated with these target parameters are again listed in the accompanying box.

Determining the Target Parameter

Parameter	Key Words or Phrases	Type of Data
μ	Mean; average	Quantitative
p	Proportion; percentage; fraction; rate	Qualitative
σ^2	Variance; variability; spread	Quantitative

6.2 Formulating Hypotheses and Setting Up the Rejection Region

In Section 6.1 we learned that the null and alternative hypotheses form the basis for inference using a test of hypothesis. The null and alternative hypotheses may take one of several forms. In the sewer pipe example, we tested the null hypothesis that the population mean strength of the pipe is less than or equal to 2,400 pounds per linear foot against the alternative hypothesis that the mean strength exceeds 2,400—that is, we tested

$$H_0: \mu \leq 2,400 \quad \text{(Pipe does not meet specifications.)}$$
$$H_a: \mu > 2,400 \quad \text{(Pipe meets specifications.)}$$

This is a **one-tailed** (or **one-sided**) **statistical test** because the alternative hypothesis specifies that the population parameter (the population mean μ in this example) is strictly greater than a specified value (2,400 in this example). If the null hypothesis had been $H_0: \mu \geq 2,400$ and the alternative hypothesis had been $H_a: \mu < 2,400$, the test would still be one-sided because the parameter is still specified to be on "one side" of the null hypothesis value. Some statistical investigations seek to show that the population parameter is *either larger or smaller* than some specified value. Such an alternative hypothesis is called a **two-tailed** (or **two-sided**) **hypothesis.**

While alternative hypotheses are always specified as strict inequalities, such as $\mu < 2,400$, $\mu > 2,400$, or $\mu \neq 2,400$, *null hypotheses are usually specified as equalities*, such as $\mu = 2,400$. Even when the null hypothesis is an inequality, such as $\mu \leq 2,400$, we specify $H_0: \mu = 2,400$, reasoning that if sufficient evidence exists to show that $H_a: \mu > 2,400$ is true when tested against $H_0: \mu = 2,400$, then surely sufficient evidence exists to reject $\mu < 2,400$ as well. Therefore, the null hypothesis is specified as the value of μ closest to a one-sided alternative hypothesis and as the only value *not* specified in a two-tailed alternative hypothesis. The steps for selecting the null and alternative hypotheses are summarized in the following box.

Steps for Selecting the Null and Alternative Hypotheses

1. Select the *alternative hypothesis* as that which the sampling experiment is intended to establish. The alternative hypothesis will assume one of three forms:
   a. One-tailed, **upper-tailed**   (e.g., $H_a: \mu > 2,400$)
   b. One-tailed, **lower-tailed**   (e.g., $H_a: \mu < 2,400$)
   c. Two-tailed   (e.g., $H_a: \mu \neq 2,400$)

2. Select the *null hypothesis* as the status quo, that which will be presumed true unless the sampling experiment conclusively establishes the alternative hypothesis. The null hypothesis will be specified as that parameter value closest to the alternative in one-tailed tests and as the complementary (or only unspecified) value in two-tailed tests.

$$\text{(e.g., } H_0: \mu = 2,400)$$

A **one-tailed test** of hypothesis is one in which the alternative hypothesis is directional and includes the symbol "<" or " > ."

A **two-tailed test** of hypothesis is one in which the alternative hypothesis does not specify departure from H_0 in a particular direction and is written with the symbol " $\neq$."

Example 6.1

Formulating H_0 and H_a for a Test of a Population Mean

Problem A metal lathe is checked periodically by quality control inspectors to determine whether it is producing machine bearings with a mean diameter of .5 inch. If the mean diameter of the bearings is larger or smaller than .5 inch, then the process is out of control and must be adjusted. Formulate the null and alternative hypotheses for a test to determine whether the bearing production process is out of control.

Solution The hypotheses must be stated in terms of a population parameter. Here, we define μ as the true mean diameter (in inches) of all bearings produced by the metal lathe. If either $\mu > .5$ or $\mu < .5$, then the lathe's production process is out of control. Because the inspectors want to be able to detect either possibility (indicating that the process is in need of adjustment), these values of μ represent the alternative (or research) hypothesis. Alternatively, because $\mu = .5$ represents an in-control process (the status quo), this represents the null hypothesis. Therefore, we want to conduct the two-tailed test:

$$H_0: \mu = .5 \text{ (i.e., the process is in control)}$$
$$H_a: \mu \neq .5 \text{ (i.e., the process is out of control)}$$

Look Back Here, the alternative hypothesis is not necessarily the hypothesis that the quality control inspectors desire to support. However, they will make adjustments to the metal lathe settings only if there is strong evidence to indicate that the process is out of control. Consequently, $\mu \neq .5$ must be stated as the alternative hypothesis.

Now Work Exercise 6.11a

Example 6.2

Formulating H_0 and H_a for a Test of a Population Proportion

Problem Cigarette advertisements are required by federal law to carry the following statement: "Warning: The surgeon general has determined that cigarette smoking is dangerous to your health." However, this warning is often located in inconspicuous corners of the advertisements and printed in small type. Suppose the Federal Trade Commission (FTC) claims that 80% of cigarette consumers fail to see the warning. A marketer for a large tobacco firm wants to gather evidence to show that the FTC's claim is too high, i.e., that fewer than 80% of cigarette consumers fail to see the warning. Specify the null and alternative hypotheses for a test of the FTC's claim.

Solution The marketer wants to make an inference about p, the true proportion of all cigarette consumers who fail to see the surgeon general's warning. In particular, the marketer wants to collect data to show that fewer than 80% of cigarette consumers fail to see the warning, i.e., $p < .80$. Consequently, $p < .80$. represents the alternative hypothesis and $p = .80$ (the claim made by the FTC) represents the null hypothesis. That is, the marketer desires the one-tailed (lower-tailed) test:

$$H_0: p = .80 \text{ (i.e., the FTC's claim is true)}$$
$$H_a: p < .80 \text{ (i.e., the FTC's claim is false)}$$

Look Back Whenever a claim is made about the value of a particular population parameter and the researcher wants to test the claim, believing that it is false, the claimed value will represent the null hypothesis.

Now Work Exercise 6.15

The rejection region for a two-tailed test differs from that for a one-tailed test. When we are trying to detect departure from the null hypothesis in *either* direction, we must establish a rejection region in both tails of the sampling distribution of the test statistic. Figures 6.4a and 6.4b show the one-tailed rejection regions for lower- and

Figure 6.4
Rejection regions corresponding to one- and two-tailed tests

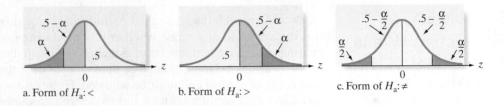

a. Form of H_a: <   b. Form of H_a: >   c. Form of H_a: ≠

upper-tailed tests, respectively. The two-tailed rejection region is illustrated in Figure 6.4c. Note that a rejection region is established in each tail of the sampling distribution for a two-tailed test.

The rejection regions corresponding to typical values selected for α are shown in Table 6.2 for one- and two-tailed tests. Note that the smaller α you select, the more evidence (the larger z) you will need before you can reject H_0.

Table 6.2	Rejection Regions for Common Values of α		
	Alternative Hypotheses		
	Lower-Tailed	Upper-Tailed	Two-Tailed
$\alpha = .10$	$z < -1.28$	$z > 1.28$	$z < -1.645$ or $z > 1.645$
$\alpha = .05$	$z < -1.645$	$z > 1.645$	$z < -1.96$ or $z > 1.96$
$\alpha = .01$	$z < -2.33$	$z > 2.33$	$z < -2.575$ or $z > 2.575$

Example 6.3

Setting Up a Hypothesis Test For μ—Mean Drug Response Time

Problem The effect of drugs and alcohol on the nervous system has been the subject of considerable research. Suppose a research neurologist is testing the effect of a drug on response time by injecting 100 rats with a unit dose of the drug, subjecting each rat to a neurological stimulus, and recording its response time. The neurologist knows that the mean response time for rats not injected with the drug (the "control" mean) is 1.2 seconds. She wishes to test whether the mean response time for drug-injected rats differs from 1.2 seconds. Set up the test of hypothesis for this experiment, using $\alpha = .01$.

Solution The key word *mean* in the statement of the problem implies that the target parameter is μ, the mean response time for all drug-injected rats. Since the neurologist wishes to detect whether μ differs from the control mean of 1.2 seconds in *either* direction—that is, $\mu < 1.2$ or $\mu > 1.2$—we conduct a two-tailed statistical test. Following the procedure for selecting the null and alternative hypotheses, we specify as the alternative hypothesis that the mean differs from 1.2 seconds, since determining whether the drug-injected mean differs from the control mean is the purpose of the experiment. The null hypothesis is the presumption that drug-injected rats have the same mean response time as control rats unless the research indicates otherwise. Thus,

H_0: $\mu = 1.2$ (Mean response time is 1.2 seconds)

H_a: $\mu \neq 1.2$ (Mean response time is less than 1.2 or greater than 1.2 seconds)

The test statistic measures the number of standard deviations between the observed value of $\bar{x}$ and the null-hypothesized value $\mu = 1.2$:

$$\text{Test statistic:} \quad z = \frac{\bar{x} - 1.2}{\sigma_{\bar{x}}}$$

The rejection region must be designated to detect a departure from $\mu = 1.2$ in *either* direction, so we will reject H_0 for values of z that are either too small (negative) or too large (positive). To determine the precise values of z that constitute the rejection region, we first select α, the probability that the test will lead to incorrect rejection of the null hypothesis. Then we divide α equally between the lower and upper tail of the distribution of z, as shown in Figure 6.5. In this example, $\alpha = .01$, so $\alpha/2 = .005$ is placed in each tail. The areas in the tails correspond to $z = -2.575$ and $z = 2.575$, respectively (from Table 6.2), so

$$\textit{Rejection region: } z < -2.575 \textit{ or } z > 2.575 \qquad \text{(see Figure 6.5)}$$

Assumptions: Since the sample size of the experiment is large enough ($n > 30$), the Central Limit Theorem will apply, and no assumptions need be made about the population of response time measurements. The sampling distribution of the sample mean response of 100 rats will be approximately normal, regardless of the distribution of the individual rats' response times.

Figure 6.5
Two-tailed rejection region:
$\alpha = .01$

Look Back Note that the test is set up *before* the sampling experiment is conducted. The data are not used to develop the test. Evidently, the neurologist wants to conclude that the mean response time for the drug-injected rats differs from the control mean only when the evidence is very convincing, because the value of α has been set quite low at .01. If the experiment results in the rejection of H_0, she can confidently conclude that the mean response time of the drug-injected rats differs from the control mean because there is only a .01 probability of a Type I error.

Now Work Exercise 6.11b and c

Once the test is set up, the neurologist of Example 6.3 is ready to perform the sampling experiment and conduct the test. The hypothesis test is performed in the next section.

Statistics IN Action | **Revisited** | ### Identifying the Key Elements of a Hypothesis Test Relevant to the KLEENEX® Survey

In Kimberly-Clark Corporation's survey of people with colds, each of 250 customers was asked to keep count of his or her use of KLEENEX® tissues in diaries. One goal of the company was to determine how many tissues to package in a cold-care box of KLEENEX®; consequently, the total number of tissues used was recorded for each person surveyed. Since number of tissues is a quantitative variable, the parameter of interest is either μ, the mean number of tissues used by all customers with colds, or σ^2, the variance of the number of tissues used.

Recall that in 2000, the company increased from 60 to 70 the number of tissues it packages in a cold-care box of KLEENEX® tissues. This decision was based on

a claim made by marketing experts that the average number of times a person will blow his or her nose during a cold exceeds the previous mean of 60. The key word *average* implies that the target parameter is μ, and the marketers are claiming that $\mu > 60$. In order to test the claim we set up the following null and alternative hypotheses:

$$H_0: \mu = 60 \qquad H_a: \mu > 60$$

We'll conduct this test in the next Statistics in Action Revisited, on p. 315.

Exercises 6.1–6.19

Understanding the Principles

6.1 Which hypothesis, the null or the alternative, is the status quo hypothesis? Which is the research hypothesis?

6.2 Which element of a test of hypothesis is used to decide whether to reject the null hypothesis in favor of the alternative hypothesis?

6.3 What is the level of significance of a test of hypothesis?

6.4 What is the difference between Type I and Type II errors in hypothesis testing? How do α and β relate to Type I and Type II errors?

6.5 List the four possible results of the combinations of decisions and true states of nature for a test of hypothesis.

6.6 We (generally) reject the null hypothesis when the test statistic falls into the rejection region, but we do not accept the null hypothesis when the test statistic does not fall into the rejection region. Why?

6.7 If you test a hypothesis and reject the null hypothesis in favor of the alternative hypothesis, does your test prove that the alternative hypothesis is correct? Explain.

Learning the Mechanics

6.8 Consider a test of $H_0: \mu = 4$. In each of the following cases, give the rejection region for the test in terms of the z-statistic:
  a. $H_a: \mu > 4, \alpha = .05$
  b. $H_a: \mu > 4, \alpha = .10$
  c. $H_a: \mu < 4, \alpha = .05$
  d. $H_a: \mu \neq 4, \alpha = .05$

6.9 For each of the following rejection regions, sketch the sampling distribution for z and indicate the location of the rejection region.
  a. $z > 1.96$
  b. $z > 1.645$
  c. $z > 2.575$
  d. $z < -1.28$
  e. $z < -1.645$ or $z > 1.645$
  f. $z < -2.575$ or $z > 2.575$
  g. For each of the rejection regions specified in parts **a–f**, what is the probability that a Type I error will be made?

Applet Exercise 6.1

Use the applet entitled *Hypotheses Test for a Mean* to investigate the frequency of Type I and Type II errors. For this exercise, use $n = 100$ and the normal distribution with mean 50 and standard deviation 10.

  a. Set the null mean equal to 50 and the alternative to *not equal*. Run the applet one time. How many times was the null hypothesis rejected at level .05? In this case, the null hypothesis is true. Which type of error occurred each time the true null hypothesis was rejected? What is the probability of rejecting a true null hypothesis at level .05? How does the proportion of times the null hypothesis was rejected compare with this probability?

  b. Clear the applet, then set the null mean equal to 47, and keep the alternative at *not equal*. Run the applet one time. How many times was the null hypothesis *not* rejected at level .05? In this case, the null hypothesis is false. Which type of error occurred each time the

null hypothesis was *not* rejected? Run the applet several more times without clearing. Based on your results, what can you conclude about the probability of failing to reject the null hypothesis for the given conditions?

Applying the Concepts—Basic

6.10 **Infants' listening time.** Researchers writing in *Analysis of Verbal Behavior* (Dec. 2007) reported that the mean listening time of 16-month-old infants exposed to nonmeaningful monosyllabic words (e.g., "giff," "cham," "gack") is 8 seconds. Set up the null and alternative hypotheses for testing the claim.

6.11 **Play Golf America program.** The Professional Golf Association (PGA) and *Golf Digest* have developed the Play Golf America program, in which teaching professionals at participating golf clubs provide a free 10-minute lesson to new customers. According to *Golf Digest* (July 2008), golf facilities that participate in the program gain, on average, \$2,400 in green fees, lessons, or equipment expenditures. A teaching professional at a golf club believes that the average gain in green fees, lessons, or equipment expenditures for participating golf facilities exceeds \$2,400.
  a. In order to support the claim made by the teaching professional, what null and alternative hypothesis should you test?
  b. Suppose you select $\alpha = .05$. Interpret this value in the words of the problem.
  c. For $\alpha = .05$, specify the rejection region of a large-sample test.

6.12 **Calories in school lunches.** A University of Florida economist conducted a study of Virginia elementary school lunch menus. During the state-mandated testing period, school lunches averaged 863 calories (*National Bureau of Economic Research*, November 2002). The economist claimed that after the testing period end, the average caloric content of Virginia school lunches dropped significantly. Set up the null and alternative hypothesis to test the economist's claim.

6.13 **A camera that detects liars.** According to *New Scientist* (January 2, 2002), a new thermal imaging camera that detects small temperature changes is now being used as a polygraph device. The United States Department of Defense Polygraph Institute (DDPI) claims that the camera can detect liars correctly 75% of the time by monitoring the temperatures of their faces. Give the null hypothesis for testing the claim made by the DDPI.

6.14 **Effectiveness of online courses.** The Sloan Survey of Online Learning, "Making the Grade: Online Education in the United States, 2006," reported that 60% of college presidents believe that their online education courses are as good as or superior to courses that utilize traditional face-to-face instruction (*Inside Higher Ed*, Nov. 2006). Give the null hypothesis for testing the claim made by the Sloan Survey.

6.15 **DNA-reading tool for quick identification of species.** A biologist and a zoologist at the University of Florida were the first scientists to test the effectiveness of a high-tech handheld device designed to instantly identify the DNA of an animal species (*PLOS Biology*, Dec. 2005). They used the DNA-reading device on tissue samples collected from mollusks with brightly colored shells. The scientists discovered that the error rate of the device is less than 5 percent. Set up the null and alternative hypotheses if you want to support the findings.

Applying the Concepts—Intermediate

6.16 **Virtual reality hypnosis for pain.** The *International Journal of Clinical and Experimental Hypnosis* (Vol. 58, 2010) investigated using virtual reality hypnosis (VRH) to reduce the pain of trauma patients. Patients reported their pain intensity on the Graphic Rating Scale (GRS) both prior to and one hour after a VRH session. The researchers reported that hypnosis reduced the pain intensity of trauma patients by an average of $\mu = 10$ points on the GRS. Suppose you want to test whether trauma patients who receive normal analgesic care (i.e., medication, but no hypnosis) will have a smaller reduction in average pain intensity than the VRH trauma patients.
   a. Set up H_0 and H_a for the test.
   b. Describe a Type I error for this test.
   c. Describe a Type II error for this test.

6.17 **Mercury levels in wading birds.** According to a University of Florida wildlife ecology and conservation researcher, the average level of mercury uptake in wading birds in the Everglades has declined over the past several years. (*UF News*, December 15, 2000). Ten years ago, the average level was 15 parts per million.
   a. Give the null and alternative hypotheses for testing whether the average level today is less than 15 ppm.
   b. Describe a Type I error for this test.
   c. Describe a Type II error for this test.

Applying the Concepts—Advanced

6.18 **Jury trial outcomes.** Sometimes, the outcome of a jury trial defies the "commonsense" expectations of the general public (e.g., the 1995 O. J. Simpson verdict and the 2011 Casey Anthony verdict). Such a verdict is more acceptable if we understand that the jury trial of an accused murderer is analogous to the statistical hypothesis-testing process. The null hypothesis in a jury trial is that the accused is innocent. (The status quo hypothesis in the U.S. system of justice is innocence, which is assumed to be true until proven *beyond a reasonable doubt*.) The alternative hypothesis is guilt, which is accepted only when sufficient evidence exists to establish its truth. If the vote of the jury is unanimous in favor of guilt, the null hypothesis of innocence is rejected and the court concludes that the accused murderer is guilty. Any vote other than a unanimous one for guilt results in a "not guilty" verdict. The court never accepts the null hypothesis; that is, the court never declares the accused "innocent." A "not guilty" verdict (as in the O. J. Simpson case) implies that the court could not find the defendant guilty *beyond a reasonable doubt*.
   a. Define Type I and Type II errors in a murder trial.
   b. Which of the two errors is the more serious? Explain.
   c. The court does not, in general, know the values of α and β, but ideally, both should be small. One of these probabilities is assumed to be smaller than the other in a jury trial. Which one, and why?
   d. The court system relies on the belief that the value of α is made very small by requiring a unanimous vote before guilt is concluded. Explain why this is so.
   e. For a jury prejudiced against a guilty verdict as the trial begins, will the value of α increase or decrease? Explain.
   f. For a jury prejudiced against a guilty verdict as the trial begins, will the value of β increase or decrease? Explain.

6.19 **Intrusion detection systems.** Refer to the *Journal of Research of the National Institute of Standards and Technology* (November–December 2003) study of a computer intrusion detection system (IDS), presented in Exercise 3.90 (p. 146). Recall that an IDS is designed to provide an alarm whenever unauthorized access (e.g., an intrusion) to a computer system occurs. The probability of the system giving a false alarm (i.e., providing a warning when, in fact, no intrusion occurs) is defined by the symbol α, while the probability of a missed detection (i.e., no warning given, when, in fact, an intrusion occurs) is defined by the symbol β. These symbols are used to represent Type I and Type II error rates, respectively, in a hypothesis-testing scenario.
   a. What is the null hypothesis H_0?
   b. What is the alternative hypothesis H_a?
   c. According to actual data on the EMERALD system collected by the Massachusetts Institute of Technology Lincoln Laboratory, only 1 in 1,000 computer sessions with no intrusions resulted in a false alarm. For the same system, the laboratory found that only 500 of 1,000 intrusions were actually detected. Use this information to estimate the values of α and β.

6.3 Test of Hypothesis about a Population Mean: Normal (z) Statistic

When testing a hypothesis about a population mean μ, the test statistic we use will depend on whether the sample size n is large (say, $n \geq 30$) or small, and whether or not we know the value of the population standard deviation, σ. In this section, we consider the large-sample case.

Because the sample size is large, the Central Limit Theorem guarantees that the sampling distribution of $\bar{x}$ is approximately normal. Consequently, the test statistic for a test based on large samples will be based on the normal z-statistic. Although the z-statistic requires that we know the true population standard deviation σ, this is rarely, if ever, the case. However, we established in Chapter 5 that when n is large, the sample standard deviation s provides a good approximation to σ, and the z-statistic can be approximated as follows:

$$z = \frac{\bar{x} - \mu_0}{\sigma_{\bar{x}}} = \frac{\bar{x} - \mu_0}{\sigma/\sqrt{n}} \approx \frac{\bar{x} - \mu_0}{s/\sqrt{n}}$$

where μ_0 represents the value of μ specified in the null hypothesis.

The setup of a large-sample test of hypothesis about a population mean is summarized in the following boxes. Both the one- and two-tailed tests are shown.

Large-Sample Test of Hypothesis about μ Based on a Normal (z) Statistic

One-Tailed Test	Two-Tailed Test
H_0: $\mu = \mu_0$	H_0: $\mu = \mu_0$
H_a: $\mu < \mu_0$ (or H_a: $\mu > \mu_0$)	H_a: $\mu \neq \mu_0$

Test Statistic (σ Known): $z = \dfrac{\bar{x} - \mu_0}{\sigma/\sqrt{n}}$ *Test statistic (σ Known):* $z = \dfrac{\bar{x} - \mu_0}{\sigma/\sqrt{n}}$

Test Statistic (σ unknown): $z \approx \dfrac{\bar{x} - \mu_0}{s/\sqrt{n}}$ *Test statistic (σ unknown):* $z \approx \dfrac{\bar{x} - \mu_0}{s/\sqrt{n}}$

Rejection region: $z < -z_\alpha$ *Rejection region:* $|z| > z_{\alpha/2}$
      (or $z > z_\alpha$ when H_a: $\mu > \mu_0$)   where $z_{\alpha/2}$ is chosen so that
where z_α is chosen so that $P(|z| > z_{\alpha/2}) = \alpha/2$
        $P(z > z_\alpha) = \alpha$

Note: The symbol for the numerical value assigned to μ under the null hypothesis is μ_0.

Conditions Required for a Valid Large-Sample Hypothesis Test for μ

1. A random sample is selected from the target population.
2. The sample size n is large (i.e., $n \geq 30$). (Due to the Central Limit Theorem, this condition guarantees that the test statistic will be approximately normal regardless of the shape of the underlying probability distribution of the population.)

Once the test has been set up, the sampling experiment is performed and the test statistic calculated. The next box contains possible conclusions for a test of hypothesis, depending on the result of the sampling experiment.

Possible Conclusions for a Test of Hypothesis

1. If the calculated test statistic falls in the rejection region, reject H_0 and conclude that the alternative hypothesis H_a is true. State that you are rejecting H_0 at the α level of significance. Remember that the confidence is in the testing *process*, not the particular result of a single test.
2. If the test statistic does not fall in the rejection region, conclude that the sampling experiment does not provide sufficient evidence to reject H_0 at the α level of significance. [Generally, we will not "accept" the null hypothesis unless the probability β of a Type II error has been estimated.]

Example 6.4

Carrying Out a Hypothesis Test for μ—Mean Drug Response Time

Problem Refer to the neurological response-time test set up in Example 6.3 (p. 302). The sample of 100 drug-injected rats yielded the results (in seconds) shown in Table 6.3. At $\alpha = .01$, use these data to conduct the test of hypothesis,

$$H_0: \mu = 1.2$$
$$H_a: \mu \neq 1.2$$

Table 6.3		Drug Response Times for 100 Rats, Example 6.4							
1.90	2.17	0.61	1.17	0.66	1.86	1.41	1.30	0.70	0.56
2.00	1.27	0.98	1.55	0.64	0.60	1.55	0.93	0.48	0.39
0.86	1.19	0.79	1.37	1.31	0.85	0.71	1.21	1.23	0.89
1.84	0.80	0.64	1.08	0.74	0.93	1.71	1.05	1.44	0.42
0.70	0.54	1.40	1.06	0.54	0.17	0.98	0.89	1.28	0.68
0.98	1.14	1.16	1.64	1.16	1.01	1.09	0.77	1.58	0.99
0.57	0.27	0.51	1.27	1.81	0.88	0.31	0.92	0.93	1.66
0.21	0.79	0.94	0.45	1.19	1.60	0.14	0.99	1.08	1.57
0.55	1.65	0.81	1.00	2.55	1.96	1.31	1.88	1.51	1.48
0.61	0.05	1.21	0.48	1.63	1.45	0.22	0.49	1.29	1.40

Data Set: DRUGRAT

Solution To carry out the test, we need to find the values of $\bar{x}$ and s. (In this study, σ is obviously unknown. So we will use s to estimate σ.) These values, $\bar{x} = 1.05$ and $s = .5$, are shown (highlighted) on the MINITAB printout of Figure 6.6. Now we substitute these sample statistics into the test statistic and obtain

$$z = \frac{\bar{x} - 1.2}{\sigma_{\bar{x}}} = \frac{\bar{x} - 1.2}{\sigma/\sqrt{n}} \approx \frac{1.05 - 1.2}{.5/\sqrt{100}} = -3.0$$

The implication is that the sample mean, 1.05, is (approximately) three standard deviations below the null-hypothesized value of 1.2 in the sampling distribution of $\bar{x}$.

Recall from Example 6.3 that the rejection region for the test at $\alpha = .01$ is

Rejection region: $|z| > 2.575$

From Figure 6.5 (p. 303), you can see that $z = -3.0$ falls in the lower-tail rejection region, which consists of all values of $z < -2.575$. Therefore, this sampling experiment provides sufficient evidence to reject H_0 and conclude, at the $\alpha = .01$ level of significance, that the mean response time for drug-injected rats differs from the control mean of 1.2 seconds. It appears that the rats receiving an injection of the drug have a mean response time that is less than 1.2 seconds.

Descriptive Statistics: TIME

Variable	N	Mean	StDev	Minimum	Q1	Median	Q3	Maximum
TIME	100	1.0517	0.4982	0.0500	0.6650	0.9950	1.4000	2.5500

Figure 6.6
MINITAB descriptive statistics for fill amounts, Example 6.4

Look Back Three points about the test of hypothesis in this example apply to all statistical tests:

1. Since z is less than -2.575, it is tempting to state our conclusion at a significance level lower than $\alpha = .01$. We resist this temptation because the level of α is determined *before* the sampling experiment is performed. If we decide that we are willing to tolerate a 1% Type I error rate, the result of the sampling experiment should have no effect on that decision. *In general, the same data should not be used both to set up and to conduct the test.*

2. When we state our conclusion at the .01 level of significance, we are referring to the failure rate of the *procedure*, not the result of this particular test. We know that the test procedure will lead to the rejection of the null hypothesis only 1% of the time when in fact $\mu = 1.2$. *Therefore, when the test statistic falls into the rejection region, we infer that the alternative $\mu \neq 1.2$ is true and express our confidence in the procedure by quoting either the α level of significance or the $100(1 - \alpha)\%$ confidence level.*

3. Although a test may lead to a "statistically significant" result (i.e., rejecting H_0 at significance level α, as in the preceding test), it may not be "practically significant." For example, suppose the neurologist tested $n = 100,000$ drug-injected rats, resulting in

$\bar{x} = 1.1995$ and $s = .05$. Now a two-tailed hypothesis test of $H_0: \mu = 1.2$ results in a test statistic of

$$z = \frac{(1.1995 - 1.2)}{.05/\sqrt{100,000}} = -3.16$$

This result at $\alpha = .01$ leads us to "reject H_0" and conclude that the mean μ is "statistically different" from 1.2. However, for all practical purposes, the sample mean $\bar{x} = 1.1995$ and the hypothesized mean $\mu = 1.2$ are the same. Because the result is not "practically significant," the neurologist is not likely to consider a unit dose of the drug as an inhibitor to response time in rats. Consequently, *not all "statistically significant" results are "practically significant."*

Now Work Exercise 6.28

Exercises 6.20–6.37

Understanding the Principles

6.20 Explain the difference between a one-tailed and a two-tailed test.

6.21 What conditions are required for a valid large-sample test for μ?

6.22 For what values of the test statistic do you reject H_0? fail to reject H_0?

Learning the Mechanics

6.23 A random sample of 100 observations from a population with standard deviation 60 yielded a sample mean of 110.
  a. Test the null hypothesis that $\mu = 100$ against the alternative hypothesis that $\mu > 100$, using $\alpha = .05$. Interpret the results of the test.
  b. Test the null hypothesis that $\mu = 100$ against the alternative hypothesis that $\mu \neq 100$, using $\alpha = .05$. Interpret the results of the test.
  c. Compare the results of the two tests you conducted. Explain why the results differ.

6.24 A random sample of 64 observations produced the following summary statistics: $\bar{x} = .323$ and $s^2 = .034$.
  a. Test the null hypothesis that $\mu = .36$ against the alternative hypothesis that $\mu < .36$, using $\alpha = .10$.
  b. Test the null hypothesis that $\mu = .36$ against the alternative hypothesis that $\mu \neq .36$, using $\alpha = .10$. Interpret the result.

6.25 Suppose you are interested in conducting the statistical test of $H_0: \mu = 200$ against $H_a: \mu > 200$, and you have decided to use the following decision rule: Reject H_0 if the sample mean of a random sample of 100 items is more than 215. Assume that the standard deviation of the population is 80.
  a. Express the decision rule in terms of z.
  b. Find α, the probability of making a Type I error, by using this decision rule.

⊙ Applet Exercise 6.2

Use the applet entitled *Hypotheses Test for a Mean* to investigate the effect of the underlying distribution on the proportion of Type I errors. For this exercise, take $n = 100$, mean $= 50$, standard deviation $= 10$, null mean $= 50$, and alternative $<$.
  a. Select the normal distribution and run the applet several times without clearing. What happens to the

proportion of times the null hypothesis is rejected at the .05 level as the applet is run more and more times?
  b. Clear the applet and then repeat part **a**, using the right-skewed distribution. Do you get similar results? Explain.
  c. Describe the effect that the underlying distribution has on the probability of making a Type I error.

⊙ Applet Exercise 6.3

Use the applet entitled *Hypotheses Test for a Mean* to investigate the effect of the underlying distribution on the proportion of Type II errors. For this exercise, take $n = 100$, mean $= 50$, standard deviation $= 10$, null mean $= 52$, and alternative $<$.
  a. Select the normal distribution and run the applet several times without clearing. What happens to the proportion of times the null hypothesis is rejected at the .01 level as the applet is run more and more times? Is this what you would expect? Explain.
  b. Clear the applet and then repeat part **a**, using the right-skewed distribution. Do you get similar results? Explain.
  c. Describe the effect that the underlying distribution has on the probability of making a Type II error.

⊙ Applet Exercise 6.4

Use the applet entitled *Hypotheses Test for a Mean* to investigate the effect of the null mean on the probability of making a Type II error. For this exercise, take $n = 100$, mean $= 50$, standard deviation $= 10$, and alternative $<$ with the normal distribution. Set the null mean to 55 and run the applet several times without clearing. Record the proportion of Type II errors that occurred at the .01 level. Clear the applet and repeat for null means of 54, 53, 52, and 51. What can you conclude about the probability of a Type II error as the null mean gets closer to the actual mean? Can you offer a reasonable explanation for this behavior?

Applying the Concepts—Basic

6.26 **Speeding and young drivers.** Teenage drivers will inevitably be tempted to drive faster, even to exceed the speed limit, by their friends. Did you resist such temptations when you first began to drive? To gain insight into this phenomenon, psychologists from the United Kingdom conducted a survey of 258 student drivers and reported the results in the *British Journal of Educational Psychology* (Vol. 80,

2010). One of the variables of interest was the response to the question, "Are you confident that you can resist your friends' persuasion to drive faster?" Each response was measured on a 7-point scale, from 1 = "definitely no" to 7 = "definitely yes." The data were collected 5 months after the students had attended a safe-driver presentation. The psychologists reported a sample mean response of 4.98 and a sample standard deviation of 1.62. Suppose it is known that the true mean response of students who *do not* attend a safe-driver presentation is $\mu = 4.7$.

a. Set up the null and alternative hypotheses for testing whether the true mean student-driver response 5 months after a safe-driver presentation is larger than 4.7.

b. Calculate the test statistic for the hypothesis test.

c. Find the rejection region for the hypothesis test, using $\alpha = .05$.

d. State the appropriate conclusion, in the words of the problem.

e. Do the test results indicate that the safe-driver presentation was effective in helping students feel more confident that they can resist their friends' persuasion to drive faster? Explain.

f. The distribution of response scores (on a 7-point scale) for all student drivers is unlikely to be normal. Does this impact the validity of the hypothesis test? Why or why not?

6.27 **Teacher perceptions of child behavior.** *Developmental Psychology* (Mar. 2003) published a study on teacher perceptions of the behavior of elementary school children. Teachers rated the aggressive behavior of a sample of 11,160 New York City public school children by responding to the statement "This child threatens or bullies others in order to get his/her own way." Responses were measured on a scale ranging from 1 (*never*) to 5 (*always*). Summary statistics for the sample of 11,160 children were reported as $\bar{x} = 2.15$ and $s = 1.05$. Let μ represent the mean response for the population of all New York City public school children. Suppose you want to test $H_0: \mu = 3$ against $H_a: \mu \neq 3$.

a. In the words of the problem, define a Type I error and a Type II error.

b. Use the sample information to conduct the test at a significance level of $\alpha = .05$.

c. Conduct the test from part **b** at a significance level of $\alpha = .10$.

6.28 **Latex allergy in health care workers.** Refer to the *Current Allergy & Clinical Immunology* (March 2004) study of $n = 46$ hospital employees who were diagnosed with a latex allergy from exposure to the powder on latex gloves, presented in Exercise 5.13 (p. 251). The number of latex gloves used per week by the sampled workers is summarized as follows: $\bar{x} = 19.3$ and $s = 11.9$. Let μ represent the mean number of latex gloves used per week by all hospital employees. Consider testing $H_0: \mu = 20$ against $H_a: \mu < 20$.

a. Give the rejection region for the test at a significance level of $\alpha = .01$.

b. Calculate the value of the test statistic.

c. Use the results from parts **a** and **b** to draw the appropriate conclusion.

6.29 **Accounting and Machiavellianism.** *Behavioral Research in Accounting* (Jan. 2008) published a study of Machiavellian traits in accountants. (*Machiavellian* describes negative character traits that include manipulation, cunning, duplicity, deception, and bad faith.) A Machiavellian ("Mach")

rating score was determined for each in a sample of 122 purchasing managers with the following results: $\bar{x} = 99.6$, $s = 12.6$. (*Note:* Scores range from a low of 40 to a high of 160, with the theoretical neutral Mach rating score of 100.) A director of purchasing at a major firm claims that the true mean Mach rating score of all purchasing managers is 85.

a. Specify the null and alternative hypothesis for a test of the director's claim.

b. Define a Type I error for this test.

c. Interpret the value, $\alpha = .10$.

d. Give the rejection region for the test using $\alpha = .10$.

e. Find the value of the test statistic.

f. Use the result, part **e,** to make the appropriate conclusion.

g. Do you need to make any assumptions about the distribution of Mach rating scores for the population of all purchasing managers? Explain.

6.30 **Heart rate during laughter.** Laughter is often called "the best medicine," since studies have shown that laughter can reduce muscle tension and increase oxygenation of the blood. In the *International Journal of Obesity* (Jan. 2007), researchers at Vanderbilt University investigated the physiological changes that accompany laughter. Ninety subjects (18–34 years old) watched film clips designed to evoke laughter. During the laughing period, the researchers measured the heart rate (beats per minute) of each subject, with the following summary results: $\bar{x} = 73.5, s = 6$. It is well known that the mean resting heart rate of adults is 71 beats per minute.

a. Setup H_0 and H_a for testing whether the true mean heart rate during laughter exceeds 71 beats per minute.

b. If $\alpha = 05$, find the rejection region for the test.

c. Calculate the value of the test statistic.

d. Make the appropriate conclusion.

Applying the Concepts—Intermediate

6.31 **Dating and disclosure.** Refer to the *Journal of Adolescence* (April 2010) study of adolescents' disclosure of their dating and romantic relationships, Exercise 1.29 (p. 21). Recall that a sample of 222 high school students was recruited to participate in the study. One of the variables of interest was the level of disclosure to an adolescent's mother (measured on a 5-point scale, where 1 = "never tell," 2 = "rarely tell," 3 = "sometimes tell," 4 = "almost always tell," and 5 = "always tell"). The sampled high school students had a mean disclosure score of 3.26 and a standard deviation of .93. The researchers hypothesize that the true mean disclosure score of all adolescents will exceed 3. Do you believe the researchers? Conduct a formal test of hypothesis using $\alpha = .01$.

6.32 **Identifying type of urban land cover.** Geographers use remote-sensing data from satellite pictures to identify urban land cover as either grassland, commercial, or residential. In *Geographical Analysis* (Oct. 2006), researchers from Arizona State, Florida State, and Louisiana State Universities collaborated on a new method for analyzing remote-sensing data. A satellite photograph of an urban area was divided into 4 × 4-meter areas (called pixels). Of interest is a numerical measure of the distribution of the sizes of gaps, or holes, in the pixel, a property called *lacunarity*. The mean and standard deviation of the lacunarity measurements for a sample of 100 pixels randomly selected from a specific urban area are 225 and 20, respectively. It is known that the mean lacunarity measurement for all grassland pixels is 220. Do the data suggest that the area sampled is grassland? Test at $\alpha = .01$.

6.33 Bone fossil study. Humerus bones from the same species of animal tend to have approximately the same length-to-width ratios. When fossils of humerus bones are discovered, archeologists can often determine the species of animal by examining the length-to-width ratios of the bones. It is known that species A exhibits a mean ratio of 8.5. Suppose 41 fossils of humerus bones were unearthed at an archeological site in East Africa, where species A is believed to have lived. (Assume that the unearthed bones were all from the same unknown species.) The length-to-width ratios of the bones are listed in the following table and saved in the **BONES** file.

10.73	8.89	9.07	9.20	10.33	9.98	9.84	9.59
8.48	8.71	9.57	9.29	9.94	8.07	8.37	6.85
8.52	8.87	6.23	9.41	6.66	9.35	8.86	9.93
8.91	11.77	10.48	10.39	9.39	9.17	9.89	8.17
8.93	8.80	10.02	8.38	11.67	8.30	9.17	12.00
9.38							

a. Test whether the population mean ratio of all bones of this particular species differs from 8.5. Use $\alpha = .01$.
b. What are the practical implications of the test you conducted in part **a**?

6.34 Cooling method for gas turbines. During periods of high demand for electricity—especially in the hot summer months—the power output from a gas turbine engine can drop dramatically. One way to counter this drop in power is by cooling the inlet air to the turbine. An increasingly popular cooling method uses high-pressure inlet fogging. The performance of a sample of 67 gas turbines augmented with high-pressure inlet fogging was investigated in the *Journal of Engineering for Gas Turbines and Power* (Jan. 2005). One measure of performance is heat rate (kilojoules per kilowatt per hour). Heat rates for the 67 gas turbines, saved in the **GASTURBINE** file, are listed in the next table. Suppose that a standard gas turbine has, on average, a heat rate of 10,000 kJ/kWh.

a. Conduct a test to determine whether the mean heat rate of gas turbines augmented with high-pressure inlet fogging exceeds 10,000 kJ/kWh. Use $\alpha = .05$.
b. Identify a Type I error for this study. Identify a Type II error.

14622	13196	11948	11289	11964	10526	10387	10592	10460	10086
14628	13396	11726	11252	12449	11030	10787	10603	10144	11674
11510	10946	10508	10604	10270	10529	10360	14796	12913	12270
11842	10656	11360	11136	10814	13523	11289	11183	10951	9722
10481	9812	9669	9643	9115	9115	11588	10888	9738	9295
9421	9105	10233	10186	9918	9209	9532	9933	9152	9295
16243	14628	12766	8714	9469	11948	12414			

6.35 Salaries of postgraduates. The *Economics of Education Review* (Vol. 21, 2002) published a paper on the relationship between education level and earnings. The data for the research were obtained from the National Adult Literacy Survey of over 25,000 respondents. The survey revealed that males with a postgraduate degree had a mean salary of $61,340 (with standard error $s_{\bar{x}} = \$2,185$) while females with a postgraduate degree had a mean of $32,227 (with standard error $s_{\bar{x}} = \$932$).

a. The article reports that a 95% confidence interval for μ_M, the population mean salary of all males with postgraduate degrees, is ($57,050, $65,631). Based on this interval, is there evidence that μ_M differs from $60,000? Explain.
b. Use the summary information to test the hypothesis that the true mean salary of males with postgraduate degrees differs from $60,000. Take $\alpha = .05$. (*Note:* $s_{\bar{x}} = s/\sqrt{n}$)
c. Explain why the inferences in parts **a** and **b** agree.
d. The article reports that a 95% confidence interval for μ_F, the population mean salary of all females with postgraduate degrees, is ($30,396, $34,058). Based on this interval, is there evidence that μ_F differs from $33,000? Explain.
e. Use the summary information to test the hypothesis that the true mean salary of females with postgraduate degrees differs from $33,000. Take $\alpha = .05$ (*Note:* $s_{\bar{x}} = s/\sqrt{n}$.)
f. Explain why the inferences in parts **d** and **e** agree.

6.36 Dissolved organic compound in lakes. The level of dissolved oxygen in the surface water of a lake is vital to maintaining the lake's ecosystem. Environmentalists from the University of Wisconsin monitored the dissolved oxygen levels over time for a sample of 25 lakes in the state (*Aquatic Biology*, May 2010). To ensure a representative sample, the environmentalists focused on several lake characteristics, including dissolved organic compound (DOC). The DOC data (measured in grams per cubic meters) for the 25 lakes are listed in the accompanying table and saved in the **WISCLAKES** file. The population of Wisconsin lakes has a mean DOC value of 15 grams/m³. Use a hypothesis test (at $\alpha = .10$) to make an inference about whether the sample is representative of all Wisconsin lakes for the characteristic, dissolved organic compound.

Lake	DOC
Allequash	9.6
Big Muskellunge	4.5
Brown	13.2
Crampton	4.1
Cranberry Bog	22.6
Crystal	2.7
EastLong	14.7
Helmet	3.5
Hiawatha	13.6
Hummingbird	19.8
Kickapoo	14.3
Little Arbor Vitae	56.9
Mary	25.1
Muskellunge	18.4
Northgate Bog	2.7
Paul	4.2
Peter	30.2
Plum	10.3
Reddington Bog	17.6
Sparkling	2.4
Tenderfoot	17.3
Trout Bog	38.8
Trout Lake	3.0
Ward	5.8
West Long	7.6

Based on Langman, O. C., et al. "Control of dissolved oxygen in northern temperate lakes over scales ranging from minutes to days." *Aquatic Biology*, Vol. 9, May 2010 (Table 1).

Applying the Concepts—Advanced

6.37 **Social interaction of mental patients.** The *Community Mental Health Journal* (Aug. 2000) presented the results of a survey of over 6,000 clients of the Department of Mental Health and Addiction Services (DMHAS) in Connecticut. One of the many variables measured for each mental health patient was frequency of social interaction (on a 5-point scale, where 1 = very infrequently, 3 = occasionally, and 5 = very frequently). The 6,681 clients who were evaluated had a mean social interaction score of 2.95 with a standard deviation of 1.10.

a. Conduct a hypothesis test (at $\alpha = .01$) to determine whether the true mean social interaction score of all Connecticut mental health patients differs from 3.

b. Examine the results of the study from a practical view, and then discuss why "statistical significance" does not always imply "practical significance."

c. Because the variable of interest is measured on a 5-point scale, it is unlikely that the population of ratings will be normally distributed. Consequently, some analysts may perceive the test from part **a** to be invalid and search for alternative methods of analysis. Defend or refute this position.

6.4 Observed Significance Levels: *p*-Values

According to the statistical test procedure described in Section 6.2, the rejection region and, correspondingly, the value of α are selected prior to conducting the test and the conclusions are stated in terms of rejecting or not rejecting the null hypothesis. A second method of presenting the results of a statistical test reports the extent to which the test statistic disagrees with the null hypothesis and leaves to the reader the task of deciding whether to reject the null hypothesis. This measure of disagreement is called the *observed significance level* (or *p-value*) for the test.

> The **observed significance level**, or **p-value**, for a specific statistical test is the probability (assuming that H_0 is true) of observing a value of the test statistic that is at least as contradictory to the null hypothesis, and supportive of the alternative hypothesis, as the actual one computed from the sample data.

Recall testing H_0: $\mu = 2{,}400$ versus H_a: $\mu > 2{,}400$, where μ is the mean breaking strength of sewer pipe (Section 6.1). The value of the test statistic computed for the sample of $n = 50$ sections of sewer pipe was $z = 2.12$. Since the test is one tailed—that is, the alternative (research) hypothesis of interest is H_a: $\mu > 2{,}400$—values of the test statistic even more contradictory to H_0 than the one observed would be values larger than $z = 2.12$. Therefore, the observed significance level (*p*-value) for this test is

$$p\text{-value} = P(z \geq 2.12)$$

or, equivalently, the area under the standard normal curve to the right of $z = 2.12$. (See Figure 6.7.)

The area A in Figure 6.7 is given in Table III in Appendix A as .4830. Therefore, the upper-tail area corresponding to $z = 2.12$ is

$$p\text{-value} = .5 - .4830 = .0170$$

Consequently, we say that these test results are "statistically significant"; that is, they disagree (rather strongly) with the null hypothesis H_0: $\mu = 2{,}400$ and favor H_a: $\mu > 2{,}400$. Hence, the probability of observing a z value as large as 2.12 is only .0170 if in fact the true value of μ is 2,400.

Figure 6.7

Finding the *p*-value for an upper-tailed test when $z = 2.12$

If you are inclined to select $\alpha = .05$ for this test, then you would reject the null hypothesis because the p-value for the test, .0170, is less than .05. In contrast, if you choose $\alpha = .01$, you would not reject the null hypothesis, because the p-value for the test is larger than .01. Thus, the use of the observed significance level is identical to the test procedure described in the preceding sections, except that the choice of α is left to you.

The steps for calculating the p-value corresponding to a test statistic for a population mean are given in the following box:

Steps for Calculating the p-Value for a Test of Hypothesis

1. Determine the value of the test statistic z corresponding to the result of the sampling experiment.

a. If the test is one-tailed, the p-value is equal to the tail area beyond z in the same direction as the alternative hypothesis. Thus, if the alternative hypothesis is of the form $>$, the p-value is the area to the right of, or above, the observed z-value. Conversely, if the alternative is of the form $<$, the p-value is the area to the left of, or below, the observed z-value. (See Figure 6.8.)

b. If the test is two tailed, the p-value is equal to twice the tail area beyond the observed z-value in the direction of the sign of z. That is, if z is positive, the p-value is twice the area to the right of, or above, the observed z value. Conversely, if z is negative, the p-value is twice the area to the left of, or below, the observed z-value. (See Figure 6.9.)

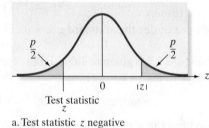

Figure 6.8

Finding the p-value for a one-tailed test

a. Lower–tailed test, $H_a: \mu < \mu_0$

b. Upper–tailed test, $H_a: \mu > \mu_0$

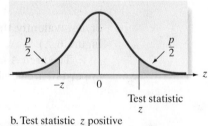

Figure 6.9

Finding the p-value for a two-tailed test: p-value $= 2(p/2)$

a. Test statistic z negative

b. Test statistic z positive

Example 6.5

Computing a p-Value—Test on Mean Drug Response Time

Problem Find the observed significance level for the test of the mean response time for drug-injected rats in Examples 6.3 and 6.4.

Solution Example 6.3 presented a two-tailed test of the hypothesis

$$H_0: \mu = 1.2 \text{ seconds}$$

against the alternative hypothesis

$$H_a: \mu \neq 1.2 \text{ seconds}$$

The observed value of the test statistic in Example 6.4 was $z = -3.0$, and any value of z less than -3.0 or greater than $+3.0$ (because this is a two-tailed test) would be even more contradictory to H_0. Therefore, the observed significance level for the test is

$$p\text{-value} = P(z < -3.0 \text{ or } z > +3.0) = P(|z| > 3.0)$$

Thus, we calculate the area below the observed z-value, $z = -3.0$, and double it. Consulting Table IV in Appendix A, we find that $P(z < -3.0) = .5 - .4987 = .0013$. Therefore, the p-value for this two-tailed test is

$$2P(z < -3.0) = 2(.0013) = .0026$$

This p-value can also be obtained with statistical software. The p-value is shown (highlighted) on the SAS printout of Figure 6.10. [*Note:* The value differs slightly due to rounding in our calculation of the test statistic.]

One Sample Z Test for a Mean

Sample Statistics for TIME

N	Mean	Std. Dev.	Std. Error
100	1.05	0.50	0.05

Hypothesis Test

Null hypothesis: Mean of TIME = 1.2
Alternative: Mean of TIME ^= 1.2

With a specified known standard deviation of 0.5

Z Statistic	Prob > Z
-2.966	0.0030

Figure 6.10
SAS test of mean response time, Example 6.5

Look Back We can interpret this p-value as a strong indication that the mean reaction time of drug-injected rats differs from the control mean ($\mu \neq 1.2$), since we would observe a test statistic this extreme or more extreme only 26 in 10,000 times if the drug-injected mean were equal to the control mean ($\mu = 1.2$). The extent to which the mean differs from 1.2 could be better determined by calculating a confidence interval for μ.

Now Work Exercise 6.44

When publishing the results of a statistical test of hypothesis in journals, case studies, reports, etc., many researchers make use of p-values. Instead of selecting α beforehand and then conducting a test, as outlined in this chapter, the researcher computes (usually with the aid of a statistical software package) and reports the value of the appropriate test statistic and its associated p-value. It is left to the reader of the report to judge the significance of the result (i.e., the reader must determine, on the basis of the reported p-value, whether to reject the null hypothesis in favor of the alternative hypothesis. Usually, the null hypothesis is rejected if the observed significance level is *less than* the fixed significance level α chosen by the reader. The inherent advantage of reporting test results in this manner is twofold: (1) Readers are permitted to select the maximum value of α that they would be willing to tolerate if they actually carried out a standard test of hypothesis in the manner outlined in this chapter, and (2) a measure of the degree of significance of the result (i.e., the p-value) is provided.

Ethics IN Statistics

Selecting the value of α *after* computing the observed significance level (p-value) in order to guarantee a preferred conclusion is considered *unethical statistical practice.*

Reporting Test Results as *p*-Values: How to Decide Whether to Reject H_0

a. Choose the maximum value of α that you are willing to tolerate.

b. If the observed significance level (p-value) of the test is less than the chosen value of α, reject the null hypothesis. Otherwise, do not reject the null hypothesis.

Example 6.6
Using *p*-Values—Test of Mean Hospital Length of Stay

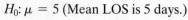

Problem The lengths of stay (in days) for 100 randomly selected hospital patients, first presented in Table 5.1, are reproduced in Table 6.4. Suppose we want to test the hypothesis that the true mean length of stay (LOS) at the hospital is less than 5 days; that is,

$H_0: \mu = 5$ (Mean LOS is 5 days.)

$H_a: \mu < 5$ (Mean LOS is less than 5 days.)

Assuming that $\sigma = 3.68$, use the data in the table to conduct the test at $\alpha = .05$.

Table 6.4		Lengths of Stay for 100 Hospital Patients							
2	3	8	6	4	4	6	4	2	5
8	10	4	4	4	2	1	3	2	10
1	3	2	3	4	3	5	2	4	1
2	9	1	7	17	9	9	9	4	4
1	1	1	3	1	6	3	3	2	5
1	3	3	14	2	3	9	6	6	3
5	1	4	6	11	22	1	9	6	5
2	2	5	4	3	6	1	5	1	6
17	1	2	4	5	4	4	3	2	3
3	5	2	3	3	2	10	2	4	2

⊙ *Data Set:* HOSPLOS

Solution The data were entered into a computer and MINITAB was used to conduct the analysis. The MINITAB printout for the **lower-tailed test** is displayed in Figure 6.11. Both the test statistic, $z = -1.28$, and the *p*-value of the test, $p = .101$, are highlighted on the MINITAB printout. Since the *p*-value exceeds our selected α value, $\alpha = .05$, we cannot reject the null hypothesis. Hence, there is insufficient evidence (at $\alpha = .05$) to conclude that the true mean LOS at the hospital is less than 5 days.

Figure 6.11
MINITAB lower-tailed test of mean LOS, Example 6.6

```
One-Sample Z: LOS

Test of mu = 5 vs < 5
The assumed standard deviation = 3.68

                                      95% Upper
Variable    N   Mean  StDev  SE Mean    Bound      Z      P
LOS       100  4.530  3.678    0.368    5.135  -1.28  0.101
```

Look Back A hospital administrator, desirous of a mean length of stay less than 5 days, may be tempted to select an α level that leads to a rejection of the null hypothesis *after* determining *p*-value $= .101$. There are two reasons one should resist this temptation. First, the administrator would need to select $\alpha > .101$ (say, $\alpha = .15$) in order to conclude that $H_a: \mu < 5$ is true. A Type I error rate of 15% is considered too large by most researchers. Second, and more importantly, such a strategy is considered unethical statistical practice (See marginal note, p. 313.).

Now Work Exercise 6.51

Note: Some statistical software packages (e.g., SPSS) will conduct only two-tailed tests of hypothesis. For these packages, you obtain the *p*-value for a one-tailed test as shown in the next box.

Converting a Two-Tailed *p*-Value from a Printout to a One-Tailed *p*-Value:

$$p = \frac{\text{Reported } p\text{-value}}{2} \text{ if } \begin{cases} H_a \text{ is of form } > \text{ and } z \text{ is positive} \\ H_a \text{ is of form } < \text{ and } z \text{ is negative} \end{cases}$$

$$p = 1 - \left(\frac{\text{Reported } p\text{-value}}{2}\right) \text{ if } \begin{cases} H_a \text{ is of form } > \text{ and } z \text{ is negative} \\ H_a \text{ is of form } < \text{ and } z \text{ is positive} \end{cases}$$

Statistics IN Action | **Revisited** | Testing a Population Mean in the KLEENEX® Survey

Refer to Kimberly-Clark Corporation's survey of 250 people who kept a count of their use of KLEENEX® tissues in diaries (p. 303). We want to test the claim made by marketing experts that μ, the average number of tissues used by people with colds, is greater than 60 tissues. That is, we want to test

$$H_0: \mu = 60 \qquad H_a: \mu > 60$$

We will select $\alpha = .05$ as the level of significance for the test.

The survey results for the 250 sampled KLEENEX® users are stored in the **TISSUES** data file. A MINITAB

analysis of the data yielded the printout displayed in Figure SIA6.1.

The observed significance level of the test, highlighted on the printout, is *p*-value = 000. Since this *p*-value is less than $\alpha = .05$, we have sufficient evidence to reject H_0; therefore, we conclude that the mean number of tissues used by a person with a cold is greater than 60 tissues. This result supports the company's decision to put 70 tissues in a standard-sized box of KLEENEX.

Data Set: TISSUES

Figure SIA6.1

MINITAB test of $\mu = 60$ for KLEENEX survey

```
Test of mu = 60 vs > 60

                                         95% Lower
Variable    N    Mean   StDev  SE Mean    Bound     T      P
NUMUSED    250  68.68   25.03   1.58      66.06   5.48   0.000
```

Exercises 6.38–6.56

Understanding the Principles

6.38 How does the observed significance level (*p*-value) of a test differ from the value of α?

6.39 In general, do large *p*-values or small *p*-values support the alternative hypothesis H_a?

6.40 If a hypothesis test using $\alpha = .05$ were conducted, for which of the following *p*-values would the null hypothesis be rejected?
- **a.** .06
- **b.** .10
- **c.** .01
- **d.** .001
- **e.** .251
- **f.** .042

6.41 For each pair consisting of α and an observed significance level (*p*-value), indicate whether the null hypothesis would be rejected.
- **a.** $\alpha = .05, p\text{-value} = .10$
- **b.** $\alpha = .10, p\text{-value} = .05$
- **c.** $\alpha = .01, p\text{-value} = .001$
- **d.** $\alpha = .025, p\text{-value} = .05$
- **e.** $\alpha = .10, p\text{-value} = .45$

Learning the Mechanics

6.42 An analyst tested the null hypothesis $\mu \geq 20$ against the alternative hypothesis $\mu < 20$. The analyst reported a *p*-value of .06. What is the smallest value of α for which the null hypothesis would be rejected?

6.43 In a test of $H_0: \mu = 100$ against $H_a: \mu > 100$, the sample data yielded the test statistic $z = 2.17$. Find the *p*-value for the test.

6.44 In a test of $H_0: \mu = 100$ against $H_a: \mu \neq 100$, the sample data yielded the test statistic $z = 2.17$. Find the *p*-value for the test.

6.45 In a test of the hypothesis $H_0: \mu = 50$ versus $H_a: \mu > 50$, a sample of $n = 100$ observations possessed mean $\bar{x} = 49.4$ and standard deviation $s = 4.1$. Find and interpret the *p*-value for this test.

6.46 In a test of the hypothesis $H_0: \mu = 10$ versus $H_a: \mu \neq 10$, a sample of $n = 50$ observations possessed mean $\bar{x} = 10.7$ and standard deviation $s = 3.1$. Find and interpret the *p*-value for this test.

6.47 Consider a test of $H_0: \mu = 75$ performed with the computer. SPSS reports a two-tailed *p*-value of .1032. Make the appropriate conclusion for each of the following situations:
- **a.** $H_a: \mu < 75, z = -1.63, \alpha = .05$
- **b.** $H_a: \mu < 75, z = 1.63, \alpha = .10$
- **c.** $H_a: \mu > 75, z = 1.63, \alpha = .10$
- **d.** $H_a: \mu \neq 75, z = -1.63, \alpha = .01$

Applying the Concepts—Basic

6.48 **Speeding and young drivers.** Refer to the *British Journal of Educational Psychology* (Vol. 80, 2010) study of teenage drivers' temptation to exceed the speed limit, Exercise 6.26 (p. 308). Recall that in a survey of 258 student drivers taken 5 months after the students had attended a safe-driver presentation, responses (measured on a 7-point scale) to the question "Are you confident that you can resist your friends' persuasion to drive faster?" had a mean of 4.98 and a standard deviation of 1.62. A MINITAB printout for testing $H_0: \mu = 4.7$ against $H_a: \mu > 4.7$ is shown below.

a. Locate the p-value of the test on the printout. What does the p-value measure?

b. Give the appropriate conclusion for $\alpha = .05$. Does this conclusion agree with your answer to Exercise 6.26d?

```
Test of mu = 4.7 vs > 4.7

                           95% Lower
  N    Mean  StDev  SE Mean   Bound    T      P
 258   4.980  1.620  0.101    4.814   2.78  0.003
```

6.49 Teacher perceptions of child behavior. Refer to the *Developmental Psychology* (Mar. 2003) study on the aggressive behavior of elementary school children, presented in Exercise 6.27 (p. 309). Recall that you tested $H_0: \mu = 3$ against $H_a: \mu \neq 3$, where μ is the mean level of aggressiveness for the population of all New York City public school children, as perceived by their teachers and based on the summary statistics $n = 11,160, \bar{x} = 2.15$, and $s = 1.05$.

a. Compute the p-value of the test.

b. Compare the p-value with $\alpha = .10$ and make the appropriate conclusion.

6.50 Latex allergy in health care workers. Refer to the *Current Allergy & Clinical Immunology* (March 2004) study of latex allergy in health care workers, presented in Exercise 6.28 (p. 309). Recall that you tested $H_0: \mu = 20$ against $H_a: \mu < 20$, where μ is the mean number of latex gloves used per week by all hospital employees, based on the summary statistics $n = 46, \bar{x} = 19.3$, and $s = 11.9$.

a. Compute the p-value of the test.

b. Compare the p-value with $\alpha = .01$ and make the appropriate conclusion.

6.51 Bone fossil study. In Exercise 6.33 (p. 310), you tested **NW** $H_0: \mu = 8.5$ versus $H_a: \mu \neq 8.5$, where μ is the population mean length-to-width ratio of humerus bones of a particular species of animal. A SAS printout for the hypothesis test is shown below. Locate the p-value on the printout and interpret it.

```
Sample Statistics for LWRATIO

    N        Mean      Std. Dev.      Std. Error
-------------------------------------------------
   41        9.26        1.20            0.19

Hypothesis Test

  Null hypothesis:    Mean of LWRATIO =  8.5
  Alternative:        Mean of LWRATIO ^= 8.5

With a specified known standard deviation of 1.2

         Z Statistic      Prob > Z
         -----------      --------
            4.042          <.0001
```

6.52 Emotional empathy in young adults. According to a theory in psychology, young female adults show more emotional empathy towards others than do males. The *Journal of Moral Education* (June 2010) tested this theory by examining the attitudes of a sample of 30 female college students. Each student completed the Ethic of Care Interview, which consisted of a series of statements on empathy attitudes. For the statement on emotional empathy (e.g., "I often have tender, concerned feelings for people less fortunate than me"), the sample mean response was 3.28. Assume the population standard deviation for females is .5. [*Note:* Empathy scores ranged from 0 to 4, where $0 =$ "never" and $4 =$ "always".] Suppose it is known that male college students have an average emotional empathy score of $\mu = 3$.

a. Specify the null and alternative hypothesis for testing whether female college students score higher than 3.0 on the emotional empathy scale.

b. Compute the test statistic.

c. Find the observed significance level (p-value) of the test.

d. At $\alpha = .01$, what is the appropriate conclusion?

e. How small of an α-value can you choose and still have sufficient evidence to reject the null hypothesis?

Applying the Concepts—Intermediate

6.53 Buy-side vs. sell-side analysts' earnings forecasts. The *Financial Analysts Journal* (Jul/Aug 2008) presented a study of earnings forecasts of buy-side and sell-side analysts. Buy-side analysts differ from sell-side analysts on a variety of factors, including scope of industry coverage, sources of information used, and target audience. Data were collected on 3,526 forecasts made by buy-side analysts and 58,562 forecasts made by sell-side analysts and the relative absolute forecast error was determined for each. A positive forecast error indicates that the analyst is overestimating earnings, while a negative forecast error implies that the analyst is underestimating earnings. Summary statistics for the forecast errors in the two samples are reproduced in the following table.

	Buy-Side Analysts	Sell-Side Analysts
Mean	0.85	−0.05
Standard Deviation	1.93	0.85

Based on Groysberg, B., Healy, P., and Chapman, C. "Buy-side vs. sell-side analysts' earning forecasts." *Financial Analysts Journal*, Vol. 64, No. 4, Jul/Aug 2008.

a. Conduct a test (at $\alpha = .01$) to determine if the true mean forecast error for buy-side analysts is positive. Use the observed significance level (p-value) of the test to make your decision and state your conclusion in the words of the problem.

b. Conduct a test (at $\alpha = .01$) to determine if the true mean forecast error for sell-side analysts is negative. Use the observed significance level (p-value) of the test to make your decision and state your conclusion in the words of the problem.

6.54 Colored string preferred by chickens. Refer to the *Applied Animal Behaviour Science* (October 2000) study of domestic chickens exposed to a pecking stimulus, presented in Exercise 5.21 (p. 252). Recall that the average number of pecks a chicken takes at a white string over a specified time interval is known to be $\mu = 7.5$ pecks. In an experiment in which 72 chickens were exposed to blue string, the average number of pecks was $\bar{x} = 1.13$ pecks, with a standard deviation of $s = 2.21$ pecks.

a. On average, are chickens more apt to peck at white string than at blue string? Conduct the appropriate test of hypothesis, using $\alpha = .05$.

b. Compare your answer to part **a** with your answer to Exercise 5.21**b**.

c. Find the *p*-value for the test and interpret it.

6.55 **Feminizing human faces.** Research published in *Nature* (August 27, 1998) revealed that people are more attracted to "feminized" faces, regardless of gender. In one experiment, 50 human subjects viewed both a Japanese female and a Caucasian male face on a computer. Using special computer graphics, each subject could morph the faces (by making them more feminine or more masculine) until they attained the "most attractive" face. The level of feminization x (measured as a percentage) was measured.

a. For the Japanese female face, $\bar{x} = 10.2\%$ and $s = 31.3\%$. The researchers used this sample information to test the null hypothesis of a mean level of feminization equal to 0%. Verify that the test statistic is equal to 2.3.

b. Refer to part **a**. The researchers reported the *p*-value of the test as $p \approx .02$. Verify and interpret this result.

c. For the Caucasian male face, $\bar{x} = 15.0\%$ and $s = 25.1\%$. The researchers reported the test statistic (for the test of the null hypothesis stated in part **a**) as 4.23, with an associated *p*-value of approximately 0. Verify and interpret these results.

Applying the Concepts—Advanced

6.56 **Ages of cable TV shoppers.** In a paper presented at the 2000 Conference of the International Association for Time Use Research, professor Margaret Sanik of Ohio State University reported the results of her study on American cable TV viewers who purchase items from one of the home-shopping channels. She found that the average age of these cable TV shoppers was 51 years. Suppose you want to test the null hypothesis $H_0: \mu = 51$, using a sample of $n = 50$ cable TV shoppers.

a. Find the *p*-value of a two-tailed test if $\bar{x} = 52.3$ and $s = 7.1$.

b. Find the *p*-value of an upper-tailed test if $\bar{x} = 52.3$ and $s = 7.1$.

c. Find the *p*-value of a two-tailed test if $\bar{x} = 52.3$ and $s = 10.4$.

d. For each of the tests in parts **a–c**, give a value of α that will lead to a rejection of the null hypothesis.

e. If $\bar{x} = 52.3$, give a value of s that will yield a *p*-value of .01 or less for a one-tailed test.

6.5 Test of Hypothesis about a Population Mean: Student's *t*-Statistic

Most water-treatment facilities monitor the quality of their drinking water on an hourly basis. One variable monitored is pH, which measures the degree of alkalinity or acidity in the water. A pH below 7.0 is acidic, one above 7.0 is alkaline, and a pH of 7.0 is neutral. One water-treatment plant has a target pH of 8.5. (Most try to maintain a slightly alkaline level.) The mean and standard deviation of 1 hour's test results, based on 17 water samples at this plant, are

$$\bar{x} = 8.42 \qquad s = .16$$

Does this sample provide sufficient evidence that the mean pH level in the water differs from 8.5?

This inference can be placed in a test-of-hypothesis framework. We establish the target pH as the null hypothesized value and then utilize a two-tailed alternative that the true mean pH differs from the target:

$$H_0: \mu = 8.5 \ (\text{Mean pH level is 8.5.})$$

$$H_a: \mu \neq 8.5 \ (\text{Mean pH level differs from 8.5.})$$

Recall from Section 5.3 that when we are faced with making inferences about a population mean from the information in a small sample, two problems emerge:

1. The normality of the sampling distribution for $\bar{x}$ does not follow from the Central Limit Theorem when the sample size is small. We must assume that the distribution of measurements from which the sample was selected is approximately normally distributed in order to ensure the approximate normality of the sampling distribution of $\bar{x}$.

2. If the population standard deviation σ is unknown, as is usually the case, then we cannot assume that s will provide a good approximation for σ when the sample size is small. Instead, we must use the t-distribution rather than the standard normal z-distribution to make inferences about the population mean μ.

Therefore, as the test statistic of a small-sample test of a population mean, we use the t-statistic:

$$\text{Test statistic: } t = \frac{\bar{x} - \mu_0}{s/\sqrt{n}} = \frac{\bar{x} - 8.5}{s/\sqrt{n}}$$

In this equation, μ_0 is the null-hypothesized value of the population mean μ. In the example here, $\mu_0 = 8.5$.

To find the rejection region, we must specify the value of α, the probability that the test will lead to rejection of the null hypothesis when it is true, and then consult the t table (Table IV of Appendix A). With $\alpha = .05$, the two-tailed rejection region is

$$\text{Rejection region: } t_{\alpha/2} = t_{.025} = 2.120 \text{ with } n - 1 = 16 \text{ degrees of freedom}$$
$$\text{Reject } H_0 \text{ if } t < -2.120 \text{ or } t > 2.120$$

The rejection region is shown in Figure 6.12.

We are now prepared to calculate the test statistic and reach a conclusion:

$$t = \frac{\bar{x} - \mu_0}{s/\sqrt{n}} = \frac{8.42 - 8.50}{.16/\sqrt{17}} = \frac{-.08}{.039} = -2.05$$

Figure 6.12

Two-tailed rejection region for small-sample t-test

Since the calculated value of t does not fall into the rejection region (Figure 6.12), we cannot reject H_0 at the $\alpha = .05$ level of significance. Thus, the water-treatment plant should not conclude that the mean pH differs from the 8.5 target on the basis of the sample evidence.

It is interesting to note that the calculated t value, -2.05, is *less than* the .05-level z-value, -1.96. The implication is that if we had *incorrectly* used a z statistic for this test, we would have rejected the null hypothesis at the .05 level and concluded that the mean pH level differs from 8.5. The important point is that the statistical procedure to be used must always be closely scrutinized and all the assumptions understood. Many statistical lies are the result of misapplications of otherwise valid procedures.

The technique for conducting a small-sample test of hypothesis about a population mean is summarized in the following boxes:

Small-Sample Test of Hypothesis about μ Based on a Student's t-Statistic

One-Tailed Test	**Two-Tailed Test**
$H_0: \mu = \mu_0$	$H_0: \mu = \mu_0$
$H_a: \mu < \mu_0 \,(\text{or } H_a: \mu > \mu_0)$	$H_a: \mu \neq \mu_0$
Test statistic: $t = \dfrac{\bar{x} - \mu_0}{s/\sqrt{n}}$	Test statistic: $t = \dfrac{\bar{x} - \mu_0}{s/\sqrt{n}}$
Rejection region: $t < -t_\alpha$ (or $t > t_\alpha$ when $H_a: \mu > \mu_0$)	Rejection region: $t < -t_{\alpha/2}$ or $t > t_{\alpha/2}$

where t_α and $t_{\alpha/2}$ are based on $(n-1)$ degrees of freedom

---

> **Conditions Required for a Valid Small-Sample Hypothesis Test for μ**
>
> 1. A random sample is selected from the target population.
>
> 2. The population from which the sample is selected has a distribution that is approximately normal.

Example 6.7

A Small-Sample Test for μ— Does a New Engine Meet Air-Pollution Standards?

Problem A major car manufacturer wants to test a new engine to determine whether it meets new air-pollution standards. The mean emission μ of all engines of this type must be less than 20 parts per million of carbon. Ten engines are manufactured for testing purposes, and the emission level of each is determined. The data (in parts per million) are listed in Table 6.5. Do the data supply sufficient evidence to allow the manufacturer to conclude that this type of engine meets the pollution standard? Assume that the manufacturer is willing to risk a Type I error with probability $\alpha = .01$.

Table 6.5		Emission Levels for Ten Engines							
15.6	16.2	22.5	20.5	16.4	19.4	19.6	17.9	12.7	14.9

Data Set: EMISSIONS

Solution The manufacturer wants to support the research hypothesis that the mean emission level μ for all engines of this type is less than 20 parts per million. The elements of this small-sample one-tailed test are as follows:

$$H_0: \mu = 20 \text{ (Mean emission level is 20 ppm.)}$$

$$H_a: \mu < 20 \text{ (Mean emission level is less than 20 ppm.)}$$

$$\text{Test statistic: } t = \frac{\bar{x} - 20}{s/\sqrt{n}}$$

Rejection region: For $\alpha = .01$ and df $= n - 1 = 9$, the one-tailed rejection region (see Figure 6.13) is $t < -t_{.01} = -2.821$.

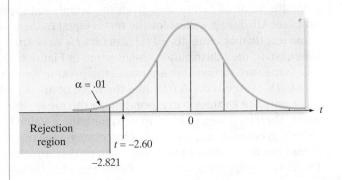

Figure 6.13

A *t*-distribution with 9 df and the rejection region for Example 6.7

Assumption: The relative frequency distribution of the population of emission levels for all engines of this type is approximately normal. Based on the shape of the MINITAB stem-and-leaf display of the data shown in Figure 6.14, this assumption appears to be reasonably satisfied.

To calculate the test statistic, we analyzed the **EMISSIONS** data with MINITAB. The MINITAB printout is shown at the bottom of Figure 6.14. From the printout, we

Stem-and-Leaf Display: E-LEVEL

```
Stem-and-leaf of E-LEVEL  N = 10
Leaf Unit = 1.0

    1    1  2
    3    1  45
   (3)   1  667
    4    1  99
    2    2  0
    1    2  2
```

One-Sample T: E-LEVEL

```
Test of mu = 20 vs < 20
```

					95% Upper		
Variable	N	Mean	StDev	SE Mean	Bound	T	P
E-LEVEL	10	17.5700	2.9522	0.9336	19.2814	-2.60	0.014

Figure 6.14
MINITAB analysis of 10 emission levels, Example 6.7

obtain $\bar{x} = 17.57$ and $s = 2.95$. Substituting these values into the test statistic formula and rounding, we get

$$t = \frac{\bar{x} - 20}{s/\sqrt{n}} = \frac{17.57 - 20}{2.95/\sqrt{10}} = -2.60$$

Since the calculated t falls outside the rejection region (see Figure 6.13), the manufacturer cannot reject H_0. There is insufficient evidence to conclude that $\mu < 20$ parts per million and that the new type of engine meets the pollution standard.

Look Back Are you satisfied with the reliability associated with this inference? The probability is only $\alpha = .01$ that the test would support the research hypothesis if in fact it were false.

Now Work Exercise 6.63a–b

Example 6.8
The *p*-Value for a Small-Sample Test of μ

Problem Find the observed significance level for the test described in Example 6.7. Interpret the result.

Solution The test of Example 6.7 was a lower-tailed test: H_0: $\mu = 20$ versus H_a: $\mu < 20$. Since the value of t computed from the sample data was $t = -2.60$, the observed significance level (or *p*-value) for the test is equal to the probability that t would assume a value less than or equal to -2.60 if in fact H_0 were true. This is equal to the area in the lower tail of the t-distribution (highlighted in Figure 6.15).

One way to find this area (i.e., the *p*-value for the test) is to consult the t-table (Table IV in Appendix A). Unlike the table of areas under the normal curve, Table IV gives only the t values corresponding to the areas .100, .050, .025, .010, .005, .001, and .0005. Therefore, we can only approximate the *p*-value for the test. Since the observed t value was based on nine degrees of freedom, we use the df $= 9$ row in Table IV and move across the row until we reach the t values that are closest to the observed $t = -2.60$. [*Note:* We ignore the minus sign.] The t values corresponding to *p*-values of .010 and .025

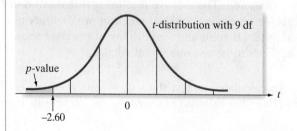

Figure 6.15
The observed significance level for the test of Example 6.7

are 2.821 and 2.262, respectively. Since the observed t value falls between $t_{.010}$ and $t_{.025}$, the p-value for the test lies between .010 and .025. In other words, $.010 < p$-value $< .025$. Thus, we would reject the null hypothesis $H_0: \mu = 20$ parts per million for any value of α larger than .025 (the upper bound of the p-value).

A second, more accurate, way to obtain the p-value is to use a statistical software package to conduct the test of hypothesis. Both the test statistic (-2.60) and the p-value $(.014)$ are highlighted on the MINITAB printout of Figure 6.14.

You can see that the actual p-value of the test falls within the bounds obtained from Table IV. Thus, the two methods agree and we will reject $H_0: \mu = 20$ in favor of $H_a: \mu < 20$ for any α level larger than .025.

Now Work Exercise 6.63c

Small-sample inferences typically require more assumptions and provide less information about the population parameter than do large-sample inferences. Nevertheless, the *t*-test is a method of testing a hypothesis about a population mean of a normal distribution when only a small number of observations is available.

What Can Be Done If the Population Relative Frequency Distribution Departs Greatly from Normal?

Answer: Use the nonparametric statistical method described in optional Section 6.8

Exercises 6.57–6.76

Understanding the Principles

6.57 In what ways are the distributions of the z-statistic and t-statistic alike? How do they differ?

6.58 Under what circumstances should you use the t-distribution in testing a hypothesis about a population mean?

Learning the Mechanics

6.59 For each of the following rejection regions, sketch the sampling distribution of t and indicate the location of the rejection region on your sketch:
a. $t > 1.440$, where df $= 6$
b. $t < -1.782$, where df $= 12$
c. $t < -2.060$ or $t > 2.060$, where df $= 25$

6.60 For each of the rejection regions defined in Exercise 6.59, what is the probability that a Type I error will be made?

6.61 A random sample of n observations is selected from a normal population to test the null hypothesis that $\mu = 10$. Specify the rejection region for each of the following combinations of H_a, α, and n:
a. $H_a: \mu \neq 10; \alpha = .05; n = 14$
b. $H_a: \mu > 10; \alpha = .01; n = 24$
c. $H_a: \mu > 10; \alpha = .10; n = 9$
d. $H_a: \mu < 10; \alpha = .01; n = 12$
e. $H_a: \mu \neq 10; \alpha = .10; n = 20$
f. $H_a: \mu < 10; \alpha = .05; n = 4$

6.62 The following sample of six measurements was randomly selected from a normally distributed population: $1, 3, -1, 5, 1, 2$.
a. Test the null hypothesis that the mean of the population is 3 against the alternative hypothesis, $\mu < 3$. Use $\alpha = .05$.

b. Test the null hypothesis that the mean of the population is 3 against the alternative hypothesis, $\mu \neq 3$. Use $\alpha = .05$.
c. Find the observed significance level for each test.

6.63 A sample of five measurements, randomly selected from a normally distributed population, resulted in the following summary statistics: $\bar{x} = 4.8$, $s = 1.3$.
a. Test the null hypothesis that the mean of the population is 6 against the alternative hypothesis, $\mu < 6$. Use $\alpha = .05$.
b. Test the null hypothesis that the mean of the population is 6 against the alternative hypothesis, $\mu \neq 6$. Use $\alpha = .05$.
c. Find the observed significance level for each test.

6.64 Suppose you conduct a t-test for the null hypothesis $H_0: \mu = 1{,}000$ versus the alternative hypothesis $H_a: \mu > 1{,}000$, based on a sample of 17 observations. The test results are $t = 1.89$, p-value $= .038$.
a. What assumptions are necessary for the validity of this procedure?
b. Interpret the results of the test.
c. Suppose the alternative hypothesis had been the two-tailed $H_a: \mu \neq 1{,}000$. If the t-statistic were unchanged, what would the p-value be for this test? Interpret the p-value for the two-tailed test.

Applying the Concepts—Basic

6.65 **Lobster trap placement.** Refer to the *Bulletin of Marine Science* (April 2010) observational study of lobster trap placement by teams fishing for the red spiny lobster in Baja California Sur, Mexico, Exercise 5.35 (p. 261). Trap spacing measurements (in meters) for a sample of seven teams of red spiny lobster fishermen are reproduced in the accompanying table (and saved in the **TRAPSPACE** file). Let μ represent the average of the trap spacing measurements for the population of red spiny lobster fishermen fishing in Baja California Sur, Mexico. In Exercise 5.35 you computed the mean and standard deviation of the sample measurements to be $\bar{x} = 89.9$ meters and $s = 11.6$ meters, respectively. Suppose you want to determine if the true value of μ differs from 95 meters.

93	99	105	94	82	70	86

From Shester, G. G. "Explaining catch variation among Baja California lobster fishers through spatial analysis of trap-placement decisions." *Bulletin of Marine Science*, Vol. 86, No. 2, April 2010 (Table 1), pp. 479–498. Reprinted with permission from the University of Miami – *Bulletin of Marine Science*.

a. Specify the null and alternative hypothesis for this test.
b. Since $\bar{x} = 89.9$ is less than 95, a fisherman wants to reject the null hypothesis. What are the problems with using such a decision rule?
c. Compute the value of the test statistic.
d. Find the approximate *p*-value of the test.
e. Select a value of α, the probability of a Type I error. Interpret this value in the words of the problem.
f. Give the appropriate conclusion, based on the results of parts **d** and **e**.
g. What conditions must be satisfied for the test results to be valid?
h. In Exercise 5.35 you found a 95% confidence interval for μ. Does the interval support your conclusion in part **f**?

6.66 **Reading Japanese books.** Refer to the *Reading in a Foreign Language* (April 2004) experiment to improve the Japanese reading comprehension levels of University of Hawaii students presented in Exercise 2.33 (p. 46). Recall that 14 students participated in a 10-week extensive reading program in a second-semester Japanese course. The data on number of books read by each student are saved in the **JAPANESE** file. A MINITAB printout of the data analysis is shown below.

One-Sample T: BOOKS

Test of mu = 25 vs > 25

Variable	N	Mean	StDev	SE Mean	95% Lower Bound	T	P
BOOKS	14	31.64	10.49	2.80	26.68	2.37	0.017

a. State the null and alternative hypotheses for determining whether the average number of books read by all students who participated in the extensive reading program exceeds 25.
b. Find the rejection region for the test, using $\alpha = .05$.
c. Compute the test statistic.
d. State the appropriate conclusion for the test.
e. What conditions are required for the test results to be valid?
f. Locate the *p*-value on the MINITAB printout and use it to test the hypothesis. Your conclusion should agree with your answer in part **d**.

6.67 **Dental anxiety study.** Refer to the *BMC Oral Health* (Vol. 9, 2009) study of adults who completed the Dental Anxiety Scale, presented in Exercise 4.81 (p. 200). Recall

that scores range from 0 (no anxiety) to 25 (extreme anxiety). Summary statistics for the scores of 15 adults who completed the questionnaire are $\bar{x} = 10.7$ and $s = 3.6$. Conduct a test of hypothesis to determine whether the mean Dental Anxiety Scale score for the population of college students differs from $\mu = 11$. Use $\alpha = .05$.

6.68 **Crab spiders hiding on flowers.** Refer to the *Behavioral Ecology* (Jan. 2005) experiment on crab spiders' use of camouflage to hide from predators (e.g., birds) on flowers, presented in Exercise 2.38 (p. 47). Researchers at the French Museum of Natural History collected a sample of 10 adult female crab spiders, each sitting on the yellow central part of a daisy, and measured the chromatic contrast between each spider and the flower. The data (for which higher values indicate a greater contrast, and, presumably, an easier detection by predators) are shown in the accompanying table and saved in the **SPIDER** file. The researchers discovered that a contrast of 70 or greater allows birds to see the spider. Of interest is whether or not the true mean chromatic contrast of crab spiders on daisies is less than 70.

57	75	116	37	96	61	56	2	43	32

Based on Thery, M., et al. "Specific color sensitivities of prey and predator explain camouflage in different visual systems." *Behavioral Ecology*, Vol. 16, No. 1, Jan. 2005 (Table 1).

a. Define the parameter of interest, μ
b. Set up the null and alternative hypotheses of interest.
c. Find $\bar{x}$ and s for the sample data, and then use these values to compute the test statistic.
d. Give the rejection region for $\alpha = .10$.
e. State the appropriate conclusion in the words of the problem.

6.69 **Cheek teeth of extinct primates.** Refer to the *American Journal of Physical Anthropology* (Vol. 142, 2010) study of the characteristics of cheek teeth (e.g., molars) in an extinct primate species, Exercise 2.34 (p. 46). Recall that the researchers recorded the dentary depth of molars (in millimeters) for a sample of 18 cheek teeth extracted from skulls. These depth measurements, saved in the **CHEEKTEETH** file, are listed in the accompanying table. Anthropologists know that the mean dentary depth of molars in an extinct primate species—called Species A—is 15 millimeters. Is there evidence to indicate that the sample of 18 cheek teeth come from some other extinct primate species (i.e., some species other than Species A)? Use the SPSS printout (p. 323) to answer the question.

18.12	16.55
19.48	15.70
19.36	17.83
15.94	13.25
15.83	16.12
19.70	18.13
15.76	14.02
17.00	14.04
13.96	16.20

Based on Boyer, D. M., Evans, A. R., and Jernvall, J. "Evidence of dietary differentiation among Late Paleocene–Early Eocene Plesiadapids (Mammalia, Primates)." *American Journal of Physical Anthropology*, Vol. 142, © 2010.

SPSS Output for Exercise 6.69

One-Sample Statistics

	N	Mean	Std. Deviation	Std. Error Mean
M2Depth	18	16.4994	1.97042	.46443

One-Sample Test

	Test Value = 15					
					95% Confidence Interval of the Difference	
	t	df	Sig. (2-tailed)	Mean Difference	Lower	Upper
M2Depth	3.229	17	.005	1.49944	.5196	2.4793

Applying the Concepts — Intermediate

6.70 **Pitch memory of amusiacs.** Amusia is a congenital disorder that adversely impacts one's perception of music. Refer to the *Advances in Cognitive Psychology* (Vol. 6, 2010) study of the pitch memory of individuals diagnosed with amusia, Exercise 5.41 (p. 262). Recall that each in a sample of 17 amusiacs listened to a series of tone pairs and was asked to determine if the tones were the same or different. In the first trial, the tones were separated by 1 second; in a second trial, the tones were separated by 5 seconds. The difference in accuracy scores for the two trials was determined for each amusiac (where the difference is the score on the first trial minus the score on the second trial). The mean score difference was .11 with a standard deviation of .19.

   a. In theory, the longer the delay between tones, the less likely one is to detect a difference between the tones. Consequently, the true mean score difference should exceed 0. Set up the null and alternative hypotheses for testing the theory.

   b. Carry out the test, part **a,** using $\alpha = .05$. Is there evidence to support the theory?

6.71 **Radioactive lichen.** Refer to the Lichen Radionuclide Baseline Research project to monitor the level of radioactivity in lichen, presented in Exercise 5.39 (p. 262). Recall that University of Alaska researchers collected nine lichen specimens and measured the amount of the radioactive element cesium-137 (in microcuries per milliliter) in each specimen. (The natural logarithms of the data values are saved in the **LICHEN** file.) Assume that in previous years the mean cesium amount in lichen was $\mu = .003$ microcurie per milliliter. Is there sufficient evidence to indicate that the mean amount of cesium in lichen specimens differs from this value? Use the SAS printout below to conduct a complete test of hypothesis at $\alpha = .10$.

```
Sample Statistics for CESIUM

    N        Mean      Std. Dev.      Std. Error
-----------------------------------------------------
    9       0.0090      0.0049         0.0016

Hypothesis Test

   Null hypothesis:     Mean of CESIUM =  0.003
   Alternative:         Mean of CESIUM ^= 0.003

    t Statistic      Df      Prob > t
-----------------------------------------------------
      3.725           8       0.0058
```

6.72 **A new dental bonding agent.** When bonding teeth, orthodontists must maintain a dry field. A new bonding adhesive (called "Smartbond") has been developed to eliminate the necessity of a dry field. However, there is concern that the new bonding adhesive is not as strong as the current standard, a composite adhesive. (*Trends in Biomaterials & Artificial Organs*, Jan. 2003.) Tests on a sample of 10 extracted teeth bonded with the new adhesive resulted in a mean breaking strength (after 24 hours) of $\bar{x} = 5.07$ Mpa and a standard deviation of $s = .46$ Mpa, where Mpa = megapascal, a measure of force per unit area. Orthodontists want to know if the true mean breaking strength of the new bonding adhesive is less than 5.70 Mpa, the mean breaking strength of the composite adhesive. Conduct the appropriate analysis for the orthodontists. Use $\alpha = .01$.

6.73 **Minimizing tractor skidding distance.** Refer to the *Journal of Forest Engineering* (July 1999) study of minimizing tractor skidding distances along a new road in a European forest, presented in Exercise 5.44 (p. 263). The skidding distances (in meters) were measured at 20 randomly selected road sites. The data (saved in the **SKIDDING** file) are repeated in the accompanying table. Recall that a logger working on the road claims that the mean skidding distance is at least 425 meters. Is there sufficient evidence to refute this claim? Use $\alpha = .10$.

| 488 | 350 | 457 | 199 | 285 | 409 | 435 | 574 | 439 | 546 |
| 385 | 295 | 184 | 261 | 273 | 400 | 311 | 312 | 141 | 425 |

Based on Tujek, J. and Pacola, E. "Algorithms for skidding distance modeling on a raster Digital Terrain Model." *Journal of Forest Engineering*, Vol. 10, No. 1, July 1999 (Table 1).

6.74 **Mongolian desert ants.** Refer to the *Journal of Biogeography* (December 2003) study of ants in Mongolia (central Asia), presented in Exercise 2.68 (p. 59). Recall that botanists placed seed baits at 11 study sites and observed the number of ant species attracted to each site. A portion of the data (saved in the **GOBIANTS** file) is provided in the next table (p. 324). Do these data indicate that the average number of ant species at Mongolian desert sites differs from 5? Conduct the appropriate test at $\alpha = .05$. Are the conditions required for a valid test satisfied?

Data for Exercise 6.74

Site	Region	Number of Ant Species
1	Dry Steppe	3
2	Dry Steppe	3
3	Dry Steppe	52
4	Dry Steppe	7
5	Dry Steppe	5
6	Gobi Desert	49
7	Gobi Desert	5
8	Gobi Desert	4
9	Gobi Desert	4
10	Gobi Desert	5
11	Gobi Desert	4

Based on Pfeiffer, M., et al. "Community organization and species richness of ants in Mongolia along an ecological gradient from steppe to Gobi desert." *Journal of Biogeography,* Vol. 30, No. 12, Dec. 2003 (Tables 1 and 2).

6.75 **Walking straight into circles.** When people get lost in unfamiliar terrain, do they really walk in circles, as is commonly believed? To answer this question, researchers conducted a field experiment and reported the results in *Current Biology* (Sept. 29, 2009). Fifteen volunteers were blindfolded and asked to walk as straight as possible in a certain direction in a large field. Walking trajectories were monitored every second for 50 minutes using GPS and the average directional bias (degrees per second) recorded for each walker. The data, saved in the **CIRCLES** file, are shown in the table (next column). A strong tendency to veer consistently in the same direction will cause walking in circles. A mean directional bias of 0 indicates that walking trajectories were random. Consequently, the researchers tested whether the true mean bias differed significantly from 0. A MINITAB printout of the analysis is shown below.

a. Interpret the results of the hypothesis test for the researchers. Use $\alpha = .10$.

b. Although most volunteers showed little overall bias, the researchers produced maps of the walking paths showing that each occasionally made several small circles during the walk. Ultimately, the researchers supported the "walking into circles" theory. Explain why the data in the table is insufficient for testing whether an individual walks into circles.

-4.50	-1.00	-0.50	-0.15	0.00	0.01	0.02	0.05	0.15
0.20	0.50	0.50	1.00	2.00	3.00			

Based on Souman, J. L., Frissen, I., Sreenivasa, M. N., and Ernst, M. O. "Walking straight into circles." *Current Biology,* Vol. 19, No. 18, Sept. 29, 2009 (Figure 2).

Applying the Concepts—Advanced

6.76 **Lengths of great white sharks.** One of the most feared predators in the ocean is the great white shark. It is known that the white shark grows to a mean length of 21 feet; however, one marine biologist believes that great white sharks off the Bermuda coast grow much longer owing to unusual feeding habits. To test this claim, some full-grown great white sharks were captured off the Bermuda coast, measured, and then set free. However, because the capture of sharks is difficult, costly, and very dangerous, only three specimens were sampled. Their lengths were 24, 20, and 22 feet. Do these data support the marine biologist's claim at $\alpha = .10$?

MINITAB Output for Exercise 6.75

One-Sample T: BIAS

Test of mu = 0 vs not = 0

Variable	N	Mean	StDev	SE Mean	95% CI	T	P
BIAS	15	0.085	1.603	0.414	(-0.802, 0.973)	0.21	0.840

6.6 Large-Sample Test of Hypothesis about a Population Proportion

Inferences about population proportions (or percentages) are often made in the context of the probability p of "success" for a binomial distribution. We saw how to use large samples from binomial distributions to form confidence intervals for p in Section 5.4. We now consider tests of hypotheses about p.

Consider, for example, a method currently used by doctors to screen women for breast cancer. The method fails to detect cancer in 20% of the women who actually have the disease. Suppose a new method has been developed that researchers hope will detect cancer more accurately. This new method was used to screen a random sample of 140 women known to have breast cancer. Of these, the new method failed to detect cancer in 12 women. Does this sample provide evidence that the failure rate of the new method differs from the one currently in use?

We first view this experiment as a binomial one with 140 screened women as the trials and failure to detect breast cancer as "Success" (in binomial terminology). Let p represent the probability that the new method fails to detect the breast cancer; on the one hand, if the new method is no better than the current one, then the failure rate is $p = .2$. On the other hand, if the new method is either better or worse than the current method, then the failure rate is either smaller or larger than 20%; that is, $p \neq .2$.

We can now place the problem in the context of a test of hypothesis:

$$H_0: p = .2 \text{ (Cancer detection rate} = .2.)$$
$$H_a: p \neq .2 \text{ (Cancer detection rate differs from .2.)}$$

Recall that the sample proportion $\hat{p}$ is really just the sample mean of the outcomes of the individual binomial trials and, as such, is approximately normally distributed (for large samples) according to the Central Limit Theorem. Thus, for large samples we can use the standard normal z as the test statistic:

$$\textit{Test statistic:} \quad z = \frac{\text{Sample proportion} - \text{Null hypothesized proportion}}{\text{Standard deviation of sample proportion}}$$

$$= \frac{\hat{p} - p_0}{\sigma_{\hat{p}}}$$

Note that we used the symbol p_0 to represent the null hypothesized value of p.

Rejection region: We use the standard normal distribution to find the appropriate rejection region for the specified value of α. Using $\alpha = .05$, the two-tailed rejection region is

$$z < -z_{\alpha/2} = -z_{.025} = -1.96 \quad \text{or} \quad z > z_{\alpha/2} = z_{.025} = 1.96$$

(See Figure 6.16.)

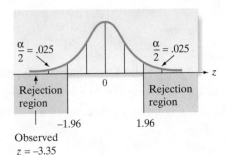

Figure 6.16

Rejection region for breast cancer example

We are now prepared to calculate the value of the test statistic. Before doing so, however, we want to be sure that the sample size is large enough to ensure that the normal approximation for the sampling distribution of $\hat{p}$ is reasonable. Recall from Section 5.4 that we require both np and nq to be at least 15. Because the null hypothesized value, p_0, is assumed to be the true value of p until our test procedure indicates otherwise, then we require $np_0 \geq 15$ and $nq_0 \geq 15$ (where $q_0 = 1 - p_0$). Since $p_0 = .2$ for our test, we have

$$np_0 = (140)(.2) = 28$$

and

$$nq_0 = (140)(.8) = 112$$

Therefore, the normal distribution will provide a reasonable approximation for the sampling distribution of $\hat{p}$.

Returning to the hypothesis test at hand, the proportion of the screenings that failed to detect breast cancer is

$$\hat{p} = \frac{12}{140} = .086$$

Finally, we calculate the number of standard deviations (the z value) between the sampled and hypothesized values of the binomial proportion:

$$z = \frac{\hat{p} - p_0}{\sigma_{\hat{p}}} = \frac{\hat{p} - p_0}{\sqrt{p_0 q_0 / n}} = \frac{.086 - .2}{\sqrt{(.2)(.8)/140}} = \frac{-.114}{.034} = -3.35$$

The implication is that the observed sample proportion is (approximately) 3.35 standard deviations below the null-hypothesized proportion, .2 (Figure 6.16). Therefore, we reject the null hypothesis, concluding at the .05 level of significance that the true failure rate of the new method for detecting breast cancer differs from .20. Since $\hat{p} = .086$, it appears that the new method is better (i.e., has a smaller failure rate) than the method currently in use. (To estimate the magnitude of the failure rate for the new method, a confidence interval can be constructed.)

The test of hypothesis about a population proportion p is summarized in the next box. Note that the procedure is entirely analogous to that used for conducting large-sample tests about a population mean.

Large-Sample Test of Hypothesis about p: Normal (z) Statistic

One-Tailed Test	**Two-Tailed Test**
$H_0: p = p_0$	$H_0: p = p_0$
$H_a: p < p_0 \,(\text{or } H_a: p > p_0)$	$H_a: p \neq p_0$

Test statistic: $z = \dfrac{\hat{p} - p_0}{\sigma_{\hat{p}}}$ Test statistic: $z = \dfrac{\hat{p} - p_0}{\sigma_{\hat{p}}}$

where p_0 = hypothesized value of p, $\sigma_{\hat{p}} = \sqrt{p_0 q_0/n}$, and $q_0 = 1 - p_0$

Rejection region: $z < -z_\alpha$ Rejection region: $z < -z_{\alpha/2}$
(or $z > z_\alpha$ when $H_a: p > p_0$)   or $z > z_{\alpha/2}$

Conditions Required for a Valid Large-Sample Hypothesis Test for p

1. A random sample is selected from a binomial population.
2. The sample size n is large. (This condition will be satisfied if np_0 and nq_0 are both at least 15.)

Example 6.9

A Hypothesis Test for p— Proportion of Defective Batteries

Problem The reputations (and hence sales) of many businesses can be severely damaged by shipments of manufactured items that contain a large percentage of defectives. For example, a manufacturer of alkaline batteries may want to be reasonably certain that less than 5% of its batteries are defective. Suppose 300 batteries are randomly selected from a very large shipment; each is tested and 10 defective batteries are found. Does this outcome provide sufficient evidence for the manufacturer to conclude that the fraction defective in the entire shipment is less than .05? Use $\alpha = .01$.

Solution The objective of the sampling is to determine whether there is sufficient evidence to indicate that the fraction defective, p, is less than .05. Consequently, we will test the null hypothesis that $p = .05$ against the alternative hypothesis that $p < .05$. The elements of the test are

$$H_0: p = .05 \text{ (Fraction of defective batteries equals .05.)}$$
$$H_a: p < .05 \text{ (Fraction of defective batteries is less than .05.)}$$

$$\text{Test statistic: } z = \frac{\hat{p} - p_0}{\sigma_{\hat{p}}}$$

$$\text{Rejection region: } z < -z_{.01} = -2.33 \qquad \text{(see Figure 6.17)}$$

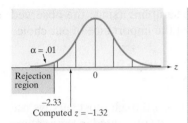

Figure 6.17
Rejection region for
Example 6.9

Before conducting the test, we check to determine whether the sample size is large enough to use the normal approximation to the sampling distribution of $\hat{p}$. Since $np_0 = (300)(.05) = 15$ and $nq_0 = (300)(.95) = 285$ are both at least 15, the normal approximation will be adequate.

We now calculate the test statistic:

$$z = \frac{\hat{p} - .05}{\sigma_{\hat{p}}} = \frac{(10/300) - .05}{\sqrt{p_0 q_0 / n}} = \frac{.03333 - .05}{\sqrt{p_0 q_0 / 300}}$$

Notice that we use p_0 to calculate $\sigma_{\hat{p}}$ because, in contrast to calculating $\sigma_{\hat{p}}$ for a confidence interval, the test statistic is computed on the assumption that the null hypothesis is true—that is, $p = p_0$. Therefore, substituting the values for $\hat{p}$ and p_0 into the z statistic, we obtain

$$z \approx \frac{-.01667}{\sqrt{(.05)(.95)/300}} = \frac{-.01667}{.0126} = -1.32$$

As shown in Figure 6.17, the calculated z-value does not fall into the rejection region. Therefore, there is insufficient evidence at the .01 level of significance to indicate that the shipment contains less than 5% defective batteries.

Now Work Exercise 6.80a–b

Example 6.10

Finding the *p*-Value for a Test about a Population Proportion *p*

Problem In Example 6.9, we found that we did not have sufficient evidence at the $\alpha = .01$ level of significance to indicate that the fraction defective, p, of alkaline batteries was less than $p = .05$. How strong was the weight of evidence favoring the alternative hypothesis (H_a: $p < .05$)? Find the observed significance level (*p*-value) for the test.

Solution The computed value of the test statistic z was $z = -1.32$. Therefore, for this lower-tailed test, the observed significance level is

$$p\text{-value} = P(z \le -1.32)$$

This lower-tail area is shown in Figure 6.18. The area between $z = 0$ and $z = 1.32$ is given in Table III in Appendix A as .4066. Therefore, the observed significance level is $.5 - .4066 = .0934$.

Note: The *p*-value can also be obtained with statistical software. The MINITAB printout shown in Figure 6.19 gives the *p*-value (highlighted).

Look Back Although we did not reject H_0: $p = .05$ at $\alpha = .01$, the probability of observing a z-value as small as or smaller than -1.32 is only .093 if in fact H_0 is true. Therefore,

Figure 6.18
The observed significance level
for Example 6.10

Test and CI for One Proportion

Test of p = 0.05 vs p < 0.05

Sample	X	N	Sample p	95% Upper Bound	Z-Value	P-Value
1	10	300	0.033333	0.050380	-1.32	0.093

Using the normal approximation.

Figure 6.19
MINITAB lower-tailed test of
p, Example 6.10

we would reject H_0 if we had selected $\alpha = .10$ prior to sampling (since the observed significance level is less than .10). This point reminds us of the importance of our choice of α prior to collecting the sample data.

Now Work Exercise 6.80c

Small-sample test procedures are also available for p, although most surveys use samples that are large enough to employ the large-sample tests presented in this section. A test of proportions that can be applied to small samples is discussed in Chapter 8.

Statistics IN Action | Revisited — Testing a Population Proportion in the KLEENEX® Survey

In the previous "Statistics in Action Revisited" (p. 315), we investigated Kimberly-Clark Corporation's assertion that the company should put more than 60 tissues in a cold-care box of KLEENEX® tissues. We did this by testing the claim that the mean number of tissues used by a person with a cold is $\mu = 60$, using data collected from a survey of 250 KLEENEX® users. Another approach to the problem is to consider the proportion of KLEENEX® users who use more than 60 tissues when they have a cold. Now the population parameter of interest is p, the proportion of all KLEENEX® users who use more than 60 tissues when they have a cold.

Kimberly-Clark Corporation's belief that the company should put more than 60 tissues in a cold-care box will be supported if over half of the KLEENEX® users surveyed use more than 60 tissues (i.e., if $p > .5$). Is there evidence to indicate that the population proportion exceeds .5? To answer this question, we set up the following null and alternative hypotheses:

$$H_0: p = .5 \quad H_a: p > .5$$

Recall that the survey results for the 250 sampled KLEENEX® users are stored in the **TISSUES** data file. In addition to the number of tissues used by each person, the file contains a qualitative variable—called USED60—representing whether the person used fewer or more than 60 tissues. (The values of USED60 in the data set are "LESS" or "MORE.") A

MINITAB analysis of this variable yielded the printout displayed in Figure SIA6.2.

On the MINITAB printout, x represents the number of the 250 people with colds that used more than 60 tissues. Note that $x = 154$. This value is used to compute the test statistic $z = 3.67$, highlighted on the printout. The p-value of the test, also highlighted on the printout, is p-value = .000. Since this value is less than $\alpha = .05$, there is sufficient evidence (at $\alpha = .05$) to reject H_0; we conclude that the proportion of all KLEENEX® users who use more than 60 tissues when they have a cold exceeds .5. This conclusion supports again, the company's decision to put more than 60 tissues in a pack of KLEENEX.

Data Set: TISSUES

Test and CI for One Proportion: USED60

```
Test of p = 0.5 vs p > 0.5

Event = MORE

                                    95% Lower
Variable    X    N  Sample p         Bound   Z-Value  P-Value
USED60    154  250  0.616000      0.565404      3.67    0.000

Using the normal approximation.
```

Figure SIA6.2
MINITAB test of $p = .5$ for KLEENEX survey

Exercises 6.77–6.95

Understanding the Principles

6.77 What type of data, quantitative or qualitative, is typically associated with making inferences about a population proportion p?

6.78 What conditions are required for a valid large-sample test for p?

Learning the Mechanics

6.79 For the binomial sample sizes and null-hypothesized values of p in each part, determine whether the sample size is large enough to use the normal approximation methodology presented in this section to conduct a test of the null hypothesis $H_0: p = p_0$.
  a. $n = 500, p_0 = .05$
  b. $n = 100, p_0 = .99$
  c. $n = 50, p_0 = .2$
  d. $n = 20, p_0 = .2$
  e. $n = 10, p_0 = .4$

6.80 Suppose a random sample of 100 observations from a binomial population gives a value of $\hat{p} = .69$ and you wish to
NW

test the null hypothesis that the population parameter p is equal to .75 against the alternative hypothesis that p is less than .75.

a. Noting that $\hat{p} = .69$, what does your intuition tell you? Does the value of $\hat{p}$ appear to contradict the null hypothesis?

b. Use the large-sample z-test to test $H_0: p = .75$ against the alternative hypothesis $H_a: p < .75$. Use $\alpha = .05$. How do the test results compare with your intuitive decision from part **a?**

c. Find and interpret the observed significance level of the test you conducted in part **b.**

6.81 Suppose the sample in Exercise 6.80 has produced $\hat{p} = .84$ and we wish to test $H_0: p = .9$ against the alternative $H_a: p < .9$.

a. Calculate the value of the z statistic for this test.

b. Note that the numerator of the z statistic ($\hat{p} - p_0 = .84 - .90 = -.06$) is the same as for Exercise 6.80. Considering this, why is the absolute value of z for this exercise larger than that calculated in Exercise 6.80?

c. Complete the test, using $\alpha = .05$, and interpret the result.

d. Find the observed significance level for the test and interpret its value.

6.82 A random sample of 100 observations is selected from a binomial population with unknown probability of success, p. The computed value of $\hat{p}$ is equal to .74.

a. Test $H_0: p = .65$ against $H_a: p > .65$. Use $\alpha = .01$.

b. Test $H_0: p = .65$ against $H_a: p > .65$. Use $\alpha = .10$.

c. Test $H_0: p = .90$ against $H_a: p \neq .90$. Use $\alpha = .05$.

d. Form a 95% confidence interval for p.

e. Form a 99% confidence interval for p.

6.83 Refer to Exercise 5.54 (p. 269), in which 50 consumers taste-tested a new snack food. The data are saved in the **SNACK** file.

a. Test $H_0: p = .5$ against $H_a: p > .5$, where p is the proportion of customers who do not like the snack food. Use $\alpha = .10$.

b. Report the observed significance level of your test.

Applet Exercise 6.5

Use the applet entitled *Hypotheses Test for a Proportion* to investigate the relationships between the probabilities of Type I and Type II errors occurring at levels .05 and .01. For this exercise, use $n = 100$, true $p = 0.5$, and alternative *not equal.*

a. Set null $p = .5$. What happens to the proportion of times the null hypothesis is rejected at the .05 level and at the .01 level as the applet is run more and more times? What type of error has occurred when the null hypothesis is rejected in this situation? Based on your results, is this type of error more likely to occur at level .05 or at level .01? Explain.

b. Set null $p = .6$. What happens to the proportion of times the null hypothesis is *not* rejected at the .05 level and at the .01 level as the applet is run more and more times? What type of error has occurred when the null hypothesis is *not* rejected in this situation? Based on your results, is this type of error more likely to occur at level .05 or at level .01? Explain.

c. Use your results from parts **a** and **b** to make a general statement about the probabilities of Type I and Type II errors at levels .05 and .01.

Applet Exercise 6.6

Use the applet entitled *Hypotheses Test for a Proportion* to investigate the effect of the true population proportion p on the probability of a Type I error occurring. For this exercise, use $n = 100$ and alternative *not equal.*

a. Set true $p = .5$ and null $p = .5$. Run the applet several times, and record the proportion of times the null hypothesis is rejected at the .01 level.

b. Clear the applet and repeat part **a** for true $p = .1$ and null $p = .1$. Then repeat one more time for true $p = .01$ and null $p = .01$.

c. Based on your results from parts **a** and **b**, what can you conclude about the probability of a Type I error occurring as the true population proportion gets closer to 0?

Applying the Concepts — Basic

6.84 **Satellite radio in cars.** A spokesperson for the National Association of Broadcasters (NAB) claims that 80% of all satellite radio subscribers have a satellite radio receiver in their car. Recall from Exercise 1.24 (p. 20), that in a June 2007 survey of 501 satellite radio subscribers, 396 had a satellite receiver in their car. Consider a test of the NAB spokesperson's claim.

a. Define the parameter of interest to the NAB spokesperson.

b. Set up the null hypothesis for testing the claim.

c. Specify the alternative hypothesis if you believe that the spokesperson's claim is too high.

d. Compute the value of the test statistic.

e. Determine the rejection region for the test using $\alpha = .10$.

f. Compute the p-value of the test.

g. Make the appropriate conclusion. Show that the decision based on the rejection region agrees with the decision based on the p-value.

6.85 **Accuracy of price scanners at Wal-Mart.** Refer to Exercise 5.81 (p. 276) and the study of the accuracy of checkout scanners at Wal-Mart Stores in California. Recall that the National Institute for Standards and Technology (NIST) mandates that, for every 100 items scanned through the electronic checkout scanner at a retail store, no more than 2 should have an inaccurate price. A study of random items purchased at California Wal-Mart stores found that 8.3% had the wrong price (*Tampa Tribune*, Nov. 22, 2005). Assume that the study included 1,000 randomly selected items.

a. Identify the population parameter of interest in the study.

b. Set up H_0 and H_a for a test to determine whether the true proportion of items scanned at California Wal-Mart stores exceeds the 2% NIST standard.

c. Find the test statistic and rejection region (at $\alpha = .05$) for the test.

d. Give a practical interpretation of the test.

e. What conditions are required for the inference made in part **d** to be valid? Are these conditions met?

6.86 **Gummy bears: red or yellow?** *Chance* (Winter 2010) presented a lesson in hypothesis testing carried out by medical students in a biostatistics class. Students were blind-folded and then given a red-colored or yellow-colored gummy bear to chew. (Half the students were randomly assigned to receive the red gummy bear and half to receive the yellow bear. The students could not see what color gummy bear they were given.) After chewing, the students were asked to guess the color of the candy based on the flavor. Of the 121 students who participated in the study, 97 correctly identified the color of the gummy bear.

a. If there is no relationship between color and gummy bear flavor, what proportion of the population of students will correctly identify the color?

b. Specify the null and alternative hypothesis for testing whether color and flavor are related.

c. Carry out the test and give the appropriate conclusion at $\alpha = .01$. Use the *p*-value of the test to make your decision.

6.87 **Single-parent families.** Examining data collected on 835 males from the National Youth Survey (a longitudinal survey of a random sample of U.S. households), researchers at Carnegie Mellon University found that 401 of the male youths were raised in a single-parent family (*Sociological Methods & Research*, February 2001). Does this information allow you to conclude that more than 45% of male youths are raised in a single-parent family? Test at $\alpha = .05$.

6.88 **Teenagers' use of emoticons in school writing.** Refer to the *Pew Internet and American Life Project* (April 2008) survey of the writing habits of U.S. teenagers, Exercise 5.61 (p. 270). Recall that in a random sample of 700 teenagers, 448 admitted to using at least one informal element in school writing assignments. [*Note*: Emoticons, such as using the symbol ":)" to represent a smile, and abbreviations, such as writing "LOL" for "laughing out loud," are considered informal elements.] Is there evidence to indicate that less than 65% of all U.S. teenagers have used at least one informal element in school writing assignments? Test the relevant hypotheses using $\alpha = .05$.

Applying the Concepts—Intermediate

6.89 **Identifying organisms using a computer.** *National Science Education Standards* recommend that all life science students be exposed to methods of identifying unknown biological specimens. Due to certain limitations of traditional identification methods, biology professors at Slippery Rock University (SRU) developed a computer-aided system for identifying common conifers (deciduous trees) called Confir ID (*The American Biology Teacher*, May 2010). A sample of 171 life science students exposed to both a traditional method of identifying conifers and Confir ID and then asked which method they preferred. As a result, 138 students indicated their preference for Confir ID. In order to change the life sciences curriculum at SRU to include Confir ID, the biology department requires that more than 70% of the students prefer the new, computerized method. Should Confir ID be added to the curriculum at SRU? Explain your reasoning.

6.90 **Graduation rates of student–athletes.** Are student–athletes at Division I universities poorer students than nonathletes? The National Collegiate Athletic Association (NCAA) measures the academic outcomes of student–athletes with the Graduation Success Rate (GSR)—the percentage of eligible athletes who graduate within six years of entering college. According to the NCAA, the GSR for all scholarship athletes at Division I institutions is 63% (*Inside Higher Ed*, Nov. 10, 2006). It is well known that the GSR for all students at Division I colleges is 60%.

a. Suppose the NCAA report was based on a sample of 500 student–athletes, of which 315 graduated within six years. Is this sufficient information to conclude that the GSR for all scholarship athletes at Division I institutions differs from 60%? Test, using $\alpha = .01$.

b. The GSR statistics were also broken down by gender and sport. For example, men's Division I college basketball players had a GSR of 42% (compared with a known GSR of 58% for all male college students). Suppose this statistic was based on a sample of 200 male basketball players, of whom 84 graduated within six years. Is this sufficient information to conclude that the GSR for all male basketball players at Division I institutions differs from 58%? Test, using $\alpha = .01$.

6.91 **Astronomy students and the Big Bang Theory.** Indiana University professors investigated first-year college students' knowledge of astronomy (*Astronomy Education Review*, Vol. 2, 2003). One concept of interest was the Big Bang theory of the creation of the universe. In a sample of 148 freshmen students, 37 believed that the Big Bang theory accurately described the creation of planetary systems. Based on this information, would you be willing to state that more than 20% of all freshmen college students believe in the Big Bang theory? How confident are you of your decision?

6.92 **Study of lunar soil.** *Meteoritics* (March 1995) reported the results of a study of lunar soil evolution. Data were obtained from the *Apollo 16* mission to the moon, during which a 62-cm core was extracted from the soil near the landing site. Monomineralic grains of lunar soil were separated out and examined for coating with dust and glass fragments. Each grain was then classified as coated or uncoated. Of interest is the "coat index"—that is, the proportion of grains that are coated. According to soil evolution theory, the coat index will exceed .5 at the top of the core, equal .5 in the middle of the core, and fall below .5 at the bottom of the core. Use the summary data in the accompanying table to test each part of the three-part theory. Use $\alpha = .05$ for each test.

	Location (depth)		
	Top (4.25 cm)	Middle (28.1 cm)	Bottom (54.5 cm)
Number of grains sampled	84	73	81
Number coated	64	35	29

Based on Basu, A., and McKay, D. S. "Lunar soil evolution processes and Apollo 16 core 60013/60014." *Meteoritics*, Vol. 30, No. 2, Mar. 1995, p. 166 (Table 2).

6.93 **Effectiveness of skin cream.** Pond's has discontinued the production of Age-Defying Complex, a cream with alpha-hydroxy acid, with Age-Defying Towlettes. Pond's advertised that the product could reduce wrinkles and improve the skin. In a study published in *Archives of Dermatology* (June 1996), 33 middle-aged women used a product with alpha-hydroxy acid for 22 weeks. At the end of the study period, a dermatologist judged whether each woman exhibited any improvement in the condition of her skin. The results for the 33 women (where I = improved skin and N = no improvement) are listed in the accompanying table and saved in the **SKINCREAM** file.

a. Do the data provide sufficient evidence to conclude that the cream will improve the skin of more than 60% of middle-aged women? Test, using $\alpha = .05$.

b. Find and interpret the *p*-value of the test.

I	I	N	I	N	N	N	I	I	I	I	I	I
N	I	I	I	N	I	I	I	N	N	I	N	I
I	I	I	I	I	N	I	I	N				

Applying the Concepts—Advanced

6.94 Killing insects with low oxygen. A group of Australian entomological toxicologists investigated the impact of exposure to low oxygen on the mortality of insects (*Journal of Agricultural, Biological, and Environmental Statistics*, Sept. 2000). Thousands of adult rice weevils were placed in a chamber filled with wheat grain, and the chamber was exposed to nitrogen gas for 4 days. Insects were assessed as dead or alive 24 hours after exposure. At the conclusion of the experiment, 31,386 weevils were dead and 35 weevils were found alive. Previous studies had shown a 99% mortality rate in adult rice weevils exposed to carbon dioxide for 4 days. What advice would you give to the entomologists regarding the use of nitrogen or carbon dioxide for killing rice weevils?

6.95 Male fetal deaths following 9/11/2001. According to research published in *BMC Public Health* (May 25, 2010), the terrorist attacks of September 11, 2001, may have led to a spike in miscarriages of male babies. Researchers from the University of California, using data from the National Vital Statistics System, reported that the male fetal death rate in September of that year was significantly higher than expected. Suppose it is known that the miscarriage rate for pregnant women expecting a boy in the United States is 3 in 1,000. In a random sample of 2,000 pregnant women expecting a baby boy during September of 2001, how many would need to have experienced a fetal death (miscarriage) in order to support the claim (at $\alpha = .05$) that the male fetal death rate in September 2001 was higher than expected?

6.7 Test of Hypothesis about a Population Variance (Optional)

Although many practical problems involve inferences about a population mean (or proportion), it is sometimes of interest to make an inference about a population variance σ^2. To illustrate, a quality control supervisor in a cannery knows that the exact amount each can contains will vary, since there are certain uncontrollable factors that affect the amount of fill. The mean fill per can is important, but equally important is the variance of the fill. If σ^2 is large, some cans will contain too little and others too much. Suppose regulatory agencies specify that the standard deviation of the amount of fill in 16-ounce cans should be less than .1 ounce. To determine whether the process is meeting this specification, the supervisor randomly selects 10 cans and weighs the contents of each. The results are given in Table 6.6.

Table 6.6	Fill Weights (ounces) of 10 Cans			
16.00	15.95	16.10	16.02	15.99
16.06	16.04	16.05	16.03	16.02

Data Set: FILLAMOUNTS

Do these data provide sufficient evidence to indicate that the variability is as small as desired? To answer this question, we need a procedure for testing a hypothesis about σ^2.

Intuitively, it seems that we should compare the sample variance σ^2 with the hypothesized value of σ^2 (or s with σ) in order to make a decision about the population's variability. The quantity

$$\frac{(n-1)s^2}{\sigma^2}$$

is known to have a **chi-square (χ^2) distribution** when the population from which the sample is taken is *normally distributed*. (Chi-square distributions were introduced in optional Section 5.6.)

Since the distribution of $\frac{(n-1)s^2}{\sigma^2}$ is known, we can use this quantity as a test statistic

in a test of hypothesis for a population variance, as illustrated in the next example.

BIOGRAPHY **FRIEDRICH R. HELMERT (1843–1917)**
Helmert Transformations

German Friedrich Helmert studied engineering sciences and mathematics at Dresden University, where he earned his Ph.D. Then he accepted a position as a professor of geodesy—the scientific study of the earth's size and shape—at the technical school in Aachen. Helmert's mathematical solutions to geodesy problems led him to several statistics-related discoveries. His greatest statistical contribution occurred in 1876, when he was the first to prove that the sampling distribution of the sample variance s^2 is a chi-square distribution. Helmert used a series of mathematical transformations to obtain the distribution of s^2—transformations that have since been named "Helmert transformations" in his honor. Later in life, Helmert was appointed professor of advanced geodesy at the prestigious University of Berlin and director of the Prussian Geodetic Institute. ∎

Example 6.11
A Test for σ^2—Fill Weight Variance

Problem Refer to the fill weights for the sample of ten 16-ounce cans in Table 6.6. Is there sufficient evidence to conclude that the true standard deviation σ of the fill measurements of 16-ounce cans is less than .1 ounce?

Solution Here, we want to test whether $\sigma < .1$. Since the null and alternative hypotheses must be stated in terms of σ^2 (rather than σ), we want to test the null hypothesis that $\sigma^2 = (.1)^2 = .01$ against the alternative that $\sigma^2 < .01$. Therefore, the elements of the test are

H_0: $\sigma^2 = .01$ (Fill variance equals .01—i.e., process specifications are not met.)

H_a: $\sigma^2 < .01$ (Fill variance is less than .01—i.e., process specifications are met.)

Test statistic: $\chi^2 = \dfrac{(n-1)s^2}{\sigma^2}$

Assumption: The distribution of the amounts of fill is approximately normal.

Rejection region: The smaller the value of s^2 we observe, the stronger is the evidence in favor of H_a. Thus, we reject H_0 for "small values" of the test statistic. Recall that the chi-square distribution depends on $(n-1)$ degrees of freedom. With $\alpha = .05$ and $(n-1) = 9$ df, the χ^2 value for rejection is found in Table V and pictured in Figure 6.20. We will reject H_0 if $\chi^2 < 3.32511$.

Figure 6.20

Rejection region for Example 6.11

[*Note:* The area given in Table V is the area to the *right* of the numerical value in the table. Thus, to determine the lower-tail value, which has $\alpha = .05$ to its *left*, we used the $\chi^2_{.95}$ column in Table V.]

A SAS printout of the analysis is displayed in Figure 6.21. The value of s (highlighted on the printout) is $s = .0412$. Substituting into the formula for the test statistic, we have

$$\chi^2 = \frac{(n-1)s^2}{\sigma^2} = \frac{9(.0412)^2}{.01} = 1.53$$

```
Sample Statistics for FILLAMT

   N      Mean      Std. Dev.      Variance
-------------------------------------------------
  10     16.026       0.0412        0.0017


Hypothesis Test

  Null hypothesis:        Variance of FILLAMT => 0.01
  Alternative:            Variance of FILLAMT <  0.01

       Chi-square         Df          Prob
-------------------------------------------------
         1.524             9         0.0030
```

Figure 6.21

SAS test of fill amount variance, Example 6.11

Conclusion: Since the value of the test statistic is less than 3.32511, the supervisor can conclude that the variance σ^2 of the population of all amounts of fill is less than .01 (i.e., $\sigma < .1$), with probability of a Type I error equal to $\alpha = .05$. If this procedure is repeatedly used, it will incorrectly reject H_0 only 5% of the time. Thus, the quality control supervisor is confident in the decision that the cannery is operating within the desired limits of variability.

Look Back Note that both the test statistic and the one-tailed *p*-value of the test are shown on the SAS printout. Note also that the *p*-value (.003) is less than $\alpha = .05$, thus confirming our conclusion to reject H_0.

Now Work Exercise 6.105

One-tailed and two-tailed tests of hypothesis for σ^2 are given in the following box.

Test of Hypothesis about σ^2

One-Tailed Test	**Two-Tailed Test**
$H_0: \sigma^2 = \sigma_0^2$	$H_0: \sigma^2 = \sigma_0^2$
$H_a: \sigma^2 < \sigma_0^2$ (or $H_a: \sigma^2 > \sigma_0^2$)	$H_a: \sigma^2 \neq \sigma_0^2$
Test statistic: $\chi^2 = \dfrac{(n-1)s^2}{\sigma_0^2}$	*Test statistic:* $\chi^2 = \dfrac{(n-1)s^2}{\sigma_0^2}$
Rejection region: $\chi^2 < \chi^2_{(1-\alpha)}$ (or $\chi^2 > \chi^2_\alpha$ when $H_a: \sigma^2 > \sigma_0^2$)	*Rejection region:* $\chi^2 < \chi^2_{(1-\alpha/2)}$ or $\chi^2 > \chi^2_{(\alpha/2)}$

where σ_0^2 is the hypothesized variance and the distribution of χ^2 is based on $(n-1)$ degrees of freedom.

Conditions Required for a Valid Large-Sample Hypothesis Test for σ^2

1. A random sample is selected from the target population.
2. The population from which the sample is selected has a distribution that is approximately normal.

⚠ **CAUTION** The procedure for conducting a hypothesis test for σ^2 in the preceding examples requires an assumption regardless of whether the sample size *n* is large or small. We must assume that the population from which the sample is selected has an approximate normal distribution. Unlike small-sample tests for μ based on the *t*-statistic, *slight to moderate departures from normality will render the χ^2 test invalid.* ▲

Exercises 6.96–6.114

Understanding the Principles

6.96 What sampling distribution is used to make inferences about σ^2?

6.97 What conditions are required for a valid test for σ^2?

6.98 *True or False*. The null hypotheses H_0: $\sigma^2 = .25$ and H_0: $\sigma = .5$ are equivalent.

6.99 *True or False*. When the sample size n is large, no assumptions about the population are necessary to test the population variance σ^2.

Learning the Mechanics

6.100 Let χ_0^2 be a particular value of χ^2. Find the value of χ_0^2 such that
a. $P(\chi^2 > \chi_0^2) = .10$ for $n = 12$
b. $P(\chi^2 > \chi_0^2) = .05$ for $n = 9$
c. $P(\chi^2 > \chi_0^2) = .025$ for $n = 5$

6.101 A random sample of n observations is selected from a normal population to test the null hypothesis that $\sigma^2 = 25$. Specify the rejection region for each of the following combinations of H_a, α, and n:
a. H_a: $\sigma^2 \neq 25$; $\alpha = .05$; $n = 16$
b. H_a: $\sigma^2 > 25$; $\alpha = .01$; $n = 23$
c. H_a: $\sigma^2 > 25$; $\alpha = .10$; $n = 15$
d. H_a: $\sigma^2 < 25$; $\alpha = .01$; $n = 13$
e. H_a: $\sigma^2 \neq 25$; $\alpha = .10$; $n = 7$
f. H_a: $\sigma^2 < 25$; $\alpha = .05$; $n = 25$

6.102 A random sample of seven measurements gave $\bar{x} = 9.4$ and $s^2 = 1.84$.
a. What assumptions must you make concerning the population in order to test a hypothesis about σ^2?
b. Suppose the assumptions in part **a** are satisfied. Test the null hypothesis $\sigma^2 = 1$ against the alternative hypothesis $\sigma^2 > 1$. Use $\alpha = .05$.
c. Refer to part **b**. Suppose the test statistic is $\chi^2 = 14.45$. Use Table V of Appendix A to find the approximate p-value of the test.
d. Test the null hypothesis $\sigma^2 = 1$ against the alternative hypothesis $\sigma^2 \neq 1$. Use $\alpha = .05$.

6.103 Refer to Exercise 6.102. Suppose we had $n = 100$, $\bar{x} = 9.4$, and $s^2 = 1.84$.
a. Test the null hypothesis H_0: $\sigma^2 = 1$ against the alternative hypothesis H_a: $\sigma^2 > 1$.
b. Compare your test result with that of Exercise 6.102.

6.104 A random sample of $n = 7$ observations from a normal population produced the following measurements: 4, 0, 6, 3, 3, 5, 9. Do the data provide sufficient evidence to indicate that $\sigma^2 < 2$? Test, using $\alpha = .05$.

Applying the Concepts—Basic

6.105 **Latex allergy in health care workers.** Refer to the *Current Allergy & Clinical Immunology* (March 2004) study of $n = 46$ hospital employees who were diagnosed with a latex allergy from exposure to the powder on latex gloves, presented in Exercise 6.28 (p. 309). Recall that the number of latex gloves used per week by the sampled workers is summarized as follows: $\bar{x} = 19.3$ and $s = 11.9$. Let σ^2 represent the variance in the number of latex gloves used per week by all hospital employees. Consider testing H_0: $\sigma^2 = 100$ against H_a: $\sigma^2 \neq 100$.
a. Give the rejection region for the test at a significance level of $\alpha = .01$.
b. Calculate the value of the test statistic.
c. Use the results from parts **a** and **b** to draw the appropriate conclusion.

6.106 **A new dental bonding agent.** Refer to the *Trends in Biomaterials & Artificial Organs* (January 2003) study of a new bonding adhesive for teeth called "Smartbond," presented in Exercise 6.72 (p. 323). Recall that tests on a sample of 10 extracted teeth bonded with the new adhesive resulted in a mean breaking strength (after 24 hours) of $\bar{x} = 5.07$ Mpa and a standard deviation of $s = .46$ Mpa. In addition to requiring a good mean breaking strength, orthodontists are concerned about the variability in breaking strength of the new bonding adhesive.
a. Set up the null and alternative hypothesis for a test to determine whether the breaking strength variance differs from .5 Mpa.
b. Find the rejection region for the test, using $\alpha = .01$.
c. Compute the test statistic.
d. Give the appropriate conclusion for the test.
e. What conditions are required for the test results to be valid?

6.107 **Characteristics of a rockfall.** Refer to the *Environmental Geology* (Vol. 58, 2009) simulation study of how far a block from a collapsing rockwall will bounce down a soil slope, Exercise 2.59 (p. 57). Rebound lengths (in meters) were estimated for 13 rock bounces. The data are repeated in the table and saved in the **ROCKFALL** file. Descriptive statistics for the rebound lengths are shown on the SAS printout at the bottom of the page. Consider a test of hypothesis for the variation in rebound lengths for the theoretical population of rock bounces from the collapsing rockwall. In particular, a geologist wants to determine if the variance differs from 10 m².

| 10.94 | 13.71 | 11.38 | 7.26 | 17.83 | 11.92 | 11.87 | 5.44 | 13.35 |
| 4.90 | 5.85 | 5.10 | 6.77 | | | | | |

Based on Paronuzzi, P. "Rockfall-induced block propagation on a soil slope, northern Italy." *Environmental Geology*, Vol. 58, 2009 (Table 2).

SAS Output for Exercise 6.107

```
              The MEANS Procedure

         Analysis Variable : LENGTH

   N       Mean        Std Dev       Minimum        Maximum
  13     9.7169231    4.0947291     4.9000000     17.8300000
```

a. Define the parameter of interest.
b. Specify the null and alternative hypothesis.
c. Compute the value of the test statistic.
d. Determine the rejection region for the test using $\alpha = .10$.
e. Make the appropriate conclusion.
f. What condition must be satisfied in order for the inference, part **e**, to be valid?

6.108 Identifying type of urban land cover. Refer to the *Geographical Analysis* (Oct. 2006) study of a new method for analyzing remote-sensing data from satellite pixels, presented in Exercise 6.32 (p. 309). Recall that the method uses a numerical measure of the distribution of gaps, or the sizes of holes, in the pixel, called *lacunarity*. Summary statistics for the lacunarity measurements in a sample of 100 grassland pixels are $\bar{x} = 225$ and $s = 20$. As stated in Exercise 6.32, it is known that the mean lacunarity measurement for all grassland pixels is 220. The method will be effective in identifying land cover if the standard deviation of the measurements is 10% (or less) of the true mean (i.e., if the standard deviation is less than 22).
a. Give the null and alternative hypothesis for a test to determine whether, in fact, the standard deviation of all grassland pixels is less than 22.
b. A MINITAB analysis of the data is provided below. Locate and interpret the *p*-value of the test. Use $\alpha = .10$.

Test for One Standard Deviation

```
Method
Null hypothesis        Sigma = 22
Alternative hypothesis Sigma = < 22

The standard method is only for the normal distribution.

Statistics
N    StDev  Variance
100  20.0   400

Tests
Method    Chi-Square  DF  P-Value
Standard  81.82       99  0.105
```

Applying the Concepts—Intermediate

6.109 Point spreads of NFL games. During the National Football League (NFL) season, Las Vegas oddsmakers establish a point spread on each game for betting purposes. For example, the Champion Green Bay Packers were established as 2.5-point favorites over the Pittsburgh Steelers in the 2011 Super Bowl. The final scores of NFL games were compared against the final point spreads established by the oddsmakers in *Chance* (Fall 1998). The difference between the outcome of the game and the point spread (called a point-spread error) was calculated for 240 NFL games. The mean and standard deviation of the point-spread errors are $\bar{x} = -1.6$ and $s = 13.3$. Suppose the researcher wants to know whether the true standard deviation of the point-spread errors exceeds 15. Conduct the analysis using $\alpha = .05$.

6.110 Mongolian desert ants. Refer to the *Journal of Biogeography* (December 2003) study of ants in Mongolia (central Asia),

presented in Exercise 6.74 (p. 323). Data on the number of ant species attracted to 11 randomly selected desert sites are saved in the **GOBIANTS** file. Do these data indicate that the standard deviation of the number of ant species at all Mongolian desert sites exceeds 15 species? Conduct the appropriate test at $\alpha = .05$. Are the conditions required for a valid test satisfied?

6.111 Birth weights of cocaine babies. A group of researchers at the University of Texas-Houston conducted a comprehensive study of pregnant cocaine-dependent women (*Journal of Drug Issues*, Summer 1997). All the women in the study used cocaine on a regular basis (at least three times a week) for more than a year. One of the many variables measured was birth weight (in grams) of the baby delivered. For a sample of 16 cocaine-dependent women, the mean birth weight was 2,971 grams and the standard deviation was 410 grams. Test (at $\alpha = .01$) to determine whether the variance in birth weights of babies delivered by cocaine-dependent women is less than 200,000 grams^2.

6.112 Cooling method for gas turbines. Refer to the *Journal of Engineering for Gas Turbines and Power* (Jan. 2005) study of the performance of augmented gas turbine engines, presented in Exercise 6.34 (p. 310). Recall that the performance of each in a sample of 67 gas turbines was measured by heat rate (kilojoules per kilowatt per hour). The data are saved in the **GASTURBINE** file. Suppose that standard gas turbines have heat rates with a standard deviation of 1,500 kJ/kWh. Is there sufficient evidence to indicate that the heat rates of the augmented gas turbine engine are more variable than the heat rates of the standard gas turbine engine? Test, using $\alpha = 0.5$.

6.113 Jitter in a water power system. Refer to the *Journal of Applied Physics* investigation of throughput jitter in the opening switch of a prototype water power system, Exercise 5.104 (p. 282). Recall that low throughput jitter is critical to successful waterline technology. An analysis of conduction time for a sample of 18 trials of the prototype system yielded $\bar{x} = 334.8$ nanoseconds and $s = 6.3$ nanoseconds. (Conduction time is defined as the length of time required for the downstream current to equal 10% of the upstream current.) A system is considered to have low throughput jitter if the true conduction time standard deviation is less than 7 nanoseconds. Does the prototype system satisfy this requirement? Test using $\alpha = .01$.

Applying the Concepts—Advanced

6.114 Motivation of drug dealers. Refer to the *Applied Psychology in Criminal Justice* (Sept. 2009) study of the personality characteristics of convicted drug dealers, Exercise 5.15 (p. 251). A random sample of 100 drug dealers had a mean Wanting Recognition (WR) score of 39 points, with a standard deviation of 6 points. Recall that the WR score is a quantitative measure of a person's level of need for approval and sensitivity to social situations. (Higher scores indicate a greater need for approval.) A criminal psychologist claims that the range of WR scores for the population of convicted drug dealers is 42 points. Do you believe the psychologist's claim? (*Hint*: Assume the population of WR scores is normally distributed.)

6.8 A Nonparametric Test about a Population Median (Optional)

In Sections 6.2–6.4, we utilized the z- and t-statistics for testing hypotheses about a population mean. The z-statistic is appropriate for large random samples selected from "general" populations—that is, samples with few limitations on the probability distribution of the underlying population. The t-statistic was developed for small-sample tests in which the sample is selected at random from a *normal* distribution. The question is, How can we conduct a test of hypothesis when we have a small sample from a *nonnormal* distribution? The answer is: Use a *distribution-free* procedure that requires fewer or less stringent assumptions about the underlying population—called a **nonparametric method.**

> **Distribution-free tests** are statistical tests that do not rely on any underlying assumptions about the probability distribution of the sampled population.

> The branch of inferential statistics devoted to distribution-free tests is called **nonparametrics**.

The **sign test** is a relatively simple nonparametric procedure for testing hypotheses about the central tendency of a nonnormal probability distribution. Note that we used the phrase *central tendency* rather than *population mean*. This is because the sign test, like many nonparametric procedures, provides inferences about the population *median* rather than the population mean μ. Denoting the population median by the Greek letter η, we know (Chapter 2) that η is the 50th percentile of the distribution (Figure 6.22) and, as such, is less affected by the skewness of the distribution and the presence of outliers (extreme observations). Since the nonparametric test must be suitable for all distributions, not just the normal, it is reasonable for nonparametric tests to focus on the more robust (less sensitive to extreme values) measure of central tendency: the median.

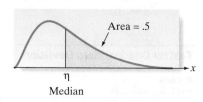

Figure 6.22
Location of the population median, η

For example, increasing numbers of both private and public agencies are requiring their employees to submit to tests for substance abuse. One laboratory that conducts such testing has developed a system with a normalized measurement scale in which values less than 1.00 indicate "normal" ranges and values equal to or greater than 1.00 are indicative of potential substance abuse. The lab reports a normal result as long as the median level for an individual is less than 1.00. Eight independent measurements of each individual's sample are made. One individual's results are shown in Table 6.7.

Table 6.7		Substance Abuse Test Results					
.78	.51	3.79	.23	.77	.98	.96	.89

Data Set: SUBABUSE

If the objective is to determine whether the *population* median (i.e., the true median level if an infinitely large number of measurements were made on the same individual sample) is less than 1.00, we establish that as our alternative hypothesis and test

$$H_0: \eta = 1.00$$
$$H_a: \eta < 1.00$$

The one-tailed sign test is conducted by counting the number of sample measurements that "favor" the alternative hypothesis—in this case, the number that are less than 1.00. If the null hypothesis is true, we expect approximately half of the measurements to fall on each side of the hypothesized median, and if the alternative is true, we expect significantly more than half to favor the alternative—that is, to be less than 1.00. Thus,

Test statistic: S = Number of measurements less than 1.00, the null hypothesized median

If we wish to conduct the test at the $\alpha = .05$ level of significance, the rejection region can be expressed in terms of the observed significance level, or p-value, of the test:

$$\text{Rejection region: } p\text{-value} \leq .05$$

In this example, $S = 7$ of the 8 measurements are less than 1.00. To determine the observed significance level associated with that outcome, we note that the number of measurements less than 1.00 is a binomial random variable (check the binomial characteristics presented in Chapter 4), and *if H_0 is true*, the binomial probability p that a measurement lies below (or above) the median 1.00 is equal to .5 (Figure 6.22). What is the probability that a result is *as contrary to or more contrary to H_0* than the one observed? That is, what is the probability that 7 *or more* of 8 binomial measurements will result in Success (be less than 1.00) if the probability of Success is .5? Binomial Table II in Appendix A (with $n = 8$ and $p = .5$) indicates that

$$P(x \geq 7) = 1 - P(x \leq 6) = 1 - .965 = .035$$

Thus, the probability that at least 7 of 8 measurements would be less than 1.00 *if the true median were 1.00* is only .035. The *p*-value of the test is therefore .035.

This *p*-value can also be obtained from a statistical software package. The MINITAB printout of the analysis is shown in Figure 6.23, with the *p*-value highlighted. Since $p = .035$ is less than $\alpha = .05$, we conclude that this sample provides sufficient evidence

---

Sign Test for Median: READING

```
Sign test of median =  1.000 versus <  1.000

            N  Below  Equal  Above       P  Median
READING     8      7      0      1  0.0352  0.8350
```

Figure 6.23

MINITAB printout of sign test

to reject the null hypothesis. The implication of this rejection is that the laboratory can conclude at the $\alpha = .05$ level of significance that the true median level for the individual tested is less than 1.00. However, we note that one of the measurements, with a value of 3.79, greatly exceeds the others and deserves special attention. This large measurement is an outlier that would make the use of a *t*-test and its concomitant assumption of normality dubious. The only assumption necessary to ensure the validity of the sign test is that the probability distribution of measurements is continuous.

The use of the sign test for testing hypotheses about population medians is summarized in the following box:

Sign Test for a Population Median η

One-Tailed Test	**Two-Tailed Test**
$H_0: \eta = \eta_0$	$H_0: \eta = \eta_0$
$H_a: \eta > \eta_0$ [or $H_a: \eta < \eta_0$]	$H_a: \eta \neq \eta_0$
Test statistic:	*Test statistic:*
$S =$ Number of sample measurements greater than η_0 [or $S =$ number of measurements less than η_0]	$S =$ Larger of S_1 and S_2, where S_1 is the number of measurements less than η_0 and S_2 is the number of measurements greater than η_0

[*Note:* Eliminate observations from the analysis that are exactly equal to the hypothesized median, η_0.]

Observed significance level:	*Observed significance level:*
p-value $= P(x \geq S)$	p-value $= 2P(x \geq S)$

where x has a binomial distribution with parameters n and $p = .5$. (Use Table II, Appendix A.)

Rejection region: Reject H_0 if p-value $\leq \alpha$

> **Conditions Required for a Valid Application of the Sign Test**
>
> The sample is selected randomly from a continuous probability distribution.
> [*Note:* No assumptions need to be made about the shape of the probability distribution.]

Recall that the normal probability distribution provides a good approximation of the binomial distribution when the sample size is large (i.e., when both $np \geq 15$ and $nq \geq 15$). For tests about the median of a distribution, the null hypothesis implies that $p = .5$, and the normal distribution provides a good approximation if $n \geq 30$. (Note that for $n = 30$ and $p = .5, np = nq = 15$.) Thus, we can use the standard normal z-distribution to conduct the sign test for large samples. The large-sample sign test is summarized in the next box.

> **Large-Sample Sign Test for a Population Median η**
>
> | **One-Tailed Test** | **Two-Tailed Test** |
> |---|---|
> | H_0: $\eta = \eta_0$ | H_0: $\eta = \eta_0$ |
> | H_a: $\eta > \eta_0$ [or H_a: $\eta < \eta_0$] | H_a: $\eta \neq \eta_0$ |
>
> *Test statistic:* $z = \dfrac{(S - .5) - .5n}{.5\sqrt{n}}$
>
> [*Note:* S is calculated as shown in the previous box. We subtract .5 from S as the "correction for continuity." The null-hypothesized mean value is $np = .5n$, and the standard deviation is
>
> $$\sqrt{npq} = \sqrt{n(.5)(.5)} = .5\sqrt{n}$$
>
> (See Chapter 4 for details on the normal approximation to the binomial distribution.)]
>
> *Rejection region:* $z > z_\alpha$ *Rejection region:* $z > z_{\alpha/2}$
>
> where tabulated z-values can be found in Table III, Appendix A.

Example 6.12

Sign Test Application—Failure Times of MP3 Players

Problem A manufacturer of MP3 players has established that the median time to failure for its players is 5,250 hours of utilization. A sample of 40 MP3 players from a competitor is obtained, and the players are tested continuously until each fails. The 40 failure times range from 5 hours (a "defective" player) to 6,575 hours, and 24 of the 40 exceed 5,250 hours. Is there evidence that the median failure time of the competitor's product differs from 5,250 hours? Use $\alpha = .10$.

Solution The null and alternative hypotheses of interest are

$$H_0: \eta = 5{,}250 \text{ hours}$$
$$H_a: \eta \neq 5{,}250 \text{ hours}$$

Since $n \geq 30$, we use the standard normal z-statistic:

$$\text{Test statistic: } z = \frac{(S - .5) - .5n}{.5\sqrt{n}}$$

Here, S is the maximum of S_1 (the number of measurements greater than 5,250) and S_2 (the number of measurements less than 5,250). Also,

$$\text{Rejection region: } \quad z > 1.645, \text{ where } z_{\alpha/2} = z_{.05} = 1.645$$

Assumptions: The probability distribution of the failure times is continuous (time is a continuous variable), but nothing is assumed about its shape.

Since the number of measurements exceeding 5,250 is $S_2 = 24$, it follows that the number of measurements less than 5,250 is $S_1 = 16$. Consequently, $S = 24$, the greater of S_1 and S_2. The calculated z-statistic is therefore

$$z = \frac{(S - .5) - .5n}{.5\sqrt{n}} = \frac{23.5 - 20}{.5\sqrt{40}} = \frac{3.5}{3.162} = 1.11$$

The value of z is not in the rejection region, so we cannot reject the null hypothesis at the $\alpha = .10$ level of significance.

Look Back The manufacturer should not conclude, on the basis of this sample, that its competitor's MP3 players have a median failure time that differs from 5,250 hours. The manufacturer will not "accept H_0," however, since the probability of a Type II error is unknown.

Now Work Exercise 6.119

The one-sample nonparametric sign test for a median provides an alternative to the t-test for small samples from nonnormal distributions. However, if the distribution is approximately normal, the t-test provides a more powerful test about the central tendency of the distribution.

Exercises 6.115–6.129

Understanding the Principles

6.115 Under what circumstances is the sign test preferred to the t-test for making inferences about the central tendency of a population?

6.116 What is the probability that a randomly selected observation exceeds the
  a. Mean of a normal distribution?
  b. Median of a normal distribution?
  c. Mean of a nonnormal distribution?
  d. Median of a nonnormal distribution?

Learning the Mechanics

6.117 Use Table II of Appendix A to calculate the following binomial probabilities:
  a. $P(x \geq 6)$ when $n = 7$ and $p = .5$
  b. $P(x \geq 5)$ when $n = 9$ and $p = .5$
  c. $P(x \geq 8)$ when $n = 8$ and $p = .5$
  d. $P(x \geq 10)$ when $n = 15$ and $p = .5$. Also, use the normal approximation to calculate this probability, and then compare the approximation with the exact value.
  e. $P(x \geq 15)$ when $n = 25$ and $p = .5$. Also, use the normal approximation to calculate this probability, and then compare the approximation with the exact value.

6.118 Consider the following sample of 10 measurements, saved in the **LM6_118** file.

8.4	16.9	15.8	12.5	10.3	4.9	12.9	9.8	23.7	7.3

Use these data, the binomial tables (Table II, Appendix A), and $\alpha = .05$ to conduct each of the following sign tests:
  a. $H_0: \eta = 9$ versus $H_a: \eta > 9$
  b. $H_0: \eta = 9$ versus $H_a: \eta \neq 9$
  c. $H_0: \eta = 20$ versus $H_a: \eta < 20$
  d. $H_0: \eta = 20$ versus $H_a: \eta \neq 20$
  e. Repeat each of the preceding tests, using the normal approximation to the binomial probabilities. Compare the results.
  f. What assumptions are necessary to ensure the validity of each of the preceding tests?

6.119 Suppose you wish to conduct a test of the research hypothesis that the median of a population is greater than 80. You randomly sample 25 measurements from the population and determine that 16 of them exceed 80. Set up and conduct the appropriate test of hypothesis at the .10 level of significance. Be sure to specify all necessary assumptions.

Applying the Concepts—Basic

6.120 **Caffeine in Starbucks coffee.** Scientists at the University of Florida College of Medicine investigated the level of caffeine in 16-ounce cups of Starbucks coffee (*Journal of Analytical Toxicology*, Oct. 2003). In one phase of the experiment, cups of Starbucks Breakfast Blend (a mix of Latin American coffees) were purchased on six consecutive days from a single specialty coffee shop. The amount of caffeine in each of the six cups (measured in milligrams) is provided in the following table and saved in the **STARBUCKS** file.

564	498	259	303	300	307

  a. Suppose the scientists are interested in determining whether the median amount of caffeine in Breakfast Blend coffee exceeds 300 milligrams. Set up the null and alternative hypotheses of interest.
  b. How many of the cups in the sample have a caffeine content that exceeds 300 milligrams?
  c. Assuming that $p = .5$, use the binomial table in Appendix A to find the probability that at least 4 of the 6 cups have caffeine amounts that exceed 300 milligrams.
  d. On the basis of the probability you found in part **c,** what do you conclude about H_0 and H_a? (Use $\alpha = .05$.)

6.121 **Cheek teeth of extinct primates.** Refer to the *American Journal of Physical Anthropology* (Vol. 142, 2010) study of the characteristics of cheek teeth (e.g., molars) in an extinct primate species, Exercise 2.34 (p. 46). Recall that the researchers measured the dentary depth of molars (in millimeters) for 18 cheek teeth extracted from skulls. These depth measurements are reproduced in the next table (p. 340) and saved in the **CHEEKTEETH** file. The researchers are interested in the median molar depth of all cheek teeth from this extinct primate species. In particular, they want to know if the population median differs from 15 mm.
  a. Specify the null and alternative hypotheses of interest of the researchers.
  b. Explain why the sign test is appropriate to apply in this case.

Data for Exercise 6.121

18.12	19.48	19.36	15.94	15.83	19.70	15.76	17.00	16.20
13.96	16.55	15.70	17.83	13.25	16.12	18.13	14.02	14.04

Based on Boyer, D. M., Evans, A. R., and Jernvall, J. "Evidence of dietary differentiation among Late Paleocene–Early Eocene Plesiadapids (Mammalia, Primates)." *American Journal of Physical Anthropology*, Vol. 142, © 2010 (Table A3).

c. A MINITAB printout of the analysis is shown below. Locate the test statistic on the printout.

d. Find the *p*-value on the printout, and use it to draw a conclusion. Test using $\alpha = .05$.

Sign Test for Median: M2Depth

```
Sign test of median = 15.00 versus not = 15.00

          N  Below  Equal  Above      P  Median
M2Depth  18      4      0     14  0.0309   16.16
```

6.122 Quality of white shrimp. In *The American Statistician* (May 2001), the nonparametric sign test was used to analyze data on the quality of white shrimp. One measure of shrimp quality is cohesiveness. Since freshly caught shrimp are usually stored on ice, there is concern that cohesiveness will deteriorate after storage. For a sample of 20 newly caught white shrimp, cohesiveness was measured both before and after storage on ice for two weeks. The difference in the cohesiveness measurements (before minus after) was obtained for each shrimp. If storage has no effect on cohesiveness, the population median of the differences will be 0. If cohesiveness deteriorates after storage, the population median of the differences will be positive.

a. Set up the null and alternative hypotheses to test whether cohesiveness will deteriorate after storage.

b. In the sample of 20 shrimp, there were 13 positive differences. Use this value to find the *p*-value of the test.

c. Make the appropriate conclusion (in the words of the problem) if $\alpha = .05$.

6.123 Emotional empathy in young adults. Refer to the *Journal of Moral Education* (June 2010) study of emotional empathy in young adults, Exercise 6.52 (p. 316). Recall that psychologists theorize that young female adults show more emotional empathy towards others than do males. To test the theory, each in a sample of 30 female college students responded to the following statement on emotional empathy: "I often have tender, concerned feelings for people less fortunate than me." Responses (i.e., empathy scores) ranged from 0 to 4, where 0 = "never" and 4 = "always." Suppose it is known that male college students have a median emotional empathy score of $\eta = 2.8$.

a. Specify the null and alternative hypothesis for testing whether female college students have a median emotional empathy scale score higher than 2.8.

b. Suppose that distribution of emotional empathy scores for the 30 female students is as shown in the table. Use this information to compute the test statistic.

c. Find the observed significance level (*p*-value) of the test.

d. At $\alpha = .01$, what is the appropriate conclusion?

Response (empathy score)	Number of Females
0	1
1	3
2	5
3	12
4	9

6.124 Crab spiders hiding on flowers. Refer to the *Behavioral Ecology* (Jan. 2005) field study on the natural camouflage of crab spiders, presented in Exercise 2.38 (p. 47). Ecologists collected a sample of 10 adult female crab spiders, each sitting on the yellow central part of a daisy, and measured the chromatic contrast between each spider and the flower. The contrast values for the 10 crab spiders are reproduced in the table and saved in the **SPIDER** file. (*Note:* The lower the contrast, the more difficult it is for predators to see the crab spider on the flower.) Recall that a contrast of 70 or greater allows bird predators to see the spider. Consider a test to determine whether the population median chromatic contrast of spiders on flowers is less than 70.

57	75	116	37	96	61	56	2	43	32

Based on Thery, M., et al. "Specific color sensitivities of prey and predator explain camouflage in different visual systems." *Behavioral Ecology*, Vol. 16, No. 1, Jan. 2005 (Table 1).

a. State the null and alternative hypotheses for the test of interest.

b. Calculate the value of the test statistic.

c. Find the *p*-value for the test.

d. At $\alpha = .10$, what is the appropriate conclusion? State your answer in the words of the problem.

Applying the Concepts—Intermediate

6.125 Lobster trap placement. Refer to the *Bulletin of Marine Science* (April 2010) observational study of lobster trap placement by teams fishing for the red spiny lobster in Baja California Sur, Mexico, Exercise 6.65 (p. 321). Trap spacing measurements (in meters) for a sample of seven teams of red spiny lobster fishermen are reproduced in the accompanying table (and saved in the **TRAPSPACE** file). In Exercise 6.65, you tested whether the average of the trap spacing measurements for the population of red spiny lobster fishermen fishing in Baja California Sur, Mexico, differs from 95 meters.

93	99	105	94	82	70	86

From Shester, G. G. "Explaining catch variation among Baja California lobster fishers through spatial analysis of trap-placement decisions." *Bulletin of Marine Science*, Vol. 86, No. 2, April 2010 (Table 1). Reprinted with permission from the University of Miami – *Bulletin of Marine Science*.

a. There is concern that the trap spacing data do not follow a normal distribution. If so, how will this impact the test you conducted in Exercise 6.65?

b. Propose an alternative nonparametric test to analyze the data.

c. Compute the value of the test statistic for the nonparametric test.

d. Find the *p*-value of the test.

e. Use the value of α you selected in Exercise 6.65 and give the appropriate conclusion.

6.126 Characteristics of a rockfall. Refer to the *Environmental Geology* (Vol. 58, 2009) simulation study of how far a block from a collapsing rockwall will bounce down a soil slope, Exercise 2.59 (p. 57). Recall that the variable of interest was *rebound length* (measured in meters) of the falling block. Based on the depth, location, and angle of block-soil impact marks left on the slope from an actual rockfall, the 13 rebound lengths shown in the table (in the next column) were estimated. (These data are saved in the **ROCKFALL** file.) Consider the following statement: "In all similar rockfalls, half of the rebound lengths will exceed 10 meters." Is this statement supported by the sample data? Test using $\alpha = .10$.

10.94	13.71	11.38	7.26	17.83	11.92	11.87	5.44	13.35
4.90	5.85	5.10	6.77					

Based on Paronuzzi, P. "Rockfall-induced block propagation on a soil slope, northern Italy." *Environmental Geology*, Vol. 58, 2009 (Table 2).

6.127 Freckling of superalloy ingots. Refer to the *Journal of Metallurgy* (Sept. 2004) study of freckling of superalloy ingots, presented in Exercise 2.185 (p. 100). Recall that freckles are defects that sometimes form during the solidification of the ingot. The freckle index for each of $n = 18$ superalloy ingots is shown in the next table and saved in the **FRECKLE** file. In the population of superalloy ingots, is there evidence to say that 50% of the ingots have a freckle index of 10 or higher? Test, using $\alpha = .01$.

30.1	22.0	14.6	16.4	12.0	2.4	22.2	10.0	15.1
12.6	6.8	4.1	2.5	1.4	33.4	16.8	8.1	3.2

Based on Yang, W. H., et al. "A freckle criterion for the solidification of superalloys with a tilted solidification front." *JOM: Journal of the Minerals, Metals and Materials Society*, Vol. 56, No. 9, Sept. 2004 (Table IV).

6.128 Study of guppy migration. To improve survival and reproductive success, many species of fish have an evolved migration history. In one migration study of guppy populations (*Zoological Science*, Vol. 6, 1989), adult female guppies were placed in the left compartment of an experimental aquarium tank divided in half by a glass plate. After the plate was removed, the numbers of fish passing through the slit from the left compartment to the right one, and vice versa, were monitored every minute for 30 minutes. If an equilibrium is reached (which is optimal for survival), the zoologists would expect about half the guppies to remain in the left compartment and half to remain in the right compartment. Consequently, if 80 guppies were placed in the aquarium, the median number of fish remaining in the left compartment should be 40. Data for a similar 30-minute experiment involving 80 guppies is shown in the table below and saved in the **GUPPY** file. (Each measurement represents the number of guppies in the left compartment at the end of a 1-minute interval.) Use the large-sample sign test to determine whether the median is less than 40. Test using $\alpha = .05$.

32	21	24	30	28	32	35	30	26	30	28	28	33	26	34
34	29	43	36	38	34	34	41	47	36	38	42	34	42	33

Based on Terami, H., and Watanabe, M. "Excessive transitory migration of guppy populations III: Analysis of perception of swimming space and a mirror effect." *Zoological Science*, Vol. 6, 1989.

6.129 Minimizing tractor skidding distance. Refer to the *Journal of Forest Engineering* (July 1999) study of minimizing tractor skidding distances along a new road in a European forest, presented in Exercise 6.73 (p. 323). The skidding distances (in meters) were measured at 20 randomly selected road sites. The data are repeated in the accompanying table and saved in the **SKIDDING** file. In Exercise 6.73, you conducted a test of hypothesis for the population mean skidding distance. Now conduct a test to determine whether the population median skidding distance is more than 400 meters. Use $\alpha = .10$.

488	350	457	199	285	409	435	574	439	546
385	295	184	261	273	400	311	312	141	425

Based on Tujek, J., and Pacola, E. "Algorithms for skidding distance modeling on a raster Digital Terrain Model." *Journal of Forest Engineering*, Vol. 10, No. 1, July 1999 (Table 1).

CHAPTER NOTES

Key Terms

Note: Starred (*) terms are from the optional sections in this chapter.

Alternative (research)
  hypothesis 295
*Chi-square (χ^2)
  distribution 331
Conclusion 298
*Distribution-free test 336
Hypothesis 294
Level of significance 299
Lower-tailed test 300
*Nonparametrics 336
Null hypothesis 295

Observed significance level
  (*p*-value) 311
One-tailed (one-sided)
  statistical test 300
Rejection region 296
*Sign test 336
Test of hypothesis 294
Test statistic 295
Two-tailed (two-sided)
  hypothesis 300
Type I error 296
Type II error 297
Upper-tailed test 300

Key Symbols

μ	Population mean
p	Population proportion, $P(\text{Success})$, in binomial trial
σ^2	Population variance
$\bar{x}$	Sample mean (estimator of μ)
$\hat{p}$	Sample proportion (estimator of p)
s^2	Sample variance (estimator of σ^2)
H_0	Null hypothesis
H_a	Alternative hypothesis
α	Probability of a Type I error
β	Probability of a Type II error
χ^2	Chi-square (sampling distribution of s^2 for normal data)
S	*Test statistic for sign test

Key Ideas

Key Words for Identifying the Target Parameter

μ—Mean, Average
p—Proportion, Fraction, Percentage, Rate, Probability
σ^2—Variance, Variability, Spread

Elements of a Hypothesis Test

1. *Null hypothesis* (H_o)
2. *Alternative hypothesis* (H_a)
3. *Test statistic* (z, t, *or* X^2)
4. *Significance level* (α)
5. *p-value*
6. *Conclusion*

Forms of Alternative Hypothesis

Lower tailed: H_a: $\mu_0 < 50$
Upper tailed: H_a: $\mu_0 > 50$
Two tailed: H_a: $\mu_0 \neq 50$

Type I Error $=$ Reject H_0 when H_0 is true (occurs with probability α)

Type II Error $=$ Accept H_0 when H_0 is false (occurs with probability β)

Power of a Test $= P$ (Reject H_0 when H_0 is false)
$= 1 - \beta$

Using *p*-values to Decide

1. Choose significance level (α)
2. Obtain *p*-value of the test
3. If $\alpha > p$-value, Reject H_0

Distribution-free (Nonparametric) Tests

Do not rely on assumptions about the distribution of the sampled population

Guide to Selecting a One-Sample Hypothesis Test

Supplementary Exercises 6.130–6.161

Note: List the assumptions necessary for the valid implementation of the statistical procedures you use in solving all these exercises. Starred (*) exercises refer to the optional sections in this chapter.

Understanding the Principles

6.130 Specify the differences between a large-sample and small-sample test of hypothesis about a population mean μ. Focus on the assumptions and test statistics.

6.131 *Complete the following statement:* The smaller the p-value associated with a test of hypothesis, the stronger is the support for the _____ hypothesis. Explain your answer.

6.132 Which of the elements of a test of hypothesis can and should be specified *prior* to analyzing the data that are to be utilized to conduct the test?

6.133 *Complete the following statement:* The larger the p-value associated with a test of hypothesis, the stronger is the support for the _____ hypothesis. Explain your answer.

Learning the Mechanics

*6.134 In a sample of $n = 10$ measurements randomly selected from a nonnormal population, seven of the measurements exceed 150.
  a. Find the p-value, for testing $H_0: \eta = 150$ against $H_a: \eta > 150$.
  b. Make thw appropriate conclusion at $\alpha = .05$.

6.135 A random sample of $n = 200$ observations from a binomial population yields $\hat{p} = .29$.
  a. Test $H_0: p = .35$ against $H_a: p < .35$. Use $\alpha = .05$.
  b. Test $H_0: p = .35$ against $H_a: p \neq .35$. Use $\alpha = .05$.

6.136 A random sample of 175 measurements possessed a mean of $\bar{x} = 8.2$ and a standard deviation of $s = .79$.
  a. Form a 95% confidence interval for μ.
  b. Test $H_0: \mu = 8.3$ against $H_a: \mu \neq 8.3$. Use $\alpha = .05$.
  c. Test $H_0: \mu = 8.4$ against $H_a: \mu \neq 8.4$. Use $\alpha = .05$.

*6.137 A random sample of 41 observations from a normal population possessed a mean of $\bar{x} = 88$ and a standard deviation of $s = 6.9$.
  a. Test $H_0: \sigma^2 = 30$ against $H_a: \sigma^2 > 30$. Use $\alpha = .05$.
  b. Test $H_0: \sigma^2 = 30$ against $H_a: \sigma^2 \neq 30$. Use $\alpha = .05$.

6.138 A t-test is conducted for the null hypothesis $H_0: \mu = 10$ versus the alternative hypothesis $H_a: \mu > 10$ for a random sample of $n = 17$ observations. The test results are $t = 1.174$ and p-value $= .1288$.
  a. Interpret the p-value.
  b. What assumptions are necessary for the validity of this test?
  c. Calculate and interpret the p-value, assuming that the alternative hypothesis was instead $H_a: \mu \neq 10$.

Applying the Concepts—Basic

6.139 **Use of herbal therapy.** According to the *Journal of Advanced Nursing* (January 2001), 45% of senior women (i.e., women over the age of 65) use herbal therapies to prevent or treat health problems. Also, senior women who use herbal therapies use an average of 2.5 herbal products in a year.
  a. Give the null hypothesis for testing the first claim by the journal.
  b. Give the null hypothesis for testing the second claim by the journal.

6.140 **FDA mandatory new-drug testing.** When a new drug is formulated, the pharmaceutical company must subject it to lengthy and involved testing before receiving the necessary permission from the Food and Drug Administration (FDA) to market the drug. The FDA requires the pharmaceutical company to provide substantial evidence that the new drug is safe for potential consumers.
  a. If the new-drug testing were to be placed in a test-of- hypothesis framework, would the null hypothesis be that the drug is safe or unsafe? the alternative hypothesis?
  b. Given the choice of null and alternative hypotheses in part **a,** describe Type I and Type II errors in terms of this application. Define α and β in terms of this application.
  c. If the FDA wants to be very confident that the drug is safe before permitting it to be marketed, is it more important that α or β be small? Explain.

6.141 **Sleep deprivation study.** In a British study, 12 healthy college students deprived of one night's sleep received an array of tests intended to measure their thinking time, fluency, flexibility, and originality of thought. The overall test scores of the sleep-deprived students were compared with the average score expected from students who received their accustomed sleep. Suppose the overall scores of the 12 sleep-deprived students had a mean of $\bar{x} = 63$ and a standard deviation of 17. (Lower scores are associated with a decreased ability to think creatively.)
  a. Test the hypothesis that the true mean score of sleep-deprived subjects is less than 80, the mean score of subjects who received sleep prior to taking the test. Use $\alpha = .05$.
  b. What assumption is required for the hypothesis test of part **a** to be valid?

6.142 **Cell phone use by drivers.** Refer to the U.S. Department of Transportation study of the level of cell phone use by drivers while they are in the act of driving a motor passenger vehicle, presented in Exercise 5.119 (p. 286). Recall that in a random sample of 1,165 drivers selected across the country, 35 were found using their cell phone.
  a. Conduct a test (at $\alpha = .05$) to determine whether p, the true driver cell phone use rate, differs from .02.
  b. Does the conclusion, you drew in part **a** agree with the inference you derived from the 95% confidence interval for p in Exercise 5.119? Explain why or why not.

*6.143 **Radioactive lichen.** Refer to the Lichen Radionuclide Baseline Research project to monitor the level of radioactivity in lichen, Exercise 6.71 (p. 323). Recall that University of Alaska researchers collected 9 lichen specimens and measured the amount of the radioactive element cesium-137 (in microcuries per milliliter) in each specimen. (The natural logarithms of the data values, saved in the **LICHEN** file, are listed in the table on the next page.) In Exercise 6.71, you used the t-statistic to test whether the mean cesium amount in lichen differs from $\mu = .003$ microcurie per milliliter. Use the MINITAB printout (page 344) to conduct an alternative nonparametric test at $\alpha = .10$. Does the result agree with that of the t-test from Exercise 6.71? [*Note:* The values in the table were converted back to microcuries per milliliter to perform the analysis.]

Data for Exercise 6.143

Location			
Bethel	−5.50	−5.00	
Eagle Summit	−4.15	−4.85	
Moose Pass	−6.05		
Turnagain Pass	−5.00		
Wickersham Dome	−4.10	−4.50	−4.60

Based on Lichen Radionuclide Baseline Research Project, 2003, p. 25. *Orion*, University of Alaska–Fairbanks.

MINITAB output for Exercise 6.143

Sign Test for Median: CESIUM

Sign test of median = 0.00300 versus not = 0.00300

```
          N  Below  Equal  Above    P   Median
CESIUM    9    1      0      8   0.0391  0.00783
```

6.144 The "Pepsi challenge." "Take the Pepsi Challenge" was a marketing campaign used by the Pepsi-Cola Company. Coca-Cola drinkers participated in a blind taste test in which they tasted unmarked cups of Pepsi and Coke and were asked to select their favorite. Pepsi claimed that "in recent blind taste tests, more than half the Diet Coke drinkers surveyed said they preferred the taste of Diet Pepsi." Suppose 100 Diet Coke drinkers took the Pepsi Challenge and 56 preferred the taste of Diet Pepsi. Test the hypothesis that more than half of all Diet Coke drinkers will select Diet Pepsi in a blind taste test. Use $\alpha = .05$.

6.145 Masculinizing human faces. Refer to the *Nature* (August 27, 1998) study of facial characteristics that are deemed attractive, presented in Exercise 6.55 (p. 317). In another experiment, 67 human subjects viewed side by side an image of a Caucasian male face and the same image 50% masculinized. Each subject was asked to select the facial image they deemed more attractive. Fifty-eight of the 67 subjects felt that masculinization of face shape decreased attractiveness of the male face. The researchers used this sample information to test whether the subjects showed a preference for either the unaltered or the morphed male face.
a. Set up the null and alternative hypotheses for this test.
b. Compute the test statistic.
c. The researchers reported a *p*-value ≈ 0 for the test. Do you agree?
d. Make the appropriate conclusion in the words of the problem. Use $\alpha = .01$.

*6.146 Time taken to solve a math programming problem. Refer to the *IEEE Transactions* study of a new hybrid algorithm for solving polynomial 0/1 mathematical programs, presented in Exercise 5.131 (p. 287). (Data on solution times are saved in the **MATHCPU** file.) A SAS printout giving

descriptive statistics for the sample of 52 solution times is reproduced at the bottom of the page. Use this information to determine whether the variance of the solution times differs from 2. Use $\alpha = .05$.

6.147 Alkalinity of river water. In Exercise 4.187 (p. 233), you learned that the mean alkalinity level of water specimens collected from the Han River in Seoul, Korea, is 50 milligrams per liter (*Environmental Science & Engineering*, September 1, 2000). Consider a random sample of 100 water specimens collected from a tributary of the Han River. Suppose the mean and standard deviation of the alkalinity levels for the sample are, respectively, $\bar{x} = 67.8$ mpl and $s = 14.4$ mpl. Is there sufficient evidence (at $\alpha = .01$) to indicate that the population mean alkalinity level of water in the tributary exceeds 50 mpl?

Applying the Concepts—Intermediate

6.148 Errors in medical tests. Medical tests have been developed to detect many serious diseases. A medical test is designed to minimize the probability that it will produce a "false positive" or a "false negative." A false positive is a positive test result for an individual who does not have the disease, whereas a false negative is a negative test result for an individual who does have the disease.
a. If we treat a medical test for a disease as a statistical test of hypothesis, what are the null and alternative hypotheses for the medical test?
b. What are the Type I and Type II errors for the test? Relate each to false positives and false negatives.
c. Which of these errors has graver consequences? Considering this error, is it more important to minimize α or β? Explain.

6.149 Post-traumatic stress of POWs. *Psychological Assessment* (March 1995) published the results of a study of World War II aviators captured by German forces after having been shot down. Having located a total of 239 World War II aviator POW survivors, the researchers asked each veteran to participate in the study; 33 responded to the letter of invitation. Each of the 33 POW survivors was administered the Minnesota Multiphasic Personality Inventory, one component of which measures level of post-traumatic stress disorder (PTSD). [*Note:* The higher the score, the higher is the level of PTSD.] The aviators produced a mean PTSD score of $\bar{x} = 9.00$ and a standard deviation of $s = 9.32$. Conduct a test to determine if the true mean PTSD score for all World war II aviator POWS is less than 16. [*Note:* The value 16 represents the mean PTSD score established for Vietnam POWS.] Use $\alpha = .10$.

6.150 Cracks in highway pavement. Using van-mounted state-of-the-art video technology, the Mississippi Department of Transportation collected data on the number of cracks (called *crack intensity*) in an undivided two-lane highway.

SAS output for Exercise 6.146

The MEANS Procedure

Analysis Variable : CPU

Mean	Std Dev	Variance	N	Minimum	Maximum
0.8121923	1.5047603	2.2643035	52	0.0360000	8.7880000

(*Journal of Infrastructure Systems*, March 1995). The mean number of cracks found in a sample of eight 50-meter sections of the highway was $\bar{x} = .210$, with a variance of $s^2 = .011$. Suppose the American Association of State Highway and Transportation Officials (AASHTO) recommends a maximum mean crack intensity of .100 for safety purposes.

a. Test the hypothesis that the true mean crack intensity of the Mississippi highway exceeds the AASHTO recommended maximum. Use $\alpha = .01$.

b. Define a Type I error and a Type II error for this study.

6.151 Inbreeding of tropical wasps. Refer to the *Science* study of inbreeding in tropical swarm-founding wasps, presented in Exercise 5.129 (p. 287). A sample of 197 wasps, captured, frozen, and subjected to a series of genetic tests, yielded a sample mean inbreeding coefficient of $\bar{x} = .044$ with a standard deviation of $s = .884$. Recall that if the wasp has no tendency to inbreed, the true mean inbreeding coefficient μ for the species will equal 0.

a. Test the hypothesis that the true mean inbreeding coefficient μ for this species of wasp exceeds 0. Use $\alpha = .05$.

b. Compare the inference you made in part **a** with the inference you obtained in Exercise 5.129, using a confidence interval. Do the inferences agree? Explain.

*6.152 Weights of parrot fish. A marine biologist wishes to use parrot fish for experimental purposes due to the belief that their weight is fairly stable (i.e., the variability in weights among parrot fish is small). The biologist randomly samples 10 parrot fish and finds that their mean weight is 4.3 pounds and the standard deviation is 1.4 pounds. The biologist will use the parrot fish only if there is evidence that the variance of their weights is less than 4.

a. Is there sufficient evidence for the biologist to claim that the variability in weights among parrot fish is small enough to justify their use in the experiment? Test at $\alpha = .05$.

b. State any assumptions that are needed for the test mentioned in part **a** to be valid.

6.153 PCB in plant discharge. The EPA sets a limit of 5 parts per million (ppm) on PCB (polychlorinated biphenyl, a dangerous substance) in water. A major manufacturing firm producing PCB for electrical insulation discharges small amounts from the plant. The company management, attempting to control the PCB in its discharge, has given instructions to halt production if the mean amount of PCB in the effluent exceeds 3 ppm. A random sample of 50 water specimens produced the following statistics: $\bar{x} = 3.1$ ppm and $s = .5$ ppm.

a. Do these statistics provide sufficient evidence to halt the production process? Use $\alpha = .01$.

b. If you were the plant manager, would you want to use a large or a small value for α for the test in part **a**?

*6.154 Fluoride in drinking water. Many water treatment facilities supplement the natural fluoride concentration with hydrofluosilicic acid in order to reach a target concentration of fluoride in drinking water. Certain levels are thought to enhance dental health, but very high concentrations can be dangerous. Suppose that one such treatment plant targets .75 milligram per liter (mg/L) for its water. The plant tests 25 samples each day to determine whether the median level differs from the target.

a. Set up the null and alternative hypotheses.

b. Set up the test statistic and rejection region, using $\alpha = .10$.

c. Explain the implication of a Type I error in the context of this application. A Type II error.

d. Suppose that one day's samples result in 18 values that exceed .75 mg/L. Conduct the test and state the appropriate conclusion in the context of this application.

e. When it was suggested to the plant's supervisor that a *t*-test should be used to conduct the daily test, she replied that the probability distribution of the fluoride concentrations was "heavily skewed to the right." Show graphically what she meant by this, and explain why this is a reason to prefer the sign test to the *t*-test.

6.155 Choosing portable grill displays. Refer to the *Journal of Consumer Research* (March 2003) experiment on influencing the choices of others by offering undesirable alternatives, presented in Exercise 3.27 (p. 120). Recall that each of 124 college students selected three portable grills out of five to display on the showroom floor. The students were instructed to include Grill #2 (a smaller-sized grill) and select the remaining two grills in the display to maximize purchases of Grill #2. If the six possible grill display combinations (1–2–3, 1–2–4, 1–2–5, 2–3–4, 2–3–5, and 2–4–5) are selected at random, then the proportion of students selecting any display will be $1/6 = .167$. One theory tested by the researcher is that the students will tend to choose the three-grill display so that Grill #2 is a compromise between a more desirable and a less desirable grill. Of the 124 students, 85 students selected a three-grill display that was consistent with that theory. Use this information to test the theory proposed by the researcher at $\alpha = .05$.

6.156 Federal civil trial appeals. Refer to the *Journal of the American Law and Economics Association* (Vol. 3, 2001) study of appeals of federal civil trials, presented in Exercise 3.59 (p. 132). A breakdown of 678 civil cases that were originally tried in front of a judge and appealed by either the plaintiff or the defendant is reproduced in the accompanying table. Do the data provide sufficient evidence to indicate that the percentage of civil cases appealed that are actually reversed is less than 25%? Test, using $\alpha = .01$.

Outcome of Appeal	Number of Cases
Plaintiff trial win—reversed	71
Plaintiff trial win—affirmed/dismissed	240
Defendant trial win—reversed	68
Defendant trial win—affirmed/dismissed	299
Total	678

Applying the Concepts—Advanced

6.157 Polygraph test error rates. In a classic study reported in *Discover* magazine, a group of physicians subjected the *polygraph* (or *lie detector*) to the same careful testing given to medical diagnostic tests. They found that if 1,000 people were subjected to the polygraph and 500 told the truth and 500 lied, the polygraph would indicate that approximately 185 of the truth tellers were liars and that approximately 120 of the liars were truth tellers.

a. In the application of a polygraph test, an individual is presumed to be a truth teller (H_0) until "proven" a liar (H_a). In this context, what is a Type I error? A Type II error?

b. According to the study, what is the probability (approximately) that a polygraph test will result in a Type I error? A Type II error?

6.158 Parents who condone spanking. In Exercise 4.182 (p. 233) you read about a nationwide survey which claimed that 60% of parents with young children condone spanking their child as a regular form of punishment (*Tampa Tribune*, October 5, 2000). In a random sample of 100 parents with young children, how many parents would need to say that they condone spanking as a form of punishment in order to refute the claim?

6.159 Solar joint inspections. X-rays and lasers are used to inspect solder-joint defects on printed circuit boards (PCBs). A particular manufacturer of laser-based inspection equipment claims that its product can inspect at least 10 solder joints per second, on average, when the joints are spaced .1 inch apart. The equipment was tested by a potential buyer on 48 different PCBs. In each case, the equipment was operated for exactly 1 second. The numbers of solder joints inspected on each run are listed in the table. These data are saved in the **PCB** file.

10	9	10	10	11	9	12	8	8	9	6	10
7	10	11	9	9	13	9	10	11	10	12	8
9	9	9	7	12	6	9	10	10	8	7	9
11	12	10	0	10	11	12	9	7	9	9	10

a. The potential buyer doubts the manufacturer's claim. Do you agree?

b. Assume that the standard deviation of the number of solder joints inspected on each run is 1.2, and the true mean number of solder joints that can be inspected is really equal to 9.5. How likely is the buyer to correctly conclude that the claim is false?

Critical Thinking Challenges

6.160 The Hot Tamale caper. "Hot Tamales" are chewy, cinnamon-flavored candies. A bulk vending machine is known to dispense, on average, 15 Hot Tamales per bag. *Chance* (Fall 2000) published an article on a classroom project in which students were required to purchase bags of Hot Tamales from the machine and count the number of candies per bag. One student group claimed it purchased five bags that had the following candy counts: 25, 23, 21, 21, and 20. There was some question as to whether the students had fabricated the data. Use a hypothesis test to gain insight into whether or not the data collected by the students were fabricated. Use a level of significance that gives the benefit of the doubt to the students.

6.161 Verifying voter petitions. To get their names on the ballot of a local election, political candidates often must obtain petitions bearing the signatures of a minimum number of registered voters. According to the *St. Petersburg Times*, in Pinellas County, Florida, a certain political candidate obtained petitions with 18,200 signatures. To verify that the names on the petitions were signed by actual registered voters, election officials randomly sampled 100 of the signatures and checked each for authenticity. Only 2 were invalid signatures.

a. Is 98 out of 100 verified signatures sufficient to believe that more than 17,000 of the total 18,200 signatures are valid? Use $\alpha = .01$.

b. Repeat part **a** if only 16,000 valid signatures are required.

Activity   Challenging a Claim: Tests of Hypotheses

Use the Internet or a newspaper or magazine to find an example of a claim made by a political or special-interest group about some characteristic (e.g., favor gun control) of the U.S. population. In this activity, you represent a rival group that believes the claim may be false.

1. In your example, what kinds of evidence might exist which would cause one to suspect that the claim might be false and therefore worthy of a statistical study? Be specific. If the claim were false, how would consumers be hurt?

2. Describe what data are relevant and how that data might be collected.

3. Explain the steps necessary to reject the group's claim at level α. State the null and alternative hypotheses. If you reject the claim, does it mean that the claim is false?

4. If you reject the claim when the claim is actually true, what type of error has occurred? What is the probability of this error occurring?

5. If you were to file a lawsuit against the group based on your rejection of its claim, how might the group use your results to defend itself?

References

Daniel, W. W. *Applied Nonparametric Statistics*, 2nd ed. Boston: PWS-Kent, 1990.

Snedecor, G. W., and Cochran, W. G. *Statistical Methods*, 7th ed. Ames, IA: Iowa State University Press, 1980.

Hollander, M., and Wolfe, D. A. *Nonparametric Statistical Methods*. 2nd ed. New York: Wiley, 1999.

Wackerly, D., Mendenhall, W., and Scheaffer, R. *Mathematical Statistics with Applications*, 7th ed. Belmont, CA: Thomson, Brooks/Cole, 2008.

Using Technology

MINITAB: Tests of Hypotheses

Testing μ

Step 1 Access the MINITAB data worksheet that contains the sample data.

Step 2 Click on the "Stat" button on the MINITAB menu bar and then click on "Basic Statistics" and "1-Sample t," as shown in Figure 8.M.1.

Figure 6.M.1
MINITAB menu options for testing a mean

Step 3 On the resulting dialog box (shown in Figure 6.M.2), click on "Samples in Columns" and then specify the quantitative variable of interest in the open box.

Figure 6.M.2
MINITAB 1-sample *t* test dialog box

Step 4 Check "Perform hypothesis test" and then specify the value of μ_0 for the null hypothesis in the "Hypothesized mean" box.

Step 5 Click on the "Options" button at the bottom of the dialog box and specify the form of the alternative hypothesis, as shown in Figure 6.M.3.

Figure 6.M.3
MINITAB 1-sample t test options

Step 6 Click "OK" to return to the "1-Sample t" dialog box and then click "OK" again to produce the hypothesis test.

Note: If you want to produce a test for the mean from summary information (e.g., the sample mean, sample standard deviation, and sample size), click on "Summarized data" in the "1-Sample t" dialog box, enter the values of the summary statistics and μ_0, and then click "OK."

Important: The MINITAB one-sample *t*-procedure uses the *t*-statistic to generate the hypothesis test. When the sample size *n* is small, this is the appropriate method. When the sample size *n* is large, the *t*-value will be approximately equal to the large-sample *z*-value, and the resulting test will still be valid. If you have a large sample and you know the value of the population standard deviation σ (which is rarely the case), select "1-sample Z" from the "Basic Statistics" menu options (see Figure 6.M.1) and make the appropriate selections.

Testing p

Step 1 Access the MINITAB data worksheet that contains the sample data.

Step 2 Click on the "Stat" button on the MINITAB menu bar and then click on "Basic Statistics" and "1 Proportion" (see Figure 6.M.1).

Step 3 On the resulting dialog box (shown in Figure 6.M.4), click on "Samples in Columns," and then specify the qualitative variable of interest in the open box.

Figure 6.M.4
MINITAB 1-proportion test dialog box

Step 4 Check "Perform hypothesis test" and then specify the null hypothesis value p_0 in the "Hypothesized proportion" box.

Step 5 Click "Options," then specify the form of the alternative hypothesis in the resulting dialog box, as shown in Figure 6.M.5. Also, check the "Use test and interval based on normal distribution" box at the bottom.

Figure 6.M.5
MINITAB 1-proportion test options

Step 6 Click "OK" to return to the "1-Proportion" dialog box and then click "OK" again to produce the test results.

Note: If you want to produce a confidence interval for a proportion from summary information (e.g., the number of successes and the sample size), click on "Summarized data" in the "1 Proportion" dialog box (see Figure 6.M.4). Enter the value for the number of trials (i.e., the sample size) and the number of events (i.e., the number of successes), and then click "OK."

Testing σ^2

Step 1 Access the MINITAB data worksheet that contains the sample data set.

Step 2 Click on the "Stat" button on the MINITAB menu bar and and then click on "Basic Statistics" and "1 Variance" (see Figure 6.M.1).

Step 3 Once the resulting dialog box appears (see Figure 6.M.6), click on "Samples in Columns" and then specify the quantitative variable of interest in the open box.

Figure 6.M.6
MINITAB 1-variance test dialog box

Step 4 Check "Perform hypothesis test" and specify the null hypothesis value of the standard deviation σ_0 in the open box.

Step 5 Click on the "Options" button at the bottom of the dialog box and specify the form of the alternative hypothesis (similar to Figure 6.M.3).

Step 6 Click "OK" twice to produce the hypothesis test.

Note: If you want to produce a test for the variance from summary information (e.g., the sample standard deviation and sample size), click on "Summarized data" in the "1 Variance" dialog box (Figure 6.M.6) and enter the values of the summary statistics.

Sign Test

Step 1 Access the MINITAB worksheet file with the sample data. It should contain a single quantitative variable.

Step 2 Click on the "Stat" button on the MINITAB menu bar, then click on "Nonparametrics" and "1-Sample Sign," as shown in Figure 6.M.7.

Figure 6.M.7
MINITAB nonparametric menu options

Step 3 On the resulting dialog box (see Figure 6.M.8), enter the quantitative variable to be analyzed in the "Variables" box.

Figure 6.M.8
MINITAB 1-sample sign dialog box

Step 4 Select the "Test median" option and specify the hypothesized value of the median and the form of the alternative hypothesis ("not equal", "less than", or "greater than").

Step 5 Click "OK" to generate the MINITAB printout.

TI-83/TI-84 Plus Graphing Calculator: Tests of Hypotheses

Note: The TI-83/TI-84 plus graphing calculator cannot currently conduct a test for a population variance.

Hypothesis Test for a Population Mean (Large Sample Case)

Step 1 *Enter the data (Skip to Step 2 if you have summary statistics, not raw data.)*

- Press **STAT** and select **1:Edit**

Note: If the list already contains data, clear the old data. Use the up **ARROW** to highlight "**L1.**"

- Press **CLEAR ENTER**
- Use the **ARROW** and **ENTER** keys to enter the data set into **L1**

Step 2 *Access the statistical tests menu*

- Press **STAT**
- Arrow right to **TESTS**
- Press **ENTER** to select either **Z-Test** (if large sample and known) or **T-Test** (if unknown)

Step 3 *Choose "**Data**" or "**Stats**" ("Data" is selected when you have entered the raw data into a List. "Stats" is selected when you are given only the mean, standard deviation, and sample size.)*

- Press **ENTER**

If you selected "Data," enter the values for the hypothesis test where μ_0 = the value for μ in the null hypothesis, σ = assumed value of the population standard deviation.

- Set **List** to **L1**
- Set **Freq** to **1**
- Use the **ARROW** to highlight the appropriate alternative hypothesis
- Press **ENTER**
- Arrow down to "**Calculate**"
- Press **ENTER**

If you selected "Stats," enter the values for the hypothesis test where μ_0 = the value for μ in the null hypothesis, σ = assumed value of the population standard deviation.

- Enter the sample mean, sample standard deviation, and sample size
- Use the **ARROW** to highlight the appropriate alternative hypothesis
- Press **ENTER**
- Arrow down to "**Calculate**"
- Press **ENTER**

```
Z-Test
 μ<10000
 z=-1.549516053
 p=.0606288707
 x̄=9755
 n=40
```

The chosen test will be displayed as well as the z (or t) test statistic, the p-value, the sample mean, and the sample size.

Hypothesis Test for a Population Proportion (Large Sample Case)

Step 1 *Enter the data (Skip to Step 2 if you have summary statistics, not raw data.)*

- Press **STAT** and select **1:Edit**

Note: If the list already contains data, clear the old data. Use the up **ARROW** to highlight "**L1.**"

- Press **CLEAR ENTER**
- Use the **ARROW** and **ENTER** keys to enter the data set into **L1**

Step 2 *Access the statistical tests menu*

- Press **STAT**
- Arrow right to **TESTS**
- Press **ENTER** after selecting **1-Prop Z Test**

Step 3 *Enter the hypothesized proportion p_0, the number of success x, and the sample size n*

- Use the **ARROW** to highlight the appropriate alternative hypothesis
- Press **ENTER**
- Arrow down to "**Calculate**"
- Press **ENTER**

The chosen test will be displayed as well as the z-test statistic and the p-value.

7 Comparing Population Means

CONTENTS

Where We've Been

- Explored two methods for making statistical inferences: *confidence intervals* and *tests of hypotheses*
- Studied confidence intervals and tests for a single population mean μ, a single population proportion p, and a single population variance σ^2
- Learned how to select the sample size necessary to estimate a population parameter with a specified margin of error

Where We're Going

- Learn how to identify the target parameter for comparing population means (7.1)
- Learn how to compare two means by using confidence intervals and tests of hypotheses (7.2–7.3)
- Determine the sizes of the samples necessary to estimate the difference between two population means with a specified margin of error (7.4)
- Present nonparametric tests for comparing two populations (7.5–7.6)
- Learn how to compare three or more population means using a designed experiment (7.7)

On the Trail of the Cockroach:
Do Roaches Travel at Random?

Entomologists have long established that insects such as ants, bees, caterpillars, and termites use chemical or "odor" trails for navigation. These trails are used as highways between sources of food and the insects' nest. Until recently, however, "bug" researchers believed that the navigational behavior of cockroaches scavenging for food was random and not linked to a chemical trail.

One of the first researchers to challenge the "random-walk" theory about cockroaches was professor and entomologist Dini Miller of Virginia Tech University. According to Miller, "The idea that roaches forage randomly means that they would have to come out of their hiding places every night and bump into food and water by accident. But roaches never seem to go hungry." Since cockroaches had never before been evaluated for trail-following behavior, Miller designed an experiment to test a cockroach's ability to follow a trail of their fecal material (*Journal of Economic Entomology*, Aug. 2000).

First, Miller developed a methanol extract from roach feces—called a pheromone. She theorized that "pheromones are communication devices between cockroaches. If you have an infestation and have a lot of fecal material around, it advertises, 'Hey, this is a good cockroach place.'" Then she created a chemical trail with the pheromone on a strip of white chromatography paper and placed the paper at the bottom of a plastic, V-shaped container, 122 square centimeters in area. German cockroaches were released into the container at the beginning of the trail, one at a time, and a video surveillance camera was used to monitor the roaches' movements.

In addition to the trail containing the fecal extract (the treatment), a trail using methanol only was created. This second trail served as a control against which the treated trail could be compared. Because Miller also wanted to determine whether trail-following ability differed among cockroaches of different age, sex, and reproductive status, four roach groups were utilized in the experiment: adult males, adult females, gravid (pregnant) females, and nymphs (immatures). Twenty roaches of each type were randomly assigned to the treatment trail, and 10 of each type were randomly assigned to the control trail. Thus, a total of 120 roaches were used in the experiment.

The movement pattern of each cockroach tested was translated into *xy*-coordinates every one-tenth of a second by the Dynamic Animal Movement Analyzer (DAMA) program. Miller measured the perpendicular distance of each *xy*-coordinate from the trail and then averaged all the distances, or deviations, for each cockroach. The average trail deviations (measured in pixels, where 1 pixel equals approximately 2 centimeters) for each of the 120 cockroaches in the study are stored in the data file named **ROACH**.

We apply the statistical methodology presented in this chapter to the cockroach data in several Statistics in Action Revisited sections.

Data Set: ROACH

Statistics IN Action Revisited

- Comparing Roach Trails—Mean Treatment versus Mean Control (p. 362)
- Comparing Treatment and Control Roach Trails Using a Nonparametric Test (p. 387)
- A One-Way Analysis of the Cockroach Data (p. 409)

7.1 Identifying the Target Parameter

Many experiments involve a comparison of population means. For instance, a sociologist may want to estimate the difference in mean life expectancy between inner-city and suburban residents. Or a consumer group may want to test whether two major brands of food freezers differ in the average amount of electricity they use. Or a professional golfer might be interested in comparing the mean driving distances of several competing brands of golf balls when struck with the same club. In this chapter, we consider techniques for using two (or more) samples to compare the populations from which they were selected.

The same procedures that are used to estimate and test hypotheses about a single population mean can be modified to make inferences about two or more population means. As in Chapters 5 and 6, the methodology used will depend on the sizes of the samples and the parameter of interest (i.e., the *target parameter*). Some key words and the type of data associated with the parameters covered in this chapter are listed in the following box.

Determining the Target Parameter

Parameter(s)	Key Words or Phrases	Type of Data	Number of Samples
$\mu_1 - \mu_2$	Difference between means; difference in averages	Quantitative	2 independent samples
μ_d	Mean of paired differences	Quantitative	1 paired sample
$\mu_1, \mu_2, \ldots \mu_k$	Compare multiple means (averages)	Quantitative	k independent samples

You can see that the key words *difference* and *compare* help identify the fact that populations are to be compared. In the previous examples, the words *mean* in *mean life expectancy* and *average* in *average amount of electricity* imply that the target parameter is the difference in two population means, $\mu_1 - \mu_2$. We will learn that if the sample consists of paired differences, then the parameter is μ_d, the mean of the differences. Finally, you may want to compare three or more population means, μ_1, μ_2, μ_3, etc.

As with inferences about a single population mean, the type of data collected on the samples will be quantitative in nature.

7.2 Comparing Two Population Means: Independent Sampling

In this section, we develop both large-sample and small-sample methodologies for comparing two population means. In the large-sample case, we use the z-statistic; in the small-sample case, we use the *t*-statistic.

Large Samples

Example 7.1

A Large-Sample Confidence Interval for $(\mu_1 - \mu_2)$—Comparing Mean Weight Loss for Two Diets

Problem A dietitian has developed a diet that is low in fats, carbohydrates, and cholesterol. Although the diet was initially intended to be used by people with heart disease, the dietitian wishes to examine the effect this diet has on the weights of obese people. Two random samples of 100 obese people each are selected, and one group of 100 is placed on the low-fat diet. The other 100 are placed on a diet that contains approximately the same quantity of food, but is not as low in fats, carbohydrates, and cholesterol. For each person, the amount of weight lost (or gained) in a three-week period is recorded. The data, saved in the **DIETSTUDY** file, are listed in Table 7.1. Form a 95% confidence interval for the difference between the population mean weight losses for the two diets. Interpret the result.

Solution Recall that the general form of a large-sample confidence interval for a single mean μ is $\bar{x} \pm z_{\alpha/2}\sigma_{\bar{x}}$. That is, we add and subtract $z_{\alpha/2}$ standard deviations of the sample estimate $\bar{x}$ to and from the value of the estimate. We employ a similar procedure to form the confidence interval for the difference between two population means.

Let μ_1 represent the mean of the conceptual population of weight losses for all obese people who could be placed on the low-fat diet. Let μ_2 be similarly defined for the other diet. We wish to form a confidence interval for $(\mu_1 - \mu_2)$. An intuitively appealing estimator for $(\mu_1 - \mu_2)$ is the difference between the sample means, $(\bar{x}_1 - \bar{x}_2)$. Thus, we will form the confidence interval of interest with

$$(\bar{x}_1 - \bar{x}_2) \pm z_{\alpha/2}\sigma_{(\bar{x}_1 - \bar{x}_2)}$$

Assuming that the two samples are independent, we write the standard deviation of the difference between the sample means (i.e., the *standard error* of $\bar{x}_1 - \bar{x}_2$) as

$$\sigma_{(\bar{x}_1 - \bar{x}_2)} = \sqrt{\frac{\sigma_1^2}{n_1} + \frac{\sigma_2^2}{n_2}}$$

| Table 7.1 | Diet Study Data, Example 7.1 |

Weight Losses for Low-Fat Diet

8	10	10	12	9	3	11	7	9	2
21	8	9	2	2	20	14	11	15	6
13	8	10	12	1	7	10	13	14	4
8	12	8	10	11	19	0	9	10	4
11	7	14	12	11	12	4	12	9	2
4	3	3	5	9	9	4	3	5	12
3	12	7	13	11	11	13	12	18	9
6	14	14	18	10	11	7	9	7	2
16	16	11	11	3	15	9	5	2	6
5	11	14	11	6	9	4	17	20	10

Weight Losses for Regular Diet

6	6	5	5	2	6	10	3	9	11
14	4	10	13	3	8	8	13	9	3
4	12	6	11	12	9	8	5	8	7
6	2	6	8	5	7	16	18	6	8
13	1	9	8	12	10	6	1	0	13
11	2	8	16	14	4	6	5	12	9
11	6	3	9	9	14	2	10	4	13
8	1	1	4	9	4	1	1	5	6
14	0	7	12	9	5	9	12	7	9
8	9	8	10	5	8	0	3	4	8

Data Set: DIETSTUDY

Typically (as in this example), the population variances σ_1^2 and σ_2^2 are unknown. Since the samples are both large ($n_1 = n_2 = 100$), the sample variances s_1^2 and s_2^2 will be good estimators of their respective population variances. Thus, the estimated standard error is

$$\sigma_{(\bar{x}_1 - \bar{x}_2)} \approx \sqrt{\frac{s_1^2}{n_1} + \frac{s_2^2}{n_2}}$$

Summary statistics for the diet data are displayed at the top of the SPSS printout shown in Figure 7.1. Note that $\bar{x}_1 = 9.31, \bar{x}_2 = 7.40, s_1 = 4.67$, and $s_2 = 4.04$. Using these values and observing that $\alpha = .05$ and $z_{.025} = 1.96$, we find that the 95% confidence interval is, approximately,

$$(9.31 - 7.40) \pm 1.96 \sqrt{\frac{(4.67)^2}{100} + \frac{(4.04)^2}{100}} = 1.91 \pm (1.96)(.62) = 1.91 \pm 1.22$$

or (.69, 3.13). This interval (rounded) is highlighted in Figure 7.1.

Using this estimation procedure over and over again for different samples, we know that approximately 95% of the confidence intervals formed in this manner will enclose the difference in population means ($\mu_1 - \mu_2$). Therefore, we are highly confident that the mean weight loss for the low-fat diet is between .69 and 3.13 pounds more

Group Statistics

	DIET	N	Mean	Std. Deviation	Std. Error Mean
WTLOSS	LOWFAT	100	9.31	4.668	.467
	REGULAR	100	7.40	4.035	.404

Independent Samples Test

		Levene's Test for Equality of Variances		t-test for Equality of Means						95% Confidence Interval of the Difference	
		F	Sig.	t	df	Sig. (2-tailed)	Mean Difference	Std. Error Difference	Lower	Upper
WTLOSS	Equal variances assumed	1.367	.244	3.095	198	.002	1.910	.617	.693	3.127
	Equal variances not assumed			3.095	193.940	.002	1.910	.617	.693	3.127

Figure 7.1
SPSS analysis of diet study data.

than the mean weight loss for the other diet. With this information, the dietitian better understands the potential of the low-fat diet as a weight-reduction diet.

Look Back If the confidence interval for $(\mu_1 - \mu_2)$ contains 0 [e.g., $(-2.5, 1.3)$], then it is possible for the difference between the population means to be 0 (i.e., $\mu_1 - \mu_2 = 0$). In this case, we could not conclude that a significant difference exists between the mean weight losses for the two diets.

Now Work Exercise 7.6a

The justification for the procedure used in Example 7.1 to estimate $(\mu_1 - \mu_2)$ relies on the properties of the sampling distribution of $(\bar{x}_1 - \bar{x}_2)$. The performance of the estimator in repeated sampling is pictured in Figure 7.2, and its properties are summarized in the following box:

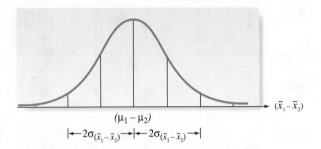

Figure 7.2
Sampling distribution of $(\bar{x}_1 - \bar{x}_2)$

Properties of the Sampling Distribution of $(\bar{x}_1 - \bar{x}_2)$

1. The mean of the sampling distribution of $(\bar{x}_1 - \bar{x}_2)$ is $(\mu_1 - \mu_2)$.

2. If the two samples are independent, the standard deviation of the sampling distribution is

$$\sigma_{(\bar{x}_1 - \bar{x}_2)} = \sqrt{\frac{\sigma_1^2}{n_1} + \frac{\sigma_2^2}{n_2}}$$

where σ_1^2 and σ_2^2 are the variances of the two populations being sampled and n_1 and n_2 are the respective sample sizes. We also refer to $\sigma_{(\bar{x}_1 - \bar{x}_2)}$ as the **standard error of the statistic** $(\bar{x}_1 - \bar{x}_2)$.

3. By the Central Limit Theorem, the sampling distribution of $(\bar{x}_1 - \bar{x}_2)$ is approximately normal *for large samples*.

In Example 7.1, we noted the similarity in the procedures for forming a large-sample confidence interval for one population mean and a large-sample confidence interval for the difference between two population means. When we are testing hypotheses, the procedures are again similar. The general large-sample procedures for forming confidence intervals and testing hypotheses about $(\mu_1 - \mu_2)$ are summarized in the following boxes:

Large, Independent Samples Confidence Interval for $(\mu_1 - \mu_2)$: Normal (z) Statistic

σ_1^2 and σ_2^2 known: $(\bar{x}_1 - \bar{x}_2) \pm z_{\alpha/2}\, \sigma_{(\bar{x}_1 - \bar{x}_2)} = (\bar{x}_1 - \bar{x}_2) \pm z_{\alpha/2}\sqrt{\dfrac{\sigma_1^2}{n_1} + \dfrac{\sigma_2^2}{n_2}}$

σ_1^2 and σ_2^2 unknown: $(\bar{x}_1 - \bar{x}_2) \pm z_{\alpha/2}\, \sigma_{(\bar{x}_1 - \bar{x}_2)} \approx (\bar{x}_1 - \bar{x}_2) \pm z_{\alpha/2}\sqrt{\dfrac{s_1^2}{n_1} + \dfrac{s_2^2}{n_2}}$

Large, Independent Samples Test of Hypothesis for $(\mu_1 - \mu_2)$: Normal (z) Statistic

One-Tailed Test

H_0: $(\mu_1 - \mu_2) = D_0$

H_a: $(\mu_1 - \mu_2) < D_0$

  [or H_a: $(\mu_1 - \mu_2) > D_0$]

Two-Tailed Test

H_0: $(\mu_1 - \mu_2) = D_0$

H_a: $(\mu_1 - \mu_2) \neq D_0$

where D_0 = Hypothesized difference between the means (this difference is often hypothesized to be equal to 0)

Test statistic:

$$z = \frac{(\bar{x}_1 - \bar{x}_2) - D_0}{\sigma_{(\bar{x}_1 - \bar{x}_2)}} \quad \text{where} \quad \sigma_{(\bar{x}_1 - \bar{x}_2)} = \sqrt{\frac{\sigma_1^2}{n_1} + \frac{\sigma_2^2}{n_2}} \text{ if both } \sigma_1^2 \text{ and } \sigma_2^2 \text{ are known}$$

$$\approx \sqrt{\frac{s_1^2}{n_1} + \frac{s_2^2}{n_2}} \text{ if } \sigma_1^2 \text{ and } \sigma_2^2 \text{ are unknown}$$

Rejection region: $z < -z_\alpha$

  [or $z > z_\alpha$ when

  H_a: $(\mu_1 - \mu_2) > D_0$]

Rejection region: $|z| > z_{\alpha/2}$

Conditions Required for Valid Large-Sample Inferences about $(\mu_1 - \mu_2)$

1. The two samples are randomly selected in an independent manner from the two target populations.

2. The sample sizes, n_1 and n_2, are both large (i.e., $n_1 \geq 30$ and $n_2 \geq 30$). (By the Central Limit Theorem, this condition guarantees that the sampling distribution of $(\bar{x}_1 - \bar{x}_2)$ will be approximately normal, regardless of the shapes of the underlying probability distributions of the populations. Also, s_1^2 and s_2^2 will provide good approximations to σ_1^2 and σ_2^2 when both samples are large.)

Example 7.2

A Large-Sample Test for $(\mu_1 - \mu_2)$— Comparing Mean Weight Loss for Two Diets

Problem Refer to the study of obese people on a low-fat diet and a regular diet presented in Example 7.1. Another way to compare the mean weight losses for the two different diets is to conduct a test of hypothesis. Use the information on the SPSS printout shown in Figure 7.1 to conduct the test. Take $\alpha = .05$.

Solution Again, we let μ_1 and μ_2 represent the population mean weight losses of obese people on the low-fat diet and regular diet, respectively. If one diet is more effective in reducing the weights of obese people, then either $\mu_1 < \mu_2$ or $\mu_2 < \mu_1$; that is, $\mu_1 \neq \mu_2$.

Thus, the elements of the test are as follows:

H_0: $(\mu_1 - \mu_2) = 0$ (i.e., $\mu_1 = \mu_2$; note that $D_0 = 0$ for this hypothesis test)

H_a: $(\mu_1 - \mu_2) \neq 0$ (i.e., $\mu_1 \neq \mu_2$)

Test statistic: $z = \dfrac{(\bar{x}_1 - \bar{x}_2) - D_0}{\sigma_{(\bar{x}_1 - \bar{x}_2)}} = \dfrac{\bar{x}_1 - \bar{x}_2 - 0}{\sigma_{(\bar{x}_1 - \bar{x}_2)}}$

Rejection region: $z < -z_{\alpha/2} = -1.96$ or $z > z_{\alpha/2} = 1.96$ (see Figure 7.3)

Substituting the summary statistics given in Figure 7.1 into the test statistic, we obtain

$$z = \frac{(\bar{x}_1 - \bar{x}_2) - 0}{\sigma_{(\bar{x}_1 - \bar{x}_2)}} = \frac{9.31 - 7.40}{\sqrt{\dfrac{\sigma_1^2}{n_1} + \dfrac{\sigma_2^2}{n_2}}}$$

Figure 7.3

Rejection region for
Example 7.2

Now, since σ_1^2 and σ_2^2 are unknown, we approximate the test statistic value as follows:

$$z \approx \frac{9.31-7.40}{\sqrt{\dfrac{s_1^2}{n_1} + \dfrac{s_2^2}{n_2}}} = \frac{1.91}{\sqrt{\dfrac{(4.67)^2}{100} + \dfrac{(4.04)^2}{100}}} = \frac{1.91}{.617} = 3.09$$

[*Note:* The value of the test statistic is highlighted in the SPSS printout of Figure 7.1.]
    As you can see in Figure 7.3, the calculated z-value clearly falls into the rejection region. Therefore, the samples provide sufficient evidence, at $\alpha = .05$, for the dietitian to conclude that the mean weight losses for the two diets differ.

Look Back This conclusion agrees with the inference drawn from the 95% confidence interval in Example 7.1. However, the confidence interval provides more information on the mean weight losses. From the hypothesis test, we know only that the two means differ; that is, $\mu_1 \neq \mu_2$. From the confidence interval in Example 7.1, we found that the mean weight loss μ_1 of the low-fat diet was between .69 and 3.13 pounds more than the mean weight loss μ_2 of the regular diet. In other words, the test tells us that the means differ, but the confidence interval tells us how large the difference is. Both inferences are made with the same degree of reliability—namely, 95% confidence (or at $\alpha = .05$).

Example 7.3
The *p*-Value for a Test of $(\mu - \mu_2)$

Problem Find the observed significance level for the test in Example 7.2. Interpret the result.

Solution The alternative hypothesis in Example 7.2, H_a: $\mu_1 - \mu_2 \neq 0$, required a two-tailed test using

$$z = \frac{\bar{x}_1 - \bar{x}_2}{\sigma_{(\bar{x}_1 - \bar{x}_2)}}$$

as a test statistic. Since the z-value calculated from the sample data was 3.09, the observed significance level (p-value) for the two-tailed test is the probability of observing a value of z at least as contradictory to the null hypothesis as $z = 3.09$; that is,

$$p\text{-value} = 2 \cdot P(z \geq 3.09)$$

This probability is computed under the assumption that H_0 is true and is equal to the highlighted area shown in Figure 7.4.
    The tabulated area corresponding to $z = 3.09$ in Table III of Appendix A is .4990. Therefore,

$$P(z \geq 3.09) = .5 - .4990 = .0010$$

and the observed significance level for the test is

$$p\text{-value} = 2(.001) = .002$$

Since our selected α value, .05, exceeds this p-value, we have sufficient evidence to reject H_0: $\mu_1 - \mu_2 = 0$.

Figure 7.4

The observed significance level
for Example 7.2

Look Back The p-value of the test is more easily obtained from a statistical software package. The p-value is highlighted at the bottom of the SPSS printout shown in Figure 7.1. This value agrees with our calculated p-value.

Now Work Exercise 7.6b

Small Samples

In comparing two population means with small samples (say, $n_1 < 30$ and $n_2 < 30$), the methodology of the previous three examples is invalid. The reason? When the sample sizes are small, estimates of σ_1^2 and σ_2^2 are unreliable and the Central Limit

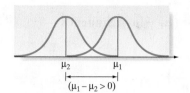

Figure 7.5

Assumptions for the two-sample t: (1) normal populations; (2) equal variances

Theorem (which guarantees that the z statistic is normal) can no longer be applied. But as in the case of a single mean (Section 6.5), we use the familiar Student's t-distribution.

To use the t-distribution, both sampled populations must be approximately normally distributed with equal population variances, and the random samples must be selected independently of each other. The assumptions of normality and equal variances imply relative frequency distributions for the populations that would appear as shown in Figure 7.5.

Since we assume that the two populations have equal variances ($\sigma_1^2 = \sigma_2^2 = \sigma^2$), it is reasonable to use the information contained in both samples to construct a **pooled sample estimator σ^2** for use in confidence intervals and test statistics. Thus, if s_1^2 and s_2^2 are the two sample variances (each estimating the variance σ^2 common to both populations), the pooled estimator of σ^2, denoted as s_p^2, is

$$s_p^2 = \frac{(n_1 - 1)s_1^2 + (n_2 - 1)s_2^2}{(n_1 - 1) + (n_2 - 1)} = \frac{(n_1 - 1)s_1^2 + (n_2 - 1)s_2^2}{n_1 + n_2 - 2}$$

or

$$s_p^2 = \frac{\overbrace{\sum(x_1 - \bar{x}_1)^2}^{\text{From sample 1}} + \overbrace{\sum(x_2 - \bar{x}_2)^2}^{\text{From sample 2}}}{n_1 + n_2 - 2}$$

where x_1 represents a measurement from sample 1 and x_2 represents a measurement from sample 2. Recall that the term *degrees of freedom* was defined in Section 5.3 as 1 less than the sample size. Thus, in this case, we have $(n_1 - 1)$ degrees of freedom for sample 1 and $(n_2 - 1)$ degrees of freedom for sample 2. Since we are pooling the information on σ^2 obtained from both samples, the number of degrees of freedom associated with the pooled variance s_p^2 is equal to the sum of the numbers of degrees of freedom for the two samples, namely, the denominator of s_p^2; that is, $(n_1 - 1) + (n_2 - 1) = n_1 + n_2 - 2$.

Note that the second formula given for s_p^2 shows that the pooled variance is simply a *weighted average* of the two sample variances s_1^2 and s_2^2. The weight given each variance is proportional to its number of degrees of freedom. If the two variances have the same number of degrees of freedom (i.e., *if the sample sizes are equal*), *then the pooled variance is a simple average of the two sample variances*. The result is an average, or "pooled," variance that is a better estimate of σ^2 than either s_1^2 or s_2^2 alone.

BIOGRAPHY BRADLEY EFRON (1938–present)

The Bootstrap Method

Bradley Efron was raised in St. Paul, Minnesota, the son of a truck driver who was the amateur statistician for his bowling and baseball leagues. Efron received a B.S. in mathematics from the California Institute of Technology in 1960, but, by his own admission, had no talent for modern abstract math. His interest in the science of statistics developed after he read a book by Harold Cramer from cover to cover. Efron went to the University of Stanford to study statistics, and he earned his Ph.D there in 1964. He has been a faculty member in Stanford's Department of Statistics since 1966. Over his career, Efron has received numerous awards and prizes for his contributions to modern statistics, including the MacArthur Prize Fellow (1983), the American Statistical Association Wilks Medal (1990), and the Parzen Prize for Statistical Innovation (1998). In 1979, Efron invented a method—called the *bootstrap*—of estimating and testing population parameters in situations in which either the sampling distribution is unknown or the assumptions are violated. The method involves repeatedly taking samples of size n (with replacement) from the original sample and calculating the value of the point estimate. Efron showed that the sampling distribution of the estimator is simply the frequency distribution of the bootstrap estimates. ∎

Both the confidence interval and the test-of-hypothesis procedures for comparing two population means with small samples are summarized in the following boxes:

Small, Independent Samples Confidence Interval for $(\mu_1 - \mu_2)$: Student's t-Statistic

$$(\bar{x}_1 - \bar{x}_2) \pm t_{\alpha/2}\sqrt{s_p^2\left(\frac{1}{n_1} + \frac{1}{n_2}\right)}$$

where $s_p^2 = \dfrac{(n_1 - 1)s_1^2 + (n_2 - 1)s_2^2}{n_1 + n_2 - 2}$

and $t_{\alpha/2}$ is based on $(n_1 + n_2 - 2)$ degrees of freedom.

[*Note*: $s_p^2 = \dfrac{s_1^2 + s_2^2}{2}$ when $n_1 = n_2$]

Small, Independent Samples Test of Hypothesis for $(\mu_1 - \mu_2)$: Student's t-Statistic

One-Tailed Test

H_0: $(\mu_1 - \mu_2) = D_0$
H_a: $(\mu_1 - \mu_2) < D_0$
[or H_a: $(\mu_1 - \mu_2) > D_0$]

Two-Tailed Test

H_0: $(\mu_1 - \mu_2) = D_0$
H_a: $(\mu_1 - \mu_2) \neq D_0$

Test statistic: $t = \dfrac{(\bar{x}_1 - \bar{x}_2) - D_0}{\sqrt{s_p^2\left(\dfrac{1}{n_1} + \dfrac{1}{n_2}\right)}}$

Rejection region: $t < -t_\alpha$
or $t > t_\alpha$ when
H_a: $(\mu_1 - \mu_2) > D_0$]

Rejection region: $|t| > t_{\alpha/2}$

where t_α and $t_{\alpha/2}$ are based on $(n_1 + n_2 - 2)$ degrees of freedom.

Conditions Required for Valid Small-Sample Inferences about $(\mu_1 - \mu_2)$

1. The two samples are randomly selected in an independent manner from the two target populations.
2. Both sampled populations have distributions that are approximately normal.
3. The population variances are equal (i.e., $\sigma_1^2 = \sigma_2^2$).

Example 7.4

A Small–Sample Confidence Interval for $(\mu_1 - \mu_2)$— Comparing Two Methods of Teaching

Problem Suppose you wish to compare a new method of teaching reading to "slow learners" with the current standard method. You decide to base your comparison on the results of a reading test given at the end of a learning period of six months. Of a random sample of 22 "slow learners," 10 are taught by the new method and 12 are taught by the standard method. All 22 children are taught by qualified instructors under similar conditions for the designated six-month period. The results of the reading test at the end of this period are given in Table 7.2.

Table 7.2	Reading Test Scores for Slow Learners						
New Method				**Standard Method**			
80	80	79	81	79	62	70	68
76	66	71	76	73	76	86	73
70	85			72	68	75	66

Data Set: READING

a. Use the data in the table to estimate the true mean difference between the test scores for the new method and the standard method. Use a 95% confidence interval.

b. Interpret the interval you found in part **a.**

c. What assumptions must be made in order that the estimate be valid? Are they reasonably satisfied?

Solution

a. For this experiment, let μ_1 and μ_2 represent the mean reading test scores of "slow learners" taught with the new and standard methods, respectively. Then the objective is to obtain a 95% confidence interval for $(\mu_1 - \mu_2)$.

The first step in constructing the confidence interval is to obtain summary statistics (e.g., $\bar{x}$ and s) on reading test scores for each method. The data of Table 7.2 were entered into a computer, and SAS was used to obtain these descriptive statistics. The SAS printout appears in Figure 7.6. Note that $\bar{x}_1 = 76.4$, $s_1 = 5.8348$, $\bar{x}_2 = 72.333$, and $s_2 = 6.3437$.

Two Sample t-test for the Means of SCORE within METHOD

Sample Statistics

Group	N	Mean	Std. Dev.	Std. Error
NEW	10	76.4	5.8348	1.8451
STD	12	72.33333	6.3437	1.8313

Hypothesis Test

Null hypothesis: Mean 1 - Mean 2 = 0
Alternative: Mean 1 - Mean 2 ^= 0

If Variances Are	t statistic	Df	Pr > t
Equal	1.552	20	0.1364
Not Equal	1.564	19.77	0.1336

95% Confidence Interval for the Difference between Two Means

Lower Limit	Upper Limit
-1.40	9.53

Figure 7.6
SAS printout for Example 7.4

Next, we calculate the pooled estimate of variance to obtain

$$s_p^2 = \frac{(n_1 - 1)s_1^2 + (n_2 - 1)s_2^2}{n_1 + n_2 - 2}$$

$$= \frac{(10 - 1)(5.8348)^2 + (12 - 1)(6.3437)^2}{10 + 12 - 2} = 37.45$$

where s_p^2 is based on $(n_1 + n_2 - 2) = (10 + 12 - 2) = 20$ degrees of freedom. Also, we find $t_{\alpha/2} = t_{.025} = 2.086$ (based on 20 degrees of freedom) from Table IV of Appendix A.

Finally, the 95% confidence interval for $(\mu_1 - \mu_2)$, the difference between mean test scores for the two methods, is

$$(\bar{x}_1 - \bar{x}_2) \pm t_{\alpha/2}\sqrt{s_p^2\left(\frac{1}{n_1} + \frac{1}{n_2}\right)} = (76.4 - 72.33) \pm t_{.025}\sqrt{37.45\left(\frac{1}{10} + \frac{1}{12}\right)}$$

$$= 4.07 \pm (2.086)(2.62)$$

$$= 4.07 \pm 5.47$$

or $(-1.4, 9.54)$. This interval agrees (except for rounding) with the one shown at the bottom of the SAS printout of Figure 7.6.

b. The interval can be interpreted as follows: With a confidence coefficient equal to .95, we estimate that the difference in mean test scores between using the new method of teaching and using the standard method falls into the interval from -1.4 to 7.54. In other words, we estimate (with 95% confidence) the mean test score for the new method to be anywhere from 1.4 points less than, to 7.54 points more than, the mean test score for the standard method. Although the sample means seem to suggest that the new method is associated with a higher mean test score, there is insufficient evidence to indicate that $(\mu_1 - \mu_2)$ differs from 0 because the interval includes 0 as a possible value for $(\mu_1 - \mu_2)$. To demonstrate a difference in mean test scores (if it exists), you could increase the sample size and thereby narrow the width of the confidence interval for $(\mu_1 - \mu_2)$. Alternatively, you can design the experiment differently. This possibility is discussed in the next section.

c. To use the small-sample confidence interval properly, the following assumptions must be satisfied:

1. The samples are randomly and independently selected from the populations of "slow learners" taught by the new method and the standard method.

2. The test scores are normally distributed for both teaching methods.

3. The variance of the test scores is the same for the two populations; that is, $\sigma_1^2 = \sigma_2^2$.

On the basis of the information provided about the sampling procedure in the description of the problem, the first assumption is satisfied. To check the plausibility of the remaining two assumptions, we resort to graphical methods. Figure 7.7 is a MINITAB printout that gives normal probability plots for the test scores of the two samples of "slow learners." The near straight-line trends on both plots indicate

Figure 7.7

MINITAB normal probability plots for Example 7.4

Figure 7.8
MINITAB box plots for
Example 7.4

that the distributions of the scores are approximately mound shaped and symmetric. Consequently, each sample data set appears to come from a population that is approximately normal.

One way to check the third assumption is to examine box plots of the sample data. Figure 7.8 is a MINITAB printout that shows side-by-side vertical box plots of the test scores in the two samples. Recall from Section 2.9 that the box plot represents the "spread" of a data set. The two box plots appear to have about the same spread; thus, the samples appear to come from populations with approximately the same variance.

Look Back All three assumptions, then, appear to be reasonably satisfied for this application of the small-sample confidence interval.

Now Work Exercise 7.9

The two-sample *t*-statistic is a powerful tool for comparing population means when the assumptions are satisfied. It has also been shown to retain its usefulness when the sampled populations are only approximately normally distributed. And when the sample sizes are equal, the assumption of equal population variances can be relaxed. That is, if $n_1 = n_2$, then σ_1^2 and σ_2^2 can be quite different, and the test statistic will still possess, approximately, a Student's *t*-distribution. In the case where $\sigma_1^2 \neq \sigma_2^2$ and $n_1 \neq n_2$, an approximate small-sample confidence interval or test can be obtained by modifying the number of degrees of freedom associated with the *t*-distribution.

The next box gives the approximate small-sample procedures to use when the assumption of equal variances is violated. The test for the case of "unequal sample sizes" is based on Satterthwaite's (1946) approximation.

Approximate Small-Sample Procedures when $\sigma_1^2 \neq \sigma_2^2$

1. Equal Sample Sizes ($n_1 = n_2 = n$)

Confidence interval: $\qquad\qquad (\bar{x}_1 - \bar{x}_2) \pm t_{\alpha/2}\sqrt{(s_1^2 + s_2^2)/n}$

Test statistic for H_0: $(\mu_1 - \mu_2) = 0$: $\quad t = (\bar{x}_1 - \bar{x}_2)/\sqrt{(s_1^2 + s_2^2)/n}$

where *t* is based on $\nu = n_1 + n_2 - 2 = 2(n - 1)$ degrees of freedom.

2. Unequal Sample Sizes ($n_1 \neq n_2$)

Confidence interval: $(\bar{x}_1 - \bar{x}_2) \pm t_{\alpha/2}\sqrt{(s_1^2/n_1) + (s_2^2/n_2)}$

Test statistic for $H_0: (\mu_1 - \mu_2) = 0$: $\quad t = (\bar{x}_1 - \bar{x}_2)/\sqrt{(s_1^2/n_1) + (s_2^2/n_2)}$

where t is based on degrees of freedom equal to

$$\nu = \frac{(s_1^2/n_1 + s_2^2/n_2)^2}{\dfrac{(s_1^2/n_1)^2}{n_1 - 1} + \dfrac{(s_2^2/n_2)^2}{n_2 - 1}}$$

Note: The value of ν will generally not be an integer. Round ν down to the nearest integer to use the t-table.

When the assumptions are not clearly satisfied, you can select larger samples from the populations or you can use other available **nonparametric statistical tests.**

What Should You Do if the Assumptions Are Not Satisfied?

Answer: If you are concerned that the assumptions are not satisfied, use the nonparametric Wilcoxon rank sum test for independent samples to test for a shift in population distributions. (See Optional Section 7.5.)

Statistics IN Action | **Revisited**

Comparing Roach Trails—Mean Treatment versus Mean Control

Consider the experiment designed to investigate the trail-following ability of German cockroaches (p. 351). Recall that an entomologist created a chemical trail with either a methanol extract from roach feces or just methanol (control). Cockroaches were then released into a container at the beginning of the trail, one at a time, and a video surveillance camera was used to monitor the roach's movements. The movement pattern of each cockroach was measured by its average trail deviation (in pixels) and the data stored in the **ROACH** file.

One of the entomologist's theories is that cockroaches, on average, will tend to follow the feces extract trail closer than they would follow the control trail. If we let μ_1 represent the population mean trail deviation of cockroaches assigned to the control trail and μ_2 represent the population mean trail deviation of cockroaches following the feces extract trail, then the speculation is that $\mu_1 > \mu_2$. Of interest, then, is the one-tailed test, $H_0:(\mu_1 - \mu_2) = 0$ versus $H_a:(\mu_1 - \mu_2) > 0$. The experiment was designed so that both samples are large ($n_1 = 40$ roaches and $n_2 = 80$ roaches); thus, the large-sample z-test is appropriate and no assumptions about the trail deviation distributions are required. We conducted this test using MINITAB. The output is shown in Figure SIA7.1.

Two-Sample T-Test and CI: DEVIATE, TRAIL

```
Two-sample T for DEVIATE

TRAIL     N   Mean   StDev   SE Mean
Control   40  64.5   37.7    6.0
Extract   80  22.8   23.9    2.7


Difference = mu (Control) - mu (Extract)
Estimate for difference:  41.73
95% CI for difference:  (28.63, 54.84)
T-Test of difference = 0 (vs not =): T-Value = 6.38   P-Value = 0.000   DF = 55
```

Figure SIA7.1
MINITAB Comparison of Means for Two Cockroach Trails

Statistics IN Action
(continued)

The two-tailed *p*-value of the test (.000) is highlighted on Figure SIA7.1. Consequently, the one-tailed *p*-value is .000/2 = 0. Therefore, at α = .05, there is sufficient evidence to conclude that the mean trail deviation of cockroaches following the control is significantly greater than the mean trail deviation of cockroaches following the feces extract trail. Thus, there is strong support for the entomologist's theory. In fact, a 95% confidence interval for $\mu_1 - \mu_2$ (highlighted on the MINITAB printout) is (28.6, 54.8). This implies that the feces extract trail mean is anywhere from 28.6 to 54.8 pixels less than the control mean (with 95% confidence).

Exercises 7.1–7.27

Understanding the Principles

7.1 Describe the sampling distribution of $(\bar{x}_1 - \bar{x}_2)$ when the samples are large.

7.2 To use the *t*-statistic to test for a difference between the means of two populations, what assumptions must be made about the two populations? About the two samples?

7.3 Two populations are described in each of the cases that follow. In which cases would it be appropriate to apply the small-sample *t*-test to investigate the difference between the population means?
  a. Population 1: Normal distribution with variance σ_1^2 Population 2: Skewed to the right with variance $\sigma_2^2 = \sigma_1^2$
  b. Population 1: Normal distribution with variance σ_1^2 Population 2: Normal distribution with variance $\sigma_2^2 \neq \sigma_1^2$
  c. Population 1: Skewed to the left with variance σ_1^2 Population 2: Skewed to the left with variance $\sigma_2^2 = \sigma_1^2$
  d. Population 1: Normal distribution with variance σ_1^2 Population 2: Normal distribution with variance $\sigma_2^2 = \sigma_1^2$
  e. Population 1: Uniform distribution with variance σ_1^2 Population 2: Uniform distribution with variance $\sigma_2^2 = \sigma_1^2$

7.4 A confidence interval for $(\mu_1 - \mu_2)$ is (−10, 4). Which of the following inferences is correct?
  a. $\mu_1 > \mu_2$
  b. $\mu_1 < \mu_2$
  c. $\mu_1 = \mu_2$
  d. no significant difference between means

7.5 A confidence interval for $(\mu_1 - \mu_2)$ is (−10, −4). Which of the following inferences is correct?
  a. $\mu_1 > \mu_2$
  b. $\mu_1 < \mu_2$
  c. $\mu_1 = \mu_2$
  d. no significant difference between means

Learning the Mechanics

7.6 In order to compare the means of two populations, independent random samples of 400 observations are selected from each population, with the following results:

Sample 1	Sample 2
$\bar{x}_1 = 5,275$	$\bar{x}_2 = 5,240$
$s_1 = 150$	$s_2 = 200$

  a. Use a 95% confidence interval to estimate the difference between the population means $(\mu_1 - \mu_2)$. Interpret the confidence interval.

b. Test the null hypothesis H_0: $(\mu_1 - \mu_2) = 0$ versus the alternative hypothesis H_a: $(\mu_1 - \mu_2) \neq 0$. Give the *p*-value of the test, and interpret the result.

c. Suppose the test in part **b** were conducted with the alternative hypothesis H_a: $(\mu_1 - \mu_2) > 0$. How would your answer to part **b** change?

d. Test the null hypothesis H_0: $(\mu_1 - \mu_2) = 25$ versus the alternative H_a: $(\mu_1 - \mu_2) \neq 25$. Give the *p*-value, and interpret the result. Compare your answer with that obtained from the test conducted in part **b**.

e. What assumptions are necessary to ensure the validity of the inferential procedures applied in parts **a–d**?

7.7 Independent random samples of 100 observations each are chosen from two normal populations with the following means and standard deviations:

Population 1	Population 2
$\mu_1 = 14$	$\mu_2 = 10$
$\sigma_1 = 4$	$\sigma_2 = 3$

Let $\bar{x}_1$ and $\bar{x}_2$ denote the two sample means.
  a. Give the mean and standard deviation of the sampling distribution of $\bar{x}_1$.
  b. Give the mean and standard deviation of the sampling distribution of $\bar{x}_2$.
  c. Suppose you were to calculate the difference $(\bar{x}_1 - \bar{x}_2)$ between the sample means. Find the mean and standard deviation of the sampling distribution of $(\bar{x}_1 - \bar{x}_2)$.
  d. Will the statistic $(\bar{x}_1 - \bar{x}_2)$ be normally distributed? Explain.

7.8 Assume that $\sigma_1^2 = \sigma_2^2 = \sigma^2$. Calculate the pooled estimator of σ^2 for each of the following cases:
  a. $s_1^2 = 200, s_2^2 = 180, n_1 = n_2 = 25$
  b. $s_1^2 = 25, s_2^2 = 40, n_1 = 20, n_2 = 10$
  c. $s_1^2 = .20, s_2^2 = .30, n_1 = 8, n_2 = 12$
  d. $s_1^2 = 2,500, s_2^2 = 1,800, n_1 = 16, n_2 = 17$
  e. Note that the pooled estimate is a weighted average of the sample variances. To which of the variances does the pooled estimate fall nearer in each of cases **a–d**?

7.9 Independent random samples from normal populations produced the following results: (saved in the **LM7_9** file).

Sample 1	Sample 2
1.2	4.2
3.1	2.7
1.7	3.6
2.8	3.9
3.0	

a. Calculate the pooled estimate of σ^2.

b. Do the data provide sufficient evidence to indicate that $\mu_2 > \mu_1$? Test, using $\alpha = .10$.

c. Find a 90% confidence interval for $(\mu_1 - \mu_2)$.

d. Which of the two inferential procedures, the test of hypothesis in part **b** or the confidence interval in part **c,** provides more information about $(\mu_1 - \mu_2)$?

7.10 Two independent random samples have been selected, 100 observations from population 1 and 100 from population 2. Sample means $\bar{x}_1 = 70$ and $\bar{x}_2 = 50$ were obtained. From previous experience with these populations, it is known that the variances are $\sigma_1^2 = 100$ and $\sigma_2^2 = 64$.

a. Find $\sigma_{(\bar{x}_1 - \bar{x}_2)}$.

b. Sketch the approximate sampling distribution $(\bar{x}_1 - \bar{x}_2)$, assuming that $(\mu_1 - \mu_2) = 5$.

c. Locate the observed value of $(\bar{x}_1 - \bar{x}_2)$ on the graph you drew in part **b.** Does it appear that this value contradicts the null hypothesis $H_0: (\mu_1 - \mu_2) = 5$?

d. Use the z-table to determine the rejection region for the test of $H_0: (\mu_1 - \mu_2) = 5$ against $H_a: (\mu_1 - \mu_2) \neq 5$. Use $\alpha = .05$.

e. Conduct the hypothesis test of part **d** and interpret your result.

f. Construct a 95% confidence interval for $(\mu_1 - \mu_2)$. Interpret the interval.

g. Which inference provides more information about the value of $(\mu_1 - \mu_2)$, the test of hypothesis in part **e** or the confidence interval in part **f?**

Applying the Concepts—Basic

7.11 **Effectiveness of teaching software.** Educational software—ranging from video-game-like programs played on Sony PlayStations to rigorous drilling exercises used on computers—has become very popular in school districts across the country. The U.S. Department of Education (DOE) recently conducted a national study of the effectiveness of educational software. In one phase of the study, a sample of 1,516 first-grade students in classrooms that used educational software was compared to a sample of 1,103 first-grade students in classrooms that did not use the technology. In its *Report to Congress* (March 2007), the DOE concluded that "[mean] test scores [of students on the SAT reading test] were not significantly higher in classrooms using reading ... software products" than in classrooms that did not use educational software.

a. Identify the parameter of interest to the DOE.

b. Specify the null and alternative hypotheses for the test conducted by the DOE.

c. The p-value for the test was reported as .62. Based on this value, do you agree with the conclusion of the DOE? Explain.

7.12 **Cognitive impairment of schizophrenics.** A study of the differences in cognitive function between normal individuals and patients diagnosed with schizophrenia was published in the *American Journal of Psychiatry* (April 2010). The total time (in minutes) a subject spent on the Trail Making Test (a standard psychological test) was used as a measure of cognitive function. The researchers theorize that the mean time on the Trail Making Test for schizophrenics will be larger than the corresponding mean for normal subjects. The data for independent random samples of 41 schizophrenics and 49 normal individuals yielded the following results:

	Schizophrenia	Normal
Sample size	41	49
Mean time	104.23	62.24
Standard deviation	45.45	16.34

Based on Perez-Iglesias, R., et al. "White matter integrity and cognitive impairment in first-episode psychosis." *American Journal of Psychiatry*, Vol. 167, No. 4, April 2010 (Table 1).

a. Define the parameter of interest to the researchers.

b. Set up the null and alternative hypothesis for testing the researchers' theory.

c. The researchers conducted the test, part **b,** and reported a p-value of .001. What conclusions can you draw from this result? (Use $\alpha = .01$.)

d. Find a 99% confidence interval for the target parameter. Interpret the result. Does your conclusion agree with that of part **c?**

7.13 **Children's recall of TV ads.** Marketing professors at Robert Morris and Kent State Universities examined children's recall and recognition of television advertisements (*Journal of Advertising*, Spring 2006). Two groups of children were shown a 60-second commercial for Sunkist FunFruit Rock-n-Roll Shapes. One group (the A/V group) was shown the ad with both audio and video; the second group (the video-only group) was shown only the video portion of the commercial. Following the viewing, the children were asked to recall 10 specific items from the ad. The number of items recalled correctly by each child is summarized in the accompanying table. The researchers theorized that "children who receive an audiovisual presentation will have the same level of mean recall of ad information as those who receive only the visual aspects of the ad."

Video-Only Group	A/V Group
$n_1 = 20$	$n_2 = 20$
$\bar{x}_1 = 3.70$	$\bar{x}_2 = 3.30$
$s_1 = 1.98$	$s_2 = 2.13$

Based on Maher, J. K., Hu, M. Y., and Kolbe, R. H. "Children's recall of television ad elements." *Journal of Advertising*, Vol. 35, No. 1, Spring 2006 (Table 1).

a. Set up the appropriate null and alternative hypotheses to test the researchers' theory.

b. Find the value of the test statistic.

c. Give the rejection region for $\alpha = .10$.

d. Make the appropriate inference. What can you say about the researchers' theory?

e. The researchers reported the p-value of the test as p-value $= .62$. Interpret this result.

f. What conditions are required for the inference to be valid?

7.14 **Index of Biotic Integrity.** The Ohio Environmental Protection Agency used the Index of Biotic Integrity (IBI) to measure the biological condition, or "health," of an aquatic region. The IBI is the sum of metrics that measure the presence, abundance, and health of fish in the region. (Higher values of the IBI correspond to healthier fish populations.) Researchers collected IBI measurements for sites located in different Ohio river basins (*Journal of Agricultural, Biological, and Environmental Sciences*,

June 2005). Summary data for two river basins, Muskingum and Hocking, are given in the accompanying table.

River Basin	Sample Size	Mean	Standard Deviation
Muskingum	53	.035	1.046
Hocking	51	.340	.960

Based on Boone, E. L., Keying, Y., and Smith, E. P. "Evaluating the relationship between ecological and habitat conditions using hierarchical models." *Journal of Agricultural, Biological, and Environmental Sciences*, Vol. 10, No. 2, June 2005 (Table 01).

a. Use a 90% confidence interval to compare the mean IBI values of the two river basins. Interpret the interval.

b. Conduct a test of hypothesis (at $\alpha = .10$) to compare the mean IBI values of the two river basins. Explain why the result will agree with the inference you derived from the 90% confidence interval in part **a**.

7.15 **Reading Japanese books.** Refer to the *Reading in a Foreign Language* (Apr. 2004) experiment to improve the Japanese reading comprehension levels of University of Hawaii students, presented in Exercise 2.33 (p. 46). Recall that 14 students participated in a 10-week extensive reading program in a second-semester Japanese course. The numbers of books read by each student and the student's course grade are repeated in the following table and saved in the **JAPANESE** file.

Number of Books	Course Grade	Number of Books	Course Grade
53	A	30	A
42	A	28	B
40	A	24	A
40	B	22	C
39	A	21	B
34	A	20	B
34	A	16	B

Source: Hitosugi, C. I., and Day, R. R. "Extensive reading in Japanese." *Reading in a Foreign Language,* Vol. 16, No. 1, Apr. 2004 (Table 4). Reprinted with permissions from the National Foreign Language Resource Center, University of Hawaii.

a. Consider two populations of students who participate in the reading program prior to taking a second-semester Japanese course: those who earn an A grade and those who earn a B or C grade. Of interest is the difference in the mean number of books read by the two populations

of students. Identify the parameter of interest in words and in symbols.

b. Form a 95% confidence interval for the target parameter identified in part **a**.

c. Give a practical interpretation of the confidence interval you formed in part **b**.

d. Compare the inference in part **c** with the inference you derived from stem-and-leaf plots in Exercise 2.33b.

7.16 **Lobster trap placement.** Refer to the *Bulletin of Marine Science* (April 2010) study of lobster trap placement, Exercise 5.35 (p. 261). Recall that the variable of interest was the average distance separating traps—called *trap spacing*—deployed by teams of fishermen fishing for the red spiny lobster in Baja California Sur, Mexico. The trap spacing measurements (in meters) for a sample of 7 teams from the Bahia Tortugas (BT) fishing cooperative are repeated in the table. In addition, trap spacing measurements for 8 teams from the Punta Abreojos (PA) fishing cooperative are listed. (All these data are saved in the **TRAPSPACE** file). For this problem, we are interested in comparing the mean trap spacing measurements of the two fishing cooperatives.

BT Cooperative:	93	99	105	94	82	70	86	
PA Cooperative:	118	94	106	72	90	66	153	98

Based on Shester, G. G. "Explaining catch variation among Baja California lobster fishers through spatial analysis of trap-placement decisions." *Bulletin of Marine Science*, Vol. 86, No. 2, April 2010 (Table 1), pp. 479–498.

a. Identify the target parameter for this study.

b. Compute a point estimate of the target parameter.

c. What is the problem with using the normal (z) statistic to find a confidence interval for the target parameter?

d. Find a 90% confidence interval for the target parameter.

e. Use the interval, part **d**, to make a statement about the difference in mean trap spacing measurements of the two fishing cooperatives.

f. What conditions must be satisfied for the inference, part **e**, to be valid?

7.17 **Bulimia study.** The "fear of negative evaluation" (FNE) scores for 11 female students known to suffer from the eating disorder bulimia and 14 female students with normal eating habits, first presented in Exercise 2.40 (p. 48), are reproduced in the next table and saved in the **BULIMIA** file. (Recall that the higher the score, the greater is the fear of a negative evaluation.)

MINITAB Output for Exercise 7.17

Two-Sample T-Test and CI: FNESCORE, GROUP

```
Two-sample T for FNESCORE

GROUP     N   Mean  StDev  SE Mean
Bulimic  11  17.82   4.92      1.5
Normal   14  14.14   5.29      1.4


Difference = mu (Bulimic) - mu (Normal)
Estimate for difference:  3.68
95% CI for difference:  (-0.60, 7.95)
T-Test of difference = 0 (vs not =): T-Value = 1.78  P-Value = 0.089  DF = 23
Both use Pooled StDev = 5.1303
```

Table for Exercise 7.17

Bulimic	
students:	21 13 10 20 25 19 16 21 24 13 14
Normal	
students:	13 6 16 13 8 19 23 18 11 19 7 10 15 20

Based on Randles, R. H. "On neutral responses (zeros) in the sign test and ties in the Wilcoxon-Mann-Whitney test." *The American Statistician,* Vol. 55, No. 2, May 2001 (Figure 3).

a. Locate a 95% confidence interval for the difference between the population means of the FNE scores for bulimic and normal female students on the MINITAB printout shown at the bottom of page 365. Interpret the result.

b. What assumptions are required for the interval of part **a** to be statistically valid? Are these assumptions reasonably satisfied? Explain.

Applying the Concepts—Intermediate

7.18 **Do video game players have superior visual attention skills?** Researchers at Griffin University (Australia) conducted a study to determine whether video game players have superior visual attention skills compared to non–video game players (*Journal of Articles in Support of the Null Hypothesis,* Vol. 6, 2009). Two groups of male psychology students—32 video game players (VGP group) and 28 nonplayers (NVGP group)—were subjected to a series of visual attention tasks that included the *attentional blink* test. A test for the difference between two means yielded $t = -.93$ and p-$value = .358$. Consequently, the researchers reported that "no statistically significant differences in the mean test performances of the two groups were found." Summary statistics for the comparison are provided in the table. Do you agree with the researchers conclusion?

	VGP	NVGP
Sample size	32	28
Mean score	84.81	82.64
Standard deviation	7.56	8.43

Based on Murphy, K., and Spencer, A. "Playing video games does not make for better visual attention skills." *Journal of Articles in Support of the Null Hypothesis,* Vol. 6, No. 1, 2007.

7.19 **Drug content assessment.** Refer to Exercise 4.108 (p. 208) and the *Analytical Chemistry* (Dec. 15, 2009) study in which scientists used high-performance liquid chromatography to determine the amount of drug in a tablet. Twenty-five tablets were produced at each of two different, independent sites. Drug concentrations (measured as a percentage) for the tablets produced at the two sites are listed in the accompanying table and saved in the **DRUGCON** file. The scientists want to know whether there is any difference between the mean drug concentration in tablets produced at Site 1 and the corresponding mean at Site 2. Use the MINITAB printout (bottom of page) to help the scientists draw a conclusion.

Site 1
91.28 92.83 87.35 91.90 82.85 94.83 87.83 87.00 84.62
86.96 88.32 91.17 83.86 87.74 92.24 92.59 84.21 87.36
90.96 92.85 89.39 89.82 87.91 92.16 88.67

Site 2
87.35 86.51 87.04 91.82 93.02 88.32 88.76 87.26 90.36
87.16 91.74 86.12 92.10 83.33 87.61 88.20 92.78 86.35
93.84 91.20 93.44 86.77 83.77 93.19 81.79

Based on Borman, P. J., Marion, J. C., Damjanov, I., and Jackson, P. "Design and analysis of method equivalence studies." *Analytical Chemistry,* Vol. 81, No. 24, December 15, 2009 (Table 3).

7.20 **Patent infringement case.** *Chance* (Fall 2002) described a lawsuit charging Intel Corp. with infringing on a patent for an invention used in the automatic manufacture of computer chips. In response, Intel accused the inventor of adding material to his patent notebook after the patent was witnessed and granted. The case rested on whether a patent witness's signature was written on top of or under key text in the notebook. Intel hired a physicist who used an X-ray beam to measure the relative concentrations of certain elements (e.g., nickel, zinc, potassium) at several spots on the notebook page. The zinc measurements for three notebook locations—on a text line, on a witness line, and on the intersection of the witness and text line—are provided in the following table and saved in the **PATENT** file.

Text line:	.335	.374	.440			
Witness line:	.210	.262	.188	.329	.439	.397
Intersection:	.393	.353	.285	.295	.319	

a. Use a test or a confidence interval (at $\alpha = .05$) to compare the mean zinc measurement for the text line with the mean for the intersection.

b. Use a test or a confidence interval (at $\alpha = .05$) to compare the mean zinc measurement for the witness line with the mean for the intersection.

MINITAB Output for Exercise 7.19

```
Two-Sample T-Test and CI: Content, Site

Two-sample T for Content

Site   N   Mean   StDev   SE Mean
1      25  89.55  3.07    0.61
2      25  89.03  3.34    0.67


Difference = mu (1) - mu (2)
Estimate for difference:  0.515
95% CI for difference:  (-1.308, 2.338)
T-Test of difference = 0 (vs not =): T-Value = 0.57  P-Value = 0.573  DF = 48
Both use Pooled StDev = 3.2057
```

c. From the results you obtained in parts **a** and **b,** what can you infer about the mean zinc measurements at the three notebook locations?

d. What assumptions are required for the inferences to be valid? Are they reasonably satisfied?

7.21 **How do you choose to argue?** Educators frequently lament weaknesses in students' oral and written arguments. In *Thinking and Reasoning* (April 2007), researchers at Columbia University conducted a series of studies to assess the cognitive skills required for successful arguments. One study focused on whether students would choose to argue by weakening the opposing position or by strengthening the favored position. (For example, suppose you are told you would do better at basketball than soccer, but you like soccer. An argument that weakens the opposing position is "You need to be tall to play basketball." An argument that strengthens the favored position is "With practice, I can become really good at soccer.") A sample of 52 graduate students in psychology was equally divided into two groups. Group 1 was presented with 10 items such that the argument always attempts to strengthens the favored position. Group 2 was presented with the same 10 items, but in this case the argument always attempts to weaken the nonfavored position. Each student then rated the 10 arguments on a five-point scale from very weak (1) to very strong (5). The variable of interest was the sum of the 10 item scores, called the *total rating.* Summary statistics for the data are shown in the accompanying table. Use the methodology of this chapter to compare the mean total ratings for the two groups at $\alpha = .05$. Give a practical interpretation of the results in the words of the problem.

	Group 1 (support favored position)	Group 2 (weaken opposing position)
Sample size	26	26
Mean	28.6	24.9
Standard deviation	12.5	12.2

Based on Kuhn, D., and Udell, W. "Coordinating own and other perspectives in argument." *Thinking and Reasoning,* October 2006.

7.22 **Pig castration study.** Two methods of castrating male piglets were investigated in *Applied Animal Behaviour Science* (Nov. 1, 2000). Method 1 involved an incision in the spermatic cords, while Method 2 involved pulling and severing the cords. Forty-nine male piglets were randomly allocated to one of the two methods. During castration, the researchers measured the number of high-frequency vocal responses (squeals) per second over a 5-second period. The data are summarized in the accompanying table. Conduct a test of hypothesis to determine whether the population mean number of high-frequency vocal responses differs for piglets castrated by the two methods. Use $\alpha = .05$.

	Method 1	Method 2
Sample size	24	25
Mean number of squeals	.74	.70
Standard deviation	.09	.09

Based on Taylor, A. A., and Weary, D. M. "Vocal responses of piglets to castration: Identifying procedural sources of pain." *Applied Animal Behaviour Science,* Vol. 70, No. 1, November 1, 2000.

7.23 **Mongolian desert ants.** Refer to the *Journal of Biogeography* (Dec. 2003) study of ants in Mongolia (central Asia), presented in Exercise 2.68 (p. 59). Recall that botanists placed seed baits at 5 sites in the Dry Steppe region and 6 sites in the Gobi Desert and observed the number of ant species attracted to each site. These data are listed in the next table and saved in the **GOBIANTS** file. Is there evidence to conclude that a difference exists between the average number of ant species found at sites in the two regions of Mongolia? Draw the appropriate conclusion, using $\alpha = .05$.

Site	Region	Number of Ant Species
1	Dry Steppe	3
2	Dry Steppe	3
3	Dry Steppe	52
4	Dry Steppe	7
5	Dry Steppe	5
6	Gobi Desert	49
7	Gobi Desert	5
8	Gobi Desert	4
9	Gobi Desert	4
10	Gobi Desert	5
11	Gobi Desert	4

Based on Pfeiffer, M., et al. "Community organization and species richness of ants in Mongolia along an ecological gradient from steppe to Gobi desert." *Journal of Biogeography,* Vol. 30, No. 12, Dec. 2003 (Tables 1 and 2).

7.24 **Does rudeness really matter in the workplace?** Studies have established that rudeness in the workplace can lead to retaliatory and counterproductive behavior. However, there has been little research on how rude behaviors influence a victim's task performance. Such a study was conducted and the results published in the *Academy of Management Journal* (Oct. 2007). College students enrolled in a management course were randomly assigned to one of two experimental conditions: rudeness condition (45 students) and control group (53 students). Each student was asked to write down as many uses for a brick as possible in five minutes; this value (total number of uses) was used as a performance measure for each student. For those students in the rudeness condition, the facilitator displayed rudeness by berating the students in general for being irresponsible and unprofessional (due to a late-arriving confederate). No comments were made about the late-arriving confederate for students in the control group. The number of different uses of a brick for each of the 98 students was recorded and the data saved in the **RUDE** file, shown in the next table. Conduct a statistical analysis (at $\alpha = .01$) to determine if the true mean performance level for students in the rudeness condition is lower than the true mean performance level for students in the control group. Use the

Control Group:

1	24	5	16	21	7	20	1	9	20	19	10	23	16	0	4	9	13
17	13	0	2	12	11	7	1	19	9	12	18	5	21	30	15	4	2
12	11	10	13	11	3	6	10	13	16	12	28	19	12	20	3	11	

Rudeness Condition:

4	11	18	11	9	6	5	11	9	12	7	5	7	3	11	1	9	11
10	7	8	9	10	7	11	4	13	5	4	7	8	3	8	15	9	16
10	0	7	15	13	9	2	13	10									

results shown on the accompanying SAS printout to draw your conclusion .

```
Sample Statistics

    Group       N     Mean     Std. Dev.   Std. Error
    ------------------------------------------------------
    Control     53   11.81132    7.3826       1.0141
    Rude        45    8.511111   3.9922       0.5951

Hypothesis Test

    Null hypothesis:      Mean 1 - Mean 2 <= 0
    Alternative:          Mean 1 - Mean 2 > 0

    If Variances Are     t statistic      Df      Pr > t
    ------------------------------------------------------
    Equal                   2.683         96      0.0043
    Not Equal               2.807       82.43     0.0031

95% Confidence Interval for the Difference between Two Means

        Lower Limit      Upper Limit
        -----------      -----------
            0.86             5.74
```

7.25 Masculinity and crime. The *Journal of Sociology* (July 2003) published a study on the link between the level of masculinity and criminal behavior in men. Using a sample of newly incarcerated men in Nebraska, the researcher identified 1,171 violent events and 532 events in which violence was avoided that the men were involved in. (A violent event involved the use of a weapon, throwing of objects, punching, choking, or kicking. An event in which violence was avoided included pushing, shoving, grabbing, or threats of violence that did not escalate into a violent event.) Each of the sampled men took the Masculinity–Femininity Scale (MFS) test to determine his level of masculinity, based on common male stereotyped traits. MFS scores ranged from 0 to 56 points, with lower scores indicating a more masculine orientation. One goal of the research was to compare the mean MFS scores for two groups of men: those involved in violent events and those who avoided violent events.

a. Identify the target parameter for this study.

b. The sample mean MFS score for the violent-event group was 44.50, while the sample mean MFS score for the avoided-violent-event group was 45.06. Is this sufficient information to make the comparison desired by the researcher? Explain.

c. In a large-sample test of hypothesis to compare the two means, the test statistic was computed to be $z = 1.21$. Compute the two-tailed p-value of the test.

d. Make the appropriate conclusion, using $\alpha = .10$.

7.26 Detection of rigged school milk prices. Each year, the state of Kentucky invites bids from dairies to supply half-pint containers of fluid milk products for its school districts. In several school districts in northern Kentucky (called the "tricounty" market), two suppliers—Meyer Dairy and Trauth Dairy—were accused of price-fixing—that is, conspiring to allocate the districts so that the winning bidder was predetermined and the price per pint was set above the competitive price. These two dairies were the only two bidders on the milk contracts in the tricounty market for eight consecutive years. (In contrast, a large number of different dairies won the milk contracts for school districts in the remainder of the northern Kentucky market, called the "surrounding" market.) Did Meyer and Trauth conspire to rig their bids in the tricounty market? Economic theory states that, if so, the mean winning price in the rigged tricounty market will be higher than the mean winning price in the competitive surrounding market. Data on all bids received from the dairies competing for the milk contracts during the time period in question are saved in the **MILK** file. A MINITAB printout of the comparison of mean prices bid for whole white milk for the two Kentucky milk markets is shown below. Is there support for the claim that the dairies in the tricounty market participated in collusive practices? Explain in detail.

Applying the Concepts—Advanced

7.27 Ethnicity and pain perception. An investigation of ethnic differences in reports of pain perception was presented at the annual meeting of the American Psychosomatic Society (March 2001). A sample of 55 blacks and 159 whites participated in the study. Subjects rated (on a 13-point scale) the intensity and unpleasantness of pain felt when a bag of ice was placed on their foreheads for two minutes. (Higher ratings correspond to higher pain intensity.) A summary of the results is provided in the following table:

	Blacks	Whites
Sample size	55	159
Mean pain intensity	8.2	6.9

a. Why is it dangerous to draw a statistical inference from the summarized data? Explain.

b. Give values of the missing sample standard deviations that would lead you to conclude (at $\alpha = .05$) that blacks, on average, have a higher pain intensity rating than whites.

c. Give values of the missing sample standard deviations that would lead you to an inconclusive decision (at $\alpha = .05$) regarding whether blacks or whites have a higher mean intensity rating.

MINITAB Output for Exercise 7.26

Two-Sample T-Test and CI: WWBID, Market

```
Two-sample T for WWBID

Market       N    Mean    StDev   SE Mean
SURROUND   254   0.1331   0.0158   0.00099
TRI-COUNTY 100   0.1431   0.0133   0.0013


Difference = mu (SURROUND) - mu (TRI-COUNTY)
Estimate for difference:  -0.009970
95% upper bound for difference:  -0.007232
T-Test of difference = 0 (vs <): T-Value = -6.02  P-Value = 0.000  DF = 213
```

7.3 Comparing Two Population Means: Paired Difference Experiments

In Example 7.4, we compared two methods of teaching reading to "slow learners" by means of a 95% confidence interval. Suppose it is possible to measure the "reading IQs" of the "slow learners" *before* they are subjected to a teaching method. Eight pairs of "slow learners" with similar reading IQs are found, and one member of each pair is randomly assigned to the standard teaching method while the other is assigned to the new method. The data are given in Table 7.3. Do the data support the hypothesis that

Table 7.3	Reading Test Scores for Eight Pairs of "Slow Learners"	
Pair	New Method (1)	Standard Method (2)
1	77	72
2	74	68
3	82	76
4	73	68
5	87	84
6	69	68
7	66	61
8	80	76

Data Set: PAIREDSCORES

the population mean reading test score for "slow learners" taught by the new method is greater than the mean reading test score for those taught by the standard method?

We want to test

$$H_0: (\mu_1 - \mu_2) = 0$$
$$H_a: (\mu_1 - \mu_2) > 0$$

Many researchers mistakenly use the t statistic for two independent samples (Section 7.2) to conduct this test. This invalid analysis is shown on the MINITAB printout of Figure 7.7. The test statistic, $t = 1.26$, and the p-value of the test, $p = .115.$, are highlighted on the printout. At $\alpha = .10$, the p-value exceeds α. Thus, from *this* analysis, we might conclude that we do not have sufficient evidence to infer a difference in the mean test scores for the two methods.

Figure 7.9

MINITAB printout of an invalid analysis of reading test scores in Table 7.3

```
Two-sample T for NEW vs STANDARD

           N    Mean   StDev  SE Mean
NEW        8   76.00    6.93      2.4
STANDARD   8   71.63    7.01      2.5


Difference = mu (NEW) - mu (STANDARD)
Estimate for difference:   4.38
95% lower bound for difference:   -1.76
T-Test of difference = 0 (vs >): T-Value = 1.26   P-Value = 0.115   DF = 14
Both use Pooled StDev = 6.9687
```

If you examine the data in Table 7.3 carefully, however, you will find this result difficult to accept. The test score of the new method is larger than the corresponding test score for the standard method *for every one of the eight pairs of "slow learners."* This, in itself, seems to provide strong evidence to indicate that μ_1 exceeds μ_2. Why, then, did the t-test fail to detect the difference? The answer is, *the independent samples t-test is not a valid procedure to use with this set of data.*

The t-test is inappropriate because the assumption of independent samples is invalid. We have randomly chosen *pairs of test scores;* thus, once we have chosen the sample for the new method, we have *not* independently chosen the sample for the standard method. The dependence between observations within pairs can be seen by examining the pairs of test scores, which tend to rise and fall together as we go from pair to pair. This pattern provides strong visual evidence of a violation of the assumption of independence required for the two-sample t-test of Section 7.2. Note also that

$$s_p^2 = \frac{(n_1 - 1)s_1^2 + (n_2 - 1)s_2^2}{n_1 + n_2 - 2} = \frac{(8 - 1)(6.93)^2 + (8 - 1)(7.01)^2}{8 + 8 - 2} = 48.58$$

Hence, there is a *large variation within samples* (reflected by the large value of s_p^2) in comparison to the relatively *small difference between the sample means*. Because s_p^2 is so large, the t-test of Section 7.2 is unable to detect a difference between μ_1 and μ_2.

We now consider a valid method of analyzing the data of Table 7.3. In Table 7.4, we add the column of differences between the test scores of the pairs of "slow learners." We can regard these differences in test scores as a random sample of differences for all pairs (matched on reading IQ) of "slow learners," past and present. Then we can use this sample to make inferences about the mean of the population of differences, μ_d, which is equal to the difference $(\mu_1 - \mu_2)$. That is, the mean of the population (and sample) of differences equals the difference between the population (and sample) means. Thus, our test becomes

$$H_0: \mu_d = 0 \quad (\mu_1 - \mu_2 = 0)$$
$$H_a: \mu_d > 0 \quad (\mu_1 - \mu_2 > 0)$$

Table 7.4   Differences in Reading Test Scores

Pair	New Method	Standard Method	Difference (New Method − Standard Method)
1	77	72	5
2	74	68	6
3	82	76	6
4	73	68	5
5	87	84	3
6	69	68	1
7	66	61	5
8	80	76	4

The test statistic is a one-sample t (Section 6.5), since we are now analyzing a single sample of differences for small n. Thus,

$$\text{Test statistic: } t = \frac{\bar{x}_d - 0}{s_d/\sqrt{n_d}}$$

where

$$\bar{x}_d = \text{Sample mean difference}$$
$$s_d = \text{Sample standard deviation of differences}$$
$$n_d = \text{Number of differences} = \text{Number of pairs}$$

Assumptions: The population of differences in test scores is approximately normally distributed. The sample differences are randomly selected from the population differences. [*Note:* We do not need to make the assumption that $\sigma_1^2 = \sigma_2^2$.]

Rejection region: At significance level $\alpha = .05$, we will reject H_0 if $t > t_{.05}$, where $t_{.05}$ is based on $(n_d - 1)$ degrees of freedom.

Referring to Table IV in Appendix A, we find the t-value corresponding to $\alpha = .05$ and $n_d - 1 = 8 - 1 = 7$ df to be $t_{.05} = 1.895$. Then we will reject the null hypothesis if $t > 1.895$. (See Figure 7.10.) Note that the number of degrees of freedom decreases from $n_1 + n_2 - 2 = 14$ to 7 when we use the paired difference experiment rather than the two independent random samples design.

Summary statistics for the $n_d = 8$ differences are shown in the MINITAB printout of Figure 7.11. Note that $\bar{x}_d = 4.375$ and $s_d = 1.685$. Substituting these values into the formula for the test statistic, we have

$$t = \frac{\bar{x}_d - 0}{s_d/\sqrt{n_d}} = \frac{4.375}{1.685/\sqrt{8}} = 7.34$$

t-distribution with 7 df

$\alpha = .05$

Rejection region

$t = 1.895$

Figure 7.10

Rejection region for Example 7.4

Because this value of t falls into the rejection region, we conclude (at $\alpha = .05$) that the population mean test score for "slow learners" taught by the new method exceeds the population mean score for those taught by the standard method. We can reach the same conclusion by noting that the p-value of the test, highlighted in Figure 7.11, is much smaller than $\alpha = .05$.

```
Paired T for NEW - STANDARD

              N    Mean   StDev   SE Mean
NEW           8   76.00    6.93      2.45
STANDARD      8   71.63    7.01      2.48
Difference    8    4.375   1.685     0.596


95% lower bound for mean difference: 3.246
T-Test of mean difference = 0 (vs > 0): T-Value = 7.34   P-Value = 0.000
```

Figure 7.11

MINITAB paired difference analysis of reading test scores

Now Work Exercises 7.33ab

This kind of experiment, in which observations are paired and the differences are analyzed, is called a **paired difference experiment.** In many cases, a paired difference experiment can provide more information about the difference between population means than an independent samples experiment can. The idea is to compare population means by comparing the differences between pairs of experimental units (objects, people, etc.) that were similar prior to the experiment. The differencing removes sources of variation that tend to inflate σ^2. For example, when two children are taught to read by two different methods, the observed difference in achievement may be due to a difference in the effectiveness of the two teaching methods, *or* it may be due to differences in the initial reading levels and IQs of the two children (random error). To reduce the effect of differences in the children on the observed differences in reading achievement, the two methods of reading are imposed on two children who are more likely to possess similar intellectual capacity, namely, children with nearly equal IQs. The effect of this pairing is to remove the larger source of variation that would be present if children with different abilities were randomly assigned to the two samples. Making comparisons within groups of similar experimental units is called , and the paired difference experiment is a simple example of a **randomized block experiment.** In our example, pairs of children with matching IQ scores represent the blocks.

Some other examples for which the paired difference experiment might be appropriate are the following:

1. Suppose you want to estimate the difference $(\mu_1 - \mu_2)$ in mean price per gallon between two major brands of premium gasoline. If you choose two independent random samples of stations for each brand, the variability in price due to geographic location may be large. To eliminate this source of variability, you could choose pairs of stations of similar size, one station for each brand, in close geographic proximity and use the sample of differences between the prices of the brands to make an inference about $(\mu_1 - \mu_2)$.

2. Suppose a college placement center wants to estimate the difference $(\mu_1 - \mu_2)$ in mean starting salaries for men and women graduates who seek jobs through the center. If it independently samples men and women, the starting salaries may vary because of their different college majors and differences in grade point averages. To eliminate these sources of variability, the placement center could match male and female job seekers according to their majors and grade point averages. Then the differences between the starting salaries of each pair in the sample could be used to make an inference about $(\mu_1 - \mu_2)$.

3. Suppose you wish to estimate the difference $(\mu_1 - \mu_2)$ in mean absorption rate into the bloodstream for two drugs that relieve pain. If you

independently sample people, the absorption rates might vary because of age, weight, sex, blood pressure, etc. In fact, there are many possible sources of nuisance variability, and pairing individuals who are similar in all the possible sources would be quite difficult. However, it may be possible to obtain two measurements *on the same person*. First, we administer one of the two drugs and record the time until absorption. After a sufficient amount of time, the other drug is administered and a second measurement on absorption time is obtained. The differences between the measurements for each person in the sample could then be used to estimate $(\mu_1 - \mu_2)$. This procedure would be advisable only if the amount of time allotted between drugs is sufficient to guarantee little or no carry-over effect. Otherwise, it would be better to use different people matched as closely as possible on the factors thought to be most important.

Now Work Exercise 7.31

The hypothesis-testing procedures and the method of forming confidence intervals for the difference between two means in a paired difference experiment are summarized in the following boxes for both large and small *n*:

Paired Difference Confidence Interval for $\mu_d = \mu_1 - \mu_2$

Large Sample, Normal (z) Statistic

$$\bar{x}_d \pm z_{\alpha/2} \frac{\sigma_d}{\sqrt{n_d}} \approx \bar{x}_d \pm z_{\alpha/2} \frac{s_d}{\sqrt{n_d}}$$

Small Sample, Student's t-Statistic

$$\bar{x}_d \pm t_{\alpha/2} \frac{s_d}{\sqrt{n_d}}$$

where $t_{\alpha/2}$ is based on $(n_d - 1)$ degrees of freedom

---

Paired Difference Test of Hypothesis for $\mu_d = \mu_1 - \mu_2$

One-Tailed Test	**Two-Tailed Test**
$H_0: \mu_d = D_0$	$H_0: \mu_d = D_0$
$H_a: \mu_d < D_0$	$H_a: \mu_d \neq D_0$
[or $H_a: \mu_d > D_0$]	

Large Sample, Normal (z) Statistic

Test statistic: $z = \dfrac{\bar{x}_d - D_0}{\sigma_d/\sqrt{n_d}} \approx \dfrac{\bar{x}_d - D_0}{s_d/\sqrt{n_d}}$

Rejection region: $z < -z_\alpha$ *Rejection region:* $|z| > z_{\alpha/2}$
[or $z > z_\alpha$ when $H_a: \mu_d > D_0$]

Small Sample, Student's t-Statistic

Test statistic: $t = \dfrac{\bar{x}_d - D_0}{s_d/\sqrt{n_d}}$

Rejection region: $t < -t_\alpha$ *Rejection region:* $|t| > t_{\alpha/2}$
[or $t > t_\alpha$ when $H_a: \mu_d > D_0$]

where t_α and $t_{\alpha/2}$ are based on $(n_d - 1)$ degrees of freedom

---

Conditions Required for Valid Large-Sample Inferences about μ_d

1. A random sample of differences is selected from the target population of differences.
2. The sample size n_d is large (i.e., $n_d \geq 30$). (By the Central Limit Theorem, this condition guarantees that the test statistic will be approximately normal, regardless of the shape of the underlying probability distribution of the population.)

Conditions Required for Valid Small-Sample Inferences about μ_d

1. A random sample of differences is selected from the target population of differences.
2. The population of differences has a distribution that is approximately normal.

Example 7.5

Confidence Interval For μ_d—Comparing Mean Salaries of Males and Females

Problem An experiment is conducted to compare the starting salaries of male and female college graduates who find jobs. Pairs are formed by choosing a male and a female with the same major and similar grade point averages (GPAs). Suppose a random sample of 10 pairs is formed in this manner and the starting annual salary of each person is recorded. The results are shown in Table 7.5. Compare the mean starting salary μ_1 for males with the mean starting salary μ_2 for females, using a 95% confidence interval. Interpret the results.

Table 7.5	Data on Annual Salaries for Matched Pairs of College Graduates		
Pair	Male	Female	Difference Male − Female
1	$29,300	$28,800	$ 500
2	41,500	41,600	−100
3	40,400	39,800	600
4	38,500	38,500	0
5	43,500	42,600	900
6	37,800	38,000	−200
7	69,500	69,200	300
8	41,200	40,100	1,100
9	38,400	38,200	200
10	59,200	58,500	700

Data Set: GRADPAIRS

Solution Since the data on annual salary are collected in pairs of males and females matched on GPA and major, a paired difference experiment is performed. To conduct the analysis, we first compute the differences between the salaries, as shown in Table 7.5. Summary statistics for these $n = 10$ differences are displayed at the top of the SAS printout shown in Figure 7.12. The 95% confidence interval for $\mu_d = (\mu_1 - \mu_2)$ for this small sample is

$$\bar{x}_d \pm t_{\alpha/2} \frac{s_d}{\sqrt{n_d}}$$

where $t_{\alpha/2} = t_{.025} = 2.262$ (obtained from Table IV, Appendix A) is based on $n_d - 1 = 9$ degrees of freedom. Substituting the values of $\bar{x}_d$ and s_d shown on the printout, we obtain

$$\bar{x}_d \pm 2.262 \frac{s_d}{\sqrt{n_d}} = 400 \pm 2.262 \left(\frac{434.613}{\sqrt{10}} \right)$$

$$= 400 \pm 310.88 \approx 400 \pm 311 = (\$89, \$711)$$

```
                        The MEANS Procedure

                    Analysis Variable : DIFF

        Mean        Std Dev     N       Minimum     Maximum
    ---------------------------------------------------------------
        400.00      434.61      10      -200.00     1100.00
    ---------------------------------------------------------------


        Two Sample Paired t-test for the Means of MALE and FEMALE

    Sample Statistics

        Group       N       Mean    Std. Dev.   Std. Error
        -----------------------------------------------------------
        MALE        10      43930       11665       3688.8
        FEMALE      10      43530       11617       3673.6


    Hypothesis Test

        Null hypothesis:    Mean of (MALE - FEMALE) =  0
        Alternative:        Mean of (MALE - FEMALE) ^= 0

            t Statistic     Df      Prob > t
        ------------------------------------------
            2.910           9       0.0173


    95% Confidence Interval for the Difference between Two Paired Means

            Lower Limit         Upper Limit
            -----------         -----------
            89.10               710.90
```

Figure 7.12

SAS analysis of salary
differences

[*Note:* This interval is also shown highlighted at the bottom of the SAS printout of Figure 7.12.] Our interpretation is that the true mean difference between the starting salaries of males and females falls between $89 and $711, with 95% confidence. Since the interval falls above 0, we infer that $\mu_1 - \mu_2 > 0$; that is, the mean salary for males exceeds the mean salary for females.

Look Back   Remember that $\mu_d = \mu_1 - \mu_2$. So if $\mu_d > 0$, then $\mu_1 > \mu_2$. Alternatively, if $\mu_d < 0$, then $\mu_1 < \mu_2$.

Now Work Exercise 7.40

To measure the amount of information about $(\mu_1 - \mu_2)$ gained by using a paired difference experiment in Example 7.5 rather than an independent samples experiment, we can compare the relative widths of the confidence intervals obtained by the two methods. A 95% confidence interval for $(\mu_1 - \mu_2)$ obtained from a paired difference experiment is, from Example 7.5, ($89, $711). If we mistakenly analyzed the same data as though this were an independent samples experiment,* we would first obtain the descriptive statistics shown in the SAS printout of Figure 7.13. Then we substitute the sample means and standard deviations shown on the printout into the formula for a 95% confidence interval for $(\mu_1 - \mu_2)$ using independent samples. The result is

$$(\bar{x}_1 - \bar{x}_2) \pm t_{.025}\sqrt{s_p^2\left(\frac{1}{n_1} + \frac{1}{n_2}\right)}$$

where

$$s_p^2 = \frac{(n_1 - 1)s_1^2 + (n_2 - 1)s_2^2}{n_1 + n_2 - 2}$$

*This is done only to provide a measure of the increase in the amount of information obtained by a paired design in comparison to an unpaired design. Actually, if an experiment were designed that used pairing, an unpaired analysis would be invalid because the assumption of independent samples would not be satisfied.

Group Statistics

	GENDER	N	Mean	Std. Deviation	Std. Error Mean
SALARY	M	10	43930.00	11665.148	3688.844
	F	10	43530.00	11616.946	3673.601

Independent Samples Test

		Levene's Test for Equality of Variances		t-test for Equality of Means						95% Confidence Interval of the Difference	
		F	Sig.	t	df	Sig. (2-tailed)	Mean Difference	Std. Error Difference	Lower	Upper
SALARY	Equal variances assumed	.000	.991	.077	18	.940	400.000	5206.046	-10537.496	11337.496
	Equal variances not assumed			.077	18.000	.940	400.000	5206.046	-10537.509	11337.509

Figure 7.13
SPSS analysis of salaries, assuming independent samples

SPSS performed these calculations and obtained the interval ($-10,537.50$, $11,337.50$), highlighted in Figure 7.13.

Notice that the independent samples interval includes 0. Consequently, if we were to use this interval to make an inference about $(\mu_1 - \mu_2)$, we would incorrectly conclude that the mean starting salaries of males and females do not differ! You can see that the confidence interval for the independent sampling experiment is about 35 times wider than for the corresponding paired difference confidence interval. Blocking out the variability due to differences in majors and grade point averages significantly increases the information about the difference in males' and females' mean starting salaries by providing a much more accurate (a smaller confidence interval for the same confidence coefficient) estimate of $(\mu_1 - \mu_2)$.

You may wonder whether a paired difference experiment is always superior to an independent samples experiment. The answer is, most of the time, but not always. We sacrifice half the degrees of freedom in the t-statistic when a paired difference design is used instead of an independent samples design. This is a loss of information, and unless that loss is more than compensated for by the reduction in variability obtained by blocking (pairing), the paired difference experiment will result in a net loss of information about $(\mu_1 - \mu_2)$. Thus, we should be convinced that the pairing will significantly reduce variability before performing a paired difference experiment. Most of the time, this will happen.

One final note: The pairing of the observations is determined *before* the experiment is performed (i.e., by the *design* of the experiment). A paired difference experiment is *never* obtained by pairing the sample observations *after* the measurements have been acquired.

Ethics IN Statistics

In a two-group analysis, intentionally pairing observations after the data have been collected in order to produce a desired result is considered *unethical statistcal practice.*

> **What Do You Do When the Assumption of a Normal Distribution for the Population of Differences Is Not Satisfied?**
>
> *Answer:* Use the nonparametric Wilcoxon signed rank test for the paired difference design (Optional Section 7.6).

Exercises 7.28–7.48

Understanding the Principles

7.28 What are the advantages of using a paired difference experiment over an independent samples design?

7.29 In a paired difference experiment, when should the observations be paired, before or after the data are collected?

7.30 What conditions are required for valid large-sample inferences about μ_d? small-sample inferences?

Learning the Mechanics

7.31 A paired difference experiment yielded n_d pairs of obser-
NW vations. In each case, what is the rejection region for testing H_0: $\mu_d = 2$ against H_a: $\mu_d > 2$?
   a. $n_d = 10, \alpha = .05$
   b. $n_d = 20, \alpha = .10$
   c. $n_d = 5, \alpha = .025$
   d. $n_d = 9, \alpha = .01$

7.32 A paired difference experiment produced the following data:

$$n_d = 16 \quad \bar{x}_1 = 143 \quad \bar{x}_2 = 150 \quad \bar{x}_d = -7 \quad s_d^2 = 64$$

a. Determine the values of t for which the null hypothesis $\mu_1 - \mu_2 = 0$ would be rejected in favor of the alternative hypothesis $\mu_1 - \mu_2 < 0$. Use $\alpha = .10$.

b. Conduct the paired difference test described in part **a.** Draw the appropriate conclusions.

c. What assumptions are necessary so that the paired difference test will be valid?

d. Find a 90% confidence interval for the mean difference μ_d.

e. Which of the two inferential procedures, the confidence interval of part **d** or the test of hypothesis of part **b,** provides more information about the difference between the population means?

7.33 The data for a random sample of six paired observations are shown in the following table and saved in the **LM7_33** file.

Pair	Sample from Population 1	Sample from Population 2
1	7	4
2	3	1
3	9	7
4	6	2
5	4	4
6	8	7

a. Calculate the difference between each pair of observations by subtracting observation 2 from observation 1. Use the differences to calculate $\bar{x}_d$ and s_d^2.

b. If μ_1 and μ_2 are the means of populations 1 and 2, respectively, express μ_d in terms of μ_1 and μ_2.

c. Form a 95% confidence interval for μ_d.

d. Test the null hypothesis $H_0: \mu_d = 0$ against the alternative hypothesis $H_a: \mu_d \neq 0$. Use $\alpha = .05$.

7.34 The data for a random sample of 10 paired observations are shown in the following table and saved in the **LM9_34** file.

Pair	Population 1	Population 2
1	19	24
2	25	27
3	31	36
4	52	53
5	49	55
6	34	34
7	59	66
8	47	51
9	17	20
10	51	55

a. If you wish to test whether these data are sufficient to indicate that the mean for population 2 is larger than that for population 1, what are the appropriate null and alternative hypotheses? Define any symbols you use.

b. Conduct the test from part **a,** using $\alpha = .10$. What is your decision?

c. Find a 90% confidence interval for μ_d. Interpret this interval.

d. What assumptions are necessary to ensure the validity of the preceding analysis?

7.35 A paired difference experiment yielded the following results:

$$n_d = 40, \quad \bar{x}_d = 11.7, \quad s_d = 6.$$

a. Test $H_0: \mu_d = 10$ against $H_a: \mu_d \neq 10$, where $\mu_d = (\mu_1 - \mu_2)$. Use $\alpha = .05$.

b. Report the p-value for the test you conducted in part **a.** Interpret the p-value.

Applying the Concepts—Basic

7.36 **Summer weight-loss camp.** Camp Jump Start is an 8-week summer camp for overweight and obese adolescents. Counselors develop a weight-management program for each camper that centers on nutrition education and physical activity. In a study published in *Pediatrics* (April 2010), the body mass index (BMI) was measured for each of 76 campers both at the start and end of camp. Summary statistics on BMI measurements are shown in the table.

	Mean	Standard Deviation
Starting BMI	34.9	6.9
Ending BMI	31.6	6.2
Paired Differences	3.3	1.5

Based on Huelsing, J., Kanafani, N., Mao, J., and White, N. H. "Camp Jump Start: Effects of a residential summer weight-loss camp for older children and adolescents." *Pediatrics*, Vol. 125, No. 4, April 2010 (Table 3).

a. Give the null and alternative hypothesis for determining whether the mean BMI at the end of camp is less than the mean BMI at the start of camp.

b. How should the data be analyzed, as an independent-samples t-test or as a paired-difference t-test? Explain.

c. Calculate the test statistic using the formula for an independent-samples t-test. (*Note:* This is *not* how the test should be conducted.)

d. Calculate the test statistic using the formula for a paired-difference t-test.

e. Compare the test statistics, parts **c** and **d.** Which test statistic provides more evidence in support of the alternative hypothesis?

f. The p-value of the test, part **d,** was reported as $p < .0001$. Interpret this result assuming $\alpha = .01$.

g. Do the differences in BMI values need to be normally distributed in order for the inference, part **f,** to be valid? Explain.

h. Find a 99% confidence interval for the true mean change in BMI for Camp Jump Start campers. Interpret the result.

7.37 **Healing potential of handling museum objects.** Does handling a museum object have a positive impact on a sick patient's well-being? To answer this question, researchers at the University College London collected data from 32 sessions with hospital patients (*Museum & Society*, Nov. 2009). Each patient's health status (measured on a 100-point scale) was recorded both before and after handling museum objects such as archaeological artifacts and brass etchings. The data (simulated) are listed in the accompanying table and saved in the **MUSEUM** file.

Session	Before	After	Session	Before	After
1	52	59	17	65	65
2	42	54	18	52	63
3	46	55	19	39	50
4	42	51	20	59	69
5	43	42	21	49	61
6	30	43	22	59	66
7	63	79	23	57	61
8	56	59	24	56	58
9	46	53	25	47	55
10	55	57	26	61	62
11	43	49	27	65	61
12	73	83	28	36	53
13	63	72	29	50	61
14	40	49	30	40	52
15	50	49	31	65	70
16	50	64	32	59	72

a. Explain why the data should be analyzed as paired differences.

b. Compute the difference between the "before" and "after" measurements for each session.

c. Find the mean and standard deviation of the differences, part **b**.

d. Use the summary statistics, part **c**, to find a 90% confidence interval for the true mean difference ("before" minus "after") in health status scale measurements.

e. Interpret the interval, part **d**. Does handling a museum object have a positive impact on a sick patient's well-being?

7.38 Laughter among deaf signers. The *Journal of Deaf Studies and Deaf Education* (Fall 2006) published an article on vocalized laughter among deaf users of American Sign Language (ASL). In videotaped ASL conversations among deaf participants, 28 laughed at least once. The researchers wanted to know if they laughed more as speakers (while signing) or as audience members (while listening). For each of the 28 deaf participants, the number of laugh episodes as a speaker and the number of laugh episodes as an audience member were determined. One goal of the research was to compare the mean numbers of laugh episodes of speakers and audience members.

a. Explain why the data should be analyzed as a paired difference experiment.

b. Identify the study's target parameter.

c. The study yielded a sample mean of 3.4 laughter episodes for speakers and a sample mean of 1.3 laughter episodes for audience members. Is this sufficient evidence to conclude that the population means are different? Explain.

d. A paired difference *t*-test resulted in $t = 3.14$ and *p*-value < .01. Interpret the results in the words of the problem.

7.39 The placebo effect and pain. According to research published in *Science* (Feb. 20, 2004), the mere belief that you are receiving an effective treatment for pain can reduce the pain you actually feel. Researchers from the University of Michigan and Princeton University tested

this placebo effect on 24 volunteers as follows: Each volunteer was put inside a magnetic resonance imaging (MRI) machine for two consecutive sessions. During the first session, electric shocks were applied to their arms and the blood oxygen level–dependent (BOLD) signal (a measure related to neural activity in the brain) was recorded during pain. The second session was identical to the first, except that, prior to applying the electric shocks, the researchers smeared a cream on the volunteer's arms. The volunteers were informed that the cream would block the pain when, in fact, it was just a regular skin lotion (i.e., a placebo). If the placebo is effective in reducing the pain experience, the BOLD measurements should be higher, on average, in the first MRI session than in the second.

a. Identify the target parameter for this study.

b. What type of design was used to collect the data?

c. Give the null and alternative hypotheses for testing the placebo effect theory.

d. The differences between the BOLD measurements in the first and second sessions were computed and summarized in the study as follows: $n_d = 24$, $\bar{x}_d = .21$, $s_d = .47$. Use this information to calculate the test statistic.

e. The *p*-value of the test was reported as *p*-value = .02. Make the appropriate conclusion at $\alpha = .05$.

7.40 NHTSA new car crash tests. Refer to the National Highway Traffic Safety Administration (NHTSA) crash test data on new cars, saved in the **CRASH** file. Crash test dummies were placed in the driver's seat and front passenger's seat of a new car model, and the car was steered by remote control into a head-on collision with a fixed barrier while traveling at 35 miles per hour. Two of the variables measured for each of the 98 new cars in the data set are (1) the severity of the driver's chest injury and (2) the severity of the passenger's chest injury. (The more points assigned to the chest injury rating, the more severe the injury is.) Suppose the NHTSA wants to determine whether the true mean driver chest injury rating exceeds the true mean passenger chest injury rating and, if so, by how much.

a. State the parameter of interest to the NHTSA.

b. Explain why the data should be analyzed as matched pairs.

c. Find a 99% confidence interval for the true difference between the mean chest injury ratings of drivers and front-seat passengers.

d. Interpret the interval you found in part **c**. Does the true mean driver chest injury rating exceed the true mean passenger chest injury rating? If so, by how much?

e. What conditions are required for the analysis to be valid? Do these conditions hold for these data?

Applying the Concepts—Intermediate

7.41 Acidity of mouthwash. Acid has been found to be a primary cause of dental caries (cavities). It is theorized that oral mouthwashes contribute to the development of caries due to the antiseptic agent oxidizing into acid over time. This theory was tested in the *Journal of Dentistry, Oral Medicine and Dental Education* (Vol. 3, 2009). Three bottles of mouthwash, each of a different brand, were randomly selected from a drugstore. The pH level (where lower pH levels indicate higher acidity) of each bottle was measured on the date of purchase and after 30 days. The data, saved in the **MOUTHWASH** file, are shown in the table. Conduct an analysis to determine if the mean initial pH level of mouthwash differs significantly from the mean pH level after 30 days. Use $\alpha = .05$ as your level of significance.

Mouthwash Brand	Initial pH	Final pH
LMW	4.56	4.27
SMW	6.71	6.51
RMW	5.65	5.58

Based on Chunhye, K. L., and Schmitz, B. C., "Determination of pH, total acid, and total ethanol in oral health products: Oxidation of ethanol and recommendations to mitigate its association with dental caries." *Journal of Dentistry, Oral Medicine and Dental Education*, Vol. 3, No. 1, 2009 (Table 1).

7.42 **Visual search and memory study.** In searching for an item (e.g., a roadside traffic sign, a lost earring, or a tumor in a mammogram), common sense dictates that you will not reexamine items previously rejected. However, researchers at Harvard Medical School found that a visual search has no memory (*Nature*, Aug. 6, 1998). In their experiment, nine subjects searched for the letter "T" mixed among several letters "L." Each subject conducted the search under two conditions: random and static. In the random condition, the locations of the letters were changed every 111 milliseconds; in the static condition, the locations of the letters remained unchanged. In each trial, the reaction time in milliseconds (i.e., the amount of time it took the subject to locate the target letter) was recorded.

a. One goal of the research was to compare the mean reaction times of subjects in the two experimental conditions. Explain why the data should be analyzed as a paired difference experiment.

b. If a visual search has no memory, then the main reaction times in the two conditions will not differ. Specify H_0 and H_a for testing the "no-memory" theory.

c. The test statistic was calculated as $t = 1.52$ with p-value $= .15$. Draw the appropriate conclusion.

7.43 **Linking dementia and leisure activities.** Does participation in leisure activities in your youth reduce the risk of Alzheimer's disease and other forms of dementia? To answer this question, a group of university researchers studied a sample of 107 same-sex Swedish pairs of twins (*Journal of Gerontology: Psychological Sciences and Social Sciences*, Sept. 2003). Each pair of twins was discordant for dementia; that is, one member of each pair was diagnosed with Alzheimer's disease while the other member (the control) was nondemented for at least five years after the sibling's onset of dementia. The level of overall leisure activity (measured on an 80-point scale, where higher values indicate higher levels of leisure activity) of each twin of each pair 20 years prior to the onset of dementia was obtained from the Swedish Twin Registry database. The leisure activity scores (simulated on the basis of summary information presented in the journal article) are saved in the **DEMENTIA** file. The first five and last five observations are shown in the following table:

Pair	Control	Demented
1	27	13
2	57	57
3	23	31
4	39	46
5	37	37
⋮	⋮	⋮
103	22	14
104	32	23
105	33	29
106	36	37
107	24	1

a. Explain why the data should be analyzed as a paired difference experiment.

b. Conduct the appropriate analysis, using $\alpha = .05$. Make an inference about which member of the pair, the demented or control (nondemented) twin, had the largest average level of leisure activity.

7.44 **Ethical sensitivity of teachers towards racial intolerance.** Many high schools have education programs that encourage teachers to embrace racial tolerance, recognize ethnic diversity, and overcome racial stereotypes. To gauge the effectiveness of one such program that utilizes two videos of teachers engaging in racial stereotypes of their students, researchers from New York University and City University of New York recruited 238 high school professionals (including teachers and counselors) to participate in a study (*Journal of Moral Education*, March 2010). Teachers watched the first video, then were given a pretest—the Quick-REST Survey—designed to measure ethical sensitivity towards racial intolerance. The teachers next participated in an all-day workshop on cultural competence. At the end of the workshop, the teachers watched the second video and again were given the Quick-REST Survey (the posttest). To determine whether the program was effective, the researchers compared the mean scores on the Quick-REST Survey using a paired-difference t-test. (*Note:* The higher the score on the Quick-REST Survey, the greater the level of racial tolerance.)

a. The researchers reported the sample means for the pretest and posttest as 75.85 and 80.35, respectively. Why is it dangerous to gauge the effectiveness of the program based only on these summary statistics?

b. The paired-difference t-test (posttest minus pretest) was reported as $t = 4.50$ with an associated observed significance level of p-value $< .001$. Interpret this result.

c. What assumptions, if any, are necessary for the validity of the inference, part **b**?

7.45 **Impact of red light cameras on car crashes.** To combat red-light-running crashes—the phenomenon of a motorist entering an intersection after the traffic signal turns red and causing a crash—many states are adopting photo-red enforcement programs. In these programs, red light cameras installed at dangerous intersections photograph the license plates of vehicles that run the red light. How effective are photo-red enforcement programs in reducing red-light-running crash incidents at intersections? The Virginia Department of Transportation (VDOT) conducted a comprehensive study of its newly adopted photo-red enforcement program and published the results in a

June 2007 report. In one portion of the study, the VDOT provided crash data both before and after installation of red light cameras at several intersections. The data (measured as the number of crashes caused by red light running per intersection per year) for 13 intersections in Fairfax County, Virginia, are given in the table and saved in the **REDLIGHT** file. Analyze the data for the VDOT. What do you conclude?

Intersection	Before Camera	After Camera
1	3.60	1.36
2	0.27	0
3	0.29	0
4	4.55	1.79
5	2.60	2.04
6	2.29	3.14
7	2.40	2.72
8	0.73	0.24
9	3.15	1.57
10	3.21	0.43
11	0.88	0.28
12	1.35	1.09
13	7.35	4.92

Based on Virginia Transportation Research Council, "Research report: The impact of red light cameras (photo-red enforcement) on crashes in Virginia." June 2007.

7.46 **Light-to-dark transition of genes.** *Synechocystis,* a type of cyanobacterium that can grow and survive under a wide range of conditions, is used by scientists to model DNA behavior. In the *Journal of Bacteriology* (July 2002), scientists isolated genes of the bacterium responsible for photosynthesis and respiration and investigated the sensitivity of the genes to light. Each gene sample was grown to midexponential phase in a growth incubator in "full light." The lights were then extinguished, and any growth of the sample was measured after 24 hours in the dark ("full dark"). The lights were then turned back on for 90 minutes ("transient light"), followed immediately by an additional 90 minutes in the dark ("transient dark"). Standardized growth measurements in each light–dark condition were obtained for 103 genes. The complete data set is saved in the **GENEDARK** file. Data on the first 10 genes are shown in the following table:

Gene ID	FULL-DARK	TR-LIGHT	TR-DARK
SLR2067	−0.00562	1.40989	−1.28569
SLR1986	−0.68372	1.83097	−0.68723
SSR3383	−0.25468	−0.79794	−0.39719
SLL0928	−0.18712	−1.20901	−1.18618
SLR0335	−0.20620	1.71404	−0.73029
SLR1459	−0.53477	2.14156	−0.33174
SLL1326	−0.06291	1.03623	0.30392
SLR1329	−0.85178	−0.21490	0.44545
SLL1327	0.63588	1.42608	−0.13664
SLL1325	−0.69866	1.93104	−0.24820

Based on Gill, R. T., et al. "Genome-wide dynamic transcriptional profiling of the light to dark transition in *Synechocystis Sp.* PCC6803." *Journal of Bacteriology,* Vol. 184, No. 13, July 2002.

a. Treat the data for the first 10 genes as a random sample collected from the population of 103 genes, and test the hypothesis that there is no difference

between the mean standardized growth of genes in the full-dark condition and genes in the transient-light condition. Use $\alpha = .01$.

b. Use a statistical software package to compute the mean difference in standardized growth of the 103 genes in the full-dark condition and the transient-light condition. Did the test you carried out in part **a** detect this difference?

c. Repeat parts **a** and **b** for a comparison of the mean standardized growth of genes in the full-dark condition and genes in the transient-dark condition.

d. Repeat parts **a** and **b** for a comparison of the mean standardized growth of genes in the transient-light condition and genes in the transient-dark condition.

Applying the Concepts—Advanced

7.47 **Homophone confusion in Alzheimer's patients.** A *homophone* is a word whose pronunciation is the same as that of another word having a different meaning and spelling (e.g., *nun* and *none, doe* and *dough,* etc.). *Brain and Language* (Apr. 1995) reported on a study of homophone spelling in patients with Alzheimer's disease. Twenty Alzheimer's patients were asked to spell 24 homophone pairs given in random order. Then the number of homophone confusions (e.g., spelling *doe* given the context, *bake bread dough*) was recorded for each patient. One year later, the same test was given to the same patients. The data for the study are provided in the next table and saved in the **HOMOPHONE** file. The researchers posed the following question: "Do Alzheimer's patients show a significant increase in mean homophone confusion errors over time?" Perform an analysis of the data to answer the researchers' question. What assumptions are necessary for the procedure used to be valid? Are they satisfied?

Patient	Time 1	Time 2
1	5	5
2	1	3
3	0	0
4	1	1
5	0	1
6	2	1
7	5	6
8	1	2
9	0	9
10	5	8
11	7	10
12	0	3
13	3	9
14	5	8
15	7	12
16	10	16
17	5	5
18	6	3
19	9	6
20	11	8

Based on Neils, J., Roeltgen, D. P., and Constantinidou, F. "Decline in homophone spelling associated with loss of semantic influence on spelling in Alzheimer's disease." *Brain and Language,* Vol. 49, No. 1, pp. 27–47.

7.48 **Alcoholic fermentation in wines.** Determining alcoholic fermentation in wine is critical to the wine-making process. Must/wine density is a good indicator of the fermentation point, since the density value decreases as sugars are converted into alcohol. For decades, winemakers have measured must/wine density with a hydrometer. Although accurate, the hydrometer employs a manual process that is very time consuming. Consequently, large wineries are searching for more rapid measures of density measurement. An alternative method utilizes the hydrostatic balance instrument (similar to the hydrometer, but digital). A winery in Portugal collected must/wine density measurements on white wine samples randomly selected from the fermentation process for a recent harvest. For each sample, the density of the wine at 20°C was measured with both the hydrometer and the hydrostatic balance. The densities for 40 wine samples are saved in the **WINE40** file. The first five and last five observations are shown in the accompanying table. The winery will use the alternative method of mea-suring wine density only if it can be demonstrated that the mean difference between the density measurements of the two methods does not exceed .002.

Perform the analysis for the winery. Provide the winery with a written report of your conclusions.

Sample	Hydrometer	Hydrostatic
1	1.08655	1.09103
2	1.00270	1.00272
3	1.01393	1.01274
4	1.09467	1.09634
5	1.10263	1.10518
⋮	⋮	⋮
36	1.08084	1.08097
37	1.09452	1.09431
38	0.99479	0.99498
39	1.00968	1.01063
40	1.00684	1.00526

Based on Cooperative Cellar of Borba (*Adega Cooperativ a de Borba*), Portugal.

7.4 Determining the Sample Size

You can find the appropriate sample size to estimate the difference between a pair of means with a specified sampling error (SE) and degree of reliability by using the method described in Section 5.5. That is, to estimate the difference between a pair of means correct to within SE units with confidence level $(1 - \alpha)$, let $z_{\alpha/2}$ standard deviations of the sampling distribution of the estimator equal SE. Then solve for the sample size. To do this, you have to solve the problem for a specific ratio between n_1 and n_2. Most often, you will want to have equal sample sizes—that is, $n_1 = n_2 = n$. We will illustrate the procedure with two examples.

Example 7.6
Finding the Sample Sizes for Estimating $(\mu_1 - \mu_2)$— Comparing Mean Crop Yields

Problem New fertilizer compounds are often advertised with the promise of increased crop yields. Suppose we want to compare the mean yield μ_1 of wheat when a new fertilizer is used with the mean yield μ_2 from a fertilizer in common use. The estimate of the difference in mean yield per acre is to be correct to within .25 bushel with a confidence coefficient of .95. If the sample sizes are to be equal, find $n_1 = n_2 = n$, the number of 1-acre plots of wheat assigned to each fertilizer.

Solution To solve the problem, you need to know something about the variation in the bushels of yield per acre. Suppose that, from past records, you know that the yields of wheat possess a range of approximately 10 bushels per acre. You could then approximate $\sigma_1 = \sigma_2 = \sigma$ by letting the range equal 4σ. Thus,

$$4\sigma \approx 10 \text{ bushels}$$
$$\sigma \approx 2.5 \text{ bushels}$$

The next step is to solve the equation

$$z_{\alpha/2}\sigma_{(\bar{x}_1-\bar{x}_2)} = \text{SE, or } z_{\alpha/2}\sqrt{\frac{\sigma_1^2}{n_1} + \frac{\sigma_2^2}{n_2}} = \text{SE}$$

for n, where $n = n_1 = n_2$. Since we want our estimate to lie within SE = .25 of $(\mu_1 - \mu_2)$ with confidence coefficient equal to .95, we have $z_{\alpha/2} = z_{.025} = 1.96$. Then, letting $\sigma_1 = \sigma_2 = 2.5$ and solving for n, we get

$$1.96\sqrt{\frac{(2.5)^2}{n} + \frac{(2.5)^2}{n}} = .25$$

$$1.96\sqrt{\frac{2(2.5)^2}{n}} = .25$$

$$n = 768.32 \approx 769 \text{ (rounding up)}$$

Consequently, you will have to sample 769 acres of wheat for each fertilizer to estimate the difference in mean yield per acre to within .25 bushel.

Look Back Since $n = 769$ would necessitate extensive and costly experimentation, you might decide to allow a larger sampling error (say, SE = .50 or SE = 1) in order to reduce the sample size, or you might decrease the confidence coefficient. The point is that we can obtain an idea of the experimental effort necessary to achieve a specified precision in our final estimate by determining the approximate sample size *before* the experiment is begun.

Now Work Exercise 7.52

Example 7.7

Finding the Sample Sizes for Estimating μ_d: Comparing Two Measuring Instruments

Problem A laboratory manager wishes to compare the difference in the mean reading of two instruments, A and B, designed to measure the potency (in parts per million) of an antibiotic. To conduct the experiment, the manager plans to select n_d specimens of the antibiotic from a vat and to measure each specimen with both instruments. The difference ($\mu_A - \mu_B$) will be estimated based on the n_d paired differences ($x_A - x_B$) obtained in the experiment. If preliminary measurements suggest that the differences will range between plus or minus 10 parts per million, how many differences will be needed to estimate ($\mu_A - \mu_B$) correct to within 1 part per million with confidence equal to .99?

Solution The estimator for ($\mu_A - \mu_B$), based on a paired difference experiment, is $\bar{x}_d = (\bar{x}_A - \bar{x}_B)$ and

$$\sigma_{\bar{x}_d} = \frac{\sigma_d}{\sqrt{n_d}}$$

Thus, the number n_d of pairs of measurements needed to estimate ($\mu_A - \mu_B$) to within 1 part per million can be obtained by solving for n_d in the equation

$$z_{\alpha/2}\left(\frac{\sigma_d}{\sqrt{n_d}}\right) = SE$$

where $z_{.005} = 2.58$ and $SE = 1$. To solve this equation for n_d, we need to have an approximate value for σ_d.

We are given the information that the differences are expected to range from -10 to 10 parts per million. Letting the range equal $4\sigma_d$, we find

$$\text{Range} = 20 \approx 4\sigma_d$$

$$\sigma_d \approx 5$$

Substituting $\sigma_d = 5$, $SE = 1$, and $z_{.005} = 2.58$ into the equation and solving for n_d, we obtain

$$2.58\left(\frac{5}{\sqrt{n_d}}\right) = 1$$

$$n_d = [(2.58)(5)]^2$$

$$= 166.41$$

Therefore, it will require $n_d = 167$ pairs of measurements to estimate ($\mu_A - \mu_B$) correct to within 1 part per million using the paried difference experiment.

The box summarizes the procedures for determining the sample sizes necessary for estimating ($\mu_1 - \mu_2$) for the case $n_1 = n_2$ and for estimating μ_d.

> **Determination of Sample Size for Comparing Two Means**
>
> **Independent Random Samples**
>
> To estimate $(\mu_1 - \mu_2)$ to within a given sampling error *SE* and with confidence level $(1 - \alpha)$, use the following formula to solve for equal sample sizes that will achieve the desired reliability:
>
> $$n_1 = n_2 = \frac{(z_{\alpha/2})^2(\sigma_1^2 + \sigma_2^2)}{(SE)^2}$$
>
> You will need to substitute estimates for the values of σ_1^2 and σ_2^2 before solving for the sample size. These estimates might be sample variances s_1^2 and s_2^2 from prior sampling (e.g., a pilot study), or from an educated (and conservatively large) guess based on the range—that is, $s \approx R/4$.
>
> **Paired Difference Experiment**
>
> To estimate μ_d to within a given sampling error SE and with confidence level $(1 - \alpha)$, use the following formula to solve for n_d that will achieve the desired reliability:
>
> $$n_d = \frac{(z_{\alpha/2})^2 \sigma_d^2}{(SE)^2}$$
>
> You will need to substitute an estimate of σ_d^2 before solving for the sample size. This estimate might be the sample variance s_d^2 from prior sampling (e.g., a pilot study), or from an educated (and conservatively large) guess based on the range—that is, $s_d = R/4$.

Exercises 7.49.–7.60

Understanding the Principles

7.49 When determining the sample sizes for estimating $\mu_1 - \mu_2$, how do you obtain estimates of the population variances $(\sigma_1)^2$ and $(\sigma_2)^2$ used in the calculations?

7.50 If the sample size calculation yields a value of n that is too large to be practical, how should you proceed?

7.51 When determining the sample size for estimating μ_d, how do you obtain an estimates of the population variance σ_d^2 used in the calculations?

Learning the Mechanics

7.52 Find the appropriate values of n_1 and n_2 (assume that
[NW] $n_1 = n_2$) needed to estimate $(\mu_1 - \mu_2)$ with
   a. A sampling error equal to 3.2 with 95% confidence. From prior experience, it is known that $\sigma_1 \approx 15$ and $\sigma_2 \approx 17$.
   b. A sampling error equal to 8 with 99% confidence. The range of each population is 60.
   c. A 90% confidence interval of width 1.0. Assume that $\sigma_1^2 \approx 5.8$ and $\sigma_2^2 \approx 7.5$.

7.53 Suppose you want to estimate the difference between two population means correct to within 2.2 with probability .95. If prior information suggests that the population variances are approximately equal to $\sigma_1^2 = \sigma_2^2 = 15$ and you want to select independent random samples of equal size from the populations, how large should the sample sizes, n_1 and n_2, be?

7.54 Enough money has been budgeted to collect $n = 100$ paired observations from populations 1 and 2 in order to estimate $\mu_d = (\mu_1 - \mu_2)$. Prior information indicates that

$\sigma_d = 12$. Have sufficient funds been allocated to construct a 90% confidence interval for μ_d of width 5 or less? Justify your answer.

Applying the Concepts—Basic

7.55 **Bulimia study.** Refer to the *American Statistician* (May 2001) study comparing the "fear of negative evaluation" (FNE) scores for bulimic and normal female students, presented in Exercise 7.17 (p. 365). Suppose you want to estimate $(\mu_B - \mu_N)$, the difference between the population means of the FNE scores for bulimic and normal female students, using a 95% confidence interval with a sampling error of two points. Find the sample sizes required to obtain such an estimate. Assume equal sample sizes of $\sigma_B^2 = \sigma_N^2 = 25$.

7.56 **Laughter among deaf signers.** Refer to the *Journal of Deaf Studies and Deaf Education* (Fall 2006) paired difference study on vocalized laughter among deaf users of sign language, presented in Exercise 7.38 (p. 377). Suppose you want to estimate $\mu_d = (\mu_S - \mu_A)$, the difference between the population mean number of laugh episodes of deaf speakers and deaf audience members, using a 90% confidence interval with a sampling error of .75. Find the number of pairs of deaf people required to obtain such an estimate, assuming that the variance of the paired differences is $\sigma_d^2 = 3$.

7.57 **Incomes of high school graduates.** A high school counselor wants to estimate the difference in mean income per day between high school graduates who have a college education and those who have not gone on to college. Suppose it is decided to compare the daily incomes of 30-year-olds, and the range of daily incomes for both groups is approximately $200 per day. How many people from each group should be sampled in order to estimate the true difference between mean daily incomes correct to within $10 per day with probability .9? Assume that $n_1 = n_2$.

Applying the Concepts—Intermediate

7.58 Study of inflamed ear lobes. In a *Journal of Ethnopharmacology* (June 1995) study, mice with inflamed ear lobes were randomly divided into two groups; one group was treated with bear bile while the other was treated with normal saline. Assuming equal-sized groups, how many mice should be included in each group in order to estimate the difference in the mean degree of swelling to within 2 milligrams with 95% confidence? Use the fact that the degree of swelling standard deviations for the bear bile and normal saline groups are approximately 4 milligrams and 3 milligrams, respectively.

7.59 Do video game players have superior visual attention skills? Refer to the *Journal of Articles in Support of the Null Hypothesis* (Vol. 6, 2009) study comparing the visual attention skill of video game and non-video game players,

Exercise 7.18 (p. 366). Recall that there was no significant difference between the mean score on the *attentional blink* test of video game players and the corresponding mean for non–video game players. It is possible that selecting larger samples would yield a significant difference. How many video game and non–video game players would need to be selected in order to estimate the difference in mean score for the two groups to within 5 points with 95% confidence? (Assume equal sample sizes will be selected from the two groups and that the score standard deviation for both groups is $\sigma \approx 9$.)

7.60 Scouting an NFL free agent. In seeking a free-agent NFL running back, a general manager is looking for a player with high mean yards gained per carry and a small standard deviation. Suppose the GM wishes to compare the mean yards gained per carry for two free agents, on the basis of independent random samples of their yards gained per carry. Data from last year's pro football season indicate that $\sigma_1 = \sigma_2 \approx 5$ yards. If the GM wants to estimate the difference in means correct to within 1 yard with a confidence level of .90, how many runs would have to be observed for each player? (Assume equal sample sizes.)

7.5 A Nonparametric Test for Comparing Two Populations: Independent Samples (Optional)

Suppose two independent random samples are to be used to compare two populations, but the *t*-test of Section 7.2 is inappropriate for making the comparison. We may be unwilling to make assumptions about the form of the underlying population probability distributions, or we may be unable to obtain exact values of the sample measurements. If the data can be ranked in order of magnitude in either of these cases, the **Wilcoxon rank sum test** (developed by Frank Wilcoxon) can be used to test the hypothesis that the probability distributions associated with the two populations are equivalent.

For example, consider an experimental psychologist who wants to compare reaction times for adult males under the influence of drug A with reaction times for those under the influence of drug B. Experience has shown that the populations of reaction-time measurements often possess probability distributions that are skewed to the right, as shown in Figure 7.14. Consequently, a *t*-test should not be used to compare the mean reaction times for the two drugs, because the normality assumption that is required for the *t*-test may not be valid.

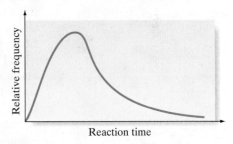

Figure 7.14

Typical probability distribution of reaction times

Suppose the psychologist randomly assigns seven subjects to each of two groups, one group to receive drug A and the other to receive drug B. The reaction time for each subject is measured at the completion of the experiment. These data (with the exception of the measurement for one subject in group A who was eliminated from the experiment for personal reasons) are shown in Table 7.6.

The population of reaction times for either of the drugs—say, drug A—is that which could conceptually be obtained by giving drug A to all adult males. To compare the probability distributions for populations A and B, *we first rank the sample observations as though they were all drawn from the same population*. That is, we pool the measurements from both samples and then rank all the measurements from the smallest (a rank of 1) to the largest (a rank of 13). The results of this ranking process are also shown in Table 7.6.

FRANK WILCOXON (1892–1965)

Wilcoxon Rank Tests

Frank Wilcoxon was born in Ireland, where his wealthy American parents were vacationing. He grew up in the family home in Catskill, New York, and then spent time working as an oil worker and tree surgeon in the back country of West Virginia. At age 25, Wilcoxon's parents sent him to Pennsylvania Military College, but he dropped out due to the death of his twin sister. Later, Wilcoxon earned degrees in chemistry from Rutgers (master's) and Cornell University (Ph.D.). After receiving his doctorate, Wilcoxon began work as a chemical researcher at the Boyce Thompson Institute for Plant Research. There, he began studying R. A. Fisher's (p. 401) newly issued *Statistical Methods for Research Workers*. In a now-famous 1945 paper, Wilcoxon presented the idea of replacing the actual sample data in Fisher's tests by their ranks and called the tests the rank sum test and signed-rank test. These tests proved to be inspirational to the further development of nonparametrics. After retiring from industry, Wilcoxon accepted a Distinguished Lectureship position at the newly created Department of Statistics at Florida State University. ∎

Table 7.6 **Reaction Times of Subjects under the Influence of Drug A or B**

Drug A		Drug B	
Reaction Time (seconds)	Rank	Reaction Time (seconds)	Rank
1.96	4	2.11	6
2.24	7	2.43	9
1.71	2	2.07	5
2.41	8	2.71	11
1.62	1	2.50	10
1.93	3	2.84	12
		2.88	13

Data Set: DRUGS

If, on the one hand, the two populations were identical, we would expect the ranks to be *randomly mixed* between the two samples. If, on the other hand, one population tends to have longer reaction times than the other, we would expect the larger ranks to be mostly in one sample and the smaller ranks mostly in the other. Thus, the test statistic for the Wilcoxon test is based on the totals of the ranks for each of the two samples— that is, on the **rank sums.** When the sample sizes are equal, the greater the difference in the rank sums, the greater will be the weight of evidence indicating a difference between the probability distributions of the populations.

In the reaction-times example, we denote the rank sum for drug A by T_1 and that for drug B by T_2. Then

$$T_1 = 4 + 7 + 2 + 8 + 1 + 3 = 25$$
$$T_2 = 6 + 9 + 5 + 11 + 10 + 12 + 13 = 66$$

The sum of T_1 and T_2 will always equal $n(n + 1)/2$, where $n = n_1 + n_2$. So, for this example, $n_1 = 6, n_2 = 7$, and

$$T_1 + T_2 = \frac{13(13 + 1)}{2} = 91$$

Since $T_1 + T_2$ is fixed, a small value for T_1 implies a large value for T_2 (and vice versa) and a large difference between T_1 and T_2. Therefore, the smaller the value of one of the rank sums, the greater is the evidence indicating that the samples were selected from different populations.

The test statistic for this test is the rank sum for the smaller sample; or, in the case where $n_1 = n_2$, either rank sum can be used. Values that locate the rejection region for this rank sum are given in Table VI of Appendix A, a partial reproduction of which is shown in Table 7.7. The columns of the table represent n_1, the first sample size, and the rows represent n_2, the second sample size. *The T_L and T_U entries in the table are the boundaries of the lower and upper regions, respectively, for the rank sum associated with the sample that has fewer measurements.* If the sample sizes n_1 and n_2 are the same, either rank sum may be used as the

Table 7.7 **Reproduction of Part of Table VI in Appendix A: Critical Values for the Wilcoxon Rank Sum Test**

$\alpha = .025$ one-tailed; $\alpha = .05$ two-tailed

n_2 \ n_1	3		4		5		6		7		8		9		10	
	T_L	T_U	T_L	T_U	T_L	T_U	T_L	T_U	T_L	T_U	T_L	T_U	T_L	T_U	T_L	T_U
3	5	16	6	18	6	21	7	23	7	26	8	28	8	31	9	33
4	6	18	11	25	12	28	12	32	13	35	14	38	15	41	16	44
5	6	21	12	28	18	37	19	41	20	45	21	49	22	53	24	56
6	7	23	12	32	19	41	26	52	28	56	29	61	31	65	32	70
7	7	26	13	35	20	45	28	56	37	68	39	73	41	78	43	83
8	8	28	14	38	21	49	29	61	39	73	49	87	51	93	54	98
9	8	31	15	41	22	53	31	65	41	78	51	93	63	108	66	114
10	9	33	16	44	24	56	32	70	43	83	54	98	66	114	79	131

test statistic. To illustrate, suppose $n_1 = 8$ and $n_2 = 10$. For a two-tailed test with $\alpha = .05$, we consult the table and find that the null hypothesis will be rejected if the rank sum of sample 1 (the sample with fewer measurements), T, is less than or equal to $T_L = 54$ *or* greater than or equal to $T_U = 98$. The Wilcoxon rank sum test is summarized in the next two boxes.

Wilcoxon Rank Sum Test: Independent Samples*

Let D_1 and D_2 represent the probability distributions for populations 1 and 2, respectively.

One-Tailed Test

H_0: D_1 and D_2 are identical
H_a: D_1 is shifted to the right of D_2
[or H_a: D_1 is shifted to the left of D_2]

Test statistic:

T_1, if $n_1 < n_2$; T_2, if $n_2 < n_1$ (Either rank sum can be used if $n_1 = n_2$.)

Rejection region:

T_1: $T_1 \geq T_U$ [or $T_1 \leq T_L$]
T_2: $T_2 \leq T_L$ [or $T_2 \geq T_U$]

Two-Tailed Test

H_0: D_1 and D_2 are identical
H_a: D_1 is shifted either to the left or to the right of D_2

Test statistic:

T_1, if $n_1 < n_2$; T_2, if $n_2 < n_1$ (Either rank sum can be used if $n_1 = n_2$.)
We will denote this rank sum as T.

Rejection region:

$T \leq T_L$ or $T \geq T_U$

where T_L and T_U are obtained from Table VI of Appendix A.

Ties: Assign tied measurements the average of the ranks they would receive if they were unequal, but occurred in successive order. For example, if the third-ranked and fourth-ranked measurements are tied, assign each a rank of $(3 + 4)/2 = 3.5$.

Conditions Required for a Valid Rank Sum Test

1. The two samples are random and independent.
2. The two probability distributions from which the samples are drawn are continuous.

Note that the assumptions necessary for the validity of the Wilcoxon rank sum test do not specify the shape or type of probability distribution. However, the distributions are assumed to be continuous so that the probability of tied measurements is 0 (see Chapter 4) and each measurement can be assigned a unique rank. In practice, however, rounding of continuous measurements will sometimes produce ties. As long as the number of ties is small relative to the sample sizes, the Wilcoxon test procedure will still have an approximate significance level of α. The test is not recommended to compare discrete distributions, for which many ties are expected.

Example 7.8

Applying The Rank Sum Test— Comparing Reaction Times of Two Drugs

Problem Do the data given in Table 7.6 provide sufficient evidence to indicate a shift in the probability distributions for drugs A and B—that is, that the probability distribution corresponding to drug A lies either to the right or to the left of the probability distribution corresponding to drug B? Test at the .05 level of significance.

---

*Another statistic used to compare two populations on the basis of independent random samples is the *Mann–Whitney U-statistic*, a simple function of the rank sums. It can be shown that the Wilcoxon rank sum test and the Mann–Whitney U-test are equivalent.

Solution

H_0: The two populations of reaction times corresponding to drug A and drug B have the same probability distribution.

H_a: The probability distribution for drug A is shifted to the right or left of the probability distribution for drug B.*

Test statistic: Since drug A has fewer subjects than drug B, the test statistic is T_1, the rank sum of drug A's reaction times.

Rejection region: Since the test is two sided, we consult part a of Table VI for the rejection region corresponding to $\alpha = .05$. We will reject H_0 for $T_1 \leq T_L$ or $T_1 \geq T_U$. Thus, we will reject H_0 if $T_1 \leq 28$ or $T_1 \geq 56$.

Since T_1, the rank sum of drug A's reaction times in Table 7.6, is 25, it is in the rejection region. (See Figure 7.15.)[†] Therefore, there is sufficient evidence to reject H_0. This same conclusion can be reached with a statistical software package. The SAS printout of the analysis is shown in Figure 7.16. Both the test statistic ($T_1 = 25$) and one-tailed

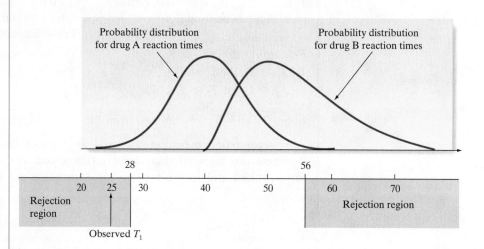

Figure 7.15
Alternative hypothesis and rejection region for Example 7.8.

The NPAR1WAY Procedure

Wilcoxon Scores (Rank Sums) for Variable REACTIME
Classified by Variable DRUG

DRUG	N	Sum of Scores	Expected Under H0	Std Dev Under H0	Mean Score
A	6	25.0	42.0	7.0	4.166667
B	7	66.0	49.0	7.0	9.428571

Wilcoxon Two-Sample Test

Statistic (S)	25.0000

Normal Approximation
Z	-2.4286
One-Sided Pr < Z	0.0076
Two-Sided Pr > |Z|	0.0152

t Approximation
One-Sided Pr < Z	0.0159
Two-Sided Pr > |Z|	0.0318

Exact Test
One-Sided Pr <= S	0.0070
Two-Sided Pr >= |S - Mean|	0.0140

Kruskal-Wallis Test

Chi-Square	5.8980
DF	1
Pr > Chi-Square	0.0152

Figure 7.16
SAS printout for Example 7.8

*The alternative hypotheses in this chapter will be stated in terms of a difference in the *location* of the distributions. However, since the shapes of the distributions may also differ under H_a, some of the figures (e.g., Figure 7.15) depicting the alternative hypothesis will show probability distributions with different shapes.

[†]Figure 7.15 depicts only one side of the two-sided alternative hypothesis. The other would show the distribution for drug A shifted to the right of the distribution for drug B.

p-value ($p = .007$) are highlighted on the printout. The one-tailed p-value is less than $\alpha = .05$, leading us to reject H_0.

Look Back Our conclusion is that the probability distributions for drugs A and B are not identical. In fact, it appears that drug B tends to be associated with reaction times that are larger than those associated with drug A (because T_1 falls into the lower tail of the rejection region).

Now Work Exercise 7.64

Table VI in Appendix A gives values of T_L and T_U for values of n_1 and n_2 less than or equal to 10. When both sample sizes, n_1 and n_2, are 10 or larger, the sampling distribution of T_1 can be approximated by a normal distribution with mean

$$E(T_1) = \frac{n_1(n_1 + n_2 + 1)}{2}$$

and variance

$$\sigma^2_{T_1} = \frac{n_1 n_2(n_1 + n_2 + 1)}{12}$$

Therefore, for $n_1 \geq 10$ and $n_2 \geq 10$, we can conduct the Wilcoxon rank sum test using the familiar z-test of Section 7.2. The test is summarized in the following box:

The Wilcoxon Rank Sum Test for Large Samples ($n_1 \geq 10$ and $n_2 \geq 10$)

Let D_1 and D_2 represent the probability distributions for populations 1 and 2, respectively.

One-Tailed Test	**Two-Tailed Test**
H_0: D_1 and D_2 are identical	H_0: D_1 and D_2 are identical
H_a: D_1 is shifted to the right of D_2	H_a: D_1 is shifted to the right or to the left
(or H_a: D_1 is shifted to the left of D_2)	of D_2

$$\text{Test statistic: } z = \frac{T_1 - \dfrac{n_1(n_1 + n_2 + 1)}{2}}{\sqrt{\dfrac{n_1 n_2(n_1 + n_2 + 1)}{12}}}$$

Rejection region:	Rejection region:
$z > z_\alpha$ (or $z < -z_\alpha$)	$\|z\| > z_{\alpha/2}$

Statistics IN Action | **Revisited**   **Comparing Treatment and Control Roach Trails Using a Nonparametric Test**

We return to the study of the trail-following ability of German cockroaches (p. 351) and the entomologist's theory that cockroaches, on average, will tend to follow a feces extract trail closer than they would follow a methanol (control) trail. Recall that four roach groups were utilized in the experiment: adult males, adult females, gravid (pregnant) females, and nymphs (immatures). In this application, we test the theory for adult males only.

Now let μ_1 represent the population mean trail deviation of adult male cockroaches assigned to the control trail and μ_2 represent the population mean trail deviation of adult male cockroaches following the feces extract trail. Again, we want to know whether $\mu_1 > \mu_2$; that is, we want to conduct the

Statistics IN Action
(continued)

one-tailed test, $H_0: (\mu_1 - \mu_2) = 0$ versus $H_a: (\mu_1 - \mu_2) > 0$.

Focusing only on the data for adult males in the **ROACH** file, we find that $n_1 = 10$ and $n_2 = 20$; thus, both samples are small. In order to apply the small-sample *t*-test, we must assume that both trail deviation distributions are approximately normal. Figure SIA7.2 shows MINITAB histograms for the two samples. Neither appear to come from an approximate normal distribution. Consequently, we will apply the appropriate nonparametric test—the Wilcoxon rank sum test.

We conducted this test using MINITAB. The output is shown in Figure SIA7.3. (*Note*: MINITAB conducts the equivalent Mann-Whitney *U*-test.) The one-tailed *p*-value of the test (.0000) is highlighted on Figure SIA7.3. Therefore, at $\alpha = .05$, there is sufficient evidence to conclude that the trail deviation distribution of adult male cockroaches following the control is shifted significantly above the trail deviation distribution of adult male cockroaches following the feces extract trail. Consequently, the entomologist's theory is supported for adult male cockroaches.

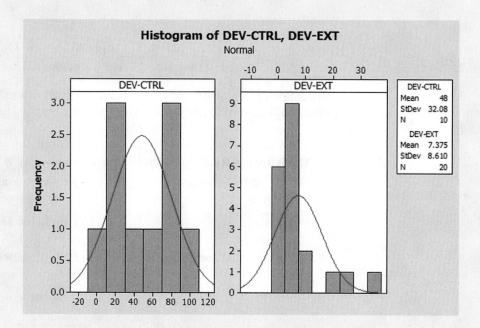

Figure SIA7.2
MINITAB histograms of trail deviations for two cockroach trails

Figure SIA7.3
MINITAB nonparametric comparison of two cockroach trails for adult males

Mann-Whitney Test and CI: DEV-CTRL, DEV-EXT

```
          N  Median
DEV-CTRL  10   46.15
DEV-EXT   20    4.50


Point estimate for ETA1-ETA2 is 39.70
95.5 Percent CI for ETA1-ETA2 is (13.11,71.69)
W = 244.0
Test of ETA1 = ETA2 vs ETA1 > ETA2 is significant at 0.0000
The test is significant at 0.0000 (adjusted for ties)
```

Exercises 7.61–7.77

Understanding the Principles

7.61 *True or False.* If the rank sum for sample 1 is much larger than the rank sum for sample 2 when $n_1 = n_2$, then the distribution of population 1 is likely to be shifted to the right of the distribution of population 2.

7.62 What conditions are required for a valid application of the Wilcoxon rank sum test?

Learning the Mechanics

7.63 Specify the rejection region for the Wilcoxon rank sum test for independent samples in each of the following situations:

a. H_0: Two probability distributions, 1 and 2, are identical H_a: The probability distribution for population 1 is shifted to the right or left of the probability distribution for population 2

$n_1 = 7, n_2 = 8, \alpha = .10$

b. H_0: Two probability distributions, 1 and 2, are identical H_a: The probability distribution for population 1 is shifted to the right of the probability distribution for population 2

$n_1 = 6, n_2 = 6, \alpha = .05$

c. H_0: Two probability distributions, 1 and 2, are identical H_a: The probability distribution for population 1 is shifted to the left of the probability distribution for population 2

$n_1 = 7, n_2 = 10, \alpha = .025$

d. H_0: Two probability distributions, 1 and 2, are identical H_a: The probability distribution for population 1 is shifted to the right or left of the probability distribution for population 2

$n_1 = 20, n_2 = 20, \alpha = .05$

7.64 Suppose you want to compare two treatments, A and B. In particular, you wish to determine whether the distribution for population B is shifted to the right of the distribution for population A. You plan to use the Wilcoxon rank sum test.

a. Specify the null and alternative hypotheses you would test.

b. Suppose you obtained the following independent random samples of observations on experimental units subjected to the two treatments. These data are saved in the **LM7_64** file.

Sample A	37,	40,	33,	29,	42,	33,	35,	28,	34,
Sample B	65,	35,	47,	52					

Conduct a test of the hypotheses you specified in part **a.** Test, using $\alpha = .05$.

7.65 Suppose you wish to compare two treatments, A and B, on the basis of independent random samples of 15 observations selected from each of the two populations. If $T_1 = 173$, do the data provide sufficient evidence to

indicate that distribution A is shifted to the left of distribution B? Test, using $\alpha = .05$.

7.66 Random samples of sizes $n_1 = 16$ and $n_2 = 12$ were drawn from populations 1 and 2, respectively. The measurements obtained are listed in the next table and saved in the **LM7_66** file.

Sample 1				Sample 2		
9.0	15.6	25.6	31.1	10.1	11.1	13.5
21.1	26.9	24.6	20.0	12.0	18.2	10.3
24.8	16.5	26.0	25.1	9.2	7.0	14.2
17.2	30.1	18.7	26.1	15.8	13.6	13.2

a. Conduct a hypothesis test to determine whether the probability distribution for population 2 is shifted to the left of the probability distribution for population 1. Use $\alpha = .05$.

b. What is the approximate p-value of the test of part **a**?

7.67 Independent random samples are selected from two populations. The data are shown in the following table and saved in the **LM7_67** file.

Sample 1		Sample 2		
15	16	5	9	5
10	13	12	8	10
12	8	9	4	

a. Use the Wilcoxon rank sum test to determine whether the data provide sufficient evidence to indicate a shift in the locations of the probability distributions of the sampled populations. Test, using $\alpha = .05$.

b. Do the data provide sufficient evidence to indicate that the probability distribution for population 1 is shifted to the right of the probability distribution for population 2? Use the Wilcoxon rank sum test with $\alpha = .05$.

Applying the Concepts—Basic

7.68 **Short Message Service for cell phones.** Short Message Service (SMS) is the formal name for the communication service that allows the interchange of short text messages between mobile telephone devices. About 75% of mobile phone subscribers worldwide send or receive SMS text messages. Consequently, SMS provides a opportunity for direct marketing. In *Management Dynamics* (2007), marketing researchers investigated the perceptions of college students towards SMS marketing. For one portion of the study, the researchers applied the Wilcoxon rank sum test to compare the distributions of the number of text messages sent and received during peak time for two groups of cell phone users: those on an annual contract and those with a pay-as-you-go option.

a. Specify the null hypothesis tested in the words of the problem.

b. Give the formula for the large-sample test statistic if there were 25 contract users and 40 pay-as-you-go users in the sample.

c. The Wilcoxon test results led the researchers to conclude "that contract users sent and received significantly more SMS's during peak time than pay-as-you-go users." Based on this information, draw a graph that is representative of the two SMS usage rate populations.

7.69 **The X-Factor in golf performance.** Many golf teaching professionals believe that a greater hip-to-shoulder differential angle during the early downswing—dubbed the "X-Factor"—leads to improved golf performance. The *Journal of Quantitative Analysis in Sports* (Vol. 5, 2009) published an article on the X-Factor and its relationship to golfing performance. The study involved 15 male golfers with a player handicap of 20 strokes or less. The golfers were divided into two groups: 8 golfers with a handicap of 10 strokes or less (low-handicapped group) and 7 golfers with a handicap between 12 and 20 strokes (high-handicapped group). The X-Factor, i.e., the hip-to-shoulder differential angle (in degrees), was measured for each golfer at the top of the backswing during his tee shot. The researchers hypothesized that low-handicapped golfers will tend to have higher X-factors than high-handicapped golfers. The researchers also discovered that the sample data were not normally distributed. Consequently, they applied a nonparametric test.

a. What nonparametric test is appropriate for analyzing these data?

b. Specify the null and alternative hypotheses of interest in the words of the problem.

c. Give the rejection region for this test, using $\alpha = .05$.

d. The researchers reported a *p*-value of .487. Use this result to draw a conclusion.

7.70 **Bursting strength of bottles.** Polyethylene terephthalate (PET) bottles are used for carbonated beverages. A critical property of PET bottles is their bursting strength (i.e., the pressure at which bottles filled with water burst when pressurized). In the *Journal of Data Science* (May 2003), researchers measured the bursting strength of PET bottles made from two different designs: an old design and a new design. The data (in pounds per square inch) for 10 bottles of each design are shown in the accompanying table and saved in the **PET** file. Suppose you want to compare the distributions of bursting strengths for the two designs.

| Old Design | 210 | 212 | 211 | 211 | 190 | 213 | 212 | 211 | 164 | 209 |
| New Design | 216 | 217 | 162 | 137 | 219 | 216 | 179 | 153 | 152 | 217 |

a. Rank all 20 observed pressures from smallest to largest, and assign ranks from 1 to 20.

b. Sum the ranks of the observations from the old design.

c. Sum the ranks of the observations from the new design.

d. Compute the Wilcoxon rank sum statistic.

e. Carry out a nonparametric test (at $\alpha = .05$) to compare the distribution of bursting strengths for the two designs.

7.71 **Research on eating disorders.** Refer to the study on eating disorders, Exercise 7.17 (p. 365). The "fear of negative evaluation" (FNE) scores for 11 female students known

to suffer from the eating disorder bulimia and 14 female students with normal eating habits, are reproduced in the table at the bottom of the page. (Recall that the higher the score, the greater is the fear of a negative evaluation.) These data are saved in the **BULIMIA** file. Suppose you want to determine whether the distribution of the FNE scores for bulimic female students is shifted above the corresponding distribution for female students with normal eating habits.

a. Specify H_0 and H_a for the test.

b. Rank all 25 FNE scores in the data set from smallest to largest.

c. Sum the ranks of the 11 FNE scores for bulimic students.

d. Sum the ranks of the 14 FNE scores for students with normal eating habits.

e. Give the rejection region for a nonparametric test of the data if $\alpha = .10$.

f. Conduct the test and give the conclusion in the words of the problem.

7.72 **Children's recall of TV ads.** Refer to the *Journal of Advertising* (Spring 2006) study of children's recall of television advertisements, presented in Exercise 7.13 (p. 364). Two groups of children were shown a 60-second commercial for Sunkist Fun Fruit Rock-n-Roll Shapes. One group (the A/V group) was shown both the audio and video portions of the ad; the other group (the video-only group) was shown only the video portion of the commercial. The number out of 10 specific items from the ad recalled correctly by each child is shown in the accompanying table. (These data are saved in the **FUNFRUIT** file.) Recall that the researchers theorized that children who receive an audiovisual presentation will have the same level of recall as those who receive only the visual aspects of the ad. Consider testing the researchers' theory, using the Wilcoxon rank sum test.

| A/V group: | 0 4 6 6 1 2 2 6 6 4 1 2 6 1 3 0 2 5 4 5 |
| Video-only group: | 6 3 6 2 2 4 7 6 1 3 6 2 3 1 3 2 5 2 4 6 |

Based on Maher, J. K., Hu, M. Y., and Kolbe, R. H. "Children's recall of television ad elements." *Journal of Advertising*, Vol. 35, No. 1, Spring 2006 (Table 1).

a. Set up the appropriate null and alternative hypotheses for the test.

b. Find the value of the test statistic.

c. Give the rejection region for $\alpha = .10$.

d. Make the appropriate inference. What can you say about the researchers' theory?

Applying the Concepts–Intermediate

7.73 **Is honey a cough remedy?** Refer to the *Archives of Pediatrics and Adolescent Medicine* (Dec. 2007) study of honey as a children's cough remedy, Exercise 2.32 (p. 45). Recall that 70 children who were ill with an upper respiratory tract infection were given either a dosage of dextromethorphan (DM)—an over-the-counter cough medicine—or a similar dose of honey. Parents then rated their children's cough symptoms and the improvement in total cough symptoms score was determined for each child. The data (improvement scores) are reproduced in the

| Bulimic Students | 21 | 13 | 10 | 20 | 25 | 19 | 16 | 21 | 24 | 13 | 14 | | | |
| Normal Students | 13 | 6 | 16 | 13 | 8 | 19 | 23 | 18 | 11 | 19 | 7 | 10 | 15 | 20 |

accompanying table and saved in the **HONEYCOUGH** file. The researchers concluded that "honey may be a preferable treatment for the cough and sleep difficulty associated with childhood upper respiratory tract infection." Use the nonparametric method presented in this section to analyze the data (use $\alpha = .05$). Do you agree with the researchers?

Honey Dosage:	12 11 15 11 10 13 10 4 15 16 9 14 10
	6 10 8 11 12 12 8 12 9 11 15 10 15
	9 13 8 12 10 8 9 5 12
DM Dosage:	4 6 9 4 7 7 7 9 12 10 11 6 3
	4 9 12 7 6 8 12 12 4 12 13 7 10
	13 9 4 4 10 15 9

Based on Paul, I. M., et al. "Effect of honey, dextromethorphan, and no treatment on nocturnal cough and sleep quality for coughing children and their parents." *Archives of Pediatrics and Adolescent Medicine*, Vol. 161, No. 12, Dec. 2007 (data simulated).

7.74 **Does rudeness really matter in the workplace?** Refer to the *Academy of Management Journal* (Oct. 2007) study on rudeness in the workplace, Exercise 7.24 (p. 367). Recall that 98 college students enrolled in a management course were randomly assigned to one of two experimental conditions: rudeness condition (where the students were berated by a facilitator for being irresponsible and unprofessional) and control group (no facilitator comments). Each student was asked to write down as many uses for a brick as possible in five minutes. The data, saved in the **RUDE** file, are reproduced below.

Control Group:
1 24 5 16 21 7 20 1 9 20 19 10 23 16 0 4 9 13
17 13 0 2 12 11 7 1 19 9 12 18 5 21 30 15 4 2
12 11 10 13 11 3 6 10 13 16 12 28 19 12 20 3 11

Rudeness Condition:
4 11 18 11 9 6 5 11 9 12 7 5 7 3 11 1 9 11
10 7 8 9 10 7 11 4 13 5 4 7 8 3 8 15 9 16
10 0 7 15 13 9 2 13 10

a. Show that although the data for the rudeness condition are approximately normally distributed, the control group data are skewed.
b. Conduct the appropriate nonparametric test (at $\alpha = .01$) to determine if the true median performance level for students in the rudeness condition is lower than the true median performance level for students in the control group.
c. Explain why the parametric two-sample test conducted in Exercise 7.24 is appropriate even though the data for both groups are not normally distributed. (Note that the nonparametric and parametric tests yield the same conclusions.)

7.75 **Computer-mediated communication study.** Computer-mediated communication (CMC) is a form of interaction that heavily involves technology (e.g., instant messaging, e-mail). A study was conducted to compare relational intimacy in people interacting via CMC with people meeting face-to-face (FTF) (*Journal of Computer-Mediated Communication*, Apr. 2004). Participants were 48 undergraduate students, of whom half were randomly assigned to the CMC group and half to the FTF group. Each group was given a task that required communication among its group members. Those in the CMC group communicated

via the "chat" mode of instant-messaging software; those in the FTF group met in a conference room. The variable of interest, relational intimacy score, was measured (on a seven-point scale) for each participant after each of three different meetings. Scores for the first meeting are given in the accompanying table and saved in the **INTIMACY** file. The researchers hypothesized that the relational intimacy scores for participants in the CMC group will tend to be lower than the relational intimacy scores for participants in the FTF group.

| CMC group: | 4 3 3 4 3 3 3 3 4 4 3 4 3 3 3 2 4 2 4 5 4 4 4 5 3 |
| FTF group: | 5 4 4 4 3 3 3 4 3 3 3 3 4 4 4 4 4 3 3 3 4 4 2 4 |

Note: Data simulated from descriptive statistics provided in article.

a. Which nonparametric procedure should be used to test the researchers' hypothesis?
b. Specify the null and alternative hypotheses of the test.
c. Give the rejection region for the test, using $\alpha = .10$.
d. Conduct the test and give the appropriate conclusion in the context of the problem.

7.76 **Brood-parasitic birds.** The term *brood-parasitic intruder* is used to describe a bird that searches for and lays eggs in a nest built by a bird of another species. For example, the brown-headed cowbird is known to be a brood parasite of the smaller willow flycatcher. Ornithologists theorize that those flycatchers which recognize, but do not vocally react to, cowbird calls are more apt to defend their nests and less likely to be found and parasitized. In a study published in *The Condor* (May 1995), each of 13 active flycatcher nests was categorized as parasitized (if at least one cowbird egg was present) or nonparasitized. Cowbird songs were taped and played back while the flycatcher pairs were sitting in the nest prior to incubation. The vocalization rate (number of calls per minute) of each flycatcher pair was recorded. The data for the two groups of flycatchers are given in the table and saved in the **COWBIRD** file. Do the data suggest (at $\alpha = .05$) that the vocalization rates of parasitized flycatchers are higher than those of nonparasitized flycatchers?

Parasitized	Not Parasitized
2.00	1.00
1.25	1.00
8.50	0
1.10	3.25
1.25	1.00
3.75	.25
5.50	

Based on Uyehara, J. C., and Narins, P. M. "Nest defense by Willow Flycatchers to brood-parasitic intruders." *The Condor*, Vol. 97, No. 2, May 1995, p. 364 (Figure 1).

7.77 **Family involvement in homework.** A study of the impact of the interactive Teachers Involve Parents in Schoolwork (TIPS) program was published in the *Journal of Educational Research* (July/Aug. 2003). A sample of 128 middle school students were assigned to complete TIPS homework assignments, while 98 students were assigned traditional, noninteractive homework assignments (ATIPS). At the end of the study, all students reported on the level of family involvement in their homework on a five-point scale (0 = Never, 1 = Rarely, 2 = Sometimes, 3 = Frequently, 4 = Always). The data for the science, math, and language arts homework are saved in the **HWSTUDY** file. (The first five and last five observations in the data set are reproduced in the accompanying table.)

a. Why might a nonparametric test be the most appropriate test to apply in order to compare the levels of family involvement in homework assignments of TIPS and ATIPS students?
b. Conduct a nonparametric analysis to compare the involvement in science homework assignments of TIPS and ATIPS students. Use $\alpha = .05$.
c. Repeat part **b** for mathematics homework assignments.
d. Repeat part **b** for language arts homework assignments.

Homework Condition	Science	Math	Language
ATIPS	1	0	0
ATIPS	0	1	1
ATIPS	0	1	0
ATIPS	1	2	0
ATIPS	1	1	2
⋮	⋮	⋮	⋮
TIPS	2	3	2
TIPS	1	4	2
TIPS	2	4	2
TIPS	4	0	3
TIPS	2	0	1

Source: Van Voorhis, F. L. "Interactive homework in middle school: Effects on family involvement and science achievement." *Journal of Educational Research,* 96(6), 2003, pp. 323–338. Reprinted with permission from Frances Van Voorhis.

7.6 A Nonparametric Test for Comparing Two Populations: Paired Difference Experiment

Nonparametric techniques may also be employed to compare two probability distributions when a paired difference design is used. For example, consumer preferences for two competing products are often compared by having each of a sample of consumers rate both products. Thus, the ratings have been paired on each consumer. Following is an example of this type of experiment.

For some paper products, softness is an important consideration in determining consumer acceptance. One method of determining softness is to have judges give a sample of the products a softness rating. Suppose each of 10 judges is given a sample of two products that a company wants to compare. Each judge rates the softness of each product on a scale from 1 to 20, with higher ratings implying a softer product. The results of the experiment are shown in Table 7.8.

Since this is a paired difference experiment, we analyze the differences between the measurements. (See Section 7.3.) However, a nonparametric approach developed by Wilcoxon requires that we calculate the ranks of the absolute values of the differences between the measurements (i.e., the ranks of the differences after removing any minus signs). *Note that tied absolute differences (e.g., the two differences of 4) are assigned the average of the ranks they would receive if they were unequal, but successive, measurements (e.g., 4.5, the average of the ranks 4 and 5).* After the absolute differences are ranked, the sum of the ranks of the positive differences of the original measurements, T_+, and the

Table 7.8	Softness Ratings of Paper				
	Product		Difference		
Judge	A	B	(A − B)	Absolute Value of Difference	Rank of (A − B) Absolute Value
1	12	8	4	4	4.5
2	16	10	6	6	7
3	8	9	−1	1	1
4	10	8	2	2	2
5	19	12	7	7	8
6	14	17	−3	3	3
7	12	4	8	8	9
8	10	6	4	4	4.5
9	12	17	−5	5	6
10	16	4	12	12	10

T_+ = Sum of positive ranks = 45
T_- = Sum of negative ranks = 10

Data Set: SOFTPAPER

sum of the ranks of the negative differences of the original measurements, T_-, are computed. (The ranks of the negative differences are highlighted in Table 7.8.)

We are now prepared to test the nonparametric hypotheses:

H_0: The probability distributions of the ratings for products A and B are identical.

H_a: The probability distributions of the ratings differ (in location) for the two products. (Note that this is a two-sided alternative and that it implies a two-tailed test.)

Test statistic: $T =$ Smaller of the positive and negative rank sums T_+ and T_-

The smaller the value of T, the greater is the evidence indicating that the two probability distributions differ in location. The rejection region for T can be determined by consulting Table VII in Appendix A, (part of which is shown in Table 7.9). This table gives a value T_0 for both one-tailed and two-tailed tests for each value of n, the number of matched pairs. For a two-tailed test with $\alpha = .05$, we will reject H_0 if $T \leq T_0$. You can see in Table 7.9 that the value of T_0 which locates the boundary of the rejection region for the judges' ratings for $\alpha = .05$ and $n = 10$ pairs of observations is 8. Thus, the rejection region for the test (see Figure 7.17) is

Rejection region: $T \leq 8$ for $\alpha = .05$

Table 7.9	**Reproduction of Part of Table VII of Appendix A: Critical Values for the Wilcoxon Paired Difference Signed Rank Test**						
One-Tailed	Two-Tailed	$n = 5$	$n = 6$	$n = 7$	$n = 8$	$n = 9$	$n = 10$
$\alpha = .05$	$\alpha = .10$	1	2	4	6	8	11
$\alpha = .025$	$\alpha = .05$		1	2	4	6	8
$\alpha = .01$	$\alpha = .02$			0	2	3	5
$\alpha = .005$	$\alpha = .01$				0	2	3
		$n = 11$	$n = 12$	$n = 13$	$n = 14$	$n = 15$	$n = 16$
$\alpha = .05$	$\alpha = .10$	14	17	21	26	30	36
$\alpha = .025$	$\alpha = .05$	11	14	17	21	25	30
$\alpha = .01$	$\alpha = .02$	7	10	13	16	20	24
$\alpha = .005$	$\alpha = .01$	5	7	10	13	16	19
		$n = 17$	$n = 18$	$n = 19$	$n = 20$	$n = 21$	$n = 22$
$\alpha = .05$	$\alpha = .10$	41	47	54	60	68	75
$\alpha = .025$	$\alpha = .05$	35	40	46	52	59	66
$\alpha = .01$	$\alpha = .02$	28	33	38	43	49	56
$\alpha = .005$	$\alpha = .01$	23	28	32	37	43	49
		$n = 23$	$n = 24$	$n = 25$	$n = 26$	$n = 27$	$n = 28$
$\alpha = .05$	$\alpha = .10$	83	92	101	110	120	130
$\alpha = .025$	$\alpha = .05$	73	81	90	98	107	117
$\alpha = .01$	$\alpha = .02$	62	69	77	85	93	102
$\alpha = .005$	$\alpha = .01$	55	61	68	76	84	92

Since the smaller rank sum for the paper data, $T_- = 10$, does not fall within the rejection region, the experiment has not provided sufficient evidence indicating that the two paper products differ with respect to their softness ratings at the $\alpha = .05$ level.

Note that if a significance level of $\alpha = .10$ had been used, the rejection region would have been $T \leq 11$ and we would have rejected H_0. In other words, the samples do provide evidence that the probability distributions of the softness ratings differ at the $\alpha = .10$ significance level.

The **Wilcoxon signed rank test** is summarized in the next box. Note that the difference measurements are assumed to have a continuous probability distribution so that

Figure 7.17

Rejection region for paired difference experiment

the absolute differences will have unique ranks. Although tied (absolute) differences can be assigned ranks by averaging, in order to ensure the validity of the test, the number of ties should be small relative to the number of observations.

Wilcoxon Signed Rank Test for a Paired Difference Experiment

Let D_1 and D_2 represent the probability distributions for populations 1 and 2, respectively.

One-Tailed Test	**Two-Tailed Test**
H_0: D_1 and D_2 are identical	H_0: D_1 and D_2 are identical
H_a: D_1 is shifted to the right of D_2	H_a: D_1 is shifted either to the
[or H_a: D_1 is shifted to the left of D_2]	left or to the right of D_2

Calculate the difference within each of the n matched pairs of observations. Then rank the absolute value of the n differences from the smallest (rank 1) to the highest (rank n), and calculate the rank sum T_- of the negative differences and the rank sum T_+ of the positive differences. [*Note:* Differences equal to 0 are eliminated, and the number n of differences is reduced accordingly.]

Test statistic:	*Test statistic:*
T_-, the rank sum of the negative differences [or T_+, the rank sum of the positive differences]	T, the smaller of T_+ or T_-
Rejection region:	*Rejection region:*
$T_- \leq T_0$ [or $T_+ \leq T_0$]	$T \leq T_0$

where T_0 is given in Table VII in Appendix A.

Ties: Assign tied absolute differences the average of the ranks they would receive if they were unequal, but occurred in successive order. For example, if the third-ranked and fourth-ranked differences are tied, assign both a rank of $(3 + 4)/2 = 3.5$.

Conditions Required for a Valid Signed Rank Test

1. The sample of differences is randomly selected from the population of differences.
2. The probability distribution from which the sample of paired differences is drawn is continuous.

Example 7.9

Applying the Signed Rank Test— Comparing Two Crime Prevention Plans

Problem Suppose the police commissioner in a small community must choose between two plans for patrolling the town's streets. Plan A, the less expensive plan, uses voluntary citizen groups to patrol certain high-risk neighborhoods. In contrast, plan B would utilize police patrols. As an aid in reaching a decision, both plans are examined by 10 trained criminologists, each of whom is asked to rate the plans on a scale from 1 to 10. (High ratings imply a more effective crime prevention plan.) The city will adopt plan B (and hire extra police) only if the data provide sufficient evidence that criminologists tend to rate plan B more effective than plan A. The results of the survey are shown in Table 7.10. Do the data provide evidence at the $\alpha = .05$ level that the distribution of ratings for plan B lies above that for plan A? Use the Wilcoxon signed rank test to answer the question.

Solution The null and alternative hypotheses are as follows:

H_0: The two probability distributions of effectiveness ratings are identical

H_a: The effectiveness ratings of the more expensive plan (B) tend to exceed those of plan A

Table 7.10 Effectiveness Ratings by 10 Qualified Crime Prevention Experts

Crime Prevention Expert	Plan A	Plan B	Difference (A − B)	Rank of Absolute Difference
1	7	9	− 2	4.5
2	4	5	− 1	2
3	8	8	0	(Eliminated)
4	9	8	1	2
5	3	6	− 3	6
6	6	10	− 4	7.5
7	8	9	− 1	2
8	10	8	2	4.5
9	9	4	5	9
10	5	9	− 4	7.5

Positive rank sum = $T_+ = 15.5$

Data Set: CRIMEPLAN

Observe that the alternative hypothesis is one sided (i.e., we only wish to detect a shift in the distribution of the B ratings to the right of the distribution of A ratings); therefore, it implies a one-tailed test of the null hypothesis. (See Figure 7.18.) If the alternative hypothesis is true, the B ratings will tend to be larger than the paired A ratings, more negative differences in pairs will occur, T_- will be large, and T_+ will be small. Because Table VII is constructed to give lower-tail values of T_0, we will use T_+ as the test statistic and reject H_0 for $T_+ \leq T_0$.

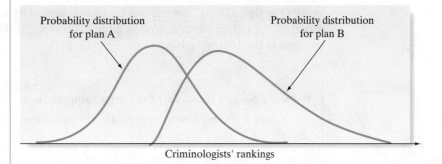

Figure 7.18

The alternative hypothesis for Example 7.9

Because a paired difference design was used (both plans were evaluated by the same criminologist), we first calculate the difference between the rating for each expert. The differences in ratings for the pairs (A − B) are shown in Table 7.10. Note that one of the differences equals 0. Consequently, we eliminate this pair from the ranking and reduce the number of pairs to $n = 9$. Looking in Table VII, we have $T_0 = 8$ for a one-tailed test with $\alpha = .05$ and $n = 9$. Therefore, the test statistic and rejection region for the test are

Test statistic: T_+, the positive rank sum

Rejection region: $T_+ \leq 8$

Summing the ranks of the positive differences (highlighted) in Table 7.10, we find that $T_+ = 15.5$. Since this value exceeds the critical value, $T_0 = 8$, we conclude that the sample provides insufficient evidence at the $\alpha = .05$ level to support the alternative hypothesis. The commissioner *cannot* conclude that the plan utilizing police patrols tends to be rated higher than the plan using citizen volunteers. That is, on the basis of this study, extra police will not be hired.

Look Back An SPSS printout of the analysis, shown in Figure 7.19, confirms the preceding conclusion. Both the test statistic and two-tailed *p*-value are highlighted on the printout. Since the one-tailed *p*-value, $.404/2 = .202$, exceeds $\alpha = .05$, we fail to reject H_0.

Wilcoxon Signed Ranks Test

Ranks

		N	Mean Rank	Sum of Ranks
A - B	Negative Ranks	6[a]	4.92	29.50
	Positive Ranks	3[b]	5.17	15.50
	Ties	1[c]		
	Total	10		

a. A < B

b. A > B

c. A = B

Test Statistics[b]

	A - B
Z	-.834[a]
Asymp. Sig. (2-tailed)	.404

a. Based on positive ranks.

b. Wilcoxon Signed Ranks Test

Figure 7.19

SPSS printout for Example 7.9

Now Work Exercise 7.81

As is the case for the rank sum test for independent samples, the sampling distribution of the signed rank statistic can be approximated by a normal distribution when the number n of paired observations is large (say, $n \geq 25$). The large-sample z-test is summarized in the following box:

Wilcoxon Signed Rank Test for Large Samples (n $\geq$ 25)

Let D_1 and D_2 represent the probability distributions for populations 1 and 2, respectively.

One-Tailed Test

H_0: D_1 and D_2 are identical
H_a: D_1 is shifted to the right of D_2 [or
H_a: D_1 is shifted to the left of D_2]

Two-Tailed Test

H_0: D_1 and D_2 are identical
H_a: D_1 is shifted either to the left or to the right of D_2

$$\text{Test statistic: } z = \frac{T_+ - [n(n + 1)/4]}{\sqrt{[n(n + 1)(2n + 1)]/24}}$$

Rejection region:
$z > z_\alpha$ [or $z < -z_\alpha$]

Rejection region:
$|z| > z_{\alpha/2}$

Assumptions: The sample size n is greater than or equal to 25. Differences equal to 0 are eliminated and the number n of differences is reduced accordingly. Tied absolute differences receive ranks equal to the average of the ranks they would have received had they not been tied.

Exercises 7.78–7.94

Understanding the Principles

7.78 Explain the difference between the one- and two-tailed versions of the Wilcoxon signed rank test for the paired difference experiment.

7.79 In order to conduct the Wilcoxon signed rank test, why do we need to assume that the probability distribution of differences is continuous?

Learning the Mechanics

7.80 Specify the test statistic and the rejection region for the Wilcoxon signed rank test for the paired difference design in each of the following situations:

a. H_0: Two probability distributions, A and B, are identical
H_a: The probability distribution for population A is shifted to the right or left of the probability distribution for population B
$n = 20, \alpha = .10$

b. H_0: Two probability distributions, A and B, are identical
H_a: The probability distribution for population A is shifted to the right of the probability distribution for population B
$n = 39, \alpha = .05$

c. H_0: Two probability distributions, A and B, are identical
H_a: The probability distribution for population A is shifted to the left of the probability distribution for population B
$n = 7, \alpha = .005$

7.81 Suppose you want to test a hypothesis that two treatments, A and B, are equivalent against the alternative hypothesis that the responses for A tend to be larger than those for B. You plan to use a paired difference experiment and to analyze the resulting data with the Wilcoxon signed rank test.
a. Specify the null and alternative hypotheses you would test.
b. Suppose the paired difference experiment yielded the data in the accompanying table. (These data are saved in the **LM7_81** file.) Conduct the test, of part **a**. Test using $\alpha = .025$.

Pair	A	B	Pair	A	B
1	54	45	6	77	75
2	60	45	7	74	63
3	98	87	8	29	30
4	43	31	9	63	59
5	82	71	10	80	82

7.82 Suppose you wish to test a hypothesis that two treatments, A and B, are equivalent against the alternative that the responses for A tend to be larger than those for B.
a. If the number of pairs equals 25, give the rejection region for the large-sample Wilcoxon signed rank test for $\alpha = .05$.
b. Suppose that $T_+ = 273$. State your test conclusions.
c. Find the p-value for the test and interpret it.

7.83 A paired difference experiment with $n = 30$ pairs yielded $T_+ = 354$.
a. Specify the null and alternative hypotheses that should be used in conducting a hypothesis test to determine

whether the probability distribution for population A is located to the right of that for population B.
b. Conduct the test of part **a**, using $\alpha = .05$.
c. What is the approximate p-value of the test of part **b**?
d. What assumptions are necessary to ensure the validity of the test you performed in part **b**?

Applying the Concepts—Basic

7.84 Treating psoriasis with the "Doctorfish of Kangal." Refer to the *Evidence-Based Research in Complementary and Alternative Medicine* (Dec. 2006) study of treating psoriasis with ichthyotherapy, presented in Exercise 2.133 (p. 85). (Recall that the therapy is also known as the "Doctorfish of Kangal," since it uses fish from the hot pools of Kangal, Turkey, to feed on skin scales.) In the study, 67 patients diagnosed with psoriasis underwent three weeks of ichthyotherapy. The Psoriasis Area Severity Index (PASI) of each patient was measured both before and after treatment. (The lower the PASI score, the better is the skin condition.) Before- and after-treatment PASI scores were compared with the use of the Wilcoxon signed rank test.

a. Explain why the PASI scores should be analyzed with a test for paired differences.
b. Refer to the box plots shown in Exercise 2.133. Give a reason that the researchers opted to use a nonparametric test to compare the PASI scores.
c. The p-value for the Wilcoxon signed ranks test was reported as $p < .0001$. Interpret this result, and comment on the effectiveness of ichthyotherapy in treating psoriasis.

7.85 Healing potential of handling museum objects. Refer to the *Museum & Society* (Nov. 2009) study of the healing potential of handling museum objects, Exercise 7.37 (p. 376). Recall that the health status of each of 32 hospital patients was recorded both before and after handling a museum object (such as an archaeological artifact or brass etching). The simulated data (measured on a 100-point scale) are reproduced in the next table and saved in the **MUSEUM** file. The Wilcoxon signed rank test was applied to the data, with the results shown in the SPSS printout (p. 398).

Session	Before	After	Session	Before	After
1	52	59	17	65	65
2	42	54	18	52	63
3	46	55	19	39	50
4	42	51	20	59	69
5	43	42	21	49	61
6	30	43	22	59	66
7	63	79	23	57	61
8	56	59	24	56	58
9	46	53	25	47	55
10	55	57	26	61	62
11	43	49	27	65	61
12	73	83	28	36	53
13	63	72	29	50	61
14	40	49	30	40	52
15	50	49	31	65	70
16	50	64	32	59	72

SPSS Output for Exercise 7.85

Wilcoxon Signed Ranks Test

Ranks

		N	Mean Rank	Sum of Ranks
AFTER - BEFORE	Negative Ranks	3[a]	3.83	11.50
	Positive Ranks	28[b]	17.30	484.50
	Ties	1[c]		
	Total	32		

a. AFTER < BEFORE

b. AFTER > BEFORE

c. AFTER = BEFORE

Test Statistics[b]

	AFTER - BEFORE
Z	-4.638[a]
Asymp. Sig. (2-tailed)	.000

a. Based on negative ranks.

b. Wilcoxon Signed Ranks Test

a. Use the information in the printout to find the large-sample Wilcoxon signed rank test statistic.

b. Does handling a museum object have a positive impact on a sick patient's well-being? Test using $\alpha = .01$.

7.86 Impact of red light cameras on car crashes. Refer to the June 2007 Virginia Department of Transportation (VDOT) study of a newly adopted photo-red-light enforcement program, Exercise 7.45 (p. 378). Recall that the VDOT provided crash data both before and after installation of red light cameras at several intersections. The data (measured as the number of crashes caused by red light running per intersection per year) for 13 intersections in Fairfax County, Virginia, are reproduced in the next table and saved in the **REDLIGHT** file. The VDOT wants to determine if the photo-red enforcement program is effective in reducing red-light-running crash incidents at intersections. Use the nonparametric Wilcoxon singed rank test (and the accompanying MINITAB printout, top right) to analyze the data for the VDOT.

Intersection	Before Camera	After Camera
1	3.60	1.36
2	0.27	0
3	0.29	0
4	4.55	1.79
5	2.60	2.04
6	2.29	3.14
7	2.40	2.72
8	0.73	0.24
9	3.15	1.57
10	3.21	0.43
11	0.88	0.28
12	1.35	1.09
13	7.35	4.92

Based on Virginia Transportation Research Council, "Research Report: The Impact of Red Light Cameras (Photo-Red Enforcement) on Crashes in Virginia," June 2007.

MINITAB Output for Exercise 7.86

Wilcoxon Signed Rank Test: Difference

```
Test of median = 0.000000 versus median > 0.000000

                  N for   Wilcoxon           Estimated
            N     Test    Statistic     P      Median
Difference  13    13          79.0   0.011    0.9650
```

7.87 Reading comprehension strategies of elementary school children. An investigation of the reading comprehension strategies employed by good and average elementary school readers was the topic of research published in *The Reading Matrix* (April 2004). Both good and average readers were recruited on the basis of their scores on a midterm language test. Each group was evaluated on how often its members employed each of eight different reading strategies. The accompanying table (saved in the **READSTRAT** file) gives the proportion of times the reading group used each strategy (called the Factor Specificity Index, or FSI score). The researchers conducted a Wilcoxon signed rank test to compare the FSI score distributions of good and average readers.

	FSI Scores	
Strategy	Good Readers	Average Readers
Word meaning	.38	.32
Words in context	.29	.25
Literal comprehension	.42	.25
Draw inference from single string	.60	.26
Draw inference from multiple string	.45	.31
Interpretation of metaphor	.32	.14
Find salient or main idea	.21	.03
Form judgment	.73	.80

Based on Ahmed, S., and Asraf, R. M. "Making sense of text: Strategies used by good and average readers." *The Reading Matrix*, Vol. 4, No. 1, April 2004 (Table 2).

a. State H_0 and H_a for the desired test of hypothesis.

b. For each strategy, compute the difference between the FSI scores of good and average readers.

c. Rank the absolute values of the differences.

d. Calculate the value of the signed rank test statistic.

e. Find the rejection region for the test, using $\alpha = .05$.

f. Make the appropriate inference in the words of the problem.

7.88 NHTSA new car crash tests. Refer to the National Highway Traffic Safety Administration (NHTSA) new-car crash test data saved in the **CRASH** file. In Exercise 7.40 (p. 377), you compared the chest injury ratings of drivers and front-seat passengers by using the Student's *t*-procedure for matched pairs. Suppose you want to make the comparison for only those cars which have a driver's rating of five stars (the highest rating). The data for these 18 cars are listed in the table (p. 399) and saved in the **CRASH** file. Now consider analyzing the data by using the Wilcoxon signed rank test.

Data for Exercise 7.88

Chest Injury Rating			Chest Injury Rating		
Car	Driver	Passenger	Car	Driver	Passenger
1	42	35	10	36	37
2	42	35	11	36	37
3	34	45	12	43	58
4	34	45	13	40	42
5	45	45	14	43	58
6	40	42	15	37	41
7	42	46	16	37	41
8	43	58	17	44	57
9	45	43	18	42	42

a. State the null and alternative hypotheses.
b. Use a statistical software package to find the signed rank test statistic.
c. Give the rejection region for the test, using $\alpha = .01$.
d. State the conclusion in practical terms. Report the *p*-value of the test.

Applying the Concepts—Intermediate

7.89 **Ethical sensitivity of teachers towards racial intolerance.** Refer to the *Journal of Moral Education* (March 2010) study of the effectiveness of a program to encourage teachers to embrace racial tolerance, Exercise 7.44 (p. 378). Recall that the level of racial tolerance was measured for each teacher before (pretest) and after (posttest) the teachers participated in an all-day workshop on cultural competence. The original sample included 238 high school teachers. The table below lists the pretest and posttest scores for a smaller sample of 10 high school teachers. These data are saved in the **TOLERANCE** file. The researchers conducted a paired-difference test to gauge the effectiveness of the program. Use the smaller sample to conduct the appropriate nonparametric test at $\alpha = .01$. What do you conclude?

Teacher	Pretest	Posttest
1	53	74
2	73	80
3	70	94
4	72	78
5	77	78
6	81	84
7	73	71
8	87	88
9	61	63
10	76	83

7.90 **Sea turtles and beach nourishment.** According to the National Oceanic and Atmospheric Administration's Office of Protected Species, sea turtle nesting rates have declined in all parts of the southeastern United States over the past 10 years. Environmentalists theorize that beach nourishment may improve the nesting rates of these turtles. (Beach nourishment involves replacing the sand on the beach in order to extend the high-water line seaward.) A study was undertaken to investigate the effect of beach nourishment on sea turtle nesting rates in Florida (Aubry Hershorin, unpublished doctoral dissertation, University of Florida, 2010). For one part of the study, eight beach zones were sampled in Jacksonville, Florida. Each beach zone was nourished by the Florida Fish and Wildlife Conservation Commission between 2000 and 2008. Nesting densitites (measured as nests per linear meter) were recorded both before and after nourishing at each of the eight beach zones. The data (saved in the **NESTDEN** file) are listed in the following table. Conduct a Wilcoxon signed rank test to compare the sea turtle nesting densities before and after beach nourishing. Use $\alpha = .05$.

Beach Zone	Before Nourishing	After Nourishing
401	0	0.003595
402	0.001456	0.007278
403	0	0.003297
404	0.002868	0.003824
405	0	0.002198
406	0	0.000898
407	0.000626	0
408	0	0

7.91 **Concrete pavement response to temperature.** Civil engineers at West Virginia University have developed a three-dimensional model to predict the response of jointed concrete pavement to temperature variations (*The International Journal of Pavement Engineering*, Sept. 2004). To validate the model, its predictions were compared with field measurements of key concrete stress variables taken at a newly constructed highway. One variable measured was slab top transverse strain (i.e., change in length per unit length per unit time) at a distance of 1 meter from the longitudinal joint. The 5-hour changes (8:20 P.M. to 1:20 A.M.) in slab top transverse strain for six days are listed in the accompanying table and saved in the **SLABSTRAIN** file. Analyze the data, using a nonparametric test. Is there a shift in the change in transverse strain distributions between field measurements and the model? Test, using $\alpha = .05$.

Day	Change in Temperature (°C)	Change in Transverse Strain	
		Field Measurement	3D Model
Oct. 24	−6.3	−58	−52
Dec. 3	13.2	69	59
Dec. 15	3.3	35	32
Feb. 2	−14.8	−32	−24
Mar. 25	1.7	−40	−39
May. 24	−.2	−83	−71

Based on Shoukry, S., William, G., and Riad, M. "Validation of 3DFE model of jointed concrete pavement response to temperature variations." *International Journal of Pavement Engineering*, Vol. 5, No. 3, Sept. 2004 (Table IV).

7.92 **Neurological impairment of POWs.** Eleven prisoners of war during the war in Croatia were evaluated for neurological impairment after their release from a Serbian detention camp (*Collegium Antropologicum*, June 1997). All 11 experienced blows to the head and neck and/or loss of consciousness during imprisonment. Neurological impairment was assessed by measuring the amplitude of the visual evoked potential (VEP) in both eyes at two points in time: 157 days and 379 days after their release. (The higher the VEP value, the greater the neurological impairment.) The data on the 11 POWs are shown in the accompanying table and saved in the **POWVEP** file. Determine whether the VEP measurements of POWs 379 days after their release tend to be greater than the VEP measurements of POWs 157 days after their release. Test, using $\alpha = .05$.

POW	157 Days after Release	379 Days after Release
1	2.46	3.73
2	4.11	5.46
3	3.93	7.04
4	4.51	4.73
5	4.96	4.71
6	4.42	6.19
7	1.02	1.42
8	4.30	8.70
9	7.56	7.37
10	7.07	8.46
11	8.00	7.16

Based on Vrca, A., et al. "The use of visual evoked potentials to follow-up prisoners of war after release from detention camps." *Collegium Antropologicum,* Vol. 21, No. 1, June 1997, p. 232. (Data simulated from information provided in Table 3.)

7.93 **Treatment for tendon pain.** In a *British Journal of Sports Medicine* (Feb. 1, 2004) study of chronic Achilles tendon pain, each in a sample of 25 patients with chronic Achilles tendinosis was treated with heavy-load eccentric calf muscle training. Tendon thickness (in millimeters) was measured both before and following the treatment of each patient. The experimental data are reproduced in the next table and saved in the **TENDON** file. Use a nonparametric test to determine whether the treatment for tendonitis tends to reduce the thickness of tendons. Test using $\alpha = .10$.

Patient	Before Thickness (millimeters)	After Thickness (millimeters)
1	11.0	11.5
2	4.0	6.4
3	6.3	6.1
4	12.0	10.0
5	18.2	14.7
6	9.2	7.3
7	7.5	6.1
8	7.1	6.4
9	7.2	5.7
10	6.7	6.5
11	14.2	13.2

Patient	Before Thickness (millimeters)	After Thickness (millimeters)
12	7.3	7.5
13	9.7	7.4
14	9.5	7.2
15	5.6	6.3
16	8.7	6.0
17	6.7	7.3
18	10.2	7.0
19	6.6	5.3
20	11.2	9.0
21	8.6	6.6
22	6.1	6.3
23	10.3	7.2
24	7.0	7.2
25	12.0	8.0

Based on Ohberg, L., et al. "Eccentric training in patients with chronic Achilles tendinosis: Normalized tendon structure and decreased thickness at follow up." *British Journal of Sports Medicine*, Vol. 38, No. 1, Feb. 1, 2004 (Table 2).

Applying the Concepts—Advanced

7.94 **Bowler's hot hand.** Is the probability of a bowler rolling a strike higher after he has thrown four consecutive strikes? An investigation into the phenomenon of a "hot hand" in bowling was published in *The American Statistician* (Feb. 2004). Frame-by-frame results were collected on 43 professional bowlers from the 2002–2003 Professional Bowlers Association (PBA) season. For each bowler, the researchers calculated the proportion of strikes rolled after bowling four consecutive strikes and the proportion after bowling four consecutive nonstrikes. The data on 4 of the 43 bowlers, saved in the **HOTBOWLER** file, are shown in the following table.

Bowler	Proportion of Strikes	
	After Four Strikes	After Four Nonstrikes
Paul Fleming	.683	.432
Bryon Smith	.684	.400
Mike DeVaney	.632	.421
Dave D'Entremont	.610	.529

Source: Dorsey-Palmateer, R., and Smith, G. "Bowlers' hot hands." *American Statistician,* Vol. 58, No. 1, Feb. 2004 (Table 3). Reprinted with permission from *The American Statistician*. Copyright 2004 by the American Statistical Association. All rights reserved.

a. Do the data on the sample of four bowlers provide support for the "hot hand" theory in bowling? Explain.

b. When the data on all 43 bowlers are used, the *p*-value for the hypothesis test is approximately 0. Interpret this result.

7.7 Comparing Three or More Population Means: Analysis of Variance (Optional)

Suppose we are interested in comparing the means of three or more populations. For example, we may want to compare the mean SAT scores of seniors at three different high schools. Or, we could compare the mean income per house-hold of residents in four census districts. Since the methods of Sections 7.2–7.6 apply to two populations only, we require an alternative technique. In this optional section, we discuss a method for comparing two or more populations based on independent random samples, called an **analysis of variance (ANOVA).**

BIOGRAPHY SIR RONALD A. FISHER (1890–1962)—THE FOUNDER OF MODERN STATISTICS

At a young age, Ronald Fisher demonstrated special abilities in mathematics, astronomy, and biology. (Fisher's biology teacher once divided all his students for "sheer brilliance" into two groups—Fisher and the rest.) Fisher graduated from prestigious Cambridge University in London in 1912 with a B.A. degree in astronomy, and, after several years teaching mathematics, he found work at the Rothamsted Agricultural Experiment station. There, Fisher began his extraordinary career as a statistician. Many consider Fisher to be the leading founder of modern statistics. His contributions to the field include the notion of unbiased statistics, the development of *p*-values for hypothesis tests, the invention of analysis of variance for designed experiments, the maximum likelihood estimation theory, and the mathematical distributions of several well-known statistics. Fisher's book, *Statistical Methods for Research Workers* (written in 1925), revolutionized applied statistics, demonstrating how to analyze data and interpret the results with very readable and practical examples. In 1935, Fisher wrote *The Design of Experiments*, where he first described his famous experiment on the "lady tasting tea." (Fisher showed, through a designed experiment, that the lady really could determine whether tea poured into milk tastes better than milk poured into tea.) Before his death, Fisher was elected a Fellow of the Royal Statistical Society, was awarded numerous medals, and was knighted by the Queen of England. ■

In the jargon of ANOVA, "treatments" represent the groups or populations of interest. Thus, the primary objective of an analysis of variance is to compare the treatment (or population) means for the quantitative (or dependent) variable of interest. If we denote the true means of the k treatments as $\mu_1, \mu_2, \ldots, \mu_k$, then we will test the null hypothesis that the treatment means are all equal against the alternative that at least two of the treatment means differ:

$$H_0: \mu_1 = \mu_2 = \cdots = \mu_k$$
$$H_a: \text{At least two of the } k \text{ treatment means differ}$$

The μ's might represent the means of *all* female and male high school seniors' SAT scores or the means of *all* households' income in each of the four census districts.

To conduct a statistical test of these hypotheses, we will use the means of the independent random samples selected from the treatment populations in a completely randomized design. That is, we compare the k sample means $\bar{x}_1, \bar{x}_2, \ldots, \bar{x}_k$.

For example, suppose you select independent random samples of five female and five male high school seniors and record their SAT scores. The data are shown in Table 7.11. A MINITAB analysis of the data, shown in Figure 7.20, reveals that the sample mean SAT scores (shaded) are 590 for females and 550 for males. Can we conclude that the population of female high school students scores 40 points higher, on average, than the population of male students?

Table 7.11

SAT Scores for High School Students

Females	Males
530	490
560	520
590	550
620	580
650	610

Data Set: TAB7_11

Figure 7.20

MINITAB descriptive statistics for data in Table 7.11

Descriptive Statistics: Females, Males

Variable	N	Mean	StDev	Variance	Minimum	Maximum
Females	5	590.0	47.4	2250.0	530.0	650.0
Males	5	550.0	47.4	2250.0	490.0	610.0

To answer this question, we must consider the amount of sampling variability among the experimental units (students). The SAT scores in Table 7.11 are depicted in the dot plot shown in Figure 7.21. Note that the difference between the sample means is small relative to the sampling variability of the scores within the treatments, namely, Female

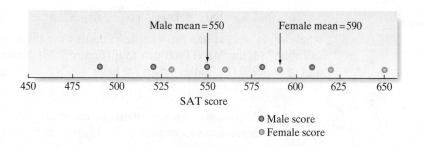

Figure 7.21

Dot plot of SAT scores: difference between means dominated by sampling variability and Male. We would be inclined not to reject the null hypothesis of equal population means in this case.

In contrast, if the data are as depicted in the dot plot of Figure 7.22, then the sampling variability is small relative to the difference between the two means. In this case, we would be inclined to favor the alternative hypothesis that the population means differ.

Figure 7.22

Dot plot of SAT scores: difference between means large relative to sampling variability

Now Work Exercise 7.99a

You can see that the key is to compare the difference between the treatment means with the amount of sampling variability. To conduct a formal statistical test of the hypothesis requires numerical measures of the difference between the treatment means and the sampling variability within each treatment. The variation between the treatment means is measured by the **sum of squares for treatments** (SST), which is calculated by squaring the distance between each treatment mean and the overall mean of *all* sample measurements, then multiplying each squared distance by the number of sample measurements for the treatment, and, finally, adding the results over all treatments. For the data in Table 7.11, the overall mean is 570. Thus, we have:

$$\text{SST} = \sum_{i=1}^{k} n_i(\bar{x}_i - \bar{x})^2 = 5(550 - 570)^2 + 5(590 - 570)^2 = 4{,}000$$

In this equation, we use $\bar{x}$ to represent the overall mean response of all sample measurements—that is, the mean of the combined samples. The symbol n_i is used to denote the sample size for the ith treatment. You can see that the value of SST is 4,000 for the two samples of five female and five male SAT scores depicted in Figures 7.21 and 7.22.

Next, we must measure the sampling variability within the treatments. We call this the **sum of squares for error** (SSE), because it measures the variability around the treatment means that is attributed to sampling error. The value of SSE is computed by summing the squared distance between each response measurement and the corresponding treatment mean and then adding the squared differences over all measurements in the entire sample:

$$\text{SSE} = \sum_{j=1}^{n_1}(x_{1j} - \bar{x}_1)^2 + \sum_{j=1}^{n_2}(x_{2j} - \bar{x}_2)^2 + \ldots \sum_{j=1}^{n_k}(x_{kj} - \bar{x}_k)^2$$

Here, the symbol x_{1j} is the jth measurement in sample 1, x_{2j} is the jth measurement in sample 2, and so on. This rather complex-looking formula can be simplified by recalling the formula for the sample variance s^2 given in Chapter 2:

$$s^2 = \sum_{i=1}^{n} \frac{(x_i - \bar{x})^2}{n - 1}$$

Note that each sum in SSE is simply the numerator of s^2 for that particular treatment. Consequently, we can rewrite SSE as

$$\text{SSE} = (n_1 - 1)s_1^2 + (n_2 - 1)s_2^2 + \cdots + (n_k - 1)s_k^2$$

where $s_1^2, s_2^2, \ldots, s_k^2$ are the sample variances for the k treatments. For the SAT scores in Table 7.11, the MINITAB printout (Figure 7.20) shows that $s_1^2 = 2{,}250$ (for females) and $s_2^2 = 2{,}250$ (for males); then we have

$$\text{SSE} = (5 - 1)(2{,}250) + (5 - 1)(2{,}250) = 18{,}000$$

To make the two measurements of variability comparable, we divide each by the number of degrees of freedom in order to convert the sums of squares to mean squares.

First, the **mean square for treatments** (MST), which measures the variability *among* the treatment means, is equal to

$$MST = \frac{SST}{k - 1} = \frac{4,000}{2 - 1} = 4,000$$

where the number of degrees of freedom for the k treatments is $(k - 1)$. Next, the **mean square for error** (MSE), which measures the sampling variability *within* the treatments, is

$$MSE = \frac{SSE}{n - k} = \frac{18,000}{10 - 2} = 2,250$$

Finally, we calculate the ratio of MST to MSE—an **F-statistic:**

$$F = \frac{MST}{MSE} = \frac{4,000}{2,250} = 1.78$$

These quantities—MST, MSE, and F—are shown (highlighted) on the MINITAB printout displayed in Figure 7.23.

Values of the F-statistic near 1 indicate that the two sources of variation, between treatment means and within treatments, are approximately equal. In this case, the difference between the treatment means may well be attributable to sampling error, which provides little support for the alternative hypothesis that the population treatment means differ. Values of F well in excess of 1 indicate that the variation among treatment means well exceeds that within means and therefore support the alternative hypothesis that the population treatment means differ.

When does F exceed 1 by enough to reject the null hypothesis that the means are equal? This depends on the sampling distribution of F. An **F-distribution**, which depends on $v_1 = (k - 1)$ numerator degrees of freedom and $v_2 = (n - k)$ denominator degrees of freedom, is shown in Figure 7.24. As you can see, the distribution is skewed to the right since $F = MST/MSE$ cannot be less than 0 and can increase without bound.

The rejection region for the F-test is taken from a table (see Tables VIII through XI of Appendix A) with $v_1 = (k - 1)$ degrees of freedom in the numerator and $v_2 = (n - k)$ degrees of freedom in the denominator and corresponding to a Type I error probability of α. For the example of the SAT scores, the F-statistic has $v_1 = (2 - 1)$ numerator degrees of freedom and $v_2 = (10 - 2) = 8$ denominator degrees of freedom. Thus, for $\alpha = .05$, we find (from Table IX of Appendix A) that

$$F_{.05} = 5.32$$

The implication is that MST would have to be 5.32 times greater than MSE before we could conclude, at the .05 level of significance, that the two population treatment means differ. Since the data yielded $F = 1.78$, our initial impressions of the dot plot in Figure 7.21 are confirmed: There is insufficient information to conclude that the mean SAT scores differ for the populations of female and male high school seniors. The rejection region and the calculated F value are shown in Figure 7.24.

In contrast, consider the dot plot in Figure 7.22. The SAT scores depicted in this dot plot are listed in Table 7.12, followed by MINITAB descriptive statistics in Figure 7.25. Note that the sample means for females and males, 590 and 550, respectively, are the

One-way ANOVA: Females, Males

Source	DF	SS	MS	F	P
Factor	1	4000	4000	1.78	0.219
Error	8	18000	2250		
Total	9	22000			

Figure 7.23

MINITAB printout with ANOVA results for data in Table 7.11

Table 7.12

SAT Scores for High School Students Shown in Figure 7.22

Females	Males
580	540
585	545
590	550
595	555
600	560

Data Set: TAB7_12

Figure 7.24

Rejection region and calculated F-values for SAT score samples same as in the previous example. Consequently, the variation between the means is the same, namely, MST = 4,000. However, the variation within the two treatments appears to be considerably smaller. In fact, Figure 7.25 shows that $s_1^2 = 62.5$ and $s_2^2 = 62.5$. Thus, the variation within the treatments is measured by

$$\text{SSE} = (5 - 1)(62.5) + (5 - 1)(62.5) = 500$$

$$\text{MSE} = \frac{\text{SSE}}{n - k} = \frac{500}{8} = 62.5 \ \text{(shaded on Figure 7.25)}$$

Descriptive Statistics: Females, Males

```
Variable  N    Mean   StDev  Variance  Minimum  Maximum
Females   5  590.00    7.91    62.50    580.00   600.00
Males     5  550.00    7.91    62.50    540.00   560.00
```

One-way ANOVA: Females, Males

```
Source  DF     SS      MS      F      P
Factor   1  4000.0  4000.0  64.00  0.000
Error    8   500.0    62.5
Total    9  4500.0
```

Figure 7.25

MINITAB descriptive statistics and ANOVA results for data in Table 7.12

Then the *F*-ratio is

$$F = \frac{\text{MST}}{\text{MSE}} = \frac{4,000}{62.5} = 64.0 \ \text{(shaded on Figure 7.25)}$$

Again, our visual analysis of the dot plot is confirmed statistically: $F = 64.0$ well exceeds the table's *F* value, 5.32, corresponding to the .05 level of significance. We would therefore reject the null hypothesis at that level and conclude that the SAT mean score of males differs from that of females.

Now Work Exercise 7.99b–h

The **analysis of variance *F*-test** for comparing treatment means is summarized in the following box:

ANOVA *F*-Test to Compare *k* Treatment Means: Completely Randomized Design

$H_0: \mu_1 = \mu_2 = \cdots = \mu_k$

H_a: At least two treatment means differ.

Test statistic: $F = \dfrac{\text{MST}}{\text{MSE}}$

Rejection region: $F > F_\alpha$, where F_α is based on $v_1 = (k - 1)$ numerator degrees of freedom (associated with MST) and $v_2 = (n - k)$ denominator degrees of freedom (associated with MSE), or, $\alpha > p$-value

---

Conditions Required for a Valid ANOVA *F*-Test: Completely Randomized Design

1. The samples are randomly selected in an independent manner from the *k* treatment populations. (This can be accomplished by randomly assigning the experimental units to the treatments.)

2. All *k* sampled populations have distributions that are approximately normal.

3. The *k* population variances are equal (i.e., $\sigma_1^2 = \sigma_2^2 = \sigma_3^2 = \cdots = \sigma_k^2$).

Computational formulas for MST and MSE are given in Appendix B. We will rely on statistical software to compute the F-statistic, concentrating on the interpretation of the results rather than their calculation.

Example 7.10

Conducting an Anova F-Test— Comparing Golf Ball Brands

Problem Suppose the USGA wants to compare the mean distances reached of four different brands of golf balls struck with a driver. An independent samples design is employed, with Iron Byron, the USGA's robotic golfer, using a driver to hit a random sample of 10 balls of each brand in a random sequence. The distance is recorded for each hit, and the results are shown in Table 7.13, organized by brand.

a. Set up the test to compare the mean distances for the four brands. Use $\alpha = .10$.

b. Use statistical software to obtain the test statistic and p-value. Give the appropriate conclusion.

Table 7.13	Results of Independent Samples Design: Iron Byron Driver			
	Brand A	Brand B	Brand C	Brand D
	251.2	263.2	269.7	251.6
	245.1	262.9	263.2	248.6
	248.0	265.0	277.5	249.4
	251.1	254.5	267.4	242.0
	260.5	264.3	270.5	246.5
	250.0	257.0	265.5	251.3
	253.9	262.8	270.7	261.8
	244.6	264.4	272.9	249.0
	254.6	260.6	275.6	247.1
	248.8	255.9	266.5	245.9
Sample means	250.8	261.1	270.0	249.3

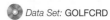 *Data Set: GOLFCRD*

Solution

a. To compare the mean distances of the $k = 4$ brands, we first specify the hypotheses to be tested. Denoting the population mean of the ith brand by μ_i, we test

$$H_0: \mu_1 = \mu_2 = \mu_3 = \mu_4$$
$$H_a: \text{The mean distances differ for at least two of the brands.}$$

The test statistic compares the variation among the four treatment (Brand) means with the sampling variability within each of the treatments:

$$\textit{Test statistic:} \quad F = \frac{\text{MST}}{\text{MSE}}$$

$$\textit{Rejection region:} \quad F > F_\alpha = F_{.10} \text{ with } v_1 = (k - 1) = 3 \text{ df}$$
$$\text{and } v_2 = (n - k) = 36 \text{ df}$$

From Table VIII of Appendix A, we find that $F_{.10} \approx 2.25$ for 3 and 36 df. Thus, we will reject H_0 if $F > 2.25$. (See Figure 7.26.)

The assumptions necessary to ensure the validity of the test are as follows:

1. The samples of 10 golf balls for each brand are selected randomly and independently.

2. The probability distributions of the distances for each brand are normal.

3. The variances of the distance probability distributions for each brand are equal.

Figure 7.26

F-test for completely randomized design: golf ball experiment

b. The MINITAB printout for the data in Table 7.13 resulting from this completely randomized design is given in Figure 7.27. The values, MST = 931.5, MSE = 21.2, and F = 43.99, are highlighted on the printout. Since $F > 2.25$, we reject H_0. Also, the p-value of the test (.000) is highlighted on the printout. Since $\alpha = .10$ exceeds the p-value, we draw the same conclusion: Reject H_0. Therefore, at the .10 level of significance, we conclude that at least two of the brands differ with respect to mean distance traveled when struck by the driver.

Figure 7.27

MINITAB ANOVA for completely randomized design

One-way ANOVA: DISTANCE versus BRAND

```
Source  DF    SS      MS      F     P
BRAND    3  2794.4  931.5  43.99  0.000
Error   36   762.3   21.2
Total   39  3556.7
```

Look Ahead Now that we know that mean distances differ, a logical follow-up question is: "Which ball brand travels farther, on average, when hit with a driver?" We discuss a method for ranking treatment means in an ANOVA later in this section.

Now Work Exercise 7.106abc

The results of an **analysis of variance (ANOVA)** can be summarized in a simple tabular format similar to that obtained from the MINITAB program in Example 7.10. The general form of the table is shown in Table 7.14, where the symbols df, SS, and MS stand for degrees of freedom, sum of squares, and mean square, respectively. Note that the two sources of variation, Treatments and Error, add to the total sum of squares, SS(Total). The ANOVA summary table for Example 7.10 is given in Table 7.15, and the partitioning of the total sum of squares into its two components is illustrated in Figure 7.28.

Table 7.14	General ANOVA Summary Table for a Completely Randomized Design			
Source	df	SS	MS	F
Treatments	$k - 1$	SST	$\text{MST} = \dfrac{\text{SST}}{k - 1}$	$\dfrac{\text{MST}}{\text{MSE}}$
Error	$n - k$	SSE	$\text{MSE} = \dfrac{\text{SSE}}{n - k}$	
Total	$n - 1$	SS(Total)		

Table 7.15	ANOVA Summary Table for Example 7.10				
Source	df	SS	MS	F	p-Value
Brands	3	2,794.39	931.46	43.99	.000
Error	36	762.30	21.18		
Total	39	3,556.69			

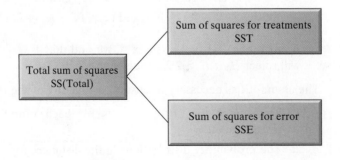

Figure 7.28

Partitioning of the total sum of squares for the completely randomized design

Example 7.11
Checking the
ANOVA Assumptions

Problem Refer to the completely randomized ANOVA design conducted in Example 7.10. Are the assumptions required for the test approximately satisfied?

Solution The assumptions for the test are repeated as follows:

1. The samples of golf balls for each brand are selected randomly and independently.
2. The probability distributions of the distances for each brand are normal.
3. The variances of the distance probability distributions for each brand are equal.

Since the sample consisted of 10 randomly selected balls of each brand, and since the robotic golfer Iron Byron was used to drive all the balls, the first assumption of independent random samples is satisfied. To check the next two assumptions, we will employ two graphical methods presented in Chapter 2: histograms and box plots. A MINITAB histogram of driving distances for each brand of golf ball is shown in Figure 7.29, and SAS box plots are shown in Figure 7.30.

Figure 7.29
MINITAB histograms for golf ball driving distances

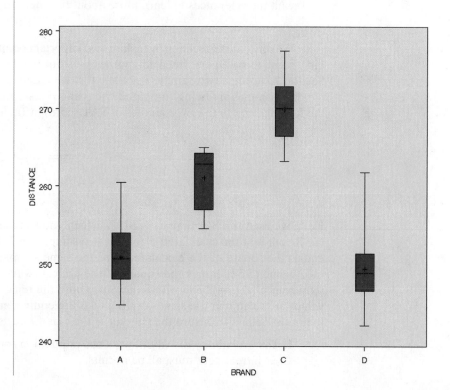

Figure 7.30
SAS box plots for golf ball distances

The normality assumption can be checked by examining the histograms in Figure 7.29. With only 10 sample measurements for each brand, however, the displays are not very informative. More data would need to be collected for each brand before we could assess whether the distances come from normal distributions. Fortunately, analysis of variance has been shown to be a very **robust method** when the assumption of normality is not satisfied exactly. That is, *moderate departures from normality do not have much effect on the significance level of the ANOVA F-test or on confidence coefficients.* Rather than spend the time, energy, or money to collect additional data for this experiment in order to verify the normality assumption, we will rely on the robustness of the ANOVA methodology.

Box plots are a convenient way to obtain a rough check on the assumption of equal variances. With the exception of a possible outlier for Brand D, the box plots in Figure 7.30 show that the spread of the distance measurements is about the same for each brand. Since the sample variances appear to be the same, the assumption of equal population variances for the brands is probably satisfied. Although robust with respect to the normality assumption, ANOVA is *not robust* with respect to the equal-variances assumption. Departures from the assumption of equal population variances can affect the associated measures of reliability (e.g., *p*-values and confidence levels). Fortunately, the effect is slight when the sample sizes are equal, as in this experiment.

Now Work Exercise 7.106d

Although graphs can be used to check the ANOVA assumptions as in Example 7.11, no measures of reliability can be attached to these graphs. When you have a plot that is unclear as to whether or not an assumption is satisfied, you can use formal statistical tests that are beyond the scope of this text. Consult the references at the end of the chapter for information on these tests. When the validity of the ANOVA assumptions is in doubt, nonparametric statistical methods are useful.

> **What Do You Do When the Assumptions Are Not Satisfied for the Analysis of Variance for a Completely Randomized Design?**
>
> *Answer:* Use a nonparametric statistical method such as the Kruskal–Wallis *H*-Test. Consult the references to learn more about this method.

We conclude this section by making two important points about an analysis of variance. First, recall that we performed a hypothesis test for the difference between two means in Section 7.2 using a two-sample *t*-statistic for two independent samples. When two independent samples are being compared, the *t*- and *F*-tests are equivalent. To see this, apply the formula for *t* to the two samples of SAT scores in Table 7.12.

$$t = \frac{\bar{x}_1 - \bar{x}_2}{\sqrt{s_p^2\left(\dfrac{1}{n_1} + \dfrac{1}{n_2}\right)}} = \frac{590 - 550}{\sqrt{(62.5)\left(\dfrac{1}{5} + \dfrac{1}{5}\right)}} = \frac{40}{5} = 8$$

Here, we used the fact that $s_p^2 = \text{MSE}$, which you can verify by comparing the formulas. Recall that the calculated *F* for the two samples in Table 7.12 is $F = 64$. This value equals the square of the calculated *t* for the same samples ($t = 8$). Likewise, the critical *F*-value (5.32) equals the square of the critical *t*-value at the two-sided .05 level of significance ($t_{.025} = 2.306$ with 8 df). Since both the rejection region and the calculated values are related in the same way, the tests are equivalent. Moreover, the assumptions that must be met to ensure the validity of the *t*- and *F*-tests are the same:

1. The probability distributions of the populations of responses associated with each treatment must all be normal.

2. The probability distributions of the populations of responses associated with each treatment must have equal variances.

3. The samples of experimental units selected for the treatments must be random and independent.

In fact, the only real difference between the tests is that the *F*-test can be used to compare *more than two* treatment means, whereas the *t*-test is applicable to two samples only.

For our second point, refer to Example 7.10. Our conclusion that at least two of the brands of golf balls have different mean distances traveled when struck with a driver leads naturally to the following questions: Which of the brands differ? and How are the brands ranked with respect to mean distance?

One way to obtain this information is to construct a confidence interval for the difference between the means of any pair of treatments, using the method of Section 7.2. For example, if a 95% confidence interval for $\mu_A - \mu_C$ in Example 7.10 is found to be $(-24, -13)$, we are confident that the mean distance for Brand C exceeds the mean for Brand A (since all differences in the interval are negative). Constructing these confidence intervals for all possible pairs of brands allows you to rank the brand means. A method for conducting these *multiple comparisons*—one that controls for Type I errors—is beyond the scope of this introductory text. Consult the references to learn more about this methodology.

Statistics in Action | **Revisited**   A One–Way Analysis of the Cockroach Data

We return to the experiment designed to investigate the trail-following ability of German cockroaches (p. 351). Recall that an entomologist created a chemical trail with either a methanol extract from roach feces or just methanol (the control). Cockroaches were then released into a container at the beginning of the trail, one at a time, and a video surveillance camera was used to monitor the roaches' movements. The movement pattern of each cockroach was measured by its average deviation (in pixels) from the extract trail and the data stored in the **ROACH** file.

For this application, consider only the cockroaches assigned to the fecal extract trail. Four roach groups were utilized in the experiment—adult males, adult females, gravid females, and nymphs—with 20 roaches of each type independently and randomly selected. Is there sufficient evidence to say that the ability to follow the extract trail differs among cockroaches of different age, sex, and reproductive status? In other words, is there evidence to suggest that the mean trail deviation μ differs for the four roach groups?

To answer this question, we conduct an analysis of variance on the **ROACH** data. The dependent (response) variable of interest is deviation from the extract trail, while the treatments are the four different roach groups. Thus, we want to test the null hypothesis:

$$H_0: \mu_{\text{Male}} = \mu_{\text{Female}} = \mu_{\text{Gravid}} = \mu_{\text{Nymph}}$$

A MINITAB printout of the ANOVA is displayed in Figure SIA7.4. The *p*-value of the test (highlighted on the printout) is 0. Since this value is less than, say, $\alpha = .05$, we reject the null hypothesis and conclude (at the .05 level of significance) that the mean deviation from the extract trail differs among the populations of adult male, adult female, gravid, and nymph cockroaches.

The sample means for the four cockroach groups are also highlighted in Figure SIA7.4. Note that adult males have the smallest sample mean deviation (7.38) while gravids have the largest sample mean deviation (44.03). A multiple comparisons of means procedure shows that the gravid mean is statistically larger than the means for the other three groups.

Data Set: ROACH

One-way ANOVA: DEVIATE versus GROUP

```
Source   DF     SS    MS      F      P
GROUP     3  14164  4721  11.61  0.000
Error    76  30918   407
Total    79  45083

S = 20.17    R-Sq = 31.42%    R-Sq(adj) = 28.71%


                        Individual 95% CIs For Mean Based on
                        Pooled StDev
Level    N   Mean  StDev  -+---------+---------+---------+--------
Female  20  21.07  26.13                  (-----*-----)
Gravid  20  44.03  24.84                            (-----*-----)
Male    20   7.38   8.61  (-----*-----)
Nymph   20  18.73  15.92            (-----*-----)
                         -+---------+---------+---------+--------
                          0        15        30        45

Pooled StDev = 20.17
```

Figure SIA7.4
MINITAB ANOVA for deviation from extract trail

Exercises 7.95–7.112

Understanding the Principles

7.95 Explain how to collect the data for a one-way ANOVA.

7.96 What conditions are required for a valid ANOVA F-test?

7.97 *True or False.* The ANOVA method is robust when the assumption of normality is not exactly satisfied.

Learning the Mechanics

7.98 Use Tables VIII, IX, X, and XI of Appendix A to find each of the following F values:
  a. $F_{.05}, v_1 = 3, v_2 = 4$
  b. $F_{.01}, v_1 = 3, v_2 = 4$
  c. $F_{.10}, v_1 = 20, v_2 = 40$
  d. $F_{.025}, v_1 = 12, v_2 = 9$

7.99 Consider dot plots A and B (shown below). Assume that
[NW] the two samples represent independent random samples corresponding to two treatments in a completely randomized design.

a. In which dot plot is the difference between the sample means small relative to the variability within the sample observations? Justify your answer.

b. Calculate the treatment means (i.e., the means of samples 1 and 2) for both dot plots.

c. Use the means to calculate the sum of squares for treatments (SST) for each dot plot.

d. Calculate the sample variance for each sample and use these values to obtain the sum of squares for error (SSE) for each dot plot.

e. Calculate the total sum of squares [SS(Total)] for the two dot plots by adding the sums of squares for treatment and error. What percentage of SS(Total) is accounted for by the treatments—that is, what percentage of the total sum of squares is the sum of squares for treatment—in each case?

f. Convert the sums of squares for treatment and error to mean squares by dividing each by the appropriate number of degrees of freedom. Calculate the F-ratio of the mean square for treatment (MST) to the mean square for error (MSE) for each dot plot.

Plot A:

Plot B:

● Treatment (Sample) 1
● Treatment (Sample) 2

g. Use the F-ratios to test the null hypothesis that the two samples are drawn from populations with equal means. Take $\alpha = .05$.

h. What assumptions must be made about the probability distributions corresponding to the responses for each treatment in order to ensure the validity of the F-tests conducted in part **g**?

i. Conduct a two-sample t-test (Section 7.2) of the null hypothesis that the two treatment means are equal for each dot plot. Use $\alpha = .05$ and two-tailed tests. Verify that the F-test and t-test results are equivalent.

j. Complete the following ANOVA table for each of the two dot plots:

Source	df	SS	MS	F
Treatments				
Error				
Total				

7.100 The data in the following table (saved in the **LM7_100** file) resulted from an experiment that utilized a completely randomized design:

Treatment 1	Treatment 2	Treatment 3
3.9	5.4	1.3
1.4	2.0	.7
4.1	4.8	2.2
5.5	3.8	
2.3	3.5	

a. Use statistical software (or the formulas in Appendix B) to complete the following ANOVA table:

Source	df	SS	MS	F
Treatments				
Error				
Total				

b. Test the null hypothesis that $\mu_1 = \mu_2 = \mu_3$, where μ_i represents the true mean for treatment i, against the alternative that at least two of the means differ. Use $\alpha = .01$.

Applying the Concepts—Basic

7.101 Treating cancer with yoga. According to a study funded by the National Institutes of Health, yoga classes can help cancer survivors sleep better. The study results were presented at the June 2010 American Society of Clinical Oncology's annual meeting. Researchers randomly assigned 410 cancer patients (who had finished cancer therapy) to receive either their usual follow-up care or attend a 75-minute yoga class twice per week. After four weeks, the researchers measured the level of fatigue and sleepiness experienced by each cancer survivor. Those who took yoga were less fatigued than those who did not.

a. Assume the patients are numbered 1 through 410. Use the random number generator of a statistical software package to randomly assign each patient to either receive the usual follow-up care or to attend yoga classes. Assign 205 patients to each treatment.

b. Consider the following treatment assignment scheme. The patients are ranked according to severity of cancer and the most severe patients are assigned to the yoga class while the others are assigned to receive their usual follow-up care. Comment on the validity of the results obtained from such an assignment.

7.102 Whales entangled in fishing gear. Entanglement of marine mammals (e.g., whales) in fishing gear is considered a significant threat to the species. A study published in *Marine Mammal Science* (April 2010) investigated the type of net most likely to entangle a certain species of whale inhabiting the East Sea of Korea. A sample of 207 entanglements of whales in the area formed the data for the study. These entanglements were caused by one of three types of fishing gear: set nets, pots, and gill nets. One of the variables investigated was body length (in meters) of the entangled whale.

a. Set up the null and alternative hypotheses for determining whether the average body length of entangled whales differs for the three types of fishing gear.

b. An ANOVA F-test yielded the following results: $F = 34.81, p - \text{value} < .0001$. Interpret the results for $\alpha = .05$.

7.103 College tennis recruiting with a team Web site. Most university athletic programs have a Web site with information on individual sports and a Prospective Student Athlete Form that allows high school athletes to submit information about their academic and sports achievements directly to the college coach. *The Sport Journal* (Winter 2004) published a study of how important team Web sites are to the recruitment of college tennis players. A survey was conducted of National Collegiate Athletic Association (NCAA) tennis coaches, of which 53 were from Division I schools, 20 were from Division II schools, and 53 were from Division III schools. Coaches were asked to respond to a series of statements, including "The Prospective Student Athlete Form on the Web site contributes very little to the recruiting process." Responses were measured on a seven-point scale (where 1 = strongly disagree and 7 = strongly agree). In order to compare the mean responses of tennis coaches from the three NCAA divisions, the data were analyzed with an independent samples ANOVA design.

a. Identify the experimental unit, the dependent (response) variable, and the treatments in this study.

b. Give the null and alternative hypothesis for the ANOVA F-test.

c. The observed significance level of the test was found to be *p-value* < .003. What conclusion can you draw if you want to test at $\alpha = .05$?

7.104 A new dental bonding agent. Refer to the *Trends in Biomaterials & Artificial Organs* (Jan. 2003) study of a new bonding adhesive for teeth, presented in Exercise 6.72 (p. 323). Recall that the new adhesive (called "Smartbond") has been developed to eliminate the necessity of a dry field. In one portion of the study, 30 extracted teeth were bonded with Smartbond and each was randomly assigned one of three different bonding times: 1 hour, 24 hours, or 48 hours. At the end of the bonding period, the breaking strength (in Mpa) of each tooth was determined. The data were analyzed with the use of analysis of variance in order to determine whether the true mean breaking strength of the new adhesive varies with the bonding time.

a. Identify the experimental units, treatments, and response variable for this design.

b. Set up the null and alternative hypotheses for the ANOVA.

c. Find the rejection region using $\alpha = .01$.

d. The test results were $F = 61.62$ and p-value ≈ 0. Give the appropriate conclusion for the test.

e. What conditions are required for the test results to be valid?

7.105 Robots trained to behave like ants. Robotics researchers investigated whether robots could be trained to behave like ants in an ant colony (*Nature*, Aug. 2000). Robots were trained and randomly assigned to "colonies" (i.e., groups) consisting of 3, 6, 9, or 12 robots. The robots were assigned the tasks of foraging for "food" and recruiting another robot when they identified a resource-rich area. One goal of the experiment was to compare the mean energy expended (per robot) of the four different sizes of colonies.

a. What type of experimental design was employed?

b. Identify the treatments and the dependent variable.

c. Set up the null and alternative hypotheses of the test.

d. The following ANOVA results were reported: $F = 7.70$, numerator df $= 3$, denominator df $= 56$, p-value $< .001$. Conduct the test at a significance level of $\alpha = .05$ and interpret the result.

7.106 Most powerful business women in America. Refer to *Fortune* (Oct. 16, 2008) magazine's study of the 50 most powerful women in business in America, Exercise 2.60 (p. 57). Recall that data on age (in years) and title of each of these 50 women are stored in the **WPOWER50** file. (Some of the data set is listed in the table below.) Suppose you want to compare the average ages of the most powerful American women in three groups based on their position (title) within the firm: Group 1 (CEO, CEO/Chairman, or CEO/President), Group 2 (Chairman, President, COO, or COO/President), and Group 3 (Director, EVP, EVP/COO, EVP/President, Executive, SVP, or VP).

a. Give the null and alternative hypotheses to be tested.

b. An SPSS analysis-of-variance printout for the test you stated in part **a** is shown in the next column. The sample means for the 3 groups appear at the bottom of the printout. Why is it insufficient to make a decision about the null hypothesis based solely on these sample means?

c. Locate the test statistic and p-value on the printout. Use this information to make the appropriate conclusion at $\alpha = .10$.

ANOVA

AGE

	Sum of Squares	df	Mean Square	F	Sig.
Between Groups	215.541	2	107.770	2.784	.072
Within Groups	1819.439	47	38.711		
Total	2034.980	49			

AGE

GROUP	Mean	N	Std. Deviation
1	52.50	18	5.044
2	49.58	19	5.357
3	47.23	13	8.516
Total	50.02	50	6.444

d. Use the data in the **WPOWER50** file to determine whether the ANOVA assumptions are reasonably satisfied.

Applying the Concepts—Intermediate

7.107 Study of recall of TV commercials. Do TV shows with violence and sex impair memory for commercials? To answer this question, Iowa St. researchers conducted a designed experiment in which 324 adults were randomly assigned to one of three viewer groups of 108 participants each (*Journal of Applied Psychology*, June 2002). One group watched a TV program with a violent content code (V) rating, the second group viewed a show with a sex content code (S) rating, and the last group watched a neutral TV program with neither a V nor an S rating. Nine commercials were embedded into each TV show. After viewing the program, each participant was scored on his or her recall of the brand names in the commercial messages, with scores ranging from 0 (no brands recalled) to 9 (all brands recalled). The data (simulated from information provided in the article) are saved in the **TVADRECALL** file. The researchers compared the mean recall scores of the three viewing groups with an analysis of variance.

a. Identify the experimental units in the study.

b. Identify the dependent (response) variable in the study.

c. Identify the factor and treatments in the study.

d. The sample mean recall scores for the three groups were $\bar{x}_v = 2.08$, $\bar{x}_s = 1.71$, and $\bar{x}_{Neutral} = 3.17$. Explain why one should not draw an inference about differences in the population mean recall scores on the basis of only these summary statistics.

Rank	Name	Age	Company	Title
1	Indra Nooyi	52	PepsiCo	CEO/Chairman
2	Irene Rosenfeld	55	Kraft Foods	CEO/Chairman
3	Pat Woertz	55	Archer Daniels Midland	CEO/Chairman
4	Anne Mulcahy	55	Xerox	CEO/Chairman
5	Angela Braley	47	Wellpoint	CEO/President
⋮	⋮	⋮	⋮	⋮
49	Cathie Black	64	Hearst Magazines	President
50	Marissa Mayer	33	Google	VP

Based on "50 most powerful women in business." *Fortune*, October 16, 2008.

e. An ANOVA on the data in the **TVADRECALL** file yielded the results shown in the accompanying MINITAB printout. Locate the test statistic and *p*-value on the printout.

f. Interpret the results from part **e**, using $\alpha = 0.01$. What can the researchers conclude about the three groups of TV ad viewers?

One-way ANOVA: VIOLENT, SEX, NEUTRAL

```
Source   DF      SS     MS      F      P
Factor    2  123.27  61.63  20.45  0.000
Error   321  967.35   3.01
Total   323 1090.62

S = 1.736   R-Sq = 11.30%   R-Sq(adj) = 10.75%
```

7.108 Restoring self-control when intoxicated. Does coffee or some other form of stimulation (e.g., an incentive to stop when seeing a flashing red light on a car) really allow a person suffering from alcohol intoxication to "sober up"? Psychologists from the University of Waterloo investigated the matter in *Experimental and Clinical Psychopharmacology* (February 2005). A sample of 44 healthy male college students participated in the experiment. Each student was asked to memorize a list of 40 words (20 words on a green list and 20 words on a red list). The students were then randomly assigned to one of four different treatment groups (11 students in each group). Students in three of the groups were each given two alcoholic beverages to drink prior to performing a word completion task. Students in Group A received only the alcoholic drinks. Participants in Group AC had caffeine powder dissolved in their drinks. Group AR participants received a monetary award for correct responses on the word completion task. Students in Group P (the placebo group) were told that they would receive alcohol, but instead received two drinks containing a carbonated beverage (with a few drops of alcohol on the surface to provide an alcoholic scent). After consuming their drinks and resting for 25 minutes, the students performed the word completion task. Their scores (simulated on the basis of summary information from the article) are reported in the table below and saved in the **DRINKERS** file.

AR	AC	A	P
.51	.50	.16	.58
.58	.30	.10	.12
.52	.47	.20	.62
.47	.36	.29	.43
.61	.39	−.14	.26
.00	.22	.18	.50
.32	.20	−.35	.44
.53	.21	.31	.20
.50	.15	.16	.42
.46	.10	.04	.43
.34	.02	−.25	.40

Based on Grattan-Miscio, K.E., and Vogel-Sprott, M. "Alcohol, intentional control, and inappropriate behavior: Regulation by caffeine or an incentive." *Experimental and Clinical Psychopharmacology*, Vol. 13, No. 1, February 2005 (Table 1).

(*Note*: A task score represents the difference between the proportion of correct responses on the green list of words and the proportion of incorrect responses on the red list of words.)

a. What type of experimental design is employed in this study?

b. Analyze the data for the researchers, using $\alpha = .05$. Are there differences among the mean task scores for the four groups?

c. What assumptions must be met in order to ensure the validity of the inference you made in part **b**?

7.109 Is honey a cough remedy? Pediatric researchers at Pennsylvania State University carried out a designed study to test whether a teaspoon of honey before bed calms a child's cough and published their results in *Archives of Pediatrics and Adolescent Medicine* (Dec. 2007). (This experiment was first described in Exercise 2.32, p. 45). A sample of 105 children who were ill with an upper respiratory tract infection and their parents participated in the study. On the first night, the parents rated their children's cough symptoms on a scale from 0 (no problems at all) to 6 (extremely severe) in five different areas. The total symptoms score (ranging from 0 to 30 points) was the variable of interest for the 105 patients. On the second night, the parents were instructed to give their sick child a dosage of liquid "medicine" prior to bedtime. Unknown to the parents, some were given a dosage of dextromethorphan (DM)—an over-the-counter cough medicine—while others were given a similar dose of honey. Also, a third group of parents (the control group) gave their sick children no dosage at all. Again, the parents rated their children's cough symptoms, and the improvement in total cough symptoms score was determined for each child. The data (improvement scores) for the study are shown in the accompanying table and saved in the **HONEYCOUGH** file. The goal of the researchers was to compare the mean improvement scores for the three treatment groups.

a. Identify the type of experimental design employed. What are the treatments?

b. Conduct an analysis of variance on the data and interpret the results.

Honey Dosage:	12 11 15 11 10 13 10 4 15 16 9 14 10 6 10 8 11 12 12 8 12 9 11 15 10 15 9 13 8 12 10 8 9 5 12
DM Dosage:	4 6 9 4 7 7 7 9 12 10 11 6 3 4 9 12 7 6 8 12 12 4 12 13 7 10 13 9 4 4 10 15 9
No Dosage (Control):	5 8 6 1 0 8 12 8 7 7 1 6 7 7 12 7 9 7 9 5 11 9 5 6 8 8 6 7 10 9 4 8 7 3 1 4 3

Based on Paul, I. M., et al. "Effect of honey, dextromethorphan, and no treatment on nocturnal cough and sleep quality for coughing children and their parents." *Archives of Pediatrics and Adolescent Medicine*, Vol. 161, No. 12, Dec. 2007 (data simulated).

7.110 The "name game." Psychologists at Lancaster University (United Kingdom) evaluated three methods of name retrieval in a controlled setting (*Journal of Experimental Psychology—Applied*, June 2000). A sample of 139 students was randomly divided into three groups, and each group of students used a different method to learn the names of the other students in the group. Group 1 used the "simple name game," in which the first student states his or her full name, the second student announces his or her name and the name of the first student, the third student says his or her name and the names of the first two students, etc. Group 2 used the "elaborate name game," a modification of the simple name game such that the students state not only their names, but also their favorite activity (e.g., sports). Group 3 used "pairwise introductions," according to which students are divided into pairs and each student must introduce the other member of the pair. One year later, all subjects were sent pictures of the students in their group and asked to state the full name of each. The researchers measured the percentage of names recalled by each student respondent. The data (simulated on the basis of summary statistics provided in the research article) are shown in the table and saved in the **NAMEGAME** file. Conduct an analysis of variance to determine whether the mean percentages of names recalled differ for the three name-retrieval methods. Use $\alpha = .05$.

Simple Name Game

24	43	38	65	35	15	44	44	18	27	0	38	50	31
7	46	33	31	0	29	0	0	52	0	29	42	39	26
51	0	42	20	37	51	0	30	43	30	99	39	35	19
24	34	3	60	0	29	40	40						

Elaborate Name Game

39	71	9	86	26	45	0	38	5	53	29	0	62	0
1	35	10	6	33	48	9	26	83	33	12	5	0	0
25	36	39	1	37	2	13	26	7	35	3	8	55	50

Pairwise Introductions

5	21	22	3	32	29	32	0	4	41	0	27	5	9
66	54	1	15	0	26	1	30	2	13	0	2	17	14
5	29	0	45	35	7	11	4	9	23	4	0	8	2
18	0	5	21	14									

Based on Morris, P. E., and Fritz, C. O. "The name game: Using retrieval practice to improve the learning of names." *Journal of Experimental Psychology—Applied*, Vol. 6, No. 2, June 2000 (data simulated from Figure 1).

7.111 Estimating the age of glacial drifts. Refer to the *American Journal of Science* (Jan. 2005) study of the chemical make-up of buried tills (glacial drifts) in Wisconsin, presented in Exercise 2.136 (p. 86). The ratio of the elements aluminum (Al) and beryllium (Be) in sediment is related to the duration of burial. Recall the Al/Be ratios for a sample of 26 buried till specimens were determined and are saved in the **TILLRATIO** file. The till specimens were obtained from five different boreholes (labeled UMRB-1, UMRB-2, UMRB-3, SWRA, and SD). The data are shown in the next table. Conduct an analysis of variance of the data. Is there sufficient evidence to indicate differences among the mean Al/Be ratios for the five boreholes? Test, using $\alpha = .10$.

UMRB-1:	3.75	4.05	3.81	3.23	3.13	3.30	3.21
UMRB-2:	3.32	4.09	3.90	5.06	3.85	3.88	
UMRB-3:	4.06	4.56	3.60	3.27	4.09	3.38	3.37
SWRA:	2.73	2.95	2.25				
SD:	2.73	2.55	3.06				

Based on *American Journal of Science*, Vol. 305, No. 1, Jan. 2005, p. 16 (Table 2).

Applying the Concepts—Advanced

7.112 Animal-assisted therapy for heart patients. Refer to the *American Heart Association Conference* (Nov. 2005) study to gauge whether animal-assisted therapy can improve the physiological responses of heart failure patients, presented in Exercise 2.106 (p. 73). Recall that 76 heart patients were randomly assigned to one of three groups. Each patient in group T was visited by a human volunteer accompanied by a trained dog, each patient in group V was visited by a volunteer only, and the patients in group C were not visited at all. The anxiety level of each patient was measured (in points) both before and after the visits. The accompanying table gives summary statistics for the drop in anxiety level for patients in the three groups. The mean drops in anxiety levels of the three groups of patients were compared with the use of an analysis of variance. Although the ANOVA table was not provided in the article, sufficient information is given to reconstruct it.

	Sample Size	Mean Drop	Std. Dev.
Group T: Volunteer + Trained Dog	26	10.5	7.6
Group V: Volunteer only	25	3.9	7.5
Group C: Control group (no visit)	25	1.4	7.5

Based on Cole, K., et al. "Animal assisted therapy decreases hemodynamics, plasma epinephrine and state anxiety in hospitalized heart failure patients." *American Journal of Critical Care*, 2007, 16: 575–585.

a. Compute SST for the ANOVA, using the formula (on p. 402)

$$SST = \sum_{i=1}^{3} n_i(\bar{x}_i - \bar{x})^2$$

where $\bar{x}$ is the overall mean drop in anxiety level of all 76 subjects. [*Hint:* $\bar{x} = (\Sigma_{i=1}^{3} n_i (\bar{x}_i)/76$.]

b. Recall that SSE for the ANOVA can be written as

$$SSE = (n_1 - 1)s_1^2 + (n_2 - 1)s_2^2 + (n_3 - 1)s_3^2$$

where s_1^2, s_2^2, and s_3^2 are the sample variances associated with the three treatments. Compute SSE for the ANOVA.

c. Use the results from parts **a** and **b** to construct the ANOVA table.

d. Is there sufficient evidence (at $\alpha = .01$) of differences among the mean drops in anxiety levels by the patients in the three groups?

e. Comment on the validity of the ANOVA assumptions. How might this affect the results of the study?

GUIDE TO COMPARING POPULATION MEANS

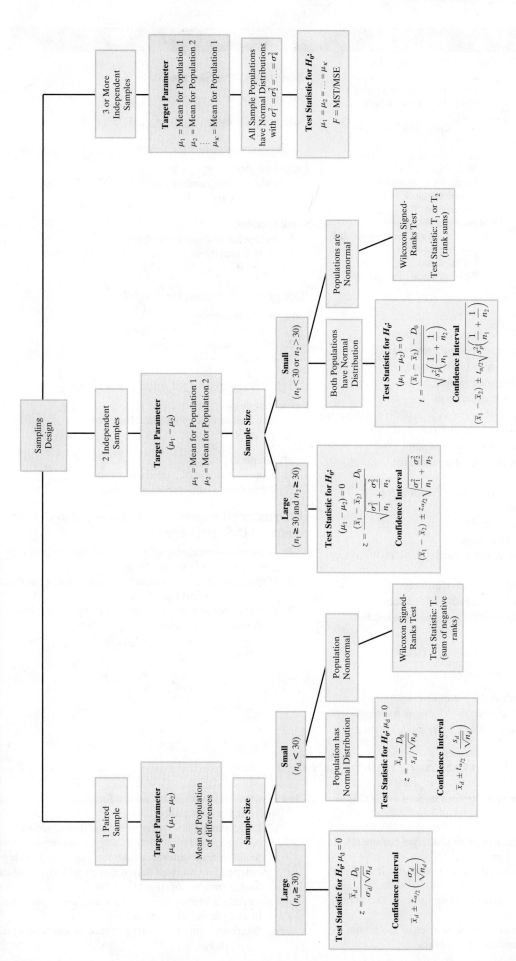

CHAPTER NOTES

Key Terms

Note: Starred (*) terms are from the optional sections in this chapter.

*Analysis of variance (ANOVA) 400
Analysis of variance F-test 404
F-distribution* 403
F-statistic 403
Nonparametric statistical tests 362
*Mean Square for Error (MSE) 403
*Mean Square for treatments (MST) 403
*One-way ANOVA 410

*Sum of Squares for Errors (SSE) 402
Sum of Squares for Treatments (SST) 402
Paired difference experiment 371
Pooled sample estimator σ^2 357
Randomized block experiment 371
Rank sums 384
Standard error of the statistic 354
Robust method 408
*Wilcoxon rank sum test 383
*Wilcoxon signal rank test 393

Key Symbols

$\mu_1 - \mu_2$	Difference between population means
μ_d	Paired difference in population means
D_0	Hypothesized value of difference
$\bar{x}_1 - \bar{x}_2$	Difference between sample means
$\bar{x}_d$	Mean of sample differences
$\sigma_{(\bar{x}_1 - \bar{x}_2)}$	Standard error for $\bar{x}_1 - \bar{x}_2$
σ_d	Standard deviation of differences, d
F_α	*Critical value for F-distribution
v_1	*Numerator degrees of freedom for F-distribution
v_2	*Denominator degrees of freedom for F-distribution
SE	Sampling error in estimation
T_i	*Sum of ranks in sample i
T_L, T_U	*Lower and upper critical values at Wilcoxon rank sum test
T_+	*Sum of ranks of positive differences
T_-	*Sum of ranks of negative differences
T_0	*Critical value of Wilcoxon signed-rank test

Key Ideas

Key Words for Identifying the Target Parameter

$\mu_1 - \mu_2$	Difference in Means or Averages
μ_d	Paired Difference in Means or Averages
$\mu_1, \mu_2, \ldots, \mu_n$	Comparing three or more means (averages)

Determining the Sample Size

Estimating $\mu_1 - \mu_2$: $n_1 = n_2 = (z_{a/2})^2(\sigma_1^2 + \sigma_2^2)/(\text{SE})^2$

Estimating μ_d: $n_d = (z_{\alpha/2})^2 \sigma_d^2/(\text{SE})^2$

Conditions Required for Inferences about $\mu_1 - \mu_2$

Large samples:
1. Independent random samples
2. $n_1 \geq 30, n_2 \geq 30$

Small samples:
1. Independent random samples
2. Both populations normal
3. $\sigma_1^2 = \sigma_2^2$

*Conditions Required for ANOVA

Large or small samples:
1. Independent random samples
2. All k populations normal
3. $\sigma_1^2 = \sigma_2^2 = \cdots = \sigma_k^2$

Conditions Required for Inferences about μ_d

Large samples:
1. Random sample of paired differences
2. $n_d \geq 30$

Small samples:
1. Random sample of paired differences
2. Population of differences is normal

Using a Confidence Interval for $(\mu_1 - \mu_2)$ to Determine whether a Difference Exists

1. If the confidence interval includes all *positive* numbers $(+, +)$: $\rightarrow$ Infer $\mu_1 > \mu_2$
2. If the confidence interval includes all *negative* numbers $(-, -)$ $\rightarrow$ Infer $\mu_1 < \mu_2$
3. If the confidence interval includes 0 $(-, +)$: $\rightarrow$ Infer "no evidence of a difference."

Supplementary Exercises 7.113–7.142

Note: Starred (*) exercises refer to the optional sections in this chapter.

Understanding the Principles

7.113 For each of the following, identify the target parameter as $\mu_1 - \mu_2$, $p_1 - p_2$, or σ_1^2/σ_2^2.
  a. Comparison of average SAT scores of males and females.
  b. Difference between mean waiting times at two super-market checkout lanes
  c. Comparison of mean starting salaries of mathematics psychology and engineering students.

7.114 List the assumptions necessary for each of the following inferential techniques:
  a. Large-sample inferences about the difference $(\mu_1 - \mu_2)$ between population means, using a two-sample z-statistic
  b. Small-sample inferences about $(\mu_1 - \mu_2)$, using an independent samples design and a two-sample t-statistic
  c. Small-sample inferences about $(\mu_1 - \mu_2)$, using a paired difference design and a single-sample t-statistic to analyze the differences
  d. Inference on comparing three treatment means, μ_1, μ_2 and μ_3.

Learning the Mechanics

7.115 Independent random samples were selected from two normally distributed populations with means μ_1 and μ_2, respectively. The sample sizes, means, and variances are shown in the following table:

Sample 1	Sample 2
$n_1 = 12$	$n_2 = 14$
$\bar{x}_1 = 17.8$	$\bar{x}_2 = 15.3$
$s_1^2 = 74.2$	$s_2^2 = 60.5$

    a. Test $H_0: (\mu_1 - \mu_2) = 0$ against $H_a: (\mu_1 - \mu_2) > 0$. Use $\alpha = .05$.

    b. Form a 99% confidence interval for $(\mu_1 - \mu_2)$.

    c. How large must n_1 and n_2 be if you wish to estimate $(\mu_1 - \mu_2)$ to within two units with 99% confidence? Assume that $n_1 = n_2$.

*7.116 An independent samples design is utilized to compare four treatment means. The data are shown in the accompanying table and saved in the **LM7_116** file.

    a. Given that SST = 36.95 and SS(Total) = 62.55, complete an ANOVA table for this experiment.

    b. Is there evidence that the treatment means differ? Use $\alpha = .10$.

Treatment 1	Treatment 2	Treatment 3	Treatment 4
8	6	9	12
10	9	10	13
9	8	8	10
10	8	11	11
11	7	12	11

7.117 Two independent random samples are taken from two populations. The results of these samples are summarized in the following table:

Sample 1	Sample 2
$n_1 = 135$	$n_2 = 148$
$\bar{x}_1 = 12.2$	$\bar{x}_2 = 8.3$
$s_1^2 = 2.1$	$s_2^2 = 3.0$

    a. Form a 90% confidence interval for $(\mu_1 - \mu_2)$.

    b. Test $H_0: (\mu_1 - \mu_2) = 0$ against $H_a: (\mu_1 - \mu_2) \neq 0$. Use $\alpha = .01$.

    c. What sample size would be required if you wish to estimate $(\mu_1 - \mu_2)$ to within .2 with 90% confidence? Assume that $n_1 = n_2$.

7.118 A random sample of five pairs of observations was selected, one observation from a population with mean μ_1, the other from a population with mean μ_2. The data are shown in the following table and saved in the **LM7_118** file.

Pair	Value from Population 1	Value from Population 2
1	28	22
2	31	27
3	24	20
4	30	27
5	22	20

    a. Test the null hypothesis $H_0: \mu_d = 0$ against $H_a: \mu_d \neq 0$, where $\mu_d = \mu_1 - \mu_2$. Use $\alpha = .05$.

    b. Form a 95% confidence interval for μ_d.

    c. When are the procedures you used in parts **a** and **b** valid?

*7.119 A random sample of nine pairs of observations is recorded on two variables x and y. The data are shown in the following table and saved in the **LM7_119** file. Do the data provide sufficient evidence to indicate that the probability distribution for x is shifted to the right of that for y? Test, using $\alpha = .05$.

Pair	x	y
1	19	12
2	27	19
3	15	7
4	35	25
5	13	11
6	29	10
7	16	16
8	22	10
9	16	18

*7.120 Two independent random samples produced the measurements listed in the next table. Do the data (saved in the **LM7_120** file) provide sufficient evidence to conclude that there is a difference between the locations of the probability distributions for the sampled populations? Test, using $\alpha = .05$.

Sample 1		Sample 2	
1.2	1.0	1.5	1.9
1.9	1.8	1.3	2.7
.7	1.1	2.9	3.5
2.5			

Applying the Concepts—Basic

7.121 **Oil spill impact on seabirds.** Refer to the *Journal of Agricultural, Biological, and Environmental Statistics* (Sept. 2000) study of the impact of a tanker oil spill on the seabird population in Alaska, presented in Exercise 2.187 (p. 100). Recall that for each of 96 shoreline locations (called transects), the number of seabirds found, the length (in kilometers) of the transect, and whether or not the transect was in an oiled area were recorded. (The data are saved in the **EVOS** file.) *Observed seabird density* is defined as the observed count divided by the length of the transect. A comparison of the mean densities of oiled and unoiled transects is displayed in the MINITAB printout on p. 418. Use this information to make an inference about the difference in the population mean seabird densities of oiled and unoiled transects.

7.122 **Rating service at five-star hotels.** A study published in the *Journal of American Academy of Business, Cambridge* (March 2002) examined whether the perception of the quality of service at five-star hotels in Jamaica differed by gender. Hotel guests were randomly selected from the lobby and restaurant areas and asked to rate 10 service-related items (e.g., "the personal attention you received from our employees"). Each item was rated on a five-point scale (1 = "much worse than I expected," 5 = "much better than I expected"),

MINITAB output for Exercise 7.121

Two-Sample T-Test and CI: Density, Oil

```
Two-sample T for Density

Oil  N   Mean   StDev   SE Mean
no   36  3.27   6.70    1.1
yes  60  3.50   5.97    0.77


Difference = mu (no ) - mu (yes)
Estimate for difference:  -0.221165
95% CI for difference:  (-2.927767, 2.485436)
T-Test of difference = 0 (vs not =): T-Value = -0.16  P-Value = 0.871  DF = 67
```

---

and the sum of the items for each guest was determined. A summary of the guest scores is provided in the following table:

Gender	Sample Size	Mean Score	Standard Deviation
Males	127	39.08	6.73
Females	114	38.79	6.94

a. Construct a 90% confidence interval for the difference between the population mean service-rating scores given by male and female guests at Jamaican five-star hotels.

b. Use the interval you constructed in part **a** to make an inference about whether the perception of the quality of service at five-star hotels in Jamaica differs by gender.

7.123 Effect of altitude on climbers. Dr. Philip Lieberman, a neuroscientist at Brown University, conducted a field experiment to gauge the effect of high altitude on a person's ability to think critically (*New York Times*, Aug. 23, 1995). The subjects of the experiment were five males who took part in an American expedition climbing Mount Everest. At the base camp, Lieberman read sentences to the climbers while they looked at simple pictures in a book. The length of time (in seconds) it took for each climber to match the picture with a sentence was recorded. Using a radio, Lieberman repeated the task when the climbers reached a camp 5 miles above sea level. At this altitude, he noted that the climbers took 50% longer to complete the task.

a. What is the variable measured in this experiment?

b. What are the experimental units?

c. Discuss how the data should be analyzed.

*7.124 Reading Japanese books. Refer to the *Reading in a Foreign Language* (Apr. 2004) experiment to improve the Japanese reading comprehension levels of University of Hawaii students, presented in Exercise 7.15 (p. 365). Recall that 14 students participated in a 10-week extensive reading program in a second-semester Japanese course. The number of books read by each student and the student's course grade are repeated in the accompanying table and saved in the **JAPANESE** file. Consider a comparison of the distributions of number of books read by students who earn an "A" grade and those who earn a "B" or "C" grade.

Number of Books	Course Grade
53	A
42	A
40	A
40	B
39	A
34	A
34	A
30	A
28	B
24	A
22	C
21	B
20	B
16	B

Source: Hitosugi, C. I., and Day, R. R. "Extensive reading in Japanese." *Reading in a Foreign Language*, Vol. 16, No. 1, Apr. 2004 (Table 4). Reprinted with permission from the National Foreign Language Resource Center, University of Hawaii.

a. Rank all 14 observations from smallest to largest, and assign ranks from 1 to 14.

b. Sum the ranks of the observations for students with an "A" grade.

c. Sum the ranks of the observations for students with either a "B" or "C" grade.

d. Compute the Wilcoxon rank sum statistic.

e. Carry out a nonparametric test (at $\alpha = .10$) to compare the distribution of the number of books read by the two populations of students.

7.125 Heights of grade school repeaters. Are children who repeat a grade in elementary school shorter, on average, than their peers? To answer this question, researchers compared the heights of Australian schoolchildren who repeated a grade with the heights of those who did not (*Archives of Disease in Childhood*, Apr. 2000). All height measurements were standardized with the use of z-scores. A summary of the results, by gender, is shown in the following table:

Gender/Status	Sample Size	Mean	Std.Dev.
Girls/Repeat	43	.26	.94
Girls/Never Repeat	1,366	.22	1.04
Boys/Repeat	86	−.04	1.17
Boys/Never Repeat	1,349	.30	.97

Based on Wake, M., Coghlan, D., and Hesketh, K. "Does height influence progression through primary school grades?" *The Archives of Disease in Childhood*, Vol. 82, No. 4, April 2000, pp. 297–301 (Table 2).

a. Set up the null and alternative hypothesis for determining whether the average height of Australian boys who repeated a grade is less than the average height of boys who never repeated.

b. Conduct the test you set up in part **a,** using $\alpha = .05$.

c. Repeat parts **a** and **b** for Australian girls.

*d. Within each grade, the children were divided into equal thirds (tertiles) based on age (youngest third, middle third, and oldest third). The researchers also compared the average heights of the three groups using an ANOVA. For boys, the F-test resulted in p-value=.01. Interpret the results at $\alpha = .05$.

7.126 Environmental impact study. Some power plants are located near rivers or oceans so that the available water can be used to cool the condensers. Suppose that, as part of an environmental impact study, a power company wants to estimate the difference in mean water temperature between the discharge of its plant and the offshore waters. How many sample measurements must be taken at each site in order to estimate the true difference between means to within .2°C with 95% confidence? Assume that the range in readings will be about 4°C at each site and that the same number of readings will be taken at each site.

7.127 Animal-assisted therapy for heart patients. Refer to the *American Heart Association Conference* (Nov. 2005) study to gauge whether animal-assisted therapy can improve the physiological responses of heart failure patients, presented in Exercise 2.106 (p. 73). Recall that a sample of $n = 26$ heart patients was visited by a human volunteer accompanied by a trained dog; the anxiety level of each patient was measured (in points) both before and after the visits. The drop (before minus after) in anxiety level for patients is summarized as follows: $\bar{x}_d = 10.5, s_d = 7.6$. Does animal-assisted therapy significantly reduce the mean anxiety level of heart failure patients? Support your answer with a 95% confidence interval.

7.128 Life expectancy of Oscar winners. Does winning an Academy of Motion Picture Arts and Sciences award lead to long-term mortality for movie actors? In an article in the *Annals of Internal Medicine* (May 15, 2001), researchers sampled 762 Academy Award winners and matched each one with another actor of the same sex who was in the same winning film and was born in the same era. The life expectancies (ages) of the pairs of actors were compared.

a. Explain why the data should be analyzed as a paired difference experiment.

b. Set up the null hypothesis for a test to compare the mean life expectancies of Academy Award winners and nonwinners.

c. The sample mean life expectancies of Academy Award winners and nonwinners were reported as 79.7 years and 75.8 years, respectively. The p-value for comparing the two population means was reported as $p = .003$. Interpret this value in the context of the problem.

Applying the Concepts—Intermediate

7.129 Reading tongue twisters. According to *Webster's New World Dictionary*, a tongue twister is "a phrase that is hard to speak rapidly." Do tongue twisters have an effect on the length of time it takes to read silently? To answer this question, 42 undergraduate psychology students participated in a reading experiment (*Memory & Cognition*, Sept. 1997). Two lists, each composed of 600 words, were constructed. One list contained a series of tongue twisters, and the other list (called the *control*) did not contain any tongue twisters. Each student read both lists, and the length of time (in minutes) required to complete the lists was recorded. The researchers used a test of hypothesis to compare the mean reading response times for the tongue-twister and control lists.

a. Set up the null hypothesis for the test.

b. For each student, the researchers computed the difference between the reading response times for the tongue-twister and control lists. The mean difference was .25 minute with a standard deviation of .78 minute.* Use the information to find the test statistic and p-value of the test.

c. Give the appropriate conclusion. Use $\alpha = .05$.

*Based on Robinson, D. H., and Katayama, A. D. "At-lexical, articulatory interference in silent reading: The 'upstream' tongue-twister effect." *Memory & Cognition*, Vol. 25, No. 5, Sept. 1997, p. 663.

7.130 Mating habits of snails. Hermaphrodites are animals that possess the reproductive organs of both sexes. *Genetical Research* (June 1995) published a study of the mating systems of hermaphroditic snail species. The mating habits of the snails were classified into two groups: (1) self-fertilizing (selfing) snails that mate with snails of the same sex and (2) cross-fertilizing (outcrossing) snails that mate with snails of the opposite sex. One variable of interest in the study was the effective population size of the snail species. The means and standard deviations of the effective population size for independent random samples of 17 outcrossing snail species and 5 selfing snail species are given in the accompanying table.

Snail Mating System	Effective Population Size		
	Sample Size	Mean	Standard Deviation
Outcrossing	17	4,894	1,932
Selfing	5	4,133	1,890

Based on Jarne, P. "Mating system, bottlenecks, and genetic polymorphism in hermaphroditic animals." *Genetical Research*, Vol. 65, No. 3, June 1995, p. 197 (Table 4).

a. Compare the mean effective population sizes of the two types of snail species with a 90% confidence interval. Interpret the result.

b. Geneticists are often more interested in comparing the variation in population size of the two types of mating systems. Conduct this analysi for the researcher. Interpret the result.

7.131 Children's use of pronouns. The *Journal of Communication Disorders* (Mar. 1995) published a study of specifically language-impaired (SLI) children. The data on deviation intelligence quotient (DIQ) for 10 SLI children and 10 younger, normally developing children are listed in the table (p. 420) and saved in the **SLI** file.

a. Use a *t*-test to compare the mean DIQ of the two groups of children. (Use $\alpha = .10$.) What do you conclude?

*__b.__ Use a nonparametric test to compare the DIQ distributions of the two groups of children.

SLI Children			YND Children		
86	87	84	110	90	105
94	86	107	92	92	96
89	98	95	86	100	92
110			90		

7.132 Identical twins reared apart. Because they share an identical genotype, twins make ideal subjects for investigating the degree to which various environmental conditions affect personality. The classical method of studying this phenomenon, nomenon, and the subject of an interesting book by Susan Farber (*Identical Twins Reared Apart*, New York: Basic Books, 1981), is the study of identical twins separated early in life and reared apart. Much of Farber's discussion focuses on a comparison of IQ scores. The data for this analysis appear in the accompanying table and are saved in the **TWINSIQ** file. One member (A) of each of the $n = 32$ pairs of twins was reared by a natural parent; the other member (B) was reared by a relative or some other person. Is there a significant difference between the average IQ scores of identical twins when one member of the pair is reared by the natural parents and the other member of the pair is not? Use $\alpha = .05$ to draw your conclusion.

Pair ID	Twin A	Twin B	Pair ID	Twin A	Twin B
112	113	109	228	100	88
114	94	100	232	100	104
126	99	86	236	93	84
132	77	80	306	99	95
136	81	95	308	109	98
148	91	106	312	95	100
170	111	117	314	75	86
172	104	107	324	104	103
174	85	85	328	73	78
180	66	84	330	88	99
184	111	125	338	92	111
186	51	66	342	108	110
202	109	108	344	88	83
216	122	121	350	90	82
218	97	98	352	79	76
220	82	94	416	97	98

Based on Farber, S. L. *Identical Twins Reared Apart*, © 1981 by Basic Books, Inc.

7.133 Treatments for panic disorder. Inositol is a complex cyclic alcohol found to be effective against clinical depression. Medical researchers believe that inositol may also be used to treat panic disorder. To test this theory, a double-blind, placebo-controlled study of 21 patients diagnosed with panic disorder was conducted (*American Journal of Psychiatry*, July 1995). Patients completed diaries recording the occurrence of their panic attacks. The data (saved in the **INOSITOL** file) for a week in which patients received a glucose placebo and for a week when they were treated with inositol are provided in the next table. [*Note:*

Neither the patients nor the treating physicians knew which week the placebo was given.] Analyze the data and interpret the results. Comment on the validity of the assumptions.

Patient	Placebo	Inositol	Patient	Placebo	Inositol
1	0	0	12	0	2
2	2	1	13	3	1
3	0	3	14	3	1
4	1	2	15	3	4
5	0	0	16	4	2
6	10	5	17	6	4
7	2	0	18	15	21
8	6	4	19	28	8
9	1	1	20	30	0
10	1	0	21	13	0
11	1	3			

Based on Benjamin, J. "Double-blind, placebo-controlled, crossover trial of inositol treatment for panic disorder." *American Journal of Psychiatry*, Vol. 152, No. 7, July 1995, p. 1085 (Table 1).

7.134 Personalities of cocaine abusers. Do cocaine abusers have radically different personalities than nonabusing college students? This was one of the questions researched in *Psychological Assessment* (June 1995). Zuckerman–Kuhlman's Personality Questionnaire (ZKPQ) was administered to a sample of 450 cocaine abusers and a sample of 589 college students. The ZKPQ yields scores (measured on a 20-point scale) on each of five dimensions: impulsive–sensation seeking, sociability, neuroticism–anxiety, aggression–hostility, and activity. The results are summarized in the accompanying table. Compare the mean ZKPQ scores of the two groups on each dimension, using a statistical test of hypothesis. Interpret the results at $\alpha = .01$.

ZKPQ Dimension	Cocaine Abusers ($n = 450$)		College Students ($n = 589$)	
	Mean	Std. Dev.	Mean	Std. Dev.
Impulsive–sensation seeking	9.4	4.4	9.5	4.4
Sociability	10.4	4.3	12.5	4.0
Neuroticism–anxiety	8.6	5.1	9.1	4.6
Aggression–hostility	8.6	3.9	7.3	4.1
Activity	11.1	3.4	8.0	4.1

Source: Ball, S. A. "The validity of an alternative five-factor measure of personality in cocaine abusers." *Psychological Assessment*, Vol. 7, No. 2, June 1995, p. 150 (Table 1). Copyright © 1995 by the American Psychological Association. Reprinted with permission.

*__7.135__ **Facial expression study.** What do people infer from facial expressions of emotion? This was the research question of interest in an article published in the *Journal of Nonverbal Behavior* (Fall 1996). A sample of 36 introductory psychology students was randomly divided into six groups. Each group was assigned to view one of six slides showing a person making a facial expression. The six expressions were (1) angry, (2) disgusted, (3) fearful, (4) happy, (5) sad, and (6) neutral. After viewing the slides, the students rated the degree of dominance they inferred from the facial expression (on a scale ranging from -15 to $+15$). The data (simulated from summary information provided in the article) are listed in the accompanying table and saved in the **FACES** file. Conduct an analysis of variance to determine whether the mean dominance ratings differ among the six facial expressions. Use $\alpha = .10$.

Angry	Disgusted	Fearful	Happy	Sad	Neutral
2.10	.40	.82	1.71	.74	1.69
.64	.73	-2.93	$-.04$	-1.26	$-.60$
.47	$-.07$	$-.74$	1.04	-2.27	$-.55$
.37	$-.25$	.79	1.44	$-.39$	.27
1.62	.89	$-.77$	1.37	-2.65	$-.57$
$-.08$	1.93	-1.60	.59	$-.44$	-2.16

7.136 Swim maze study. Merck Research Labs used the single-T swim maze to conduct an experiment to evaluate the effect of a new drug. Nineteen impregnated dam rats were allocated a dosage of 12.5 milligrams of the drug. One male and one female rat pup were randomly selected from each resulting litter to perform in the swim maze. Each pup was placed in the water at one end of the maze and allowed to swim until it escaped at the opposite end. If the pup failed to escape after a certain period of time, it was placed at the beginning of the maze and given another chance. The experiment was repeated until each pup accomplished three successful escapes. The accompanying table (saved in the **RATPUPS** file) reports the number of swims required by each pup to perform three successful escapes.

Litter	Male	Female	Litter	Male	Female
1	8	5	11	6	5
2	8	4	12	6	3
3	6	7	13	12	5
4	6	3	14	3	8
5	6	5	15	3	4
6	6	3	16	8	12
7	3	8	17	3	6
8	5	10	18	6	4
9	4	4	19	9	5
10	4	4			

Based on Thomas E. Bradstreet, "Favorite Data Sets from Early Phases of Drug Research—Part 2." *Proceedings of the Section on Statistical Education of the American Statistical Association.*

a. Is there sufficient evidence of a difference between the mean number of swims required by male and female pups? Conduct the t-test (at $\alpha = .10$). Comment on the assumptions required for the test to be valid.

*b. Conduct the appropriate nonparametric test on the data. Interpret the results using $\alpha = .10$.

7.137 Rating music teachers. Students enrolled in music classes at the University of Texas (Austin) participated in a study to compare the observations and teacher evaluations of music education majors and nonmusic majors (*Journal of Research in Music Education*, Winter 1991). Independent random samples of 100 music majors and 100 nonmajors rated the overall performance of their teacher, using a six-point scale, where 1 was the lowest rating and 6 the highest. Use the information in the accompanying table to compare the mean teacher ratings of the two groups of music students with a 95% confidence interval. Interpret the result.

	Music Majors	Nonmusic Majors
Sample size	100	100
Mean "overall" rating	4.26	4.59
Standard deviation	.81	.78

Based on Duke, R. A., and Blackman, M. D. "The relationship between observers' recorded teacher behavior and evaluation of music instruction." *Journal of Research in Music Education,* Vol. 39, No. 4, Winter 1991 (Table 2).

7.138 Testing electronic circuits. Japanese researchers have developed a compression–depression method of testing electronic circuits based on Huffman coding (*IEICE Transactions on Information & Systems*, Jan. 2005). The new method is designed to reduce the time required for input decompression and output compression—called the compression ratio. Experimental results were obtained by testing a sample of 11 benchmark circuits (all of different sizes) from a SUN Blade 1000 workstation. Each circuit was tested with the standard compression–depression method and the new Huffman-based coding method and the compression ratio recorded. The data are given below and saved in the **CIRCUITS** file.

a. Compare the two methods with a 95% confidence interval. Which method has the smaller mean compression ratio?

b. How many circuits need to be sampled in order to estimate the mean difference in compression ratio to within .03 with 95% confidence.

Circuit	Standard Method	Huffman Coding Method
1	.80	.78
2	.80	.80
3	.83	.86
4	.53	.53
5	.50	.51
6	.96	.68
7	.99	.82
8	.98	.72
9	.81	.45
10	.95	.79
11	.99	.77

Based on Ichihara, H., Shintani, M., and Inoue, T. "Huffman-based test response coding." *IEICE Transactions on Information & Systems,* Vol. E88-D, No. 1, Jan. 2005 (Table 3).

*7.139 Agent Orange and Vietnam Vets. Agent Orange, the code name for a herbicide developed for the U.S. armed forces in the 1960s, was found to be extremely contaminated with TCDD, or dioxin. During the Vietnam War, an estimated 19 million gallons of Agent Orange was used to destroy the dense plant and tree cover of the Asian jungle. As a result of this exposure, many Vietnam veterans have dangerously high levels of TCDD in their blood and adipose (fatty) tissue. A study published in *Chemosphere* (Vol. 20, 1990) reported on the TCDD levels of 20

Massachusetts Vietnam vets who were possibly exposed to Agent Orange. The TCDD amounts (measured in parts per trillion) in both plasma and fat tissue of the 20 vets are listed in the accompanying table. The data are saved in the **TCDD** file. Medical researchers are interested in comparing the TCDD levels in fat tissue and plasma for Vietnam veterans. Specifically, they want to determine whether the distribution of TCDD levels in fat is shifted above or below the distribution of TCDD levels in plasma. Conduct this analysis (at $\alpha = .05$) and make the appropriate inference.

Vet	Fat	Plasma
1	4.9	2.5
2	6.9	3.5
3	10.0	6.8
4	4.4	4.7
5	4.6	4.6
6	1.1	1.8
7	2.3	2.5
8	5.9	3.1
9	7.0	3.1
10	5.5	3.0
11	7.0	6.9
12	1.4	1.6
13	11.0	20.0
14	2.5	4.1
15	4.4	2.1
16	4.2	1.8
17	41.0	36.0
18	2.9	3.3
19	7.7	7.2
20	2.5	2.0

Based on Schecter, A., et al. "Partitioning of 2,3,7,8-chlorinated dibenzo-*p*-dioxins and dibenzofurans between adipose tissue and plasma lipid of 20 Massachusetts Vietnam veterans." *Chemosphere,* Vol. 20, Nos. 7–9, 1990, pp. 954–955 (Tables I and II).

Applying the Concepts—Advanced

7.140 Feeding habits of sea urchins. The *Florida Scientist* (Summer/Autumn 1991) reported on a study of the feeding habits of sea urchins. A sample of 20 urchins was captured from Biscayne Bay (Miami), placed in marine aquaria, and then starved for 48 hours. Each sea urchin was then fed a 5-cm blade of turtle grass. Ten of the urchins received only green blades, while the other half received only decayed blades. (Assume that the two samples of 10 sea urchins each were randomly and independently selected.) The ingestion time, measured from the time the blade first made contact with the urchin's mouth to the time the urchin had finished ingesting the blade, was recorded. A summary of the results is provided in the following table:

	Green Blades	Decayed Blades
Number of sea urchins	10	10
Mean ingestion time (hours)	3.35	2.36
Standard deviation (hours)	.79	.47

From "Laboratory measurement of ingestion rate for the sea urchin Lytechinus variegatus" by Dr. Jeremy Montague. *Florida Scientist,* Vol. 54, Nos. 3/4, Summer/Autumn 1991. Reprinted with permission from the Florida Academy of Sciences.

According to the researchers, "The difference in rates at which the urchins ingested the blades suggest that green, unblemished turtle grass may not be a particularly palatable food compared with decayed turtle grass. If so, urchins in the field may find it more profitable to selectively graze on decayed portions of the leaves." Do the results support this conclusion?

Critical Thinking Challenges

7.141 Self-managed work teams and family life. To improve quality, productivity, and timeliness, more and more American industries are utilizing self-managed work teams (SMWTs). A team typically consists of 5 to 15 workers who are collectively responsible for making decisions and performing all tasks related to a particular project. Researchers L. Stanley-Stevens (Tarleton State University), D. E. Yeatts, and R. R. Seward (both from the University of North Texas) investigated the connection between SMWTs, work characteristics, and workers' perceptions of positive spillover into family life (*Quality Management Journal,* Summer 1995). Survey data were collected from 114 AT&T employees who worked on 1 of 15 SMWTs at an AT&T technical division. The workers were divided into two groups: (1) those who reported a positive spillover of work skills to family life and (2) those who did not report any such positive work spillover. The two groups were compared on a variety of job and demographic characteristics, several of which are shown in the table (next column). All but the demographic characteristics were measured on a seven-point scale, ranging from 1 = "strongly disagree" to 7 = "strongly agree"; thus, the larger the number, the more the characteristic was indicated. The file named **SPILLOVER** includes the values of the variables listed in the table for each of the 114 survey participants. The researchers' objectives were to compare the two groups of workers on each characteristic. In particular, they wanted to know which job-related characteristics are most highly associated with positive work spillover. Conduct a complete analysis of the data for the researchers.

Characteristic	Variable
Information Flow	Use of creative ideas (seven-point scale)
Information Flow	Utilization of information (seven-point scale)
Decision Making	Participation in decisions regarding personnel matters (seven-point scale)
Job	Good use of skills (seven-point scale)
Job	Task identity (seven-point scale)
Demographic	Age (years)
Demographic	Education (years)
Comparison	Group (positive spillover or no spillover)

7.142 MS and exercise study. A study published in *Clinical Kinesiology* (Spring 1995) was designed to examine the metabolic and cardiopulmonary responses during exercise of persons diagnosed with multiple sclerosis (MS). Leg-cycling and arm-cranking exercises were performed by 10 MS patients and 10 healthy (non-MS) control subjects. Each member of the control group was selected on the basis of gender, age, height, and weight to match (as closely as possible) with one member of the MS group. Consequently, the researchers compared the MS and non-MS groups by matched-pairs t-tests on such outcome variables as oxygen uptake, carbon dioxide output, and peak aerobic power. The data on the matching variables used in the experiment are shown in the table below and saved in the **MSSTUDY** file. Have the researchers successfully matched the MS and non-MS subjects?

Matched Pair	MS Subjects				Non-MS Subjects			
	Gender	Age (years)	Height (cm)	Weight (kg)	Gender	Age (years)	Height (cm)	Weight (kg)
1	M	48	171.0	80.8	M	45	173.0	76.3
2	F	34	158.5	75.0	F	34	158.0	75.6
3	F	34	167.6	55.5	F	34	164.5	57.7
4	M	38	167.0	71.3	M	34	161.3	70.0
5	M	45	182.5	90.9	M	39	179.0	96.0
6	F	42	166.0	72.4	F	42	167.0	77.8
7	M	32	172.0	70.5	M	34	165.8	74.7
8	F	35	166.5	55.3	F	43	165.1	71.4
9	F	33	166.5	57.9	F	31	170.1	60.4
10	F	46	175.0	79.9	F	43	175.0	77.9

From "Maximal aerobic exercise of individuals with multiple sclerosis using three modes of ergometry." *Clinical Kinesiology,* Vol. 49, No. 1, Spring 1995, p. 7. Reprinted with permission from W. Jeffrey Armstrong.

Activity | Paired vs. Unpaired Experiments

We have now discussed two methods of collecting data to compare two population means. In many experimental situations, a decision must be made either to collect two independent samples or to conduct a paired difference experiment. The importance of this decision cannot be overemphasized, since the amount of information obtained and the cost of the experiment are both directly related to the method of experimentation that is chosen.

Choose two populations (pertinent to your school major) that have unknown means and for which you could both collect two independent samples and collect paired observations. Before conducting the experiment, state which method of sampling you think will provide more information (and why). Compare the two methods, first performing the independent sampling procedure by collecting 10 observations from each population (a total of 20 measurements) and then performing the paired difference experiment by collecting 10 pairs of observations.

Construct two 95% confidence intervals, one for each experiment you conduct. Which method provides the narrower confidence interval and hence more information on this performance of the experiment? Does your result agree with your preliminary expectations?

References

Cochran, W. G., and Cox, G. M. *Experimental Designs*, 2nd ed. New York: Wiley, 1957.

Conover, W. J. *Practical Nonparametric Statistics*, 2nd ed. New York: Wiley, 1980.

Daniel, W. W. *Applied Nonparametric Statistics*, 2nd ed. Boston: PWS-Kent, 1990.

Dunn, O. J. "Multiple comparisons using rank sums." *Technometrics*, Vol. 6, 1964.

Friedman, M. "The use of ranks to avoid the assumption of normality implicit in the analysis of variance." *Journal of the American Statistical Association*, Vol. 32, 1937.

Freedman, D., Pisani, R., and Purves, R. *Statistics*. New York: W. W. Norton and Co., 1978.

Gibbons, J. D. *Nonparametric Statistical Inference*, 4th ed. Boca Raton, FL: CRC Press, 2003.

Hsu, J. C. *Multiple Comparisons: Theory and Methods*. London: Chapman & Hall, 1996.

Hollander, M., and Wolfe, D. A. *Nonparametric Statistical Methods*. 2nd ed. New York: Wiley, 1999.

Kramer, C. Y. "Extension of multiple range tests to group means with unequal number of replications." *Biometrics*, Vol. 12, 1956, pp. 307–310.

Kruskal, W. H., and Wallis, W. A. "Use of ranks in one-criterion variance analysis." *Journal of the American Statistical Association*, Vol. 47, 1952.

Kutner, M., Nachtsheim, C., Neter, J., and Li, W. *Applied Linear Statistical Models*, 5th ed. New York: McGraw-Hill/Irwin, 2005.

Lehmann, E. L. *Nonparametrics: Statistical Methods Based on Ranks*. (revised). New York: Springer, 2006.

Marascuilo, L. A., and McSweeney, M. *Nonparametric and Distribution-Free Methods for the Social Sciences*. Monterey, CA: Brooks/Cole, 1977.

Mason, R. L., Gunst, R. F., and Hess, J. L. *Statistical Design and Analysis of Experiments*. New York: Wiley, 1989.

Mendenhall, W., Beaver, R. J., and Beaver, B. *Introduction to Probability and Statistics*, 13th ed. Belmont, CA: Brooks/Cole, 2009.

Mendenhall, W. *Introduction to Linear Models and the Design and Analysis of Experiments*. Belmont, CA: Wadsworth, 1968.

Miller, R. G., Jr. *Simultaneous Statistical Inference*. New York: Springer-Verlag, 1981.

Satterthwaite, F. W. "An approximate distribution of estimates of variance components." *Biometrics Bulletin*, Vol. 2, 1946, pp. 110–114.

Scheffé, H. "A method for judging all contrasts in the Analysis of Variance," *Biometrica*, Vol. 40, 1953, pp. 87–104.

Scheffé, H. *The Analysis of Variance*. New York: Wiley, 1959.

Snedecor, G. W., and Cochran, W. G. *Statistical Methods*, 7th ed. Ames, IA: Iowa State University Press, 1980.

Steele, R. G. D., and Torrie, J. H. *Principles and Procedures of Statistics: A Biometrical Approach*, 2nd ed. New York: McGraw-Hill, 1980.

Tukey, J. "Comparing individual means in the Analysis of Variance," *Biometrics*, Vol. 5, 1949, pp. 99–114.

Wilcoxon, F., and Wilcox, R. A. "Some rapid approximate statistical procedures." The American Cyanamid Co., 1964.

USING TECHNOLOGY

MINITAB: Comparing Means

MINITAB can be used to make both parametric and nonparametric inferences about $\mu_1 - \mu_2$ for independent samples, μ_d for paired samples, and $\mu_1, \mu_2, \cdots, \mu_k$ in an analysis of variance.

Comparing Two Means with Independent Samples

Step 1 Access the MINITAB worksheet that contains the sample data.

Step 2 Click on the "Stat" button on the MINITAB menu bar and then click on "Basic Statistics" and "2-Sample t," as shown in Figure 7.M.1. The resulting dialog box appears as shown in Figure 7.M.2.

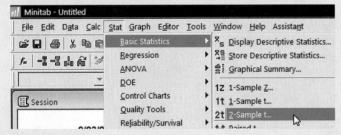

Figure 7.M.1
MINITAB menu options for comparing two means

Step 3a If the worksheet contains data for one quantitative variable (which the means will be computed on) and one qualitative variable (which represents the two groups or populations), select "Samples in one column" and then specify the quantitative variable in the "Samples" area and the qualitative variable in the "Subscripts" area. (See Figure 7.M.2.)

Step 3b If the worksheet contains the data for the first sample in one column and the data for the second sample in another column, select "Samples in different columns" and then specify the "First" and "Second" variables. Alternatively, if you have only summarized data (i.e., sample sizes, sample means, and sample standard deviations), select "Summarized data" and enter these summarized values in the appropriate boxes.

Step 4 Click the "Options" button on the MINITAB "2-Sample T" dialog box. Specify the confidence level for a confidence interval, the null-hypothesized value of the difference, $\mu_1 - \mu_2$,

Figure 7.M.2
MINITAB 2-sample *t* dialog box

and the form of the alternative hypothesis (lower tailed, two tailed, or upper tailed) in the resulting dialog box, as shown in Figure 7.M.3.

Figure 7.M.3
MINITAB options dialog box

Step 5 Click "OK" to return to the "2-Sample T" dialog box and then click "OK" again to generate the MINITAB printout.

Important Note: The MINITAB two-sample *t*-procedure uses the *t*-statistic to conduct the test of hypothesis. When the sample sizes are small, check the "Assume equal variances" box in Figure 7.M.2. When the sample sizes are large, leave the "Assume equal variances" box unchecked; the *t*-value will be approximately equal to the large-sample *z*-value, and the resulting test will still be valid.

Comparing Two Means with Paired Samples

Step 1 Access the MINITAB worksheet that contains the sample data. The data file should contain two quantitative variables—one with the data values for the first group (or population) and one with the data values for the second group. (*Note:* The sample size should be the same for each group.)

Step 2 Click on the "Stat" button on the MINITAB menu bar and then click on "Basic Statistics" and "Paired t" (see Figure 7.M.1).

Step 3 On the resulting dialog box, select the "Samples in columns" option and specify the two quantitative variables of interest in the "First sample" and "Second sample" boxes, as shown in Figure 7.M.4. [Alternatively, if you have only summarized data of the paired differences, select the "Summarized data (differences)" option and enter the sample size, sample mean, and sample standard deviation in the appropriate boxes.]

Figure 7.M.4
MINITAB paired-samples *t* dialog box

Step 4 Click the "Options" button and specify the confidence level for a confidence interval, the null-hypothesized value of the difference μ_d, and the form of the alternative hypothesis (lower tailed, two tailed, or upper tailed) in the resulting dialog box. (See Figure 7.M.3.)

Figure 7.M.5
MINITAB nonparametric menu options

Step 5 Click "OK" to return to the "Paired t" dialog box and then click "OK" again to generate the MINITAB printout.

Rank Sum Test

Step 1 Access the MINITAB worksheet file with the sample data. It should contain two quantitative variables, one for each of the two samples being compared.

Step 2 Click on the "Stat" button on the MINITAB menu bar, then click on "Nonparametrics" and "Mann-Whitney" (see Figure 7.M.5).

Step 3 On the resulting dialog box (see Figure 7.M.6), specify the variable for the first sample in the "First Sample" box and the variable for the second sample in the "Second Sample" box.

Figure 7.M.6
MINITAB Mann-Whitney (rank sum) test dialog box

Step 4 Specify the form of the alternative hypothesis ("not equal," "less than," or "greater than").

Step 5 Click "OK" to generate the MINITAB printout.

Signed Rank Test

Step 1 Access the MINITAB worksheet file with the matched-pairs data. It should contain two quantitative variables, one for each of the two groups being compared.

Step 2 Compute the difference between these two variables and save it in a column on the worksheet. (Use the "Calc" button on the MINITAB menu bar.)

Step 3 Click on the "Stat" button on the MINITAB menu bar, then click on "Nonparametrics" and "1-Sample Wilcoxon" (see Figure 7.M.1).

Step 4 On the resulting dialog box (see Figure 7.M.7), enter the variable representing the paired differences in the "Variables" box.

Step 5 Select the "Test median" option and specify the hypothesized value of the median as "0." Select the form of the alternative hypothesis ("not equal," "less than," or "greater than").

Step 6 Click "OK" to generate the MINITAB printout.

Figure 7.M.7
MINITAB 1-sample Wilcoxon (signed rank) test dialog box

ANOVA *F*-test

Step 1 Access the MINITAB worksheet file that contains the sample data. The data file should contain one quantitative variable (the response, or dependent, variable) and one factor variable with at least two levels.

Step 2 Click on the "Stat" button on the MINITAB menu bar and then click on "ANOVA" and "One-Way," as shown in Figure 7.M.8.

Figure 7.M.8
MINITAB menu options for one-way ANOVA

Step 3 On the resulting dialog screen Figure 7.M.9, specify the response variable. in the "Response" box and the factor variable in the "Factor" box.

Figure 7.M.9
MINITAB one-way ANOVA dialog box

Step 4 Click "OK" to generate the MINITAB printout.

TI-83/TI-84 Plus Graphing Calculator: Comparing Means

The TI-83/ TI-84 plus graphing calculator can be used to conduct tests and form confidence intervals for the difference between two means with independent samples, the difference between two means with matched pairs, and an analysis of variance.

Confidence Interval for $\mu_1 - \mu_2$

Step 1 *Enter the data (Skip to Step 2 if you have summary statistics, not raw data)*

- Press **STAT** and select **1:Edit**

Note: If the lists already contain data, clear the old data. Use the up **ARROW** to highlight "**L1**."

- Press **CLEAR ENTER**
- Use the up **ARROW** to highlight "**L2**"
- Press **CLEAR ENTER**
- Use the **ARROW** and **ENTER** keys to enter the first data set into **L1**
- Use the **ARROW** and **ENTER** keys to enter the second data set into **L2**

Step 2 *Access the statistical tests menu*

- Press **STAT**
- Arrow right to **TESTS**
- Arrow down to **2-SampTInt**
- Press **ENTER**

Step 3 *Choose "Data" or "Stats" ("Data" is selected when you have entered the raw data into the Lists. "Stats" is selected when you are given only the means, standard deviations, and sample sizes)*

- Press **ENTER**
- If you selected "Data," set **List1** to **L1** and **List2** to **L2**
- Set **Freq1** to **1** and set **Freq2** to **1**
- Set **C-Level** to the confidence level
- If you are assuming that the two populations have equal variances, select **Yes** for **Pooled**

- If you are not assuming equal variances, select **No**
- Press **ENTER**
- Arrow down to "**Calculate**"
- Press **ENTER**
- If you selected "Stats," enter the means, standard deviations, and sample sizes
- Set **C-Level** to the confidence level
- If you are assuming that the two populations have equal variances, select **Yes** for **Pooled**
- If you are not assuming equal variances, select **No**

- Press **ENTER**
- Arrow down to "Calculate"
- Press **ENTER**

(The accompanying screen is set up for an example with a mean of 100, a standard deviation of 10, and a sample size of 15 for the first data set and a mean of 105, a standard deviation of 12, and a sample size of 18 for the second data set.)

The confidence interval will be displayed with the number of degrees of freedom, the sample statistics, and the pooled standard deviation (when appropriate).

Hypothesis Test for $\mu_1 - \mu_2$

Step 1 *Enter the data (Skip to Step 2 if you have summary statistics, not raw data)*

- Press **STAT** and select **1:Edit**

Note: If the lists already contain data, clear the old data. Use the up **ARROW** to highlight "**L1**."

- Press **CLEAR ENTER**
- Use the up **ARROW** to highlight "**L2**"
- Press **CLEAR ENTER**
- Use the **ARROW** and **ENTER** keys to enter the first data set into **L1**
- Use the **ARROW** and **ENTER** keys to enter the second data set into **L2**

Step 2 *Access the statistical tests menu*

- Press **STAT**
- Arrow right to **TESTS**
- Arrow down to **2-SampTTest**
- Press **ENTER**

Step 3 *Choose "Data" or "Stats" ("Data" is selected when you have entered the raw data into the Lists. "Stats" is selected when you are given only the means, standard deviations, and sample sizes)*

- Press **ENTER**
- If you selected "Data," set **List1** to **L1** and **List2** to **L2**
- Set **Freq1** to **1** and set **Freq2** to **1**
- Use the **ARROW** to highlight the appropriate alternative hypothesis
- Press **ENTER**
- If you are assuming that the two populations have equal variances, select **Yes** for **Pooled**
- If you are not assuming equal variances, select **No**
- Press **ENTER**
- Arrow down to "Calculate"
- Press **ENTER**
- If you selected "Stats," enter the means, standard deviations, and sample sizes
- Use the **ARROW** to highlight the appropriate alternative hypothesis

- Press **ENTER**
- If you are assuming that the two populations have equal variances, select **Yes** for **Pooled**
- If you are not assuming equal variances, select **No**
- Press **ENTER**
- Arrow down to "Calculate"
- Press **ENTER**

(The screen that follows is set up for an example with a mean of 100, a standard deviation of 10, and a sample size of 15 for the first data set and a mean of 120, a standard deviation of 12, and a sample size of 18 for the second data set.)

The results of the hypothesis test will be displayed with the p-value, the number of degrees of freedom, the sample statistics, and the pooled standard deviation (when appropriate).

Confidence Interval for a Paired Difference Mean

Note: There is no paired difference option on the calculator. These instructions demonstrate how to calculate the differences and then use the 1-sample *t*-interval.

Step 1 *Enter the data and calculate the differences*

- Press **STAT** and select **1:Edit**

Note: If the lists already contain data, clear the old data. Use the up **ARROW** to highlight "**L1**."

- Press **CLEAR ENTER**
- Use the up **ARROW** to highlight "**L2**"
- Press **CLEAR ENTER**
- Use the **ARROW** and **ENTER** keys to enter the first data set into **L1**
- Use the **ARROW** and **ENTER** keys to enter the second data set into **L2**
- The differences will be calculated in **L3**
- Use the up **ARROW** to highlight "**L3**"
- Press **CLEAR**—This will clear any old data, but **L3** will remain highlighted
- To enter the equation L3 = L1 − L2, use the following keystrokes:
- Press **2ND "1"** (this will enter L1)
- Press the **MINUS** button
- Press **2ND "2"** (this will enter L2)

(Notice the equation at the bottom of the screen.)

- Press **ENTER** (the differences should be calculated in L3)

Step 2 *Access the statistical tests menu*

- Press **STAT**
- Arrow right to **TESTS**

- Arrow down to **TInterval** (even for large sample case)
- Press **ENTER**

Step 3 *Choose "Data"*

- Press **ENTER**
- Set **List** to L3
- Set **Freq** to 1
- Set **C-Level** to the confidence level
- Arrow down to "**Calculate**"
- Press **ENTER**

The confidence interval will be displayed with the mean, standard deviation, and sample size of the differences.

Hypothesis Test for a Paired Difference Mean

Note: There is no paired difference option on the calculator. These instructions demonstrate how to calculate the differences and then use the 1-sample *t*-test.

Step 1 *Enter the data and calculate the differences*

- Press **STAT** and select **1:Edit**

Note: If the lists already contain data, clear the old data. Use the up **ARROW** to highlight "**L1.**"

- Press **CLEAR ENTER**
- Use the up **ARROW** to highlight "**L2**"
- Press **CLEAR ENTER**
- Use the **ARROW** and **ENTER** keys to enter the first data set into **L1**
- Use the **ARROW** and **ENTER** keys to enter the second data set into **L2**
- The differences will be calculated in **L3**
- Use the up **ARROW** to highlight "**L3**"
- Press **CLEAR**—This will clear any old data, but **L3** will remain highlighted
- To enter the equation $L3 = L1 - L2$, use the following keystrokes:
- Press **2ND "1"** (this will enter L1)
- Press the **MINUS** button
- Press **2ND "2"** (this will enter L2)

(Notice the equation at the bottom of the screen.)

- Press **ENTER** (the differences should be calculated in L3)

L1	L2	**L3** 3
65	62	3
58	53	5
78	36	42
60	34	26
68	56	12
69	50	19
66	42	24

L3 =L1-L2

Step 2 *Access the statistical tests menu*

- Press **STAT**
- Arrow right to **TESTS**
- Arrow down to **T-Test (even for a large-sample case)**
- Press **ENTER**

Step 3 *Choose "Data"*

- Press **ENTER**
- Enter the values for the hypothesis test, where $\mu_0 =$ the value for μ_d in the null hypothesis
- Set **List** to L3
- Set **Freq** to 1
- Use the **ARROW** to highlight the appropriate alternative hypothesis
- Press **ENTER**
- Arrow down to "**Calculate**"
- Press **ENTER**

The test statistic and the *p*-value will be displayed, as will the sample mean, standard deviation, and sample size of the differences.

ANOVA *F*-test

Step 1 *Enter each data set into its own list (i.e., sample 1 into L1, sample 2 into L2, sample 3 into L3, etc.)*

Step 2 *Access the statistical test menu*

- Press **STAT**
- Arrow right to **TESTS**
- Arrow down to **ANOVA**
- Press **ENTER**
- Type in each List name, separated by commas (*e.g., L1, L2, L3, L4*)
- Press **ENTER**

ANOVA(L₁,L₂,L₃,L
4)

Step 3 *View display*

The calculator will display the *F*-test statistic, as well as the *p*-value, the factor degrees of freedom, sum of squares, mean square, and by arrowing down, the Error degrees of freedom, sum of squares, mean square, and the pooled standard deviation.

One-way ANOVA
F=2.25
p=.1597672711
Factor
 df=3
 SS=54
↓ MS=18

One-way ANOVA
↑ MS=18
 Error
  df=8
  SS=64
  MS=8
 Sxp=2.82842712

8 Comparing Population Proportions

CONTENTS

Where We've Been

- Presented both parametric and nonparametric methods for comparing two or more population means

Where We're Going

- Discuss methods for comparing two population proportions. (8.1–8.2)
- Present a *chi-square* hypothesis test for comparing the category proportions associated with a single qualitative variable—called a *one-way analysis* (8.3)
- Present a chi-square hypothesis test for relating two qualitative variables—called a *two-way analysis.* (8.4)

Traditionally, a common social activity on American college campuses is drinking alcohol. Despite laws on underage drinking, fraternities, sororities, and other campus groups often have alcohol available at their weekend parties. For some students, this activity leads to binge drinking and excessive alcohol use, often resulting in academic failure, physical violence, accidental injury, and even death. In fact, the Journal of Studies on Alcohol (Vol. 63, 2002) recently reported that about 1,400 alcohol-related deaths occur each year on American college campuses.

To gain insight into the alcohol consumption behavior of college students, professors Soyeon Shim (University of Arizona) and Jennifer Maggs (Pennsylvania State University) designed a study and reported their results in *Family and Consumer Sciences Research Journal* (Mar. 2005). Among the researchers' main objectives were (1) to segment college students on the basis of their rates of alcohol consumption and (2) to establish a statistical link between the frequency of drinking and the amount of alcohol consumed. They collected survey data from undergraduate students enrolled in a variety of courses at the University of Arizona, a large state university in the Southwest. To increase the likelihood of obtaining a representative sample, the researchers balanced the sample with both lower and upper division students, as well as students with majors in the social sciences, humanities, business, engineering, and the natural sciences. A total of 657 students completed usable surveys.

The survey consisted of a six-page booklet that took approximately 10 minutes to complete. Two of the many questions on the survey (and the subject of this *Statistics in Action*) pertained to the frequency with which the student drank alcohol (beer, wine, or liquor) during the previous one-month period and the average number of drinks the student consumed per occasion. From this information, the researchers categorized students according to Type of drinker. Responses for the three variables of interest were classified qualitatively as shown in Table SIA8.1. The data for the 657 students are saved in the **COLLDRINKS** file.

In an attempt to help the researchers achieve their objectives, we apply the statistical methodology presented in this chapter to this data set in two *Statistics in Action Revisited* examples.

Statistics IN Action Revisited

- Testing Category Proportions for Type of College Drinker (p. 444)

- Testing whether Frequency of Drinking Is Related to Amount of Alcohol Consumed (p. 456)

Table SIA8.1 Qualitative Variables Measured in the Drinking Study	
Variable Name	Levels (possible values)
AMOUNT	None, 1 drink, 2–3 drinks, 4–6 drinks, 7–9 drinks, 10 or more drinks
FREQUENCY	None, Once a month, Once or twice per week, More
TYPE	Non/Seldom, Social, Typical binge, Heavy binge

 Data Set: COLLDRINKS

Many experiments are conducted in the biological, physical, and social sciences to compare two or more proportions. Those conducted in business and the social sciences to sample the opinions of people are called **sample surveys**. For example, a state government might wish to estimate the difference between the proportions of people in two regions of the state who would qualify for a new welfare program. Or, after an innovative process change, an engineer might wish to determine whether the proportion of defective items produced by a manufacturing process was less than the proportion of defectives produced before the change. Or, a sports psychologist might want to determine if the proportions of professional hockey players differ depending on their highest level of education attained (some high school, high school diploma, some college, college undergraduate degree, college graduate degree). In Section 8.1 we show you how to test hypotheses about the difference between two population proportions based on independent random sampling. We will also show how to find a confidence interval for the difference. Then, in Section 8.3 we will compare more than two population proportions, and in Section 8.4 we will present a related problem.

8.1 Comparing Two Population Proportions: Independent Sampling

Suppose a presidential candidate wants to compare the preferences of registered voters in the northeastern United States with those in the southeastern United States. Such a comparison would help determine where to concentrate campaign efforts. The candidate hires a professional pollster to randomly choose 1,000 registered voters in the northeast and 1,000 in the southeast and interview each to learn her or his voting preference. The objective is to use this sample information to make an inference about the difference $(p_1 - p_2)$ between the proportion p_1 of *all* registered voters in the northeast and the proportion p_2 of *all* registered voters in the southeast who plan to vote for the presidential candidate.

Table 8.1	Results of Poll
Northeast	Southeast
$n_1 = 1{,}000$	$n_2 = 1{,}000$
$x_1 = 546$	$x_2 = 475$

The two samples represent independent binomial experiments. (See Section 4.3 for the characteristics of binomial experiments.) The binomial random variables are the numbers x_1 and x_2 of the 1,000 sampled voters in each area who indicate that they will vote for the candidate. The results are summarized in Table 8.1.

We can now calculate the sample proportions $\hat{p}_1$ and $\hat{p}_2$ of the voters in favor of the candidate in the northeast and southeast, respectively:

$$\hat{p} = \frac{x_1}{n_1} = \frac{546}{1{,}000} = .546 \quad \hat{p}_2 = \frac{x_2}{n_2} = \frac{475}{1{,}000} = .475$$

The difference between the sample proportions $(\hat{p}_1 - \hat{p}_2)$ makes an intuitively appealing point estimator of the difference between the population $(p_1 - p_2)$. For this example, the estimate is

$$(\hat{p}_1 - \hat{p}_2) = .546 - .475 = .071$$

To judge the reliability of the estimator $(\hat{p}_1 - \hat{p}_2)$, we must observe its performance in repeated sampling from the two populations. That is, we need to know the sampling distribution of $(\hat{p}_1 - \hat{p}_2)$. The properties of the sampling distribution are given in the next box. Remember that $\hat{p}_1$ and $\hat{p}_2$ can be viewed as means of the number of successes per trial in the respective samples, so the Central Limit Theorem applies when the sample sizes are large.

Properties of the Sampling Distribution of $(\hat{p}_1 - \hat{p}_2)$

1. The mean of the sampling distribution of $(\hat{p}_1 - \hat{p}_2)$ is $(p_1 - p_2)$; that is,

$$E(\hat{p}_1 - \hat{p}_2) = p_1 - p_2$$

   Thus, $(\hat{p}_1 - \hat{p}_2)$ is an unbiased estimator of $(p_1 - p_2)$.

2. The standard deviation of the sampling distribution of $(\hat{p}_1 - \hat{p}_2)$ is

$$\sigma_{(\hat{p}_1 - \hat{p}_2)} = \sqrt{\frac{p_1 q_1}{n_1} + \frac{p_2 q_2}{n_2}}$$

3. If the sample sizes n_1 and n_2 are large (see Section 5.4 for a guideline), the sampling distribution of $(\hat{p}_1 - \hat{p}_2)$ is approximately normal.

Since the distribution of $(\hat{p}_1 - \hat{p}_2)$ in repeated sampling is approximately normal, we can use the z-statistic to derive confidence intervals for $(p_1 - p_2)$ or to test a hypothesis about $(p_1 - p_2)$.

For the voter example, a 95% confidence interval for the difference $(p_1 - p_2)$ is

$$(\hat{p}_1 - \hat{p}_2) \pm 1.96 \sigma_{(\hat{p}_1 - \hat{p}_2)}, \text{ or } (\hat{p}_1 - \hat{p}_2) \pm 1.96 \sqrt{\frac{p_1 q_1}{n_1} + \frac{p_2 q_2}{n_2}}$$

The quantities $p_1 q_1$ and $p_2 q_2$ must be estimated in order to complete the calculation of the standard deviation $\sigma_{(\hat{p}_1 - \hat{p}_2)}$ and, hence, the calculation of the confidence interval. In Section 5.4, we showed that the value of pq is relatively insensitive to the value chosen to approximate p. Therefore, $\hat{p}_1 \hat{q}_1$ and $\hat{p}_2 \hat{q}_2$ will provide satisfactory approximations of $p_1 q_1$ and $p_2 q_2$, respectively. Then

$$\sqrt{\frac{p_1 q_1}{n_1} + \frac{p_2 q_2}{n_2}} \approx \sqrt{\frac{\hat{p}_1 \hat{q}_1}{n_1} + \frac{\hat{p}_2 \hat{q}_2}{n_2}}$$

and we will approximate the 95% confidence interval by

$$(\hat{p}_1 - \hat{p}_2) \pm 1.96 \sqrt{\frac{\hat{p}_1 \hat{q}_1}{n_1} + \frac{\hat{p}_2 \hat{q}_2}{n_2}}$$

Substituting the sample quantities yields

$$(.546 - .475) \pm 1.96 \sqrt{\frac{(.546)(.454)}{1,000} + \frac{(.475)(.525)}{1,000}}$$

or $.071 \pm .044$. Thus, we are 95% confident that the interval from .027 to .115 contains $(p_1 - p_2)$.

We infer that there are between 2.7% and 11.5% more registered voters in the northeast than in the southeast who plan to vote for the presidential candidate. It seems that the candidate should direct a greater campaign effort in the southeast than in the northeast.

Now Work Exercise 8.9

The general form of a confidence interval for the difference $(p_1 - p_2)$ between population proportions is given in the following box:

Large-Sample $100(1 - \alpha)\%$ Confidence Interval for $(p_1 - p_2)$: Normal (z) Statistic

$$(\hat{p}_1 - \hat{p}_2) \pm z_{\alpha/2} \sigma_{(\hat{p}_1 - \hat{p}_2)} = (\hat{p}_1 - \hat{p}_2) \pm z_{\alpha/2} \sqrt{\frac{p_1 q_1}{n_1} + \frac{p_2 q_2}{n_2}}$$

$$\approx (\hat{p}_1 - \hat{p}_2) \pm z_{\alpha/2} \sqrt{\frac{\hat{p}_1 \hat{q}_1}{n_1} + \frac{\hat{p}_2 \hat{q}_2}{n_2}}$$

The z-statistic,

$$z = \frac{(\hat{p}_1 - \hat{p}_2) - (p_1 - p_2)}{\sigma_{(\hat{p}_1 - \hat{p}_2)}}$$

is used to test the null hypothesis that $(p_1 - p_2)$ equals some specified difference, say, D_0. For the special case where $D_0 = 0$—that is, where we want to test the null hypothesis $H_0: (p_1 - p_2) = 0$ (or, equivalently, $H_0: p_1 = p_2$)—the best estimate of $p_1 = p_2 = p$ is obtained by dividing the total number of successes $(x_1 + x_2)$ for the two samples by the total number of observations $(n_1 + n_2)$, that is,

$$\hat{p} = \frac{x_1 + x_2}{n_1 + n_2}, \quad \text{or} \quad \hat{p} = \frac{n_1 \hat{p}_1 + n_2 \hat{p}_2}{n_1 + n_2}$$

The second equation shows that $\hat{p}$ is a weighted average of $\hat{p}_1$ and $\hat{p}_2$, with the larger sample receiving more weight. If the sample sizes are equal, then $\hat{p}$ is a simple average of the two sample proportions of successes.

We now substitute the weighted average $\hat{p}$ for both p_1 and p_2 in the formula for the standard deviation of $(\hat{p}_1 - \hat{p}_2)$:

$$\sigma_{(\hat{p}_1 - \hat{p}_2)} = \sqrt{\frac{p_1 q_1}{n_1} + \frac{p_2 q_2}{n_2}} \approx \sqrt{\frac{\hat{p}\hat{q}}{n_1} + \frac{\hat{p}\hat{q}}{n_2}} = \sqrt{\hat{p}\hat{q}\left(\frac{1}{n_1} + \frac{1}{n_2}\right)}$$

The test is summarized in the following boxes:

Conditions Required for Valid Large-Sample Inferences about $(p_1 - p_2)$

1. The two samples are randomly selected in an independent manner from the two target populations.

2. The sample sizes, n_1 and n_2, are both large, so the sampling distribution of $(\hat{p}_1 - \hat{p}_2)$ will be approximately normal. (This condition will be satisfied if both $n_1 \hat{p}_1 \geq 15, n_1 \hat{q}_1 \geq 15$, and $n_2 \hat{p}_2 \geq 15, n_2 \hat{q}_2 \geq 15$.)

Large-Sample Test of Hypothesis about $(p_1 - p_2)$: Normal (z) Statistic

One-Tailed Test	**Two-Tailed Test**
H_0: $(p_1 - p_2) = 0*$	H_0: $(p_1 - p_2) = 0$
H_a: $(p_1 - p_2) < 0$	H_a: $(p_1 - p_2) \neq 0$
[or H_a: $(p_1 - p_2) > 0$]	

$$\text{Test statistic: } z = \frac{(\hat{p}_1 - \hat{p}_2)}{\sigma_{(\hat{p}_1 - \hat{p}_2)}}$$

Rejection region: $z < -z_\alpha$ Rejection region: $|z| > z_{\alpha/2}$
[or $z > z_\alpha$ when H_a: $(p_1 - p_2) > 0$]

$$\text{Note: } \sigma_{(\hat{p}_1 - \hat{p}_2)} = \sqrt{\frac{p_1 q_1}{n_1} + \frac{p_2 q_2}{n_2}} \approx \sqrt{\hat{p}\hat{q}\left(\frac{1}{n_1} + \frac{1}{n_2}\right)} \text{ where } \hat{p} = \frac{x_1 + x_2}{n_1 + n_2}$$

Example 8.1

A Large-Sample Test about $(p_1 - p_2)$—Comparing Fractions of Smokers for Two Years

Problem In the past decade, intensive antismoking campaigns have been sponsored by both federal and private agencies. Suppose the American Cancer Society randomly sampled 1,500 adults in 2000 and then sampled 1,750 adults in 2010 to determine whether there was evidence that the percentage of smokers had decreased. The results of the two sample surveys are shown in Table 8.2, where x_1 and x_2 represent the numbers of smokers in the 2000 and 2010 samples, respectively. Do these data indicate that the fraction of smokers decreased over this 10-year period? Use $\alpha = .05$.

Solution If we define p_1 and p_2 as the true proportions of adult smokers in 2000 and 2010, respectively, then the elements of our test are

$$H_0: (p_1 - p_2) = 0$$
$$H_a: (p_1 - p_2) > 0$$

(The test is one tailed, since we are interested only in determining whether the proportion of smokers *decreased*.)

Table 8.2
Results of Smoking Survey

2000	2010
$n_1 = 1,500$	$n_2 = 1,750$
$x_1 = 555$	$x_2 = 578$

$$\text{Test statistic: } z = \frac{(\hat{p}_1 - \hat{p}_2) - 0}{\sigma_{(\hat{p}_1 - \hat{p}_2)}}$$

Rejection region using $\alpha = .05$: $z > z_\alpha = z_{.05} = 1.645$ (see Figure 8.1)

We now calculate the sample proportions of smokers:

$$\hat{p}_1 = \frac{555}{1,500} = .37 \qquad \hat{p}_2 = \frac{578}{1,750} = .33$$

Then

$$z = \frac{(\hat{p}_1 - \hat{p}_2) - 0}{\sigma_{(\hat{p}_1 - \hat{p}_2)}} \approx \frac{(\hat{p}_1 - \hat{p}_2)}{\sqrt{\hat{p}\hat{q}\left(\frac{1}{n_1} + \frac{1}{n_2}\right)}}$$

where

$$\hat{p} = \frac{x_1 + x_2}{n_1 + n_2} = \frac{555 + 578}{1,500 + 1,750} = .349$$

Note that $\hat{p}$ is a weighted average of $\hat{p}_1$ and $\hat{p}_2$, with more weight given to the larger (2010) sample. Thus, the computed value of the test statistic is

$$z = \frac{.37 - .33}{\sqrt{(.349)(.651)\left(\frac{1}{1,500} + \frac{1}{1,750}\right)}} = \frac{.040}{.0168} = 2.38$$

Figure 8.1
Rejection region for Example 8.1

---

*The test can be adapted to test for a difference $D_0 \neq 0$. Because most applications call for a comparison of p_1 and p_2, implying that $D_0 = 0$, we will confine our attention to this case.

There is sufficient evidence at the $\alpha = .05$ level to conclude that the proportion of adults who smoke has decreased over the 10-year period.

Look Back We could place a confidence interval on $(p_1 - p_2)$ if we were interested in estimating the extent of the decrease.

Now Work Exercise 8.12

Example 8.2
Finding The Observed Significance Level of a Test for $(p_1 - p_2)$

Problem Use a statistical software package to conduct the test presented in Example 8.1. Find and interpret the p-value of the test.

Solution We entered the sample sizes (n_1 and n_2) and numbers of successes (x_1 and x_2) into MINITAB and obtained the printout shown in Figure 8.2. The test statistic for this one-tailed test, $z = 2.37$, as well as the p-value of the test, are highlighted on the printout. Note that p-value $= .009$ is smaller than $\alpha = .05$. Consequently, we have strong evidence to reject H_0 and conclude that p_1 exceeds p_2.

Figure 8.2
MINITAB output for test of two proportions

```
Test and CI for Two Proportions

Sample    X     N    Sample p
1        555  1500   0.370000
2        578  1750   0.330286


Difference = p (1) - p (2)
Estimate for difference:  0.0397143
95% lower bound for difference:  0.0121024
Test for difference = 0 (vs > 0):  Z = 2.37  P-Value = 0.009

Fisher's exact test: P-Value = 0.010
```

Exercises 8.1–8.21

Understanding the Principles

8.1 What conditions are required for valid large-sample inferences about $p_1 - p_2$?

8.2 What is the problem with using the z-statistic to make inferences about $p_1 - p_2$ when the sample sizes are both small?

8.3 Consider making an inference about $p_1 - p_2$, where there are x_1 successes in n_1 binomial trials and x_2 successes in n_2 binomial trials.
  a. Describe the distributions of x_1 and x_2.
  b. For large samples, describe the sampling distribution of $(\hat{p}_1 - \hat{p}_2)$.

Learning the Mechanics

8.4 In each case, determine whether the sample sizes are large enough to conclude that the sampling distribution of $(\hat{p}_1 - \hat{p}_2)$ is approximately normal.
  a. $n_1 = 10, n_2 = 12, \hat{p}_1 = .50, \hat{p}_2 = .50$
  b. $n_1 = 10, n_2 = 12, \hat{p}_1 = .10, \hat{p}_2 = .08$
  c. $n_1 = n_2 = 30, \hat{p}_1 = .20, \hat{p}_2 = .30$
  d. $n_1 = 100, n_2 = 200, \hat{p}_1 = .05, \hat{p}_2 = .09$
  e. $n_1 = 100, n_2 = 200, \hat{p}_1 = .95, \hat{p}_2 = .91$

8.5 Construct a 95% confidence interval for $(p_1 - p_2)$ in each of the following situations:
  a. $n_1 = 400, \hat{p}_1 = .65; n_2 = 400, \hat{p}_2 = .58$
  b. $n_1 = 180, \hat{p}_1 = .31; n_2 = 250, \hat{p}_2 = .25$
  c. $n_1 = 100, \hat{p}_1 = .46; n_2 = 120, \hat{p}_2 = .61$

8.6 Independent random samples, each containing 800 observations, were selected from two binomial populations. The samples from populations 1 and 2 produced 320 and 400 successes, respectively.
  a. Test $H_0: (p_1 - p_2) = 0$ against $H_a: (p_1 - p_2) \neq 0$. Use $\alpha = .05$.
  b. Test $H_0: (p_1 - p_2) = 0$ against $H_a: (p_1 - p_2) \neq 0$. Use $\alpha = .01$.
  c. Test $H_0: (p_1 - p_2) = 0$ against $H_a: (p_1 - p_2) < 0$. Use $\alpha = .01$.
  d. Form a 90% confidence interval for $(p_1 - p_2)$.

8.7 Random samples of size $n_1 = 50$ and $n_2 = 60$ were drawn from populations 1 and 2, respectively. The samples yielded $\hat{p}_1 = .4$ and $\hat{p}_2 = .2$. Test $H_0: (p_1 - p_2) = .1$ against $H_a: (p_1 - p_2) > .1$, using $\alpha = .05$.

8.8 Sketch the sampling distribution of $(\hat{p}_1 - \hat{p}_2)$ based on independent random samples of $n_1 = 100$ and $n_2 = 200$ observations from two binomial populations with probabilities of success $p_1 = .1$ and $p_2 = .5$, respectively.

Applying the Concepts—Basic

8.9 **Bullying behavior study.** School bullying is a form of aggressive behavior that occurs when a student is exposed repeatedly to negative actions (e.g., name-calling, hitting, kicking, spreading slander) from another student. In order to study the effectiveness of an antibullying policy at Dutch elementary schools, a survey of over 2,000 elementary school children was conducted (*Health Education Research*, Feb. 2005). Each student was asked if he or she ever bullied another student. In a sample of 1,358 boys, 746 claimed they had never bullied another student. In a sample of 1,379 girls, 967 claimed they had never bullied another student.

a. Estimate the true proportion of Dutch boys who have never bullied another student.

b. Estimate the true proportion of Dutch girls who have never bullied another student.

c. Estimate the difference in the proportions with a 90% confidence interval.

d. Make a statement about how likely the interval you used in part **c** contains the true difference in proportions.

e. Which group is more likely to bully another student, Dutch boys or Dutch girls?

8.10 **Is steak your favorite barbecue food?** July is National Grilling Month in the United States. On July 1, 2008, *The Harris Poll #70* reported on a survey of Americans' grilling preferences. When asked about their favorite food prepared on a barbeque grill, 662 of 1,250 randomly sampled Democrats preferred steak, as compared to 586 of 930 randomly sampled Republicans.

a. Give a point estimate for the proportion of all Democrats who prefer steak as their favorite barbeque food.

b. Give a point estimate for the proportion of all Republicans who prefer steak as their favorite barbeque food.

c. Give a point estimate for the difference between the proportions of all Democrats and all Republicans who prefer steak as their favorite barbeque food.

d. Construct a 95% confidence interval for the difference between the proportions of all Democrats and all Republicans who prefer steak as their favorite barbeque food.

e. Give a practical interpretation of the interval, part **d.**

f. Explain the meaning of the phrase "95% confident" in your answer to part **e.**

8.11 **Treating depression with St. John's wort.** The *Journal of the American Medical Association* (April 18, 2001) published a study of the effectiveness of using extracts of the herbal medicine St. John's wort in treating major depression. In an eight-week randomized, controlled trial, 200 patients diagnosed with major depression were divided into two groups, one of which ($n_1 = 98$) received St. John's wort extract while the other ($n_2 = 102$) received a placebo (no drug). At the end of the study period, 14 of the St. John's wort patients were in remission, compared with 5 of the placebo patients.

a. Compute the proportion of the St. John's wort patients who were in remission.

b. Compute the proportion of the placebo patients who were in remission.

c. If St. John's wort is effective in treating major depression, then the proportion of St. John's wort patients in remission will exceed the proportion of placebo patients in remission. At $\alpha = .01$, is St. John's wort effective in treating major depression?

d. Repeat part **c,** but use $\alpha = .10$.

e. Explain why the choice of α is critical for this study.

8.12 **Planning-habits survey.** *American Demographics* (Jan. 2002) [NW] reported the results of a survey on the planning habits of men and women. In response to the question "What is your preferred method of planning and keeping track of meetings,

appointments, and deadlines?" 56% of the men and 46% of the women answered "I keep them in my head." A nationally representative sample of 1,000 adults participated in the survey; therefore, assume that 500 were men and 500 were women.

a. Set up the null and alternative hypotheses for testing whether the percentage of men who prefer keeping track of appointments in their head is larger than the corresponding percentage of women.

b. Compute the test statistic for the test.

c. Give the rejection region for the test, using $\alpha = .01$.

d. Find the *p*-value for the test.

e. Draw the appropriate conclusion.

8.13 **Racial profiling by the LAPD.** *Racial profiling* is a term used to describe any police action that relies on ethnicity rather than behavior to target suspects engaged in criminal activities. Does the Los Angeles Police Department (LAPD) invoke racial profiling in stops and searches of Los Angeles drivers? This question was addressed in *Chance* (Spring 2006).

a. Data on stops and searches of both African-Americans and white drivers from January through June 2005 are summarized in the accompanying table. Conduct a test (at $\alpha = .05$) to determine whether there is a disparity in the proportions of African-American and white drivers who are searched by the LAPD after being stopped.

Race	Number Stopped	Number Searched	Number of "Hits"
African-American	61,688	12,016	5,134
White	106,892	5,312	3,006

Based on Khadjavi, L. S. "Driving while black in the City of Angels." *Chance*, Vol. 19, No. 2, Spring 2006 (Tables 1 and 2), pp. 43–46.

b. The LAPD defines a "hit rate" as the proportion of searches that result in a discovery of criminal activity. Use the data in the table to estimate the disparity in the hit rates for African-American and white drivers under a 95% confidence interval. Interpret the results.

Applying the Concepts—Intermediate

8.14 **Angioplasty's benefits challenged.** Each year, more than 1 million heart patients undergo an angioplasty. The benefits of an angioplasty were challenged in a recent study of 2,287 patients (2007 Annual Conference of the American College of Cardiology, New Orleans). All the patients had substantial blockage of the arteries, but were medically stable. All were treated with medication such as aspirin and beta-blockers. However, half the patients were randomly assigned to get an angioplasty and half were not. After five years, the researchers found that 211 of the 1,145 patients in the angioplasty group had subsequent heart attacks, compared with 202 of 1,142 patients in the medication-only group. Do you agree with the study's conclusion that "There was no significant difference in the rate of heart attacks for the two groups"? Support your answer with a 95% confidence interval.

8.15 **Killing insects with low oxygen.** Refer to the *Journal of Agricultural, Biological, and Environmental Statistics* (Sept. 2000) study of the mortality of rice weevils exposed to low oxygen, presented in Exercise 6.94 (p. 331). Recall that 31,386 of 31,421 rice weevils were found dead after exposure to nitrogen gas for 4 days. In a second experiment, 23,516 of 23,676 rice weevils were found dead after exposure to nitrogen gas for 3.5 days. Conduct a test of hypothesis to compare the mortality rates of adult rice weevils exposed to nitrogen at the two exposure times. Is there a significant difference (at $\alpha = .10$) in the mortality rates?

8.16 **Effectiveness of drug tests of Olympic athletes.** Erythropoietin (EPO) is a banned drug used by athletes to increase the oxygen-carrying capacity of their blood. New tests for EPO were first introduced prior to the 2000 Olympic Games held in Sydney, Australia. *Chance* (Spring 2004) reported that of a sample of 830 world-class athletes, 159 did not compete in the 1999 World Championships (a year prior to the introduction of the new EPO test). Similarly, 133 of 825 potential athletes did not compete in the 2000 Olympic games. Was the new test effective in deterring an athlete from participating in the 2000 Olympics? If so, then the proportion of nonparticipating athletes in 2000 will be more than the proportion of nonparticipating athletes in 1999. Conduct the analysis (at $\alpha = .10$) and draw the proper conclusion.

8.17 **"Tip-of-the-tongue" study.** Trying to think of a word you know, but can't instantly retrieve, is called the "tip-of-the-tongue" phenomenon. *Psychology and Aging* (Sept. 2001) published a study of this phenomenon in senior citizens. The researchers compared 40 people between 60 and 72 years of age with 40 between 73 and 83 years of age. When primed with the initial syllable of a missing word (e.g., seeing the word *include* to help recall the word *incisor*), the younger seniors had a higher recall rate. Suppose 31 of the 40 seniors in the younger group could recall the word when primed with the initial syllable, while only 22 of the 40 seniors could recall the word. Compare the recall rates of the two groups, using $\alpha = .05$. Does one group of elderly people have a significantly higher recall rate than the other?

8.18 **Detection of rigged school milk prices (cont'd).** Refer to the investigation of collusive bidding in the northern Kentucky school milk market, presented in Exercise 7.26 (p. 368). Market allocation is a common form of collusive behavior in bid-rigging conspiracies. Under collusion, the same dairy usually controls the same school districts year after year. The *incumbency rate* for a market is defined as the proportion of school districts that are won by the vendor that won the previous year. Past experience with milk bids in a competitive environment reveals that a typical incumbency rate is .7. That is, 70% of the school districts are expected to purchase their milk from the dairy that won the previous year. Incumbency rates of .9 or higher are strong indicators of collusive bidding. Over the years, when bid collusion was alleged to have occurred in northern Kentucky, there were 51 potential vendor transitions (i.e., changes in milk supplier from one year to the next in a district) in the tricounty market and 134 potential vendor transitions in the surrounding market. These values represent the sample sizes ($n_1 = 134$ and $n_2 = 51$) for calculating incumbency rates. Examining the data saved in the **MILK** file, you'll find that in 50 of the 51 potential vendor transitions for the tricounty market, the winning dairy from the previous year won the bid the next year; similarly, you'll find that in 91 of the 134 potential vendor transitions for the surrounding area, the same dairy won the bid the next year.

a. Estimate the incumbency rates for the tricounty and surrounding milk markets.

b. A MINITAB printout comparing the two incumbency rates is shown below. Give a practical interpretation of the results. Do they show further support for the bid collusion theory?

Test and CI for Two Proportions

```
Sample   X    N    Sample p
1        91  134   0.679104
2        50   51   0.980392


Difference = p (1) - p (2)
Estimate for difference:  -0.301288
95% upper bound for difference:  -0.227669
Test for difference = 0 (vs < 0):  Z = -4.30  P-Value = 0.000
```

8.19 **Does sleep improve mental performance?** Are creativity and problem solving linked to adequate sleep? This question was the subject of research conducted by German scientists at the University of Lübeck (*Nature*, Jan. 22, 2004). One hundred volunteers were divided into two equal-sized groups. Each volunteer took a math test that involved transforming strings of eight digits into a new string that fit a set of given rules, as well as a third, hidden rule. Prior to taking the test, one group received eight hours of sleep, while the other group stayed awake all night. The scientists monitored the volunteers to determine whether and when they figured out the third rule. Of the volunteers who slept, 39 discovered the third rule; of the volunteers who stayed awake all night, 15 discovered the third rule. From the study results, what can you infer about the proportions of volunteers in the two groups who discover the third rule? Support your answer with a 90% confidence interval.

Applying the Concepts—Advanced

8.20 **Religious symbolism in TV commercials.** Gonzaga University professors conducted a study of television commercials and published their results in the *Journal of Sociology, Social Work and Social Welfare* (Vol. 2, 2008). The key research question was: "Do television advertisers use religious symbolism to sell goods and services?" In a sample of 797 TV commercials collected in 1998, only 16 commercials used religious symbolism. Of the sample of 1,499 TV commercials examined in the more recent study, 51 commercials used religious symbolism. Conduct an analysis to determine if the percentage of TV commercials that use religious symbolism has changed since the 1998 study. If you detect a change, estimate the magnitude of the difference and attach a measure of reliability to the estimate.

8.21 **Teeth defects and stress in prehistoric Japan.** Linear enamel hypoplasia (LEH) defects are pits or grooves on the tooth surface that are typically caused by malnutrition, chronic infection, stress, and trauma. A study of LEH defects in prehistoric Japanese cultures was published in the *American Journal of Physical Anthropology* (May 2010). Three groups of Japanese people were studied: Yayoi farmers (early agriculturists), eastern Jomon foragers (broad-based economy), and western Jomon foragers (wet rice economy). LEH defect prevalence was determined from skulls of individuals obtained from each of the three cultures. Of the 182 Yayoi farmers in the study, 63.1% had at least one LEH defect; of the 164 Eastern Jomon foragers, 48.2% had at least one LEH defect; and, of the 122 Western Jomon foragers, 64.8% had at least one LEH defect. Two theories were tested. Theory 1 states that foragers with a broad-based economy will have a lower LEH defect prevalence than early agriculturists. Theory 2 states that foragers with a wet rice economy will not differ in LEH defect prevalence from early agriculturists. Use the results to test both theories, each at $\alpha = .01$.

Based on Temple, D. H. "Patterns of systemic stress during the agricultural transition in prehistoric Japan." *American Journal of Physical Anthropology*, Vol. 142, No. 1, May 2010.

8.2 Determining the Sample Size

The sample sizes n_1 and n_2 required to compare two population proportions can be found in a manner similar to the method described in Section 7.4 for comparing two population means. We will assume equal sized samples (i.e., $n_1 = n_2 = n$) and then choose n so that $(\hat{p}_1 - \hat{p}_2)$ will differ from $(p_1 - p_2)$ by no more than a sampling error SE with a specified level of confidence. We will illustrate the procedure with an example.

Example 8.3

Finding The Sample Sizes For Estimating $(p_1 - p_2)$: Comparing Defect Rates of Two Machines

Problem A production supervisor suspects that a difference exists between the proportions p_1 and p_2 of defective items produced by two different machines. Experience has shown that the proportion defective for each of the two machines is in the neighborhood of .03. If the supervisor wants to estimate the difference in the proportions to within .005, using a 95% confidence interval, how many items must be randomly sampled from the output produced by each machine? (Assume that the supervisor wants $n_1 = n_2 = n$.)

Solution In this sampling problem, the sampling error $SE = .005$, and for the specified level of reliability, $z_{\alpha/2} = z_{0.25} = 1.96$. Then, letting $p_1 = p_2 = .03$ and $n_1 = n_2 = n$, we find the required sample size per machine by solving the following equation for n:

$$z_{\alpha/2}\sigma_{(\hat{p}_1 - \hat{p}_2)} = SE$$

or

$$z_{\alpha/2}\sqrt{\frac{p_1 q_1}{n_1} + \frac{p_2 q_2}{n_2}} = SE$$

$$1.96\sqrt{\frac{(0.3)(.97)}{n} + \frac{(0.3)(.97)}{n}} = .005$$

$$1.96\sqrt{\frac{2(.03)(.97)}{n}} = .005$$

$$n = 8{,}943.2$$

Look Back This large n will likely result in a tedious sampling procedure. If the supervisor insists on estimating $(p_1 - p_2)$ correct to within .005 with 95% confidence, approximately 9,000 items will have to be inspected for each machine.

Now Work Exercise 8.24a

You can see from the calculations in Example 8.3 that $\sigma_{(\hat{p}_1 - \hat{p}_2)}$ (and hence the solution, $n_1 = n_2 = n$) depends on the actual (but unknown) values of p_1 and p_2. In fact, the required sample size $n_1 = n_2 = n$ is largest when $p_1 = p_2 = .5$. Therefore, if you have no prior information on the approximate values of p_1 and p_2, use $p_1 = p_2 = .5$ in the formula for $\sigma_{(\hat{p}_1 - \hat{p}_2)}$. If p_1 and p_2 are in fact close to .5, then the values of n_1 and n_2 that you have calculated will be correct. If p_1 and p_2 differ substantially from .5, then your solutions for n_1 and n_2 will be larger than needed. Consequently, using $p_1 = p_2 = .5$ when solving for n_1 and n_2 is a conservative procedure because the sample sizes n_1 and n_2 will be at least as large as (and probably larger than) needed.

The procedure for determining sample sizes necessary for estimating $(p_1 - p_2)$ for the case $n_1 = n_2$ is given in the following box:

Determination of Sample Size for Estimating $p_1 - p_2$

To estimate $(p_1 - p_2)$ to within a given sampling error SE and with confidence level $(1 - \alpha)$, use the following formula to solve for equal sample sizes that will achieve the desired reliability:

$$n_1 = n_2 = \frac{(z_{\alpha/2})^2(p_1 q_1 + p_2 q_2)}{(SE)^2}$$

You will need to substitute estimates for the values of p_1 and p_2 before solving for the sample size. These estimates might be based on prior samples, obtained from educated guesses or, most conservatively, specified as $p_1 = p_2 = .5$.

Exercises 8.22–8.30

Understanding the Principles

8.22 In determining the sample sizes for estimating $p_1 - p_2$, how do you obtain estimates of the binomial proportions p_1 and p_2 used in the calculations?

8.23 If the sample-size calculation yields a value of n that is too large to be practical, how should you proceed?

Learning the Mechanics

8.24 Assuming that $n_1 = n_2$, find the sample sizes needed to
[NW] estimate $(p_1 - p_2)$ for each of the following situations:
  a. SE = .01 with 99% confidence. Assume that $p_1 \approx .4$ and $p_2 \approx .7$.
  b. A 90% confidence interval of width .05. Assume there is no prior information available with which to obtain approximate values of p_1 and p_2.
  c. SE = .03 with 90% confidence. Assume that $p_1 \approx .2$ and $p_2 \approx .3$.

8.25 Enough money has been budgeted to collect independent random samples of size $n_1 = n_2 = 100$ from populations 1 and 2 in order to estimate $(p_1 - p_2)$. Prior information indicates that $p_1 = p_2 \approx .6$. Have sufficient funds been allocated to construct a 90% confidence interval for $(p_1 - p_2)$ of width .1 or less? Justify your answer.

Applying the Concepts—Basic

8.26 **Size of a political poll.** A pollster wants to estimate the difference between the proportions of men and women who favor a particular national candidate using a 90% confidence interval of width .04. Suppose the pollster has no prior information about the proportions. If equal numbers of men and women are to be polled, how large should the sample sizes be?

8.27 **Angioplasty's benefits challenged.** Refer to the study of patients with substantial blockage of the arteries presented at the 2007 Annual Conference of the American College of Cardiology, Exercise 8.14 (p. 435). Recall that half the patients were randomly assigned to get an angioplasty and half were not. The researchers compared the proportion of patients with subsequent heart attacks for the two groups and reported no significant difference between the two proportions. Although the study involved over 2,000 patients, the sample size may have been too small to detect a difference in heart attack rates.

a. How many patients must be sampled in each group in order to estimate the difference in heart attack rates to within .015 with 95% confidence? (Use summary data from Exercise 8.14 in your calculation.)

b. Comment on the practicality of carrying out the study with the sample sizes determined in part **a**.

c. Comment on the practical significance of the difference detected in the confidence interval for the study, part **a**.

Applying the Concepts—Intermediate

8.28 **Cable-TV home shoppers.** All cable television companies carry at least one home-shopping channel. Who uses these home-shopping services? Are the shoppers primarily men or women? Suppose you want to estimate the difference in the percentages of men and women who say they have used or expect to use televised home shopping. You want an 80% confidence interval of width .06 or less.
  a. Approximately how many people should be included in your samples?
  b. Suppose you want to obtain individual estimates for the two percentages of interest. Will the sample size found in part **a** be large enough to provide estimates of each percentage correct to within .02 with probability equal to .90? Justify your response.

8.29 **Buyers of TVs.** A manufacturer of large-screen televisions wants to compare with a competitor the proportions of its best sets that need repair within 1 year. If it is desired to estimate the difference between proportions to within .05 with 90% confidence, and if the manufacturer plans to sample twice as many buyers (n_1) of its sets as buyers (n_2) of the competitor's sets, how many buyers of each brand must be sample? Assume that the propotion of sets that need repair will be about .2 for both brands.

8.30 **Rat damage in sugarcane.** Poisons are used to prevent rat damage in sugarcane fields. The U.S. Department of Agriculture is investigating whether the rat poison should be located in the middle of the field or on the outer perimeter. One way to answer this question is to determine where the greater amount of damage occurs. If damage is measured by the proportion of cane stalks that have been damaged by rats, how many stalks from each section of the field should be sampled in order to estimate the true difference between proportions of stalks damaged in the two sections, to within .02 with 95% confidence?

8.3 Testing Category Probabilities: Multinomial Experiment

Recall from Section 1.4 (p. 9) that observations on a qualitative variable can only be categorized. For example, consider the highest level of education attained by a professional hockey player. Level of education is a qualitative variable with several categories, including some high school, high school diploma, some college, college undergraduate degree, and graduate degree. If we were to record education level for all professional hockey players, the result of the categorization would be a count of the numbers of players falling into the respective categories.

When the qualitative variable of interest results in one of two responses (e.g., yes or no, success or failure, favor or do not favor), the data—called *counts*—can be analyzed with the binomial probability distribution discussed in Section 4.3. However, qualitative variables, such as level of education, that allow for more than two categories for a response are much more common, and these must be analyzed by a different method.

Qualitative data with more than two levels often result from a **multinomial experiment**. The characteristics for a multinomial experiment with k outcomes are described in the next box. You can see that the binomial experiment of Chapter 4 is a multinomial experiment with $k = 2$.

Properties of the Multinomial Experiment

1. The experiment consists of n identical trials.

2. There are k possible outcomes to each trial. These outcomes are sometimes called **classes, categories,** or **cells.**

3. The probabilities of the k outcomes, denoted by $p_1, p_2, \ldots, p_k$, where $p_1 + p_2 + \cdots + p_k = 1$, remain the same from trial to trial.

4. The trials are independent.

5. The random variables of interest are the **cell counts** $n_1, n_2, \ldots, n_k$ of the number of observations that fall into each of the k categories.

Example 8.4

Identifying a Multinomial Experiment

Problem Consider the problem of determining the highest level of education attained by each of a sample of $n = 40$ National Hockey League (NHL) players. Suppose we categorize level of education into one of five categories—some high school, high school diploma, some college, college undergraduate degree, and graduate degree—and count the number of the 40 players that fall into each category. Is this a multinomial experiment, to a reasonable degree of approximation?

Solution Checking the five properties of a multinomial experiment shown in the box, we have the following:

1. The experiment consists of $n = 40$ identical trials, each of which is undertaken to determine the education level of an NHL player.

2. There are $k = 5$ possible outcomes to each trial, corresponding to the five education-level responses.

3. The probabilities of the $k = 5$ outcomes p_1, p_2, p_3, p_4, and p_5, where p_i represents the true probability that an NHL player attains level-of-education category i, remain the same from trial to trial (to a reasonable degree of approximation).

4. The trials are independent; that is, the education level attained by one NHL player does not affect the level attained by any other player.

5. We are interested in the count of the number of hockey players who fall into each of the five education-level categories. These five cell counts are denoted $n_1, n_2, n_3, n_4,$ and n_5.

Thus, the properties of a multinomial experiment are satisfied.

In this section, we consider a multinomial experiment with k outcomes that correspond to the categories of a *single* qualitative variable. The results of such an experiment are summarized in a **one-way table.** The term *one-way* is used because only one variable is classified. Typically, we want to make inferences about the true percentages that occur in the k categories on the basis of the sample information in the one-way table.

To illustrate, suppose three political candidates are running for the same elective position. Prior to the election, we conduct a survey to determine the voting preferences of a random sample of 150 eligible voters. The qualitative variable of interest is *preferred candidate*, which has three possible outcomes: candidate 1, candidate 2, and candidate 3. Suppose the number of voters preferring each candidate is tabulated and the resulting count data appear as in Table 8.3.

Table 8.3	Results of Voter Preference Survey	
	Candidate	
1	**2**	**3**
61 votes	53 votes	36 votes

Note that our voter preference survey satisfies the properties of a multinomial experiment for the qualitative variable, preferred candidate. The experiment consists of randomly sampling $n = 150$ voters from a large population of voters containing an unknown proportion p_1 that favors candidate 1, a proportion p_2 that favors candidate 2, and a proportion p_3 that favors candidate 3. Each voter sampled represents a single trial that can result in one of three outcomes: The voter will favor candidate 1, 2, or 3 with probabilities $p_1, p_2,$ and p_3, respectively. (Assume that all voters will have a preference.) The voting preference of any single voter in the sample does not affect the preference of any other; consequently, the trials are independent. Finally, you can see that the recorded data are the numbers of voters in each of the three preference categories. Thus, the voter preference survey satisfies the five properties of a multinomial experiment.

In this survey, and in most practical applications of the multinomial experiment, the k outcome probabilities $p_1, p_2, \ldots, p_k$ are unknown and we want to use the survey data to make inferences about their values. The unknown probabilities in the voter preference survey are

$$p_1 = \text{Proportion of all voters who favor candidate 1}$$
$$p_2 = \text{Proportion of all voters who favor candidate 2}$$
$$p_3 = \text{Proportion of all voters who favor candidate 3}$$

To decide whether the voters, in total, have a preference for any one of the candidates, we will test the null hypothesis that the candidates are equally preferred (i.e., $p_1 = p_2 = p_3 = \frac{1}{3}$) against the alternative hypothesis that one candidate is preferred (i.e., at least one of the probabilities $p_1, p_2,$ and p_3 exceeds $\frac{1}{3}$). Thus, we want to test

$H_0: p_1 = p_2 = p_3 = \frac{1}{3}$ (no preference)

$H_a:$ At least one of the proportions exceeds $\frac{1}{3}$ (a preference exists)

If the null hypothesis is true and $p_1 = p_2 = p_3 = \frac{1}{3}$, then the expected value (mean value) of the number of voters who prefer candidate 1 is given by

$$E_1 = np_1 = (n)\frac{1}{3} = (150)\frac{1}{3} = 50$$

Similarly, $E_2 = E_3 = 50$ if the null hypothesis is true and no preference exists.

KARL PEARSON (1895–1980)
The Father of Statistics

While attending college, London-born Karl Pearson exhibited a wide range of interests, including mathematics, physics, religion, history, socialism, and Darwinism. After earning a law degree at Cambridge University and a Ph.D. in political science at the University of Heidelberg (Germany), Pearson became a professor of applied mathematics at University College in London. His 1892 book, *The Grammar of Science*, illustrated his conviction that statistical data analysis lies at the foundation of all knowledge; consequently, many consider Pearson to be the "father of statistics." Among Pearson's many contributions to the field are introducing the term *standard deviation* and its associated symbol (σ); developing the distribution of the correlation coefficient; cofounding and editing the prestigious statistics journal *Biometrika*; and (what many consider his greatest achievement) creating the first chi-square "goodness-of-fit" test. Pearson inspired his students (including his son, Egon, and William Gossett) with his wonderful lectures and enthusiasm for statistics. ■

The **chi-square test** measures the degree of disagreement between the data and the null hypothesis:

$$\chi^2 = \frac{[n_1 - E_1]^2}{E_1} + \frac{[n_2 - E_2]^2}{E_2} + \frac{[n_3 - E_3]^2}{E_3}$$
$$= \frac{(n_1 - 50)^2}{50} + \frac{(n_2 - 50)^2}{50} + \frac{(n_3 - 50)^2}{50}$$

Note that the farther the observed numbers n_1, n_2, and n_3 are from their expected value (50), the larger χ^2 will become. That is, large values of χ^2 imply that the null hypothesis is false.

We have to know the distribution of χ^2 in repeated sampling before we can decide whether the data indicate that a preference exists. When H_0 is true, χ^2 can be shown to have (approximately) a **chi-square distribution**. The shape of the chi-square distribution is nonsymmetric (see Figure 8.3), and for this one-way classification, the χ^2 distribution has $(k - 1)$ degrees of freedom.* Critical values of χ^2 are provided in Table V of Appendix A, a portion of which is shown in Table 8.4.

To illustrate, the rejection region for the voter preference survey for $\alpha = .05$ and $k - 1 = 3 - 1 = 2$ df is

Rejection region: $\chi^2 > \chi^2_{.05}$

This value of $\chi^2_{.05}$ (found in Table V) is 5.99147. (See Figure 8.3.) The computed value of the test statistic is

$$\chi^2 = \frac{(n_1 - 50)^2}{50} + \frac{(n_2 - 50)^2}{50} + \frac{(n_3 - 50)^2}{50}$$
$$= \frac{(61 - 50)^2}{50} + \frac{(53 - 50)^2}{50} + \frac{(36 - 50)^2}{50} = 6.52$$

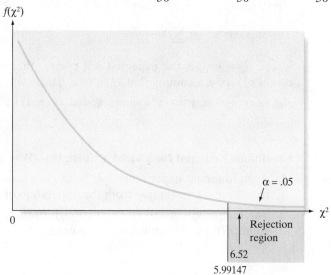

Figure 8.3
Rejection region for voter preference survey

Since the computed $\chi^2 = 6.52$ exceeds the critical value of 5.99147, we conclude at the $\alpha = .05$ level of significance that there does exist a voter preference for one or more of the candidates.

Now that we have evidence to indicate that the proportions p_1, p_2, and p_3 are unequal, we can use the methods of Section 5.4 to make inferences concerning their individual values. [*Note:* We cannot use the methods of Section 7.2 to compare two

---

*The derivation of the number of degrees of freedom for χ^2 involves the number of linear restrictions imposed on the count data. In the present case, the only constraint is that $\Sigma n_i = n$, where n (the sample size) is fixed in advance. Therefore, df $= k - 1$. For other cases, we will give the number of degrees of freedom for each usage of χ^2 and refer the interested reader to the references for more detail.

Table 8.4   Reproduction of Part of Table V in Appendix A

Degrees of Freedom	$\chi^2_{.100}$	$\chi^2_{.050}$	$\chi^2_{.025}$	$\chi^2_{.010}$	$\chi^2_{.005}$
1	2.70554	3.84146	5.02389	6.63490	7.87944
2	4.60517	5.99147	7.37776	9.21034	10.5966
3	6.25139	7.81473	9.34840	11.3449	12.8381
4	7.77944	9.48773	11.1433	13.2767	14.8602
5	9.23635	11.0705	12.8325	15.0863	16.7496
6	10.6446	12.5916	14.4494	16.8119	18.5476
7	12.0170	14.0671	16.0128	18.4753	20.2777

proportions, because the cell counts are dependent random variables.] The general form for a test of hypothesis concerning multinomial probabilities is shown in the following box:

A Test of a Hypothesis about Multinomial Probabilities: One-Way Table

H_0: $p_1 = p_{1,0}$, $p_2 = p_{2,0}$, $\ldots$, $p_k = p_{k,0}$

where $p_{1,0}, p_{2,0}, \ldots, p_{k,0}$ represent the hypothesized values of the multinomial probabilities

H_a: At least one of the multinomial probabilities does not equal its hypothesized value

Test statistic: $\chi^2 = \sum \dfrac{[n_i - E_i]^2}{E_i}$

where $E_i = np_{i,0}$ is the **expected cell count**—that is, the expected number of outcomes of type i, assuming that H_0 is true. The total sample size is n.

Rejection region: $\chi^2 > \chi^2_\alpha$, where χ^2_α has $(k - 1)$ df

---

Conditions Required for a Valid χ^2 Test: One-Way Table

1. A multinomial experiment has been conducted. This is generally satisfied by taking a random sample from the population of interest.

2. The sample size n will be large enough so that, for every cell, the expected cell count $E(n_i)$ will be equal to 5 or more.*

Example 8.5

A One-Way χ^2 Test—Effectiveness of a TV Program on Marijuana

Problem Suppose an educational television station has broadcast a series of programs on the physiological and psychological effects of smoking marijuana. Now that the series is finished, the station wants to see whether the citizens within the viewing area have changed their minds about how the possession of marijuana should be considered legally. Before the series was shown, it was determined that 7% of the citizens favored legalization, 18% favored decriminalization, 65% favored the existing law (an offender could be fined or imprisoned), and 10% had no opinion.

*The assumption that all expected cell counts are at least 5 is necessary in order to ensure that the χ^2 approximation is appropriate. Exact methods for conducting the test of hypothesis exist and may be used for small expected cell counts, but these methods are beyond the scope of this text.

A summary of the opinions (after the series was shown) of a random sample of 500 people in the viewing area is given in Table 8.5. Test at the $\alpha = .01$ level to see whether these data indicate that the distribution of opinions differs significantly from the proportions that existed before the educational series was aired.

Table 8.5	Distribution of Opinions about Marijuana Possession		
Legalization	Decriminalization	Existing Laws	No Opinion
39	99	336	26

Data Set: MARIJUANA

Solution Define the proportions after the airing to be

p_1 = Proportion of citizens favoring legalization
p_2 = Proportion of citizens favoring decriminalization
p_3 = Proportion of citizens favoring existing laws
p_4 = Proportion of citizens with no opinion

Then the null hypothesis representing no change in the distribution of percentages is

$$H_0: p_1 = .07, p_2 = .18, p_3 = .65, p_4 = .10$$

and the alternative is

H_a: At least one of the proportions differs from its null hypothesized value

Thus, we have

$$\text{Test statistic: } \chi^2 = \sum \frac{[n_i - E_i]^2}{E_i}$$

where

$$E_1 = np_{1,0} = 500(.07) = 35$$
$$E_2 = np_{2,0} = 500(.18) = 90$$
$$E_3 = np_{3,0} = 500(.65) = 325$$
$$E_4 = np_{4,0} = 500(.10) = 50$$

Since all these values are larger than 5, the χ^2 approximation is appropriate. Also, if the citizens in the sample were randomly selected, then the properties of the multinomial probability distribution are satisfied.

Rejection region: For $\alpha = .01$ and df $= k - 1 = 3$, reject H_0 if $\chi^2 > \chi^2_{.01}$, where (from Table V in Appendix A) $\chi^2_{.01} = 11.3449$.

We now calculate the test statistic:

$$\chi^2 = \frac{(39 - 35)^2}{35} + \frac{(99 - 90)^2}{90} + \frac{(336 - 325)^2}{325} + \frac{(26 - 50)^2}{50} = 13.249$$

Since this value exceeds the table value of χ^2 (11.3449), the data provide sufficient evidence ($\alpha = .01$) that the opinions on the legalization of marijuana have changed since the series was aired.

The χ^2 test can also be conducted with the use of an available statistical software package. Figure 8.4 is an SPSS printout of the analysis of the data in Table 8.5. The test statistic and *p*-value of the test are highlighted on the printout. Since $\alpha = .01$ exceeds $p = .004$, there is sufficient evidence to reject H_0.

OPINION

	Observed N	Expected N	Residual
LEGAL	39	35.0	4.0
DECRIM	99	90.0	9.0
EXISTLAW	336	325.0	11.0
NONE	26	50.0	-24.0
Total	500		

Test Statistics

	OPINION
Chi-Square[a]	13.249
df	3
Asymp. Sig.	.004

a. 0 cells (.0%) have expected frequencies less than 5. The minimum expected cell frequency is 35.0.

Figure 8.4

SPSS analysis of data in Table 8.5

Look Back If the conclusion for the χ^2 test is "fail to reject H_0," then there is insufficient evidence to conclude that the distribution of opinions differs from the proportions stated in H_0. Be careful not to "accept H_0" and conclude that $p_1 = .07, p_2 = .18, p_3 = .65$, and $p_4 = .10$. The probability (β) of a Type II error is unknown.

Now Work Exercise 8.39

If we focus on one particular outcome of a multinomial experiment, we can use the methods developed in Section 5.4 for a binomial proportion to establish a confidence interval for any one of the multinomial probabilities.* For example, if we want a 95% confidence interval for the proportion of citizens in the viewing area who have no opinion about the issue, we calculate

$$\hat{p}_4 \pm 1.96\sigma_{\hat{p}_4}$$

where

$$\hat{p}_4 = \frac{n_4}{n} = \frac{26}{500} = .052 \quad \text{and} \quad \sigma_{\hat{p}_4} \approx \sqrt{\frac{\hat{p}_4(1 - \hat{p}_4)}{n}}$$

Thus, we get

$$.052 \pm 1.96\sqrt{\frac{(.052)(.948)}{500}} = .052 \pm .019$$

or (.033, .071). Consequently, we estimate that between 3.3% and 7.1% of the citizens now have no opinion on the issue of the legalization of marijuana. The series of programs may have helped citizens who formerly had no opinion on the issue to form an opinion, since it appears that the proportion of "no opinions" is now less than 10%.

Statistics IN Action | **Revisited** | **Testing Category Proportions for Type of College Drinker**

In the *Family and Consumer Sciences Research Journal* (Mar. 2005) study of college students and drinking (p. 430), one of the researchers' main objectives was to segment college students according to their rates of alcohol consumption. A segmentation was developed on the basis of the students' responses to the questions on frequency of drinking and average number of drinks per occasion. Four types, or groups, of college drinkers emerged: non/seldom drinkers, social drinkers, typical binge drinkers, and heavy binge drinkers. What are the proportions of students in each

*Note that focusing on one outcome has the effect of lumping the other $(k - 1)$ outcomes into a single group. Thus, we obtain, in effect, two outcomes—or a binomial experiment.

Statistics IN Action (continued)	of these groups, and are these proportions statistically different?

of these groups, and are these proportions statistically different?

To answer these questions, we used SPSS to analyze the type-of-drinker variable in the **COLLDRINKS** file. Figure SIA8.1 shows summary statistics and a graph describing the four categories. From the summary table at the top of the printout, you can see that 118 (or 18%) of the students are non/seldom

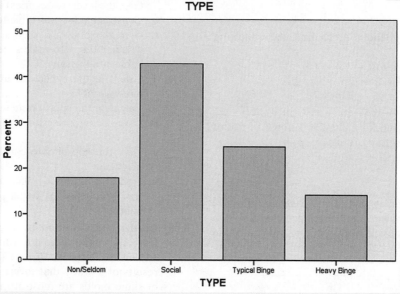

TYPE

		Frequency	Percent	Valid Percent	Cumulative Percent
Valid	Non/Seldom	118	18.0	18.0	18.0
	Social	282	42.9	42.9	60.9
	Typical Binge	163	24.8	24.8	85.7
	Heavy Binge	94	14.3	14.3	100.0
	Total	657	100.0	100.0	

Figure SIA8.1
SPSS descriptive statistics and graph for type of drinker

drinkers, 282 (or 43%) are social drinkers, 163 (or 25%) are typical binge drinkers, and 94 (or 14%) are heavy binge drinkers. These sample percentages are illustrated in the bar graph in Figure SIA8.1. In this sample of students, the largest percentage (43%) consists of social drinkers.

Is this sufficient evidence to indicate that the true proportions in the population of college students are different? Letting p_1, p_2, p_3, and p_4 represent the true proportions for non/seldom, social, typical binge, and heavy binge drinkers, respectively, we tested $H_0: p_1 = p_2 = p_3 = p_4 = .25$, using the chi-square test in SPSS. The printout is displayed in Figure SIA8.2. The cell frequencies and expected numbers are shown in the top table of the figure, while the chi-square test statistic (127.5) and p-value (.000) are shown in the bottom table. At any reasonably selected α-level (say, $\alpha = .01$), the small p-value indicates that there is sufficient evidence to reject the null hypothesis; thus, we conclude that the true proportions associated with the four type-of-drinker categories are indeed statistically different.

Chi-Square Test

Frequencies

TYPE

	Observed N	Expected N	Residual
Non/Seldom	118	164.3	-46.3
Social	282	164.3	117.8
Typical Binge	163	164.3	-1.3
Heavy Binge	94	164.3	-70.3
Total	657		

Test Statistics

	TYPE
Chi-Square[a]	127.493
df	3
Asymp. Sig.	.000

a. 0 cells (.0%) have expected frequencies less than 5. The minimum expected cell frequency is 164.3.

Figure SIA8.2
SPSS chi-square test for type-of-drinker categories

Exercises 8.31–8.49

Understanding the Principles

8.31 What are the characteristics of a multinomial experiment? Compare the characteristics with those of a binomial experiment.

8.32 What conditions must n satisfy to make the χ^2 test for a one-way table valid?

Learning the Mechanics

8.33 Use Table V of Appendix A to find each of the following χ^2 values:
a. $\chi^2_{.05}$ for df = 10
b. $\chi^2_{.990}$ for df = 50
c. $\chi^2_{.10}$ for df = 16
d. $\chi^2_{.005}$ for df = 50

8.34 Use Table V of Appendix A to find the following probabilities:
a. $P(\chi^2 \leq 1.063623)$ for df = 4
b. $P(\chi^2 > 30.5779)$ for df = 15
c. $P(\chi^2 \geq 82.3581)$ for df = 100
d. $P(\chi^2 < 18.4926)$ for df = 30

8.35 Find the rejection region for a one-dimensional χ^2-test of a null hypothesis concerning $p_1, p_2, \ldots, p_k$ if
a. $k = 3; \alpha = .05$
b. $k = 5; \alpha = .10$
c. $k = 4; \alpha = .01$

8.36 A multinomial experiment with $k = 3$ cells and $n = 320$ produced the data shown in the accompanying table. Do these data provide sufficient evidence to contradict the null hypothesis that $p_1 = .25, p_2 = .25$, and $p_3 = .50$? Test, using $\alpha = .05$.

	Cell		
	1	2	3
n_i	78	60	182

8.37 A multinomial experiment with $k = 4$ cells and $n = 205$ produced the data shown in the following table:

	Cell			
	1	2	3	4
n_i	43	56	59	47

a. Is there sufficient evidence to conclude that the multinomial probabilities differ? Test, using $\alpha = .05$.
b. What are the Type I and Type II errors associated with the test of part **a?**
c. Construct a 95% confidence interval for the multinomial probability associated with cell 3.

8.38 A multinomial experiment with $k = 4$ cells and $n = 400$ produced the data shown in the accompanying table. Do these data provide sufficient evidence to contradict the null hypothesis that $p_1 = .2, p_2 = .4, p_3 = .1$, and $p_4 = .3$? Test, using $\alpha = .05$.

	Cell			
	1	2	3	4
n_i	70	196	46	88

Applying the Concepts—Basic

8.39 **Jaw dysfunction study.** A report on dental patients with temporomandibular (jaw) joint dysfunction (TMD) was published in *General Dentistry* (Jan/Feb. 2004). A random sample of 60 patients was selected for an experimental treatment of TMD. Prior to treatment, the patients filled out a survey on two nonfunctional jaw habits—bruxism (teeth grinding) and teeth clenching—that have been linked to TMD. Of the 60 patients, 3 admitted to bruxism, 11 admitted to teeth clenching, 30 admitted to both habits, and 16 claimed they had neither habit.
a. Describe the qualitative variable of interest in the study. Give the levels (categories) associated with the variable.
b. Construct a one-way table for the sample data.
c. Give the null and alternative hypotheses for testing whether the percentages associated with the admitted habits are the same.
d. Calculate the expected numbers for each cell of the one-way table.
e. Calculate the appropriate test statistic.
f. Give the rejection region for the test at $\alpha = .05$
g. Give the appropriate conclusion in the words of the problem.
h. Find and interpret a 95% confidence interval for the true proportion of dental patients who admit to both habits.

8.40 **Beetles and slime molds.** Myxomycetes are mushroom-like slime molds that are a food source for insects. The *Journal of Natural History* (May 2010) published the results of a study that investigated which of six species of slime molds are most attractive to beetles inhabiting an Atlantic rain forest. A sample of 19 beetles feeding on slime mold was obtained and the species of slime mold was determined for each beetle. The numbers of beetles captured on each of the six species are given in the accompanying table. These data are saved in the **SLIMEMOLD** file. The researchers want to know if the relative frequency of occurrence of beetles differs for the six slime mold species.

Slime mold species:	LE	TM	AC	AD	HC	HS
Number of beetles:	3	2	7	3	1	3

a. Identify the categorical variable (and its levels) of interest in this study.
b. Set up the null and alternative hypotheses of interest to the researchers.
c. Find the test statistic and corresponding p-value.
d. The researchers found "no significant differences in the relative frequencies of occurrence" using $\alpha = .05$. Do you agree?
e. Comment on the validity of the inference, part **d**. (Determine the expected cell counts.)

8.41 **Excavating ancient pottery.** Refer to the *Chance* (Fall 2000) study of ancient Greek pottery, presented in Exercise 2.14 (p. 36). Recall that 837 pottery pieces were uncovered at the excavation site. The table describing the types of pottery found is reproduced in the next table and the information saved in the **POTTERY** file.

Pot Category	Number Found
Burnished	133
Monochrome	460
Painted	183
Other	61
Total	837

Based on Berg, I., and Bliedon, S. "The pots of phyiakopi: Applying statistical techniques to archaeology." *Chance*, Vol. 13, No. 4, Fall 2000.

a. Describe the qualitative variable of interest in the study. Give the levels (categories) associated with the variable.

b. Assume that the four types of pottery occur with equal probability at the excavation site. What are the values of p_1, p_2, p_3, and p_4, the probabilities associated with the four pottery types?

c. Give the null and alternative hypotheses for testing whether one type of pottery is more likely to occur at the site than any of the other types.

d. Find the test statistic for testing the hypotheses stated in part **c.**

e. Find and interpret the *p*-value of the test. State the conclusion in the words of the problem if you use $\alpha = .10$.

8.42 Museum management. Refer to the *Museum Management and Curatorship* (June 2010) worldwide survey of 30 leading museums of contemporary art, Exercise 2.19 (p. 37). Recall that each museum manager was asked to provide the performance measure used most often for internal evaluation. A summary of the results is provided in the table and saved in the **MUSEUM2** file. The data were analyzed using a chi-square test for a multinomial experiment. The results are shown in the MINITAB printout (bottom of the page).

a. Is there evidence to indicate that one performance measure is used more often than any of the others? Test using $\alpha = .10$.

b. Find a 90% confidence interval for the proportion of museums worldwide that use total visitors as their performance measure. Interpret the result.

Performance Measure	Number of Museums
Total visitors	8
Paying visitors	5
Big shows	6
Funds raised	7
Members	4

Applying the Concepts—Intermediate

8.43 Gender in two-child families. Refer to the *Human Biology* (Feb. 2009) study on the gender of children in two-child families, Exercise 4.29 (p. 175). The article reported on the results of the National Health Interview Survey (NHIS) of 42,888 two-child families. The table below (saved in the **BOYGIRL** file) gives the number of families with each gender configuration.

Gender Configuration	Number of Families
Girl-girl (GG)	9,523
Boy-girl (BG)	11,118
Girl-boy (GB)	10,913
Boy-boy (BB)	11,334

a. If it is just as likely to have a boy as a girl, find the probability of each of the gender configurations for a two-child family.

b. Use the probabilities, part **a,** to determine the expected number of families for each gender configuration.

c. Compute the chi-square test statistic for testing the hypothesis that it is just as likely to have a boy as a girl.

d. Interpret the result, part **c,** if you conduct the test using $\alpha = .10$.

e. Recent research indicates that the ratio of boys to girls in the world population is not 1 to 1, but instead higher (e.g., 1.06 to 1). Using a ratio of 1.06 to 1, the researchers showed that the probabilities of the different gender configurations are: GG—.23795, BG—.24985, GB—.24985, and BB—.26235. Repeat parts **b–d** using these probabilities.

8.44 Sociology fieldwork methods. Refer to the *Teaching Sociology* (July 2006) study of the fieldwork methods used by qualitative sociologists, presented in Exercise 2.16 (p. 36). Recall that fieldwork methods can be categorized as follows: Interview, Observation plus Participation, Observation Only, and Grounded Theory. The accompanying table (saved in the **FIELDWORK** file) shows the number of papers published over the past seven years in each category. Suppose a sociologist claims that 70%, 15%, 10%, and 5% of the fieldwork methods involve interview, observation plus participation, observation only, and grounded theory, respectively. Do the data support or refute the claim? Explain.

Fieldwork Method	Number of Papers
Interview	5,079
Observation + Participation	1,042
Observation Only	848
Grounded Theory	537

Based on Hood, J. C. "Teaching against the text: The case of qualitative methods." *Teaching Sociology*, Vol. 34, July 2006 (Exhibit 2).

MINITAB Output for Exercise 8.42

Chi-Square Goodness-of-Fit Test for Observed Counts in Variable: NUMBER

Using category names in PERFORM

Category	Observed	Test Proportion	Expected	Contribution to Chi-Sq
Total visitors	8	0.2	6	0.666667
Paying visitors	5	0.2	6	0.166667
Big shows	6	0.2	6	0.000000
Funds raised	7	0.2	6	0.166667
Members	4	0.2	6	0.666667

N	DF	Chi-Sq	P-Value
30	4	1.66667	0.797

8.45 **Do you believe in the Bible?** Refer to the General Social Survey (GSS) and the question pertaining to a person's belief in the Bible, presented in Exercise 2.18 (p. 36). Recall that approximately 2,800 Americans selected from one of the following answers: (1) The Bible is the actual word of God and is to be taken literally; (2) the Bible is the inspired word of God, but not everything is to be taken literally; (3) the Bible is an ancient book of fables; or (4) the Bible has some other origin, but is recorded by men. The variable "Bible1" in the **BIBLE** file contains the responses.

a. Summarize the responses in a one-way table.

b. State the null and alternative hypotheses for testing whether the true proportions in each category are equal.

c. Find the expected number of responses in each answer category for the test mentioned in part **b.**

d. Compute the chi-square statistic for the test.

e. Give the appropriate conclusion for the test if $\alpha = .10$.

f. A more realistic null hypothesis is that 30% of Americans believe that the Bible is the actual word of God; 50% believe that it is inspired by God, but not to be taken literally; 15% believe that it is an ancient book of fables; and 5% believe that the Bible has some other origin. Repeat parts **b–e** for this hypothesis.

8.46 **Characteristics of ice-melt ponds.** The National Snow and Ice Data Center (NSIDC) collected data on 504 ice-melt ponds in the Canadian Arctic. The data are saved in the **PONDICE** file. One variable of interest to environmental engineers studying the ponds is the type of ice observed in each. Ice type is classified as first-year ice, multiyear ice, or landfast ice. The SAS summary table for the types of ice of the 504 ice-melt ponds is reproduced below.

a. Use a 90% confidence interval to estimate the proportion of ice-melt ponds in the Canadian Arctic that have first-year ice.

b. Suppose environmental engineers hypothesize that 15% of Canadian Arctic ice-melt ponds have first-year ice, 40% have landfast ice, and 45% have multiyear ice. Test the engineers' theory, using $\alpha = .01$.

The FREQ Procedure

ICETYPE	Frequency	Percent	Cumulative Frequency	Cumulative Percent
First-year	88	17.46	88	17.46
Landfast	196	38.89	284	56.35
Multi-year	220	43.65	504	100.00

8.47 **Detecting Alzheimer's disease at an early age.** Geneticists at Australian National University are studying whether the cognitive effects of Alzheimer's disease can be detected at an early age (*Neuropsychology*, Jan. 2007). One portion of the study focused on a particular strand of DNA extracted from each in a sample of 2,097 young adults between the ages of 20 and 24. The DNA strand was classified into one of three genotypes: $E4^+/E4^+$, $E4^+/E4^-$, and $E4^-/E4^-$. The number of young adults with each genotype is shown in the next table and the data are saved in the **E4E4** file. Suppose that in adults who are not afflicted with Alzheimer's disease, the distribution of genotypes for this strand of DNA is 2% with $E4^+/E4^+$, 25% with $E4^+/E4^-$, and 73% with $E4^-/E4^-$. If differences in this distribution are detected, then this strand of DNA could lead researchers to an early test for the onset of Alzheimer's. Conduct a test (at $\alpha = .05$) to determine if the distribution of E4/E4 genotypes for the population of young adults differs from the norm.

Genotype:	$E4^+/E4^+$	$E4^+/E4^-$	$E4^-/E4^-$
Number of young adults:	56	517	1,524

Applying the Concepts—Advanced

8.48 **Political representation of religious groups.** Do those elected to the U.S. House of Representatives really "represent" their constituents demographically? This was a question of interest in *Chance* (Summer 2002). One of several demographics studied was religious affiliation. The accompanying table (saved in the **USHOUSE** file) gives the proportion of the U.S. population for several religions, as well as the number of the 435 seats in the House of Representatives affiliated with that religion. Give your opinion on whether or not the members of the House of Representatives are statistically representative of the religious affiliation of their constituents in the United States.

Religion	Proportion of U.S. Population	Number of Seats in House
Catholic	.28	117
Methodist	.04	61
Jewish	.02	30
Other	.66	227
Totals	1.00	435

8.49 **Analysis of a Scrabble game.** In the board game Scrabble™, a player initially draws a "hand" of seven tiles at random from 100 tiles. Each tile has a letter of the alphabet, and the player attempts to form a word from the letters in his or her hand. In *Chance* (Winter 2002), scientist C. J. Robinove

Letter	Relative Frequency in Board Game	Frequency in Electronic Game
A	.09	39
B	.02	18
C	.02	30
D	.04	30
E	.12	31
F	.02	21
G	.03	35
H	.02	21
I	.09	25
J	.01	17
K	.01	27
L	.04	18
M	.02	31
N	.06	36
O	.08	20
P	.02	27
Q	.01	13
R	.06	27
S	.04	29
T	.06	27
U	.04	21
V	.02	33
W	.02	29
X	.01	15
Y	.02	32
Z	.01	14
# (blank)	.02	34
Total		700

Source: Robinove, C. J. "Letter-frequency bias in an electronic scrabble game." *Chance*, Vol. 15, No. 1, Winter 2002, p. 31 (Table 3). Reprinted with permission from Chance. Copyright 2002 by the American Statistical Association. All rights reserved.

investigated whether a handheld electronic version of the game, called ScrabbleExpress™, produces too few vowels in the 7-letter draws. For each of the 26 letters (and "blank" for any letter), the table (p. 448) gives the true relative frequency of the letter in the board game, as well as the frequency of occurrence of the letter in a sample of 700 tiles (i.e., 100 "hands") randomly drawn in the electronic game. These data are saved in the **SCRABBLE** file.

a. Do the data support the scientist's contention that ScrabbleExpress™ "presents the player with unfair word selection opportunities" that are not the same as in the Scrabble™ board game? Test, using $\alpha = .05$.

b. Use a 95% confidence interval to estimate the true proportion of letters drawn in the electronic game that are vowels. Compare the results with the true relative frequency of a vowel in the board game.

8.4 Testing Categorical Probabilities: Two-Way (Contingency) Table

In Section 8.3, we introduced the multinomial probability distribution and considered data classified according to a single criterion. We now consider multinomial experiments in which the data are classified according to two criteria—that is, *classification with respect to two qualitative factors*.

Consider a study similar to one in the *Journal of Marketing* on the impact of using celebrities in television advertisements. The researchers investigated the relationship between the gender of a viewer and the viewer's brand awareness. Three hundred TV viewers were randomly selected and each asked to identify products advertised by male celebrity spokespersons. The data are summarized in the **two-way table** shown in Table 8.6. This table, called a **contingency table,** presents multinomial count data classified on two scales, or **dimensions, of classification:** gender of viewer and brand awareness.

Table 8.6   Contingency Table for Marketing Example

Brand Awareness		Gender		
		Male	Female	Totals
	Could Identify Product	95	41	136
	Could Not Identify Product	50	114	164
	Totals	145	155	300

Data Set: CELEBRITY

The symbols representing the cell counts for the multinomial experiment in Table 8.6 are shown in Table 8.7a, and the corresponding cell, row, and column probabilities are shown in Table 8.7b. Thus, n_{11} represents the number of viewers who are male and could identify the brand, and p_{11} represents the corresponding cell probability. Note the symbols for the row and column totals and also the symbols for the probability totals. The latter are called **marginal probabilities** for each row and column.

Table 8.7a   Observed Counts for Contingency Table 8.6

Brand Awareness		Gender		
		Male	Female	Totals
	Could Identify Product	n_{11}	n_{12}	R_1
	Could Not Identify Product	n_{21}	n_{22}	R_2
	Totals	C_1	C_2	n

Table 8.7b   Probabilities for Contingency Table 8.6

Brand Awareness		Gender		
		Male	Female	Totals
	Could Identify Product	p_{11}	p_{12}	p_{r1}
	Could Not Identify Product	p_{21}	p_{22}	p_{r2}
	Totals	p_{c1}	p_{c2}	1

The marginal probability p_{r1} is the probability that a TV viewer identifies the product; the marginal probability p_{c1} is the probability that a TV viewer is male. Thus,

$$p_{r1} = p_{11} + p_{12} \text{ and } p_{c1} = p_{11} + p_{21}$$

We can see, then, that this really is a multinomial experiment with a total of 300 trials, $(2)(2) = 4$ cells or possible outcomes, and probabilities for each cell as shown in Table 8.7b. Since the 300 TV viewers are randomly chosen, the trials are considered independent and the probabilities are viewed as remaining constant from trial to trial.

Suppose we want to know whether the two classifications of gender and brand awareness are dependent. That is, if we know the gender of the TV viewer, does that information give us a clue about the viewer's brand awareness? In a probabilistic sense, we know (Chapter 3) that the independence of events A and B implies that $P(AB) = P(A)P(B)$. Similarly, in the contingency table analysis, if the **two classifications are independent,** the probability that an item is classified into any particular cell of the table is the product of the corresponding marginal probabilities. Thus, under the hypothesis of independence, in Table 8.7b we must have

$$p_{11} = p_{r1}p_{c1} \quad p_{12} = p_{r1}p_{c2}$$
$$p_{21} = p_{r2}p_{c1} \quad p_{22} = p_{r2}p_{c2}$$

To test the hypothesis of independence, we use the same reasoning employed in the one-dimensional tests of Section 8.3. First, we calculate the *expected*, or *mean, count in each cell*, assuming that the null hypothesis of independence is true. We do this by noting that the expected count in a cell of the table is just the total number of multinomial trials, n, times the cell probability. Recall that n_{ij} represents the **observed count** in the cell located in the ith row and jth column. Then the expected cell count for the upper left-hand cell (first row, first column) is

$$E_{11} = np_{11}$$

or, when the null hypothesis (the classifications are independent) is true,

$$E_{11} = np_{r1}p_{c1}$$

Since these true probabilities are not known, we estimate p_{r1} and p_{c1} by the same proportions $\hat{p}_{r1} = R_1/n$ and $\hat{p}_{c1} = C_1/n$. Thus, the estimate of the expected value E_{11} is

$$\hat{E}_{11} = n\left(\frac{R_1}{n}\right)\left(\frac{C_1}{n}\right) = \frac{R_1 C_1}{n}$$

Similarly, for each i, j,

$$\hat{E}_{ij} = \frac{(\text{Row total})(\text{Column total})}{\text{Total sample size}}$$

Hence,

$$\hat{E}_{12} = \frac{R_1 C_2}{n}$$

$$\hat{E}_{21} = \frac{R_2 C_1}{n}$$

$$\hat{E}_{22} = \frac{R_2 C_2}{n}$$

Finding Expected Cell Counts for a Two-Way Contingency Table

The estimate of the expected number of observations falling into the cell in row i and column j is given by

$$\hat{E}_{ij} = \frac{R_i C_j}{n}$$

where R_i = total for row i, C_j = total for column j, and n = sample size.

Using the data in Table 8.6, we find that

$$\hat{E}_{11} = \frac{R_1 C_1}{n} = \frac{(136)(145)}{300} = 65.73$$

$$\hat{E}_{12} = \frac{R_1 C_2}{n} = \frac{(136)(155)}{300} = 70.27$$

$$\hat{E}_{21} = \frac{R_2 C_1}{n} = \frac{(164)(145)}{300} = 79.27$$

$$\hat{E}_{22} = \frac{R_2 C_2}{n} = \frac{(164)(155)}{300} = 84.73$$

These estimated expected values are more easily obtained using computer software. Figure 8.5 is a MINITAB printout of the analysis, with the expected values highlighted.

```
Tabulated statistics: AWARE, GENDER

Using frequencies in NUMBER


Rows: AWARE    Columns: GENDER

            Male   Female    All

ID-Product    95       41    136
            65.7     70.3  136.0

No-ID         50      114    164
            79.3     84.7  164.0

All          145      155    300
           145.0    155.0  300.0

Cell Contents:      Count
                    Expected count
```

Figure 8.5

MINITAB contingency table analysis of data in Table 8.6

```
Pearson Chi-Square = 46.135, DF = 1, P-Value = 0.000
Likelihood Ratio Chi-Square = 47.362, DF = 1, P-Value = 0.000
```

We now use the χ^2 statistic to compare the observed and expected (estimated) counts in each cell of the contingency table:

$$\chi^2 = \frac{[n_{11} - \hat{E}_{11}]^2}{\hat{E}_{11}} + \frac{[n_{12} - \hat{E}_{12}]^2}{\hat{E}_{12}} + \frac{[n_{21} - \hat{E}_{21}]^2}{\hat{E}_{21}} + \frac{[n_{22} - \hat{E}_{22}]^2}{\hat{E}_{22}}$$

$$= \sum \frac{[n_{ij} - \hat{E}_{ij}]^2}{\hat{E}_{ij}}$$

(*Note:* The use of $\sum$ in the context of a contingency table analysis refers to a sum over all cells in the table.)

Substituting the data of Table 8.6 and the expected values into this expression, we get

$$\chi^2 = \frac{(95 - 65.73)^2}{65.73} + \frac{(41 - 70.27)^2}{70.27} + \frac{(50 - 79.27)^2}{79.27} + \frac{(114 - 84.73)^2}{84.73} = 46.14$$

Note that this value is also shown (highlighted) in Figure 8.5.

Large values of χ^2 imply that the observed counts do not closely agree and hence that the hypothesis of independence is false. To determine how large χ^2 must be before it is too large to be attributed to chance, we make use of the fact that the sampling distribution of χ^2 is approximately a χ^2 probability distribution when the classifications are independent.

When testing the null hypothesis of independence in a two-way contingency table, the appropriate degrees of freedom will be $(r - 1)(c - 1)$, where r is the number of rows and c is the number of columns in the table. For the brand awareness example, the number of degrees of freedom for χ^2 is $(r - 1)(c - 1) = (2 - 1)(2 - 1) = 1$. Then, for $\alpha = .05$, we reject the hypothesis of independence when

$$\chi^2 > \chi^2_{.05} = 3.84146$$

Since the computed $\chi^2 = 46.14$ exceeds the value 3.84146, we conclude that viewer gender and brand awareness are dependent events. This result may also be obtained by noting that the p-value of the test (highlighted on Figure 8.5) is approximately 0.

The pattern of **dependence** can be seen more clearly by expressing the data as percentages. We first select one of the two classifications to be used as the base variable. In the preceding example, suppose we select gender of the TV viewer as the classificatory variable to be the base. Next, we represent the responses for each level of the second categorical variable (brand awareness here) as a percentage of the subtotal for the base variable. For example, from Table 8.6, we convert the response for males who identify the brand (95) to a percentage of the total number of male viewers (145). That is,

$$\left(^{95}\!/_{145}\right)100\% = 65.5\%$$

All of the entries in Table 8.6 are similarly converted, and the values are shown in Table 8.8. The value shown at the right of each row is the row's total, expressed as a percentage of the total number of responses in the entire table. Thus, the percentage of TV viewers who identify the product is $\left(\frac{136}{300}\right)100\% = 45.3\%$ (rounded to the nearest percent).

Table 8.8	Percentage of TV Viewers Who Identify Brand, by Gender			
		Gender		
		Male	Female	Totals
Brand Awareness	**Could Identify Product**	65.5	26.5	45.3
	Could Not Identify Product	34.5	73.5	54.7
	Totals	100	100	100

If the gender and brand awareness variables are independent, then the percentages in the cells of the table are expected to be approximately equal to the corresponding row percentages. Thus, we would expect the percentage of viewers who identify the brand for each gender to be approximately 45% if the two variables are independent. The extent to which each gender's percentage departs from this value determines the dependence of the two classifications, with greater variability of the row percentages meaning a greater degree of dependence. A plot of the percentages helps summarize the observed pattern. In the SPSS bar graph in Figure 8.6, we show the gender of the viewer (the base variable) on the horizontal axis and the percentage of TV viewers who identify the brand (green bars) on the vertical axis. The "expected" percentage under the assumption of independence is shown as a horizontal line.

Figure 8.6 clearly indicates the reason that the test resulted in the conclusion that the two classifications in the contingency table are dependent. The percentage of male TV viewers who identify the brand promoted by a male celebrity is more than twice as high as the percentage of female TV viewers who identify the brand. Statistical measures of the degree of dependence and procedures for making comparisons of pairs of levels for classifications are beyond the scope of this text, but can be found in the references. We will utilize descriptive summaries such as Figure 8.6 to examine the degree of dependence exhibited by the sample data.

The general form of a two-way contingency table containing r rows and c columns (called an $r \times c$ contingency table) is shown in Table 8.9. Note that the observed count in the ijth cell is denoted by n_{ij}, the ith row total is r_i, the jth column total is c_j, and the total sample size is n. Using this notation, we give the general form of the contingency table test for independent classifications in the box.

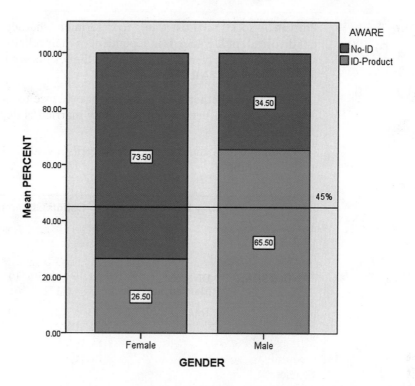

Figure 8.6
SPSS bar graph showing percent of viewers who identify TV product

Table 8.9   General $r \times c$ Contingency Table

		Column				
		1	2	...	c	Row Totals
Row	**1**	n_{11}	n_{12}	...	n_{1c}	R_1
	2	n_{21}	n_{22}	...	n_{2c}	R_2
	$\vdots$	$\vdots$	$\vdots$		$\vdots$	$\vdots$
	r	n_{r1}	n_{r2}	...	n_{rc}	R_r
Column Totals		C_1	C_2	...	C_c	n

General Form of a Two-Way (Contingency) Table Analysis: A Test for Independence

H_0: The two classifications are independent

H_a: The two classifications are dependent

Test statistic: $\chi^2 = \sum \dfrac{[n_{ij} - \hat{E}_{ij}]^2}{\hat{E}_{ij}}$

where $\hat{E}_{ij} = \dfrac{R_i C_j}{n}$

Rejection region: $\chi^2 > \chi_\alpha^2$, where χ_α^2 has $(r - 1)(c - 1)$ df

---

Conditions Required for a Valid χ^2 Test: Contingency Tables

1. The n observed counts are a random sample from the population of interest. We may then consider this to be a multinomial experiment with $r \times c$ possible outcomes.

2. The sample size n will be large enough so that, for every cell, the expected count $\hat{E}(n_{ij})$ will be equal to 5 or more.

Example 8.6

Conducting a Two-Way Analysis—Marital Status and Religion

Problem A social scientist wants to determine whether the marital status (divorced or not divorced) of U.S. men is independent of their religious affiliation (or lack thereof). A sample of 500 U.S. men is surveyed, and the results are tabulated as shown in Table 8.10 and saved in the **MARREL** file.

a. Test to see whether there is sufficient evidence to indicate that the marital status of men who have been or are currently married is dependent on religious affiliation. Take $\alpha = .01$.

b. Graph the data and describe the patterns revealed. Is the result of the test supported by the graph?

Table 8.10   Survey Results (Observed Counts), Example 8.6

		Religious Affiliation					
		A	B	C	D	None	Totals
Marital Status	**Divorced**	39	19	12	28	18	116
	Married, never divorced	172	61	44	70	37	384
	Totals	211	80	56	98	55	500

Data Set: MARREL

Solution

a. The first step is to calculate estimated expected cell frequencies under the assumption that the classifications are independent. Rather than compute these values by hand, we resort to a computer. The SAS printout of the analysis of Table 8.10 is displayed in Figure 8.7, each cell of which contains the observed (top) and expected (bottom) frequency in that cell. Note that $\hat{E}_{11}$, the estimated expected count for the Divorced, A cell, is 48.952. Similarly, the estimated expected count for the Divorced, B cell, is $\hat{E}_{12} = 18.56$. Since all the estimated expected cell frequencies are greater than 5, the χ^2 approximation for the test statistic is appropriate. Assuming that the men chosen were randomly selected from all married or previously married American men, the characteristics of the multinomial probability distribution are satisfied.

```
                          The FREQ Procedure

                     Table of MARITAL by RELIGION

     MARITAL        RELIGION

     Frequency|
     Expected |A        |B        |C        |D        |NONE     |  Total

     DIVORCED |      39 |      19 |      12 |      28 |      18 |    116
              |  48.952 |   18.56 |  12.992 |  22.736 |   12.76 |

     NEVER    |     172 |      61 |      44 |      70 |      37 |    384
              |  162.05 |   61.44 |  43.008 |  75.264 |   42.24 |

     Total          211        80        56        98        55       500


             Statistics for Table of MARITAL by RELIGION

        Statistic                       DF       Value       Prob

        Chi-Square                       4        7.1355      0.1289
        Likelihood Ratio Chi-Square      4        6.9854      0.1367
        Mantel-Haenszel Chi-Square       1        6.4943      0.0108
        Phi Coefficient                           0.1195
        Contingency Coefficient                   0.1186
        Cramer's V                                0.1195


                      Fisher's Exact Test

              Table Probability (P)     6.936E-06
              Pr <= P                      0.1251

                   Sample Size = 500
```

Figure 8.7

SAS contingency table printout for Example 8.6

The null and alternative hypotheses we want to test are

H_0: The marital status of U.S. men and their religious affiliation are independent

H_a: The marital status of U.S. men and their religious affiliation are dependent

The test statistic, $\chi^2 = 7.135$, is highlighted at the bottom of the printout, as is the observed significance level (*p*-value) of the test. Since $\alpha = .01$ is less than $p = .129$, we fail to reject H_0; that is, we cannot conclude that the marital status of U.S. men depends on their religious affiliation. (Note that we could not reject H_0 even with $\alpha = .10$.)

b. The marital status frequencies can be expressed as percentages of the number of men in each religious affiliation category. The expected percentage of divorced men under the assumption of independence is $(^{116}/_{500})100\% = 23\%$. A SAS graph of the percentages is shown in Figure 8.8. Note that the percentages of divorced men (see

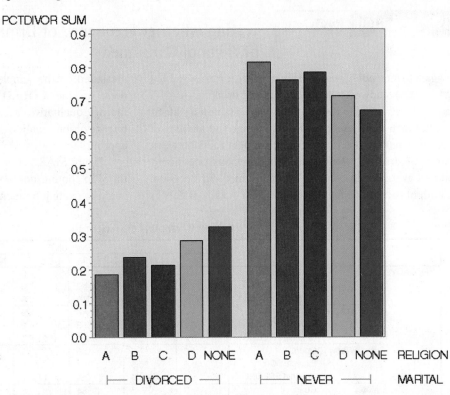

Figure 8.8

SAS side-by-side bar graphs showing percentage of divorced and never divorced males by religion

the bars in the "DIVORCED" block of the SAS graph) deviate only slightly from that expected under the assumption of independence, supporting the result of the test in part **a.** That is, neither the descriptive bar graph nor the statistical test provides evidence that the male divorce rate depends on (varies with) religious affiliation.

Now Work Exercise 8.59

Contingency Tables with Fixed Marginals

In the *Journal of Marketing* study on celebrities in TV ads, a single random sample was selected from the target population of all TV viewers and the outcomes—values of gender and brand awareness—were recorded for each viewer. For this type of study, the researchers had no a *priori* knowledge of how many observations would fall into the categories of the qualititative variables. In other words, prior to obtaining the sample, the researchers did not know how many males or how many brand identifiers would make up the sample. Often times, it is advantageous to select a random sample from each of the levels of one of the qualitative variables.

For example, in the *Journal of Marketing* study, the researchers may want to be sure of an equivalent number of males and females in their sample. Consequently, they will select independent random samples of 150 males and 150 females. (In fact, this was the sampling plan for the actual study.) Summary data for this type of study yield a **contingency table**

with fixed marginals since the column totals for one qualitative variable (e.g., gender) are known in advance.* The goal of the analysis does not change—determine whether the two qualitative variables (e.g., gender and brand awareness) are dependent.

The procedure for conducting a chi-square analysis for a contingency table with fixed marginals is identical to the one outlined above, since it can be shown (proof omitted) that the χ^2 test statistic for this type of sampling also has an approximate chi-square distribution with $(r - 1)(c - 1)$ degrees of freedom. One reason why you might choose this alternative sampling plan is to obtain sufficient observations in each cell of the contingency table to ensure that the chi-square approximation is valid. Remember, this will usually occur when the expected cell counts are all greater than or equal to 5. By selecting a large sample (150 observations) for each gender in the *Journal of Marketing* study, the researchers improved the odds of obtaining large expected cell counts in the contingency table.

Statistics IN Action | **Revisited**

Testing whether Frequency of Drinking Is Related to Amount of Alcohol Consumed

Refer again to the *Family and Consumer Sciences Research Journal* (Mar. 2005) study of college students and drinking (p. 430). A second objective of the researchers was to establish a statistical link between the frequency of drinking and the amount of alcohol consumed. That is, the researchers sought a link between frequency of drinking alcohol over the previous one-month period and average number of drinks consumed per occasion. Since both of these variables (FREQUENCY and AMOUNT)

measured on the sample of 657 students in the **COLLDRINKS** file are qualitative, a contingency table analysis is appropriate.

Figure SIA8.3 shows the SPSS contingency table analyses relating frequency

FREQNCY * AMOUNT Crosstabulation

| | | | \multicolumn{6}{c}{AMOUNT} | |
			None	1 drink	2-3 drinks	4-6 drinks	7-9 drinks	More	Total
FREQNCY	None	Count	118	0	0	0	0	0	118
		Expected Count	21.2	11.0	51.2	17.8	11.5	5.4	118.0
		% within FREQNCY	100.0%	.0%	.0%	.0%	.0%	.0%	100.0%
	OnceMonth	Count	0	30	43	20	0	0	93
		Expected Count	16.7	8.6	40.3	14.0	9.1	4.2	93.0
		% within FREQNCY	.0%	32.3%	46.2%	21.5%	.0%	.0%	100.0%
	TwiceWeek	Count	0	26	158	46	35	11	276
		Expected Count	49.6	25.6	119.7	41.6	26.9	12.6	276.0
		% within FREQNCY	.0%	9.4%	57.2%	16.7%	12.7%	4.0%	100.0%
	More	Count	0	5	84	33	29	19	170
		Expected Count	30.5	15.8	73.7	25.6	16.6	7.8	170.0
		% within FREQNCY	.0%	2.9%	49.4%	19.4%	17.1%	11.2%	100.0%
Total		Count	118	61	285	99	64	30	657
		Expected Count	118.0	61.0	285.0	99.0	64.0	30.0	657.0
		% within FREQNCY	18.0%	9.3%	43.4%	15.1%	9.7%	4.6%	100.0%

Chi-Square Tests

	Value	df	Asymp. Sig. (2-sided)
Pearson Chi-Square	756.606[a]	15	.000
Likelihood Ratio	706.412	15	.000
Linear-by-Linear Association	322.813	1	.000
N of Valid Cases	657		

a. 1 cells (4.2%) have expected count less than 5. The minimum expected count is 4.25.

Figure SIA8.3
SPSS contingency table analysis: frequency of drinking vs. average amount

---

*Data from this type of study are also known as *product binomial data.*

Statistics IN Action
(continued)

of drinking to average amount of alcohol consumed. The null hypothesis for the test is H_0: Frequency and Amount are independent. The chi-square test statistic (756.6) and the p-value of the test (.000) are highlighted on the printout. If we conduct the test at $\alpha = .01$, there is sufficient evidence to reject H_0. That is, the data provide evidence indicating that, for college students, the average amount of alcohol consumed per occasion is associated with the frequency of drinking.

The row percentages highlighted in the contingency table of Figure SIA8.3 reveal the differences in drinking amounts for the different levels of drinking frequency. For frequency of drinking "None" and "Once a month", 0% drink heavily (7–9 or 10 or more drinks per occasion). However, for frequency of drinking "Twice a week" and "More," 12.7% and 17.1%, respectively, have 7–9 drinks per occasion, while 4.0% and 11.2%, respectively, have 10 or more drinks per occasion. These results led the researchers to report that "The frequent drinkers were more likely to consume more [alcohol] on each occasion, a tendency that clearly makes them heavy drinkers."

Data Set: COLLDRINKS

Exercises 8.50–8.74

Understanding the Principles

8.50 What is a two-way (contingency) table?

8.51 What is a contingency table with fixed marginals?

8.52 *True or False.* One goal of a contingency table analysis is to determine whether the two classifications are independent or dependent.

8.53 What conditions are required for a valid chi-square test of data from a contingency table?

Learning the Mechanics

8.54 Find the rejection region for a test of independence of two classifications for which the contingency table contains r rows and c columns and
a. $r = 5$, $c = 5$, $\alpha = .05$
b. $r = 3$, $c = 6$, $\alpha = .10$
c. $r = 2$, $c = 3$, $\alpha = .01$

8.55 Consider the following 2×3 (i.e., $r = 2$ and $c = 3$) contingency table:

		Column 1	Column 2	Column 3
Row	**1**	9	34	53
	2	16	30	25

a. Specify the null and alternative hypotheses that should be used in testing the independence of the row and column classifications.
b. Specify the test statistic and the rejection region that should be used in conducting the hypothesis test of part **a**. Use $\alpha = .01$.
c. Assuming that the row classification and the column classification are independent, find estimates for the expected cell counts.
d. Conduct the hypothesis test of part **a**. Interpret your result.

8.56 Refer to Exercise 8.55.
a. Convert the frequency responses to percentages by calculating the percentage of each column total falling in each row. Also, convert the row totals to percentages of the total number of responses. Display the percentages in a table.
b. Create a bar graph with row 1 percentage on the vertical axis and column number on the horizontal axis. Show the row 1 total percentage as a horizontal line on the graph.
c. What pattern do you expect to see if the rows and columns are independent? Does the plot support the result of the test of independence in Exercise 8.55?

8.57 Test the null hypothesis of independence of the two classifications A and B of the 3×3 contingency table shown here. Use $\alpha = .05$.

		B		
		B_1	B_2	B_3
A	A_1	40	72	42
	A_2	63	53	70
	A_3	31	38	30

8.58 Refer to Exercise 8.57. Convert the responses to percentages by calculating the percentage of each B class total falling into each A classification. Also, calculate the percentage of the total number of responses that constitute each of the A classification totals.
a. Create a bar graph with row A_1 percentage on the vertical axis and B classification on the horizontal axis. Does the graph support the result of the test of hypothesis in Exercise 8.57? Explain.
b. Repeat part **a** for the row A_2 percentages.
c. Repeat part **a** for the row A_3 percentages.

Applying the Concepts—Basic

8.59 **Children's perceptions of their neighborhood.** In *Health Education Research* (Feb. 2005), nutrition scientists at Deakin University (Australia) investigated children's perceptions of their environments. Each in a sample of 147 ten-year-old children drew maps of their home and neighborhood environment. The researchers examined the maps for certain themes (e.g., presence of a pet, television in the bedroom, opportunities for physical activity). The results, broken down by gender, for two themes (presence of a dog and TV in the bedroom) are shown in the tables below and saved in the **MAPDOG** and **MAPTV** files, respectively.

a. Find the sample proportion of boys who drew a dog on their maps.

b. Find the sample proportion of girls who drew a dog on their maps.

c. Compare the proportions you found in parts **a** and **b**. Does it appear that the likelihood of drawing a dog on the neighborhood map depends on gender?

d. Give the null hypothesis for testing whether the likelihood of a drawing a dog on the neighborhood map depends on gender.

e. Use the MINITAB printout (below) to conduct the test mentioned in part **d** at $\alpha = .05$.

f. Conduct a test to determine whether the likelihood of drawing a TV in the bedroom is different for boys and girls. Use $\alpha = .05$.

Presence of a Dog	Number of Boys	Number of Girls
Yes	6	11
No	71	59
Total	77	70

Presence of TV in Bedroom	Number of Boys	Number of Girls
Yes	11	9
No	66	61
Total	77	70

Based on Hume, C., Salmon, J., and Ball, K. "Children's perceptions of their home and neighborhood environments, and their association with objectively measured physical activity: A qualitative and quantitative study," *Health Education Research*, Vol. 20, No. 1, February 2005 (Table III).

MINITAB output for Exercise 8.59

Tabulated statistics: DOG, GENDER

```
Using frequencies in NUMBER


Rows: DOG    Columns: GENDER

         Boy    Girl    All

No        71      59     130
        68.10   61.90  130.00

Yes        6      11      17
         8.90    8.10   17.00

All       77      70     147
        77.00   70.00  147.00

Cell Contents:      Count
                    Expected count


Pearson Chi-Square = 2.250, DF = 1, P-Value = 0.134
Likelihood Ratio Chi-Square = 2.268, DF = 1, P-Value = 0.132
```

8.60 Eyewitnesses and mug shots. *Applied Psychology in Criminal Justice* (April 2010) published a study of mug shot choices by eyewitnesses to a crime. Each in a sample of 96 college students was shown a video of a simulated theft, then asked to select the mug shot that most closely resembled the thief. The students were randomly assigned to view either 3, 6, or 12 mug shots at a time, with 32 students in each group. The number of students in the 3-, 6-, or 12-photos-per-page groups who selected the target mugshot were 19, 19, and 15, respectively.

a. For each photo group, compute the proportion of students who selected the target mug shot. Which group yielded the lowest proportion?

b. Create a contingency table for these data, with photo group in the rows and whether or not the target mug shot was selected in the columns.

c. Refer to, part **b**. Are there differences in the proportions who selected the target mug shot among the three photo groups? Test using $\alpha = .10$.

8.61 Stereotyping deceptive and authentic news stories. Major newspapers lose their credibility (and subscribers) when they are found to have published deceptive or misleading news stories. In *Journalism and Mass Communication Quarterly* (Summer 2007), University of Texas researchers investigated whether certain stereotypes (e.g., negative references to certain nationalities) occur more often in deceptive news stories than in authentic news stories. The researchers analyzed 183 news stories that were proven to be deceptive in nature and 128 news stories that were considered authentic. Specifically, the researchers determined whether each story was negative, neutral, or positive in tone. The accompanying table (saved in the **NEWSSTORY** file) gives the number of news stories found in each tone category.

	Authentic News Stories	Deceptive News Stories
Negative Tone	59	111
Neutral Tone	49	61
Positive Tone	20	11
Total	128	183

Based on Lasorsa, D., and Dai, J. "When news reporters deceive: The production of stereotypes." *Journalism and Mass Communication Quarterly*, Vol. 84, No. 2, Summer 2007 (Table 2).

a. Find the sample proportion of negative tone news stories that is deceptive.

b. Find the sample proportion of neutral news stories that is deceptive.

c. Find the sample proportion of positive news stories that is deceptive.

d. Compare the sample proportions, parts **a–c**. Does it appear that the proportion of news stories that is deceptive depends on story tone?

e. Give the null hypothesis for testing whether the authenticity of a news story depends on tone.

f. Use the SPSS printout (p. 459) to conduct the test, part **e**. Test at $\alpha = .05$.

SPSS output for Exercise 8.61

TONE * STORY Crosstabulation

			STORY		
			Authentic	Deceptive	Total
TONE	Negative	Count	59	111	170
		Expected Count	70.0	100.0	170.0
	Neutral	Count	49	61	110
		Expected Count	45.3	64.7	110.0
	Positive	Count	20	11	31
		Expected Count	12.8	18.2	31.0
Total		Count	128	183	311
		Expected Count	128.0	183.0	311.0

Chi-Square Tests

	Value	df	Asymp. Sig. (2-sided)
Pearson Chi-Square	10.427[a]	2	.005
Likelihood Ratio	10.348	2	.006
N of Valid Cases	311		

a. 0 cells (.0%) have expected count less than 5. The minimum expected count is 12.76.

8.62 Healing heart patients with music, imagery, touch, and prayer. "Frontier medicine" is a term used to describe medical therapies (e.g., energy healing, therapeutic prayer, spiritual healing) for which there is no plausible explanation. *The Lancet* (July 16, 2005) published the results of a study designed to test the effectiveness of two types of frontier medicine—music, imagery, and touch (MIT) therapy and therapeutic prayer—in healing cardiac care patients. Patients were randomly assigned to receive one of four types of treatment: (1) prayer, (2) MIT, (3) prayer and MIT, and (4) standard care (no prayer and no MIT). Six months after therapy, the patients were evaluated for a major adverse cardiovascular event (e.g., a heart attack). The results of the study are summarized in the accompanying table and saved in the **HEALING** file.

a. Identify the two qualitative variables (and associated levels) measured in the study.

b. State H_o and H_a for testing whether a major adverse cardiovascular event depends on type of therapy.

c. Use the MINITAB printout (next column) to conduct the test mentioned in part **b** at $\alpha = .10$. On the basis of this test, what can the researchers infer about the effectiveness of music, imagery, and touch therapy and the effectiveness of healing prayer in heart patients?

Therapy	Number of Patients with Major Cardiovascular Events	Number of Patients with No Events	Total
Prayer	43	139	182
MIT	47	138	185
Prayer and MIT	39	150	189
Standard	50	142	192

Based on Krucoff, M. W., et al. "Music, imagery, touch, and prayer as adjuncts to interventional cardiac care: The Monitoring and Actualization of Noetic Trainings (MANTRA) II randomized study." *The Lancet*, Vol. 366, July 16, 2005 (Table 4).

MINITAB output for Exercise 8.62

Tabulated statistics: THERAPY, EVENT

```
Using frequencies in NUMBER


Rows: THERAPY    Columns: EVENT

              No      Yes     All

MIT          138       47     185
           140.7     44.3   185.0

Prayer       139       43     182
           138.4     43.6   182.0

Prayer&MIT   150       39     189
           143.8     45.2   189.0

Standard     142       50     192
           146.1     45.9   192.0

All          569      179     748
           569.0    179.0   748.0


Cell Contents:        Count
                      Expected count


Pearson Chi-Square = 1.828, DF = 3, P-Value = 0.609
Likelihood Ratio Chi-Square = 1.855, DF = 3, P-Value = 0.603
```

8.63 Masculinity and crime. Refer to the *Journal of Sociology* (July 2003) study on the link between the level of masculinity and criminal behavior in men, presented in Exercise 7.25 (p. 368). The researcher identified events that a sample of newly incarcerated men were involved in and classified each event as "violent" (involving the use of a weapon, the throwing of objects, punching, choking, or kicking) or "avoided-violent" (involving pushing, shoving, grabbing, or threats of violence that did not escalate into a violent event). Each man (and corresponding event) was also classified as possessing "high-risk masculinity" (scored high on the Masculinity–Femininity Scale test and low on the Traditional Outlets of Masculinity Scale test) or "low-risk masculinity." The data on 1,507 events are summarized in the following table and saved in the **HRM** file.

	Violent Events	Avoided-Violent Events	Totals
High-Risk Masculinity	236	143	379
Low-Risk Masculinity	801	327	1,128
Totals	1,037	470	1,507

Based on Krienert, J. L. "Masculinity and crime: A quantitative exploration of Messerschmidt's hypothesis." *Journal of Sociology,* Vol. 7, No. 2, July 2003 (Table 4).

a. Identify the two categorical variables measured (and their levels) in the study.

b. Identify the experimental units.

c. If the type of event (violent or avoided-violent) is independent of high- low-risk masculinity, how many of the 1,507 events would you expect to be violent and involve a high-risk-masculine man?

d. Repeat part **c** for the other combinations of event type and high- low-risk masculinity.

e. Calculate the χ^2 statistic for testing whether event type depends on high- low-risk masculinity.

f. Give the appropriate conclusion of the test mentioned in part **e**, using $\alpha = .05$.

Applying the Concepts—Intermediate

8.64 **"Cry wolf" effect in air traffic controlling.** Researchers at Alion Science Corporation and New Mexico State University collaborated on a study of how air traffic controllers respond to false alarms (*Human Factors*, Aug. 2009). The researchers theorize that the high rate of false alarms regarding midair collisions leads to the "cry wolf" effect, i.e., the tendency for air traffic controllers to ignore true alerts in the future. The investigation examined data on a random sample of 437 conflict alerts. Each alert was first classified as a "true" or "false" alert. Then, each was classified according to whether or not there was a human controller response to the alert. The number of the 437 alerts that fall into each of the combined categories is given as follows: True alert/No response–3; True alert/Response–231; False alert/No response–37; False alert/Response–166. This summary information is saved in the **ATC** file. Do the data indicate that the response rate of air traffic controllers to midair collision alarms differs for true and false alerts? Test using $\alpha = .05$. What inference can you make concerning the "cry wolf" effect?

Based on Wickens, C. D., Rice, S., Keller, D., Hutchins, S., Hughes, J., and Clayton, K., "False alerts in air traffic control conflict alerting system: Is there a 'cry wolf' effect?" Human Factors, Vol. 51, Issue 4, August 2009 (Table 2).

8.65 **IQ and mental deficiency.** A person is diagnosed with a mental deficiency if, before the age of 18, his or her score on a standard IQ test is no higher than 70 (two standard deviations below the mean of 100). Researchers at Cornell and West Virginia Universities examined the impact of rising IQ scores on diagnoses of mental deficiency (MD) (*American Psychologist*, October, 2003). IQ data were collected from different school districts across the United States, and the students were tested with either the Wechsler Intelligence Scale for Children—Revised (WISC-R) or the Wechsler Intelligence Scale for Children—Third Revision (WISC-III) IQ tests. The researchers focused on those students with IQs just above the mental deficiency cutoff (between 70 and 85), based on the original IQ test. These "borderline" MD students were then retested one year later with one of the IQ tests. The accompanying table gives the number of students diagnosed with mental deficiency on the basis of the retest. These data are saved in the **MDIQ** file. Conduct a chi-square test for independence to determine whether the proportion of students diagnosed with MD depends on the IQ test/retest method. Use $\alpha = .01$.

Test/Retest	Diagnosed with MD	Above MD Cutoff IQ	Total
WISC-R / WISC-R	25	167	192
WISC-R / WISC-III	54	103	157
WISC-III / WISC-III	36	141	177

Source: Kanaya, T., Scullin, M. H., and Ceci, S. J. "The Flynn effect and U.S. Policies." *American Psychologist*, Vol. 58, No. 10, Oct. 2003 (Figure 1). Copyright © 2003 by the American Psychological Association. Reprinted with permission.

8.66 **Creating menus to influence others.** The *Journal of Consumer Research* (Mar. 2003) published a study on influencing the choices of others by offering undesirable alter-

natives. In one experiment conducted by the researcher, 96 subjects were asked to imagine that they had just moved to an apartment with two others and that they were shopping for a new appliance (e.g., a television, a microwave oven). Each subject was asked to create a menu of three brand choices for his or her roommates; then subjects were randomly assigned (in equal numbers) to one of three different "goal" conditions: (1) Create the menu in order to influence roommates to buy a preselected brand, (2) create the menu in order to influence roommates to buy a brand of your choice, and (3) create the menu with no intent to influence roommates. The researcher theorized that the menus created to influence others would likely include undesirable alternative brands. Consequently, the number of menus in each goal condition that was consistent with the theory was determined. The data are summarized in the accompanying table and saved in the **MENU3** file. Analyze the data for the purpose of determining whether the proportion of subjects who select menus consistent with the theory depends on the goal condition. Use $\alpha = .01$.

Goal Condition	Number Consistent with Theory	Number Not Consistent with Theory	Totals
Influence/preselected brand	15	17	32
Influence/own brand	14	18	32
No influence	3	29	32

Based on Hamilton, R. W. "Why do people suggest what they do not want? Using context effects to influence others' choices." *Journal of Consumer Research*, Vol. 29, March 2003 (Table 2).

8.67 **Detecting Alzheimer's disease at an early age.** Refer to the *Neuropsychology* (Jan. 2007) study of whether the cognitive effects of Alzheimer's disease can be detected at an early age, Exercise 8.47 (p. 448). Recall that a particular strand of DNA was classified into one of three genotypes: $E4^+/E4^+$, $E4^+/E4^-$, and $E4^-/E4^-$. In addition to a sample of 2,097 young adults (20–24 years), two other age groups were studied: a sample of 2,182 middle-aged adults (40–44 years) and a sample of 2,281 elderly adults (60–64 years). The accompanying table gives a breakdown of the number of adults with the three genotypes in each age category for the total sample of 6,560 adults. (These data are saved in the **E4E4ALL** file.) The researchers concluded that "there were no significant genotype differences across the three age groups" using $\alpha = .05$. Do you agree?

Age Group	$E4^+/E4^+$ Genotype	$E4^+/E4^-$ Genotype	$E4^-/E4^-$ Genotype	Sample Size
20–24	56	517	1,524	2,097
40–44	45	566	1,571	2,182
60–64	48	564	1,669	2,281

Source: Jorm, A. F., et al. "APOE genotype and cognitive functioning in a large age-stratified population sample." *Neuropsychology*, Vol. 21, No. 1, January 2007 (Table 1). Copyright © 2007 by the American Psychological Association. Reprinted with permission.

8.68 **Trapping grain moths.** In an experiment described in the *Journal of Agricultural, Biological, and Environmental Statistics* (Dec. 2000), bins of corn were stocked with various parasites (e.g., grain moths) in late winter. In early summer (June), three bowl-shaped traps were placed on the surface of the grain in order to capture the moths.

All three traps were baited with a sex pheromone lure; however, one trap used an unmarked sticky adhesive, one was marked with a fluorescent red powder, and one was marked with a fluorescent blue powder. The traps were set on a Wednesday, and the catch was collected the following Thursday and Friday. The accompanying table (saved in the **MOTHTRAP** file) shows the number of moths captured in each trap on each day. Conduct a test (at $\alpha = .10$) to determine whether the percentages of moths caught by the three traps depends on the day of the week.

	Adhesive, No Mark	Red Mark	Blue Mark
Thursday	136	41	17
Friday	101	50	18

Based on Wileyto, E. P. et al. "Self-marking recapture models for estimating closed insect populations." *Journal of Agricultural, Biological, and Environmental Statistics*, Vol. 5, No. 4, December 2000 (Table 5A).

8.69 Classifying air threats with heuristics. The *Journal of Behavioral Decision Making* (Jan. 2007) published a study on the use of heuristics to classify the threat level of approaching aircraft. Of special interest was the use of a fast and frugal heuristic—a computationally simple procedure for making judgments with limited information—named "Take-the-Best-for-Classification" (TTB-C). The subjects were 48 men and women, some from a Canadian Forces reserve unit, others university students. Each subject was presented with a radar screen on which simulated approaching aircraft were identified with asterisks. By using the computer mouse to click on the asterisk, one could receive further information about the aircraft. The goal was to identify the aircraft as "friend" or "foe" as fast as possible. Half the subjects were given cue-based instructions for determining the type of aircraft, while the other half were given pattern-based instructions. The researcher also classified the heuristic strategy used by the subject as TTB-C, Guess, or Other. Data on the two variables Instruction type and Strategy, measured for each of the 48 subjects, are saved in the **AIRTHREAT** file. (Data on the first and last five subjects are shown in the table below.) Do the data provide sufficient evidence at $\alpha = .05$ to indicate that choice of heuristic strategy depends on type of instruction provided? How about at $\alpha = .01$?

Instruction	Strategy
Pattern	Other
Pattern	Other
Pattern	Other
Cue	TTBC
Cue	TTBC
⋮	⋮
Pattern	TTBC
Cue	Guess
Cue	TTBC
Cue	Guess
Pattern	Guess

Based on Bryant, D. J. "Classifying simulated air threats with fast and frugal heuristics." *Journal of Behavioral Decision Making*, Vol. 20, January 2007 (Appendix C).

8.70 Subarctic plant study. The traits of seed-bearing plants indigenous to subarctic Finland were studied in *Arctic, Antarctic, and Alpine Research* (May 2004). Plants were categorized according to *type* (dwarf shrub, herb, or grass), *abundance of seedlings* (no seedlings, rare seedlings, or abundant seedlings), *regenerative group* (no vegetative reproduction, vegetative reproduction possible, vegetative reproduction ineffective, or vegetative reproduction effective), *seed weight class* (0–.1, .1–.5, .5–1.0, 1.0–5.0, and > 5.0 milligrams), and *diaspore morphology* (no structure, pappus, wings, fleshy fruits, or awns/hooks). The data on a sample of 73 plants are saved in the **SEEDLING** file.

```
Tabulated statistics: Abundance, Type

Rows: Abundance   Columns: Type

          DwarfShrub   Grasses   Herbs   All

NS            3            1        1      5
SA            5           14       32     51
SR            5            2       10     17
All          13           17       43     73

Cell Contents:        Count
```

a. A contingency table for plant type and seedling abundance, produced by MINITAB, is shown above. (*Note:* NS = no seedlings, SA = seedlings abundant, and SR = seedlings rare.) Suppose you want to perform a chi-square test of independence to determine whether seedling abundance depends on plant type. Find the expected cell counts for the contingency table. Are the assumptions required for the test satisfied?

b. Reformulate the contingency table by combining the NS and SR categories of seedling abundance. Find the expected cell counts for this new contingency table. Are the assumptions required for the test satisfied?

c. Reformulate the contingency table of part **b** by combining the dwarf shrub and grasses categories of plant type. Find the expected cell counts for this contingency table. Are the assumptions required for the test satisfied?

d. Carry out the chi-square test for independence on the contingency table you came up with in part **c**, using $\alpha = .10$. What do you conclude?

8.71 Susceptibility to hypnosis. A standardized procedure for determining a person's susceptibility to hypnosis is the Stanford Hypnotic Susceptibility Scale, Form C (SHSS:C). Recently, a new method called the Computer-Assisted Hypnosis Scale (CAHS), which uses a computer as a facilitator of hypnosis, has been developed. Each scale classifies a person's hypnotic susceptibility as low, medium, high, or very high. Researchers at the University of Tennessee compared the two scales by administering both tests to each of 130 undergraduate volunteers. (*Psychological Assessment*, Mar. 1995). The hypnotic classifications are summarized in the next table (p. 462) and saved in the **HYPNOSIS** file. A contingency table analysis will be performed to determine whether CAHS level and SHSS level are independent.

Table for Excercise 8.71

		CAHS Level				
		Low	Medium	High	Very High	Totals
SHSS: C Level	**Low**	32	14	2	0	48
	Medium	11	14	6	0	31
	High	6	14	19	3	42
	Very High	0	2	4	3	9
	Totals	49	44	31	6	130

a. Check to see if the assumption of expected cell counts of 5 or more is satisfied. Should you proceed with the analysis? Explain.

b. One way to satisfy the assumption of part **a** is to combine the data for two or more categories (e.g., high and very high) in the contingency table. Form a new contingency table by combining the data for the high and very high categories in both the rows and the columns.

c. Calculate the expected cell counts in the new contingency table you formed in part **c**. Is the assumption now satisfied?

d. Perform the chi-square test on the new contingency table. Use $\alpha = .05$. Interpret the results.

8.72 **Guilt in decision making.** The effect of guilt emotion on how a decision maker focuses on the problem was investigated in the Jan. 2007 issue of the *Journal of Behavioral Decision Making*. A total of 171 volunteer students participated in the experiment, where each was randomly assigned to one of three emotional states (guilt, anger, or neutral) through a reading/writing task. Immediately after the task, the students were presented with a decision problem where the stated option has predominantly negative features (e.g., spending money on repairing a very old car). The results (number responding in each category) are summarized in the accompanying table and saved in the **GUILT** file. Is there sufficient evidence (at $\alpha = .10$) to claim that the option choice depends on emotional state?

Emotional State	Choose Stated Option	Do Not Choose Stated Option	Totals
Guilt	45	12	57
Anger	8	50	58
Neutral	7	49	56
Totals	60	111	171

Based on Gangemi, A., and Mancini, F. "Guilt and focusing in decision-making." *Journal of Behavioral Decision Making*, Vol. 20, Jan. 2007 (Table 2).

Applying the Concepts—Advanced

8.73 **Efficacy of an HIV vaccine.** New, effective AIDS vaccines are now being developed through the process of "sieving"—that is, sifting out infections with some strains of HIV. Harvard School of Public Health statistician Peter Gilbert demonstrated how to test the efficacy of an HIV vaccine in *Chance* (Fall 2000). As an example, using the 2×2 table shown next, Gilbert reported the results of VaxGen's preliminary HIV vaccine trial. The vaccine was designed to eliminate a particular strain of the virus called the "MN strain." The trial consisted of 7 AIDS patients vaccinated with the new drug and 31 AIDS patients who were treated with a placebo (no vaccination). The first table (saved in the **HIVVAC1** file) shows the number of patients who tested positive and negative for the MN strain in the trial follow-up period.

		MN Strain		
		Positive	Negative	Totals
Patient Group	**Unvaccinated**	22	9	31
	Vaccinated	2	5	7
	Totals	24	14	38

Source: Gilbert, P. "Developing an AIDS vaccine by sieving." *Chance*, Vol. 13, No. 4, Fall 2000. Reprinted with permission from Chance. Copyright 2000 by the American Statistical Association. All rights reserved.

a. Conduct a test to determine whether the vaccine is effective in treating the MN strain of HIV. Use $\alpha = .05$.

b. Are the assumptions for the test you carried out in part **a** satisfied? What are the consequences if the assumptions are violated?

c. In the case of a 2×2 contingency table, R. A. Fisher (1935) developed a procedure for computing the exact *p*-value

SAS output for Exercise 8.73

```
                    The FREQ Procedure

                Table of GROUP by MNSTRAIN

           GROUP     MNSTRAIN

           Frequency
           Expected  NEG      POS      Total
           UN        9        22       31
                     11.421   19.579
           V         5        2        7
                     2.5789   4.4211
           Total     14       24       38


        Statistics for Table of GROUP by MNSTRAIN

   Statistic                    DF    Value     Prob

   Chi-Square                   1     4.4112    0.0357
   Likelihood Ratio Chi-Square  1     4.2893    0.0384
   Continuity Adj. Chi-Square   1     2.7773    0.0956
   Mantel-Haenszel Chi-Square   1     4.2952    0.0382
   Phi Coefficient                   -0.3407
   Contingency Coefficient            0.3225
   Cramer's V                        -0.3407

   WARNING: 50% of the cells have expected counts less
            than 5. Chi-Square may not be a valid test.


              Fisher's Exact Test

      Cell (1,1) Frequency (F)        9
      Left-sided Pr <= F         0.0498
      Right-sided Pr >= F        0.9940

      Table Probability (P)      0.0438
      Two-sided Pr <= P          0.0772

            Sample Size = 38
```

for the test (called *Fisher's exact test*). The method utilizes the *hypergeometric probability distribution*. Consider the hypergeometric probability

$$\frac{\binom{7}{2}\binom{31}{22}}{\binom{38}{24}}$$

which represents the probability that 2 out of 7 vaccinated AIDS patients test positive and 22 out of 31 unvaccinated patients test positive—that is, the probability of the result shown in the table, given that the null hypothesis of independence is true. Compute this probability (called the *probability of the contingency table*).

d. Refer to part **c.** Two contingency tables (with the same marginal totals as the original table) that are more unsupportive of the null hypothesis of independence than the observed table are shown below (These data are saved in the **HIVVAC2** and **HIVVAC3** files respectively.). First, explain why these tables provide more evidence to reject H_0 than the original table does. Then compute the probability of each table, using the hypergeometric formula.

e. The *p*-value of Fisher's exact test is the probability of observing a result at least as unsupportive of the null hypothesis as is the observed contingency table, given the same marginal totals. Sum the probabilities of parts **c** and **d** to obtain the *p*-value of Fisher's exact test. (To verify your calculations, check the *p*-value labeled **Left-sided Pr < = F** at the bottom of the SAS printout

shown on p. 462. Interpret this value in the context of the vaccine trial.

8.74 **Examining the "Monty Hall Dilemma."** In Exercise 3.151 (p. 159) you solved the game show problem of whether or not to switch your choice of three doors, one of which hides a prize, after the host reveals what is behind a door that is not chosen. (Despite the natural inclination of many to keep one's first choice, the correct answer is that you should switch your choice of doors.) This problem is sometimes called the "Monty Hall Dilemma," named for Monty Hall, the host of the popular TV game show *Let's Make a Deal.* In *Thinking & Reasoning* (July, 2007), Wichita State University professors set up an experiment designed to influence subjects to switch their original choice of doors. Each subject participated in 23 trials. In trial 1, three (boxes) representing doors were presented on a computer screen; only one box hid a prize. In each subsequent trial, an additional box was presented, so that in trial 23, twenty-five boxes were presented. In each trial, after a box was selected, all of the remaining boxes except for one either (1) were shown to be empty (*Empty* condition), (2) disappeared (*Vanish* condition), (3) disappeared, and the chosen box was enlarged (*Steroids* condition), or (4) disappeared, and the remaining box not chosen was enlarged (*Steroids2* condition). Twenty-seven subjects were assigned to each condition. The number of subjects who ultimately switched boxes is tallied, by condition, in the following table for both the first trial and the last trial, These data are saved in the **MONTYHALL** file.

	MN Strain		
	Positive	Negative	Totals
Patient Group Unvaccinated	23	8	31
Vaccinated	1	6	7
Totals	24	14	38

	MN Strain		
	Positive	Negative	Totals
Patient Group Unvaccinated	24	7	31
Vaccinated	0	7	7
Totals	24	14	38

| | First Trial (1) | | Last Trial (23) | |
Condition	Switch Boxes	No Switch	Switch Boxes	No Switch
Empty	10	17	23	4
Vanish	3	24	12	15
Steroids	5	22	21	6
Steroids2	8	19	19	8

Based on Howard, J. N., Lambdin, C. G., and Datteri, D. L. "Let's make a deal: quality and availability of second-stage information as a catalyst for change." *Thinking & Reasoning*, Vol. 13, No. 3, July 2007 (Table 2).

a. For a selected trial, does the likelihood of switching boxes depend on condition?

b. For a given condition, does the likelihood of switching boxes depend on trial number?

c. On the basis of the results you obtained in parts **a** and **b**, what factors influence a subject to switch choices?

CHAPTER NOTES

Key Terms

Categories 439
Cells 439
Cell counts 439
Chi-square test 441
Chi-square distribution 441
Dependence 452
Classes 439
Contingency table 449
Contingency table with fixed
  marginals 455–456
Dependence 452
Dimensions of classification 449

Expected cell count 442
Independence of two
  classifications 450
Marginal probabilities 449
Multinomial experiment 439
Observed cell count 450
One-way table 440
Two-way table 449
Sampling distribution of
  $(\hat{p}_1 - \hat{p}_2)$ 431
Sample surveys 430

Key Symbols/Notation

$p_1 - p_2$	Difference between population propotions
$\hat{p}_1 - \hat{p}_2$	Difference between sample proportions
SE	Sampling error of estimation
$p_{i,0}$	Value of multinomial probability p_1 hypothesized in H_0
χ^2	Chi-square test statistic used in analysis of categorical data
n_i	Number of observed outcomes in cell i of a one-way table

E_i Expected number of outcomes in cell i of a one-way table

p_{ij} Probability of an outcome in row i and column j of a two-way table

n_{ij} Number of observed outcomes in row i and column j of a two-way table

E_{ij} Expected number of outcomes in row i and column j of a two-way table

R_i Total number of outcomes in row i of a two-way table

C_j Total number of outcomes in column j of a two-way table

Key Ideas

Conditions Required for Inferences about $(p_1 - p_2)$

(1) Large independent random samples
(2) $n_1 p_1 \geq 15, n_1 q_1 \geq 15$
(3) $n_2 p_2 \geq 15, n_2 q_2 \geq 15$

Determining the Sample Size for Estimating $(p_1 - p_2)$

$$n_1 = n_2 = \frac{(z_{\alpha/2})^2 (p_1 q_1 + p_2 q_2)}{(SE)^2}$$

Categorical Data Analysis Guide

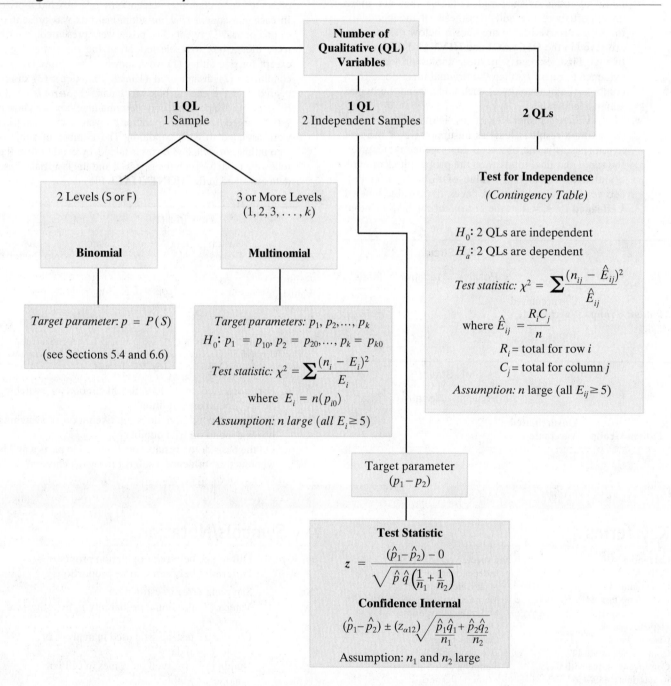

Multinomial Data

Qualitative data that fall into more than two categories (or classes)

Properties of a Multinomial Experiment

1. n identical trials
2. k possible outcomes to each trial
3. probabilities of the k outcomes $(p_1, p_2, \ldots, p_k)$ where $p_1 + p_2 + \ldots + p_k = 1$, remain the same from trial to trial
4. trials are independent
5. variables of interest: *cell counts* (i.e., number of observations falling into each outcome category), denoted $n_1, n_2, \ldots, n_k$

One-Way Table

Summary table for a *single* qualitative variable

Two-Way (Contingency) Table

Summary table for *two* qualitative variables

Chi-Square (χ^2) Statistic

used to test category probabilities in one-way and two-way tables

Chi-Square Tests for Independence

should **not** be used to *infer a causal relationship between two QLs*

Conditions Required for Valid χ^2 Tests

1. multinomial experiment
2. sample size n is large (expected cell counts are all greater than or equal to 5)

Supplementary Exercises 8.75–8.101

Understanding the Principles

8.75 *True or False.* Rejecting the null hypothesis in a chi-square test for independence implies that a causal relationship exists between the two categorical variables.

8.76 What is the difference between a one-way chi-square analysis and a two-way chi-square analysis?

Learning the Mechanics

8.77 Independent random samples were selected from two binomial populations. The size and number of observed successes for each sample are shown in the following table:

Sample 1	Sample 2
$n_1 = 200$	$n_2 = 200$
$x_1 = 110$	$x_2 = 130$

  a. Test $H_0: (p_1 - p_2) = 0$ against $H_a: (p_1 - p_2) < 0$. Use $\alpha = .10$.

  b. Form a 95% confidence interval for $(p_1 - p_2)$.

  c. What sample sizes would be required if we wish to use a 95% confidence interval of width .01 to estimate $(p_1 - p_2)$?

8.78 A random sample of 150 observations was classified into the categories shown in the following table:

	Category				
	1	2	3	4	5
n_i	28	35	33	25	29

  a. Do the data provide sufficient evidence that the categories are not equally likely? Use $\alpha = .10$.

  b. Form a 90% confidence interval for p_2, the probability that an observation will fall into category 2.

8.79 A random sample of 250 observations was classified according to the row and column categories shown in the next table.

  a. Do the data provide sufficient evidence to conclude that the rows and columns are dependent? Test, using $\alpha = .05$.

		Column		
		1	2	3
Row	1	20	20	10
	2	10	20	70
	3	20	50	30

  b. Would the analysis change if the row totals were fixed before the data were collected?

  c. Do the assumptions required for the analysis to be valid differ according to whether the row (or column) totals are fixed? Explain.

  d. Convert the table entries to percentages by using each column total as a base and calculating each row response as a percentage of the corresponding column total. In addition, calculate the row totals and convert them to percentages of all 250 observations.

  e. Create a bar graph with the percentages from row 1 on the vertical axis and the column number on the horizontal axis. Draw a horizontal line corresponding to the total percentage for row 1. Does the graph support the result of the test conducted in part **a**?

  f. Repeat part **e** for the percentages from row 2.

  g. Repeat part **e** for the percentages from row 3.

Applying the Concepts—Basic

8.80 **Oil spill impact on seabirds.** Refer to the *Journal of Agricultural, Biological, and Environmental Statistics* (Sept. 2000) study of the impact of a tanker oil spill on the seabird population in Alaska, presented in Exercise 2.187 (p. 100). Recall that for each of 96 shoreline locations (called transects), the number of seabirds found, the length (in kilometers) of the transect, and whether or not the transect was in an oiled area were recorded. (The data are saved in the **EVOS** file.) *Observed seabird density* is defined as the observed count divided by the length of the transect. A comparison of the mean densities of oiled and unoiled transects is displayed in the MINITAB printout on p. 466. Use this information to make an inference about the difference in the population mean seabird densities of oiled and unoiled transects.

MINITAB output for Exercise 8.80

Two-Sample T-Test and CI: Density, Oil

```
Two-sample T for Density

Oil  N  Mean  StDev  SE Mean
no   36  3.27  6.70    1.1
yes  60  3.50  5.97    0.77


Difference = mu (no ) - mu (yes)
Estimate for difference:  -0.221165
95% CI for difference: (-2.927767, 2.485436)
T-Test of difference = 0 (vs not =): T-Value = -0.16  P-Value = 0.871  DF = 67
```

8.81 **Location of major sports venues.** There has been a recent trend for professional sports franchises in Major League Baseball (MLB), the National Football League (NFL), the National Basketball Association (NBA), and the National Hockey League (NHL) to build new stadiums and ballparks in urban, downtown venues. An article in *Professional Geographer* (Feb. 2000) investigated whether there has been a significant suburban-to-urban shift in the location of major sport facilities. In 1985, 40% of all major sport facilities were located downtown, 30% in central cities, and 30% in suburban areas. In contrast, of the 113 major sports franchises that existed in 1997, 58 were built downtown, 26 in central cities, and 29 in suburban areas.

a. Describe the qualitative variable of interest in the study. Give the levels (categories) associated with the variable.

b. Give the null hypothesis for a test to determine whether the proportions of major sports facilities in downtown, central city, and suburban areas in 1997 are the same as in 1985.

c. If the null hypothesis of part **b** is true, how many of the 113 sports facilities in 1997 would you expect to be located in downtown, central city, and suburban areas, respectively?

d. Find the value of the chi-square statistic for testing the null hypothesis of part **b.**

e. Find the (approximate) p-value of the test, and give the appropriate conclusion in the words of the problem. Assume that $\alpha = .05$.

8.82 **Scanning Internet messages.** *Inc. Technology* (Mar. 18, 1997) reported the results of an Equifax/Harris Consumer Privacy Survey in which 328 Internet users indicated their level of agreement with the following statement: "The government needs to be able to scan Internet messages and user communications to prevent fraud and other crimes." The number of users in each response category is summarized as follows:

Agree Strongly	Agree Somewhat	Disagree Somewhat	Disagree Strongly
59	108	82	79

a. Specify the null and alternative hypotheses you would use to determine whether the opinions of Internet users are evenly divided among the four categories.

b. Conduct the test of part **a,** using $\alpha = .05$.

c. In the context of this exercise, what is a Type I error? A Type II error?

d. What assumptions must hold in order to ensure the validity of the test you conducted in part **b**?

8.83 **Risk factor for lumbar disease.** One of the most common musculoskeletal disorders is lumbar disk disease (LDD).

Medical researchers reported finding a common genetic risk factor for LDD (*Journal of the American Medical Association*, Apr. 11, 2001). The study included 171 Finnish patients diagnosed with LDD (the patient group) and 321 without LDD (the control group). Of the 171 LDD patients, 21 were discovered to have the genetic trait. Of the 321 people in the control group, 15 had the genetic trait.

a. Consider the two categorical variables group and presence/absence of genetic trait. Form a 2 × 2 contingency table for these variables.

b. Conduct a test to determine whether the genetic trait occurs at a higher rate in LDD patients than in the controls. Use $\alpha = .01$.

c. Construct a bar graph that will visually support your conclusion in part **b.**

8.84 **Late-emerging reading disabilities.** Studies of children with reading disabilities typically focus on "early-emerging" difficulties identified prior to the fourth grade. Psychologists at Haskins Laboratories recently studied children with "late-emerging" reading difficulties (i.e., children who appeared to undergo a fourth-grade "slump" in reading achievement) and published their findings in the *Journal of Educational Psychology* (June 2003). A sample of 161 children was selected from fourth and fifth graders at elementary schools in Philadelphia. In addition to recording the grade level, the researchers determined whether each child had a previously undetected reading disability. Sixty-six children were diagnosed with a reading disability. Of these children, 32 were fourth graders and 34 were fifth graders. Similarly, of the 95 children with normal reading achievement, 55 were fourth graders and 40 were fifth graders.

a. Identify the two qualitative variables (and corresponding levels) measured in the study.

b. From the information provided, form a contingency table.

c. Assuming that the two variables are independent, calculate the expected cell counts.

d. Find the test statistic for determining whether the proportions of fourth and fifth graders with reading disabilities differs from the proportions of fourth and fifth graders with normal reading skills.

e. Find the rejection region for the test if $\alpha = .10$.

f. Is there a link between reading disability and grade level? Give the appropriate conclusion of the test.

8.85 **Politics and religion.** University of Maryland professor Ted R. Gurr examined the political strategies used by ethnic groups worldwide in their fight for minority rights. (*Political Science & Politics*, June 2000). Each in a sample of 275 ethnic groups was classified according to world region

Table for Exercise 8.85

		Political Strategy		
		No Political Action	Mobilization, Mass Action	Terrorism, Rebellion, Civil War
	Latin American	24	31	7
	Post-Communist	32	23	4
World Region	**South, Southeast, East Asia**	11	22	26
	Africa/Middle East	39	36	20

Table 1 from "Nonviolence ethnopolitics: Strategies for the attainment of group rights and autonomy" by Ted Robert Gurr. Political Science & Politics, Vol. 33, No. 2, June 2000. Copyright © 2000 The American Political Science Association. Reprinted with the permission of Cambridge University Press.

and highest level of political action reported. The data are summarized in the contingency table above and saved in the **ETHNIC** file. Conduct a test at $\alpha = .10$ to determine whether political strategy of ethnic groups depends on world region. Support your answer with a graph.

8.86 **"Made in the USA" survey.** Refer to the *Journal of Global Business* (Spring 2002) study of what "Made in the USA" on product labels means to the typical consumer, presented in Exercise 2.179 (p. 99). Recall that 106 shoppers participated in the survey. Their responses, given as a percentage of U.S. labor and materials in four categories, are summarized as follows: 64 shoppers responded 100%; 20 shoppers stated 75 to 99%; 18 shoppers stated 50 to 74%; and 4 shoppers said less than 50%. Suppose a consumer advocate group claims that half of all consumers believe that "Made in the USA" means "100%" of labor and materials are produced in the United States, one-fourth believe that "75 to 99%" are produced in the United States, one-fifth believe that "50 to 74%" are produced in the United States, and 5 percent believe that "less than 50%" are produced in the United States.

a. Describe the qualitative variable of interest in the study. Give the levels (categories) associated with the variable.

b. What are the values of p_1, p_2, p_3, and p_4, the probabilities associated with the four response categories hypothesized by the consumer advocate group?

c. Give the null and alternative hypotheses for testing the consumer advocate group's claim.

d. Compute the test statistic for testing the hypotheses stated in part **c.**

e. Find the rejection region of the test at $\alpha = .10$.

f. State the conclusion in the words of the problem.

g. Find and interpret a 90% confidence interval for the true proportion of consumers who believe that "Made in the USA" means that "100%" of labor and materials are produced in the United States.

Based on "'Made in the USA': Consumer perceptions, deception and policy alternatives." *Journal of Global Business*, Vol. 13, No. 24, Spring 2002 (Table 3).

8.87 **The "winner's curse" in auction bidding.** In auction bidding, the "winner's curse" is the phenomenon of the winning (or highest) bid price being above the expected value of the item being auctioned. *The Review of Economics and Statistics* (Aug. 2001) published a study on whether experience in bidding affects the likelihood of the winner's curse occurring. Two groups of bidders in a sealed-bid auction were compared: (1) superexperienced bidders and (2) less experienced bidders. In the superexperienced group, 29 of 189 winning bids were above the item's expected value; in the less experienced group, 32 of 149 winning bids were above the item's expected value.

a. Find an estimate of p_1, the true proportion of super-experienced bidders who fall prey to the winner's curse.

b. Find an estimate of p_2, the true proportion of less experienced bidders who fall prey to the winner's curse.

c. Construct a 90% confidence interval for $p_1 - p_2$.

d. Give a practical interpretation of the confidence interval you constructed in part **c.** Make a statement about whether experience in bidding affects the likelihood of the winner's curse occurring.

Applying the Concepts—Intermediate

8.88 **Switching majors in college.** When female undergraduates switch from science, mathematics, and engineering (SME) majors into disciplines that are not based on science, are their reasons different from those of their male counterparts? This question was investigated in *Science Education* (July 1995). A sample of 335 junior/senior undergraduates—172 females and 163 males—at two large research universities were identified as "switchers"; that is, they left a declared SME major for a non-SME major. Each student listed one or more factors that contributed to the switching decision.

a. Of the 172 females in the sample, 74 listed lack or loss of interest in SME (i.e., they were "turned off" by science) as a major factor, compared with 72 of the 163 males. Conduct a test (at $\alpha = .10$) to determine whether the proportion of female switchers who give "lack of interest in SME" as a major reason for switching differs from the corresponding proportion of males.

b. Thirty-three of the 172 females in the sample indicated that they because discouraged or lost confidence because of low grades in SME during their early years, compared with 44 of 163 males. Construct a 90% confidence interval for the difference between the proportions of female and male switchers who lost confidence due to low grades in SME. Interpret the result.

8.89 **Iraq War survey.** The Pew Internet & American Life Project commissioned Princeton Survey Research Associates to develop and carry out a survey of what Americans think about the War in Iraq. Some of the results of the March 2003 survey of over 1,400 American adults are saved in the **IRAQWAR** file. Responses to the following questions were recorded:

1. Do you support or oppose the Iraq War? (1 = Support, 2 = Oppose)

2. Do you ever go online to access the Internet or World Wide Web? (1 = Yes, 2 = No)

3. Do you consider yourself a Republican, Democrat, or Independent? (1 = Rep., 2 = Dem., 3 = Ind.)

4. Have you or anyone in your household served in the U.S. military? (1 = Yes, I have; 2 = Yes, other; 3 = Yes, both; 4 = No)

5. In general, would you describe your political views as very conservative, conservative, moderate, liberal, or very liberal? (1 = Very conservative, 2 = Conservative, 3 = Moderate, 4 = Liberal, 5 = Very liberal)

6. What is your race? (1 = White, 2 = African-American, 3 = Asian, 4 = Mixed, 5 = Native American, 6 = Other)

7. What is your income range? (1 = < 10K, 2 = 10–20K, 3 = 20–30K, 4 = 30–40K, 5 = 40–50K, 6 = 50–75K, 7 = 75–100K, 8 = > 100K)

8. Do you live in a suburban, rural, or urban community? (1 = urban, 2 = suburban, 3 = rural)

Conduct a series of contingency table analyses to determine whether support for the Iraq War depends on one or more of the other categorical variables measured in the March 2003 survey.

8.90 Pig farmer study. An article in *Sociological Methods & Research* (May 2001) analyzed the data presented in the accompanying table. A sample of 262 Kansas pig farmers was classified according to their education level (college or not) and size of their pig farm (number of pigs). The data are saved in the **PIGFARM** file. Conduct a test to determine whether a pig farmer's education level has an impact on the size of the pig farm. Use $\alpha = .05$ and support your answer with a graph.

		Education Level		
		No College	College	Totals
	<1,000 pigs	42	53	95
	1,000–2,000 pigs	27	42	69
Farm Size	2,001–5,000 pigs	22	20	42
	>5,000 pigs	27	29	56
	Totals	118	144	262

Based on Agresti, A., and Liu, I. "Strategies for modeling a categorical variable allowing multiple category choices." *Sociological Methods & Research,* Vol. 29, No. 4, May 2001 (Table I).

8.91 Multiple-sclerosis drug. Interferons are proteins produced naturally by the human body that help fight infections and regulate the immune system. A drug developed from interferons, called Avonex, is now available for treating patients with multiple sclerosis (MS). In a clinical study, 85 MS patients received weekly injections of Avonex over a two-year period. The number of exacerbations (i.e., flare-ups of symptoms) was recorded for each patient and is summarized in the accompanying table. For MS patients who take a placebo (no drug) over a similar two-year period, it is known from previous studies that 26% will experience no exacerbations, 30% one exacerbation, 11% two exacerbations, 14% three exacerbations, and 19% four or more exacerbations.

Number of Exacerbations	Number of Patients
0	32
1	26
2	15
3	6
4 or more	6

Based on data from Biogen, Inc.

a. Conduct a test to determine whether the exacerbation distribution of MS patients who take Avonex differs from the percentages reported for placebo patients. Use $\alpha = .05$.

b. Find a 95% confidence interval for the true percentage of Avonex MS patients who remain free of exacerbations during a two-year period.

c. Refer to part **b.** Is there evidence that Avonex patients are more likely to have no exacerbations than placebo patients? Explain.

8.92 Flight response of geese to helicopter traffic. Offshore oil drilling near an Alaskan estuary has led to increased air traffic—mostly large helicopters—in the area. The U.S. Fish and Wildlife Service commissioned a study to investigate the impact these helicopters have on the flocks of Pacific brant geese that inhabit the estuary in the fall before migrating (*Statistical Case Studies: A Collaboration between Academe and Industry,* 1998). Two large helicopters were flown repeatedly over the estuary at different altitudes and lateral distances from the flock. The flight responses of the geese (recorded as "low" or "high"), the altitude (in hundreds of meters), and the lateral distance (also in hundreds of meters) for each of 464 helicopter overflights were recorded and are saved in the **PACGEESE** file. The data for the first 10 overflights are shown in the following table:

Overflight	Altitude	Lateral Distance	Flight Response
1	0.91	4.99	HIGH
2	0.91	8.21	HIGH
3	0.91	3.38	HIGH
4	9.14	21.08	LOW
5	1.52	6.60	HIGH
6	0.91	3.38	HIGH
7	3.05	0.16	HIGH
8	6.10	3.38	HIGH
9	3.05	6.60	HIGH
10	12.19	6.60	HIGH

Source: From Erickson, W., Nick, T., and Ward, D. "Investigating Flight Response of Pacific Brant to Helicopters at Izembek Lagoon, Alaska by Using Logistic Regression." *Statistical Case Studies: A Collaboration between Academe and Industry,* ASA-SIAM Series on Statistics and Applied Probability, 1998. Copyright © 1998 Society for Industrial and Applied Mathematics. Reprinted with permission. All rights reserved.

a. The researchers categorized altitude as follows: less than 300 meters, 300–600 meters, and 600 or more meters. Summarize the data in the **PACGEESE** file by creating a contingency table for altitude category and flight response.

b. Conduct a test to determine whether flight response of the geese depends on altitude of the helicopter. Test, using $\alpha = .01$.

c. The researchers categorized lateral distance as follows: less than 1,000 meters, 1,000–2,000 meters, 2,000–3,000 meters, and 3,000 or more meters. Summarize the data in the **PACGEESE** file by creating a contingency table for lateral distance category and flight response.

d. Conduct a test to determine whether flight response of the geese depends on lateral distance of helicopter from the flock. Test, using $\alpha = .01$.

e. The current Federal Aviation Authority (FAA) minimum altitude standard for flying over the estuary is 2,000 feet (approximately 610 meters). On the basis of the results obtained in parts **a–d,** what changes to the FAA regulations do you recommend in order to minimize the effects to Pacific brant geese?

8.93 Birds feeding on gypsy moths. A field study was conducted to identify the natural predators of the gypsy moth. (*Environmental Entomology*, June 1995). For one part of the study, 24 black-capped chickadees (common wintering birds) were captured in mist nets and individually caged. Each bird was offered a mass of gypsy moth eggs attached to a piece of bark. Half the birds were offered no other food (no choice), and half were offered a variety of other naturally occurring foods such as spruce and pine seeds (choice). The numbers of birds that did and did not feed on the gypsy moth egg mass are given in the next table and saved in the **MOTH** file. Analyze the data in the table to determine whether a relationship exists between food choice and feeding or not feeding on gypsy moth eggs. Use $\alpha = .10$.

	Fed on Egg Mass	
	Yes	No
Choice of foods	2	10
No choice	8	4

8.94 Gangs and homemade weapons. The National Gang Crime Research Center (NGCRC) has developed a six-level gang classification system for both adults and juveniles. The six categories are shown in the accompanying table. The classification system was developed as a potential predictor of a gang member's propensity for violence in prison, jail, or a correctional facility. To test the system, the NGCRC collected data on approximately 10,000 confined offenders and assigned each a score from the gang classification system. (*Journal of Gang Research*, Winter 1997). One of several other variables measured by the NGCRC was whether or not the offender had ever carried a homemade weapon (e.g., knife) while in custody. The data on gang score and homemade weapon are summarized in the table below and saved in the **GANGS** file. Conduct a test to determine whether carrying a homemade weapon in custody depends on gang classification score. (Use $\alpha = .01$.) Support your conclusion with a graph.

Gang Classification Score	Homemade Weapon Carried	
	Yes	No
0 (Never joined a gang, no close friends in a gang)	255	2,551
1 (Never joined a gang, 1–4 close friends in a gang)	110	560
2 (Never joined a gang, 5 or more friends in a gang)	151	636
3 (Inactive gang member)	271	959
4 (Active gang member, no position of rank)	175	513
5 (Active gang member, holds position of rank)	476	831

Source: From Knox, G. W., et al. "A gang classification system for corrections." *Journal of Gang Research*, Vol. 4, No. 2, Winter 1997, p. 54 (Table 4). Reprinted with permission from National Gang Crime Research Center.

8.95 Top Internet search engines. Nielsen/NetRatings is a global leader in Internet media and market research. In May 2006, the firm reported on the "search" shares (i.e., the percentage of all Internet searches) for the most popular search engines available on the Web. Google Search accounted for 50% of all searches, Yahoo! Search for 22%, MSN Search for 11%, and all other search engines for 17%. Suppose that, in a random sample of 1,000 recent Internet searches, 487 used Google Search, 245 used Yahoo! Search, 121 used MSN Search, and 147 used another search engine.
a. Do the sample data disagree with the percentages reported by Nielsen/NetRatings? Test, using $\alpha = .05$.
b. Find and interpret a 95% confidence interval for the percentage of all Internet searches that use the Google Search engine.

8.96 Kicking the cigarette habit. Can taking an antidepressant drug help cigarette smokers kick their habit? The *New England Journal of Medicine* (Oct. 23, 1997) published a study in which 615 smokers (all of whom wanted to give up smoking) were randomly assigned to receive either Zyban (an antidepressant) or a placebo (a dummy pill) for six weeks. Of the 309 patients who received Zyban, 71 were not smoking one year later. Of the 306 patients who received a placebo, 37 were not smoking one year later. Conduct a test of hypothesis (at $\alpha = .05$) to answer the research question posed in the first sentence of this exercise.

8.97 Coupon usage study. A hot topic in marketing research is the exploration of a technology-based self-service (TBSS) encounter, in which various technologies (e.g., ATMs, online banking, self-scanning at retail stores) allow the customer to perform all or part of the service. Marketing professor Dan Ladik of the University of Suffolk investigated whether there were differences in customer characteristics and customer satisfaction between users of discount coupons distributed through the mail (nontechnology users) and users of coupons distributed via the Internet (TBSS users). A questionnaire measured several qualitative variables (defined in the accompanying table) for each of 440 coupon users. The data are saved in the **COUPONS** file.

Variable Name	Levels (Possible Values)
Coupon User Type	Mail, Internet, or Both
Gender	Male or Female
Education	High School, Vo-Tech/College, 4-year College Degree, or Graduate School
Work Status	Full Time, Part Time, Not Working, Retired
Coupon Satisfaction	Satisfied, Unsatisfied, Indifferent

a. Consider the variable Coupon User Type. Conduct a test (at $\alpha = .05$) to determine whether the proportions of mail-only users, Internet-only users, and users of both media are statistically different. Illustrate the results with a graph.
b. The researcher wants to know whether there are differences in customer characteristics (i.e., Gender, Education, Work Status, and Coupon Satisfaction) among the three types of coupon users. For each characteristic, conduct a contingency table analysis (at $\alpha = .05$) to determine whether Coupon User Type is related to that characteristic. Illustrate your results with graphs.

Table for Exercise 8.100

Interval	$x < -2$	$-2 \leq x < -1$	$-1 \leq x < 0$	$0 \leq x < 1$	$1 \leq x < 2$	$x \geq 2$
Frequency	7	20	61	77	26	9

8.98 Battle simulation trials. In order to evaluate their situational awareness, fighter aircraft pilots participate in battle simulations. At a random point in the trial, the simulator is frozen and data on situational awareness are immediately collected. The simulation is then continued until, ultimately, performance (e.g., number of kills) is measured. A study reported in *Human Factors* (Mar. 1995) investigated whether temporarily stopping the simulation results in any change in pilot performance. Trials were designed so that some simulations were stopped to collect situational awareness data while others were not. Each trial was then classified according to the number of kills made by the pilot. The data for 180 trials are summarized in the accompanying contingency table and saved in the **SIMKILLS** file. Conduct a contingency table analysis and interpret the results fully.

	Number of Kills					
	0	1	2	3	4	Totals
Stops	32	33	19	5	2	91
No Stops	24	36	18	8	3	89
Totals	56	69	37	13	5	180

Applying the Concepts—Advanced

8.99 Goodness-of-fit test. A statistical analysis is to be done on a set of data consisting of 1,000 monthly salaries. The analysis requires the assumption that the sample was drawn from a normal distribution. A preliminary test, called the χ^2 *goodness-of-fit test*, can be used to help determine whether it is reasonable to assume that the sample is from a normal distribution. Suppose the mean and standard deviation of the 1,000 salaries are hypothesized to be $1,200 and $200, respectively. Using the standard normal table, we can approximate the probability of a salary being in the intervals listed in the accompanying table. The third column represents the expected number of the 1,000 salaries to be found in each interval if the sample was drawn from a normal distribution with $\mu = $1,200$ and $\sigma = 200. Suppose the last column contains the actual

Interval	Probability	Expected Frequency	Observed Frequency
Less than $800	.023	23	26
$800 < $1,000	.136	136	146
$1,000 < $1,200	.341	341	361
$1,200 < $1,400	.341	341	311
$1,400 < $1,600	.136	136	143
$1,600 or above	.023	23	13

observed frequencies in the sample. Large differences between the observed and expected frequencies cast doubt on the normality assumption.

a. Compute the χ^2 statistic on the basis of the observed and expected frequencies.

b. Find the tabulated χ^2 value when $\alpha = .05$ and there are five degrees of freedom. (There are $k - 1 = 5$ df associated with this χ^2 statistic.)

c. On the basis of the χ^2 statistic and the tabulated χ^2 value, is there evidence that the salary distribution is nonnormal?

d. Find the approximate observed significance level for the test in part **c.**

8.100 Testing normality. Suppose a random variable is hypothesized to be normally distributed with a mean of 0 and a standard deviation of 1. A random sample of 200 observations of the variable yields frequencies in the intervals listed in the table shown above. Do the data provide sufficient evidence to contradict the hypothesis that x is normally distributed with $\mu = 0$ and $\sigma = 1$? Use the technique developed in Exercise 8.99.

Critical Thinking Challenge

8.101 A "rigged" election? *Chance* (Spring 2004) presented data from a recent election held to determine the board of directors of a local community. There were 27 candidates for the board, and each of 5,553 voters was allowed to choose 6 candidates. The claim was that "a fixed vote with fixed percentages [was] assigned to each and every candidate, making it impossible to participate in an honest election." Votes were tallied in six time slots: after 600 total votes were in, after 1,200, after 2,444, after 3,444, after 4,444, and, finally, after 5,553 votes. The data on three of the candidates (Smith, Coppin, and Montes) are shown in the accompanying table and saved in the **RIGVOTE** file. A residential organization believes that "there was nothing random about the count and tallies for each time slot, and specific unnatural or rigged percentages were being assigned to each and every candidate." Give your opinion. Is the probability of a candidate receiving votes independent of the time slot, and if so, does this imply a rigged election?

Time Slot	1	2	3	4	5	6
Votes for Smith	208	208	451	392	351	410
Votes for Coppin	55	51	109	98	88	104
Votes for Montes	133	117	255	211	186	227
Total Votes	600	600	1,244	1,000	1,000	1,109

Based on Gelman, A. "55,000 residents desperately need your help!" *Chance*, Vol. 17, No. 2, Spring 2004 (Figures 1 and 5).

Activity   Binomial versus Multinomial Experiments

In this activity, you will study the difference between binomial and multinomial experiments.

1. A television station has hired an independent research group to determine whether television viewers in the area prefer its local news program to the news programs of two other stations in the same city. Explain why a multinomial experiment would be appropriate, and design a poll that satisfies the five properties of a multinomial experiment. State the null and alternative hypotheses for the corresponding χ^2 test.

2. Suppose the television station believes that a majority of local viewers prefers its news program to those of its two competitors. Explain why a binomial experiment would be appropriate to support this claim, and design a poll that satisfies the five properties of a binomial experiment. State the null and alternative hypotheses for the corresponding test.

3. Generalize the situations in Exercises 1 and 2 in order to describe conditions under which a multinomial experiment can be rephrased as a binomial experiment. Is there any advantage in doing so? Explain.

References

Agresti, A. *Categorical Data Analysis*. New York: Wiley, 1990.

Cochran, W. G. "The χ^2 test of goodness of fit." *Annals of Mathematical Statistics*, 1952, 23.

Conover, W. J. *Practical Nonparametric Statistics*, 2nd ed. New York: Wiley, 1980.

Fisher, R. A. "The logic of inductive inference (with discussion)." *Journal of the Royal Statistical Society*, Vol. 98, 1935, pp. 39–82.

Hollander, M., and Wolfe, D. A. *Nonparametric Statistical Methods*. New York: Wiley, 1973.

Savage, I. R. "Bibliography of nonparametric statistics and related topics." *Journal of the American Statistical Association*, 1953, 48.

USING TECHNOLOGY

MINITAB: Categorical Data Analyses

MINITAB can analyse the difference between two proportions and conduct chi-square tests for both one-way and two-way (contingency) tables.

Comparing Two Proportions with Large Independent Samples

Step 1 Access the MINITAB worksheet that contains the sample data.

Step 2 Click on the "Stat" button on the MINITAB menu bar and then click on "Basic Statistics" and "2 Proportions."

Step 3 On the resulting dialog box (shown in Figure 8.M.1), select the data option ("Samples in different columns" or "Summarized data") and make the appropriate menu choices. (Figure 8.M.1 shows the menu options when you select "Summarized data.")

Figure 8.M.1
MINITAB 2 proportions dialog box

Step 4 Click the "Options" button and specify the confidence level for a confidence interval, the null-hypothesized value of the difference, and the form of the alternative hypothesis (lower tailed, two tailed, or upper tailed) in the resulting dialog box, as shown in Figure 8.M.2. (If you desire a pooled estimate of p for the test, be sure to check the appropriate box.)

Figure 8.M.2
MINITAB 2 proportions options

Step 5 Click "OK" to return to the "2 Proportions" dialog box and then click "OK" again to generate the MINITAB printout.

One-Way Table

Step 1 Access the MINITAB worksheet file that contains the sample data for the qualitative variable of interest. [*Note*: The data file can have actual values (levels) of the variable for each observation, or, alternatively, two columns—one listing the levels of the qualitative variable and the other column with the observed counts for each level.]

Step 2 Click on the "Stat" button on the MINITAB menu bar, and then click on "Tables" and "Chi-Square Goodness-of-Fit

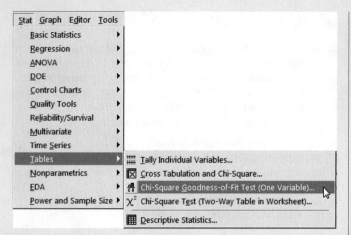

Figure 8.M.3
MINITAB menu options for a one-way chi-square analysis

Test (One Variable)," as shown in Figure 8.M.3. The resulting dialog box appears as shown in Figure 8.M.4.

Figure 8.M.4
MINITAB one-way chi-square dialog box

Step 3 If your data have one column of values for your qualitative variable, select "Categorical data" and specify the variable name (or column) in the box. If your data have summary information in two columns (see above), select "Observed counts" and specify the column with the counts and the column with the variable names in the respective boxes.

Step 4 Select "Equal proportions" for a test of equal proportions, or select "Specific proportions" and enter the hypothesized proportion next to each level in the resulting box.

Step 5 Click "OK" to generate the MINITAB printout.

Two-Way Table

Step 1 Access the MINITAB worksheet file that contains the sample data. The data file should contain two qualitative variables, with category values for each of the n observations in the data set. Alternatively, the worksheet can contain the cell counts for each of the categories of the two qualitative variables.

Step 2 Click on the "Stat" button on the MINITAB menu bar and then click on "Tables" and "Cross Tabulation and Chi-Square," (see Figure 8.M.3). The resulting dialog box appears as shown in Figure 8.M.5.

Step 3 Specify one qualitative variable in the "For rows" box and the other qualitative variable in the "For columns" box. [*Note:* If your worksheet contains cell counts for the categories, enter the variable with the cell counts in the "Frequencies are in" box.]

Figure 8.M.5
MINITAB cross tabulation dialog box

Step 4 Select the summary statistics (e.g., counts, percentages) you want to display in the contingency table.

Step 5 Click the "Chi-square" button. The resulting dialog box is shown in Figure 8.M.6.

Figure 8.M.6
MINITAB chi-square dialog box

Step 6 Select "Chi-Square analysis" and "Expected cell counts" and click "OK."

Step 7 When you return to the "Cross Tabulation" menu screen, click "OK" to generate the MINITAB printout.
Note: If your MINITAB worksheet contains only the cell counts for the contingency table in columns, click the "Chi-Square Test (Table in Worksheet)" menu option (see Figure 8.M.3) and specify the columns in the "Columns containing the table" box. Click "OK" to produce the MINITAB printout.

TI-83/TI-84 Plus Graphing Calculator: Categorical Data Analyses

The TI-83/TI-84 plus graphing calculator can be used to make inferences about $(p_1 - p_2)$ and to conduct a chi-square test for a two-way (contingency) table, but cannot conduct a chi-square test for a one-way table.

Confidence Interval for $(p_1 - p_2)$

Step 1 *Access the statistical tests menu*

- Press **STAT**
- Arrow right to **TESTS**
- Arrow down to **2-PropZInt**
- Press **ENTER**

```
EDIT CALC TESTS
6↑2-PropZTest…
7:ZInterval…
8:TInterval…
9:2-SampZInt…
0:2-SampTInt…
A:1-PropZInt…
B:2-PropZInt…
```

Step 2 *Enter the values from the sample information and* **the confidence level**

where $x_1 =$ number of successes in the first sample (e.g., 53)

$n_1 =$ sample size for the first sample (e.g., 400)

$x_2 =$ number of successes in the second sample (e.g., 78)

$n_2 =$ sample size for the second sample (e.g., 500)

```
2-PropZInt
 (-.0695,.02249)
 p̂1=.1325
 p̂2=.156
 n1=400
 n2=500
```

- Set **C-Level** to the confidence level
- Arrow down to "**Calculate**"
- Press **ENTER**

Hypothesis Test for $(p_1 - p_2)$

Step 1 *Access the statistical tests menu*

- Press **STAT**
- Arrow right to **TESTS**
- Arrow down to **2-PropZTest**
- Press **ENTER**

```
EDIT CALC TESTS
1:Z-Test…
2:T-Test…
3:2-SampZTest…
4:2-SampTTest…
5:1-PropZTest…
6:2-PropZTest…
7↓ZInterval…
```

Step 2 *Enter the values from the sample information and* **select the alternative hypothesis**

where $x_1 =$ number of successes in the first sample (e.g., 53)

$n_1 =$ sample size for the first sample (e.g., 400)

$x_2 =$ number of successes in the second sample (e.g., 78)

$n_2 =$ sample size for the second sample (e.g., 500)

- Use the **ARROW** to highlight the appropriate alternative hypothesis
- Press **ENTER**
- Arrow down to "**Calculate**"
- Press **ENTER**

```
2-PropZTest
 p1≠p2
 z=-.9933568641
 P=.3205360811
 p̂1=.1325
 p̂2=.156
 ↓p̂=.1455555556
```

Two-Way (Contingency) Table

Step 1 *Access the matrix menu to enter the observed values*

- Press **2nd x⁻¹** for **MATRX**
- Arrow right to **EDIT**
- Press **ENTER**
- Use the **ARROW** key to enter the row and column dimensions of your observed Matrix
- Use the **ARROW** key to enter your observed values into Matrix [A]

```
MATRIX[A] 2 ×5
[ 39   19   12   _
[ 172  61   44   _
```

Step 2 *Access the matrix menu to enter the expected values*

- Press **2nd x⁻¹** for **MATRX**
- Arrow right to **EDIT**
- Arrow down to **2:[B]**
- Press **ENTER**
- Use the **ARROW** key to enter the row and column dimensions of your expected matrix (The dimensions will be the same as in Matrix A)
- Use the **ARROW** key to enter your expected values into Matrix [B]

```
MATRIX[B] 2 ×5
[ 48.952  18.56  12.992 _
[ 162.05  61.44  43.008 _

2,1=162.05
```

Step 3 *Access the statistical tests menu and perform the chi-square test*

- Press **STAT**
- Arrow right to **TESTS**
- Arrow down to χ^2 **Test**
- Press **ENTER**
- Arrow down to **Calculate**
- Press **ENTER**

```
X²-Test
Observed: [A]
Expected: [B]
Calculate Draw
```

9 Simple Linear Regression

CONTENTS

Where We've Been

- Presented methods for estimating and testing population parameters (e.g., the mean, proportion, and variance) for a single sample
- Extended these methods to allow for a comparison of population parameters for multiple samples

Where We're Going

- Introduce the straight-line (*simple linear regression*) model as a means of relating one quantitative variable to another quantitative variable (9.1)
- Assess how well the simple linear regression model fits the sample data (9.2–9.4)
- Introduce the *correlation coefficient* as a means of relating one quantitative variable to another quantitative variable (9.5)
- Utilize the simple linear regression model to predict the value of one variable from a specified value of another variable (9.6, 9.7)
- Introduce a nonparametric test for rank correlation (9.8)

The act of searching for and finding underground supplies of water with the use of nothing more than a divining rod is commonly known as "dowsing." Although widely regarded among scientists as no more than a superstitious relic from medieval times, dowsing remains popular in folklore, and to this day, there are individuals who claim to have this mysterious skill.

Many dowsers in Germany claim that they respond to "earthrays" which emanate from the water source. Earthrays, say the dowsers, are a subtle form of radiation that is potentially hazardous to human health. As a result of these claims, in the mid-1980s the German government conducted a two-year experiment to investigate the possibility that dowsing is a genuine skill. If such a skill could be demonstrated, reasoned government officials, then dangerous levels of radiation in Germany could be detected, avoided, and disposed of.

A group of university physicists in Munich, Germany, was provided a grant of 400,000 marks (about $250,000) to conduct the study. Approximately 500 candidate dowsers were recruited to participate in preliminary tests of their skill. To avoid fraudulent claims, the 43 individuals who seemed to be the most successful in the preliminary tests were selected for the final, carefully controlled, experiment.

The researchers set up a 10-meter-long line on the ground floor of a vacant barn, along which a small wagon could be moved. Attached to the wagon was a short length of pipe, perpendicular to the test line, that was connected by hoses to a pump with running water. The location of the pipe along the line for each trial of the experiment was assigned by a computer-generated random number. On the upper floor of the barn, directly above the experimental line, a 10-meter test line was painted. In each trial, a dowser was admitted to this upper level and required, with his or her rod, stick, or other tool of choice, to ascertain where the pipe with running water on the ground floor was located.

Each dowser participated in at least one test series constituting a sequence of from 5 to 15 trials (typically, 10), with the pipe randomly repositioned after each trial. (Some dowsers undertook only 1 test series, whereas selected others underwent more than 10 test series.) Over the two-year experimental period, the 43 dowsers participated in a total of 843 tests. The experiment was "double blind" in that neither the observer (researcher) on the top floor nor the dowser knew the pipe's location, even after a guess was made. [*Note:* Before the experiment began, a professional magician inspected the entire arrangement for potential deception or cheating by the dowsers.]

For each trial, two variables were recorded: the actual location of the pipe (in decimeters from the beginning of the line) and the dowser's guess (also measured in decimeters). On the basis of an examination of these data, the German physicists concluded in their final report that although most dowsers did not do particularly well in the experiments, "some few dowsers, in particular tests, showed an extraordinarily high rate of success, which can scarcely if at all be explained as due to chance . . . a real core of dowser-phenomena can be regarded as empirically proven . . ." (Wagner, Betz, and König, 1990. Final Report 01 KB8602, Federal Ministry for Research and Technology).

This conclusion was critically assessed by Professor J. T. Enright of the University of California at San Diego (*Skeptical Inquirer*, Jan./Feb. 1999). In the Statistics in Action Revisited sections of this chapter, we demonstrate how Enright concluded the exact opposite of the German physicists.

Statistics IN Action Revisited

- Estimating a Straight-Line Regression Model for the Dowsing Data (p. 484)
- Assessing How Well the Straight-Line Model Fits the Dowsing Data (p. 499)
- Using the Coefficients of Correlation and Determination to Assess the Dowsing Data (p. 509)
- Using the Straight-Line Model to Predict Pipe Location for the Dowsing Data (p. 517)

In Chapters 5–7, we described methods for making inferences about population means. The mean of a population has been treated as a *constant*, and we have shown how to use sample data to estimate or to test hypotheses about this constant mean. In many applications, the mean of a population is not viewed as a constant, but rather as a variable. For example, the mean sale price of residences in a large city might be treated as a variable that depends on the number of square feet of living space in the residence. The relationship might be

$$\text{Mean sale price} = \$30,000 + \$60 \,(\text{Square feet})$$

This formula implies that the mean sale price of 1,000-square-foot homes is $90,000, the mean sale price of 2,000-square-foot homes is $150,000, and the mean sale price of 3,000-square-foot homes is $210,000.

In this chapter, we discuss situations in which the mean of the population is treated as a variable, dependent on the value of another variable. The dependence of the residential sale price on the number of square feet of living space is one illustration. Other examples include the dependence of the mean reaction time on the amount of a drug in the bloodstream, the dependence of the mean starting salary of a college graduate on the student's GPA, and the dependence of the mean number of years to which a criminal is sentenced on the number of previous convictions.

Here, we present the simplest of all models relating a populating mean to another variable: the *straight-line model*. We show how to use the sample data to estimate the straight-line relationship between the mean value of one variable, *y*, as it relates to a second variable, *x*. The methodology of estimating and using a straight-line relationship is referred to as *simple linear regression analysis*.

9.1 Probabilistic Models

An important consideration in taking a drug is how it may affect one's perception or general awareness. Suppose you want to model the length of time it takes to respond to a stimulus (a measure of awareness) as a function of the percentage of a certain drug in the bloodstream. The first question to be answered is this: "Do you think that an exact relationship exists between these two variables?" That is, do you think that it is possible to state the exact length of time it takes an individual (subject) to respond if the amount of the drug in the bloodstream is known? We think that you will agree with us that this is *not* possible, for several reasons: The reaction time depends on many variables other than the percentage of the drug in the bloodstream—for example, the time of day, the amount of sleep the subject had the night before, the subject's visual acuity, the subject's general reaction time without the drug, and the subject's age. Even if many variables are included in a model, it is still unlikely that we would be able to predict the subject's reaction time *exactly*. There will almost certainly be some variation in response times due strictly to *random phenomena* that cannot be modeled or explained.

If we were to construct a model that hypothesized an exact relationship between variables, it would be called a **deterministic model**. For example, if we believe that *y*, the reaction time (in seconds), will be exactly one-and-one-half times *x*, the amount of drug in the blood, we write

$$y = 1.5x$$

This represents a **deterministic relationship** between the variables *y* and *x*. It implies that *y* can always be determined exactly when the value of *x* is known. *There is no allowance for error in this prediction.*

If, however, we believe that there will be unexplained variation in reaction times— perhaps caused by important, but unincluded, variables or by random phenomena—we discard the deterministic model and use a model that accounts for this **random error.** Our **probabilistic model** will include both a deterministic component and a random-error component. For example, if we hypothesize that the response time *y* is related to the percentage *x* of drug by

$$y = 1.5x + \text{Random error}$$

we are hypothesizing a **probabilistic relationship** between *y* and *x*. Note that the deterministic component of this probabilistic model is 1.5*x*.

Figure 9.1a shows the possible responses for five different values of *x*, the percentage of drug in the blood, when the model is deterministic. All the responses must fall exactly on the line, because a deterministic model leaves no room for error.

Figure 9.1b shows a possible set of responses for the same values of *x* when we are using a probabilistic model. Note that the deterministic part of the model (the straight line itself) is the same. Now, however, the inclusion of a random-error component allows the response times to vary from this line. Since we know that the response time does vary randomly for a given value of *x*, the probabilistic model for *y* is more realistic than the deterministic model.

Figure 9.1
Possible reaction times y for five different drug percentages x

a. Deterministic relationship: $y = 1.5x$

b. Probabilistic relationship: $y = 1.5x + $ Random error

General Form of Probabilistic Models

$$y = \text{Deterministic component} + \text{Random error}$$

where y is the variable of interest. We always assume that the mean value of the random error equals 0. This is equivalent to assuming that the mean value of y, $E(y)$, equals the deterministic component of the model; that is,

$$E(y) = \text{Deterministic component}$$

BIOGRAPHY **FRANCIS GALTON (1822–1911)**

The Law of Universal Regression

Francis Galton was the youngest of seven children born to a middle-class English family of Quaker faith. A cousin of Charles Darwin, Galton attended Trinity College (Cambridge, England) to study medicine. Due to the death of his father, Galton was unable to obtain his degree. His competence in both medicine and mathematics, however, led Galton to pursue a career as a scientist. He made major contributions to the fields of genetics, psychology, meteorology, and anthropology. Some consider Galton to be the first social scientist for his applications of the novel statistical concepts of the time—in particular, regression and correlation. While studying natural inheritance in 1886, Galton collected data on heights of parents and adult children. He noticed the tendency for tall (or short) parents to have tall (or short) children, but that the children were not as tall (or short), on average as their parents. Galton called this phenomenon the "law of universal regression," for the average heights of adult children tended to "regress" to the mean of the population. With the help of his friend and disciple, Karl Pearson, Galton applied the straight-line model to the height data, and the term *regression model* was coined. ∎

In this chapter, we present the simplest of probabilistic models—the **straight-line model**—which gets its name from the fact that the deterministic portion of the model graphs as a straight line. Fitting this model to a set of data is an example of **regression analysis**, or **regression modeling**. The elements of the straight-line model are summarized in the following box:

A First-Order (Straight-Line) Probabilistic Model

$$y = \beta_0 + \beta_1 x + \varepsilon$$

where

$y = $ **Dependent** *or* **response variable** (variable to be modeled)

$x = $ **Independent** *or* **predictor variable** (variable used as a predictor of y)*

$\beta_0 + \beta_1 x = E(y) = $ Deterministic component

ε (epsilon) $ = $ Random error component

---

*The word *independent* should not be interpreted in a probabilistic sense as defined in Chapter 3. The phrase *independent variable* is used in regression analysis to refer to a predictor variable for the response y.

β_0 (beta zero) = **y-intercept of the line**—that is, the point at which the line intersects, or cuts through, the y-axis (see Figure 9.2)

β_1 (beta one) = **Slope of the line**—that is, the change (amount of increase or decrease) in the deterministic component of y for every one-unit increase in x.

[*Note*: A *positive* slope implies that $E(y)$ *increases* by the amount β_1. (See Figure 9.2.) A *negative* slope implies that $E(y)$ *decreases* by the amount β_1.]

Figure 9.2
The straight-line model

In the probabilistic model, the deterministic component is referred to as the **line of means**, because the mean of y, $E(y)$, is equal to the straight-line component of the model. That is,

$$E(y) = \beta_0 + \beta_1 x$$

Note that the Greek symbols β_0 and β_1 respectively represent the y-intercept and slope of the model. They are population parameters that will be known only if we have access to the entire population of (x, y) measurements. Together with a specific value of the independent variable x, they determine the mean value of y, which is just a specific point on the line of means (Figure 9.2).

The values of β_0 and β_1 will be unknown in almost all practical applications of regression analysis. The process of developing a model, estimating the unknown parameters, and using the model can be viewed as the five-step procedure shown in the following box:

Conducting a Simple Linear Regression:

Step 1 Hypothesize the deterministic component of the model that relates the mean $E(y)$ to the independent variable x (Section 9.2).

Step 2 Use the sample data to estimate unknown parameters in the model (Section 9.2).

Step 3 Specify the probability distribution of the random-error term and estimate the standard deviation of this distribution (Section 9.3).

Step 4 Statistically evaluate the usefulness of the model (Sections 9.4 and 9.5).

Step 5 When satisfied that the model is useful, use it for prediction, estimation, and other purposes (Section 9.6).

Exercises 9.1–9.10

Understanding the Principles

9.1 Why do we generally prefer a probabilistic model to a deterministic model? Give examples for which the two types of models might be appropriate.

9.2 What is the difference between a dependent variable and an independent variable in a probabilistic model?

9.3 What is the line of means?

9.4 If a straight-line probabilistic relationship relates the mean $E(y)$ to an independent variable x, does it imply that every value of the variable y will always fall exactly on the line of means? Why or why not?

Learning the Mechanics

9.5 In each case, graph the line that passes through the given points.
   a. $(1, 1)$ and $(5, 5)$
   b. $(0, 3)$ and $(3, 0)$
   c. $(-1, 1)$ and $(4, 2)$
   d. $(-6, -3)$ and $(2, 6)$

9.6 Give the slope and y-intercept for each of the lines graphed in Exercise 9.5.

9.7 The equation (deterministic) for a straight line is

$$y = \beta_0 + \beta_1 x$$

If the line passes through the point $(-2, 4)$, then $x = -2, y = 4$ must satisfy the equation; that is,

$$4 = \beta_0 + \beta_1(-2)$$

Similarly, if the line passes through the point $(4, 6)$, then $x = 4, y = 6$ must satisfy the equation; that is,

$$6 = \beta_0 + \beta_1(4)$$

Use these two equations to solve for β_0 and β_1; then find the equation of the line that passes through the points $(-2, 4)$ and $(4, 6)$.

9.8 Refer to Exercise 9.7. Find the equations of the lines that pass through the points listed in Exercise 9.5.

9.9 Plot the following lines:
   a. $y = 4 + x$
   b. $y = 5 - 2x$
   c. $y = -4 + 3x$
   d. $y = -2x$
   e. $y = x$
   f. $y = .50 + 1.5x$

9.10 Give the slope and y-intercept for each of the lines defined in Exercise 9.9.

9.2 Fitting the Model: The Least Squares Approach

After the straight-line model has been hypothesized to relate the mean $E(y)$ to the independent variable x, the next step is to collect data and to estimate the (unknown) population parameters, the y-intercept β_0 and the slope β_1.

To begin with a simple example, suppose an experiment involving five subjects is conducted to determine the relationship between the percentage of a certain drug in the bloodstream and the length of time it takes to react to a stimulus. The results are shown in Table 9.1. (The number of measurements and the measurements themselves are unrealistically simple in order to avoid arithmetic confusion in this introductory example.) This set of data will be used to demonstrate the five-step procedure of regression modeling given in the previous section. In the current section, we hypothesize the deterministic component of the model and estimate its unknown parameters (steps 1 and 2). The model's assumptions and the random-error component (step 3) are the subjects of Section 9.3, whereas Sections 9.4 and 9.5 assess the utility of the model (step 4). Finally, we use the model for prediction and estimation (step 5) in Section 9.6.

Table 9.1   Reaction Time versus Drug Percentage

Subject	Percent x of Drug	Reaction Time y (seconds)
1	1	1
2	2	1
3	3	2
4	4	2
5	5	4

Data Set: STIMULUS

Step 1   *Hypothesize the deterministic component of the probabilistic model.* As stated before, we will consider only straight-line models in this chapter. Thus, the complete model relating mean response time $E(y)$ to drug percentage x is given by

$$E(y) = \beta_0 + \beta_1 x$$

Step 2   *Use sample data to estimate unknown parameters in the model.* This step is the subject of this section—namely, how can we best use the information in the sample of five observations in Table 9.1 to estimate the unknown y-intercept β_0 and slope β_1?

Figure 9.3
Scatterplot for data in Table 9.1

To determine whether a linear relationship between y and x is plausible, it is helpful to plot the sample data in a **scatterplot** (or **scattergram**). Recall (Section 2.9) that a scatterplot locates each data point on a graph, as shown in Figure 9.3 for the five data points of Table 9.4. Note that the scatterplot suggests a general tendency for y to increase as x increases. If you place a ruler on the scatterplot, you will see that a line may be drawn through three of the five points, as shown in Figure 9.4. To obtain the equation of this visually fitted line, note that the line intersects the y-axis at $y = -1$, so the y-intercept is -1. Also, y increases exactly one unit for every one-unit increase in x, indicating that the slope is $+1$. Therefore, the equation is

$$\tilde{y} = -1 + 1(x) = -1 + x$$

where $\tilde{y}$ is used to denote the y that is predicted from the visual model.

One way to decide quantitatively how well a straight line fits a set of data is to note the extent to which the data points deviate from the line. For example, to evaluate the model in Figure 9.4, we calculate the magnitude of the *deviations* (i.e., the differences between the observed and the predicted values of y). These deviations, or **errors of prediction**, are the vertical distances between observed and predicted values (see Figure 9.4).* The observed and predicted values of y, their differences, and their squared differences are shown in Table 9.2. Note that the *sum of errors* equals 0 and the *sum of squares of the errors* (SSE), which places a greater emphasis on large deviations of the points from the line, is equal to 2.

Figure 9.4
Visual straight line fitted to the data in Figure 9.3

Table 9.2 Comparing Observed and Predicted Values for the Visual Model

x	y	$\tilde{y} = -1 + x$	$(y - \tilde{y})$	$(y - \tilde{y})^2$
1	1	0	$(1 - 0) = 1$	1
2	1	1	$(1 - 1) = 0$	0
3	2	2	$(2 - 2) = 0$	0
4	2	3	$(2 - 3) = -1$	1
5	4	4	$(4 - 4) = 0$	0
			Sum of errors $= 0$	Sum of squared errors (SSE) $= 2$

You can see by shifting the ruler around the graph that it is possible to find many lines for which the sum of errors is equal to 0, but it can be shown that there is one (and only one) line for which the SSE is a *minimum*. This line is called the **least squares line**, the **regression line**, or the **least squares prediction equation**. The methodology used to obtain that line is called the **method of least squares.**

Now Work Exercise 9.16a–d

To find the least squares prediction equation for a set of data, assume that we have a sample of n data points consisting of pairs of values of x and y, say, $(x_1, y_1), (x_2, y_2), \ldots, (x_n, y_n)$. For example, the $n = 5$ data points shown in Table 9.2 are $(1, 1), (2, 1), (3, 2), (4, 2),$ and $(5, 4)$. The fitted line, which we will calculate on the basis of the five data points, is written as

$$\hat{y} = \hat{\beta}_0 + \hat{\beta}_1 x$$

The "hats" indicate that the symbols below them are estimates: $\hat{y}$ (y-hat) is an estimator of the mean value of y, $E(y)$, and is a predictor of some future value of y; and $\hat{\beta}_0$ and $\hat{\beta}_1$ are estimators of β_0 and β_1, respectively.

For a given data point—say, the point (x_i, y_i),—the observed value of y is y_i and the predicted value of y would be obtained by substituting x_i into the prediction equation:

$$\hat{y}_i = \hat{\beta}_0 + \hat{\beta}_1 x_i$$

The deviation of the ith value of y from its predicted value is

$$(y_i - \hat{y}_i) = [y_i - (\hat{\beta}_0 + \hat{\beta}_1 x_i)]$$

*These errors of prediction are also known as **regression residuals**.

Then the sum of the squares of the deviations of the y-values about their predicted values for all the n data points is

$$\text{SSE} = \sum [y_i - (\hat{\beta}_0 + \hat{\beta}_1 x_i)]^2$$

The quantities $\hat{\beta}_0$ and $\hat{\beta}_1$ that make the SSE a minimum are called the **least squares estimates** of the population parameters β_0 and β_1, and the prediction equation $\hat{y} = \hat{\beta}_0 + \hat{\beta}_1 x$ is called the *least squares line*.

> The **least squares line** $\hat{y} = \hat{\beta}_0 + \hat{\beta}_1 x$ is the line that has the following two properties:
>
> 1. The sum of the errors equals 0, i.e., mean error of prediction = 0.
> 2. The sum of squared errors (SSE) is smaller than that for any other straight-line model

The values of $\hat{\beta}_0$ and $\hat{\beta}_1$ that minimize the SSE are given by the formulas in the following box (proof omitted):*

Formulas for the Least Squares Estimates

Slope: $\hat{\beta}_1 = \dfrac{\text{SS}_{xy}}{\text{SS}_{xx}}$

y-intercept: $\hat{\beta}_0 = \bar{y} - \hat{\beta}_1 \bar{x}$

where

$$\text{SS}_{xy} = \sum (x_i - \bar{x})(y_i - \bar{y}) = \sum x_i y_i - \frac{\left(\sum x_i\right)\left(\sum y_i\right)}{n}$$

$$\text{SS}_{xx} = \sum (x_i - \bar{x})^2 = \sum x_i^2 - \frac{\left(\sum x_i\right)^2}{n}$$

$$n = \text{Sample size}$$

Example 9.1
Applying the Method of Least Squares— Drug Reaction Data

Problem Refer to the reaction data presented in Table 9.1. Consider the straight-line model $E(y) = \beta_0 + \beta_1 x$, where y = reaction time (in seconds) and x = percent of drug received.

a. Use the method of least squares to estimate the values of β_0 and β_1.
b. Predict the reaction time when $x = 2\%$.
c. Find the SSE for the analysis.
d. Give practical interpretations of $\hat{\beta}_0$ and $\hat{\beta}_1$.

Solution

a. Preliminary computations for finding the least squares line for the drug reaction example are presented in Table 9.3. We can now calculate

$$\text{SS}_{xy} = \sum x_i y_i - \frac{\left(\sum x_i\right)\left(\sum y_i\right)}{5} = 37 - \frac{(15)(10)}{5} = 37 - 30 = 7$$

$$\text{SS}_{xx} = \sum x_i^2 - \frac{\left(\sum x_i\right)^2}{5} = 55 - \frac{(15)^2}{5} = 55 - 45 = 10$$

*Students who are familiar with calculus should note that the values of β_0 and β_1 that minimize SSE $= \Sigma (y_i - \hat{y}_i)^2$ are obtained by setting the two partial derivatives $\partial \text{SSE}/\partial\beta_0$ and $\partial \text{SSE}/\partial\beta_1$ equal to 0. The solutions of these two equations yield the formulas shown in the box. Furthermore, we denote the *sample* solutions of the equations by $\hat{\beta}_0$ and $\hat{\beta}_1$, where the "hat" denotes that these are sample estimates of the true population intercept β_0 and slope β_1.

Table 9.3 Preliminary Computations for the Drug Reaction Example

	x_i	y_i	x_i^2	$x_i y_i$
	1	1	1	1
	2	1	4	2
	3	2	9	6
	4	2	16	8
	5	4	25	20
Totals	$\sum x_i = 15$	$\sum y_i = 10$	$\sum x_i^2 = 55$	$\sum x_i y_i = 37$

Then the slope of the least squares line is

$$\hat{\beta}_1 = \frac{SS_{xy}}{SS_{xx}} = \frac{7}{10} = .7$$

and the y-intercept is

$$\hat{\beta}_0 = \bar{y} - \hat{\beta}_1 \bar{x} = \frac{\sum y_i}{5} - \hat{\beta}_1 \frac{\sum x_i}{5}$$

$$= \frac{10}{5} - (.7)\left(\frac{15}{5}\right) = 2 - (.7)(3) = 2 - 2.1 = -.1$$

The least squares line is thus

$$\hat{y} = \hat{\beta}_0 + \hat{\beta}_1 x = -.1 + .7x$$

The graph of this line is shown in Figure 9.5.

b. The predicted value of y for a given value of x can be obtained by substituting into the formula for the least squares line. Thus, when $x = 2$, we predict y to be

$$\hat{y} = -.1 + .7x = -.1 + .7(2) = 1.3$$

We show how to find a prediction interval for y in Section 9.6.

c. The observed and predicted values of y, the deviations of the y values about their predicted values, and the squares of these deviations are shown in Table 9.4. Note that the sum of the squares of the deviations, SSE, is 1.10 and (as we would expect) this is less than the SSE = 2.0 obtained in Table 9.2 for the visually fitted line.

d. The estimated y-intercept, $\hat{\beta}_0 = -.1$, appears to imply that the estimated mean reaction time is equal to $-.1$ second when the percent x of drug is equal to 0%. Since negative reaction times are not possible, this seems to make the model nonsensical. However, *the model parameters should be interpreted only within the sampled range of the independent variable*—in this case, for amounts of drug in the bloodstream between 1% and 5%. Thus, the y-intercept—which is, by definition, at $x = 0$ (0% drug)—is not within the range of the sampled values of x and is not subject to meaningful interpretation.

The slope of the least squares line, $\hat{\beta}_1 = .7$, implies that for every unit increase in x, the mean value of y is estimated to increase by .7 unit. In terms of this example, for every 1% increase in the amount of drug in the bloodstream, the mean reaction time

Figure 9.5
The line $\hat{y} = -.1 + .7x$ fitted to the data

Table 9.4 Comparing Observed and Predicted Values for the Least Squares Prediction Equation

x	y	$\hat{y} = -.1 + .7x$	$(y - \hat{y})$	$(y - \hat{y})^2$
1	1	.6	$(1 - .6) = .4$	.16
2	1	1.3	$(1 - 1.3) = -.3$	.09
3	2	2.0	$(2 - 2.0) = 0$	.00
4	2	2.7	$(2 - 2.7) = -.7$	.49
5	4	3.4	$(4 - 3.4) = .6$	.36
			Sum of errors = 0	SSE = 1.10

is estimated to increase by .7 second *over the sampled range of drug amounts from 1% to 5%*. Thus, the model does not imply that increasing the drug amount from 5% to 10% will result in an increase in mean reaction time of 3.5 seconds, because the range of x in the sample does not extend to 10% ($x = 10$). In fact, 10% might be such a high concentration that the drug would kill the subject! Be careful to interpret the estimated parameters only within the sampled range of x.

Look Back The calculations required to obtain $\hat{\beta}_0$, $\hat{\beta}_1$, and SSE in simple linear regression, although straightforward, can become rather tedious. Even with the use of a pocket calculator, the process is laborious and susceptible to error, especially when the sample size is large. Fortunately, the use of statistical computer software can significantly reduce the labor involved in regression calculations. The SAS, SPSS, and MINITAB outputs for the simple linear regression of the data in Table 9.1 are displayed in Figure 9.6a–c. The values of $\hat{\beta}_0$ and $\hat{\beta}_1$ are highlighted on the printouts. These values, $\hat{\beta}_0 = -.1$ and $\hat{\beta}_1 = .7$, agree exactly with our hand-calculated values. The value of SSE $= 1.10$ is also highlighted on the printouts.

Dependent Variable: TIME_Y

Number of Observations Read	5	
Number of Observations Used	5	

Analysis of Variance

Source	DF	Sum of Squares	Mean Square	F Value	Pr > F
Model	1	4.90000	4.90000	13.36	0.0354
Error	3	1.10000	0.36667		
Corrected Total	4	6.00000			

Root MSE	0.60553	R-Square	0.8167
Dependent Mean	2.00000	Adj R-Sq	0.7556
Coeff Var	30.27650		

Parameter Estimates

Variable	DF	Parameter Estimate	Standard Error	t Value	Pr > |t|
Intercept	1	-0.10000	0.63509	-0.16	0.8849
DRUG_X	1	0.70000	0.19149	3.66	0.0354

Figure 9.6a
SAS printout for the time–drug regression

Model Summary

Model	R	R Square	Adjusted R Square	Std. Error of the Estimate
1	.904[a]	.817	.756	.606

a. Predictors: (Constant), DRUG_X

ANOVA[b]

Model		Sum of Squares	df	Mean Square	F	Sig.
1	Regression	4.900	1	4.900	13.364	.035[a]
	Residual	1.100	3	.367		
	Total	6.000	4			

a. Predictors: (Constant), DRUG_X
b. Dependent Variable: TIME_Y

Coefficients[a]

Model		Unstandardized Coefficients		Standardized Coefficients	t	Sig.
		B	Std. Error	Beta		
1	(Constant)	-.100	.635		-.157	.885
	DRUG_X	.700	.191	.904	3.656	.035

a. Dependent Variable: TIME_Y

Figure 9.6b
SPSS printout for the time–drug regression

```
Regression Analysis: TIME_Y versus DRUG_X

The regression equation is
TIME_Y = - 0.100 + 0.700 DRUG_X


Predictor    Coef   SE Coef     T      P
Constant   -0.1000    0.6351  -0.16  0.885
DRUG_X      0.7000    0.1915   3.66  0.035


S = 0.605530   R-Sq = 81.7%   R-Sq(adj) = 75.6%


Analysis of Variance

Source          DF     SS      MS      F      P
Regression       1  4.9000  4.9000  13.36  0.035
Residual Error   3  1.1000  0.3667
Total            4  6.0000
```

Figure 9.6c

MINITAB printout for the time–drug regression

Now Work Exercise 9.23

Interpreting the Estimates of β_0 and β_1 in Simple Linear Regression

y-intercept: $\hat{\beta}_0$ represents the predicted value of y when $x = 0$. (*Caution*: This value will not be meaningful if the value $x = 0$ is nonsensical or outside the range of the sample data.)

slope: $\hat{\beta}_1$ represents the increase (or decrease) in y for every 1-unit increase in x. (*Caution:* This interpretation is valid only for x-values within the range of the sample data.)

Even when the interpretations of the estimated parameters in a simple linear regression are meaningful, we need to remember that they are only estimates based on the sample. As such, their values will typically change in repeated sampling. How much confidence do we have that the estimated slope $\hat{\beta}_1$ accurately approximates the true slope β_1? Determining this requires statistical inference, in the form of confidence intervals and tests of hypotheses, which we address in Section 9.4.

To summarize, we defined the best-fitting straight line to be the line that minimizes the sum of squared errors around it, and we called it the least squares line. We should interpret the least squares line only within the sampled range of the independent variable. In subsequent sections, we show how to make statistical inferences about the model.

Statistics IN Action | **Revisited**

Estimating a Straight–Line Regression Model for the Dowsing Data

After conducting a series of experiments in a Munich barn, a group of German physicists concluded that dowsing (i.e., the ability to find underground water with a divining rod) "can be regarded as empirically proven." This observation was based on the data collected on 3 (of the participating 500) dowsers who had particularly impressive results. All of these "best" dowsers (numbered 99, 18, and 108) performed the experiment multiple times, and the best test series (sequence of trials) for each of them was identified. These data, saved in the **DOWSING** file, are listed in Table SIA9.1.

Recall (p. 475) that for various hidden pipe locations, each dowser guessed where the pipe with running water was located. Let x = dowser's guess (in meters) and y = pipe location (in meters) for each trial. One way to determine whether the "best" dowsers are effective is to fit the straight-line model $E(y) = \beta_0 + \beta_1 x$ to the data in Table SIA9.1.

A MINITAB scatterplot of the data is shown in Figure SIA9.1. The least squares line, obtained from the MINITAB regression printout shown in Figure SIA9.2, is also displayed on the scatterplot. Although the least squares line

(continued)

(continued)

Table SIA9.1	Dowsing Trial Results: Best Series for the Three Best Dowsers		
Trial	Dowser Number	Pipe Location	Dowser's Guess
1	99	4	4
2	99	5	87
3	99	30	95
4	99	35	74
5	99	36	78
6	99	58	65
7	99	40	39
8	99	70	75
9	99	74	32
10	99	98	100
11	18	7	10
12	18	38	40
13	18	40	30
14	18	49	47
15	18	75	9
16	18	82	95
17	108	5	52
18	108	18	16
19	108	33	37
20	108	45	40
21	108	38	66
22	108	50	58
23	108	52	74
24	108	63	65
25	108	72	60
26	108	95	49

Based on Enright, J. T. "Testing dowsing: The failure of the Munich experiments." *Skeptical Inquirer*, Jan./Feb. 1999, p. 45 (Figure 6a).

Data Set: DOWSING

Figure SIA9.1
MINITAB scatterplot of dowsing data

Regression Analysis: PIPE versus GUESS

```
The regression equation is
PIPE = 30.1 + 0.308 GUESS


Predictor    Coef   SE Coef     T       P
Constant    30.07     11.41   2.63   0.015
GUESS      0.3079    0.1900   1.62   0.118


S = 26.0298   R-Sq = 9.9%   R-Sq(adj) = 6.1%


Analysis of Variance

Source           DF        SS       MS      F       P
Regression        1    1778.9   1778.9   2.63   0.118
Residual Error   24   16261.2    677.6
Total            25   18040.2
```

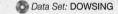

Figure SIA9.2
MINITAB simple linear regression for dowsing data

has a slight upward trend, the variation of the data points around the line is large. It does not appear that a dowser's guess (*x*) will be a very good predictor of the actual pipe location (*y*). In fact, the estimated slope (obtained from Figure SIA9.2) is $\hat{\beta}_1 = .31$. Thus, for every 1-meter increase in a dowser's guess, we estimate that the actual pipe location will increase only .31 meter. In the Statistics in Action Revisited sections that follow, we will provide a measure of reliability for this inference and investigate the phenomenon of dowsing further.

Exercises 9.11–9.31

Understanding the Principles

9.11 In regression, what is an error of prediction?

9.12 Give two properties of the line estimated with the method of least squares.

9.13 *True or False.* The estimates of β_0 and β_1 should be interpreted only within the sampled range of the independent variable, x.

Learning the Mechanics

9.14 The accompanying table is similar to Table 9.3. It is used to make the preliminary computations for finding the least squares line for the given pairs of x and y values.
a. Complete the table.   b. Find SS_{xy}.
c. Find SS_{xx}.   d. Find $\hat{\beta}_1$.
e. Find $\bar{x}$ and $\bar{y}$.   f. Find $\hat{\beta}_0$.
g. Find the least squares line.

x_i	y_i	x_i^2	$x_i y_i$
7	2	—	—
4	4	—	—
6	2	—	—
2	5	—	—
1	7	—	—
1	6	—	—
3	5	—	—
Totals $\sum x_i =$	$\sum y_i =$	$\sum x_i^2 =$	$\sum x_i y_i =$

9.15 Refer to Exercise 9.14. After the least squares line has been obtained, the following table (which is similar to Table 9.4) can be used (1) to compare the observed and the predicted values of y and (2) to compute SSE.

x	y	$\hat{y}$	$(y - \hat{y})$	$(y - \hat{y})^2$
7	2	—	—	—
4	4	—	—	—
6	2	—	—	—
2	5	—	—	—
1	7	—	—	—
1	6	—	—	—
3	5	—	—	—
			$\sum(y - \hat{y}) =$	SSE $= \sum(y - \hat{y})^2 =$

a. Complete the table.
b. Plot the least squares line on a scatterplot of the data. Plot the following line on the same graph:

$$\hat{y} = 14 - 2.5x$$

c. Show that SSE is larger for the line in part **b** than it is for the least squares line.

9.16 Construct a scatterplot of the following data.

[NW]

x	.5	1	1.5
y	2	1	3

a. Plot the following two lines on your scatterplot:

$$y = 3 - x \quad \text{and} \quad y = 1 + x$$

b. Which of these lines would you choose to characterize the relationship between x and y? Explain.
c. Show that the sum of errors for both of these lines equals 0.
d. Which of these lines has the smaller SSE?
e. Find the least squares line for the data, and compare it with the two lines described in part **a**.

9.17 Consider the following pairs of measurements, saved in the **LM9_17** file:

x	5	3	−1	2	7	6	4
y	4	3	0	1	8	5	3

a. Construct a scatterplot of these data.
b. What does the scatterplot suggest about the relationship between x and y?
c. Given that $SS_{xx} = 43.4286$, $SS_{xy} = 39.8571$, $\bar{y} = 3.4286$, and $\bar{x} = 3.7143$, calculate the least squares estimates of β_0 and β_1.
d. Plot the least squares line on your scatterplot. Does the line appear to fit the data well? Explain.
e. Interpret the y-intercept and slope of the least squares line. Over what range of x are these interpretations meaningful?

Applet Exercise 9.1

Use the applet entitled *Regression by Eye* to explore the relationship between the pattern of data in a scatterplot and the corresponding least squares model.
a. Run the applet several times. For each time, attempt to move the green line into a position that appears to minimize the vertical distances of the points from the line. Then click *Show regression line* to see the actual regression line. How close is your line to the actual line? Click *New data* to reset the applet.
b. Click the trash can to clear the graph. Use the mouse to place five points on the scatterplot that are approximately in a straight line. Then move the green line to approximate the regression line. Click *Show regression line* to see the actual regression line. How close were you this time?
c. Continue to clear the graph, and plot sets of five points with different patterns among the points. Use the green line to approximate the regression line. How close do you come to the actual regression line each time?
d. On the basis of your experiences with the applet, explain why we need to use more reliable methods of finding the regression line than just "eyeing" it.

Applying the Concepts—Basic

9.18 **Do nice guys really finish last?** In baseball, there is an old saying that "nice guys finish last." Is this true in the competitive corporate world? Researchers at Harvard University attempted to answer this question and reported their results in *Nature* (March 20, 2008). In the study, Boston-area college students repeatedly played a version of the game "prisoner's dilemma," where competitors choose cooperation, defection, or costly punishment. (Cooperation meant paying 1 unit for the opponent to receive 2 units; defection meant gaining 1 unit at a cost of

1 unit for the opponent; and punishment meant paying 1 unit for the opponent to lose 4 units.) At the conclusion of the games, the researchers recorded the average payoff and the number of times punishment was used for each player. A graph of the data is shown in the accompanying scatterplot.

a. Consider punishment use (*x*) as a predictor of average payoff (*y*). Based on the scatterplot, is there evidence of a linear trend?

b. Refer to part **a.** Is the slope of the line relating punishment use (*x*) to average payoff (*y*) positive or negative?

c. The researchers concluded that "winners don't punish"? Do you agree? Explain.

9.19 New method for blood typing. Refer to the *Analytical Chemistry* (May 2010) study in which medical researchers tested a new method of typing blood using lost cost paper, Exercise 2.151 (p. 90). The researchers applied blood drops to the paper and recorded the rate of absorption (called *blood wicking*). The table gives the wicking lengths (millimeters) for six blood drops, each at a different antibody concentration. The data are saved in the **BLOODTYPE** file. Let *y* = wicking length and *x* = antibody concentration.

Droplet	Length (mm)	Concentration
1	22.50	0.0
2	16.00	0.2
3	13.50	0.4
4	14.00	0.6
5	13.75	0.8
6	12.50	1.0

Based on Khan, M. S., et al. "Paper diagnostic for instant blood typing." *Analytical Chemistry*, Vol. 82, No. 10, May 2010 (Figure 4b).

a. Give the equation of the straight-line model relating *y* to *x*.

b. An SPSS printout of the simple linear regression analysis is shown below. Give the equation of the least squares line.

c. Give practical interpretations (if possible) of the estimated *y*-intercept and slope of the line.

Coefficients[a]

Model		Unstandardized Coefficients		Standardized Coefficients	t	Sig.
		B	Std. Error	Beta		
1	(Constant)	19.393	1.708		11.357	.000
	ABConc	-8.036	2.820	-.819	-2.849	.046

a. Dependent Variable: WickLength

9.20 Quantitative models of music. Writing in *Chance* (Fall 2004), University of Konstanz (Germany) statistics professor Jan Beran demonstrated that certain aspects of music can be described by quantitative models. For example, the information content of a musical composition (called *entropy*) can be quantified by determining how many times a certain pitch occurs. In a sample of 147 famous compositions ranging from the 13th to the 20th century, Beran computed the Z12-note entropy (*y*) and plotted it against the year of birth (*x*) of the composer. The graph is reproduced here.

a. Do you observe a trend, especially since the year 1400?

b. The least squares line for the data since 1400 is shown on the graph. Is the slope of the line positive or negative? What does this imply?

c. Explain why the line shown is not the true line of means.

9.21 Wind turbine blade stress. Mechanical engineers at the University of Newcastle (Australia) investigated the use of timber in high-efficiency small wind turbine blades (*Wind Engineering*, Jan. 2004). The strengths of two types of timber—radiata pine and hoop pine—were compared. Twenty specimens (called "coupons") of each timber blade were fatigue tested by measuring the stress (in MPa) on the blade after various numbers of blade cycles. A simple linear regression analysis of the data, one conducted for each type of timber, yielded the following results (where *y* = stress and *x* = natural logarithm of number of cycles):

Radiata Pine: $\hat{y} = 97.37 - 2.50x$

Hoop Pine: $\hat{y} = 122.03 - 2.36x$

a. Interpret the estimated slope of each line.

b. Interpret the estimated y-intercept of each line.

c. On the basis of these results, which type of timber blade appears to be stronger and more fatigue resis-tant? Explain.

9.22 Mongolian desert ants. Refer to the *Journal of Biogeography* (Dec. 2003) study of ants in Mongolia, presented in Exercise 2.155 (p. 91). Data on annual rainfall, maximum daily temperature, and number of ant species recorded at each of 11 study sites are listed in the table and saved in the **GOBIANTS** file.

Site	Region	Annual Rainfall (mm)	Max. Daily Temp. (°C)	Number of Ant Species
1	Dry Steppe	196	5.7	3
2	Dry Steppe	196	5.7	3
3	Dry Steppe	179	7.0	52
4	Dry Steppe	197	8.0	7
5	Dry Steppe	149	8.5	5
6	Gobi Desert	112	10.7	49
7	Gobi Desert	125	11.4	5
8	Gobi Desert	99	10.9	4
9	Gobi Desert	125	11.4	4
10	Gobi Desert	84	11.4	5
11	Gobi Desert	115	11.4	4

Based on Pfeiffer, M., et al. "Community organization and species richness of ants in Mongolia along an ecological gradient from steppe to Gobi desert." *Journal of Biogeography*, Vol. 30, No. 12, Dec. 2003 (Tables 1 and 2).

a. Consider a straight-line model relating annual rainfall (y) and maximum daily temperature (x). A MINITAB printout of the simple linear regression is shown below. Give the least squares prediction equation.

Regression Analysis: Rain versus Temp

```
The regression equation is
Rain = 295 - 16.4 Temp

Predictor    Coef   SE Coef      T      P
Constant   295.25     22.41  13.18  0.000
Temp       -16.364     2.346  -6.97  0.000

S = 17.5111   R-Sq = 84.4%    R-Sq(adj) = 82.7%

Analysis of Variance
Source           DF      SS     MS      F      P
Regression        1   14915  14915  48.64  0.000
Residual Error    9    2760    307
Total            10   17675
```

---

b. Construct a scatterplot for the analysis you performed in part **a**. Include the least square line on the plot. Does the line appear to be a good predictor of annual rainfall?

c. Now consider a straight-line model relating number of ant species (y) to annual rainfall (x). On the basis of the MINITAB printout below. repeat parts **a** and **b**.

Regression Analysis: AntSpecies versus Rain

```
The regression equation is
AntSpecies = 10.5 + 0.016 Rain

Predictor    Coef   SE Coef      T      P
Constant   10.52     22.03   0.48  0.644
Rain       0.0160    0.1480   0.11  0.916

S = 19.6726   R-Sq = 0.1%    R-Sq(adj) = 0.0%

Analysis of Variance
Source           DF      SS     MS      F      P
Regression        1     4.5    4.5   0.01  0.916
Residual Error    9  3483.1  387.0
Total            10  3487.6
```

9.23 Redshifts of quasi-stellar objects. Astronomers call a [NW] shift in the spectrum of galaxies a "redshift." A correlation between redshift level and apparent magnitude (i.e., brightness on a logarithmic scale) of a quasi-stellar object was discovered and reported in the *Journal of Astrophysics & Astronomy* (Mar./Jun. 2003). Physicist D. Basu (Carleton University, Ottawa) applied simple linear regression to data collected for a sample of over 6,000 quasi-stellar objects with confirmed redshifts. The analysis yielded the following results for a specific range of magnitudes: $\hat{y} = 18.13 + 6.21x$, where y = magnitude and x = redshift level.

a. Graph the least squares line. Is the slope of the line positive or negative?

b. Interpret the estimate of the y-intercept in the words of the problem.

c. Interpret the estimate of the slope in the words of the problem.

Applying the Concepts—Intermediate

9.24 Extending the life of an aluminum smelter pot. An investigation of the properties of bricks used to line aluminum smelter pots was published in *The American Ceramic Society Bulletin* (Feb. 2005). Six different commercial bricks were evaluated. The life span of a smelter pot depends on the porosity of the brick lining (the less porosity, the longer is the life); consequently, the researchers measured the apparent porosity of each brick specimen, as well as the mean pore diameter of each brick. The data are given in the next table and saved in the **SMELTPOT** file.

Brick	Apparent Porosity (%)	Mean Pore Diameter (micrometers)
A	18.8	12.0
B	18.3	9.7
C	16.3	7.3
D	6.9	5.3
E	17.1	10.9
F	20.4	16.8

Based on Bonadia, P., et al. "Aluminosilicate refractories for aluminum cell linings." *The American Ceramic Society Bulletin*, Vol. 84, No. 2, Feb. 2005 (Table II).

a. Find the least squares line relating porosity (y) to mean pore diameter (x).

b. Interpret the y-intercept of the line.

c. Interpret the slope of the line.

d. Predict the apparent percentage of porosity for a brick with a mean pore diameter of 10 micrometers.

9.25 Ranking driving performance of professional golfers. Refer to *The Sport Journal* (Winter 2007) study of a new method for ranking the total driving performance of golfers on the Professional Golf Association (PGA) tour, presented in Exercise 2.64 (p. 59). Recall that the method computes a driving performance index based on a golfer's average driving distance (yards) and driving accuracy (percent of drives that land in the fairway). Data for the top 40 PGA golfers (as ranked by the new method) are saved in the **PGADRIVER** file. (The first five and last five observations are listed in the next table.)

a. Write the equation of a straight-line model relating driving accuracy (y) to diving distance (x).

Data for Exercise 9.25

Rank	Player	Driving Distance (yards)	Driving Accuracy (%)	Driving Performance Index
1	Woods	316.1	54.6	3.58
2	Perry	304.7	63.4	3.48
3	Gutschewski	310.5	57.9	3.27
4	Wetterich	311.7	56.6	3.18
5	Hearn	295.2	68.5	2.82
⋮	⋮	⋮	⋮	⋮
36	Senden	291	66	1.31
37	Mickelson	300	58.7	1.30
38	Watney	298.9	59.4	1.26
39	Trahan	295.8	61.8	1.23
40	Pappas	309.4	50.6	1.17

Based on Frederick Wiseman, Ph.D., Mohamed Habibullah, Ph.D., and Mustafa Yilmaz, Ph.D, *Sports Journal,* Vol. 10, No. 1, Winter 2007 (Table 2).

b. Use simple linear regression to fit the model you found in part **a** to the data. Give the least squares prediction equation.

c. Interpret the estimated y-intercept of the line.

d. Interpret the estimated slope of the line.

e. In Exercise 2.157 (p. 91), you were informed that a professional golfer practicing a new swing to increase his average driving distance is concerned that his driving accuracy will be lower. Which of the two estimates, y-intercept or slope, will help you determine whether the golfer's concern is a valid one? Explain.

9.26 FCAT scores and poverty. In the state of Florida, elementary school performance is based on the average score obtained by students on a standardized exam, called the Florida Comprehensive Assessment Test (FCAT). An analysis of the link between FCAT scores and sociodemographic factors was published in the *Journal of Educational*

Elementary School	FCAT—Math	FCAT—Reading	% Below Poverty
1	166.4	165.0	91.7
2	159.6	157.2	90.2
3	159.1	164.4	86.0
4	155.5	162.4	83.9
5	164.3	162.5	80.4
6	169.8	164.9	76.5
7	155.7	162.0	76.0
8	165.2	165.0	75.8
9	175.4	173.7	75.6
10	178.1	171.0	75.0
11	167.1	169.4	74.7
12	177.0	172.9	63.2
13	174.2	172.7	52.9
14	175.6	174.9	48.5
15	170.8	174.8	39.1
16	175.1	170.1	38.4
17	182.8	181.4	34.3
18	180.3	180.6	30.3
19	178.8	178.0	30.3
20	181.4	175.9	29.6
21	182.8	181.6	26.5
22	186.1	183.8	13.8

Based on Tekwe, C. D., et al. "An empirical comparison of statistical models for value-added assessment of school performance." *Journal of Educational and Behavioral Statistics*, Vol. 29, No. 1, Spring 2004 (Table 2).

and Behavioral Statistics (Spring 2004). Data on average math and reading FCAT scores of third graders, as well as the percentage of students below the poverty level, for a sample of 22 Florida elementary schools are listed in the accompanying table and saved in the **FCAT** file.

a. Propose a straight-line model relating math score (y) to percentage (x) of students below the poverty level.

b. Use the method of least squares to fit the model to the data in the FCAT file.

c. Graph the least squares line on a scatterplot of the data. Is there visual evidence of a relationship between the two variables? Is the relationship positive or negative?

d. Interpret the estimates of the y-intercept and slope in the words of the problem.

e. Now consider a model relating reading score (y) to percentage (x) of students below the poverty level. Repeat parts **a–d** for this model.

9.27 Sound waves from a basketball. Refer to the *American Journal of Physics* (June 2010) study of sound waves in a spherical cavity, Exercise 2.37 (p. 47). The frequencies of sound waves (estimated using a mathematical formula) resulting from the first 24 resonances (echoes) after striking a basketball with a metal rod are reproduced in the following table and saved in the **BBALL** file. Recall that the researcher expects the sound wave frequency to increase as the number of resonances increases.

Resonance	Frequency
1	979
2	1572
3	2113
4	2122
5	2659
6	2795
7	3181
8	3431
9	3638
10	3694
11	4038
12	4203
13	4334
14	4631
15	4711
16	4993
17	5130
18	5210
19	5214
20	5633
21	5779
22	5836
23	6259
24	6339

Based on Russell, D. A. "Basketballs as spherical acoustic cavities." *American Journal of Physics*, Vol. 48, No. 6, June 2010 (Table I).

a. Hypothesize a model for frequency (y) as a function of number of resonances (x) that proposes a linearly increasing relationship.

b. According to the researcher's theory, will the slope of the line be positive or negative?

c. Estimate the beta parameters of the model and (if possible) give a practical interpretation of each.

9.28 **Sweetness of orange juice** The quality of the orange juice produced by a manufacturer is constantly monitored. There are numerous sensory and chemical components that combine to make the best-tasting orange juice. For example, one manufacturer has developed a quantitative index of the "sweetness" of orange juice. (The higher the index, the sweeter is the juice.) Is there a relationship between the sweetness index and a chemical measure such as the amount of water-soluble pectin (parts per million) in the orange juice? Data collected on these two variables during 24 production runs at a juice-manufacturing plant are shown in the table and saved in the **OJUICE** file. Suppose a manufacturer wants to use simple linear regression to predict the sweetness (y) from the amount of pectin (x).

a. Find the least squares line for the data.

b. Interpret $\hat{\beta}_0$ and $\hat{\beta}_1$ in the words of the problem.

c. Predict the sweetness index if the amount of pectin in the orange juice is 300 ppm. [*Note:* A measure of reliability of such a prediction is discussed in Section 9.6.]

Run	Sweetness Index	Pectin (ppm)
1	5.2	220
2	5.5	227
3	6.0	259
4	5.9	210
5	5.8	224
6	6.0	215
7	5.8	231
8	5.6	268
9	5.6	239
10	5.9	212
11	5.4	410
12	5.6	256
13	5.8	306
14	5.5	259
15	5.3	284
16	5.3	383
17	5.7	271
18	5.5	264
19	5.7	227
20	5.3	263
21	5.9	232
22	5.8	220
23	5.8	246
24	5.9	241

Note: The data in the table are authentic. For reasons of confidentiality, the name of the manufacturer cannot be disclosed.

9.29 **Ideal height of your mate.** Anthropologists theorize that humans tend to choose mates who are similar to themselves. This includes choosing mates who are similar in height. To test this theory, a study was conducted on 147 Cornell University students (*Chance*, Summer 2008). Each student was asked to select the height of his or her ideal spouse or life partner. The researchers fit the simple linear regression model, $E(y) = \beta_0 + \beta_1 x$, where y = ideal partner's height (in inches) and x = student's height (in inches). The data for the study (simulated from information provided in a scatterplot) are saved in the **IDHEIGHT** file. The next table lists selected observations from the full data set.

a. The researchers found the estimated slope of the line to be negative. Fit the model to the data in the **IDHEIGHT** file using statistical software and verify this result.

Gender	Actual Height	Ideal Height
F	59	66
F	60	70
F	60	72
F	61	65
F	61	67
⋮	⋮	⋮
M	73.5	66
M	74	67
M	74	68
M	74	69
M	74	70

Based on Lee, G., Velleman, P., and Wainer, H. "Giving the finger to dating services." *Chance*, Vol. 21, No. 3, Summer 2008 (adapted from Figure 3).

b. The negative slope was interpreted as follows: "The taller the respondent was, the shorter they felt their ideal partner ought to be." Do you agree?

c. The result, part **b**, contradicts the theory developed by anthropologists. To gain insight into this phenomenon, use a scatterplot to graph the full data set. Use a different plotting symbol for male and female students. Now focus on just the data for the female students. What trend do you observe? Repeat for male students.

d. Fit the straight-line model to the data for the female students. Interpret the estimated slope of the line.

e. Repeat part **d** for the male students.

f. Based on the results, parts **d** and **e**, comment on whether the study data support the theory developed by anthropologists.

9.30 **The "name game."** Refer to the *Journal of Experimental Psychology—Applied* (June 2000) study in which the "name game" was used to help groups of students learn the names of other students in the group, presented in Exercise 7.110 (p. 413). Recall that the "name game" requires the first student in the group to state his or her full name, the second student to say his or her name and the name of the first student, the third student to say his or her name and the names of the first two students, etc. After making their introductions, the students listened to a seminar speaker for 30 minutes. At the end of the seminar, all students were asked to remember the full name of each of the other students in their group, and the researchers measured the proportion of names recalled for each. One goal of the study was to investigate the linear trend between y = proportion of names recalled and x = position (order) of the student during the game. The data (simulated on the basis of summary statistics provided in the research article) for 144 students in the first eight positions are saved in the **NAMEGAME2** file. The first five and last five observations in the data set are listed in the table on the next page. [*Note:* Since the student in position 1 actually must recall the names of all the other students, he or she is assigned position number 9 in the data set.] Use the method of least squares to estimate the line $E(y) = \beta_0 + \beta_1 x$. Interpret the β estimates in the words of the problem.

Data for Exercise 9.30

Position	Recall
2	0.04
2	0.37
2	1.00
2	0.99
2	0.79
⋮	⋮
9	0.72
9	0.88
9	0.46
9	0.54
9	0.99

Based on Morris, P. E., and Fritz, C. O. "The name game: Using retrieval practice to improve the learning of names." *Journal of Experimental Psychology—Applied*, Vol. 6, No. 2, June 2000 (data simulated from Figure 1).

Applying the Concepts—Advanced

9.31 **Spreading rate of spilled liquid.** Refer to the *Chemical Engineering Progress* (Jan. 2005) study of the rate at which a spilled volatile liquid will spread across a surface, presented in Exercise 2.158 (p. 91). Recall that a DuPont Corp. engineer calculated the mass (in pounds) of a 50-gallon methanol spill after a period ranging from 0 to 60 minutes. Do the data shown in the accompanying table (saved in the **LIQUIDSPILL** file) indicate that the mass of the spill tends to diminish as time increases? If so, how much will the mass diminish each minute?

Time (minutes)	Mass (pounds)
0	6.64
1	6.34
2	6.04
4	5.47
6	4.94
8	4.44
10	3.98
12	3.55
14	3.15
16	2.79
18	2.45
20	2.14
22	1.86
24	1.60
26	1.37
28	1.17
30	0.98
35	0.60
40	0.34
45	0.17
50	0.06
55	0.02
60	0.00

Based on Barry, J. "Estimating rates of spreading and evaporation of volatile liquids." *Chemical Engineering Progress*, Vol. 101, No. 1, Jan. 2005, p. 38.

9.3 Model Assumptions

In Section 9.2, we assumed that the probabilistic model relating the drug reaction time y to the percentage x of drug in the bloodstream is

$$y = \beta_0 + \beta_1 x + \varepsilon$$

We also recall that the least squares estimate of the deterministic component of the model, $\beta_0 + \beta_1 x$, is

$$\hat{y} = \hat{\beta}_0 + \hat{\beta}_1 x = -.1 + .7x$$

Now we turn our attention to the random component ε of the probabilistic model and its relation to the errors in estimating β_0 and β_1. We will use a probability distribution to characterize the behavior of ε. We will see how the probability distribution of ε determines how well the model describes the relationship between the **dependent variable** y and the independent variable x.

Step 3 in a regression analysis requires us to specify the probability distribution of the random error ε. We will make four basic assumptions about the general form of this probability distribution:

Assumption 1 The mean of the probability distribution of ε is 0. That is, the average of the values of ε over an infinitely long series of experiments is 0 for each setting of the independent variable x. This assumption implies that the mean value of y, for a given value of x, is $E(y) = \beta_0 + \beta_1 x$.

Assumption 2 The variance of the probability distribution of ε is constant for all settings of the independent variable x. For our straight-line model, this assumption means that the variance of ε is equal to a constant—say, σ^2—for all values of x.

Assumption 3 The probability distribution of ε is normal.

Assumption 4 The values of ε associated with any two observed values of y are independent. That is, the value of ε associated with one value of y has no effect on any of the values of ε associated with any other y values.

The implications of the first three assumptions can be seen in Figure 9.7, which shows distributions of errors for three values of x, namely, 5, 10, and 15. Note that the relative frequency distributions of the errors are normal with a mean of 0 and a constant variance σ^2. (All of the distributions shown have the same amount of spread or variability.) The straight line shown in Figure 9.7 is the line of means; it indicates the mean value of y for a given value of x. We denote this mean value as $E(y)$. Then the line of means is given by the equation

$$E(y) = \beta_0 + \beta_1 x$$

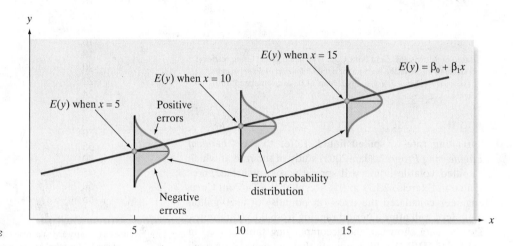

Figure 9.7

The probability distribution of ε

These assumptions make it possible for us to develop measures of reliability for the least squares estimators and to devise hypothesis tests for examining the usefulness of the least squares line. We have various techniques for checking the validity of these assumptions, and we have remedies to apply when they appear to be invalid. (These topics are beyond the scope of this text, but they are discussed in some of the chapter references.) Fortunately, the assumptions need not hold exactly in order for least squares estimators to be useful. The assumptions will be satisfied adequately for many applications encountered in practice.

It seems reasonable to assume that the greater the variability of the random error ε (which is measured by its variance σ^2), the greater will be the errors in the estimation of the model parameters β_0 and β_1 and in the error of prediction when $\hat{y}$ is used to predict y for some value of x. Consequently, you should not be surprised, as we proceed through this chapter, to find that σ^2 appears in the formulas for all confidence intervals and test statistics that we will be using.

Estimation of σ^2 for a (First-Order) Straight-Line Model

$$s^2 = \frac{\text{SSE}}{\text{Degrees of freedom for error}} = \frac{\text{SSE}}{n - 2}$$

where $\text{SSE} = \sum (y_i - \hat{y}_i)^2 = \text{SS}_{yy} - \hat{\beta}_1 \text{SS}_{xy}$

in which

$$\text{SS}_{yy} = \sum (y_i - \bar{y})^2 = \sum y_i^2 - \frac{\left(\sum y_i \right)^2}{n}$$

To estimate the standard deviation σ of ε, we calculate

$$s = \sqrt{s^2} = \sqrt{\frac{\text{SSE}}{n - 2}}$$

We will refer to s as the **estimated standard error of the regression model**.

In most practical situations, σ^2 is unknown and we must use our data to estimate its value. The best estimate of σ^2, denoted by s^2, is obtained by dividing the sum of the squares of the deviations of the y values from the prediction line, or

$$\text{SSE} = \sum (y_i - \hat{y}_i)^2$$

by the number of degrees of freedom associated with this quantity. We use 2 df to estimate the two parameters β_0 and β_1 in the straight-line model, leaving $(n - 2)$ df for the estimation of the error variance.

⚠ **CAUTION** When performing these calculations, you may be tempted to round the calculated values of SS_{yy}, $\hat{\beta}_1$, and SS_{xy}. Be certain to carry at least six significant figures for each of these quantities, to avoid substantial errors in calculating the SSE. ▲

Example 9.2
Estimating σ in Regression—Drug Reaction Data

Problem Refer to Example 9.1 and the simple linear regression of the drug reaction data in Table 9.3.

a. Compute an estimate of σ.

b. Give a practical interpretation of the estimate.

Solution

a. We previously calculated SSE = 1.10 for the least squares line $\hat{y} = -.1 + .7x$. Recalling that there were $n = 5$ data points, we have $n - 2 = 5 - 2 = 3$ df for estimating σ^2. Thus,

$$s^2 = \frac{\text{SSE}}{n - 2} = \frac{1.10}{3} = .367$$

is the estimated variance, and

$$s = \sqrt{.367} = .61$$

is the standard error of the regression model.

b. You may be able to grasp s intuitively by recalling the interpretation of a standard deviation given in Chapter 2 and remembering that the least squares line estimates the mean value of y for a given value of x. Since s measures the spread of the distribution of y values about the least squares line and these errors of prediction are assumed to be normally distributed, we should not be surprised to find that most (about 95%) of the observations lie within $2s$, or $2(.61) = 1.22$, of the least squares line. For this simple example (only five data points), all five data points fall within $2s$ of the least squares line. In Section 9.6, we use s to evaluate the error of prediction when the least squares line is used to predict a value of y to be observed for a given value of x.

Dependent Variable: TIME_Y

Number of Observations Read	5
Number of Observations Used	5

Analysis of Variance

Source	DF	Sum of Squares	Mean Square	F Value	Pr > F
Model	1	4.90000	4.90000	13.36	0.0354
Error	3	1.10000	0.36667		
Corrected Total	4	6.00000			

Root MSE	0.60553	R-Square	0.8167
Dependent Mean	2.00000	Adj R-Sq	0.7556
Coeff Var	30.27650		

Parameter Estimates

Variable	DF	Parameter Estimate	Standard Error	t Value	Pr > |t|
Intercept	1	-0.10000	0.63509	-0.16	0.8849
DRUG_X	1	0.70000	0.19149	3.66	0.0354

Figure 9.8
SAS printout for the time–drug regression

Look Back The values of s^2 and s can also be obtained from a simple linear regression print-out. The SAS printout for the drug reaction example is reproduced in Figure 9.8. The value of s^2 is highlighted on the printout (in the **Mean Square** column in the row labeled **Error**). The value $s^2 = .36667$, rounded to three decimal places, agrees with the one calculated by hand. The value of s is also highlighted in Figure 9.8 (next to the heading **Root MSE**). This value, $s = .60553$, agrees (except for rounding) with our hand-calculated value.

Now Work Exercise 9.36a–b

Interpretation of s, the Estimated Standard Deviation of ε

We expect most ($\approx 95\%$) of the observed y values to lie within $2s$ of their respective least squares predicted values, $\hat{y}$.

Exercises 9.30–9.42

Understanding the Principles

9.32 What are the four assumptions made about the probability distribution of ε in regression?

9.33 Illustrate the assumptions of Exercise 9.32 with a graph.

9.34 Visually compare the scatterplots shown below. If a least squares line were determined for each data set, which do you think would have the smallest variance s^2? Explain.

a.

b.

c.

Learning the Mechanics

9.35 Calculate SSE and s^2 for each of the following cases:
  a. $n = 20$, $SS_{yy} = 95$, $SS_{xy} = 50$, $\hat{\beta}_1 = .75$

  b. $n = 40$, $\sum y^2 = 860$, $\sum y = 50$,

  $SS_{xy} = 2{,}700$, $\hat{\beta}_1 = .2$

  c. $n = 10$, $\sum (y_i - \bar{y})^2 = 58$,

  $SS_{xy} = 91$, $SS_{xx} = 170$

9.36 Suppose you fit a least squares line to 12 data points and
NW the calculated value of SSE is .429.
  a. Find s^2, the estimator of σ^2 (the variance of the random error term ε).
  b. Find s, the estimate of σ.

  c. What is the largest deviation that you might expect between any one of the 12 points and the least squares line?

9.37 Refer to Exercises 9.14 and 9.17 (p. 486). Calculate SSE, s^2, and s for the least squares lines obtained in those exercises. Interpret the standard errors of the regression model for each.

Applying the Concepts—Basic

9.38 **Do nice guys really finish last?** Refer to the *Nature* (March 20, 2008) study of whether "nice guys finish last," Exercise 9.18 (p. 486). Recall that Boston-area college students repeatedly played a version of the game "prisoner's dilemma," where competitors choose cooperation, defection, or costly punishment. At the conclusion of the games, the researchers recorded the average payoff and the number of times punishment was used for each player. Based on a scatterplot of the data, the simple linear regression relating average payoff (y) to punishment use (x) resulted in SSE = 1.04.
  a. Assuming a sample size of $n = 28$, compute the estimated standard deviation of the error distribution, s.
  b. Give a practical interpretation of s.

9.39 **New method for blood typing.** Refer to the *Analytical Chemistry* (May 2010) study in which medical researchers tested a new method of typing blood using lost cost paper, Exercise 9.19 (p. 487). The data in the **BLOODTYPE** file were used to fit the straight-line model relating $y =$ wicking length to $x =$ antibody concentration. The SPSS printout follows.

Model Summary

Model	R	R Square	Adjusted R Square	Std. Error of the Estimate
1	.819[a]	.670	.587	2.35944

a. Predictors: (Constant), ABConc

ANOVA[b]

Model		Sum of Squares	df	Mean Square	F	Sig.
1	Regression	45.201	1	45.201	8.119	.046[a]
	Residual	22.268	4	5.567		
	Total	67.469	5			

a. Predictors: (Constant), ABConc
b. Dependent Variable: WickLength

a. Give the values of SSE, s^2, and s shown on the print-out.

b. Give a practical interpretation of s. Recall that wicking length is measured in millimeters.

9.40 **Quantitative models of music.** Refer to the *Chance* (Fall 2004) study on modeling a certain pitch of a musical composition, presented in Exercise 9.20 (p. 487). Recall that the number of times (y) a certain pitch occurs—called entropy—was modeled as a straight-line function of year of birth (x) of the composer. On the basis of the scatterplot of the data, the standard deviation σ of the model is estimated to be $s = .1$. For a given year (x), about 95% of the actual entropy values (y) will fall within d units of their predicted values. Find the value of d.

9.41 **Mongolian desert ants.** Refer to the *Journal of Biogeography* (Dec. 2003) study of ant sites in Mongolia, presented in Exercise 9.22 (p. 487). The data in the **GOBIANTS** file was used to estimate the straight-line model relating annual rainfall (y) to maximum daily temperature (x).

a. Give the values of SSE, s^2, and s, shown on the MINITAB printout (p. 488).

b. Give a practical interpretation of the value of s.

Applying the Concepts—Intermediate

9.42 **Extending the life of an aluminum smelter pot.** Refer to *The American Ceramic Society Bulletin* (Feb. 2005) study of bricks that line aluminum smelter pots, presented in Exercise 9.24 (p. 488). You fit the simple linear regression model relating brick porosity (y) to mean pore diameter (x) to the data in the **SMELTPOT** file.

a. Find an estimate of the standard deviation σ of the model.

b. In Exercise 9.24d, you predicted brick porosity percentage when $x = 10$ micrometers. Use the result of part **a** to estimate the error of prediction.

9.43 **FCAT scores and poverty.** Refer to the *Journal of Educational and Behavioral Statistics* (Spring 2004) study of scores on the Florida Comprehensive Assessment Test (FCAT), presented in Exercise 9.26 (p. 489). The data are saved in the **FCAT** file.

a. Consider the simple linear regression relating math score (y) to percentage (x) of students below the poverty level. Find and interpret the value of s for this regression.

b. Consider the simple linear regression relating reading score (y) to percentage (x) of students below the poverty level. Find and interpret the value of s for this regression.

c. Which dependent variable, math score or reading score, can be more accurately predicted by percentage (x) of students below the poverty level? Explain.

9.44 **Sweetness of orange juice.** Refer to the study of the quality of orange juice produced at a juice manufacturing plant, Exercise 9.28 (p. 490). The data are saved in the **OJUICE** file. Recall that simple linear regression was used to predict

the sweetness index (y) from the amount of pectin (x) in orange juice manufactured during a production run.

a. Give the values of SSE, s^2, and s for this regression.

b. Explain why it is difficult to give a practical interpretation to s^2.

c. Use the value of s to derive a range within which most (about 95%) of the errors of prediction of sweetness index fall.

9.45 **Ideal height of your mate.** Refer to the *Chance* (Summer 2008) study of the height of the ideal mate, Exercise 9.29 (p. 490). The data in the **IDHEIGHT** file were used to fit the simple linear regression model, $E(y) = \beta_0 + \beta_1 x$, where y = ideal partner's height (in inches) and x = student's height (in inches).

a. Fit the straight-line model to the data for the male students. Find an estimate for σ, the standard deviation of the error term, and interpret its value practically.

b. Repeat part **a** for the female students.

c. For which group, males or females, is student's height the more accurate predictor of ideal partner's height?

Applying the Concepts—Advanced

9.46 **Life tests of cutting tools.** To improve the quality of the output of any production process, it is necessary first to understand the capabilities of the process (For example, see Gitlow, H., *Quality Management Systems: A Practical Guide*, 2000.) In a particular manufacturing process, the useful life of a cutting tool is linearly related to the speed at which the tool is operated. The data in the accompanying table, saved in the **CUTTOOL** file, were derived from life tests for the two different brands of cutting tools currently used in the production process. For which brand would you feel more confident using the least squares line to predict useful life for a given cutting speed? Explain.

Cutting Speed (meters per minute)	Useful Life (hours)	
	Brand A	Brand B
30	4.5	6.0
30	3.5	6.5
30	5.2	5.0
40	5.2	6.0
40	4.0	4.5
40	2.5	5.0
50	4.4	4.5
50	2.8	4.0
50	1.0	3.7
60	4.0	3.8
60	2.0	3.0
60	1.1	2.4
70	1.1	1.5
70	.5	2.0
70	3.0	1.0

9.4 Assessing the Utility of the Model: Making Inferences about the Slope β_1

Now that we have specified the probability distribution of ε and found an estimate of the variance σ^2, we are ready to make statistical inferences about the linear model's usefulness in predicting the response y. This is step 4 in our regression modeling procedure.

Refer again to the data of Table 9.1, and suppose the reaction times are *completely unrelated* to the percentage of drug in the bloodstream. What could then be said about the values of β_0 and β_1 in the hypothesized probabilistic model

$$y = \beta_0 + \beta_1 x + \varepsilon$$

if x contributes no information for the prediction of y? The implication is that the mean of y—that is, the deterministic part of the model $E(y) = \beta_0 + \beta_1 x$—does not change as x changes. In the straight-line model, this means that the true slope, β_1, is equal to 0. (See Figure 9.9.) Therefore, to test the null hypothesis that the linear model contributes no information for the prediction of y against the alternative hypothesis that the linear model is useful in predicting y, we test

$$H_0: \beta_1 = 0$$
$$H_a: \beta_1 \neq 0$$

If the data support the alternative hypothesis, we will conclude that x does contribute information for the prediction of y with the straight-line model [although the true relationship between $E(y)$ and x could be more complicated than a straight line]. In effect, then, this is a test of the usefulness of the hypothesized model.

Figure 9.9

Graph of the straight-line model when the slope is zero, i.e., $y = \beta_0 + \varepsilon$

The appropriate test statistic is found by considering the sampling distribution of $\hat{\beta}_1$, the least squares estimator of the slope β_1, as shown in the following box:

Sampling Distribution of $\hat{\beta}_1$

If we make the four assumptions about ε (see Section 9.3), the sampling distribution of the least squares estimator $\hat{\beta}_1$ of the slope will be normal with mean β_1 (the true slope) and standard deviation

$$\sigma_{\hat{\beta}_1} = \frac{\sigma}{\sqrt{SS_{xx}}} \qquad \text{(see Figure 9.10)}$$

We estimate $\sigma_{\hat{\beta}_1}$ by $s_{\hat{\beta}_1} = \dfrac{s}{\sqrt{SS_{xx}}}$ and refer to $s_{\hat{\beta}_1}$ as the **estimated standard error of the least squares slope $\hat{\beta}_1$.**

Figure 9.10

Sampling distribution of $\hat{\beta}_1$

Since σ is usually unknown, the appropriate test statistic is a *t*-statistic, formed as:

$$t = \frac{\hat{\beta}_1 - \text{Hypothesized value of } \beta_1}{s_{\hat{\beta}_1}} \qquad \text{where} \qquad s_{\hat{\beta}_1} = \frac{s}{\sqrt{SS_{xx}}}$$

Thus,

$$t = \frac{\hat{\beta}_1 - 0}{s/\sqrt{SS_{xx}}}$$

Note that we have substituted the estimator s for σ and then formed the estimated standard error $s_{\hat{\beta}_1}$ by dividing s by $\sqrt{SS_{xx}}$. The number of degrees of freedom associated with this t statistic is the same as the number of degrees of freedom associated with s. Recall that this number is $(n - 2)$ df when the hypothesized model is a straight line. (See Section 9.3.) The setup of our test of the usefulness of the straight-line model is summarized in the following two boxes.

A Test of Model Utility: Simple Linear Regression

One-Tailed Test	**Two-Tailed Test**
$H_0: \beta_1 = 0$	$H_0: \beta_1 = 0$
$H_a: \beta_1 < 0$ (or $H_a: \beta_1 > 0$)	$H_a: \beta_1 \neq 0$

Test statistic: $t = \dfrac{\hat{\beta}_1}{s_{\hat{\beta}_1}} = \dfrac{\hat{\beta}_1}{s/\sqrt{SS_{xx}}}$

Rejection region: $t < -t_\alpha$ Rejection region: $|t| > t_{\alpha/2}$
(or $t > t_\alpha$ when $H_a: \beta_1 > 0$)
where t_α and $t_{\alpha/2}$ are based on $(n - 2)$ degrees of freedom

---

Conditions Required for a Valid Test: Simple Linear Regression

The four assumptions about ε listed in Section 9.3.

Example 9.3

Testing the Regression Slope, β_1—Drug Reaction Model

Problem Refer to the simple linear regression analysis of the drug reaction data performed in Examples 9.1 and 9.2. Conduct a test (at $\alpha = .05$) to determine whether the reaction time (y) is linearly related to the amount of drug (x).

Solution For the drug reaction example, $n = 5$. Thus, t will be based on $n - 2 = 3$ df, and the rejection region t (at $\alpha = .05$) will be

$$|t| > t_{.025} = 3.182$$

We previously calculated $\hat{\beta}_1 = .7, s = .61$, and $SS_{xx} = 10$. Thus,

$$t = \frac{\hat{\beta}_1}{s/\sqrt{SS_{xx}}} = \frac{.7}{.61/\sqrt{10}} = \frac{.7}{.19} = 3.7$$

Since this calculated t-value falls into the upper-tail rejection region (see Figure 9.11), we reject the null hypothesis and conclude that the slope β_1 is not 0. The sample evidence indicates that the percentage x of drug in the bloodstream contributes information for the prediction of the reaction time y when a linear model is used.

[*Note:* We can reach the same conclusion by using the observed significance level (p-value) of the test from a computer printout. The MINITAB printout for the drug reaction example is reproduced in Figure 9.12. The test statistic and the two-tailed p-value are highlighted on the printout. Since the p-value $= .035$ is smaller than $\alpha = .05$, we will reject H_0.]

Figure 9.11

Rejection region and calculated t value for testing $H_0: \beta_1 = 0$ versus $H_a: \beta_1 \neq 0$

Figure 9.12

MINITAB printout for the time–drug regression

```
Regression Analysis: TIME_Y versus DRUG_X

The regression equation is
TIME_Y = - 0.100 + 0.700 DRUG_X


Predictor      Coef   SE Coef      T      P
Constant    -0.1000    0.6351  -0.16  0.885
DRUG_X       0.7000    0.1915   3.66  0.035


S = 0.605530   R-Sq = 81.7%   R-Sq(adj) = 75.6%


Analysis of Variance

Source           DF      SS      MS      F      P
Regression        1  4.9000  4.9000  13.36  0.035
Residual Error    3  1.1000  0.3667
Total             4  6.0000
```

Look Back What conclusion can be drawn if the calculated t-value does not fall into the rejection region or if the observed significance level of the test exceeds α? We know from previous discussions of the philosophy of hypothesis testing that such a t-value does *not* lead us to accept the null hypothesis. That is, we do not conclude that $\beta_1 = 0$. Additional data might indicate that β_1 differs from 0, or a more complicated relationship may exist between x and y, requiring the fitting of a model other than the straight-line model.

Now Work Exercise 9.54

Interpreting p-Values for β Coefficients in Regression

Almost all statistical computer software packages report a *two-tailed* p-value for each of the β parameters in the regression model. For example, in simple linear regression, the p-value for the two-tailed test $H_0: \beta_1 = 0$ versus $H_a: \beta_1 \neq 0$ is given on the printout. If you want to conduct a *one-tailed* test of hypothesis, you will need to adjust the p-value reported on the printout as follows:

$$Upper - tailed\ test(H_a: \beta_1 > 0): \quad p\text{-value} = \begin{cases} p/2 & \text{if } t > 0 \\ 1 - p/2 & \text{if } t < 0 \end{cases}$$

$$Lower - tailed\ test(H_a: \beta_1 < 0): \quad p\text{-value} = \begin{cases} p/2 & \text{if } t < 0 \\ 1 - p/2 & \text{if } t > 0 \end{cases}$$

where p is the p-value reported on the printout and t is the value of the test statistic.

Another way to make inferences about the slope β_1 is to estimate it with a confidence interval, formed as shown in the following box:

A $100(1 - \alpha)\%$ Confidence Interval for the Simple Linear Regression Slope β_1

$$\hat{\beta}_1 \pm (t_{\alpha/2}) s_{\hat{\beta}_1}$$

where the estimated standard error of $\hat{\beta}_1$ is calculated by

$$s_{\hat{\beta}_1} = \frac{s}{\sqrt{SS_{xx}}}$$

and $t_{\alpha/2}$ is based on $(n - 2)$ degrees of freedom.

Conditions Required for a Valid Confidence Interval: Simple Linear Regression

The four assumptions about ε listed in Section 9.3.

For the simple linear regression for the drug reaction (Examples 9.1–9.3), $t_{\alpha/2}$ is based on $(n - 2) = 3$ degrees of freedom. Therefore, a 95% confidence interval for the slope β_1, the expected change in reaction time for a 1% increase in the amount of drug in the bloodstream, is

$$\hat{\beta}_1 \pm t_{.025}s_{\hat{\beta}_1} = .7 \pm 3.182\left(\frac{s}{\sqrt{SS_{xx}}}\right) = .7 \pm 3.182\left(\frac{.61}{\sqrt{10}}\right) = .7 \pm .61$$

Thus, the estimate of the interval for the slope parameter β_1 is from .09 to 1.31. [*Note:* This interval can also be obtained with statistical software and is highlighted on the SPSS printout shown in Figure 9.13.] In terms of this example, the implication is that we can be 95% confident that the *true* mean increase in reaction time per additional 1% of the drug is between .09 and 1.31 seconds. This inference is meaningful only over the sampled range of x—that is, from 1% to 5% of the drug in the bloodstream.

Figure 9.13

SPSS printout with 95% confidence intervals for the time–drug regression betas

Coefficients^a

Model		Unstandardized Coefficients B	Unstandardized Coefficients Std. Error	Standardized Coefficients Beta	t	Sig.	95% Confidence Interval for B Lower Bound	95% Confidence Interval for B Upper Bound
1	(Constant)	-.100	.635		-.157	.885	-2.121	1.921
	TIME_X	.700	.191	.904	3.656	.035	.091	1.309

a. Dependent Variable: DRUG_Y

Now Work Exercise 9.59

Since all the values in this interval are positive, it appears that β_1 is positive and that the mean of y, $E(y)$, increases as x increases. However, the rather large width of the confidence interval reflects the small number of data points (and, consequently, a lack of information) used in the experiment. We would expect a narrower interval if the sample size were increased.

We conclude this section with a comment on the other β-parameter in the straight-line model—the y-intercept, β_0. Why not conduct a test of hypothesis on β_0? For example, we could conduct the test $H_0: \beta_0 = 0$ against $H_a: \beta_0 \neq 0$. The p-value for this test appears on the printouts for SAS, SPSS, MINITAB, and most other statistical software packages. The answer lies in the interpretation of β_0. In the previous section, we learned that the y-intercept represents the mean value of y when $x = 0$. Thus, the test $H_0: \beta_0 = 0$ is equivalent to testing whether $E(y) = 0$ when $x = 0$. In the drug reaction simple linear regression, we would be testing whether the mean reaction time (y) is 0 seconds when the amount of drug in the blood (x) is 0%. The value $x = 0$ is typically not a meaningful value (as in the drug reaction example), or, $x = 0$ is typically outside the range of the sample data. In either of these cases, the test $H_0: \beta_0 = 0$ is not meaningful and should be avoided.

For those regression analyses where $x = 0$ is a meaningful value, one may desire to predict the value of y when $x = 0$. We discuss a confidence interval for such a prediction in Section 9.6.

⚠ **CAUTION** In simple linear regression, the test $H_0: \beta_0 = 0$ is only meaningful if the value $x = 0$ makes sense and is within the range of the sample data. ▲

Statistics IN Action | **Revisited**   Assessing How Well the Straight–Line Model Fits the Dowsing Data

In the previous Statistics in Action Revisited, we fit the straight-line model $E(y) = \beta_0 + \beta_1 x$, where x = dowser's guess (in meters) and y = pipe location (in meters) for each trial. The MINITAB regression printout is reproduced in Figure SIA9.3. The two-tailed p-value for testing the null hypothesis $H_0: \beta_1 = 0$ (highlighted on the printout) is p-value = .118. Even for an α-level as high as $\alpha = .10$, there is insufficient evidence to reject H_0. Consequently, the dowsing data in Table SIA9.1 provide no statistical support for the German researchers' claim that the three best dowsers have an ability to find underground water with a divining rod.

This lack of support for the dowsing theory is made clearer with a confidence interval for the slope of the line. When

Statistics IN Action
(continued)

$n = 26$, df $= (n - 2) = 24$ and $t_{.025} = 2.064$. Substituting the latter value and the relevant values shown on the MINITAB printout, we find that a 95% confidence interval for β_1 is

$$\hat{\beta}_1 \pm t_{.025}(s_{\hat{\beta}_1}) = .31 \pm (2.064)(.19)$$
$$= .31 \pm .39, \text{ or } (-.08, .70)$$

Thus, for every 1-meter increase in a dowser's guess, we estimate (with 95% confidence) that the change in the actual pipe location will range anywhere from a decrease of .08 meter to an increase of .70 meter. In other words, we're not sure whether the pipe location will increase or decrease along the 10-meter pipeline! Keep in mind also that the data in Table SIA9.1 represent the "best" performances of the three dowsers (i.e., the outcome of the dowsing experiment in its most favorable light). When the data for all trials are considered and plotted, there is not even a hint of a trend.

Regression Analysis: PIPE versus GUESS

```
The regression equation is
PIPE = 30.1 + 0.308 GUESS


Predictor    Coef   SE Coef     T      P
Constant    30.07     11.41   2.63  0.015
GUESS      0.3079    0.1900   1.62  0.118


S = 26.0298    R-Sq = 9.9%    R-Sq(adj) = 6.1%


Analysis of Variance

Source         DF       SS      MS     F      P
Regression      1   1778.9  1778.9  2.63  0.118
Residual Error 24  16261.2   677.6
Total          25  18040.2
```

Figure SIA9.3
MINITAB simple linear regression for dowsing data

Exercises 9.47–9.66

Understanding the Principles

9.47 In the equation $E(y) = \beta_0 + \beta_1 x$, what is the value of β_1 if x has no linear relationship to y?

9.48 What conditions are required for valid inferences about the β's in simple linear regression?

9.49 How do you adjust the p-value obtained from a computer printout when you perform a one-tailed test of β_1 in simple linear regression?

9.50 For each of the following 95% confidence intervals for β_1 in simple linear regression, decide whether there is evidence of a positive or negative linear relationship between y and x:
a. $(22, 58)$ **b.** $(-30, 111)$ **c.** $(-45, -7)$

Learning the Mechanics

9.51 Construct both a 95% and a 90% confidence interval for β_1 for each of the following cases:
a. $\hat{\beta}_1 = 31, s = 3, SS_{xx} = 35, n = 12$
b. $\hat{\beta}_1 = 64, SSE = 1,960, SS_{xx} = 30, n = 18$
c. $\hat{\beta}_1 = -8.4, SSE = 146, SS_{xx} = 64, n = 24$

9.52 Consider the following pairs of observations:

x	1	5	3	2	6	6	0
y	1	3	3	1	4	5	1

a. Construct a scatterplot of the data.
b. Use the method of least squares to fit a straight line to the seven data points in the table.
c. Plot the least squares line on your scatterplot of part **a.**
d. Specify the null and alternative hypotheses you would use to test whether the data provide sufficient evidence to indicate that x contributes information for the (linear) prediction of y.
e. What is the test statistic that should be used in conducting the hypothesis test of part **d**? Specify the number of degrees of freedom associated with the test statistic.
f. Conduct the hypothesis test of part **d**, using $\alpha = .05$.
g. Construct a 95% confidence interval for β_1.

9.53 Consider the following pairs of observations:

y	4	2	5	3	2	4
x	1	4	5	3	2	4

a. Construct a scatterplot of the data.
b. Use the method of least squares to fit a straight line to the six data points.
c. Graph the least squares line on the scatterplot of part **a.**
d. Compute the test statistic for determining whether x and y are linearly related.
e. Carry out the test you set up in part **d**, using $\alpha = .01$.
f. Find a 99% confidence interval for β_1.

Applying the Concepts—Basic

9.54 **English as a second language reading ability.** What are the factors that allow a native Spanish-speaking person to understand and read English? A study published in the *Bilingual Research Journal* (Summer 2006) investigated the relationship of Spanish (first-language) grammatical knowledge to English (second-language) reading. The study involved a sample of $n = 55$ native Spanish-speaking adults who were students in an English as a second language (ESL) college class. Each student took four standardized exams: Spanish grammar (SG), Spanish reading (SR), English grammar (EG), and English reading (ESLR). Simple linear regression was used to model the ESLR score (y) as a function of each of the other exam scores (x). The results are summarized in the next table (p. 501).
a. At $\alpha = .05$, is there sufficient evidence to indicate that ESLR score is linearly related to SG score?

Independent variable (x)	p-value for testing H_0: $\beta_1 = 0$
SG score	.739
SR score	.012
ER score	.022

b. At $\alpha = .05$, is there sufficient evidence to indicate that ESLR score is linearly related to SR score?

c. At $\alpha = .05$, is there sufficient evidence to indicate that ESLR score is linearly related to ER score?

9.55 Lobster fishing study. Refer to the *Bulletin of Marine Science* (April 2010) study of teams of fishermen fishing for the red spiny lobster in Baja California Sur, Mexico, Exercise 7.16 (p. 365). Two variables measured for each of 8 teams from the Punta Abreojos (PA) fishing cooperative were y = total catch of lobsters (in kilograms) during the season and x = average percentage of traps allocated per day to exploring areas of unknown catch (called *search frequency*). These data, saved in the **TRAPSPACE** file, are listed in the table.

Total Catch	Search Frequency
2,785	35
6,535	21
6,695	26
4,891	29
4,937	23
5,727	17
7,019	21
5,735	20

Source: From Shester, G. G. "Explaining catch variation among Baja California lobster fishers through spatial analysis of trap-placement decisions." *Bulletin of Marine Science*, Vol. 86, No. 2, April 2010 (Table 1). Reprinted with permission from the University of Miami - *Bulletin of Marine Science.*

a. Graph the data in a scatterplot. What type of trend, if any, do you observe?

b. A simple linear regression analysis was conducted using SAS. Find the least squares prediction equation on the accompanying SAS printout. Interpret the slope of the least squares line.

c. Give the null and alternative hypothesis for testing whether total catch (y) is negatively linearly related to search frequency (x).

d. Find the *p*-value of the test, part **c**, on the SAS printout.

e. Give the appropriate conclusion of the test, part **c**, using $\alpha = .05$.

9.56 Ranking driving performance of professional golfers. Refer to *The Sport Journal* (Winter 2007) study of a new method for ranking the total driving performance of golfers on the Professional Golf Association (PGA) tour, presented in Exercise 9.25 (p. 488). You fit a straight-line model relating driving accuracy (y) to driving distance (x) to the data saved in the **PGADRIVER** file.

a. Give the null and alternative hypotheses for testing whether driving accuracy (y) decreases linearly as driving distance (x) increases.

b. Find the test statistic and *p*-value of the test you set up in part **a**.

c. Make the appropriate conclusion at $\alpha = .01$.

9.57 FCAT scores and poverty. Refer to the *Journal of Educational and Behavioral Statistics* (Spring 2004) study of scores on the Florida Comprehensive Assessment Test (FCAT), first presented in Exercise 9.26 (p. 489). Consider the simple linear regression relating math score (y) to percentage (x) of students below the poverty level. The data are saved in the **FCAT** file.

a. Test whether y is negatively related to x. Use $\alpha = .01$.

b. Construct a 99% confidence interval for β_1. Interpret the result practically.

9.58 Ideal height of your mate. Refer to the *Chance* (Summer 2008) study of the height of the ideal mate, Exercise 9.45 (p. 495). You used the data in the **IDHEIGHT** file to fit the simple linear regression model $E(y) = \beta_0 + \beta_1 x$, where y = ideal partner's height (in inches) and x = student's height (in inches), for both males and females.

a. Find a 90% confidence interval for β_1 in the model for the male students. Give a practical interpretation of the result.

b. Repeat part **a** for the female students.

c. Which group, males or females, has the greater increase in ideal partner's height for every 1 inch increase in student's height?

9.59 Sweetness of orange juice. Refer to Exercise 9.28 (p. 490) and the simple linear regression relating the sweetness index (y) of an orange juice sample to the amount of

SAS Output for Exercise 9.55

Dependent Variable: CATCH

Number of Observations Read	8
Number of Observations Used	8

Analysis of Variance

Source	DF	Sum of Squares	Mean Square	F Value	Pr > F
Model	1	6888299	6888299	6.81	0.0402
Error	6	6071019	1011837		
Corrected Total	7	12959318			

Root MSE	1005.90086	R-Square	0.5315
Dependent Mean	5540.50000	Adj R-Sq	0.4535
Coeff Var	18.15542		

Parameter Estimates

Variable	DF	Parameter Estimate	Standard Error	t Value	Pr > |t|
Intercept	1	9658.24359	1617.76208	5.97	0.0010
SEARCHFREQ	1	-171.57265	65.75780	-2.61	0.0402

water-soluble pectin (x) in the juice. The data are saved in the **OJUICE** file. Find a 95% confidence interval for the true slope of the line. Interpret the result.

Applying the Concepts—Intermediate

9.60 **Effect of massage on boxers.** The *British Journal of Sports Medicine* (Apr. 2000) published a study of the effect of massage on boxing performance. Two variables measured on the boxers were blood lactate concentration (in mM) and the boxer's perceived recovery (on a 28-point scale). On the basis of information provided in the article, the data shown in the accompanying table (and saved in the **BOXING2** file) were obtained for 16 five-round boxing performances in which a massage was given to the boxer between rounds. Conduct a test to determine whether blood lactate level (y) is linearly related to perceived recovery (x). Use $\alpha = .10$.

Blood Lactate Level	Perceived Recovery
3.8	7
4.2	7
4.8	11
4.1	12
5.0	12
5.3	12
4.2	13
2.4	17
3.7	17
5.3	17
5.8	18
6.0	18
5.9	21
6.3	21
5.5	20
6.5	24

Based on Hemmings, B., Smith, M., Graydon, J., and Dyson, R. "Effects of massage on physiological restoration, perceived recovery, and repeated sports performance." *British Journal of Sports Medicine*, Vol. 34, No. 2, Apr. 2000 (data adapted from Table 3).

9.61 **Forest fragmentation study.** Refer to the *Conservation Ecology* (Dec. 2003) study on the causes of fragmentation of 54 South American forests, presented in Exercise 2.156 (p. 91). Recall that researchers developed two fragmentation indexes for each forest—one index for anthropogenic (human development activities) fragmentation and one for fragmentation from natural causes. Data on 5 of the 54 forests saved in the **FORFRAG** file are listed in the following table:

Ecoregion (forest)	Anthropogenic Index, y	Natural Origin Index, x
Araucaria moist forests	34.09	30.08
Atlantic Coast *restingas*	40.87	27.60
Bahia coastal forests	44.75	28.16
Bahia interior forests	37.58	27.44
Bolivian *Yungas*	12.40	16.75

Based on Wade, T. G., et al. "Distribution and causes of global forest fragmentation." *Conservation Ecology*, Vol. 72, No. 2, Dec. 2003 (Table 6).

a. Ecologists theorize that a linear relationship exists between the two fragmentation indexes. Write the model relating y to x.

b. Fit the model to the data in the **FORFRAG** file, using the method of least squares. Give the equation of the least squares prediction equation.

c. Interpret the estimates of β_0 and β_1 in the context of the problem.

d. Is there sufficient evidence to indicate that the natural origin index (x) and the anthropogenic index (y) are positively linearly related? Test, using $\alpha = .05$.

e. Find and interpret a 95% confidence interval for the change in the anthropogenic index (y) for every 1-point increase in the natural origin index (x).

9.62 **Pain empathy and brain activity.** *Empathy* refers to being able to understand and vicariously feel what others actually feel. Neuroscientists at University College of London investigated the relationship between brain activity and pain-related empathy in persons who watch others in pain (*Science*, Feb. 20, 2004). Sixteen couples participated in the experiment. The female partner watched while painful stimulation was applied to the finger of her male partner. Two variables were measured for each female: y = pain-related brain activity (measured on a scale ranging from −2 to 2) and x = score on the Empathic Concern Scale (0 to 25 points). The data are listed in the accompanying table and saved in the **BRAINPAIN** file. The research question of interest was "Do people scoring higher in empathy show higher pain-related brain activity?" Use simple linear regression analysis to answer this question.

Couple	Brain Activity (y)	Empathic Concern (x)
1	.05	12
2	−.03	13
3	.12	14
4	.20	16
5	.35	16
6	0	17
7	.26	17
8	.50	18
9	.20	18
10	.21	18
11	.45	19
12	.30	20
13	.20	21
14	.22	22
15	.76	23
16	.35	24

Based on Singer, T. et al. "Empathy for pain involves the affective but not sensory components of pain." *Science*, Vol. 303, Feb. 20, 2004. (Adapted from Figure 4.)

9.63 **Relation of eye and head movements.** How do eye and head movements relate to body movements when a person reacts to a visual stimulus? Scientists at the California Institute of Technology designed an experiment to answer this question and reported their results in *Nature* (Aug. 1998). Adult male rhesus monkeys were exposed to a visual stimulus (i.e., a panel of light-emitting diodes), and their eye, head, and body movements were electronically recorded. In one variation of the experiment, two variables were measured: active head movement (x, percent per degree) and body-plus-head rotation (y, percent per degree). The data for n = 39 trials were subjected to a simple linear regression analysis, with the following results: $\hat{\beta}_1 = .88, s_{\hat{\beta}_1} = .14$

a. Conduct a test to determine whether the two variables, active head movement x and body-plus-head rotation y, are positively linearly related. Use $\alpha = .05$.

b. Construct and interpret a 90% confidence interval for β_1.

c. The scientists want to know whether the true slope of the line differs significantly from 1. On the basis of your answer to part **b**, make the appropriate inference.

9.64 **The "name game."** Refer to the *Journal of Experimental Psychology—Applied* (June 2000) name-retrieval study, presented in Exercise 9.30 (p. 490). Recall that the goal of the study was to investigate the linear trend between proportion of names recalled (y) and position (order) of the student (x) during the "name game." Is there sufficient evidence (at $\alpha = .01$) of a linear trend? Answer the question by analyzing the data for 144 students saved in the **NAMEGAME2** file.

Applying the Concepts—Advanced

9.65 **Does elevation affect hitting performance in baseball?** Refer to the *Chance* (Winter 2006) investigation of the effects of elevation on slugging percentage in Major League Baseball, Exercise 2.148 (p. 89). Data were compiled on players' composite slugging percentages at each of 29 cities for the 2003 season, as well as on each city's elevation (feet above sea level.) The data are saved in the **MLBPARKS** file. (Selected observations are shown in the table in the next column.) Consider a straight-line model relating slugging percentage (y) to elevation (x).

a. The model was fit to the data with the use of MINITAB, with the results shown in the printout below. Locate the estimates of the model parameters on the printout.

Regression Analysis: SLUGPCT versus ELEVATION

```
The regression equation is
SLUGPCT = 0.515 + 0.000021 ELEVATION

Predictor      Coef      SE Coef      T      P
Constant       0.515140  0.007954   64.76  0.000
ELEVATION   0.00002074  0.00000719   2.89  0.008

S = 0.0369803    R-Sq = 23.6%    R-Sq(adj) = 20.7%

Analysis of Variance

Source          DF      SS        MS       F      P
Regression       1   0.011390  0.011390  8.33  0.008
Residual Error  27   0.036924  0.001368
Total           28   0.048314
```

City	Slug Pct.	Elevation
Anaheim	.480	160
Arlington	.605	616
Atlanta	.530	1,050
Baltimore	.505	130
Boston	.505	20
⋮	⋮	⋮
Denver	.625	5,277
⋮	⋮	⋮
Seattle	.550	350
San Francisco	.510	63
St. Louis	.570	465
Tampa	.500	10
Toronto	.535	566

Based on Schaffer, J., and Heiny, E. L. "The effects of elevation on slugging percentage in Major League Baseball." *Chance*, Vol. 19, No. 1, Winter 2006 (adapted from Figure 2).

b. Is there sufficient evidence (at $\alpha = .01$) of a positive linear relationship between elevation (x) and slugging percentage (y)? Use the *p*-value shown on the printout to make the inference.

c. Construct a scatterplot of the data and draw the least squares line on the graph. Locate the data point for Denver on the graph. What do you observe?

d. Recall that the Colorado Rockies, who play their home games in Denver, are annually among the league leaders in slugging percentage. Baseball experts attribute this to the "thin air" of Denver—called the Mile High city due to its elevation. Remove the data point for Denver from the data set and refit the straight-line model to the remaining data. Repeat parts **a** and **b**. What conclusions can you draw about the "thin air" theory from this analysis?

9.66 **Spreading rate of spilled liquid.** Refer to the *Chemical Engineering Progress* (Jan. 2005) study of the rate at which a spilled volatile liquid will spread across a surface, Exercise 9.31 (p. 491). Recall that the data on mass of the spill and elapsed time is saved in the **LIQUIDSPILL** file. Is there sufficient evidence (at $\alpha = .05$) to indicate that the mass of the spill tends to diminish linearly as elapsed time increases? If so, give an interval estimate (with 95% confidence) of the decrease in spill mass for each minute of elapsed time.

9.5 The Coefficients of Correlation and Determination

In this section, we present two statistics that describe the adequacy of a model: the *coefficient of correlation* and the *coefficient of determination*.

Coefficient of Correlation

Recall (from optional Section 2.9) that a **bivariate relationship** describes a relationship—or correlation—between two variables x and y. Scatterplots are used to describe a bivariate relationship graphically. In this section, we will discuss the concept of **correlation** and how it can be used to measure the linear relationship between two variables x and y. A numerical descriptive measure of correlation is provided by the *coefficient of correlation, r*.

The **coefficient of correlation**,* r, is a measure of the strength of the *linear* relationship between two variables x and y. It is computed (for a sample of n measurements on x and y) as follows:

$$r = \frac{SS_{xy}}{\sqrt{SS_{xx}\,SS_{yy}}}$$

where

$$SS_{xy} = \sum(x - \bar{x})(y - \bar{y}) = \sum xy - \frac{(\sum x)(\sum y)}{n}$$

$$SS_{xx} = \sum(x - \bar{x})^2 = \sum x^2 - \frac{(\sum x)^2}{n}$$

$$SS_{yy} = \sum(y - \bar{y})^2 = \sum y^2 - \frac{(\sum y)^2}{n}$$

Note that the computational formula for the correlation coefficient r given above involves the same quantities that were used in computing the least squares prediction equation. In fact, since the numerators of the expressions for $\hat{\beta}_1$ and r are identical, it is clear that $r = 0$ when $\hat{\beta}_1 = 0$ (the case where x contributes no information for the prediction of y) and that r is when the slope is positive and negative when the slope is negative. Unlike $\hat{\beta}_1$, the correlation coefficient r is *scaleless* and assumes a value between -1 and $+1$, regardless of the units of x and y.

A value of r near or equal to 0 implies little or no linear relationship between y and x. In contrast, the closer r comes to 1 or -1, the stronger is the linear relationship between y and x. And if $r = 1$ or $r = -1$, all the sample points fall exactly on the least squares line. Positive values of r imply a positive linear relationship between y and x; that is, y increases as x increases. Negative values of r imply a negative linear relationship between y and x; that is, y decreases as x increases. Each of these situations is portrayed in Figure 9.14.

Now Work Exercise 9.69

We use the data in Table 9.1 for the drug reaction example to demonstrate how to calculate the coefficient of correlation, r. The quantities needed to calculate r are SS_{xy}, SS_{xx}, and SS_{yy}. The first two quantities have been calculated previously and are repeated here for convenience:

$$SS_{xy} = 7, \quad SS_{xx} = 10, \quad SS_{yy} = \sum y^2 - \frac{(\sum y)^2}{n}$$

$$= 26 - \frac{(10)^2}{5} = 26 - 20 = 6$$

We now find the coefficient of correlation:

$$r = \frac{SS_{xy}}{\sqrt{SS_{xx}\,SS_{yy}}} = \frac{7}{\sqrt{(10)(6)}} = \frac{7}{\sqrt{60}} = .904$$

The fact that r is positive and near 1 indicates that the reaction time tends to increase as the amount of drug in the bloodstream increases—*for the given sample of five subjects.* This is the same conclusion we reached when we found the calculated value of the least squares slope to be positive.

*The value of r is often called the *Pearson correlation coefficient* to honor its developer, Karl Pearson. (See Biography, p. 441).

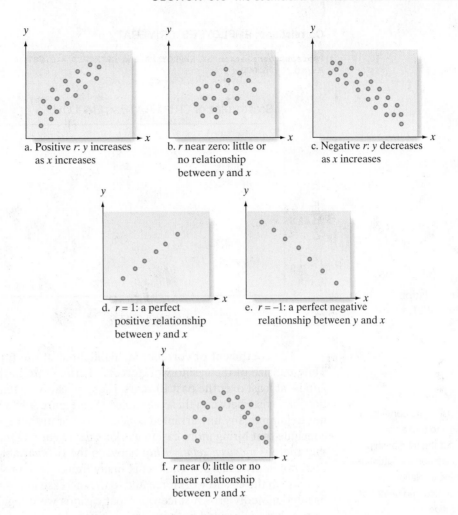

Figure 9.14
Values of r and their implications

a. Positive r: y increases as x increases

b. r near zero: little or no relationship between y and x

c. Negative r: y decreases as x increases

d. $r = 1$: a perfect positive relationship between y and x

e. $r = -1$: a perfect negative relationship between y and x

f. r near 0: little or no linear relationship between y and x

---

Example 9.4
Using the Correlation Coefficient— Relating Crime Rate and Casino Employment

Problem Legalized gambling is available on several riverboat casinos operated by a city in Mississippi. The mayor of the city wants to know the correlation between the number of casino employees and the yearly crime rate. The records for the past 10 years are examined, and the results listed in Table 9.5 are obtained. Calculate the coefficient of correlation, r, for the data. Interpret the result.

Table 9.5	**Data on Casino Employees and Crime Rate, Example 9.4**	
Year	Number x of Casino Employees (thousands)	Crime Rate y (number of crimes per 1,000 population)
2001	15	1.35
2002	18	1.63
2003	24	2.33
2004	22	2.41
2005	25	2.63
2006	29	2.93
2007	30	3.41
2008	32	3.26
2009	35	3.63
2010	38	4.15

Data Set: CASINO

Solution Rather than use the computing formula given earlier, we resort to a statistical software package. The data of Table 9.5 were entered into a computer and MINITAB was used to compute r. The MINITAB printout is shown in Figure 9.15.

Figure 9.15

MINITAB correlation printout and scatterplot for Example 9.4

Ethics IN Statistics

Intentionally using the correlation coefficient only to make an inference about the relationship between two variables in situations where a nonlinear relationship may exist is considered *unethical statistical practice*.

The coefficient of correlation, highlighted at the top of the printout, is $r = .987$. Thus, the size of the casino workforce and crime rate in this city are very highly correlated—at least over the past 10 years. The implication is that a strong positive linear relationship exists between these variables. (See Figure 9.15.) We must be careful, however, not to jump to any unwarranted conclusions. For instance, the mayor may be tempted to conclude that hiring more casino workers next year will increase the crime rate—that is, that there is a *causal relationship* between the two variables. However, high correlation does not imply causality. The fact is, many things have probably contributed both to the increase in the casino workforce and to the increase in crime rate. The city's tourist trade has undoubtedly grown since riverboat casinos were legalized, and it is likely that the casinos have expanded both in services offered and in number. *We cannot infer a causal relationship on the basis of high sample correlation. When a high correlation is observed in the sample data, the only safe conclusion is that a linear trend may exist between x and y.*

Look Back Another variable, such as the increase in tourism, may be the underlying cause of the high correlation between x and y.

Now Work Exercise 9.79a

⚠ **CAUTION** Two caveats apply in using the sample correlation coefficient r to infer the nature of the relationship between x and y: (1) A *high correlation* does not necessarily imply that a causal relationship exists between x and y—only that a linear trend may exist; (2) a *low correlation* does not necessarily imply that x and y are unrelated—only that x and y are not strongly linearly related. ▲

Keep in mind that the correlation coefficient r measures the linear correlation between x values and y values in the sample, and a similar linear coefficient of correlation exists for the population from which the data points were selected. The **population correlation coefficient** is denoted by the symbol ρ (rho). As you might expect, ρ is estimated by the corresponding sample statistic r. Or, instead of estimating ρ, we might want to test the null hypothesis $H_0: \rho = 0$ against $H_a: \rho \neq 0$; that is, we can test the hypothesis that x contributes no information for the prediction of y by using the straight-line model against the alternative that the two variables are at least linearly related.

However, we already performed this *identical* test in Section 9.4 when we tested $H_0: \beta_1 = 0$ against $H_a: \beta_1 \neq 0$. That is, the null hypothesis $H_0: \rho = 0$ is equivalent to the hypothesis $H_0: \beta_1 = 0$.* When we tested the null hypothesis $H_0: \beta_1 = 0$ in connection

*The two tests are equivalent in simple linear regression only.

with the drug reaction example, the data led to a rejection of the null hypothesis at the $\alpha = .05$ level. This rejection implies that the null hypothesis of a 0 linear correlation between the two variables (drug and reaction time) can also be rejected at the $\alpha = .05$ level. The only real difference between the least squares slope $\hat{\beta}_1$ and the coefficient of correlation, r, is the measurement scale. Therefore, the information they provide about the usefulness of the least squares model is to some extent redundant. For this reason, we will use the slope to make inferences about the existence of a positive or negative linear relationship between two variables.

For the sake of completeness, a summary of the test for linear correlation is provided in the following boxes.

A Test for Linear Correlation

One-Tailed Test

$H_0: \rho = 0$

$H_a: \rho > 0 \, (\text{or } H_a: \rho < 0)$

Two-Tailed Test

$H_0: \rho = 0$

$H_a: \rho \neq 0$

$$\text{Test statistic: } t = \frac{r\sqrt{n-2}}{\sqrt{1-r^2}} = \frac{\hat{\beta}}{s_{\hat{\beta}_1}}$$

Rejection region: $t > t_\alpha \, (\text{or } t < -t_\alpha)$ Rejection region: $|t| > t_{\alpha/2}$

where the distribution of t depends on $(n-2)$ df.

Condition Required for a Valid Test of Correlation

The sample of (x, y) values is randomly selected from a normal population.

Coefficient of Determination

Another way to measure the usefulness of a linear model is to measure the contribution of x in predicting y. To accomplish this, we calculate how much the errors of prediction of y were reduced by using the information provided by x. To illustrate, consider the sample shown in the scatterplot of Figure 9.16a. If we assume that x contributes no information for the prediction of y, the best prediction for a value of y is the sample mean $\bar{y}$, which is shown as the horizontal line in Figure 9.16b. The vertical line segments in Figure 9.16b are the deviations of the points about the mean $\bar{y}$. Note that the sum of the squares of the deviations for the prediction equation $\hat{y} = \bar{y}$ is

$$SS_{yy} = \sum(y_i - \bar{y})^2$$

Now suppose you fit a least squares line to the same set of data and locate the deviations of the points about the line, as shown in Figure 9.16c. Compare the deviations about the prediction lines in Figures 9.16b and 9.16c. You can see that

1. If x contributes little or no information for the prediction of y, the sums of the squares of the deviations for the two lines

$$SS_{yy} = \sum(y_i - \bar{y})^2 \quad \text{and} \quad SSE = \sum(y_i - \hat{y}_i)^2$$

will be nearly equal.

2. If x does contribute information for the prediction of y, the SSE will be smaller than SS_{yy}. In fact, if all the points fall on the least squares line, then SSE = 0.

Consequently, the reduction in the sum of the squares of the deviations that can be attributed to x, expressed as a proportion of SS_{yy}, is

$$\frac{SS_{yy} - SSE}{SS_{yy}}$$

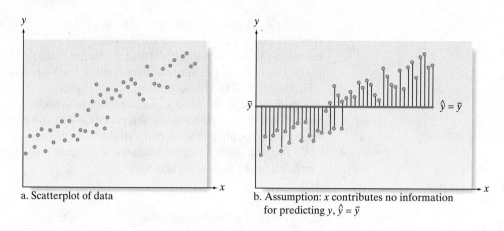

a. Scatterplot of data

b. Assumption: x contributes no information for predicting y, $\hat{y} = \bar{y}$

c. Assumption: x contributes information for predicting y, $\hat{y} = \hat{\beta}_0 + \hat{\beta}_1 x$

Figure 9.16

A comparison of the sum of squares of deviations for two models

Note that SS_{yy} is the "total sample variability" of the observations around the mean $\bar{y}$ and that SSE is the remaining "unexplained sample variability" after fitting the line $\hat{y}$. Thus, the difference $(SS_{yy} - SSE)$ is the "explained sample variability" attributable to the linear relationship with x. Thus, a verbal description of the proportion is

$$\frac{SS_{yy} - SSE}{SS_{yy}} = \frac{\text{Explained sample variability}}{\text{Total sample variability}}$$

$$= \text{Proportion of total sample variability explained by the linear relationship}$$

In simple linear regression, it can be shown that this proportion—called the *coefficient of determination*—is equal to the square of the simple linear coefficient of correlation, r.

> The **coefficient of determination** is
>
> $$r^2 = \frac{SS_{yy} - SSE}{SS_{yy}} = 1 - \frac{SSE}{SS_{yy}}$$
>
> and represents the proportion of the total sample variability around $\bar{y}$ that is explained by the linear relationship between y and x. (In simple linear regression, it may also be computed as the square of the coefficient of correlation, r.)

Note that r^2 is always between 0 and 1, because r is between -1 and $+1$. Thus, an r^2 of .60 means that the sum of the squares of the deviations of the y values about their predicted values has been reduced 60% by the use of the least squares equation $\hat{y}$, instead of $\bar{y}$, to predict y.

Example 9.5
Obtaining the Value of r^2—Drug Reaction Regression

Problem Calculate the coefficient of determination for the drug reaction example. The data are repeated in Table 9.6 for convenience. Interpret the result.

Solution From previous calculations,

$$SS_{yy} = 6 \quad \text{and} \quad SSE = \sum(y - \hat{y})^2 = 1.10$$

Then, from our earlier definition, the coefficient of determination is

$$r^2 = \frac{SS_{yy} - SSE}{SS_{yy}} = \frac{6.0 - 1.1}{6.0} = \frac{4.9}{6.0} = .817$$

Table 9.6

Percent x of Drug	Reaction Time y (seconds)
1	1
2	1
3	2
4	2
5	4

Data Set: STIMULUS

Another way to compute r^2 is to recall from Section 9.5 that $r = .904$. Then we have $r^2 = (.904)^2 = .817$. A third way to obtain r^2 is from a computer printout. Its value is highlighted on the SPSS printout in Figure 9.17. Our interpretation is as follows: We know that using the percent x of drug in the blood to predict y with the least squares line

$$\hat{y} = -.1 + .7x$$

accounts for nearly 82% of the total sum of the squares of the deviations of the five sample y values about their mean. Or, stated another way, 82% of the sample variation in reaction time (y) can be "explained" by using the percent x of drug in a straight-line model.

Figure 9.17
Portion of SPSS printout for time-drug regression

Model Summary

Model	R	R Square	Adjusted R Square	Std. Error of the Estimate
1	.904ᵃ	.817	.756	.606

a. Predictors: (Constant), DRUG_X

Now Work Exercise 9.79b

Practical Interpretation of the Coefficient of Determination, r^2

$100(r^2)$% of the sample variation in y (measured by the total sum of the squares of the deviations of the sample y values about their mean $\bar{y}$) can be explained by (or attributed to) using x to predict y in the straight-line model.

Statistics IN Action | **Revised**

Using the Coefficients of Correlation and Determination to Assess the Dowsing Data

In the previous Statistics in Action Revisited, we discovered that using a dowser's guess (x) in a straight-line model was not statistically useful in predicting actual pipe location (y). Both the coefficient of correlation and the coefficient of determination (highlighted on the MINITAB printouts in Figure SIA9.4) also support this conclusion. The value of the correlation coefficient, $r = .314$, indicates a fairly weak positive linear relationship between the variables. This value, however, is not statistically significant (p-value $= .118$). In other words, there is no evidence to indicate that the population correlation coefficient is different from 0. The coefficient of determination, $r^2 = .099$, implies that only about 10% of the sample variation in pipe location values can be explained by the simple linear model.

Statistics IN Action
(continued)

Regression Analysis: PIPE versus GUESS

```
The regression equation is
PIPE = 30.1 + 0.308 GUESS

Predictor    Coef   SE Coef    T      P
Constant    30.07    11.41   2.63   0.015
GUESS       0.3079   0.1900  1.62   0.118

S = 26.0298    R-Sq = 9.9%    R-Sq(adj) = 6.1%

Analysis of Variance

Source           DF      SS       MS      F      P
Regression        1    1778.9   1778.9   2.63   0.118
Residual Error   24   16261.2    677.6
Total            25   18040.2
```

Correlations: PIPE, GUESS

```
Pearson correlation of PIPE and GUESS = 0.314
P-Value = 0.118
```

Figure SIA9.4

MINITAB printouts with coefficients of correlation and determination for the dowsing data

Exercises 9.67–9.88

Understanding the Principles

9.67 *True or False.* The correlation coefficient is a measure of the strength of the linear relationship between x and y.

9.68 Describe the slope of the least squares line if
a. $r = .7$ **b.** $r = -.7$ **c.** $r = 0$ **d.** $r^2 = .64$

9.69 Explain what each of the following sample correlation
[NW] coefficients tells you about the relationship between the x and y values in the sample:
a. $r = 1$ **b.** $r = -1$
c. $r = 0$ **d.** $r = .90$
e. $r = .10$ **f.** $r = -.88$

9.70 *True or False.* A value of the correlation coefficient near 1 or near -1 implies a causal relationship between x and y.

Learning the Mechanics

9.71 Construct a scatterplot for each data set. Then calculate r and r^2 for each data set.

a.

x	-2	-1	0	1	2
y	-2	1	2	5	6

b.

x	-2	-1	0	1	2
y	6	5	3	2	0

c.

x	1	2	2	3	3	3	4
y	2	1	3	1	2	3	2

d.

x	0	1	3	5	6
y	0	1	2	1	0

9.72 Calculate r^2 for the least squares line in Exercise 9.14 (p. 486).

9.73 Calculate r^2 for the least squares line in Exercise 9.17 (p. 486).

Applet Exercise 9.2

Use the applet entitled *Correlation by the Eye* to explore the relationship between the pattern of data in a scatterplot and the corresponding correlation coefficient.

a. Run the applet several times. Each time, guess the value of the correlation coefficient. Then click *Show r* to see the actual correlation coefficient. How close is your value to the actual value of *r*? Click *New data* to reset the applet.

b. Click the trash can to clear the graph. Use the mouse to place five points on the scatterplot that are approximately in a straight line. Then guess the value of the correlation coefficient. Click *Show r* to see the actual correlation coefficient. How close were you this time?

c. Continue to clear the graph and plot sets of five points with different patterns among the points. Guess the value of *r*. How close do you come to the actual value of *r* each time?

d. On the basis of your experiences with the applet, explain why we need to use more reliable methods of finding the correlation coefficient than just "eyeing" it.

Applying the Concepts—Basic

9.74 **RateMyProfessors.com.** A popular Web site among college students is RateMyProfessors.com (RMP). Established over 10 years ago, RMP allows students to post quantitative ratings of their instructors. In *Practical Assessment, Research & Evaluation* (May 2007), University of Maine researchers investigated whether instructor ratings posted on RMP are correlated with the formal in-class student evaluations of teaching (SET) that all universities are required to administer at the end of the semester. Data collected for $n = 426$ University of Maine instructors yielded a correlation between RMP and SET ratings of .68.

a. Give the equation of a linear model relating SET rating (y) to RMP rating (x).

b. Give a practical interpretation of the value $r = .68$.

c. Is the estimated slope of the line, part **a**, positive or negative? Explain.

d. A test of the null hypothesis $H_0: \rho = 0$ yielded a p-value of .001. Interpret this result.

e. Compute the coefficient of determination, r^2, for the regression analysis. Interpret the result.

9.75 **Going for it on fourth down in the NFL.** Each week coaches in the National Football League (NFL) face a decision during the game. On fourth down, should the team punt the ball or go for a first down? To aid in the decision-making process, statisticians at California State University, Northridge, developed a regression model for predicting the number of points scored (y) by a team that has a first down with a given number of yards (x) from the opposing goal line (*Chance*, Winter 2009). One of the models fit to data collected on five NFL teams from a recent season was the simple linear regression model, $E(y) = \beta_0 + \beta_1 x$. The regression yielded the following results: $\hat{y} = 4.42 - .048\,x, r^2 = .18$.

a. Give a practical interpretation of the coefficient of determination, r^2.

b. Compute the value of the coefficient of correlation, r, from the value of r^2. Is the value of r positive or negative? Why?

9.76 **Lobster fishing study.** Refer to the *Bulletin of Marine Science* (April 2010) study of teams of fishermen fishing for the red spiny lobster in Baja California Sur, Mexico, Exercise 9.55 (p. 501). Recall that simple linear regression was used to model $y = $ total catch of lobsters (in kilograms) during the season as a function of $x = $ average percentage of traps allocated per day to exploring areas of unknown catch (called *search frequency*).

a. Locate and interpret the coefficient of determination, r^2, on the SAS printout shown on p. 576.

b. Note that the coefficient of correlation, r, is not shown on the SAS printout. Is there information on the printout to determine whether total catch (y) is negatively linearly related to search frequency (x)? Explain.

9.77 **Physical activity of obese young adults.** Refer to the *International Journal of Obesity* (Jan. 2007) study of the physical activity of obese young adults, presented in Exercise 4.157 (p. 228). For two groups of young adults — 13 obese and 15 of normal weight — researchers recorded the total number of registered movements (counts) of each young adult over a period of time. *Baseline* physical activity was then computed as the number of counts per minute (cpm). Four years later, physical activity measurements were taken again — called physical activity *at follow-up*.

a. For the 13 obese young adults, the researchers reported a correlation of $r = .50$ between baseline and follow-up physical activity, with an associated p-value of .07. Give a practical interpretation of this correlation coefficient and p-value.

b. Refer to part **a**. Construct a scatterplot of the 13 data points that would yield a value of $r = .50$.

c. For the 15 young adults of normal weight, the researchers reported a correlation of $r = -.12$ between baseline and follow-up physical activity, with an associated p-value of .66. Give a practical interpretation of this correlation coefficient and p-value.

d. Refer to part **c**. Construct a scatterplot of the 15 data points that would yield a value of $r = -12$.

9.78 **Wind turbine blade stress.** Refer to the *Wind Engineering* (Jan. 2004) study of two types of timber—radiata pine and hoop pine—used in high-efficiency small wind turbine blades, presented in Exercise 9.21 (p. 487). Data on stress (y) and the natural logarithm of the number of blade cycles (x) for each type of timber were analyzed by means of simple linear regression. The results are as follows, with additional information on the coefficient of determination:

Radiata Pine: $\hat{y} = 97.37 - 2.50x, r^2 = .84$
Hoop Pine: $\hat{y} = 122.03 - 2.36x, r^2 = .90$

Interpret the value of r^2 for each type of timber.

9.79 **Sports news on local TV broadcasts.** *The Sports Journal* (Winter 2004) published the results of a study conducted to assess the factors that affect the time allotted to sports news on local television news broadcasts. Information on total time (in minutes) allotted to sports and on audience ratings of the TV news broadcast (measured on a 100-point scale) was obtained from a national sample of 163 news directors. A correlation analysis of the data yielded $r = .43$.

a. Interpret the value of the correlation coefficient r.

b. Find and interpret the value of the coefficient of determination r^2.

9.80 **Redshifts of quasi-stellar objects.** Refer to the *Journal of Astrophysics & Astronomy* (Mar./Jun. 2003) study of redshifts in quasi-stellar objects presented in Exercise 9.23 (p. 488). Recall that simple linear regression was used to model the magnitude (y) of a quasi-stellar object as a function of the redshift level (x). In addition to the least squares line, $\hat{y} = 18.13 + 6.21x$, the coefficient of correlation was determined to be $r = .84$.

a. Interpret the value of r in the words of the problem.

b. What is the relationship between r and the estimated slope of the line?

c. Find and interpret the value of r^2.

Applying the Concepts—Intermediate

9.81 **Performance in online courses.** Florida State University information scientists assessed the impact of online courses on student performance (*Educational Technology & Society*, Jan. 2005). Each in a sample of 24 graduate students enrolled in an online advanced Web application course was asked, "How many courses per semester (on average) do you take online?" Each student's performance on weekly quizzes was also recorded. The information scientists found that the number of online courses and the weekly quiz grade were negatively correlated at $r = -.726$.

a. Give a practical interpretation of r.

b. The researchers concluded that there was a "significant negative correlation" between the number of online courses and the weekly quiz grade. Do you agree?

9.82   **Salary linked to height.** Are short people shortchanged when it comes to salary? According to business professors T. A. Judge (University of Florida) and D. M. Cable (University of North Carolina), tall people tend to earn more money over their career than short people earn (*Journal of Applied Psychology*, June 2004). Using data collected from participants in the National Longitudinal Surveys, the researchers computed the correlation between average earnings (in dollars) from 1985 to 2000 and height (in inches) for several occupations. The results are given in the following table:

Occupation	Correlation, r	Sample Size, n
Sales	.41	117
Managers	.35	455
Blue Collar	.32	349
Service Workers	.31	265
Professional/Technical	.30	453
Clerical	.25	358
Crafts/Forepersons	.24	250

Source: Judge, T. A., and Cable, D. M. "The effect of physical height on workplace success and income: Preliminary test of a theoretical model." *Journal of Applied Psychology*, Vol. 89, No. 3, June 2004 (Table 5). Copyright © 2004 by the American Psychological Association. Reprinted with permission.

a. Interpret the value of r for people in sales occupations.
b. Compute r^2 for people in sales occupations. Interpret the result.
c. Give H_0 and H_a for testing whether average earnings and height are positively correlated.
d. Compute the test statistic for testing H_0 and H_a in part c for people in sales occupations.
e. Use the result you obtained in part d to conduct the test at $\alpha = .01$. State the appropriate conclusion.
f. Select another occupation and repeat parts a–e.

9.83   **View of rotated objects.** *Perception & Psychophysics* (July 1998) reported on a study of how people view three-dimensional objects projected onto a rotating two-dimensional image. Each in a sample of 25 university students viewed various depth-rotated objects (e.g., a hairbrush, a duck, and a shoe) until they recognized the object. The recognition exposure time—that is, the minimum time (in milliseconds) required for the subject to recognize the object—was recorded for each object. In addition, each subject rated the "goodness of view" of the object on a numerical scale, with lower scale values corresponding to better views. The following table gives the correlation coefficient r between recognition exposure time and goodness of view for several different rotated objects:

Object	r	t
Piano	.447	2.40
Bench	−.057	.27
Motorbike	.619	3.78
Armchair	.294	1.47
Teapot	.949	14.50

a. Interpret the value of r for each object.
b. Calculate and interpret the value of r^2 for each object.
c. The table also includes the t-value for testing the null hypothesis of no correlation (i.e., for testing H_0: $\beta_1 = 0$). Interpret these results.

9.84   **Snow geese feeding trial.** Botanists at the University of Toronto conducted a series of experiments to investigate the feeding habits of baby snow geese (*Journal of Applied Ecology*, Vol. 32, 1995). Goslings were deprived of food until their guts were empty and then were allowed to feed for 6 hours on a diet of plants or Purina® Duck Chow®. For each feeding trial, the change in the weight of the gosling after 2.5 hours was recorded as a percentage of the bird's initial weight. Two other variables recorded were digestion efficiency (measured as a percentage) and amount of acid-detergent fiber in the digestive tract (also measured as a percentage). Data on 42 feeding trials are saved in the **SNOWGEESE** file. The first and last 5 observations are listed in the table below.

Feeding Trial	Diet	Weight Change (%)	Digestion Efficiency (%)	Acid-Detergent Fiber (%)
1	Plants	−6	0	28.5
2	Plants	−5	2.5	27.5
3	Plants	−4.5	5	27.5
4	Plants	0	0	32.5
5	Plants	2	0	32
⋮	⋮	⋮	⋮	⋮
38	Duck Chow	9	59	8.5
39	Duck Chow	12	52.5	8
40	Duck Chow	8.5	75	6
41	Duck Chow	10.5	72.5	6.5
42	Duck Chow	14	69	7

Based on Gadallah, F. L., and Jefferies, R. L. "Forage quality in brood rearing areas of the lesser snow goose and the growth of captive goslings." *Journal of Applied Biology*, Vol. 32, No. 2, 1995, pp. 281–282 (adapted from Figures 2 and 3).

a. The botanists were interested in the correlation between weight change (y) and digestion efficiency (x). Plot the data for these two variables in a scatterplot. Do you observe a trend?
b. Find the coefficient of correlation relating weight change y to digestion efficiency x. Interpret this value.
c. Conduct a test to determine whether weight change y is correlated with digestion efficiency x. Use $\alpha = .01$.
d. Repeat parts b and c, but exclude the data for trials that used duck chow. What do you conclude?
e. The botanists were also interested in the correlation between digestion efficiency y and acid-detergent fiber x. Repeat parts a–d for these two variables.

9.85   **Do nice guys finish first or last?** Refer to the *Nature* (March 20, 2008) study of the use of punishment in cooperation games, Exercise 9.18 (p. 486). Recall that college students repeatedly played a version of the game "prisoner's dilemma" and the researchers recorded the average payoff (y) and the number of times punishment was used (x) for each player. A negative correlation was discovered between x and y.

a. Give the null and alternative hypothesis for testing whether average payoff and punishment use are negatively correlated.
b. The test, part a, yielded a p-value of .001. Interpret this result.
c. Does the result, part b, imply that increasing punishment causes your payoff to decrease? Explain.

9.86   **The "name game."** Refer to the *Journal of Experimental Psychology—Applied* (June 2000) name-retrieval study, first presented in Exercise 9.30 (p. 490). The data for the study are saved in the **NAMEGAME2** file. Find and interpret the values of r and r^2 for the simple linear regression relating the proportion of names recalled (y) and the position (order) of the student (x) during the "name game."

9.87 **Effect of massage on boxing.** Refer to the *British Journal of Sports Medicine* (April 2000) study of the effect of massage on boxing performance, presented in Exercise 9.60 (p. 502). The data for the study are saved in the **BOXING2** file. Find and interpret the values of r and r^2 for the simple linear regression relating the blood lac-tate concentration and the boxer's perceived recovery.

Applying the Concepts—Advanced

9.88 **Pain tolerance study.** A study published in *Psychosomatic Medicine* (Mar./Apr. 2001) explored the relationship between reported severity of pain and actual pain tolerance in 337 patients who suffer from chronic pain. Each patient reported his or her severity of chronic pain on a seven-point scale ($1 = $ no pain, $7 = $ extreme pain). To obtain a pain tolerance level, a tourniquet was applied to the arm of each patient and twisted. The maximum pain level tolerated was measured on a quantitative scale.

a. According to the researchers, "Correlational analysis revealed a small but significant inverse relationship between [actual] pain tolerance and the reported severity of chronic pain." On the basis of this statement, is the value of r for the 337 patients positive or negative?

b. Suppose that the result reported in part **a** is significant at $\alpha = .05$. Find the approximate value of r for the sample of 337 patients.

9.6 Using the Model for Estimation and Prediction

If we are satisfied that a useful model has been found to describe the relationship between reaction time and percent of drug in the bloodstream, we are ready for step 5 in our regression modeling procedure: using the model for estimation and prediction.

> *The most common uses of a probabilistic model for making inferences can be divided into two categories. The first is the use of the model for estimating the mean value of y, E(y), for a specific value of x.*

For our drug reaction example, we may want to estimate the mean response time for all people whose blood contains 4% of the drug.

> *The second use of the model entails predicting a new individual y value for a given x.*

That is, we may want to predict the reaction time for a specific person who possesses 4% of the drug in the bloodstream.

In the first case, we are attempting to estimate the mean value of y for a very large number of experiments at the given x value. In the second case, we are trying to predict the outcome of a single experiment at the given x value. Which of these uses of the model—estimating the mean value of y or predicting an individual new value of y (for the same value of x)—can be accomplished with the greater accuracy?

Before answering this question, we first consider the problem of choosing an estimator (or predictor) of the mean (or a new individual) y value. We will use the least squares prediction equation

$$\hat{y} = \hat{\beta}_0 + \hat{\beta}_1 x$$

both to estimate the mean value of y and to predict a specific new value of y for a given value of x. For our example, we found that

$$\hat{y} = -.1 + .7x$$

so the estimated mean reaction time for all people when $x = 4$ (the drug is 4% of the blood content) is

$$\hat{y} = -.1 + .7(4) = 2.7 \text{ seconds}$$

The same value is used to predict a new y value when $x = 4$. That is, both the estimated mean and the predicted value of y are $\hat{y} = 2.7$ when $x = 4$, as shown in Figure 9.18.

The difference between these two uses of the model lies in the accuracies of the estimate and the prediction, best measured by the sampling errors of the least squares line when it is used as an estimator and as a predictor, respectively. These errors are reflected in the standard deviations given in the following box:

Figure 9.18

Estimated mean value and predicted individual value of reaction time y for $x = 4$

Sampling Errors for the Estimator of the Mean of y and the Predictor of an Individual New Value of y

1. The standard deviation of the sampling distribution of the estimator $\hat{y}$ of the mean value of y at a specific value of x, say x_p, is

$$\sigma_y = \sigma\sqrt{\frac{1}{n} + \frac{(x_p - \bar{x})^2}{SS_{xx}}}$$

where σ is the standard deviation of the random error ε. We refer to $\sigma_{\hat{y}}$ as the standard error of $\hat{y}$.

2. The standard deviation of the prediction error for the predictor $\hat{y}$ of an individual new y value at a specific value of x is

$$\sigma_{(y-\hat{y})} = \sigma\sqrt{1 + \frac{1}{n} + \frac{(x_p - \bar{x})^2}{SS_{xx}}}$$

where σ is the standard deviation of the random error ε. We refer to $\sigma_{(y-\hat{y})}$ as the standard error of prediction.

The true value of σ is rarely known, so we estimate σ by s and calculate the estimation and prediction intervals as shown in the next two boxes:

A $100(1 - \alpha)$% Confidence Interval for the Mean Value of y at $x = x_p$

$$\hat{y} + t_{\alpha/2}(\text{Estimated standard error of } \hat{y})$$

or

$$\hat{y} \pm t_{\alpha/2}s\sqrt{\frac{1}{n} + \frac{(x_p - \bar{x})^2}{SS_{xx}}}$$

where $t_{\alpha/2}$ is based on $(n - 2)$ degrees of freedom.

A $100(1 - \alpha)$% Prediction Interval* for an Individual New Value of y at $x = x_p$

$$\hat{y} \pm t_{\alpha/2}(\text{Estimated standard error of prediction})$$

or

$$\hat{y} \pm t_{\alpha/2}s\sqrt{1 + \frac{1}{n} + \frac{(x_p - \bar{x})^2}{SS_{xx}}}$$

where $t_{\alpha/2}$ is based on $(n - 2)$ degrees of freedom.

Example 9.6

Estimating the Mean of y—Drug Reaction Regression

Problem Refer to the simple linear regression on drug reaction. Find a 95% confidence interval for the mean reaction time when the concentration of the drug in the bloodstream is 4%.

Solution For a 4% concentration, $x = 4$ and the confidence interval for the mean value of y is

$$\hat{y} \pm t_{\alpha/2}s\sqrt{\frac{1}{n} + \frac{(x_p - \bar{x})^2}{SS_{xx}}} = \hat{y} \pm t_{.025}s\sqrt{\frac{1}{5} + \frac{(4 - \bar{x})^2}{SS_{xx}}}$$

*The term *prediction interval* is used when the interval formed is intended to enclose the value of a random variable. The term *confidence interval* is reserved for the estimation of population parameters (such as the mean).

where $t_{.025}$ is based on $n - 2 = 5 - 2 = 3$ degrees of freedom. Recall that $\hat{y} = 2.7$, $s = .61, \bar{x} = 3$, and $SS_{xx} = 10$. From Table IV in Appendix A, $t_{.025} = 3.182$. Thus, we have

$$2.7 \pm (3.182)(.61)\sqrt{\frac{1}{5} + \frac{(4-3)^2}{10}} = 2.7 \pm (3.182)(.61)(.55)$$
$$= 2.7 \pm (3.182)(.34)$$
$$= 2.7 \pm 1.1$$

Therefore, when the percentage of drug in the bloodstream is 4%, we can be 95% confident that the mean reaction time for all possible subjects will range from 1.6 to 3.8 seconds.

Look Back Note that we used a small amount of data (a small sample size) for purposes of illustration in fitting the least squares line. The interval would probably be narrower if more information had been obtained from a larger sample.

Now Work Exercise 9.94a–d

Example 9.7

Predicting an Individual Value of y— Drug Reaction Regression

Problem Refer again to the drug reaction regression. Predict the reaction time for the next performance of the experiment for a subject with a drug concentration of 4%. Use a 95% prediction interval.

Solution To predict the response time for an individual new subject for whom $x = 4$, we calculate the 95% prediction interval as

$$\hat{y} \pm t_{\alpha/2}s\sqrt{1 + \frac{1}{n} + \frac{(x_p - \bar{x})^2}{SS_{xx}}} = 2.7 \pm (3.182)(.61)\sqrt{1 + \frac{1}{5} + \frac{(4-3)^2}{10}}$$
$$= 2.7 \pm (3.182)(.61)(1.14)$$
$$= 2.7 \pm (3.182)(.70)$$
$$= 2.7 \pm 2.2$$

Therefore, when the drug concentration for an individual is 4%, we predict with 95% confidence that the reaction time for this new individual will fall into the interval from .5 to 4.9 seconds.

Look Back Like the confidence interval for the mean value of y, the prediction interval for y is quite large. This is because we have chosen a simple example (one with only five data points) to fit the least squares line. The width of the prediction interval could be reduced by using a larger number of data points.

Now Work Exercise 9.94e

Both the confidence interval for $E(y)$ and the prediction interval for y can be obtained from a statistical software package. Figure 9.19 is a MINITAB printout showing the confidence interval and prediction interval, respectively, for the data in the drug example.

The 95% confidence interval for $E(y)$ when $x = 4$, highlighted under "95% CI" in Figure 9.19, is (1.645, 3.755). The 95% prediction interval for y when $x = 4$, highlighted in Figure 9.19 under "95% PI," is (.503, 4.897). These agree with the ones computed in Examples 9.6 and 9.7.

Note that the prediction interval for an individual new value of y is *always* wider than the corresponding confidence interval for the mean value of y. Will this always be true? The answer is "Yes." The error in estimating the mean value of y, $E(y)$, for a given value of x, say, x_p, is the distance between the least squares line and the true line of means, $E(y) = \beta_0 + \beta_1 x$. This error, $[\hat{y} - E(y)]$, is shown in Figure 9.20. In contrast, *the error $(y_p - \hat{y})$ in predicting some future value of y is the sum of two errors*: the error

Predicted Values for New Observations

```
New
Obs    Fit   SE Fit      95% CI            95% PI
 1    2.700  0.332    (1.645, 3.755)    (0.503, 4.897)
```

Figure 9.19

MINITAB printout giving 95% confidence interval for $E(y)$ and 95% prediction interval for y

Values of Predictors for New Observations

```
New
Obs   DRUG_X
 1     4.00
```

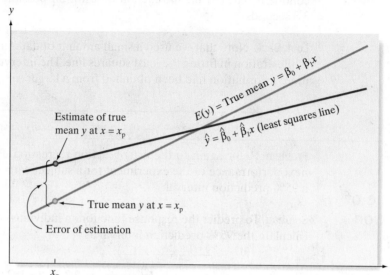

Figure 9.20

Error in estimating the mean value of y for a given value of x

in estimating the mean of y, $E(y)$, shown in Figure 9.20, plus the random error that is a component of the value of y that is to be predicted. (See Figure 9.21.) Consequently, the error in predicting a particular value of y will be larger than the error in estimating the mean value of y for a particular value of x. Note from their formulas that both the error of estimation and the error of prediction take their smallest values when $x_p = \bar{x}$. The farther x_p lies from $\bar{x}$, the larger will be the errors of estimation and prediction. You can see why this is true by noting the deviations for different values of x_p between the actual line of means $E(y) = \beta_0 + \beta_1 x$ and the predicted line of means $\hat{y} = \hat{\beta}_0 + \hat{\beta}_1 x$ shown in Figure 9.21. The deviation is larger at the extremes of the interval, where the largest and smallest values of x in the data set occur.

Both the confidence intervals for mean values and the prediction intervals for new values are depicted over the entire range of the regression line in Figure 9.22. You

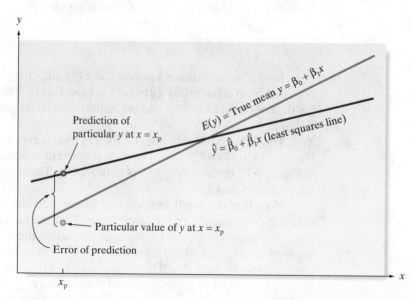

Figure 9.21

Error in predicting a future value of y for a given value of x

Figure 9.22

Confidence intervals for mean values and prediction intervals for new values

can see that the confidence interval is always narrower than the prediction interval and that they are both narrowest at the mean $\bar{x}$, increasing steadily as the distance $|x - \bar{x}|$ increases. In fact, when x is selected far enough away from $\bar{x}$ so that it falls outside the range of the sample data, it is dangerous to make any inferences about $E(y)$ or y.

⚠ **CAUTION** Using the least squares prediction equation to estimate the mean value of y or to predict a particular value of y for values of x that fall outside the range of the values of x contained in your sample data may lead to errors of estimation or prediction that are much larger than expected. Although the least squares model may provide a very good fit to the data over the range of x values contained in the sample, it could give a poor representation of the true model for values of x outside that region. ▲

The width of the confidence interval grows smaller as n is increased; thus, in theory, you can obtain as precise an estimate of the mean value of y as desired (at any given x) by selecting a large enough sample. The prediction interval for a new value of y also grows smaller as n increases, but there is a lower limit on its width. If you examine the formula for the prediction interval, you will see that the interval can get no smaller than $\hat{y} \pm z_{\alpha/2}\sigma$.* Thus, the only way to obtain more accurate predictions for new values of y is to reduce the standard deviation σ of the regression model. This can be accomplished only by improving the model, either by using a curvilinear (rather than linear) relationship with x or by adding new independent variables to the model (or both). Consult the chapter references to learn more about these methods of improving the model.

Now Work Exercise 9.94f

Using the Straight–Line Model to Predict Pipe Location for the Dowsing Data

The group of German physicists who conducted the dowsing experiments stated that the data for the three "best" dowsers empirically support the dowsing theory. If so, then the straight-line model relating a dowser's guess (x) to actual pipe location (y) should yield accurate predictions. The MINITAB printout shown in Figure SIA9.5 gives a 95% prediction interval for y when a dowser guesses $x = 50$ meters (the middle of the 100-meter-long waterpipe). The highlighted interval is $(-9.3, 100.23)$. Thus, we can be 95% confident that the actual

pipe location will fall between -9.3 meters and 100.23 meters for this guess. Since the pipe is only 100 meters long, the interval in effect ranges from 0 to 100 meters—the entire length of the pipe! This result, of course, is due to the fact that the straight-line model is not a statistically useful predictor of pipe location, a fact we discovered in the previous Statistics in Action Revisited sections.

*The result follows from the facts that, for large n, $t_{\alpha/2} \approx z_{\alpha/2}$, $s \approx \sigma$, and the last two terms under the radical in the standard error of the predictor are approximately 0.

Statistics in Action
(continued)

```
Predicted Values for New Observations

New
Obs    Fit   SE Fit      95% CI            95% PI
 1    45.47   5.15   (34.83, 56.10)   (-9.30, 100.23)


Values of Predictors for New Observations

New
Obs   GUESS
 1    50.0
```

Figure SIA9.5
MINITAB prediction interval
for dowsing data

Exercises 9.89–9.107

Understanding the Principles

9.89 Explain the difference between y and $E(y)$ for a given x.

9.90 *True or False.* For a given x, a confidence interval for $E(y)$ will always be wider than a prediction interval for y.

9.91 *True or False.* The greater the deviation between x and $\bar{x}$, the wider the prediction interval for y will be.

9.92 For each of the following, decide whether the proper inference is a prediction interval for y or a confidence interval for $E(y)$:

a. A jeweler wants to predict the selling price of a diamond stone on the basis of its size (number of carats).

b. A psychologist wants to estimate the average IQ of all patients who have a certain income level.

Learning the Mechanics

9.93 In fitting a least squares line to $n = 10$ data points, the following quantities were computed:

$$SS_{xx} = 32, \bar{x} = 3, SS_{yy} = 26, \bar{y} = 4, SS_{xy} = 28$$

a. Find the least squares line.
b. Graph the least squares line.
c. Calculate SSE.
d. Calculate s^2.
e. Find a 95% confidence interval for the mean value of y when $x_p = 2.5$.
f. Find a 95% prediction interval for y when $x_p = 4$.

9.94 Consider the following pairs of measurements saved in the [NW] LM9_94 file.

x	1	2	3	4	5	6	7
y	3	5	4	6	7	7	10

a. Construct a scatterplot of these data.
b. Find the least squares line, and plot it on your scatterplot.
c. Find s^2.
d. Find a 90% confidence interval for the mean value of y when $x = 4$. Plot the upper and lower bounds of the confidence interval on your scatterplot.
e. Find a 90% prediction interval for a new value of y when $x = 4$. Plot the upper and lower bounds of the prediction interval on your scatterplot.

f. Compare the widths of the intervals you constructed in parts **d** and **e**. Which is wider and why?

9.95 Consider the following pairs of measurements, saved in the [NW] LM9_95 file.

x	4	6	0	5	2	3	2	6	2	1
y	3	5	−1	4	3	2	0	4	1	1

For these data, $SS_{xx} = 38.900$, $SS_{yy} = 33.600$, $SS_{xy} = 32.8$, and $\hat{y} = -.414 + .843x$.

a. Construct a scatterplot of the data.
b. Plot the least squares line on your scatterplot.
c. Use a 95% confidence interval to estimate the mean value of y when $x_p = 6$. Plot the upper and lower bounds of the interval on your scatterplot.
d. Repeat part **c** for $x_p = 3.2$ and $x_p = 0$.
e. Compare the widths of the three confidence intervals you constructed in parts **c** and **d,** and explain why they differ.

9.96 Refer to Exercise 9.95.

a. Using no information about x, estimate and calculate a 95% confidence interval for the mean value of y. [*Hint:* Use the one-sample t methodology of Section 7.3.]
b. Plot the estimated mean value and the confidence interval as horizontal lines on your scatterplot.
c. Compare the confidence intervals you calculated in parts **c** and **d** of Exercise 9.95 with the one you calculated in part **a** of this exercise. Does x appear to contribute information about the mean value of y?
d. Check the answer you gave in part **c** with a statistical test of the null hypothesis $H_0: \beta_1 = 0$ against $H_a: \beta_1 \neq 0$. Use $\alpha = .05$.

Applying the Concepts—Basic

9.97 **Do nice guys finish first or last?** Refer to the *Nature* (March 20, 2008) study of the use of punishment in cooperation games, Exercise 9.18 (p. 486). Recall that simple linear regression was used to model a player's average payoff (y) as a straight-line function of the number of times punishment was used (x) by the player.

a. If the researchers want to predict average payoff for a single player who used punishment 10 times, how should they proceed?

SAS output for Exercise 9.99

Predictions

Obs	Region	Rain	Temp	Predicted Rain	Lower prediction limit of Rain	Upper prediction limit of Rain
1	DryStepp	196	5.7	201.977	179.525	224.430
2	DryStepp	196	5.7	201.977	179.525	224.430
3	DryStepp	179	7.0	180.704	163.694	197.714
4	DryStepp	197	8.0	164.340	150.594	178.085
5	DryStepp	149	8.5	156.157	143.513	168.802
6	GobiDese	112	10.7	120.156	106.038	134.274
7	GobiDese	125	11.4	108.701	92.298	125.104
8	GobiDese	99	10.9	116.883	102.172	131.595
9	GobiDese	125	11.4	108.701	92.298	125.104
10	GobiDese	84	11.4	108.701	92.298	125.104
11	GobiDese	115	11.4	108.701	92.298	125.104

b. If the researchers want to estimate the mean of the average payoffs for all players who used punishment 10 times, how should they proceed?

9.98 English as a second language reading ability. Refer to the *Bilingual Research Journal* (Summer 2006) study of the relationship of Spanish (first-language) grammatical knowledge to English (second-language) reading, presented in Exercise 9.54 (p. 500). Recall that three simple linear regressions were used to model the English reading (ESLR) score (y) as a function of Spanish grammar (SG), Spanish reading (SR), and English grammar (EG), respectively.

a. If the researchers want to predict the ELSR score (y) of a native Spanish-speaking adult who scored 50% in Spanish grammar (x), how should they proceed?

b. If the researchers want to estimate the mean ELSR score $E(y)$ of all native Spanish-speaking adults who scored 70% in Spanish grammar (x), how should they proceed?

9.99 Mongolian desert ants. Refer to the *Journal of Biogeography* (Dec. 2003) study of ant sites in Mongolia, presented in Exercise 9.22 (p. 487). You applied the method of least squares to the data in the **GOBIANTS** file to estimate the straight-line model relating annual rainfall (y) and maximum daily temperature (x). A SAS printout giving 95% prediction intervals for the amount of rainfall at each of the 11 sites is shown above. Select the interval associated with site (observation) 7 and interpret it practically.

9.100 Ranking driving performance of professional golfers. Refer to *The Sport Journal* (Winter 2007) study of a new method for ranking the total driving performance of golfers on the Professional Golf Association (PGA) tour, presented in Exercise 9.25 (p. 488). You fit a straight-line model relating diving accuracy (y) to driving distance (x) to the data saved in the **PGADRIVER** file. A MINITAB printout with prediction and confidence intervals for a driving distance of $x = 300$ yards is shown below.

```
Predicted Values for New Observations

New
Obs    Fit   SE Fit      95% CI            95% PI
  1  61.309  0.357  (60.586, 62.032)  (56.724, 65.894)


Values of Predictors for New Observations

New
Obs  DISTANCE
  1    300
```

a. Locate the 95% prediction interval for driving accuracy (y) on the printout, and give a practical interpretation of the result.

b. Locate the 95% prediction interval for mean driving accuracy (y) on the printout, and give a practical interpretation of the result.

c. If you are interested in knowing the average driving accuracy of all PGA golfers who have a driving distance of 300 yards, which of the intervals is relevant? Explain.

9.101 Sweetness of orange juice. Refer to the simple linear regression of sweetness index y and amount of pectin, x, for $n = 24$ orange juice samples, presented in Exercise 9.28 (p. 490). The SPSS printout of the analysis is shown at the top of page 520. A 90% confidence interval for the mean sweetness index $E(y)$ for each value of x is shown on the SPSS spreadsheet on the next page. Select an observation and interpret this interval.

Applying the Concepts—Intermediate

9.102 Sound waves from a basketball. Refer to the *American Journal of Physics* (June 2010) study of sound waves in a spherical cavity, Exercise 9.27 (p. 489). The frequencies of sound waves resulting from the first 24 resonances (echoes) after striking a basketball with a metal rod are saved in the **BBALL** file. You fit a straight-line model relating frequency (y) to number of resonances (x) in Exercise 9.27.

a. Use the model to predict the sound wave frequency for the 10th resonance.

b. Form a 90% confidence interval for the prediction, part **a.** Interpret the result.

c. Suppose you want to predict the sound wave frequency for the 30th resonance. What are the dangers in making this prediction with the fitted model?

9.103 Ideal height of your mate. Refer to the *Chance* (Summer 2008) study of the height of the ideal mate, Exercise 9.29 (p. 490). The data in the **IDHEIGHT** file was used to fit the simple linear regression model, $E(y) = \beta_0 + \beta_1 x$, where y = ideal partner's height (in inches) and x = student's height (in inches). One model was fitted for male students and one model was fitted for female students. Consider a student who is 66 inches tall.

a. If the student is a female, use the model to predict the height of her ideal mate. Form a 95% confidence interval for the prediction and interpret the result.

	run	sweet	pectin	lower90m	upper90m
1	1	5.2	220	5.64898	5.83848
2	2	5.5	227	5.63898	5.81613
3	3	6.0	259	5.57819	5.72904
4	4	5.9	210	5.66194	5.87173
5	5	5.8	224	5.64337	5.82560
6	6	6.0	215	5.65564	5.85493
7	7	5.8	231	5.63284	5.80379
8	8	5.6	268	5.55553	5.71011
9	9	5.6	239	5.61947	5.78019
10	10	5.9	212	5.65946	5.86497
11	11	5.4	410	5.05526	5.55416
12	12	5.6	256	5.58517	5.73592
13	13	5.8	306	5.43785	5.65219
14	14	5.5	259	5.57819	5.72904
15	15	5.3	284	5.50957	5.68213
16	16	5.3	383	5.15725	5.57694
17	17	5.7	271	5.54743	5.70434
18	18	5.5	264	5.56591	5.71821
19	19	5.7	227	5.63898	5.81613
20	20	5.3	263	5.56843	5.72031
21	21	5.9	232	5.63125	5.80075
22	22	5.8	220	5.64898	5.83848
23	23	5.8	246	5.60640	5.76091
24	24	5.9	241	5.61587	5.77454

SPSS output for Exercise 9.101

b. If the student is a male, use the model to predict the height of his ideal mate. Form a 95% confidence interval for the prediction and interpret the result.

c. Which of the two inferences, parts **a** and **b,** may be invalid? Why?

9.104 The "name game." Refer to the *Journal of Experimental Psychology—Applied* (June 2000) name-retrieval study, presented in Exercise 9.30, (p. 490). The data for the study are saved in the **NAMEGAME2** file.

a. Find a 99% confidence interval for the mean recall proportion for students in the fifth position during the "name game." Interpret the result.

b. Find a 99% prediction interval for the recall proportion of a particular student in the fifth position during the "name game." Interpret the result.

c. Compare the intervals you found in parts **a** and **b.** Which interval is wider? Will this always be the case? Explain.

9.105 Spreading rate of spilled liquid. Refer to the *Chemicial Engineering Progress* (Jan. 2005) study of the rate at which a spilled volatile liquid will spread across a surface, presented in Exercise 9.31 (p. 491). Recall that simple linear regression was used to model y = mass of the spill as a function of y = elapsed time of the spill. The data for the study are saved in the **LIQUIDSPILL** file.

a. Find a 99% confidence interval for the mean mass of all spills with an elapsed time of 15 minutes. Interpret the result.

b. Find a 99% prediction interval for the mass of a single spill with an elapsed time of 15 minutes. Interpret the result.

c. Compare the intervals you found in parts **a** and **b.** Which interval is wider? Will this always be the case? Explain.

9.106 Feeding habits of snow geese. Refer to the *Journal of Applied Ecology* feeding study of the relationship between the weight change y of baby snow geese and their digestion efficiency x, presented in Exercise 9.84 (p. 512). The data for the study are saved in the **SNOWGEESE** file.

a. Fit the simple linear regression model to the data.

b. Do you recommend using the model to predict weight change y? Explain.

c. Use the model to form a 95% confidence interval for the mean weight change of all baby snow geese with a digestion efficiency of $x = 15\%$. Interpret the interval.

Applying the Concepts—Advanced

9.107 Life tests of cutting tools. Refer to the data saved in the **CUTTOOL** file of Exercise 9.46 (p. 495).

a. Use a 90% confidence interval to estimate the mean useful life of a brand-A cutting tool when the cutting speed is 45 meters per minute. Repeat for brand B. Compare the widths of the two intervals and comment on the reasons for any difference.

b. Use a 90% prediction interval to predict the useful life of a brand-A cutting tool when the cutting speed is 45 meters per minute. Repeat for brand B. Compare the widths of the two intervals with each other and with the two intervals you calculated in part **a.** Comment on the reasons for any differences.

c. Note that the estimation and prediction you performed in parts **a** and **b** were for a value of x that was not included in the original sample. That is, the value $x = 45$ was not part of the sample. However, the value is within the range of x values in the sample, so that the regression model spans the x value for which the estimation and prediction were made. In such situations, estimation and prediction represent **interpolations.**

Suppose you were asked to predict the useful life of a brand-A cutting tool for a cutting speed of $x = 100$ meters per minute. Since the given value of x is outside the range of the sample x values, the prediction is an example of **extrapolation.** Predict the useful life of a brand-A cutting tool that is operated at 100 meters per minute, and construct a 95% confidence interval for the actual useful life of the tool. What additional assumption do you have to make in order to ensure the validity of an extrapolation?

9.7 A Complete Example

In the previous sections, we presented the basic elements necessary to fit and use a straight-line regression model. In this section, we will assemble these elements by applying them in an example with the aid of computer software.

Suppose a fire insurance company wants to relate the amount of fire damage in major residential fires to the distance between the burning house and the nearest fire station. The study is to be conducted in a large suburb of a major city; a sample of 15 recent fires in this suburb is selected. The amount of damage, y, and the distance between the fire and the nearest fire station, x, are recorded for each fire. The results are given in Table 9.7 and saved in the **FIREDAM** file.

Step 1 First, we hypothesize a model to relate fire damage, y, to the distance from the nearest fire station, x. We hypothesize a straight-line probabilistic model:

$$y = \beta_0 + \beta_1 x + \varepsilon$$

Step 2 Next, we open the **FIREDAM** file and use statistical software to estimate the unknown parameters in the deterministic component of the hypothesized model. The SAS printout for the simple linear regression analysis is shown in Figure 9.23. The least squares estimate of the slope β_1 and intercept β_0, highlighted on the printout, are

$$\hat{\beta}_1 = 4.91933$$
$$\hat{\beta}_0 = 10.27793$$

Table 9.7
Fire Damage Data

Distance from Fire Station, x (miles)	Fire Damage y (thousands of dollars)
3.4	26.2
1.8	17.8
4.6	31.3
2.3	23.1
3.1	27.5
5.5	36.0
.7	14.1
3.0	22.3
2.6	19.6
4.3	31.3
2.1	24.0
1.1	17.3
6.1	43.2
4.8	36.4
3.8	26.1

Data Set: FIREDAM

Dependent Variable: DAMAGE

Analysis of Variance

Source	DF	Sum of Squares	Mean Square	F Value	Pr > F
Model	1	841.76636	841.76636	156.89	<.0001
Error	13	69.75098	5.36546		
Corrected Total	14	911.51733			

Root MSE	2.31635	R-Square	0.9235	
Dependent Mean	26.41333	Adj R-Sq	0.9176	
Coeff Var	8.76961			

Parameter Estimates

Variable	DF	Parameter Estimate	Standard Error	t Value	Pr > |t|	95% Confidence Limits	
Intercept	1	10.27793	1.42028	7.24	<.0001	7.20960	13.34625
DISTANCE	1	4.91933	0.39275	12.53	<.0001	4.07085	5.76781

Output Statistics

Obs	DISTANCE	Dep Var DAMAGE	Predicted Value	Std Error Mean Predict	95% CL Predict		Residual
1	3.4	26.2000	27.0037	0.5999	21.8344	32.1729	-0.8037
2	1.8	17.8000	19.1327	0.8340	13.8141	24.4514	-1.3327
3	4.6	31.3000	32.9068	0.7915	27.6186	38.1951	-1.6068
4	2.3	23.1000	21.5924	0.7112	16.3577	26.8271	1.5076
5	3.1	27.5000	25.5279	0.6022	20.3573	30.6984	1.9721
6	5.5	36.0000	37.3342	1.0573	31.8334	42.8351	-1.3342
7	0.7	14.1000	13.7215	1.1766	8.1087	19.3342	0.3785
8	.3	22.3000	25.0359	0.6081	19.8622	30.2097	-2.7359
9	2.6	19.6000	23.0682	0.6550	17.8678	28.2686	-3.4682
10	4.3	31.3000	31.4311	0.7198	26.1908	36.6713	-0.1311
11	2.1	24.0000	20.6085	0.7566	15.3442	25.8729	3.3915
12	1.1	17.3000	15.6892	1.0444	10.1999	21.1785	1.6108
13	6.1	43.2000	40.2858	1.2587	34.5906	45.9811	2.9142
14	4.8	36.4000	33.8907	0.8450	28.5640	39.2175	2.5093
15	3.8	26.1000	28.9714	0.6320	23.7843	34.1585	-2.8714
16	3.5	.	27.4956	0.6043	22.3239	32.6672	.

Figure 9.23
SAS printout for fire damage regression

and the least squares equation is (rounded)

$$\hat{y} = 10.28 + 4.92x$$

This prediction equation is graphed by MINITAB in Figure 9.24, along with a plot of the data points.

Figure 9.24

MINITAB scatterplot with least squares line for fire damage regression analysis

The least squares estimate of the slope, $\hat{\beta}_1 = 4.92$, implies that the estimated mean damage increases by \$4,920 for each additional mile from the fire station. This interpretation is valid over the range of x, or from .7 to 6.1 miles from the station. The estimated y-intercept, $\hat{\beta}_0 = 10.28$, has the interpretation that a fire 0 miles from the fire station has an estimated mean damage of \$10,280. Although this would seem to apply to the fire station itself, remember that the y-intercept is meaningfully interpretable only if $x = 0$ is within the sampled range of the independent variable. Since $x = 0$ is outside the range, β_0 has no practical interpretation.

Step 3   Now we specify the probability distribution of the random-error component ε. The assumptions about the distribution are identical to those listed in Section 9.3. Although we know that these assumptions are not completely satisfied (they rarely are for practical problems), we are willing to assume that they are approximately satisfied for this example. The estimate of the standard deviation σ of ε, highlighted on the SAS printout, is

$$s = 2.31635$$

This implies that most of the observed fire damage (y) values will fall within approximately $2s = 4.64$ thousand dollars of their respective predicted values when the least squares line is used.

Step 4   We can now check the usefulness of the hypothesized model—in other words, whether x really contributes information for the prediction of y by the straight-line model. First, test the null hypothesis that the slope β_1 is 0—that is, that there is no linear relationship between fire damage and the distance from the nearest fire station—against the alternative hypothesis that fire damage increases as the distance increases. We test

$$H_0: \beta_1 = 0$$
$$H_a: \beta_1 > 0$$

The two-tailed observed significance level for testing $H_a: \beta_1 \neq 0$, highlighted on the SAS printout, is less than .0001. Thus, the p-value for our one-tailed test is less than half of this value (.00005). This small p-value leaves little doubt that mean fire damage and distance between the fire and the fire station are at least linearly related, with mean fire damage increasing as the distance increases.

We gain additional information about the relationship by forming a confidence interval for the slope β_1. A 95% confidence interval, highlighted on the

SAS printout, is (4.071, 5.768). Thus, with 95% confidence, we estimate that the interval from $4,071 to $5,768 encloses the mean increase (β_1) in fire damage per additional mile in distance from the fire station.

Another measure of the utility of the model is the coefficient of determination, r^2. The value (also highlighted on the printout) is $r^2 = .9235$, which implies that about 92% of the sample variation in fire damage (y) is explained by the distance (x) between the fire and the fire station.

The coefficient of correlation, r, that measures the strength of the linear relationship between y and x is not shown on the SAS printout and must be calculated. Using the facts that $r = \sqrt{r^2}$ in simple linear regression and that r and $\hat{\beta}_1$ have the same sign, we calculate

$$r = +\sqrt{r^2} = \sqrt{.9235} = .96$$

The high correlation confirms our conclusion that β_1 is greater than 0; it appears that fire damage and distance from the fire station are positively correlated. All signs point to a strong linear relationship between y and x.

Step 5   We are now prepared to use the least squares model. Suppose the insurance company wants to predict the fire damage if a major residential fire were to occur 3.5 miles from the nearest fire station. The predicted value (highlighted at the bottom of the SAS printout) is $\hat{y} = 27.496$, while the 95% prediction interval (also highlighted) is (22.324, 32.667). Therefore, with 95% confidence, we predict fire damage in a major residential fire 3.5 miles from the nearest station to be between $22,324 and $32,667.

⚠ CAUTION We would not use this model to make predictions for homes less than .7 mile or more than 6.1 miles from the nearest fire station. A look at the data in Table 9.7 reveals that all the x-values fall between .7 and 6.1. It is dangerous to use the model to make predictions outside the region in which the sample data fall. A straight line might not provide a good model for the relationship between the mean value of y and the value of x when stretched over a wider range of x-values. ▲

Exercises 9.108–9.109

Applying the Concepts—Intermediate

9.108 An MBA's work-life balance. The importance of having employees with a healthy work-life balance has been recognized by U.S. companies for decades. Many business schools offer courses that assist MBA students with developing good work-life balance habits and most large companies have developed work-life balance programs for their employees. In April 2005, the Graduate Management Admission Council (GMAC) conducted a survey of over 2,000 MBA alumni to explore the work-life balance issue. (For example, one question asked alumni to state their level of agreement with the statement, "My personal and work demands are overwhelming.") Based on these responses, the GMAC determined a work-life balance scale score for each MBA alumni. Scores ranged from 0 to 100, with lower scores indicating a higher imbalance between work and life. Many other variables, including average number of hours worked per week, were also measured. The data for the work-life balance study are saved in the **GMAC** file. (The first 15 observations are listed in the accompanying table.) Let x = average number of hours worked per week and y = work-life balance scale score for each MBA alumnus. Investigate the link between these two variables by conducting a complete simple linear regression analysis of the data. Summarize your findings in a professional report.

WLB Score	Hours
75.22	50
64.98	45
49.62	50
44.51	55
70.10	50
54.74	60
55.98	55
21.24	60
59.86	50
70.10	50
29.00	70
64.98	45
36.75	40
35.45	40
45.75	50

Based on "Work-life balance: An MBA alumni report." *Graduate Management Admission Council (GMAC) Research Report* (Oct. 13, 2005).

9.109 Legal advertising—does it pay? According to the American Bar Association, there are over one million lawyers competing for your business. To gain a competitive edge, these lawyers are aggressively advertising their services. In fact, Erickson Marketing, Inc., reports that "attorneys are the #1 category of advertising in the Yellow Pages." Does legal advertising really pay? To partially answer this question, consider the case of an actual law firm that specializes in personal injury (PI) cases. The firm spends thousands of dollars each month on advertising. The accompanying table shows the firm's new personal injury cases each month over a 42-month period. Also shown is the total expenditure on advertising each month, and over the previous 6 months. These data are saved in the **LEGALADV** file. Do these data provide support for the hypothesis that increased advertising expenditures are associated with more personal injury cases? Conduct a complete simple linear regression analysis of the data, letting y = new PI cases and x = 6-month cumulative advertising expenditure. Summarize your findings in a professional report.

Month	New PI Cases	6 Months Cumulative Adv. Exp.
7	11	$41,632.74
8	7	$38,227.39
9	13	$39,779.77
10	7	$37,490.22
11	9	$52,225.71
12	8	$56,249.15
13	18	$59,938.03
14	9	$65,250.59
15	25	$66,071.85
16	26	$81,765.94
17	27	$66,895.46
18	12	$71,426.16
19	14	$75,346.40
20	5	$81,589.97
21	22	$78,828.68

Month	New PI Cases	6 Months Cumulative Adv. Exp.
22	15	$78,415.73
23	12	$90,802.77
24	18	$95,689.44
25	20	$83,099.55
26	38	$82,703.75
27	13	$90,484.38
28	18	$102,084.54
29	21	$84,976.99
30	7	$95,314.29
31	16	$115,858.28
32	12	$108,557.00
33	15	$127,693.57
34	18	$122,761.67
35	30	$123,545.67
36	12	$119,388.26
37	30	$134,675.68
38	20	$133,812.93
39	19	$142,417.13
40	29	$149,956.61
41	58	$165,204.46
42	42	$156,725.72
43	24	$146,397.56
44	47	$197,792.64
45	24	$198,460.28
46	14	$206,662.87
47	31	$253,09.27
48	26	$249,496.28

Source: Info Tech, Inc., Gainesville, Florida.

9.8 A Nonparametric Test for Correlation (Optional)

When the simple linear regression assumptions (Section 9.3) are violated (e.g., the random error ε has a highly skewed distribution), an alternative method of analysis may be required. One approach is to apply a nonparametric test for correlation based on ranks.

To illustrate, suppose 10 new paintings are shown to two art critics and each critic ranks the paintings from 1 (best) to 10 (worst). We want to determine whether the critics' ranks are related. Does a correspondence exist between their ratings? If a painting is ranked high by critic 1, is it likely to be ranked high by critic 2? Or do high rankings by one critic correspond to low rankings by the other? That is, are the rankings of the critics *correlated*?

If the rankings are as shown in the "Perfect Agreement" columns of Table 9.8, we immediately notice that the critics agree on the rank of every painting. High ranks correspond to high ranks and low ranks to low ranks. This is an example of a *perfect*

Table 9.8 Rankings of 10 Paintings by Two Critics

Painting	Perfect Agreement		Perfect Disagreement	
	Critic 1	Critic 2	Critic 1	Critic 2
1	4	4	9	2
2	1	1	3	8
3	7	7	5	6
4	5	5	1	10
5	2	2	2	9
6	6	6	10	1
7	8	8	6	5
8	3	3	4	7
9	10	10	8	3
10	9	9	7	4

positive correlation between the ranks. In contrast, if the rankings appear as shown in the "Perfect Disagreement" columns of Table 9.8, then high ranks for one critic correspond to low ranks for the other. This is an example of *perfect negative correlation*.

In practice, you will rarely see perfect positive or perfect negative correlation between the ranks. In fact, it is quite possible for the critics' ranks to appear as shown in Table 9.9. Note that these rankings indicate some agreement between the critics, but not perfect agreement, thus pointing up a need for a measure of rank correlation.

Table 9.9   Rankings of Paintings: Less-than-Perfect Agreement

Painting	Critic 1	Critic 2	Difference between Rank 1 and Rank 2 d	d^2
1	4	5	−1	1
2	1	2	−1	1
3	9	10	−1	1
4	5	6	−1	1
5	2	1	1	1
6	10	9	1	1
7	7	7	0	0
8	3	3	0	0
9	6	4	2	4
10	8	8	0	0
				$\Sigma d^2 = 10$

Spearman's rank correlation coefficient, r_s, provides a measure of correlation between ranks. The formula for this measure of correlation is given in the next box. We also give a formula that is identical to r_s when there are no ties in rankings; this formula provides a good approximation to r_s when the number of ties is small relative to the number of pairs.

Note that if the ranks for the two critics are identical, as in the second and third columns of Table 9.8, the differences between the ranks will all be 0. Thus,

$$r_s = 1 - \frac{6\sum d^2}{n(n^2 - 1)} = 1 - \frac{6(0)}{10(99)} = 1$$

That is, *perfect positive correlation* between the pairs of ranks is characterized by a Spearman correlation coefficient of $r_s = 1$. When the ranks indicate perfect disagreement, as in the fourth and fifth columns of Table 9.8, $\Sigma d_i^2 = 330$ and

$$r_s = 1 - \frac{6(330)}{10(99)} = -1.$$

Thus, *perfect negative correlation* is indicated by $r_s = -1$.

BIOGRAPHY CHARLES E. SPEARMAN (1863–1945)

Spearman's Correlation

London-born Charles Spearman was educated at Leamington College before joining the British Army. After 20 years as a highly decorated officer, Spearman retired from the army and moved to Germany to begin his study of experimental psychology at the University of Leipzig. At the age of 41, he earned his Ph.D. and ultimately became one of the most influential figures in the field of psychology. Spearman was the originator of the classical theory of mental tests and developed the "two-factor" theory of intelligence. These theories were used to develop and support the "Plus-Elevens" tests in England: exams administered to British 11-year-olds that predict whether they should attend a university or a technical school. Spearman was greatly influenced by the works of Francis Galton (p. 477); consequently, he developed a strong statistical background. While conducting his research on intelligence, he proposed the rank-order correlation coefficient—now called "Spearman's correlation coefficient." During his career, Spearman spent time at various universities, including University College (London), Columbia University, Catholic University, and the University of Cairo (Egypt). ∎

Spearman's Rank Correlation Coefficient

$$r_s = \frac{SS_{uv}}{\sqrt{SS_{uu}SS_{vv}}}$$

where

$$SS_{uv} = \sum(u_i - \bar{u})(v_i - \bar{v}) = \sum u_i v_i - \frac{\left(\sum u_i\right)\left(\sum v_i\right)}{n}$$

$$SS_{uu} = \sum(u_i - \bar{u})^2 = \sum u_i^2 - \frac{\left(\sum u_i\right)^2}{n}$$

$$SS_{vv} = \sum(v_i - \bar{v})^2 = \sum v_i^2 - \frac{\left(\sum v_i\right)^2}{n}$$

u_i = Rank of the ith observation in sample 1

v_i = Rank of the ith observation in sample 2

n = Number of pairs of observations (number of observations in each sample)

Shortcut Formula for r_s*

$$r_s = 1 - \frac{6\sum d_i^2}{n(n^2 - 1)}$$

where

$d_i = u_i - v_i$ (difference in the ranks of the ith observations for samples 1 and 2)

n = number of pairs of observations (number of observations in each sample)

For the data of Table 9.9,

$$r_s = 1 - \frac{6\sum d^2}{n(n^2 - 1)} = 1 - \frac{6(10)}{10(99)} = 1 - \frac{6}{99} = .94$$

The fact that r_s is *close* to 1 indicates that the critics tend to agree, but the agreement is not perfect.

The value of r_s always falls between −1 and +1, with +1 indicating perfect positive correlation and −1 indicating a perfect negative correlation. The closer r_s falls to +1 or −1, the greater the correlation between the ranks. Conversely, the nearer r_s is to 0, the less is the correlation.

Note that the concept of correlation implies that two responses are obtained for each experimental unit. In the art critics example, each painting received two ranks (one from each critic) and the objective of the study was to determine the degree of positive correlation between the two rankings. Rank correlation methods can be used to measure the correlation between any pair of variables. If two variables are measured on each of n experimental units, we rank the measurements associated with each variable separately. Ties receive the average of the ranks of the tied observations. Then we calculate the value of r_s for the two rankings. This value measures the rank correlation between the two variables. We illustrate the procedure in Example 9.8.

---

*The shortcut formula is not exact when there are tied measurements, but it is a good approximation when the total number of ties is not large relative to n.

Example 9.8
Spearman's Rank Correlation— Smoking versus Babies' Weights

Problem A study is conducted to investigate the relationship between cigarette smoking during pregnancy and the weights of newborn infants. The 15 women smokers who make up the sample kept accurate records of the number of cigarettes smoked during their pregnancies, and the weights of their children were recorded at birth. The data are given in Table 9.10.

Table 9.10 Data and Calculations for Example 9.8

Woman	Cigarettes per Day	Rank	Baby's Weight (pounds)	Rank	d	d^2
1	12	1	7.7	5	−4	16
2	15	2	8.1	9	−7	49
3	35	13	6.9	4	9	81
4	21	7	8.2	10	−3	9
5	20	5.5	8.6	13.5	−8	64
6	17	3	8.3	11.5	−8.5	72.25
7	19	4	9.4	15	−11	121
8	46	15	7.8	6	9	81
9	20	5.5	8.3	11.5	−6	36
10	25	8.5	5.2	1	7.5	56.25
11	39	14	6.4	3	11	121
12	25	8.5	7.9	7	1.5	2.25
13	30	12	8.0	8	4	16
14	27	10	6.1	2	8	64
15	29	11	8.6	13.5	−2.5	6.25
					Total =	795

Data Set: NEWBORN

a. Calculate and interpret Spearman's rank correlation coefficient for the data.

b. Use a nonparametric test to determine whether level of cigarette smoking and weights of newborns are negatively correlated for all smoking mothers. Use $\alpha = .05$.

Solution

a. We first rank the number of cigarettes smoked per day, assigning a 1 to the smallest number (12) and a 15 to the largest (46). Note that the two ties receive the averages of their respective ranks. Similarly, we assign ranks to the 15 babies' weights. Since the number of ties is relatively small, we will use the shortcut formula to calculate r_s. The differences d between the ranks of the babies' weights and the ranks of the number of cigarettes smoked per day are shown in Table 9.10. The squares of the differences, d^2, are also given. Thus,

$$r_s = 1 - \frac{6\sum d_i^2}{n(n^2 - 1)} = 1 - \frac{6(795)}{15(15^2 - 1)} = 1 - 1.42 = -.42$$

The value of r_s can also be obtained by computer. A SAS printout of the analysis is shown in Figure 9.25. The value of r_s, highlighted on the printout, agrees (except for rounding) with our hand-calculated value, −.42. The negative correlation coefficient indicates that in this sample an increase in the number of cigarettes smoked per day is *associated with* (but is not necessarily the *cause of*) a decrease in the weight of the newborn infant.

Figure 9.25
SAS Spearman correlation printout for Example 9.8

```
            The CORR Procedure

  2  Variables:    CIGARETTES WEIGHT


Spearman Correlation Coefficients, N = 15
        Prob > |r| under H0: Rho=0

                 CIGARETTES        WEIGHT

  CIGARETTES       1.00000       -0.42473
                                  0.1145

  WEIGHT          -0.42473        1.00000
                   0.1145
```

b. If we define ρ as the **population rank correlation coefficient** [i.e., the rank correlation coefficient that could be calculated from all (x, y) values in the population], we can determine whether level of cigarette smoking and weights of newborns are negatively correlated by conducting the following test:

H_0: $\rho = 0$ (no population correlation between ranks)

H_a: $\rho < 0$ (negative population correlation between ranks)

Test statistic: r_s (the *sample* Spearman rank correlation coefficient)

To determine a rejection region, we consult Table XII in Appendix A, which is partially reproduced in Table 9.11. Note that the left-hand column gives values of n, the number of pairs of observations. The entries in the table are values for an upper-tail rejection region, since only positive values are given. Thus, for $n = 15$ and $\alpha = .05$, the value .441 is the boundary of the upper-tailed rejection region, so $P(r_s > .441) = .05$ if H_0: $\rho = 0$ is true. Similarly, for negative values of r_s, we have $P(r_s < -.441) = .05$ if $\rho = 0$. That is, we expect to see $r_s < -.441$ only 5% of the time if there is really no relationship between the ranks of the variables.

Table 9.11 **Reproduction of Part of Table XII in Appendix A: Critical Values of Spearman's Rank Correlation Coefficient**

n	$\alpha = .05$	$\alpha = .025$	$\alpha = .01$	$\alpha = .005$
5	.900	–	–	–
6	.829	.886	.943	–
7	.714	.786	.893	–
8	.643	.738	.833	.881
9	.600	.683	.783	.833
10	.564	.648	.745	.794
11	.523	.623	.736	.818
12	.497	.591	.703	.780
13	.475	.566	.673	.745
14	.457	.545	.646	.716
15	.441	.525	.623	.689
16	.425	.507	.601	.666
17	.412	.490	.582	.645
18	.399	.476	.564	.625
19	.388	.462	.549	.608
20	.377	.450	.534	.591

The lower-tailed rejection region is therefore

Rejection region ($\alpha = .05$): $r_s < -.441$

Since the calculated $r_s = -.42$ is not less than $-.441$, we cannot reject H_0 at the $\alpha = .05$ level of significance. That is, this sample of 15 smoking mothers provides insufficient evidence to conclude that a negative correlation exists between the number of cigarettes smoked and the weight of newborns for the populations of measurements corresponding to all smoking mothers. This does not, of course, mean that no relationship exists. A study using a larger sample of smokers and taking other factors into account (father's weight, sex of newborn child, etc.) would be more likely to reveal whether smoking and the weight of a newborn child are related.

Look Back The two-tailed p-value of the test (.1145) is highlighted on the SAS printout, shown in Figure 9.25. Since the lower-tailed p-value, $.1145/2 = .05725$, exceeds $\alpha = .05$, our conclusion is the same: Do not reject H_0.

Now Work Exercise 9.115

A summary of Spearman's nonparametric test for correlation is given in the following box:

Spearman's Nonparametric Test for Rank Correlation

One-Tailed Test	**Two-Tailed Test**
$H_0: \rho = 0$	$H_0: \rho = 0$
$H_a: \rho > 0 \ [\text{or } H_a: \rho < 0]$	$H_a: \rho \neq 0$

Test statistic: r_s, the sample rank correlation (see the formulas for calculating r_s)

Rejection region: $r_s > r_{s,\alpha}$ $\left[\text{or } r_s < -r_{s,\alpha} \text{ when } H_a: \rho < 0 \right]$	Rejection region: $\lvert r_s \rvert > r_{s,\alpha/2}$
where $r_{s,\alpha}$ is the value from Table XII corresponding to the upper-tail area α and n pairs of observations	where $r_{s,\alpha/2}$ is the value from Table XII corresponding to the upper-tail area $\alpha/2$ and n pairs of observations

Ties: Assign tied measurements the average of the ranks they would receive if they were unequal, but occurred in successive order. For example, if the third-ranked and fourth-ranked measurements are tied, assign each a rank of $(3 + 4)/2 = 3.5$. The number of ties should be small relative to the total number of observations.

Conditions Required for a Valid Spearman's Test

1. The sample of experimental units on which the two variables are measured is randomly selected.
2. The probability distributions of the two variables are continuous.

Exercises 9.110–9.117

Understanding the Principles

9.110 What is the value of r_s when there is perfect negative rank correlation between two variables? Perfect positive rank correlation?

9.111 What conditions are required for a valid Spearman's test?

Learning the Mechanics

9.112 Use Table XII of Appendix A to find each of the following probabilities:
   a. $P(r_s > .508)$ when $n = 22$
   b. $P(r_s > .448)$ when $n = 28$
   c. $P(r_s \leq .648)$ when $n = 10$
   d. $P(r_s < -.738 \text{ or } r_s > .738)$ when $n = 8$

9.113 Specify the rejection region for Spearman's nonparametric test for rank correlation in each of the following situations:
   a. $H_0: \rho = 0, H_a: \rho \neq 0, n = 10, \alpha = .05$
   b. $H_0: \rho = 0, H_a: \rho > 0, n = 20, \alpha = .025$
   c. $H_0: \rho = 0, H_a: \rho < 0, n = 30, \alpha = .01$

9.114 Compute Spearman's rank correlation coefficient for each of the following pairs of sample observations:

a.

x	33	61	20	19	40
y	26	36	65	25	35

b.

x	89	102	120	137	41
y	81	94	75	52	136

c.

x	2	15	4	10
y	11	2	15	21

d.

x	5	20	15	10	3
y	80	83	91	82	87

9.115 The following sample data, saved in the **LM9_115** file, [NW] were collected on variables x and y:

x	0	3	0	−4	3	0	4
y	0	2	2	0	3	1	2

   a. Specify the null and alternative hypotheses that should be used in conducting a hypothesis test to determine whether the variables x and y are correlated.
   b. Conduct the test of part **a,** using $\alpha = .05$.
   c. What is the approximate p-value of the test of part **b**?
   d. What assumptions are necessary to ensure the validity of the test of part **b**?

Applying the Concepts—Basic

9.116 **Mongolian desert ants.** Refer to the *Journal of Biogeography* (Dec. 2003) study of ants in Mongolia, presented in Exercise 9.22 (p. 487). Data on annual rainfall, maximum daily temperature, and number of ant species recorded at each of 11 study sites are reproduced in the table below and saved in the **GOBIANTS** file.

Site	Region	Annual Rainfall (mm)	Max. Daily Temp. (°C)	Number of Ant Species
1	Dry Steppe	196	5.7	3
2	Dry Steppe	196	5.7	3
3	Dry Steppe	179	7.0	52
4	Dry Steppe	197	8.0	7
5	Dry Steppe	149	8.5	5
6	Gobi Desert	112	10.7	49
7	Gobi Desert	125	11.4	5
8	Gobi Desert	99	10.9	4
9	Gobi Desert	125	11.4	4
10	Gobi Desert	84	11.4	5
11	Gobi Desert	115	11.4	4

Based on Pfeiffer, M., et al. "Community organization and species richness of ants in Mongolia along an ecological gradient from steppe to Gobi desert." *Journal of Biogeography*, Vol. 30, No. 12, Dec. 2003 (Tables 1 and 2).

a. Consider the data for the five sites in the Dry Steppe region only. Rank the five annual rainfall amounts. Then rank the five maximum daily temperature values.

b. Use the ranks from part **a** to find and interpret the rank correlation between annual rainfall (y) and maximum daily temperature (x).

c. Repeat parts **a** and **b** for the six sites in the Gobi Desert region.

d. Now consider the rank correlation between the number of ant species (y) and annual rainfall (x). Using all the data, compute and interpret Spearman's rank correlation statistic.

9.117 Extending the life of an aluminum smelter pot. Refer to the *American Ceramic Society Bulletin* (Feb. 2005) study of the lifetime of an aluminum smelter pot, presented in Exercise 9.24 (p. 488). Since the life of a smelter pot depends on the porosity of the brick lining, the researchers measured the apparent porosity and the mean pore diameter of each of six bricks. The data, saved in the **SMELTPOT** file, are reproduced in the following table:

Brick	Apparent Porosity (%)	Mean Pore Diameter (micrometers)
A	18.8	12.0
B	18.3	9.7
C	16.3	7.3
D	6.9	5.3
E	17.1	10.9
F	20.4	16.8

Based on Bonadia, P., et al. "Aluminosilicate refractories for aluminum cell linings." *American Ceramic Society Bulletin*, Vol. 84, No. 2, Feb. 2005 (Table II).

a. Rank the apparent porosity values for the six bricks. Then rank the six pore diameter values.

b. Use the ranks from part **a** to find the rank correlation between apparent porosity (y) and mean pore diameter (x). Interpret the result.

c. Conduct a test for positive rank correlation. Use $\alpha = .01$.

9.118 Lobster fishing study. Refer to the *Bulletin of Marine Science* (April 2010) study of teams of fishermen fishing for the red spiny lobster in Baja California Sur, Mexico, Exercise 9.55 (p. 501). Recall that two variables measured

for each of 8 teams from the Punta Abreojos (PA) fishing cooperative were total catch of lobsters (in kilograms) during the season and average percentage of traps allocated per day to exploring areas of unknown catch (called *search frequency*). These data, saved in the **TRAPSPACE** file, are reproduced in the table.

Total Catch	Search Frequency
2,785	35
6,535	21
6,695	26
4,891	29
4,937	23
5,727	17
7,019	21
5,735	20

From Shester, G. G. "Explaining catch variation among Baja California lobster fishers through spatial analysis of trap-placement decisions." *Bulletin of Marine Science*, Vol. 86, No. 2, April 2010 (Table 1). Reprinted with permission from the University of Miami, *Bulletin of Marine Science*.

a. Rank the total catch values from 1 to 8.

b. Rank the search frequency values from 1 to 8.

c. Use the ranks, parts **a** and **b**, to compute Spearman's rank correlation coefficient.

d. Based on the result, part **c**, is there sufficient evidence to indicate that total catch is negatively rank correlated with search frequency? Test using $\alpha = .05$.

9.119 Effect of massage on boxers. Refer to the *British Journal of Sports Medicine* (Apr. 2000) study of the effect of massaging boxers between rounds, presented in Exercise 9.60 (p. 502). Two variables measured on the boxers were blood lactate level (y) and the boxer's perceived recovery (x). The data for 16 five-round boxing performances are reproduced in the table and saved in the **BOXING2** file.

Blood Lactate Level	Perceived Recovery
3.8	7
4.2	7
4.8	11
4.1	12
5.0	12
5.3	12
4.2	13
2.4	17
3.7	17
5.3	17
5.8	18
6.0	18
5.9	21
6.3	21
5.5	20
6.5	24

Based on Hemmings, B., Smith, M., Graydon, J., and Dyson, R. "Effects of massage on physiological restoration, perceived recovery, and repeated sports performance." *British Journal of Sports Medicine*, Vol. 34, No. 2, Apr. 2000 (data adapted from Table 3).

a. Rank the values of the 16 blood lactate levels.

b. Rank the values of the 16 perceived recovery values.

c. Use the ranks from parts **a** and **b** to compute Spearman's rank correlation coefficient. Give a practical interpretation of the result.

d. Find the rejection region for a test to determine whether y and x are rank correlated. Use $\alpha = .10$.

e. What is the conclusion of the test you conducted in part **d**? State your answer in the words of the problem.

9.120 Assessment of biometric recognition methods. Biometric technologies have been developed to detect or verify an individual's identity. These methods are based on physiological characteristics (called *biometric signatures*), such as facial features, the iris of the eye, fingerprints, the voice, the shape of the hand, and the gait. In *Chance* (Winter 2004), four biometric recognition algorithms were compared. All four were applied to 1,196 biometric signatures, and "match" scores were obtained. The Spearman correlation between match scores for each possible pair of algorithms was determined. The rank correlation matrix is as follows:

Method	I	II	III	IV
I	1	.189	.592	.340
II		1	.205	.324
III			1	.314
IV				1

a. Locate the largest rank correlation and interpret its value.

b. Locate the smallest rank correlation and interpret its value.

9.121 Childhood obesity study. The *Journal of Education and Human Development* (Vol. 3, 2009) published a study of the eating patterns of families of overweight preschool children. The body mass index for each in a sample of 10 overweight children and their parents were determined. These data, saved in the **BMI** file, are reproduced in the table below.

Child	Parent
17.10	24.62
17.15	24.70
17.20	25.70
17.24	25.80
17.25	26.20
17.30	26.30
17.32	26.60
17.40	26.80
17.60	27.20
17.80	27.35

Based on Seal, N., and Seal, J. "Eating patterns of the rural families of overweight preschool children: A pilot study." *Journal of Education and Human Development*, Vol. 3, No. 1, 2009 (Figure 1).

a. Demonstrate that the rank correlation between the BMI values in the sample is $r_s = 1$.

b. Interpret the result, part **a**.

c. Why should a researcher avoid concluding that there is a perfect linear relationship between BMI values in overweight children and their parents?

Applying the Concepts—Intermediate

9.122 The "name game" Refer to the *Journal of Experimental Psychology—Applied* (June 2000) study in which the "name game" was used to help groups of students learn the names of other students in the group, presented in Exercise 9.30 (p. 490). Recall that one goal of the study was to investigate the relationship between proportion y of names recalled by a student and position (order x) of the student during the game. The data for 144 students in the first eight positions are saved in the **NAMEGAME2** file. (The first five and last five observations in the data set are listed below.) A SAS printout follows.

Position	Recall
2	0.04
2	0.37
2	1.00
2	0.99
2	0.79
⋮	⋮
9	0.72
9	0.88
9	0.46
9	0.54
9	0.99

Based on Morris, P. E., and Fritz, C. O. "The name game: Using retrieval practice to improve the learning of names." *Journal of Experimental Psychology—Applied*, Vol. 6, No. 2, June 2000 (data simulated from Figure 1).

```
          The CORR Procedure

    2  Variables:    POSITION RECALL

Spearman Correlation Coefficients, N = 144
     Prob > |r| under H0: Rho=0

                  POSITION        RECALL

  POSITION        1.00000        0.20652
                                 0.0130

  RECALL          0.20652        1.00000
                  0.0130
```

a. To properly apply the parametric test for correlation on the basis of the Pearson coefficient of correlation, r (Section 9.6), both the x and y variables must be normally distributed. Demonstrate that this assumption is violated for these data. What are the consequences of the violation?

b. Find Spearman's rank correlation coefficient on the accompanying SAS printout and interpret its value.

c. Find the observed significance level for testing for zero rank correlation on the SAS printout, and interpret its value.

d. At $\alpha = .05$, is there sufficient evidence of rank correlation between proportion y of names recalled by a student and position (order x) of the student during the game?

9.123 Study of child bipolar disorders. Psychiatric researchers at the University of Pittsburgh Medical Center have developed a new test for measuring manic symptoms in pediatric bipolar patients (*Journal of Child and Adolescent Psychopharmacology*, Dec. 2003). The new test is called the Kiddie Schedule for Affective Disorders and Schizophrenia-Mania Rating Scale (KSADS-MRS). The new test was compared with the standard test, the Clinical Global Impressions—Bipolar Scale (CGI-BP). Both tests were administered to a sample of 18 pediatric patients before and after they were treated for manic symptoms. The changes in the test scores are recorded in the accompanying table and saved in the **MANIA** file.

Patient	Change in KSADS-MRS (%)	Improvement in CGI-BP
1	80	6
2	65	5
3	20	4
4	−15	4
5	−50	4
6	20	3
7	−30	3
8	−70	3
9	−10	2
10	−25	2
11	−35	2
12	−65	2
13	−65	2
14	−70	2
15	−80	2
16	−90	2
17	−95	2
18	−90	1

Based on Axelson, D. et al. "A preliminary study of the Kiddie Schedule for Affective Disorders and Schizophrenia for School-Age Children Mania Rating Scale for children and adolescents." *Journal of Child and Adolescent Psychopharmacology,* Vol. 13, No. 4, Dec. 2003 (Figure 2).

a. The researchers used Spearman's statistic to measure the correlation between the changes in the two test scores. Compute the value of r_s.

b. Is there sufficient evidence (at $\alpha = .05$) of positive rank correlation between the two test score changes in the population of all pediatric patients with manic symptoms?

9.124 Do nice guys finish first or last? Refer to the *Nature* (March 20, 2008) study of whether the saying "nice guys finish last" applies to the competitive corporate world, Exercise 9.18 (p. 486). Recall that college students repeatedly played a version of the game "prisoner's dilemma," where competitors choose cooperation, defection, or costly punishment. At the conclusion of the games, the researchers recorded the average payoff and the number of times punishment was used for each player. The data in the table, saved in the **PUNISH**

Punish	Payoff
0	0.50
1	0.20
2	0.30
3	0.25
4	0.00
5	0.30
6	0.10
8	−0.20
10	0.15
12	−0.30
14	−0.10
16	−0.20
17	−0.25

file, are representative of the data obtained in the study. The researchers concluded that "punishers tend to have lower payoffs." Do you agree? Use Spearman's rank correlation statistic to support your conclusion.

9.125 FCAT scores and poverty. Refer to the *Journal of Educational and Behavioral Statistics* (Spring 2004) analysis of the link between Florida Comprehensive Assessment Test (FCAT) scores and sociodemographic factors, presented in Exercise 9.26 (p. 489). Data on average math and reading FCAT scores of third graders, as well as the percentage of students below the poverty level, for a sample of 22 Florida elementary schools are saved in the **FCAT** file.

a. Compute and interpret Spearman's rank correlation between FCAT math score (y) and percentage (x) of students below the poverty level.

b. Compute and interpret Spearman's rank correlation between FCAT reading score (y) and percentage (x) of students below the poverty level.

c. Determine whether the value of r_s in part **a** would lead you to conclude that FCAT math score and percent below poverty level are negatively rank correlated in the population of all Florida elementary schools. Use $\alpha = .01$ to make your decision.

d. Determine whether the value of r_s in part **b** would lead you to conclude that FCAT reading score and per cent below poverty level are negatively rank correlated in the population of all Florida elementary schools. Use $\alpha = .01$ to make your decision.

9.126 Pain empathy and brain activity. Refer to the *Science* (Feb. 20, 2004) study on the relationship between brain activity and pain-related empathy in persons who watch others in pain, presented in Exercise 9.62 (p. 502). Recall that 16 female partners watched while painful stimulation was applied to the finger of their respective male partners. The two variables of interest were y = female's pain-related brain activity (measured on a scale ranging from −2 to 2) and x = female's score on the Empathic Concern Scale (0 to 25 points). The data, saved in the **BRAINPAIN** file, are reproduced in the

Couple	Brain Activity (y)	Empathic Concern (x)
1	.05	12
2	−.03	13
3	.12	14
4	.20	16
5	.35	16
6	0	17
7	.26	17
8	.50	18
9	.20	18
10	.21	18
11	.45	19
12	.30	20
13	.20	21
14	.22	22
15	.76	23
16	.35	24

Based on Singer, T., et al. "Empathy for pain involves the affective but not sensory components of pain." *Science,* Vol. 303, Feb. 20, 2004. (Adapted from Figure 4.)

accompanying table. Use Spearman's rank correlation test to answer the research question, "Do people scoring higher in empathy show higher pain-related brain activity?"

9.127 Public perceptions of health risks. The *Journal of Experimental Psychology: Learning, Memory, and Cognition* (July 2005) published a study of the ability of people to judge the risk of an infectious disease. The researchers asked German college students to estimate the number of people infected with a certain disease in a typical year. The median estimates, as well as the actual incidence for each in a sample of 24 infections, are reproduced in the accompanying table and saved in the **INFECTION** file.

a. Use graphs to demonstrate that the variables Actual incidence and Estimated incidence are not normally distributed.

b. Recall that the researchers used regression to model the relationship between Actual incidence and Estimated incidence. How does the result you found in part **a** affect this analysis?

c. Find Spearman's correlation coefficient for the two variables. Interpret this value.

d. Refer to part **c.** At $\alpha = .01$, is there a positive association between Actual incidence and Estimated incidence?

Infection	Actual Incidence	Estimated Incidence
Polio	0.25	300
Diphtheria	1	1000
Trachoma	1.75	691
Rabbit Fever	2	200
Cholera	3	17.5
Leprosy	5	0.8
Tetanus	9	1000
Hemorrhagic Fever	10	150
Trichinosis	22	326.5
Undulant Fever	23	146.5
Well's Disease	39	370
Gas Gangrene	98	400
Parrot Fever	119	225
Typhoid	152	200
Q Fever	179	200
Malaria	936	400
Syphilis	1514	1500
Dysentery	1627	1000
Gonorrhea	2926	6000
Meningitis	4019	5000
Tuberculosis	12619	1500
Hepatitis	14889	1000
Gastroenteritis	203864	37000
Botulism	15	37500

Based on Hertwig, R., Pachur, T., and Kurzenhauser, S. "Judgments of risk frequencies: Tests of possible cognitive mechanisms." *Journal of Experimental Psychology: Learning, Memory, and Cognition*, Vol. 31, No. 4, July 2005 (Table 1).

CHAPTER NOTES

Key Terms

Note: Starred (*) items are from the optional Section in this chapter.

Bivariate relationship 503
Coefficient of correlation 504
Coefficient of determination 508
Correlation 503
Dependent (or response) variable 477
Deterministic model 476
Deterministic relationship 476
Errors of prediction 480
Estimated standard error of the least squares slope β_1 496
Estimated standard error of the regression model 492
Extrapolation 521
Independent (or predictor) variable 477
Interpolation 520
Least squares estimates 481

Least squares line (or regression line or least squares prediction equation) 481
Line of means 478
Method of least squares 480
Population correlation coefficient 506
*Population rank correlation coefficient 528
Probabilistic model 476
Probabilistic relationship 476
Random error 476
Regression analysis (modeling) 477
Regression residuals 480
Scatterplot (or scattergram) 480
Slope of the line 478
*Spearman's rank correlation coefficient 525
Straight-line (first-order) model 477
y-intercept of the line 478

Key Symbols/Notation

y	Dependent variable (variable to be predicted)
x	Independent variable (variable used to predict)
$E(y)$	Expected (mean) of y
β_0	y-intercept of true line
β_1	slope of true line
$\hat{\beta}_0$	Least squared estimate of y-intercept
$\hat{\beta}_1$	Least squares estimate of slope
ε	Random error
$\hat{y}$	Predicted value of y for a given x-value
$(y - \hat{y})$	Estimated error of prediction
SSE	Sum of squared errors of prediction
r	Coefficient of correlation
r^2	Coefficient of determination
x_p	Value of x used to predict y
$r^2 = \dfrac{SS_{yy} - SSE}{SS_{yy}}$	Coefficient of determination
$\hat{y} \pm t_{\alpha/2}s\sqrt{\dfrac{1}{n} + \dfrac{(x_p - \bar{x})^2}{SS_{xx}}}$	$(1 - \alpha)100\%$ confidence interval *for* $E(y)$ when $x = x_p$
$\hat{y} \pm t_{\alpha/2}s\sqrt{1 + \dfrac{1}{n} + \dfrac{(x_p - \bar{x})^2}{SS_{xx}}}$	$(1 - \alpha)100\%$ prediction *interval for* y when $x = x_p$
$r_s = 1 - \dfrac{6\sum d_i^2}{n(n^2 - 1)}$	*Spearman's rank correlation coefficient
d_i	*Difference between ranks of ith observations for x and y

Key Ideas

Simple Linear Regression Variables

$y =$ **Dependent** variable (quantitative)

$x =$ **Independent** variable (quantitative)

Method of Least Squares Properties

1. average error of prediction = 0
2. sum of squared errors is minimum

First-order (straight-line) Model

$$E(y) = \beta_0 + \beta_1 x$$

where $E(y)$ = mean of y

β_0 = **y-intercept** of line (point where line intercepts y-axis)

β_1 = **slope** of line (change in y for every one-unit change in x)

Practical Interpretation of *y*-intercept

Predicted y-value when $x = 0$

(no practical interpretation if $x = 0$ is either nonsensical or outside range of sample data)

Practical Interpretation of Slope

Increase (or decrease) in y for every one-unit increase in x

Coefficient of Correlation, *r*

1. ranges between -1 and $+1$
2. measures strength of *linear relationship* between y and x

Guide to Simple Linear Regression

Coefficient of Determination, r^2

1. ranges between 0 and 1
2. measures proportion of sample variation in y "explained" by the model.

Practical Interpretation of Model Standard Deviation s

Ninety-five percent of y-values fall within $2s$ of their respective predicted values

Comparing Intervals in Step 5

Width of *confidence interval for* $E(y)$ will always be **narrower** than width of *prediction interval for y.*

*Nonparametric Test for Rank Correlation

Spearman's test

Supplementary Exercises 9.128–9.154

[Note: Starred (*) exercises are from the optional section in this chapter.]

Understanding the Principles

9.128 Explain the difference between a probabilistic model and a deterministic model.

9.129 Give the general form of a straight-line model for $E(y)$.

9.130 Outline the five steps in a simple linear regression analysis.

9.131 *True or False.* In simple linear regression, about 95% of the y-values in the sample will fall within $2s$ of their respective predicted values.

Learning the Mechanics

9.132 In fitting a least squares line to $n = 15$ data points, the following quantities were computed: $SS_{xx} = 55$, $SS_{yy} = 198$, $SS_{xy} = -88$, $\bar{x} = 1.3$, and $\bar{y} = 35$.
   a. Find the least squares line.
   b. Graph the least squares line.
   c. Calculate SSE.
   d. Calculate s^2.
   e. Find a 90% confidence interval for β_1. Interpret this estimate.
   f. Find a 90% confidence interval for the mean value of y when $x = 15$.
   g. Find a 90% prediction interval for y when $x = 15$.

9.133 Consider the following sample data:

y	5	1	3
x	5	1	3

   a. Construct a scatterplot for the data.
   b. It is possible to find many lines for which $\Sigma(y - \hat{y}) = 0$. For this reason, the criterion $\Sigma(y - \hat{y}) = 0$ is not used to identify the "best-fitting" straight line. Find two lines that have $\Sigma(y - \hat{y}) = 0$.
   c. Find the least squares line.
   d. Compare the value of SSE for the least squares line with that of the two lines you found in part **b.** What principle of least squares is demonstrated by this comparison?

9.134 Consider the following 10 data points saved in the **LM9_134** file.

x	3	5	6	4	3	7	6	5	4	7
y	4	3	2	1	2	3	3	5	4	2

   a. Plot the data on a scatterplot.
   b. Calculate the values of r and r^2.
   c. Is there sufficient evidence to indicate that x and y are linearly correlated? Test at the $\alpha = .10$ level of significance.

*d. Calculate Spearman's rank correlation coefficient for the data. Is there evidence that x and y are rank correlated? Test at $\alpha = .10$.

Applying the Concepts—Basic

9.135 **Arsenic in soil.** In Denver, Colorado, environmentalists have discovered a link between high arsenic levels in soil and a crabgrass killer used in the 1950s and 1960s (*Environmental Science & Technology*, Sept. 1, 2000). The recent discovery was based, in part, on the accompanying scatterplots. The graphs plot the level of the metals cadmium and arsenic, respectively, against the distance from a former smelter plant for samples of soil taken from Denver residential properties.

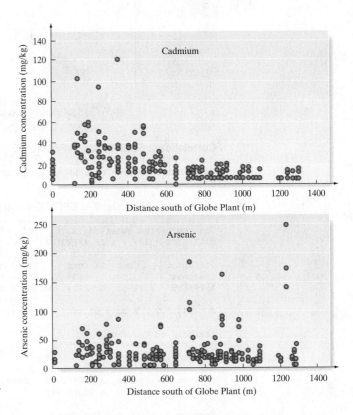

   a. Normally, the metal level in soil decreases as distance from the source (e.g., a smelter plant) increases. Propose a straight-line model relating metal level y to distance x from the plant. On the basis of the theory, would you expect the slope of the line to be positive or negative?

b. Examine the scatterplot for cadmium. Does the plot support the theory you set forth in part **a**?

c. Examine the scatterplot for arsenic. Does the plot support the theory of part **a**? (*Note:* This finding led investigators to discover the link between high arsenic levels and the use of the crabgrass killer.)

9.136 **Predicting sale prices of homes.** Real-estate investors, home buyers, and homeowners often use the appraised value of property as a basis for predicting the sale of that property. Data on sale prices and total appraised value of 78 residential properties sold recently in an upscale Tampa, Florida, neighborhood named Hunter's Green are saved in the **HUNGREEN** file. Selected observations are listed in the accompanying table.

Property	Sale Price	Appraised Value
1	$489,900	$418,601
2	1,825,000	1,577,919
3	890,000	687,836
4	250,00	191,620
5	1,275,000	1,063,901
:	:	:
74	325,000	292,702
75	516,000	407,449
76	309,300	272,275
77	370,000	347,320
78	580,000	511,359

Based on data from Hillsborough Country (Florida) Property Appraiser's Officer.

MINITAB output for Exercise 9.136

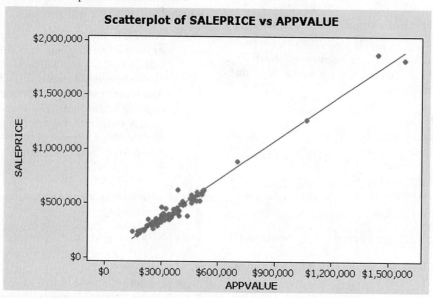

Correlations: SALEPRICE, APPVALUE

```
Pearson correlation of SALEPRICE and APPVALUE = 0.987
P-Value = 0.000
```

Regression Analysis: SALEPRICE versus APPVALUE

```
The regression equation is
SALEPRICE = 184 + 1.20 APPVALUE


Predictor     Coef   SE Coef      T      P
Constant       184      9834   0.02  0.985
APPVALUE   1.19956   0.02234  53.70  0.000


S = 44859.7   R-Sq = 97.4%   R-Sq(adj) = 97.4%

Analysis of Variance

Source          DF           SS           MS        F      P
Regression       1   5.80234E+12  5.80234E+12  2883.31  0.000
Residual Error  76   1.52942E+11   2012390103
Total           77   5.95528E+12


Predicted Values for New Observations

New
Obs     Fit  SE Fit          95% CI              95% PI
  1  480007    5105  (469839, 490175)  (390085, 569930)


Values of Predictors for New Observations

New
Obs  APPVALUE
  1    400000
```

a. Propose a straight-line model to relate the appraised property value (x) to the sale price (y) for residential properties in this neighborhood.

b. A MINITAB scatterplot of the data with the least squared line is shown on p. 536. Does it appear that a straight-line model will be an appropriate fit to the data?

c. A MINITAB simple linear regression printout is also shown (p. 536). Find the equation of the least squared line. Interpret the estimated slope and y-intercept in the words of the problem.

d. Locate the test statistic and p-value for testing H_0: $\beta_1 = 0$ against H_a: $\beta_1 > 0$. Is there sufficient evidence (at $\alpha = .01$) of a positive linear relationship between apprised property value (x) and sale price (y)?

e. Locate and interpret practically the values of r and r^2 on the printout.

f. Locate and interpret practically the 95% prediction interval for sale price (y) on the printout.

9.137 **Baseball batting averages versus wins.** Is the number of games won by a major league baseball team in a season related to the team's batting average? Consider data from the *Baseball Almanac* on the number of games won and the batting averages for the 14 teams in the American League for the 2010 Major League Baseball season. The data are listed in the next table and saved in the **ALWINS** file.

a. If you were to model the relationship between the mean (or expected) number of games won by a major league team and the team's batting average x, using a straight line, would you expect the slope of the line to be positive or negative? Explain.

b. Construct a scatterplot of the data. Does the pattern revealed by the scatterplot agree with your answer to part **a**?

c. An SAS printout of the simple linear regression is shown below. Find the estimates of the β's on the printout and write the equation of the least squares line.

d. Graph the least squares line on your scatterplot. Does your least squares line seem to fit the points on your scatterplot?

Team	Games Won	Batting Avg. (average number of hits per 1,000 at bats)
New York	95	.267
Toronto	85	.248
Baltimore	66	.259
Boston	89	.268
Tampa Bay	96	.247
Cleveland	69	.248
Detroit	81	.268
Chicago	88	.268
Kansas City	67	.274
Minnesota	94	.273
Los Angeles	80	.248
Texas	90	.276
Seattle	61	.236
Oakland	81	.256

Based on data from *Baseball Almanac*, 2010; www.mlb.com.

e. Interpret the estimates of β_0 and β_1 in the words of the problem.

f. Conduct a test (at $\alpha = .05$) to determine whether the mean (or expected) number of games won by a major league baseball team is positively linearly related to the team's batting average.

g. Find the coefficient of determination, r^2, and interpret its value.

h. Do you recommend using the model to predict the number of games won by a team during the 2010 season?

9.138 **College protests of labor exploitation.** Refer to the *Journal of World-Systems Research* (Winter 2004) study of student "sit-ins" for a "sweat-free campus" at universities, presented in Exercise 2.153 (p. 90). Recall that the **SITIN** file contains data on the duration (in days) of each sit-in, as well as the number of student arrests. The data for 5 sit-ins in which there was at least one arrest are shown in the table (p. 548). Let y = number of arrests and x = duration.

a. Give the equation of a straight-line model relating y to x.

b. SPSS was used to fit the model to the data for the 5 sitins. The printout is shown on page 538. Give the least squares prediction equation.

SAS output for Exercise 9.137

Dependent Variable: WINS

Number of Observations Read   14
Number of Observations Used   14

Analysis of Variance

Source	DF	Sum of Squares	Mean Square	F Value	Pr > F
Model	1	270.31115	270.31115	2.18	0.1660
Error	12	1491.11742	124.25979		
Corrected Total	13	1761.42857			

Root MSE	11.14719	R-Square	0.1535	
Dependent Mean	81.57143	Adj R-Sq	0.0829	
Coeff Var	13.66555			

Parameter Estimates

Variable	DF	Parameter Estimate	Standard Error	t Value	Pr > |t|
Intercept	1	-12.62556	63.93554	-0.20	0.8468
BATAVG	1	0.36269	0.24591	1.47	0.1660

SPSS output for Exercise 9.138

Model Summary

Model	R	R Square	Adjusted R Square	Std. Error of the Estimate
1	.601[a]	.361	.148	16.913

a. Predictors: (Constant), DURATION

ANOVA[b]

Model		Sum of Squares	df	Mean Square	F	Sig.
1	Regression	485.026	1	485.026	1.696	.284[a]
	Residual	858.174	3	286.058		
	Total	1343.200	4			

a. Predictors: (Constant), DURATION

b. Dependent Variable: ARRESTS

Coefficients[a]

Model		Unstandardized Coefficients		Standardized Coefficients	t	Sig.
		B	Std. Error	Beta		
1	(Constant)	2.522	16.352		.154	.887
	DURATION	7.261	5.576	.601	1.302	.284

a. Dependent Variable: ARRESTS

Sit-In	University	Duration (days)	Number of Arrests
12	Wisconsin	4	54
14	SUNY Albany	1	11
15	Oregon	3	14
17	Iowa	4	16
18	Kentucky	1	12

Based on Ross, R. J. S. "From antisweatshop to global justice to antiwar: How the new new left is the same and different from the old new left." *Journal of Word-Systems Research*, Vol. X, No. 1, Winter 2004 (Tables 1 and 3).

c. Interpret the estimates of β_0 and β_1 in the context of the problem.

d. Find and interpret the value of s on the printout.

e. Find and interpret the value of r^2 on the printout.

f. Conduct a test to determine whether number of arrests is positively linearly related to duration. (Use $\alpha = .10$.)

9.139 Feeding habits of fish. Refer to the *Brain and Behavior Evolution* (Apr. 2000) study of the feeding behavior of black-bream fish, presented in Exercise 2.150 (p. 89). Recall that the zoologists recorded the number of aggressive strikes of two black-bream fish feeding at the bottom of an aquarium in the 10-minute period following the addition of food. The table listing the weekly number of strikes and the age of the fish (in days) is reproduced in the next column. These data are saved in the **BLACKBREAM** file.

a. Write the equation of a straight-line model relating number of strikes (y) to age of fish (x).

b. Fit the model to the data by the method of least squares and give the least squares prediction equation.

c. Give a practical interpretation of the value of $\hat{\beta}_0$ if possible.

Week	Number of Strikes	Age of Fish (days)
1	85	120
2	63	136
3	34	150
4	39	155
5	58	162
6	35	169
7	57	178
8	12	184
9	15	190

Based on Shand, J., et al. "Variability in the location of the retinal ganglion cell area centralis is correlated with ontogenetic changes in feeding behavior in the Blackbream, Acanthopagrus 'butcher'." *Brain and Behavior*, Vol. 55, No. 4, Apr. 2000 (Figure H).

d. Give a practical interpretation of the value of $\hat{\beta}_1$ if possible.

e. Test $H_0: \beta_1 = 0$ versus $H_a: \beta_1 < 0$, using $\alpha = .10$. Interpret the result.

*f. Find Spearman's rank correlation coefficient relating number of strikes (y) to age (x).

*g. Test whether number of strikes (y) and age (x) are negatively rank correlated. Use $\alpha = .01$.

9.140 English as a second language reading ability. Refer to the *Bilingual Research Journal* (Summer 2006) study of the relationship of Spanish (first-language) grammatical knowledge to English (second-language) reading, presented in Exercise 9.54 (p. 500). Recall that each in a sample of $n = 55$ native Spanish-speaking adults took four standardized exams: Spanish grammar (SG), Spanish reading, (SR), English grammar (EG), and English reading (ESLR). Simple linear regressions were used to model the ESLR score (y) as a function of each of the other exam scores (x). The coefficient of determination, r^2, for each model is listed in the next table. Give a practical interpretation of each of these values.

Independent Variable (x)	r^2
SG score	.002
SR score	.099
EG score	.078

9.141 Removing metal from water. In the *Electronic Journal of Biotechnology* (Apr. 15, 2004), Egyptian scientists studied a new method for removing heavy metals from water. Metal solutions were prepared in glass vessels, and then biosorption was used to remove the metal ions. Two variables were measured for each test vessel: y = metal uptake (milligrams of metal per gram of biosorbent) and x = final concentration of metal in the solution (milligrams per liter).

a. Write a simple linear regression model relating y to x.]

b. For one metal, a simple linear regression analysis yielded r^2 = .92. Interpret this result.

Applying the Concepts—Intermediate

9.142 New method of estimating rainfall. Accurate measurements of rainfall are critical for many hydrological and meteorological projects. Two standard methods of monitoring rainfall use rain gauges and weather radar. Both, however, can be contaminated by human and environmental interference. In the *Journal of Data Science* (Apr. 2004), researchers employed artificial neural networks (i.e., computer-based mathematical models) to estimate rainfall at a meteorological station in Montreal. Rainfall estimates were made every 5 minutes over a 70-minute period by each of the three methods. The data (in millimeters) are listed in the table and saved in the **RAINFALL** file.

Time	Radar	Rain Gauge	Neural Network
8:00 a.m.	3.6	0	1.8
8:05	2.0	1.2	1.8
8:10	1.1	1.2	1.4
8:15	1.3	1.3	1.9
8:20	1.8	1.4	1.7
8:25	2.1	1.4	1.5
8:30	3.2	2.0	2.1
8:35	2.7	2.1	1.0
8:40	2.5	2.5	2.6
8:45	3.5	2.9	2.6
8:50	3.9	4.0	4.0
8:55	3.5	4.9	3.4
9:00 a.m.	6.5	6.2	6.2
9:05	7.3	6.6	7.5
9:10	6.4	7.8	7.2

Based on Hessami, M. et al. "Selection of an artificial neural network model for the post-calibration of weather radar rainfall estimation." *Journal of Data Science*, Vol. 2, No. 2, Apr. 2004. (Adapted from Figures 2 and 4.)

a. Propose a straight-line model relating rain gauge amount (y) to weather radar rain estimate (x).

b. Use the method of least squares to fit the model to the data in the **RAINFALL** file.

c. Graph the least squares line on a scatterplot of the data. Is there visual evidence of a relationship between the two variables? Is the relationship positive or negative?

d. Interpret the estimates of the y-intercept and slope in the words of the problem.

e. Find and interpret the value of s for this regression.

f. Test whether y is linearly related to x. Use α = .01.

g. Construct a 99% confidence interval for β_1. Interpret the result practically.

h. Now consider a model relating rain gauge amount (y) to the artificial neural network rain estimate (x). Repeat parts **a–g** for this model.

9.143 Are geography journals worth their cost? Refer to the *Geoforum* (Vol. 37, 2006) study of whether the price of a geography journal is correlated with quality, presented in Exercise 2.154 (p. 90). Several quantitative variables were recorded for each in a sample of 28 geography journals: cost of a one-year subscription (dollars); journal impact factor (JIF), the average number of times articles from the journal have been cited; number of citations for the journals over the past five years; and relative price index (RPI). The data for the 28 journals are saved in the **GEOJRNL** file. Selected observations are listed in the table.

Journal	Cost ($)	JIF	Citations	RPI
J. Econ. Geogr	468	3.139	207	1.16
Prog. Hum. Geog	624	2.943	544	0.77
T. I. Brit. Geogr	499	2.388	249	1.11
Econ. Geogr.	90	2.325	173	0.30
A. A. A. Geogr.	698	2.115	377	0.93
⋮	⋮	⋮	⋮	⋮
Geogr.Anal.	213	0.902	106	0.88
Geogr.J.	223	0.857	81	0.94
Appl.Geogr	646	0.853	74	3.38

Based on Blomley, N. "Is this journal worth US$1118?" *Geoforum*, Vol. 37, No. 6, p. 118 (Tables 1, 2, and 3).

a. Fit a straight-line model relating cost (y) to JIF (x). Give a practical interpretation of the estimated slope of the line.

b. Within how many dollars can you expect to predict cost?

c. Find and interpret a 95% confidence interval for the slope.

d. Repeat parts a–c for a line relating cost (y) to number of citations (x).

e. Repeat parts a–c for a line relating cost (y) to RPI (x).

9.144 Organic chemistry experiment. Chemists at Kyushu University (Japan) examined the linear relationship between the maximum absorption rate y (in nanomoles) and the Hammett substituent constant x for metacyclophane compounds (*Journal of Organic Chemistry*, July 1995). The data for variants of two compounds are given in the next table and saved in the **ORGCHEM** file. The variants of compound 1 are labeled 1a, 1b, 1d, 1e, 1f, 1g, and 1h; the variants of compound 2 are 2a, 2b, 2c, and 2d.

a. Plot the data in a scatterplot. Use two different plotting symbols for the two compounds. What do you observe?

b. Using only the data for compound 1, fit the model $E(y) = \beta_0 + \beta_1 x$.

c. Assess the adequacy of the model you fit in part **b**. Use α = .01.

d. Repeat parts **b** and **c**, using only the data for compound 2.

*e. Conduct Spearman's test for rank correlation between x and y.

Data for Exercise 9.144

Compound	Maximum Absorption y	Hammett Constant x
1a	298	0.00
1b	346	.75
1d	303	.06
1e	314	−.26
1f	302	.18
1g	332	.42
1h	302	−.19
2a	343	.52
2b	367	1.01
2c	325	.37
2d	331	.53

Source: Tsuge, A., et al. "Preparation and spectral properties of disubstituted [2-2] metacyclophanes." *Journal of Organic Chemistry,* Vol. 60, No. 15, July 1995, pp. 4390–4391 (Table 1 and Figure 1). Reprinted with permission from the American Chemical Society.

9.145 Mortality of predatory birds. Two species of predatory birds—collard flycatchers and tits—compete for nest holes during breeding season on the island of Gotland, Sweden. Frequently, dead flycatchers are found in nest boxes occupied by tits. A field study examined whether the risk of mortality to flycatchers is related to the degree of competition between the two bird species for nest sites (*The Condor,* May 1995). The table below gives data on the number y of flycatchers killed at each of 14 discrete locations (plots) on the island, as well as on the nest box tit occupancy x (i.e., the percentage of nest boxes occupied by tits) at each plot. These data are saved in the **CONDOR2** file. Consider the simple linear regression model $E(y) = \beta_0 + \beta_1 x$.

Plot	Number of Flycatchers Killed y	Nest Box Tit Occupancy x (%)
1	0	24
2	0	33
3	0	34
4	0	43
5	0	50
6	1	35
7	1	35
8	1	38
9	1	40
10	2	31
11	2	43
12	3	55
13	4	57
14	5	64

Based on Merila, J., and Wiggins, D. A. "Interspecific competition for nest holes causes adult mortality in the collard flycatcher." *The Condor,* Vol. 97, No. 2, May 1995, p. 447.

a. Plot the data in a scatterplot. Does the frequency of flycatcher casualties per plot appear to increase linearly with increasing proportion of nest boxes occupied by tits?

b. Use the method of least squares to find the estimates of β_0 and β_1. Interpret their values.

c. Test the utility of the model, using $\alpha = .05$.

d. Find r and r^2 and interpret their values.

e. Find s and interpret the result.

f. Do you recommend using the model to predict the number of flycatchers killed? Explain.

9.146 Winning marathon times. In *Chance* (Winter 2000), statistician Howard Wainer and two students compared men's and women's winning times in the Boston Marathon. One of the graphs used to illustrate gender differences is reproduced below. The scatterplot graphs the winning times (in minutes) against the year in which the race was run. Men's times are represented by purple dots and women's times by red dots.

a. Consider only the winning times for men. Is there evidence of a linear trend? If so, propose a straight-line model for predicting winning time (y) based on year (x). Would you expect the slope of this line to be positive or negative?

b. Repeat part b for women's times.

c. Which slope, the men's or the women's, will be greater in absolute value?

d. Would you recommend using the straight-line models to predict the winning time in the 2020 Boston Marathon? Why or why not?

e. Which model, the men's or the women's, is likely to have the smallest estimate of σ?

Winning Times in the Boston Marathon for Men and Women

9.147 Quantum tunneling. At temperatures approaching absolute zero (−273°C), helium exhibits traits that seem to defy many laws of Newtonian physics. An experiment has been conducted with helium in solid form at various temperatures near absolute zero. The solid helium is placed in a dilution refrigerator along with a solid impure substance, and the fraction (in weight) of the impurity passing through the solid helium is recorded. (This phenomenon of solids passing directly through solids is known as *quantum tunneling*.) The data are given in the next table and saved in the **HELIUM** file.

a. Find the least squares estimates of the intercept and slope. Interpret them.

b. Use a 95% confidence interval to estimate the slope β_1. Interpret the interval in terms of this application. Does the interval support the hypothesis that temperature contributes information about the proportion of impurity passing through helium?

c. Interpret the coefficient of determination for this model.

Data for Exercise 9.147

Temperature x (°C)	Proportion of Impurity
−262.0	.315
−265.0	.202
−256.0	.204
−267.0	.620
−270.0	.715
−272.0	.935
−272.4	.957
−272.7	.906
−272.8	.985
−272.9	.987

d. Find a 95% prediction interval for the percentage of impurity passing through solid helium at −273°C. Interpret the result.

e. Note that the value of x in part **d** is outside the experimental region. Why might this lead to an unreliable prediction?

9.148 **Dance/movement therapy.** In cotherapy, two or more therapists lead a group. An article in the *American Journal of Dance Therapy* (Spring/Summer 1995) examined the use of cotherapy in dance/movement therapy. Two of several variables measured on each of a sample of 136 professional dance/movement therapists were years x of formal training and reported success rate y (measured as a percentage) of coleading dance/movement therapy groups.

a. Propose a linear model relating y to x.

b. The researcher hypothesized that dance/movement therapists with more years in formal dance training will report higher perceived success rates in cotherapy relationships. State the hypothesis in terms of the parameter of the model you proposed in part **a**.

c. The correlation coefficient for the sample data was reported as $r = -.26$. Interpret this result.

d. Does the value of r in part **c** support the hypothesis in part **b**? Test, using $\alpha = .05$.

9.149 **Conversing with the hearing impaired.** A study was conducted to investigate how people with a hearing impairment communicate with their conversational partners. (*Journal of the Academy of Rehabilitative Audiology*, Vol. 27, 1994). Each of 13 hearing-impaired subjects, all fitted with a cochlear implant, participated in a structured communication interaction with a familiar conversational partner

Subject	Words with Familiar Partner x	Words with Unfamiliar Partner y
1	65	47
2	160	78
3	55	90
4	83	75
5	0	6
6	140	101
7	49	40
8	164	215
9	62	29
10	56	75
11	207	121
12	207	139
13	93	83

Based on Tye-Murray, N., et al. "Communication breakdowns: Partner contingencies and partner reactions." *Journal of the Academy of Rehabilitative Audiology*, Vol. 27, 1994, pp. 116–117 (Tables 6, 7).

(a family member) and with an unfamiliar conversational partner (who was instructed not to take the initiative to repair breakdowns in communication). The total number of words used by the subject in each of the two conversations is given in the accompanying table and saved in the **HEARAID** file.

a. Plot the data in a scatterplot. Is there visual evidence of a linear relationship between x and y? If so, is it positive or negative?

b. Propose a straight-line model relating y to x.

c. Use the method of least squares to find the estimates of β_0 and β_1.

d. Interpret the values of $\hat{\beta}_0$ and $\hat{\beta}_1$.

9.150 **Sports participation survey.** The Sasakawa Sports Foundation conducted a national survey to assess the physical activity patterns of Japanese adults. The next table lists the frequency (average number of days in the past year) and the amount of time (average number of minutes per single activity) Japanese adults spent participating in a sample of 11 sports activities. The data are saved in the **JAPANSPORTS** file.

a. Write the equation of a straight-line model relating duration (y) to frequency (x).

b. Find the least squares prediction equation.

c. Is there evidence of a linear relationship between y and x? Test, using $\alpha = .05$.

d. Use the least squares line to predict the amount of time Japanese adults participate in a sport that they play 25 times a year. Form a 95% confidence interval around the prediction and interpret the result.

Activity	Frequency x (days/years)	Amount of Time y (minutes)
Jogging	135	43
Cycling	68	99
Aerobics	44	61
Swimming	39	60
Volleyball	30	80
Tennis	21	100
Softball	16	91
Baseball	19	127
Skating	7	115
Skiing	10	249
Golf	5	262

Based on J. Bennett, ed. *Statistics in Sport*. London: Arnold, 1998 (adapted from Figure 11.6).

Applying the Concepts—Advanced

9.151 **Regression through the origin.** Sometimes it is known from theoretical considerations that the straight-line relationship between two variables x and y passes through the origin of the xy-plane. Consider the relationship between the total weight y of a shipment of 50-pound bags of flour and the number x of bags in the shipment. Since a shipment containing $x = 0$ bags (i.e., no shipment at all) has a total weight of $y = 0$, a straight-line model of the relationship between x and y should pass through the point $x = 0$, $y = 0$. In such a case, you could assume that $\beta_0 = 0$ and characterize the relationship between x and y with the following model:

$$y = \beta_1 x + \varepsilon$$

The least squares estimate of β_1 for this model is

$$\hat{\beta}_1 = \frac{\Sigma x_i y_i}{\Sigma x_i^2}$$

From the records of past flour shipments, 15 shipments were randomly chosen and the data shown in the following table were recorded. These data are saved in the **FLOUR** file.

Weight of Shipment	Number of 50-Pound Bags in Shipment
5,050	100
10,249	205
20,000	450
7,420	150
24,685	500
10,206	200
7,325	150
4,958	100
7,162	150
24,000	500
4,900	100
14,501	300
28,000	600
17,002	400
16,100	400

a. Find the least squares line for the given data under the assumption that $\beta_0 = 0$. Plot the least squares line on a scatterplot of the data.

b. Find the least squares line for the given data, using the model

$$y = \beta_0 + \beta_1 x + \varepsilon$$

(i.e., do not restrict β_0 to equal 0). Plot this line on the same scatterplot you constructed in part **a.**

c. Refer to part **b.** Why might $\hat{\beta}_0$ be different from 0 even though the true value of β_0 is known to be 0?

d. The estimated standard error of $\hat{\beta}_0$ is equal to

$$s\sqrt{\frac{1}{n} + \frac{\bar{x}^2}{SS_{xx}}}$$

Use the t-statistic

$$t = \frac{\hat{\beta}_0 - 0}{s\sqrt{(1/n) + (\bar{x}^2/SS_{xx})}}$$

to test the null hypothesis $H_0: \beta_0 = 0$ against the alternative $H_a: \beta_0 \neq 0$. Take $\alpha = .10$. Should you include β_0 in your model?

9.152 Long-jump "takeoff error." The long jump is a track-and-field event in which a competitor attempts to jump a maximum distance into a sandpit after a running start. At the edge of the pit is a takeoff board. Jumpers usually try to plant their toes at the front edge of this board to maximize their jumping distance. The absolute distance between the front edge of the takeoff board and the spot where the toe actually lands on the board prior to jumping is called "takeoff error." Is takeoff error in the long jump linearly related to best jumping distance? To answer this question, kinesiology researchers videotaped the performances of 18 novice long jumpers at a high school track meet. (*Journal of Applied Biomechanics*, May 1995). The average takeoff error x and the best jumping distance y (out of three jumps) for each jumper are recorded in the accompanying table and saved in the **LONGJUMP** file. If a jumper can reduce his or her average takeoff error by .1 meter, how much would you estimate the jumper's best jumping distance to change? On the basis of your answer, comment on the usefulness of the model for predicting best jumping distance.

Jumper	Best Jumping Distance y (meters)	Average Takeoff Error x (meters)
1	5.30	.09
2	5.55	.17
3	5.47	.19
4	5.45	.24
5	5.07	.16
6	5.32	.22
7	6.15	.09
8	4.70	.12
9	5.22	.09
10	5.77	.09
11	5.12	.13
12	5.77	.16
13	6.22	.03
14	5.82	.50
15	5.15	.13
16	4.92	.04
17	5.20	.07
18	5.42	.04

Based on Berg, W. P., and Greer, N. L. "A kinematic profile of the approach run of novice long jumpers." *Journal of Applied Biomechanics*, Vol. 11, No. 2, May 1995, p. 147 (Table 1).

Critical Thinking Challenges

9.153 Study of fertility rates. The fertility rate of a country is defined as the number of children a woman citizen bears, on average, in her lifetime. *Scientific American* (Dec. 1993) reported on the declining fertility rate in developing countries. The researchers found that family planning can have a great effect on fertility rate. The table (p. 543) gives the fertility rate y and contraceptive prevalence x (measured as the percentage of married women who use contraception) for each of 27 developing countries. These data are saved in the **FERTRATE** file.

a. According to the researchers, "The data reveal that differences in contraceptive prevalence explain about 90% of the variation in fertility rates." Do you concur?

b. The researchers also concluded that "if contraceptive use increases by 18 percent, women bear, on average, one fewer child." Is this statement supported by the data? Explain.

9.154 Spall damage in bricks. A recent civil suit revolved around a five-building brick apartment complex located in the Bronx, New York, which began to suffer *spalling* damage (i.e., a separation of some portion of the face of a brick from its body). The owner of the complex alleged that the bricks were manufactured defectively. The brick manufacturer countered that poor design and shoddy

Data for Exercise 9.153

Country	Contraceptive Prevalence x	Fertility Rate y
Mauritius	76	2.2
Thailand	69	2.3
Colombia	66	2.9
Costa Rica	71	3.5
Sri Lanka	63	2.7
Turkey	62	3.4
Peru	60	3.5
Mexico	55	4.0
Jamaica	55	2.9
Indonesia	50	3.1
Tunisia	51	4.3
El Salvador	48	4.5
Morocco	42	4.0
Zimbabwe	46	5.4
Egypt	40	4.5
Bangladesh	40	5.5
Botswana	35	4.8
Jordan	35	5.5
Kenya	28	6.5
Guatemala	24	5.5
Cameroon	16	5.8
Ghana	14	6.0
Pakistan	13	5.0
Senegal	13	6.5
Sudan	10	4.8
Yemen	9	7.0
Nigeria	7	5.7

Based on Robey, B., et al. "The fertility decline in developing countries." *Scientific American*, Dec. 1993, p. 62. [*Note:* The data values are estimated from a scatterplot.]

management led to the damage. To settle the suit, an estimate of the rate of damage per 1,000 bricks, called the spall rate, was required (*Chance*, Summer 1994). The owner estimated the spall rate by using several *scaffold-drop* surveys. (With this method, an engineer lowers a scaffold down at selected places on building walls and counts the number of visible spalls for every 1,000 bricks in the observation area.) The brick manufacturer conducted its own survey by dividing the walls of the complex into 83 wall segments and taking a photograph of each

one. (The number of spalled bricks that could be made out from each photo was recorded, and the sum over all 83 wall segments was used as an estimate of total spall damage.) In this court case, the jury was faced with the following dilemma: On the one hand, the scaffold-drop survey provided the most accurate estimate of spall rates in a given wall segment. Unfortunately, however, the drop areas were not selected at random from the entire complex; rather, drops were made at areas with high spall concentrations, leading to an overestimate of the total damage. On the other hand, the photo survey was complete in that all 83 wall segments in the complex were checked for spall damage. But the spall rate estimated by the photos, at least in areas of high spall concentration, was biased low (spalling damage cannot always be seen from a photo), leading to an underestimate of the total damage.

The data in the table (saved in the **BRICKS** file) are the spall rates obtained from the two methods at 11 drop locations. Use the data, as did expert statisticians who testified in the case, to help the jury estimate the true spall rate at a given wall segment. Then explain how this information, coupled with the data (not given here) on all 83 wall segments, can provide a reasonable estimate of the total spall damage (i.e., total number of damaged bricks).

Drop Location	Drop Spall Rate (per 1,000 bricks)	Photo Spall Rate (per 1,000 bricks)
1	0	0
2	5.1	0
3	6.6	0
4	1.1	.8
5	1.8	1.0
6	3.9	1.0
7	9.5	1.9
8	22.1	7.7
9	39.3	14.9
10	39.9	13.9
11	43.0	9.8

Based on Fairley, W. B., et al. "Bricks, buildings, and the Bronx: Estimating masonry deterioration." *Chance*, Vol. 7. No. 3, Summer 1994, p. 36 (Figure 3). [*Note:* The data points are estimated from the points shown on a scatterplot.]

Activity   Applying Simple Linear Regression to Your Favorite Data

Many dependent variables in all areas of research serve as the subjects of regression-modeling efforts. We list five such variables here:

1. Crime rate in various communities

2. Daily maximum temperature in your town

3. Grade point average of students who have completed one academic year at your college

4. Gross domestic product of the United States

5. Points scored by your favorite football team in a single game

Choose one of these dependent variables, or choose some other dependent variable, for which you want to construct a prediction model. There may be a large number of independent variables that should be included in a prediction equation for the dependent variable you choose. List three potentially important independent

variables, x_1, x_2, and x_3, that you think might be (individually) strongly related to your dependent variable. Next, obtain 10 data values, each of which consists of a measure of your dependent variable y and the corresponding values of x_1, x_2, and x_3.

a. Use the least squares formulas given in this chapter to fit three straight-line models—one for each independent variable—for predicting y.

b. Interpret the sign of the estimated slope coefficient $\hat{\beta}_1$ in each case, and test the utility of each model by testing $H_0: \beta_1 = 0$ against $H_a: \beta_1 \neq 0$. What assumptions must be satisfied to ensure the validity of these tests?

c. Calculate the coefficient of determination, r^2, for each model. Which of the independent variables predicts y best for the 10 sampled sets of data? Is this variable necessarily best in general (i.e., for the entire population)? Explain.

References

Chatterjee, S., and Price, B. *Regression Analysis by Example*, 2nd ed. New York: Wiley, 1991.

Conover, W. J. *Practical Nonparametric Statistics*, 2nd. ed. New York: Wiley, 1980.

Daniel, W. W., *Applied Nonparametric Statistics*, 2nd. ed. Boston: PWS-Kent, 1990.

Draper, N., and Smith, H. *Applied Regression Analysis*, 3rd ed. New York: Wiley, 1987.

Graybill, F. *Theory and Application of the Linear Model*. North Scituate, MA: Duxbury, 1976.

Kleinbaum, D., and Kupper, L. *Applied Regression Analysis and Other Multivariable Methods,* 2nd ed. North Scituate, MA: Duxbury, 1997.

Kutner, M., Nachtsheim, C., Neter, J., and Li, W. *Applied Linear Statistical Models*, 5th ed. New York: McGraw-Hill/Irwin, 2006.

Mendenhall, W. *Introduction to Linear Models and the Design and Analysis of Experiments*. Belmont, CA.: Wadsworth, 1968.

Mendenhall, W., and Sincich, T. *A Second Course in Statistics: Regression Analysis*, 7th ed. Upper Saddle River, NJ: Prentice Hall, 209.

Montgomery, D., Peck, E., and Vining, G. *Introduction to Linear Regression Analysis*, 3rd ed. New York: Wiley, 2001.

Mosteller, F., and Tukey, J. W. *Data Analysis and Regression: A Second Course in Statistics*. Reading, MA: Addison-Wesley, 1977.

Rousseeuw, P. J., and Leroy, A. M. *Robust Regression and Outlier Detection*. New York: Wiley, 1987.

Weisburg, S. *Applied Linear Regression*, 2nd ed. New York: Wiley, 1985.

USING TECHNOLOGY

MINITAB: Simple Linear Regression

Simple Linear Regression Analysis

Step 1 Access the MINITAB worksheet file that contains the two quantitative variables (dependent and independent variables).

Step 2 Click on the "Stat" button on the MINITAB menu bar, and then click on "Regression" and "Regression" again, as shown in Figure 9.M.1.

Figure 9.M.1
MINITAB menu options for regression

Step 3 On the resulting dialog box (see Figure 9.M.2), specify the dependent variable in the "Response" box and the independent variable in the "Predictors" box.

Figure 9.M.2
MINITAB regression dialog box

Step 4 To produce prediction intervals for y and confidence intervals for $E(y)$, click the "Options" button. The resulting dialog box is shown in Figure 9.M.3.

Step 5 Check "Confidence limits" and/or "Prediction limits", specify the "Confidence level," and enter the value of x in the "Prediction intervals for new observations" box.

Figure 9.M.3
MINITAB regression options

Step 6 Click "OK" to return to the main Regression dialog box and then click "OK" again to produce the MINITAB simple linear regression printout.

Correlation Analysis

Step 1 Click on the "Stat" button on the MINITAB main menu bar, then click on "Basic Statistics," and then click on "Correlation," as shown in Figure 9.M.4.

Step 2 On the resulting dialog box (see Figure 9.M.5), enter the two variables of interest in the "Variables" box.

Step 3 Click "OK" to obtain a printout of the correlation.

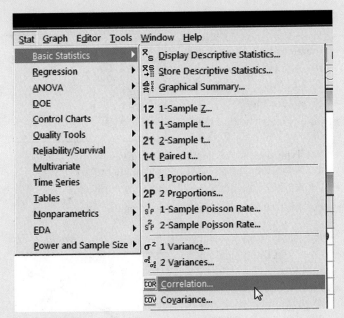

Figure 9.M.4
MINITAB menu options for correlation

Figure 9.M.5
MINITAB correlation dialog box

Rank Correlation

Step 1 To obtain Spearman's rank correlation coefficient in MINITAB, you must first rank the values of the two quantitative variables of interest. Click the "Calc" button on the MINITAB menu bar and create two additional columns, one for the ranks of the x-variable and one for the ranks of the y-variable. (Use the "Rank" function on the MINITAB calculator as shown in Figure 9.M.6.)

Step 2 Click on the "Stat" button on the main menu bar, then click on "Basic Statistics" and "Correlation." (See Figure 9.M.4.)

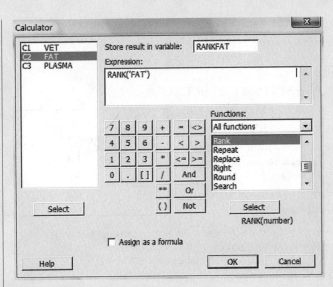

Figure 9.M.6
MINITAB calculator menu screen

Step 3 On the resulting dialog box (see Figure 9.M.7), enter the ranked variables in the "Variables" box and unselect the "Display p-values" option.

Step 4 Click "OK" to obtain the MINITAB printout. (You will need to look up the critical value of Spearman's rank correlation to conduct the test.)

Figure 9.M.7
MINITAB correlation dialog box

TI-83/TI-84 Plus Graphing Calculator: Simple Linear Regression

Finding the Least Squares Regression Equation

Step 1 *Enter the data*

- Press **STAT** and select **1:Edit**

Note: If a list already contains data, clear the old data.

- Use the up arrow to highlight the list name, "**L1**" or "**L2**"

- Press **CLEAR ENTER**
- Enter your *x*-data in **L1** and your *y*-data in **L2**

Step 2 *Find the equation*

- Press **STAT** and highlight **CALC**
- Press **4** for **LinReg(ax + b)**
- Press **ENTER**
- The screen will show the values for *a* and *b* in the equation $y = ax + b$

Finding *r* and *r²*

Use this procedure if *r* and r^2 do not already appear on the LinReg screen from part I:

Step 1 *Turn the diagnostics feature on*

- Press **2nd 0** for **CATALOG**
- Press the **ALPHA** key and x^{-1} for **D**
- Press the down **ARROW** until **DiagnosticsOn** is highlighted
- Press **ENTER** twice

Step 2 *Find the regression equation as shown in part I above*

The values for *r* and r^2 will appear on the screen as well.

Graphing the Least Squares Line with the Scatterplot

Step 1 *Enter the data as shown in part I above*

Step 2 *Set up the data plot*

- Press **Y =** and **CLEAR** all functions from the **Y** register
- Press **2ndY =** for **STAT PLOT**
- Press **1** for **Plot1**
- Set the cursor so that **ON** is flashing and press **ENTER**
- For **Type**, use the **ARROW** and **ENTER** keys to highlight and select the scatterplot (first icon in the first row)
- For **Xlist**, choose the column containing the *x*-data
- For **Ylist**, choose the column containing the *y*-data

Step 3 *Find the regression equation and store the equation in Y1*

- Press **STAT** and highlight **CALC**
- Press **4** for **LinReg(ax + b)** (*Note:* Don't press **ENTER** here because you want to store the regression equation in Y1.)
- Press **VARS**
- Use the right arrow to highlight **Y-VARS**
- Press **ENTER** to select **1:Function**
- Press **ENTER** to select **1:Y1**
- Press **ENTER**

Step 4 *View the scatterplot and regression line*

- Press **ZOOM** and then press **9** to select **9:ZoomStat**

You should see the data graphed along with the regression line.

Appendix A: Tables

548 APPENDIX A: Tables

Table I — Random Numbers

Row \ Column	1	2	3	4	5	6	7	8	9	10	11	12	13	14
1	10480	15011	01536	02011	81647	91646	69179	14194	62590	36207	20969	99570	91291	90700
2	22368	46573	25595	85393	30995	89198	27982	53402	93965	34095	52666	19174	39615	99505
3	24130	48360	22527	97265	76393	64809	15179	24830	49340	32081	30680	19655	63348	58629
4	42167	93093	06243	61680	07856	16376	39440	53537	71341	57004	00849	74917	97758	16379
5	37570	39975	81837	16656	06121	91782	60468	81305	49684	60672	14110	06927	01263	54613
6	77921	06907	11008	42751	27756	53498	18602	70659	90655	15053	21916	81825	44394	42880
7	99562	72905	56420	69994	98872	31016	71194	18738	44013	48840	63213	21069	10634	12952
8	96301	91977	05463	07972	18876	20922	94595	56869	69014	60045	18425	84903	42508	32307
9	89579	14342	63661	10281	17453	18103	57740	84378	25331	12566	58678	44947	05585	56941
10	85475	36857	53342	53988	53060	59533	38867	62300	08158	17983	16439	11458	18593	64952
11	28918	69578	88231	33276	70997	79936	56865	05859	90106	31595	01547	85590	91610	78188
12	63553	40961	48235	03427	49626	69445	18663	72695	52180	20847	12234	90511	33703	90322
13	09429	93969	52636	92737	88974	33488	36320	17617	30015	08272	84115	27156	30613	74952
14	10365	61129	87529	85689	48237	52267	67689	93394	01511	26358	85104	20285	29975	89868
15	07119	97336	71048	08178	77233	13916	47564	81056	97735	85977	29372	74461	28551	90707
16	51085	12765	51821	51259	77452	16308	60756	92144	49442	53900	70960	63990	75601	40719
17	02368	21382	52404	60268	89368	19885	55322	44819	01188	65255	64835	44919	05944	55157
18	01011	54092	33362	94904	31273	04146	18594	29852	71585	85030	51132	01915	92747	64951
19	52162	53916	46369	58586	23216	14513	83149	98736	23495	64350	94738	17752	35156	35749
20	07056	97628	33787	09998	42698	06691	76988	13602	51851	46104	88916	19509	25625	58104
21	48663	91245	85828	14346	09172	30168	90229	04734	59193	22178	30421	61666	99904	32812
22	54164	58492	22421	74103	47070	25306	76468	26384	58151	06646	21524	15227	96909	44592
23	32639	32363	05597	24200	13363	38005	94342	28728	35806	06912	17012	64161	18296	22851
24	29334	27001	87637	87308	58731	00256	45834	15398	46557	41135	10367	07684	36188	18510
25	02488	33062	28834	07351	19731	92420	60952	61280	50001	67658	32586	86679	50720	94953
26	81525	72295	04839	96423	24878	82651	66566	14778	76797	14780	13300	87074	79666	95725
27	29676	20591	68086	26432	46901	20849	89768	81536	86645	12659	92259	57102	80428	25280
28	00742	57392	39064	66432	84673	40027	32832	61362	98947	96067	64760	64584	96096	98253
29	05366	04213	25669	26422	44407	44048	37937	63904	45766	66134	75470	66520	34693	90449
30	91921	26418	64117	94305	26766	25940	39972	22209	71500	64568	91402	42416	07844	69618
31	00582	04711	87917	77341	42206	35126	74087	99547	81817	42607	43808	76655	62028	76630
32	00725	69884	62797	56170	86324	88072	76222	36086	84637	93161	76038	65855	77919	88006
33	69011	65795	95876	55293	18988	27354	26575	08625	40801	59920	29841	80150	12777	48501
34	25976	57948	29888	88604	67917	48708	18912	82271	65424	69774	33611	54262	85963	03547
35	09763	83473	73577	12908	30883	18317	28290	35797	05998	41688	34952	37888	38917	88050
36	91576	42595	27958	30134	04024	86385	29880	99730	55536	84855	29080	09250	79656	73211
37	17955	56349	90999	49127	20044	59931	06115	20542	18059	02008	73708	83517	36103	42791
38	46503	18584	18845	49618	02304	51038	20655	58727	28168	15475	56942	53389	20562	87338
39	92157	89634	94824	78171	84610	82834	09922	25417	44137	48413	25555	21246	35509	20468
40	14577	62765	35605	81263	39667	47358	56873	56307	61607	49518	89656	20103	77490	18062
41	98427	07523	33362	64270	01638	92477	66969	98420	04880	45585	46565	04102	46880	45709

(continued)

Table I (continued)

Row	1	2	3	4	5	6	7	8	9	10	11	12	13	14
42	34914	63976	88720	82765	34476	17032	87589	40836	32427	70002	70663	88863	77775	69348
43	70060	28277	39475	46473	23219	53416	94970	25832	69975	94884	19661	72828	00102	66794
44	53976	54914	06990	67245	68350	82948	11398	42878	80287	88267	47363	46634	06541	97809
45	76072	29515	40980	07391	58745	25774	22987	80059	39911	96189	41151	14222	60697	59583
46	90725	52210	83974	29992	65831	38857	50490	83765	55657	14361	31720	57375	56228	41546
47	64364	67412	33339	31926	14883	24413	59744	92351	97473	89286	35931	04110	23726	51900
48	08962	00358	31662	25388	61642	34072	81249	35648	56891	69352	48373	45578	78547	81788
49	95012	68379	93526	70765	10592	04542	76463	54328	02349	17247	28865	14777	62730	92277
50	15664	10493	20492	38391	91132	21999	59516	81652	27195	48223	46751	22923	32261	85653
51	16408	81899	04153	53381	79401	21438	83035	92350	36693	31238	59649	91754	72772	02338
52	18629	81953	05520	91962	04739	13092	97662	24822	94730	06496	35090	04822	86774	98289
53	73115	35101	47498	87637	99016	71060	88824	71013	18735	20286	23153	72924	35165	43040
54	57491	16703	23167	49323	45021	33132	12544	41035	80780	45393	44812	12512	98931	91202
55	30405	83946	23792	14422	15059	45799	22716	19792	09983	74353	68668	30429	70735	25499
56	16631	35006	85900	98275	32388	52390	16815	69290	82732	38480	73817	32523	41961	44437
57	96773	20206	42559	78985	05300	22164	24369	54224	35083	19687	11052	91491	60383	19746
58	38935	64202	14349	82674	66523	44133	00697	35552	35970	19124	63318	29686	03387	59846
59	31624	76384	17403	53363	44167	64486	64758	75366	76554	31601	12614	33072	60332	92325
60	78919	19474	23632	27889	47914	02584	37680	20801	72152	39339	34806	08930	85001	87820
61	03931	33309	57047	74211	63445	17361	62825	39908	05607	91284	68833	25570	38818	46920
62	74426	33278	43972	10110	89917	15665	52872	73823	73144	88662	88970	74492	51805	99378
63	09066	00903	20795	95452	92648	45454	09552	88815	16553	51125	79375	97596	16296	66092
64	42238	12426	87025	14267	20979	04508	64535	31355	86064	29472	47689	05974	52468	16834
65	16153	08002	26504	41744	81959	65642	74240	56302	00033	67107	77510	70625	28725	34191
66	21457	40742	29820	96783	29400	21840	15035	34537	33310	06116	95240	15957	16572	06004
67	21581	57802	02050	89728	17937	37621	47075	42080	97403	48626	68995	43805	33386	21597
68	55612	78095	83197	33732	05810	24813	86902	60397	16489	03264	88525	42786	05269	92532
69	44657	66999	99324	51281	84463	60563	79312	93454	68876	25471	93911	25650	12682	73572
70	91340	84979	46949	81973	37949	61023	43997	15263	80644	43942	89203	71795	99533	50501
71	91227	21199	31935	27022	84067	05462	35216	14486	29891	68607	41867	14951	91696	85065
72	50001	38140	66321	19924	72163	09538	12151	06878	91903	18749	34405	56087	82790	70925
73	65390	05224	72958	28609	81406	39147	25549	48542	42627	45233	57202	94617	23772	07896
74	27504	96131	83944	41575	10573	08619	64482	73923	36152	05184	94142	25299	84387	34925
75	37169	94851	39117	89632	00959	16487	65536	49071	39782	17095	02330	74301	00275	48280
76	11508	70225	51111	38351	19444	66499	71945	05422	13442	78675	84081	66938	93654	59894
77	37449	30362	06694	54690	04052	53115	62757	95348	78662	11163	81651	50245	34971	52924
78	46515	70331	85922	38329	57015	15765	97161	17869	45349	61796	66345	81073	49106	79860
79	30986	81223	42416	58353	21532	30502	32305	86482	05174	07901	54339	58861	74818	46942
80	63798	64995	46583	09785	44160	78128	83991	42865	92520	83531	80377	35909	81250	54238
81	82486	84846	99254	67632	43218	50076	21361	64816	51202	88124	41870	52689	51275	83556
82	21885	32906	92431	09060	64297	51674	64126	62570	26123	05155	59194	52799	28225	85762

(continued)

Table I (continued)

Column / Row	1	2	3	4	5	6	7	8	9	10	11	12	13	14
83	60336	98782	07408	53458	13564	59089	26445	29789	85205	41001	12535	12133	14645	23541
84	43937	46891	24010	25560	86355	33941	25786	54990	71899	15475	95434	98227	21824	19585
85	97656	63175	89303	16275	07100	92063	21942	18611	47348	20203	18534	03862	78095	50136
86	03299	01221	05418	38982	55758	92237	26759	86367	21216	98442	08303	56613	91511	75928
87	79626	06486	03574	17668	07785	76020	79924	25651	83325	88428	85076	72811	22717	50585
88	85636	68335	47539	03129	65651	11977	02510	26113	99447	68645	34327	15152	55230	93448
89	18039	14367	64337	06177	12143	46609	32989	74014	64708	00533	35398	58408	13261	47908
90	08362	15656	60627	36478	65648	16764	53412	09013	07832	41574	17639	82163	60859	75567
91	79556	29068	04142	16268	15387	12856	66227	38358	22478	73373	88732	09443	82558	05250
92	92608	82674	27072	32534	17075	27698	98204	63863	11951	34648	88022	56148	34925	57031
93	23982	25835	40055	67006	12293	02753	14827	23235	35071	99704	37543	11601	35503	85171
94	09915	96306	05908	97901	28395	14186	00821	80703	70426	75647	76310	88717	37890	40129
95	59037	33300	26695	62247	69927	76123	50842	43834	86654	70959	79725	93872	28117	19233
96	42488	78077	69882	61657	34136	79180	97526	43092	04098	73571	80799	76536	71255	64239
97	46764	86273	63003	93017	31204	36692	40202	35275	57306	55543	53203	18098	47625	88684
98	03237	45430	55417	63282	90816	17349	88298	90183	36600	78406	06216	95787	42579	90730
99	86591	81482	52667	61582	14972	90053	89534	76036	49199	43716	97548	04379	46370	28672
100	38534	01715	94964	87288	65680	43772	39560	12918	86537	62738	19636	51132	25739	56947

Table II   Binomial Probabilities

Tabulated values are $\sum_{x=0}^{k} p(x)$. *(Computations are rounded at the third decimal place.)*

a. $n = 5$

p k	.01	.05	.10	.20	.30	.40	.50	.60	.70	.80	.90	.95	.99
0	.951	.774	.590	.328	.168	.078	.031	.010	.002	.000	.000	.000	.000
1	.999	.977	.919	.737	.528	.337	.188	.087	.031	.007	.000	.000	.000
2	1.000	.999	.991	.942	.837	.683	.500	.317	.163	.058	.009	.001	.000
3	1.000	1.000	1.000	.993	.969	.913	.812	.663	.472	.263	.081	.023	.001
4	1.000	1.000	1.000	1.000	.998	.990	.969	.922	.832	.672	.410	.226	.049

b. $n = 6$

p k	.01	.05	.10	.20	.30	.40	.50	.60	.70	.80	.90	.95	.99
0	.941	.735	.531	.262	.118	.047	.016	.004	.001	.000	.000	.000	.000
1	.999	.967	.886	.655	.420	.233	.109	.041	.011	.002	.000	.000	.000
2	1.000	.998	.984	.901	.744	.544	.344	.179	.070	.017	.001	.000	.000
3	1.000	1.000	.999	.983	.930	.821	.656	.456	.256	.099	.016	.002	.000
4	1.000	1.000	1.000	.998	.989	.959	.891	.767	.580	.345	.114	.033	.001
5	1.000	1.000	1.000	1.000	.999	.996	.984	.953	.882	.738	.469	.265	.059

c. $n = 7$

p k	.01	.05	.10	.20	.30	.40	.50	.60	.70	.80	.90	.95	.99
0	.932	.698	.478	.210	.082	.028	.008	.002	.000	.000	.000	.000	.000
1	.998	.956	.850	.577	.329	.159	.063	.019	.004	.000	.000	.000	.000
2	1.000	.996	.974	.852	.647	.420	.227	.096	.029	.005	.000	.000	.000
3	1.000	1.000	.997	.967	.874	.710	.500	.290	.126	.033	.003	.000	.000
4	1.000	1.000	1.000	.995	.971	.904	.773	.580	.353	.148	.026	.004	.000
5	1.000	1.000	1.000	1.000	.996	.981	.937	.841	.671	.423	.150	.044	.002
6	1.000	1.000	1.000	1.000	1.000	.998	.992	.972	.918	.790	.522	.302	.068

(continued)

Table II	(continued)

d. $n = 8$

k \ p	.01	.05	.10	.20	.30	.40	.50	.60	.70	.80	.90	.95	.99
0	.923	.663	.430	.168	.058	.017	.004	.001	.000	.000	.000	.000	.000
1	.997	.943	.813	.503	.255	.106	.035	.009	.001	.000	.000	.000	.000
2	1.000	.994	.962	.797	.552	.315	.145	.050	.011	.001	.000	.000	.000
3	1.000	1.000	.995	.944	.806	.594	.363	.174	.058	.010	.000	.000	.000
4	1.000	1.000	1.000	.990	.942	.826	.637	.406	.194	.056	.005	.000	.000
5	1.000	1.000	1.000	.999	.989	.950	.855	.685	.448	.203	.038	.006	.000
6	1.000	1.000	1.000	1.000	.999	.991	.965	.894	.745	.497	.187	.057	.003
7	1.000	1.000	1.000	1.000	1.000	.999	.996	.983	.942	.832	.570	.337	.077

e. $n = 9$

k \ p	.01	.05	.10	.20	.30	.40	.50	.60	.70	.80	.90	.95	.99
0	.914	.630	.387	.134	.040	.010	.002	.000	.000	.000	.000	.000	.000
1	.997	.929	.775	.436	.196	.071	.020	.004	.000	.000	.000	.000	.000
2	1.000	.992	.947	.738	.463	.232	.090	.025	.004	.000	.000	.000	.000
3	1.000	.999	.992	.914	.730	.483	.254	.099	.025	.003	.000	.000	.000
4	1.000	1.000	.999	.980	.901	.733	.500	.267	.099	.020	.001	.000	.000
5	1.000	1.000	1.000	.997	.975	.901	.746	.517	.270	.086	.008	.001	.000
6	1.000	1.000	1.000	1.000	.996	.975	.910	.768	.537	.262	.053	.008	.000
7	1.000	1.000	1.000	1.000	1.000	.996	.980	.929	.804	.564	.225	.071	.003
8	1.000	1.000	1.000	1.000	1.000	1.000	.998	.990	.960	.866	.613	.370	.086

f. $n = 10$

k \ p	.01	.05	.10	.20	.30	.40	.50	.60	.70	.80	.90	.95	.99
0	.904	.599	.349	.107	.028	.006	.001	.000	.000	.000	.000	.000	.000
1	.996	.914	.736	.376	.149	.046	.011	.002	.000	.000	.000	.000	.000
2	1.000	.988	.930	.678	.383	.167	.055	.012	.002	.000	.000	.000	.000
3	1.000	.999	.987	.879	.650	.382	.172	.055	.011	.001	.000	.000	.000
4	1.000	1.000	.998	.967	.850	.633	.377	.166	.047	.006	.000	.000	.000
5	1.000	1.000	1.000	.994	.953	.834	.623	.367	.150	.033	.002	.000	.000
6	1.000	1.000	1.000	.999	.989	.945	.828	.618	.350	.121	.013	.001	.000
7	1.000	1.000	1.000	1.000	.998	.988	.945	.833	.617	.322	.070	.012	.000
8	1.000	1.000	1.000	1.000	1.000	.998	.989	.954	.851	.624	.264	.086	.004
9	1.000	1.000	1.000	1.000	1.000	1.000	.999	.994	.972	.893	.651	.401	.096

(continued)

Table II **(continued)**

g. $n = 15$

k \ p	.01	.05	.10	.20	.30	.40	.50	.60	.70	.80	.90	.95	.99
0	.860	.463	.206	.035	.005	.000	.000	.000	.000	.000	.000	.000	.000
1	.990	.829	.549	.167	.035	.005	.000	.000	.000	.000	.000	.000	.000
2	1.000	.964	.816	.398	.127	.027	.004	.000	.000	.000	.000	.000	.000
3	1.000	.995	.944	.648	.297	.091	.018	.002	.000	.000	.000	.000	.000
4	1.000	.999	.987	.838	.515	.217	.059	.009	.001	.000	.000	.000	.000
5	1.000	1.000	.998	.939	.722	.403	.151	.034	.004	.000	.000	.000	.000
6	1.000	1.000	1.000	.982	.869	.610	.304	.095	.015	.001	.000	.000	.000
7	1.000	1.000	1.000	.996	.950	.787	.500	.213	.050	.004	.000	.000	.000
8	1.000	1.000	1.000	.999	.985	.905	.696	.390	.131	.018	.000	.000	.000
9	1.000	1.000	1.000	1.000	.996	.966	.849	.597	.278	.061	.002	.000	.000
10	1.000	1.000	1.000	1.000	.999	.991	.941	.783	.485	.164	.013	.001	.000
11	1.000	1.000	1.000	1.000	1.000	.998	.982	.909	.703	.352	.056	.005	.000
12	1.000	1.000	1.000	1.000	1.000	1.000	.996	.973	.873	.602	.184	.036	.000
13	1.000	1.000	1.000	1.000	1.000	1.000	1.000	.995	.965	.833	.451	.171	.010
14	1.000	1.000	1.000	1.000	1.000	1.000	1.000	1.000	.995	.965	.794	.537	.140

h. $n = 20$

k \ p	.01	.05	.10	.20	.30	.40	.50	.60	.70	.80	.90	.95	.99
0	.818	.358	.122	.012	.001	.000	.000	.000	.000	.000	.000	.000	.000
1	.983	.736	.392	.069	.008	.001	.000	.000	.000	.000	.000	.000	.000
2	.999	.925	.677	.206	.035	.004	.000	.000	.000	.000	.000	.000	.000
3	1.000	.984	.867	.411	.107	.016	.001	.000	.000	.000	.000	.000	.000
4	1.000	.997	.957	.630	.238	.051	.006	.000	.000	.000	.000	.000	.000
5	1.000	1.000	.989	.804	.416	.126	.021	.002	.000	.000	.000	.000	.000
6	1.000	1.000	.998	.913	.608	.250	.058	.006	.000	.000	.000	.000	.000
7	1.000	1.000	1.000	.968	.772	.416	.132	.021	.001	.000	.000	.000	.000
8	1.000	1.000	1.000	.990	.887	.596	.252	.057	.005	.000	.000	.000	.000
9	1.000	1.000	1.000	.997	.952	.755	.412	.128	.017	.001	.000	.000	.000
10	1.000	1.000	1.000	.999	.983	.872	.588	.245	.048	.003	.000	.000	.000
11	1.000	1.000	1.000	1.000	.995	.943	.748	.404	.113	.010	.000	.000	.000
12	1.000	1.000	1.000	1.000	.999	.979	.868	.584	.228	.032	.000	.000	.000
13	1.000	1.000	1.000	1.000	1.000	.994	.942	.750	.392	.087	.002	.000	.000
14	1.000	1.000	1.000	1.000	1.000	.998	.979	.874	.584	.196	.011	.000	.000
15	1.000	1.000	1.000	1.000	1.000	1.000	.994	.949	.762	.370	.043	.003	.000
16	1.000	1.000	1.000	1.000	1.000	1.000	.999	.984	.893	.589	.133	.016	.000
17	1.000	1.000	1.000	1.000	1.000	1.000	1.000	.996	.965	.794	.323	.075	.001
18	1.000	1.000	1.000	1.000	1.000	1.000	1.000	.999	.992	.931	.608	.264	.017
19	1.000	1.000	1.000	1.000	1.000	1.000	1.000	1.000	.999	.988	.878	.642	.182

(continued)

Table II   (continued)

i. $n = 25$

k \ p	.01	.05	.10	.20	.30	.40	.50	.60	.70	.80	.90	.95	.99
0	.778	.277	.072	.004	.000	.000	.000	.000	.000	.000	.000	.000	.000
1	.974	.642	.271	.027	.002	.000	.000	.000	.000	.000	.000	.000	.000
2	.998	.873	.537	.098	.009	.000	.000	.000	.000	.000	.000	.000	.000
3	1.000	.966	.764	.234	.033	.002	.000	.000	.000	.000	.000	.000	.000
4	1.000	.993	.902	.421	.090	.009	.000	.000	.000	.000	.000	.000	.000
5	1.000	.999	.967	.617	.193	.029	.002	.000	.000	.000	.000	.000	.000
6	1.000	1.000	.991	.780	.341	.074	.007	.000	.000	.000	.000	.000	.000
7	1.000	1.000	.998	.891	.512	.154	.022	.001	.000	.000	.000	.000	.000
8	1.000	1.000	1.000	.953	.677	.274	.054	.004	.000	.000	.000	.000	.000
9	1.000	1.000	1.000	.983	.811	.425	.115	.013	.000	.000	.000	.000	.000
10	1.000	1.000	1.000	.994	.902	.586	.212	.034	.002	.000	.000	.000	.000
11	1.000	1.000	1.000	.998	.956	.732	.345	.078	.006	.000	.000	.000	.000
12	1.000	1.000	1.000	1.000	.983	.846	.500	.154	.017	.000	.000	.000	.000
13	1.000	1.000	1.000	1.000	.994	.922	.655	.268	.044	.002	.000	.000	.000
14	1.000	1.000	1.000	1.000	.998	.966	.788	.414	.098	.006	.000	.000	.000
15	1.000	1.000	1.000	1.000	1.000	.987	.885	.575	.189	.017	.000	.000	.000
16	1.000	1.000	1.000	1.000	1.000	.996	.946	.726	.323	.047	.000	.000	.000
17	1.000	1.000	1.000	1.000	1.000	.999	.978	.846	.488	.109	.002	.000	.000
18	1.000	1.000	1.000	1.000	1.000	1.000	.993	.926	.659	.220	.009	.000	.000
19	1.000	1.000	1.000	1.000	1.000	1.000	.998	.971	.807	.383	.033	.001	.000
20	1.000	1.000	1.000	1.000	1.000	1.000	1.000	.991	.910	.579	.098	.007	.000
21	1.000	1.000	1.000	1.000	1.000	1.000	1.000	.998	.967	.766	.236	.034	.000
22	1.000	1.000	1.000	1.000	1.000	1.000	1.000	1.000	.991	.902	.463	.127	.002
23	1.000	1.000	1.000	1.000	1.000	1.000	1.000	1.000	.998	.973	.729	.358	.026
24	1.000	1.000	1.000	1.000	1.000	1.000	1.000	1.000	1.000	.996	.928	.723	.222

Table III   Normal Curve Areas

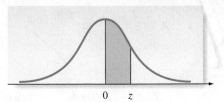

z	.00	.01	.02	.03	.04	.05	.06	.07	.08	.09
.0	.0000	.0040	.0080	.0120	.0160	.0199	.0239	.0279	.0319	.0359
.1	.0398	.0438	.0478	.0517	.0557	.0596	.0636	.0675	.0714	.0753
.2	.0793	.0832	.0871	.0910	.0948	.0987	.1026	.1064	.1103	.1141
.3	.1179	.1217	.1255	.1293	.1331	.1368	.1406	.1443	.1480	.1517
.4	.1554	.1591	.1628	.1664	.1700	.1736	.1772	.1808	.1844	.1879
.5	.1915	.1950	.1985	.2019	.2054	.2088	.2123	.2157	.2190	.2224
.6	.2257	.2291	.2324	.2357	.2389	.2422	.2454	.2486	.2517	.2549
.7	.2580	.2611	.2642	.2673	.2704	.2734	.2764	.2794	.2823	.2852
.8	.2881	.2910	.2939	.2967	.2995	.3023	.3051	.3078	.3106	.3133
.9	.3159	.3186	.3212	.3238	.3264	.3289	.3315	.3340	.3365	.3389
1.0	.3413	.3438	.3461	.3485	.3508	.3531	.3554	.3577	.3599	.3621
1.1	.3643	.3665	.3686	.3708	.3729	.3749	.3770	.3790	.3810	.3830
1.2	.3849	.3869	.3888	.3907	.3925	.3944	.3962	.3980	.3997	.4015
1.3	.4032	.4049	.4066	.4082	.4099	.4115	.4131	.4147	.4162	.4177
1.4	.4192	.4207	.4222	.4236	.4251	.4265	.4279	.4292	.4306	.4319
1.5	.4332	.4345	.4357	.4370	.4382	.4394	.4406	.4418	.4429	.4441
1.6	.4452	.4463	.4474	.4484	.4495	.4505	.4515	.4525	.4535	.4545
1.7	.4554	.4564	.4573	.4582	.4591	.4599	.4608	.4616	.4625	.4633
1.8	.4641	.4649	.4656	.4664	.4671	.4678	.4686	.4693	.4699	.4706
1.9	.4713	.4719	.4726	.4732	.4738	.4744	.4750	.4756	.4761	.4767
2.0	.4772	.4778	.4783	.4788	.4793	.4798	.4803	.4808	.4812	.4817
2.1	.4821	.4826	.4830	.4834	.4838	.4842	.4846	.4850	.4854	.4857
2.2	.4861	.4864	.4868	.4871	.4875	.4878	.4881	.4884	.4887	.4890
2.3	.4893	.4896	.4898	.4901	.4904	.4906	.4909	.4911	.4913	.4916
2.4	.4918	.4920	.4922	.4925	.4927	.4929	.4931	.4932	.4934	.4936
2.5	.4938	.4940	.4941	.4943	.4945	.4946	.4948	.4949	.4951	.4952
2.6	.4953	.4955	.4956	.4957	.4959	.4960	.4961	.4962	.4963	.4964
2.7	.4965	.4966	.4967	.4968	.4969	.4970	.4971	.4972	.4973	.4974
2.8	.4974	.4975	.4976	.4977	.4977	.4978	.4979	.4979	.4980	.4981
2.9	.4981	.4982	.4982	.4983	.4984	.4984	.4985	.4985	.4986	.4986
3.0	.4987	.4987	.4987	.4988	.4988	.4989	.4989	.4989	.4990	.4990

Based on Table I of A. Hald, *Statistical Tables and Formulas* (New York: Wiley), 1952.

Table IV   Critical Values of *t*

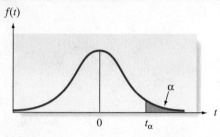

Degrees of Freedom	60 $t_{.100}$	90 $t_{.050}$	95 $t_{.025}$	98 $t_{.010}$	99 $t_{.005}$	99.8 $t_{.001}$	99.9 $t_{.0005}$
1	3.078	6.314	12.706	31.821	63.657	318.31	636.62
2	1.886	2.920	4.303	6.965	9.925	22.326	31.598
3	1.638	2.353	3.182	4.541	5.841	10.213	12.924
4	1.533	2.132	2.776	3.747	4.604	7.173	8.610
5	1.476	2.015	2.571	3.365	4.032	5.893	6.869
6	1.440	1.943	2.447	3.143	3.707	5.208	5.959
7	1.415	1.895	2.365	2.998	3.499	4.785	5.408
8	1.397	1.860	2.306	2.896	3.355	4.501	5.041
9	1.383	1.833	2.262	2.821	3.250	4.297	4.781
10	1.372	1.812	2.228	2.764	3.169	4.144	4.587
11	1.363	1.796	2.201	2.718	3.106	4.025	4.437
12	1.356	1.782	2.179	2.681	3.055	3.930	4.318
13	1.350	1.771	2.160	2.650	3.012	3.852	4.221
14	1.345	1.761	2.145	2.624	2.977	3.787	4.140
15	1.341	1.753	2.131	2.602	2.947	3.733	4.073
16	1.337	1.746	2.120	2.583	2.921	3.686	4.015
17	1.333	1.740	2.110	2.567	2.898	3.646	3.965
18	1.330	1.734	2.101	2.552	2.878	3.610	3.922
19	1.328	1.729	2.093	2.539	2.861	3.579	3.883
20	1.325	1.725	2.086	2.528	2.845	3.552	3.850
21	1.323	1.721	2.080	2.518	2.831	3.527	3.819
22	1.321	1.717	2.074	2.508	2.819	3.505	3.792
23	1.319	1.714	2.069	2.500	2.807	3.485	3.767
24	1.318	1.711	2.064	2.492	2.797	3.467	3.745
25	1.316	1.708	2.060	2.485	2.787	3.450	3.725
26	1.315	1.706	2.056	2.479	2.779	3.435	3.707
27	1.314	1.703	2.052	2.473	2.771	3.421	3.690
28	1.313	1.701	2.048	2.467	2.763	3.408	3.674
29	1.311	1.699	2.045	2.462	2.756	3.396	3.659
30	1.310	1.697	2.042	2.457	2.750	3.385	3.646
40	1.303	1.684	2.021	2.423	2.704	3.307	3.551
60	1.296	1.671	2.000	2.390	2.660	3.232	3.460
120	1.289	1.658	1.980	2.358	2.617	3.160	3.373
∞	1.282	1.645	1.960	2.326	2.576	3.090	3.291

Table V   **Critical Values of χ^2**

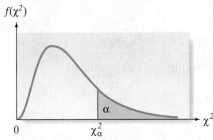

Degrees of Freedom	$\chi^2_{.995}$	$\chi^2_{.990}$	$\chi^2_{.975}$	$\chi^2_{.950}$	$\chi^2_{.900}$
1	.0000393	.0001571	.0009821	.0039321	.0157908
2	.0100251	.0201007	.0506356	.102587	.210720
3	.0717212	.114832	.215795	.351846	.584375
4	.206990	.297110	.484419	.710721	1.063623
5	.411740	.554300	.831211	1.145476	1.61031
6	.675727	.872085	1.237347	1.63539	2.20413
7	.989265	1.239043	1.68987	2.16735	2.83311
8	1.344419	1.646482	2.17973	2.73264	3.48954
9	1.734926	2.087912	2.70039	3.32511	4.16816
10	2.15585	2.55821	3.24697	3.94030	4.86518
11	2.60321	3.05347	3.81575	4.57481	5.57779
12	3.07382	3.57056	4.40379	5.22603	6.30380
13	3.56503	4.10691	5.00874	5.89186	7.04150
14	4.07468	4.66043	5.62872	6.57063	7.78953
15	4.60094	5.22935	6.26214	7.26094	8.54675
16	5.14224	5.81221	6.90766	7.96164	9.31223
17	5.69724	6.40776	7.56418	8.67176	10.0852
18	6.26481	7.01491	8.23075	9.39046	10.8649
19	6.84398	7.63273	8.90655	10.1170	11.6509
20	7.43386	8.26040	9.59083	10.8508	12.4426
21	8.03366	8.89720	10.28293	11.5913	13.2396
22	8.64272	9.54249	10.9823	12.3380	14.0415
23	9.26042	10.19567	11.6885	13.0905	14.8479
24	9.88623	10.8564	12.4011	13.8484	15.6587
25	10.5197	11.5240	13.1197	14.6114	16.4734
26	11.1603	12.1981	13.8439	15.3791	17.2919
27	11.8076	12.8786	14.5733	16.1513	18.1138
28	12.4613	13.5648	15.3079	16.9279	18.9392
29	13.1211	14.2565	16.0471	17.7083	19.7677
30	13.7867	14.9535	16.7908	18.4926	20.5992
40	20.7065	22.1643	24.4331	26.5093	29.0505
50	27.9907	29.7067	32.3574	34.7642	37.6886
60	35.5346	37.4848	40.4817	43.1879	46.4589
70	43.2752	45.4418	48.7576	51.7393	55.3290
80	51.1720	53.5400	57.1532	60.3915	64.2778
90	59.1963	61.7541	65.6466	69.1260	73.2912
100	67.3276	70.0648	74.2219	77.9295	82.3581

(continued)

Table V	(continued)				
Degrees of Freedom	$\chi^2_{.100}$	$\chi^2_{.050}$	$\chi^2_{.025}$	$\chi^2_{.010}$	$\chi^2_{.005}$
1	2.70554	3.84146	5.02389	6.63490	7.87944
2	4.60517	5.99147	7.37776	9.21034	10.5966
3	6.25139	7.81473	9.34840	11.3449	12.8381
4	7.77944	9.48773	11.1433	13.2767	14.8602
5	9.23635	11.0705	12.8325	15.0863	16.7496
6	10.6446	12.5916	14.4494	16.8119	18.5476
7	12.0170	14.0671	16.0128	18.4753	20.2777
8	13.3616	15.5073	17.5346	20.0902	21.9550
9	14.6837	16.9190	19.0228	21.6660	23.5893
10	15.9871	18.3070	20.4831	23.2093	25.1882
11	17.2750	19.6751	21.9200	24.7250	26.7569
12	18.5494	21.0261	23.3367	26.2170	28.2995
13	19.8119	22.3621	24.7356	27.6883	29.8194
14	21.0642	23.6848	26.1190	29.1413	31.3193
15	22.3072	24.9958	27.4884	30.5779	32.8013
16	23.5418	26.2962	28.8454	31.9999	34.2672
17	24.7690	27.5871	30.1910	33.4087	35.7185
18	25.9894	28.8693	31.5264	34.8053	37.1564
19	27.2036	30.1435	32.8523	36.1908	38.5822
20	28.4120	31.4104	34.1696	37.5662	39.9968
21	29.6151	32.6705	35.4789	38.9321	41.4010
22	30.8133	33.9244	36.7807	40.2894	42.7956
23	32.0069	35.1725	38.0757	41.6384	44.1813
24	33.1963	36.4151	39.3641	42.9798	45.5585
25	34.3816	37.6525	40.6465	44.3141	46.9278
26	35.5631	38.8852	41.9232	45.6417	48.2899
27	36.7412	40.1133	43.1944	46.9630	49.6449
28	37.9159	41.3372	44.4607	48.2782	50.9933
29	39.0875	42.5569	45.7222	49.5879	52.3356
30	40.2560	43.7729	46.9792	50.8922	53.6720
40	51.8050	55.7585	59.3417	63.6907	66.7659
50	63.1671	67.5048	71.4202	76.1539	79.4900
60	74.3970	79.0819	83.2976	88.3794	91.9517
70	85.5271	90.5312	95.0231	100.425	104.215
80	96.5782	101.879	106.629	112.329	116.321
90	107.565	113.145	118.136	124.116	128.299
100	118.498	124.342	129.561	135.807	140.169

| Table VI | **Critical Values of T_L and T_U for the Wilcoxon Rank Sum Test: Independent Samples** |

Test statistic is the rank sum associated with the smaller sample (if equal sample sizes, either rank sum can be used).

a. $\alpha = .025$ one-tailed; $\alpha = .05$ two-tailed

n_2 \ n_1	3		4		5		6		7		8		9		10	
	T_L	T_U	T_L	T_U	T_L	T_U	T_L	T_U	T_L	T_U	T_L	T_U	T_L	T_U	T_L	T_U
3	5	16	6	18	6	21	7	23	7	26	8	28	8	31	9	33
4	6	18	11	25	12	28	12	32	13	35	14	38	15	41	16	44
5	6	21	12	28	18	37	19	41	20	45	21	49	22	53	24	56
6	7	23	12	32	19	41	26	52	28	56	29	61	31	65	32	70
7	7	26	13	35	20	45	28	56	37	68	39	73	41	78	43	83
8	8	28	14	38	21	49	29	61	39	73	49	87	51	93	54	98
9	8	31	15	41	22	53	31	65	41	78	51	93	63	108	66	114
10	9	33	16	44	24	56	32	70	43	83	54	98	66	114	79	131

b. $\alpha = .05$ one-tailed; $\alpha = .10$ two-tailed

n_2 \ n_1	3		4		5		6		7		8		9		10	
	T_L	T_U	T_L	T_U	T_L	T_U	T_L	T_U	T_L	T_U	T_L	T_U	T_L	T_U	T_L	T_U
3	6	15	7	17	7	20	8	22	9	24	9	27	10	29	11	31
4	7	17	12	24	13	27	14	30	15	33	16	36	17	39	18	42
5	7	20	13	27	19	36	20	40	22	43	24	46	25	50	26	54
6	8	22	14	30	20	40	28	50	30	54	32	58	33	63	35	67
7	9	24	15	33	22	43	30	54	39	66	41	71	43	76	46	80
8	9	27	16	36	24	46	32	58	41	71	52	84	54	90	57	95
9	10	29	17	39	25	50	33	63	43	76	54	90	66	105	69	111
10	11	31	18	42	26	54	35	67	46	80	57	95	69	111	83	127

Based on F. Wilcoxon and R. A. Wilcox, "Some Rapid Approximate Statistical Procedures," 1964, 20–23.

Table VII	Critical Values of T_0 in the Wilcoxon Paired Difference Signed Rank Test					
One-Tailed	Two-Tailed	$n = 5$	$n = 6$	$n = 7$	$n = 8$	$n = 9$	$n = 10$
$\alpha = .05$	$\alpha = .10$	1	2	4	6	8	11
$\alpha = .025$	$\alpha = .05$		1	2	4	6	8
$\alpha = .01$	$\alpha = .02$			0	2	3	5
$\alpha = .005$	$\alpha = .01$				0	2	3
		$n = 11$	$n = 12$	$n = 13$	$n = 14$	$n = 15$	$n = 16$
$\alpha = .05$	$\alpha = .10$	14	17	21	26	30	36
$\alpha = .025$	$\alpha = .05$	11	14	17	21	25	30
$\alpha = .01$	$\alpha = .02$	7	10	13	16	20	24
$\alpha = .005$	$\alpha = .01$	5	7	10	13	16	19
		$n = 17$	$n = 18$	$n = 19$	$n = 20$	$n = 21$	$n = 22$
$\alpha = .05$	$\alpha = .10$	41	47	54	60	68	75
$\alpha = .025$	$\alpha = .05$	35	40	46	52	59	66
$\alpha = .01$	$\alpha = .02$	28	33	38	43	49	56
$\alpha = .005$	$\alpha = .01$	23	28	32	37	43	49
		$n = 23$	$n = 24$	$n = 25$	$n = 26$	$n = 27$	$n = 28$
$\alpha = .05$	$\alpha = .10$	83	92	101	110	120	130
$\alpha = .025$	$\alpha = .05$	73	81	90	98	107	117
$\alpha = .01$	$\alpha = .02$	62	69	77	85	93	102
$\alpha = .005$	$\alpha = .01$	55	61	68	76	84	92
		$n = 29$	$n = 30$	$n = 31$	$n = 32$	$n = 33$	$n = 34$
$\alpha = .05$	$\alpha = .10$	141	152	163	175	188	201
$\alpha = .025$	$\alpha = .05$	127	137	148	159	171	183
$\alpha = .01$	$\alpha = .02$	111	120	130	141	151	162
$\alpha = .005$	$\alpha = .01$	100	109	118	128	138	149
		$n = 35$	$n = 36$	$n = 37$	$n = 38$	$n = 39$	
$\alpha = .05$	$\alpha = .10$	214	228	242	256	271	
$\alpha = .025$	$\alpha = .05$	195	208	222	235	250	
$\alpha = .01$	$\alpha = .02$	174	186	198	211	224	
$\alpha = .005$	$\alpha = .01$	160	171	183	195	208	
		$n = 40$	$n = 41$	$n = 42$	$n = 43$	$n = 44$	$n = 45$
$\alpha = .05$	$\alpha = .10$	287	303	319	336	353	371
$\alpha = .025$	$\alpha = .05$	264	279	295	311	327	344
$\alpha = .01$	$\alpha = .02$	238	252	267	281	297	313
$\alpha = .005$	$\alpha = .01$	221	234	248	262	277	292
		$n = 46$	$n = 47$	$n = 48$	$n = 49$	$n = 50$	
$\alpha = .05$	$\alpha = .10$	389	408	427	446	466	
$\alpha = .025$	$\alpha = .05$	361	379	397	415	434	
$\alpha = .01$	$\alpha = .02$	329	345	362	380	398	
$\alpha = .005$	$\alpha = .01$	307	323	339	356	373	

Based on F. Wilcoxon and R. A. Wilcox, "Some Rapid Approximate Statistical Procedures," 1964, p. 28.

Table VIII   Percentage Points of the *F*-Distribution, $\alpha = .10$

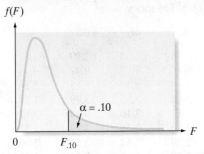

ν_1 ν_2	Numerator Degrees of Freedom								
	1	2	3	4	5	6	7	8	9
1	39.86	49.50	53.59	55.83	57.24	58.20	58.91	59.44	59.86
2	8.53	9.00	9.16	9.24	9.29	9.33	9.35	9.37	9.38
3	5.54	5.46	5.39	5.34	5.31	5.28	5.27	5.25	5.24
4	4.54	4.32	4.19	4.11	4.05	4.01	3.98	3.95	3.94
5	4.06	3.78	3.62	3.52	3.45	3.40	3.37	3.34	3.32
6	3.78	3.46	3.29	3.18	3.11	3.05	3.01	2.98	2.96
7	3.59	3.26	3.07	2.96	2.88	2.83	2.78	2.75	2.72
8	3.46	3.11	2.92	2.81	2.73	2.67	2.62	2.59	2.56
9	3.36	3.01	2.81	2.69	2.61	2.55	2.51	2.47	2.44
10	3.29	2.92	2.73	2.61	2.52	2.46	2.41	2.38	2.35
11	3.23	2.86	2.66	2.54	2.45	2.39	2.34	2.30	2.27
12	3.18	2.81	2.61	2.48	2.39	2.33	2.28	2.24	2.21
13	3.14	2.76	2.56	2.43	2.35	2.28	2.23	2.20	2.16
14	3.10	2.73	2.52	2.39	2.31	2.24	2.19	2.15	2.12
15	3.07	2.70	2.49	2.36	2.27	2.21	2.16	2.12	2.09
16	3.05	2.67	2.46	2.33	2.24	2.18	2.13	2.09	2.06
17	3.03	2.64	2.44	2.31	2.22	2.15	2.10	2.06	2.03
18	3.01	2.62	2.42	2.29	2.20	2.13	2.08	2.04	2.00
19	2.99	2.61	2.40	2.27	2.18	2.11	2.06	2.02	1.98
20	2.97	2.59	2.38	2.25	2.16	2.09	2.04	2.00	1.96
21	2.96	2.57	2.36	2.23	2.14	2.08	2.02	1.98	1.95
22	2.95	2.56	2.35	2.22	2.13	2.06	2.01	1.97	1.93
23	2.94	2.55	2.34	2.21	2.11	2.05	1.99	1.95	1.92
24	2.93	2.54	2.33	2.19	2.10	2.04	1.98	1.94	1.91
25	2.92	2.53	2.32	2.18	2.09	2.02	1.97	1.93	1.89
26	2.91	2.52	2.31	2.17	2.08	2.01	1.96	1.92	1.88
27	2.90	2.51	2.30	2.17	2.07	2.00	1.95	1.91	1.87
28	2.89	2.50	2.29	2.16	2.06	2.00	1.94	1.90	1.87
29	2.89	2.50	2.28	2.15	2.06	1.99	1.93	1.89	1.86
30	2.88	2.49	2.28	2.14	2.05	1.98	1.93	1.88	1.85
40	2.84	2.44	2.23	2.09	2.00	1.93	1.87	1.83	1.79
60	2.79	2.39	2.18	2.04	1.95	1.87	1.82	1.77	1.74
120	2.75	2.35	2.13	1.99	1.90	1.82	1.77	1.72	1.68
∞	2.71	2.30	2.08	1.94	1.85	1.77	1.72	1.67	1.63

(continued)

The row label at left reads "Denominator Degrees of Freedom."

Table VIII	(continued)								

ν_1	Numerator Degrees of Freedom								
ν_2	10	12	15	20	24	30	40	60	120	∞
1	60.19	60.71	61.22	61.74	62.00	62.26	62.53	62.79	63.06	63.33
2	9.39	9.41	9.42	9.44	9.45	9.46	9.47	9.47	9.48	9.49
3	5.23	5.22	5.20	5.18	5.18	5.17	5.16	5.15	5.14	5.13
4	3.92	3.90	3.87	3.84	3.83	3.82	3.80	3.79	3.78	3.76
5	3.30	3.27	3.24	3.21	3.19	3.17	3.16	3.14	3.12	3.10
6	2.94	2.90	2.87	2.84	2.82	2.80	2.78	2.76	2.74	2.72
7	2.70	2.67	2.63	2.59	2.58	2.56	2.54	2.51	2.49	2.47
8	2.54	2.50	2.46	2.42	2.40	2.38	2.36	2.34	2.32	2.29
9	2.42	2.38	2.34	2.30	2.28	2.25	2.23	2.21	2.18	2.16
10	2.32	2.28	2.24	2.20	2.18	2.16	2.13	2.11	2.08	2.06
11	2.25	2.21	2.17	2.12	2.10	2.08	2.05	2.03	2.00	1.97
12	2.19	2.15	2.10	2.06	2.04	2.01	1.99	1.96	1.93	1.90
13	2.14	2.10	2.05	2.01	1.98	1.96	1.93	1.90	1.88	1.85
14	2.10	2.05	2.01	1.96	1.94	1.91	1.89	1.86	1.83	1.80
15	2.06	2.02	1.97	1.92	1.90	1.87	1.85	1.82	1.79	1.76
16	2.03	1.99	1.94	1.89	1.87	1.84	1.81	1.78	1.75	1.72
17	2.00	1.96	1.91	1.86	1.84	1.81	1.78	1.75	1.72	1.69
18	1.98	1.93	1.89	1.84	1.81	1.78	1.75	1.72	1.69	1.66
19	1.96	1.91	1.86	1.81	1.79	1.76	1.73	1.70	1.67	1.63
20	1.94	1.89	1.84	1.79	1.77	1.74	1.71	1.68	1.64	1.61
21	1.92	1.87	1.83	1.78	1.75	1.72	1.69	1.66	1.62	1.59
22	1.90	1.86	1.81	1.76	1.73	1.70	1.67	1.64	1.60	1.57
23	1.89	1.84	1.80	1.74	1.72	1.69	1.66	1.62	1.59	1.55
24	1.88	1.83	1.78	1.73	1.70	1.67	1.64	1.61	1.57	1.53
25	1.87	1.82	1.77	1.72	1.69	1.66	1.63	1.59	1.56	1.52
26	1.86	1.81	1.76	1.71	1.68	1.65	1.61	1.58	1.54	1.50
27	1.85	1.80	1.75	1.70	1.67	1.64	1.60	1.57	1.53	1.49
28	1.84	1.79	1.74	1.69	1.66	1.63	1.59	1.56	1.52	1.48
29	1.83	1.78	1.73	1.68	1.65	1.62	1.58	1.55	1.51	1.47
30	1.82	1.77	1.72	1.67	1.64	1.61	1.57	1.54	1.50	1.46
40	1.76	1.71	1.66	1.61	1.57	1.54	1.51	1.47	1.42	1.38
60	1.71	1.66	1.60	1.54	1.51	1.48	1.44	1.40	1.35	1.29
120	1.65	1.60	1.55	1.48	1.45	1.41	1.37	1.32	1.26	1.19
∞	1.60	1.55	1.49	1.42	1.38	1.34	1.30	1.24	1.17	1.00

Denominator Degrees of Freedom

Table IX   Percentage Points of the *F*-Distribution, $\alpha = .05$

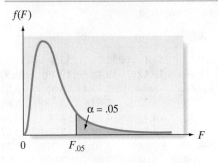

ν_1 ν_2	Numerator Degrees of Freedom								
	1	2	3	4	5	6	7	8	9
1	161.4	199.5	215.7	224.6	230.2	234.0	236.8	238.9	240.5
2	18.51	19.00	19.16	19.25	19.30	19.33	19.35	19.37	19.38
3	10.13	9.55	9.28	9.12	9.01	8.94	8.89	8.85	8.81
4	7.71	6.94	6.59	6.39	6.26	6.16	6.09	6.04	6.00
5	6.61	5.79	5.41	5.19	5.05	4.95	4.88	4.82	4.77
6	5.99	5.14	4.76	4.53	4.39	4.28	4.21	4.15	4.10
7	5.59	4.74	4.35	4.12	3.97	3.87	3.79	3.73	3.68
8	5.32	4.46	4.07	3.84	3.69	3.58	3.50	3.44	3.39
9	5.12	4.26	3.86	3.63	3.48	3.37	3.29	3.23	3.18
10	4.96	4.10	3.71	3.48	3.33	3.22	3.14	3.07	3.02
11	4.84	3.98	3.59	3.36	3.20	3.09	3.01	2.95	2.90
12	4.75	3.89	3.49	3.26	3.11	3.00	2.91	2.85	2.80
13	4.67	3.81	3.41	3.18	3.03	2.92	2.83	2.77	2.71
14	4.60	3.74	3.34	3.11	2.96	2.85	2.76	2.70	2.65
15	4.54	3.68	3.29	3.06	2.90	2.79	2.71	2.64	2.59
16	4.49	3.63	3.24	3.01	2.85	2.74	2.66	2.59	2.54
17	4.45	3.59	3.20	2.96	2.81	2.70	2.61	2.55	2.49
18	4.41	3.55	3.16	2.93	2.77	2.66	2.58	2.51	2.46
19	4.38	3.52	3.13	2.90	2.74	2.63	2.54	2.48	2.42
20	4.35	3.49	3.10	2.87	2.71	2.60	2.51	2.45	2.39
21	4.32	3.47	3.07	2.84	2.68	2.57	2.49	2.42	2.37
22	4.30	3.44	3.05	2.82	2.66	2.55	2.46	2.40	2.34
23	4.28	3.42	3.03	2.80	2.64	2.53	2.44	2.37	2.32
24	4.26	3.40	3.01	2.78	2.62	2.51	2.42	2.36	2.30
25	4.24	3.39	2.99	2.76	2.60	2.49	2.40	2.34	2.28
26	4.23	3.37	2.98	2.74	2.59	2.47	2.39	2.32	2.77
27	4.21	3.35	2.96	2.73	2.57	2.46	2.37	2.31	2.25
28	4.20	3.34	2.95	2.71	2.56	2.45	2.36	2.29	2.24
29	4.18	3.33	2.93	2.70	2.55	2.43	2.35	2.28	2.22
30	4.17	3.32	2.92	2.69	2.53	2.42	2.33	2.27	2.21
40	4.08	3.23	2.84	2.61	2.45	2.34	2.25	2.18	2.12
60	4.00	3.15	2.76	2.53	2.37	2.25	2.17	2.10	2.04
120	3.92	3.07	2.68	2.45	2.29	2.17	2.09	2.02	1.96
∞	3.84	3.00	2.60	2.37	2.21	2.10	2.01	1.94	1.88

(continued)

Table IX	(continued)								

ν_2 \ ν_1	Numerator Degrees of Freedom								
	10	12	15	20	24	30	40	60	120	∞
1	241.9	243.9	245.9	248.0	249.1	250.1	251.1	252.2	253.3	254.3
2	19.40	19.41	19.43	19.45	19.45	19.46	19.47	19.48	19.49	19.50
3	8.79	8.74	8.70	8.66	8.64	8.62	8.59	8.57	8.55	8.53
4	5.96	5.91	5.86	5.80	5.77	5.75	5.72	5.69	5.66	5.63
5	4.74	4.68	4.62	4.56	4.53	4.50	4.46	4.43	4.40	4.36
6	4.06	4.00	3.94	3.87	3.84	3.81	3.77	3.74	3.70	3.67
7	3.64	3.57	3.51	3.44	3.41	3.38	3.34	3.30	3.27	3.23
8	3.35	3.28	3.22	3.15	3.12	3.08	3.04	3.01	2.97	2.93
9	3.14	3.07	3.01	2.94	2.90	2.86	2.83	2.79	2.75	2.71
10	2.98	2.91	2.85	2.77	2.74	2.70	2.66	2.62	2.58	2.54
11	2.85	2.79	2.72	2.65	2.61	2.57	2.53	2.49	2.45	2.40
12	2.75	2.69	2.62	2.54	2.51	2.47	2.43	2.38	2.34	2.30
13	2.67	2.60	2.53	2.46	2.42	2.38	2.34	2.30	2.25	2.21
14	2.60	2.53	2.46	2.39	2.35	2.31	2.27	2.22	2.18	2.13
15	2.54	2.48	2.40	2.33	2.29	2.25	2.20	2.16	2.11	2.07
16	2.49	2.42	2.35	2.28	2.24	2.19	2.15	2.11	2.06	2.01
17	2.45	2.38	2.31	2.23	2.19	2.15	2.10	2.06	2.01	1.96
18	2.41	2.34	2.27	2.19	2.15	2.11	2.06	2.02	1.97	1.92
19	2.38	2.31	2.23	2.16	2.11	2.07	2.03	1.98	1.93	1.88
20	2.35	2.28	2.20	2.12	2.08	2.04	1.99	1.95	1.90	1.84
21	2.32	2.25	2.18	2.10	2.05	2.01	1.96	1.92	1.87	1.81
22	2.30	2.23	2.15	2.07	2.03	1.98	1.94	1.89	1.84	1.78
23	2.27	2.20	2.13	2.05	2.01	1.96	1.91	1.86	1.81	1.76
24	2.25	2.18	2.11	2.03	1.98	1.94	1.89	1.84	1.79	1.73
25	2.24	2.16	2.09	2.01	1.96	1.92	1.87	1.82	1.77	1.71
26	2.22	2.15	2.07	1.99	1.95	1.90	1.85	1.80	1.75	1.69
27	2.20	2.13	2.06	1.97	1.93	1.88	1.84	1.79	1.73	1.67
28	2.19	2.12	2.04	1.96	1.91	1.87	1.82	1.77	1.71	1.65
29	2.18	2.10	2.03	1.94	1.90	1.85	1.81	1.75	1.70	1.64
30	2.16	2.09	2.01	1.93	1.89	1.84	1.79	1.74	1.68	1.62
40	2.08	2.00	1.92	1.84	1.79	1.74	1.69	1.64	1.58	1.51
60	1.99	1.92	1.84	1.75	1.70	1.65	1.59	1.53	1.47	1.39
120	1.91	1.83	1.75	1.66	1.61	1.55	1.50	1.43	1.35	1.25
∞	1.83	1.75	1.67	1.57	1.52	1.46	1.39	1.32	1.22	1.00

Denominator degrees of freedom

Table X   Percentage Points of the *F*-Distribution, $\alpha = .025$

ν_1	Numerator Degrees of Freedom								
ν_2	1	2	3	4	5	6	7	8	9
1	647.8	799.5	864.2	899.6	921.8	937.1	948.2	956.7	963.3
2	38.51	39.00	39.17	39.25	39.30	39.33	39.36	39.37	39.39
3	17.44	16.04	15.44	15.10	14.88	14.73	14.62	14.54	14.47
4	12.22	10.65	9.98	9.60	9.36	9.20	9.07	8.98	8.90
5	10.01	8.43	7.76	7.39	7.15	6.98	6.85	6.76	6.68
6	8.81	7.26	6.60	6.23	5.99	5.82	5.70	5.60	5.52
7	8.07	6.54	5.89	5.52	5.29	5.12	4.99	4.90	4.82
8	7.57	6.06	5.42	5.05	4.82	4.65	4.53	4.43	4.36
9	7.21	5.71	5.08	4.72	4.48	4.32	4.20	4.10	4.03
10	6.94	5.46	4.83	4.47	4.24	4.07	3.95	3.85	3.78
11	6.72	5.26	4.63	4.28	4.04	3.88	3.76	3.66	3.59
12	6.55	5.10	4.47	4.12	3.89	3.73	3.61	3.51	3.44
13	6.41	4.97	4.35	4.00	3.77	3.60	3.48	3.39	3.31
14	6.30	4.86	4.24	3.89	3.66	3.50	3.38	3.29	3.21
15	6.20	4.77	4.15	3.80	3.58	3.41	3.29	3.20	3.12
16	6.12	4.69	4.08	3.73	3.50	3.34	3.22	3.12	3.05
17	6.04	4.62	4.01	3.66	3.44	3.28	3.16	3.06	2.98
18	5.98	4.56	3.95	3.61	3.38	3.22	3.10	3.01	2.93
19	5.92	4.51	3.90	3.56	3.33	3.17	3.05	2.96	2.88
20	5.87	4.46	3.86	3.51	3.29	3.13	3.01	2.91	2.84
21	5.83	4.42	3.82	3.48	3.25	3.09	2.97	2.87	2.80
22	5.79	4.38	3.78	3.44	3.22	3.05	2.93	2.84	2.76
23	5.75	4.35	3.75	3.41	3.18	3.02	2.90	2.81	2.73
24	5.72	4.32	3.72	3.38	3.15	2.99	2.87	2.78	2.70
25	5.69	4.29	3.69	3.35	3.13	2.97	2.85	2.75	2.68
26	5.66	4.27	3.67	3.33	3.10	2.94	2.82	2.73	2.65
27	5.63	4.24	3.65	3.31	3.08	2.92	2.80	2.71	2.63
28	5.61	4.22	3.63	3.29	3.06	2.90	2.78	2.69	2.61
29	5.59	4.20	3.61	3.27	3.04	2.88	2.76	2.67	2.59
30	5.57	4.18	3.59	3.25	3.03	2.87	2.75	2.65	2.57
40	5.42	4.05	3.46	3.13	2.90	2.74	2.62	2.53	2.45
60	5.29	3.93	3.34	3.01	2.79	2.63	2.51	2.41	2.33
120	5.15	3.80	3.23	2.89	2.67	2.52	2.39	2.30	2.22
∞	5.02	3.69	3.12	2.79	2.57	2.41	2.29	2.19	2.11

(continued)

Table X	(continued)									
ν_1	Numerator Degrees of Freedom									
ν_2	10	12	15	20	24	30	40	60	120	∞
1	968.6	976.7	984.9	993.1	997.2	1,001	1,006	1,010	1,014	1,018
2	39.40	39.41	39.43	39.45	39.46	39.46	39.47	39.48	39.49	39.50
3	14.42	14.34	14.25	14.17	14.12	14.08	14.04	13.99	13.95	13.90
4	8.84	8.75	8.66	8.56	8.51	8.46	8.41	8.36	8.31	8.26
5	6.62	6.52	6.43	6.33	6.28	6.23	6.18	6.12	6.07	6.02
6	5.46	5.37	5.27	5.17	5.12	5.07	5.01	4.96	4.90	4.85
7	4.76	4.67	4.57	4.47	4.42	4.36	4.31	4.25	4.20	4.14
8	4.30	4.20	4.10	4.00	3.95	3.89	3.84	3.78	3.73	3.67
9	3.96	3.87	3.77	3.67	3.61	3.56	3.51	3.45	3.39	3.33
10	3.72	3.62	3.52	3.42	3.37	3.31	3.26	3.20	3.14	3.08
11	3.53	3.43	3.33	3.23	3.17	3.12	3.06	3.00	2.94	2.88
12	3.37	3.28	3.18	3.07	3.02	2.96	2.91	2.85	2.79	2.72
13	3.25	3.15	3.05	2.95	2.89	2.84	2.78	2.72	2.66	2.60
14	3.15	3.05	2.95	2.84	2.79	2.73	2.67	2.61	2.55	2.49
15	3.06	2.96	2.86	2.76	2.70	2.64	2.59	2.52	2.46	2.40
16	2.99	2.89	2.79	2.68	2.63	2.57	2.51	2.45	2.38	2.32
17	2.92	2.82	2.72	2.62	2.56	2.50	2.44	2.38	2.32	2.25
18	2.87	2.77	2.67	2.56	2.50	2.44	2.38	2.32	2.26	2.19
19	2.82	2.72	2.62	2.51	2.45	2.39	2.33	2.27	2.20	2.13
20	2.77	2.68	2.57	2.46	2.41	2.35	2.29	2.22	2.16	2.09
21	2.73	2.64	2.53	2.42	2.37	2.31	2.25	2.18	2.11	2.04
22	2.70	2.60	2.50	2.39	2.33	2.27	2.21	2.14	2.08	2.00
23	2.67	2.57	2.47	2.36	2.30	2.24	2.18	2.11	2.04	1.97
24	2.64	2.54	2.44	2.33	2.27	2.21	2.15	2.08	2.01	1.94
25	2.61	2.51	2.41	2.30	2.24	2.18	2.12	2.05	1.98	1.91
26	2.59	2.49	2.39	2.28	2.22	2.16	2.09	2.03	1.95	1.88
27	2.57	2.47	2.36	2.25	2.19	2.13	2.07	2.00	1.93	1.85
28	2.55	2.45	2.34	2.23	2.17	2.11	2.05	1.98	1.91	1.83
29	2.53	2.43	2.32	2.21	2.15	2.09	2.03	1.96	1.89	1.81
30	2.51	2.41	2.31	2.20	2.14	2.07	2.01	1.94	1.87	1.79
40	2.39	2.29	2.18	2.07	2.01	1.94	1.88	1.80	1.72	1.64
60	2.27	2.17	2.06	1.94	1.88	1.82	1.74	1.67	1.58	1.48
120	2.16	2.05	1.94	1.82	1.76	1.69	1.61	1.53	1.43	1.31
∞	2.05	1.94	1.83	1.71	1.64	1.57	1.48	1.39	1.27	1.00

Denominator Degrees of Freedom

Table XI   Percentage Points of the *F*-Distribution, α = .01

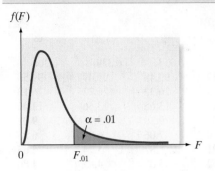

ν_1 ν_2	Numerator Degrees of Freedom								
	1	2	3	4	5	6	7	8	9
1	4,052	4,999.5	5,403	5,625	5,764	5,859	5,928	5,982	6,022
2	98.50	99.00	99.17	99.25	99.30	99.33	99.36	99.37	99.39
3	34.12	30.82	29.46	28.71	28.24	27.91	27.67	27.49	27.35
4	21.20	18.00	16.69	15.98	15.52	15.21	14.98	14.80	14.66
5	16.26	13.27	12.06	11.39	10.97	10.67	10.46	10.29	10.16
6	13.75	10.92	9.78	9.15	8.75	8.47	8.26	8.10	7.98
7	12.25	9.55	8.45	7.85	7.46	7.19	6.99	6.84	6.72
8	11.26	8.65	7.59	7.01	6.63	6.37	6.18	6.03	5.91
9	10.56	8.02	6.99	6.42	6.06	5.80	5.61	5.47	5.35
10	10.04	7.56	6.55	5.99	5.64	5.39	5.20	5.06	4.94
11	9.65	7.21	6.22	5.67	5.32	5.07	4.89	4.74	4.63
12	9.33	6.93	5.95	5.41	5.06	4.82	4.64	4.50	4.39
13	9.07	6.70	5.74	5.21	4.86	4.62	4.44	4.30	4.19
14	8.86	6.51	5.56	5.04	4.69	4.46	4.28	4.14	4.03
15	8.68	6.36	5.42	4.89	4.56	4.32	4.14	4.00	3.89
16	8.53	6.23	5.29	4.77	4.44	4.20	4.03	3.89	3.78
17	8.40	6.11	5.18	4.67	4.34	4.10	3.93	3.79	3.68
18	8.29	6.01	5.09	4.58	4.25	4.01	3.84	3.71	3.60
19	8.18	5.93	5.01	4.50	4.17	3.94	3.77	3.63	3.52
20	8.10	5.85	4.94	4.43	4.10	3.87	3.70	3.56	3.46
21	8.02	5.78	4.87	4.37	4.04	3.81	3.64	3.51	3.40
22	7.95	5.72	4.82	4.31	3.99	3.76	3.59	3.45	3.35
23	7.88	5.66	4.76	4.26	3.94	3.71	3.54	3.41	3.30
24	7.82	5.61	4.72	4.22	3.90	3.67	3.50	3.36	3.26
25	7.77	5.57	4.68	4.18	3.85	3.63	3.46	3.32	3.22
26	7.72	5.53	4.64	4.14	3.82	3.59	3.42	3.29	3.18
27	7.68	5.49	4.60	4.11	3.78	3.56	3.39	3.26	3.15
28	7.64	5.45	4.57	4.07	3.75	3.53	3.36	3.23	3.12
29	7.60	5.42	4.54	4.04	3.73	3.50	3.33	3.20	3.09
30	7.56	5.39	4.51	4.02	3.70	3.47	3.30	3.17	3.07
40	7.31	5.18	4.31	3.83	3.51	3.29	3.12	2.99	2.89
60	7.08	4.98	4.13	3.65	3.34	3.12	2.95	2.82	2.72
120	6.85	4.79	3.95	3.48	3.17	2.96	2.79	2.66	2.56
∞	6.63	4.61	3.78	3.32	3.02	2.80	2.64	2.51	2.41

(continued)

Denominator Degrees of Freedom

Table XI   (continued)

ν_2 \ ν_1	Numerator Degrees of Freedom									
	10	12	15	20	24	30	40	60	120	∞
1	6,056	6,106	6,157	6,209	6,235	6,261	6,287	6,313	6,339	6,366
2	99.40	99.42	99.43	99.45	99.46	99.47	99.47	99.48	99.49	99.50
3	27.23	27.05	26.87	26.69	26.60	26.50	26.41	26.32	26.22	26.13
4	14.55	14.37	14.20	14.02	13.93	13.84	13.75	13.65	13.56	13.46
5	10.05	9.89	9.72	9.55	9.47	9.38	9.29	9.20	9.11	9.02
6	7.87	7.72	7.56	7.40	7.31	7.23	7.14	7.06	6.97	6.88
7	6.62	6.47	6.31	6.16	6.07	5.99	5.91	5.82	5.74	5.65
8	5.81	5.67	5.52	5.36	5.28	5.20	5.12	5.03	4.95	4.86
9	5.26	5.11	4.96	4.81	4.73	4.65	4.57	4.48	4.40	4.31
10	4.85	4.71	4.56	4.41	4.33	4.25	4.17	4.08	4.00	3.91
11	4.54	4.40	4.25	4.10	4.02	3.94	3.86	3.78	3.69	3.60
12	4.30	4.16	4.01	3.86	3.78	3.70	3.62	3.54	3.45	3.36
13	4.10	3.96	3.82	3.66	3.59	3.51	3.43	3.34	3.25	3.17
14	3.94	3.80	3.66	3.51	3.43	3.35	3.27	3.18	3.09	3.00
15	3.80	3.67	3.52	3.37	3.29	3.21	3.13	3.05	2.96	2.87
16	3.69	3.55	3.41	3.26	3.18	3.10	3.02	2.93	2.84	2.75
17	3.59	3.46	3.31	3.16	3.08	3.00	2.92	2.83	2.75	2.65
18	3.51	3.37	3.23	3.08	3.00	2.92	2.84	2.75	2.66	2.57
19	3.43	3.30	3.15	3.00	2.92	2.84	2.76	2.67	2.58	2.49
20	3.37	3.23	3.09	2.94	2.86	2.78	2.69	2.61	2.52	2.42
21	3.31	3.17	3.03	2.88	2.80	2.72	2.64	2.55	2.46	2.36
22	3.26	3.12	2.98	2.83	2.75	2.67	2.58	2.50	2.40	2.31
23	3.21	3.07	2.93	2.78	2.70	2.62	2.54	2.45	2.35	2.26
24	3.17	3.03	2.89	2.74	2.66	2.58	2.49	2.40	2.31	2.21
25	3.13	2.99	2.85	2.70	2.62	2.54	2.45	2.36	2.27	2.17
26	3.09	2.96	2.81	2.66	2.58	2.50	2.42	2.33	2.23	2.13
27	3.06	2.93	2.78	2.63	2.55	2.47	2.38	2.29	2.20	2.10
28	3.03	2.90	2.75	2.60	2.52	2.44	2.35	2.26	2.17	2.06
29	3.00	2.87	2.73	2.57	2.49	2.41	2.33	2.23	2.14	2.03
30	2.98	2.84	2.70	2.55	2.47	2.39	2.30	2.21	2.11	2.01
40	2.80	2.66	2.52	2.37	2.29	2.20	2.11	2.02	1.92	1.80
60	2.63	2.50	2.35	2.20	2.12	2.03	1.94	1.84	1.73	1.60
120	2.47	2.34	2.19	2.03	1.95	1.86	1.76	1.66	1.53	1.38
∞	2.32	2.18	2.04	1.88	1.79	1.70	1.59	1.47	1.32	1.00

Denominator Degrees of Freedom

Table XII   Critical Values of Spearman's Rank Correlation Coefficient

The α values correspond to a one-tailed test of H_0: $\rho = 0$. The value should be doubled for two-tailed tests.

n	$\alpha = .05$	$\alpha = .025$	$\alpha = .01$	$\alpha = .005$	n	$\alpha = .05$	$\alpha = .025$	$\alpha = .01$	$\alpha = .005$
5	.900	–	–	–	18	.399	.476	.564	.625
6	.829	.886	.943	–	19	.388	.462	.549	.608
7	.714	.786	.893	–	20	.377	.450	.534	.591
8	.643	.738	.833	.881	21	.368	.438	.521	.576
9	.600	.683	.783	.833	22	.359	.428	.508	.562
10	.564	.648	.745	.794	23	.351	.418	.496	.549
11	.523	.623	.736	.818	24	.343	.409	.485	.537
12	.497	.591	.703	.780	25	.336	.400	.475	.526
13	.475	.566	.673	.745	26	.329	.392	.465	.515
14	.457	.545	.646	.716	27	.323	.385	.456	.505
15	.441	.525	.623	.689	28	.317	.377	.448	.496
16	.425	.507	.601	.666	29	.311	.370	.440	.487
17	.412	.490	.582	.645	30	.305	.364	.432	.478

Based on E. G. Olds, "Distribution of Sums of Squares of Rank Differences for Small Samples." *Annals of Mathematical Statistics*, p. 9, © 1938.

Appendix B: Calculation Formulas for Analysis of Variance (Independent Sampling)

$$
\begin{aligned}
\text{CM} &= \text{Correction for mean} \\
&= \frac{(\text{Total of all observations})^2}{\text{Total number of abservations}} = \frac{(\Sigma y_i)^2}{n} \\
\text{SS(Total)} &= \text{Total sum of squares} \\
&= (\text{Sum of squares of all observations}) - \text{CM} = \Sigma y_i^2 - \text{CM} \\
\text{SST} &= \text{Sum of squares for treatments} \\
&= \left(\begin{array}{c} \text{Sum of squares of treatments totals with} \\ \text{each square divided by the number of} \\ \text{observations for that treatment} \end{array} \right) - \text{CM} \\
&= \frac{T_1^2}{n_1} + \frac{T_2^2}{n_2} + \cdots + \frac{T_k^2}{n_k} - \text{CM} \\
\text{SSE} &= \text{Sum of squares for error} = \text{SS(Total)} - \text{SST} \\
\text{MST} &= \text{Mean square for treatments} = \frac{\text{SST}}{k-1} \\
\text{MSE} &= \text{Mean square for error} = \frac{\text{SSE}}{n-k} \\
F &= \text{Test statistic} = \frac{\text{MST}}{\text{MSE}}
\end{aligned}
$$

where

$$
\begin{aligned}
n &= \text{Total number of observations} \\
k &= \text{Number of treatments} \\
T_i &= \text{Total for treatment } i \ (i = 1, 2, \ldots, k)
\end{aligned}
$$

Short Answers to Selected Odd Exercises

Chapter 1

1.3 population; variable(s); sample; inference; measure of reliability **1.11** qualitative; qualitative **1.13 a.** earthquake sites
b. sample **c.** ground motion (qualitative); magnitude (quantitative); ground acceleration (quantitative) **1.15 a.** (1) qualitative;
(2) quantitative; (3) qualitative **b.** sample **1.17** descriptive **1.19** Town, Type of water supply, and Presence of hydrogen sulphide
are qualitative; all others are quantitative **1.21 a.** quantitative **b.** quantitative **c.** qualitative **d.** quantitative **e.** qualitative **f.** quantitative
g. qualitative **1.23 a.** designed experiment **b.** smokers **c.** quantitative **d.** population: all smokers in the U.S.; sample: 50,000 smokers
in trial **e.** the difference in mean age at which each of the scanning methods first detects a tumor. **1.25. a.** designed experiment
b. amateur boxers **c.** heart rate (quantitative); blood lactate level (quantitative) **d.** no difference between the two groups of boxers
e. no **1.27 a.** population; all students; sample: 155 volunteer students **b.** designed experiment **c.** higher proportion of students in
guilty-state group chose not to repair the car than those in neutral-state and anger-state groups **d.** representativeness of sample
1.29 a. quantitative: age and dating experience; qualitative: gender and willingness to tell **b.** possible nonrepresentative sample
1.31 a. survey **b.** qualitative **c.** nonrepresentative sample **1.33 a.** results valid if only eat oat bran **b.** observational (survey) **c.** only
most positive responses reported **d.** children not surveyed about their level of hunger

Chapter 2

2.5 a. X−8, Y−9, Z−3 **b.** X−.40, Y−.45, Z−.15 **2.7 a.** qualitative **b.** Unknown−5, Unworn−2, Slight−4, Light/Mod−2, Mod−3,
Mod/Heavy−1, Heavy−1 **c.** Unknown−.278, Unworn−.111, Slight−.222, Light/Mod−.111, Mod−.167, Mod/Heavy−.056,
Heavy−.056 **d.** Unknown **2.9 a.** 39/266 = .147 **b.** level 2−.286, level 3−.188, level 4−.327, level 5−.041, level 6−.011
e. level 4 **2.11 a.** relative frequencies: Black−.203; White−.637; Sumatran−.017; Javan−.003; Indian−.140 **c.** .839; .161
2.13 a. .389 **b.** yes **c.** multiyear ice is most common **2.15** LEO−most government owned (43.7%); GEO−most commercially
owned (69.1%) **2.17** 50% of sampled CEOs had advanced degrees **2.19** No **2.21 b.** public−40% contaminated; private−21.4%
contaminated **c.** bedrock−31.3% contaminated; unconsolidated−31.8% contaminated **2.27 a.** 23 **2.29** frequencies: 50, 75, 125, 100,
25, 50, 50, 25 **2.31 a.** histogram **b.** 38 **c.** .475 **d.** .1375

2.33 a.

Stem	Leaf
1	6
2	0 1 2 4 8
3	0 4 4 9
4	0 0 2
5	3

b. A students tend to read the most books **2.35. b.** .962 **2.37.** 67% of frequencies exceed 3,500 hertz **2.39.** most of the PMIs
range from 3 to 7.5

2.41 No;

Stem	Leaf
1	0000000
2	0
3	00
4	000
5	
6	
7	
8	
9	0
10	0
11	0
12	00

2.43 histogram looks similar to graph; "inside job" not likely **2.45 a.** 33 **b.** 175 **c.** 20 **d.** 71 **e.** 1,089 **2.47 a.** 6 **b.** 50 **c.** 42.8 **2.49** mean,
median, mode **2.51** sample size and variability of the data **2.53 a.** mean < median **b.** mean > median **c.** mean = median
2.55 mode = 15; mean = 14.545; median = 15 **2.57 a.** 8.5 **b.** 25 **c.** .78 **d.** 13.44 **2.59** mean = 9.72; median = 10.94
2.61 a. mean = 31.6; median = 32; mode = 34 and 40 **b.** approx. symmetric **2.63 a.** mean = −4.86; median = −4.85;
mode = −5.00 **2.65 b.** probably none **2.67 a.** 16.5; increase **b.** 16.16; no change **c.** no mode **2.69 a.** mean = −.15; median = −.11;
4 modes **c.** mean = −.12; median = −.105; 4 modes **2.71** largest value minus smallest value **2.75** more variable **2.77 a.** 5, 3.7, 1.92

b. 99, 1949.25, 44.15 **c.** 98, 1307.84, 36.16 **2.79** data set 1: 1, 1, 2, 2, 3, 3, 4, 4, 5, 5; data set 2: 1, 1, 1, 1, 1, 5, 5, 5, 5, 5 **2.81 a.** 3, 1.3, 1.14 **b.** 3, 1.3, 1.14 **c.** 3, 1.3, 1.14 **2.83 a.** 51.26 **b.** 128.57 **c.** 11.34 **d.** σ^2 and σ **2.85 a.** 2.86 **b.** 3.26 **c.** 2.94 **d.** DM; honey **2.87 a.** 6.45; increase **b.** 3.883; increase **c.** 1.97; increase **2.89 a.** 10, 7.67, 2.77 **b.** 8, 5.15, 2.27; less variable **c.** 8, 5.06, 2.25; less variable **2.91 a.** dollars; quantitative **b.** at least 3/4; at least 8/9; nothing; nothing **2.93 a.** ≈68% **b.** ≈95% **c.** ≈all **2.95** range/6 = 104.17, range/4 = 156.25; no **2.97 a.** $\bar{x}$ = 95.7, s = 4.96 **b.** (90.74, 100.66); (85.77, 105.63); (80.81, 110.59) **c.** 89.2%, 96.2%, 97.8%; agree (Chebychev) **2.99 a.** unknown **b.** ≈84% **2.101 a.** (0, 153) **b.** (0, 281) **c.** hand rubbing appears to be more effective **2.103 a.** at least 8/9 of the velocities will fall within 936 ± 30 **b.** No **2.105 a.** 19 ± 195 **b.** 7 ± 147 **c.** SAT-Math **2.107** do not purchase **2.109 a.** 25%; 75% **b.** 50%; 50% **c.** 80%; 20% **d.** 16%; 84% **2.111 a.** 2 **b.** .5 **c.** 0 **d.** −2.5 **e.** sample: a, d; population: b, c **f.** above: a, b; below: d **2.113** μ = 60, σ = 10 **2.115** 26th percentile **2.117 a.** z = −1.5 **b.** .50 **2.119 a.** 23 **b.** z = 3.28 **2.121 a.** z = 5.08 **b.** z = −.81 **c.** yes **2.123 a.** z = 2.0: 3.7; z = −1.0: 2.2; z = .5: 2.95; z = −2.5: 1.45 **b.** 1.9 **c.** z = 1.0 and 2.0; GPA = 3.2 and 3.7; mound shaped; symmetric **2.129 a.** no, z = .73 **b.** yes, z = −3.27 **c.** no, z = 1.36 **d.** yes, z = 3.73 **2.131 a.** 4 **b.** Q_U ≈ 6, Q_L ≈ 3 **c.** 3 **d.** skewed right **e.** 50%; 75% **f.** 12, 13, and 16 **2.133 a.** 10, 15, 27.5 **b.** 3.5, 5, 7.5 **c.** effective **2.135 a.** − 1.26 **b.** No **2.137 a.** 69, 73, 74, 78, 83, 84, and 86 **b.** 69, 73, 74, and 78 **c.** no; highly skewed data **2.139** outliers: 28 and 33 **2.141 b.** σ_{2005} = 67.9; σ_{2009} = 81.4 **c.** no **2.147** slight positive linear trend **2.149** positive linear trend **2.151** negative trend, non-linear **2.153 a.** no **b.** yes, slight positive **c.** reliability is suspect (only 5 data points) **2.155 a.** negative trend **b.** positive trend with both plant coverage and diversity **2.157** Yes; accuracy decreases as driving distance increases **2.165 a.** −1, 1, 2 **b.** −2, 2, 4 **c.** 1, 3, 4 **d.** .1, .3, .4 **2.167 a.** 3.12 **b.** 9.02 **c.** 9.79 **2.169 a.** $\bar{x}$ = 5.67, s^2 = 1.07, s = 1.03 **b.** $\bar{x}$ = −$1.5, s^2 = 11.5, s = $3.39 **c.** $\bar{x}$ = .413%, s^2 = .088, s = .297% **d.** 3; $10; .7375% **2.171** yes, positive **2.173** skewed to the right **2.175 b.** z = −1.06 **2.177 a.** survey **b.** quantitative **c.** about 60% **2.181 a.** no **b.** unknown; at least 84%; at least 93.75% **c.** 55%; 97.5; 100% **d.** 87.5%; 95%; 100% **2.183** Over half of the whistle types were "Type a" **2.185 a.** no outliers **2.187 a.** Seabirds—quantitative; length—quantitative; oil—qualitative **b.** transect **c.** oiled: 37.5%; unoiled: 62.5% **e.** distributions are similar **f.** 3.27 ± 13.4 **g.** 3.50 ± 11.94 **h.** unoiled **2.189 b.** yes **c.** A1775A: = 19,462.2, s = 532.29; A1775B: = 22,838.5, s = 560.98 **d.** cluster A1775A **2.191** yes; z = −2.5 **2.193 a.** median **b.** mean

Chapter 3

3.9 a. .5 **b.** .3 **c.** .6 **3.11** $P(A)$ = .55; $P(B)$ = .50; $P(C)$ = .70 **3.13 a.** 10 **b.** 20 **c.** 15,504 **3.15 a.** (R_1R_2), (R_1R_3), (R_2R_3), (R_1B_1), (R_1B_2), (R_2B_1), (R_2B_2), (R_3B_1), (R_3B_2), (B_1B_2) **b.** 1/10 for each sample point **c.** $P(A)$ = 1/10, $P(B)$ = 6/10, $P(C)$ = 3/10 **3.17 a.** Blue, orange, green, yellow, brown, and red **b.** P(blue) = .24, P(orange) = .20, P(green) = .16, P(yellow) = .14, P(brown) = .13, P(red) = .13 **c.** .13 **d.** .43 **e.** .76 **3.19 a.** .01 **b.** yes **3.21 a.** (None), (1 or 2), (3–5), (6–9), (10 or more) **b.** P(None) = .25, P(1 or 2) = .31, $P(3 - 5)$ = .25, $P(6 - 9)$ = .05, P(10 or more) = .14 **c.** .44 **3.23** .748 **3.25** .389 **3.27 a.** 6 **b.** .282, .065, .339, .032, .008, .274 **c.** .686 **3.29 a.** 28 **b.** 1/28 **3.31** 693/1,686,366 = .000411 **3.33 a.** 15 **b.** 20 **c.** 15 **d.** 6 **3.39** $P(A \cup B)$ = $P(A)$ + $P(B)$ − $P(A \cap B)$ **3.41** $P(A \cup B)$ = $P(A)$ + $P(B)$ **3.43 b.** $P(A)$ = 7/8, $P(B)$ = 1/2, $P(A \cup B)$ = 7/8, $P(A^c)$ = 1/8, $P(A \cap B)$ = 1/2 **c.** 7/8 **d.** no **3.45 a.** $^3/_4$ **b.** 13/20 **c.** 1 **d.** 2/5 **e.** $^1/_4$ **f.** 7/20 **g.** 1 **h.** 1/4 **3.47 a.** .65 **b.** .72 **c.** .25 **d.** .08 **e.** .35 **f.** .72 **g.** .65 **h.** A and C, B and C, C and D **3.49. b.** .06 **c.** .94 **3.51. a.** School laboratory, In transit, Chemical plant, Non chemical plant, Other **b.** .06, .26, .21, .35, .12 **c.** .06 **d.** .56 **e.** .74 **3.53 b.** .43 **c.** .57 **3.55 a.** .333 **b.** .351 **c.** .263 **d.** .421 **3.57 a.** .156 **b.** .617 **c.** .210 **d.** .449 **e.** yes **3.59 a.** .684 **b.** .124 **c.** no **d.** .316 **e.** .717 **f.** .091 **3.61 a.** .19 **b.** .405 **c.** .595 **3.65** $P(A|B)$ = $P(A \cap B)/P(B)$ **3.67 a.** .5 **b.** .25 **c.** no **3.69 a.** .08 **b.** .40 **c.** .52 **3.71 a.** $P(A)$ = .4; $P(B)$ = .4; $P(A \cap B)$ = .3 **b.** $P(E_1|A)$ = .25, $P(E_2|A)$ = .25, $P(E_3|A)$ = .5 **c.** .75 **3.73** no **3.75** 1/3, 0, 1/14, 1/7, 1 **3.77 a.** .26 **b.** .013 **3.79** .40 **3.81 a.** .055 **b.** .214 **3.83** .013; independent events **3.85 a.** .23 **b.** .729 **3.87 a.** 38/132 = .2879 **b.** 29/123 = .236 **3.89 a.** .406 **b.** .294 **3.91** $P(A|B)$ = 0 **3.93 a.** (A defeats B, C defeats D, A defeats C), (A defeats B, C defeats D, C defeats A), (A defeats B, D defeats C, A defeats D), (A defeats B, D defeats C, D defeats A), (B defeats A, C defeats D, B defeats C), (B defeats A, C defeats D, C defeats B), (B defeats A, D defeats C, B defeats D), (B defeats A, D defeats C, D defeats B) **b.** $P(A$ wins) = 2/8 = .25 **c.** .576 **3.95 a.** .1 **b.** .522 **c.** The person is guessing **3.97 b.** Worst—.250, 2nd worst—.200, 3rd worst—.157, 4th worst—.120, 5th worst—.089, 6th worst—.064, 7th worst—.044, 8th worst—.029, 9th worst—.018, 10th worst—.011, 11th worst—.007, 12th worst—.006, 13th worst—.005 **c.** 250/800 = .313 **d.** .297 **e.** .288 **3.103 a.** 35, 820, 200 **b.** 1/35, 820, 200 **c.** highly unlikely **3.113 a.** $A \cup B$ **b.** B^c **c.** $A \cap B$ **d.** $A^c|B$ **3.115 a.** 0 **b.** no **3.117** .5 **3.119 c.** $P(A)$ = 1/4; $P(B)$ = 1/2 **e.** $P(A^c)$ = 3/4; $P(B^c)$ = 1/2; $P(A \cap B)$ = 1/4; $P(A \cup B)$ = 1/2; $P(A|B)$ = 1/2; $P(B|A)$ = 1 **f.** no, no **3.121 a.** no **b.** .3, .1 **c.** .37 **3.123** .05 **3.125 a.** false **b.** true **c.** true **d.** false **3.127** $A = \{8th \ grader \ scores \ above \ 655\}$; $P(A^c)$ = .95 **3.129 a.** .261 **b.** Trunk—.85, Leaves—.10, Branch—.05 **3.131 a.** {Single, shore parallel; Other; Planar} **b.** 1/3, 1/3, 1/3 **c.** 2/3 **d.** {No dunes/flat; Bluff/scarp; Single dune; Not observed} **e.** 1/3, 1/6, 1/3, 1/6 **f.** 2/3 **3.133 a.** { 11, 13, 15, 17, 29, 31, 33, 35} **b.** {2, 4, 6, 8, 10, 11, 13, 15, 17, 20, 22, 24, 26, 28, 29, 31, 33, 35, 1, 3, 5, 7, 9, 19, 21, 23, 25, 27} **c.** $P(A)$ = 9/19, $P(B)$ = 9/19, $P(A \cap B)$ = 4/19, $P (A \cup B)$ = 14/19, $P(C)$ = 9/19 **d.** {11, 13, 15, 17} **e.** 14/19, no **f.** 2/19 **g.** {1, 2, 3, $\cdots$, 29, 31, 33, 35} **h.** 16/19 **3.135 a.** .64, .32, .04 **b.** .72, .22, .06 **c.** dependent **3.137 a.** .116 **b.** .728 **3.139 a.** .7127 **b.** .2873 **c.** .9639 **d.** .3078 **e.** .0361 **f.** at least 3 **3.141 a.** .006 **b.** .0007 **c.** .001 **d.** .018 **e.** .006 **3.143 a.** 1 to 2 **b.** .5 **c.** .4 **3.145 a.** .25 **b.** .0156 **c.** .4219 **d.** .001, .000000001, .9970 **3.147** .993 **3.149** no **3.151** Marilyn

Chapter 4

4.3 a. discrete **b.** continuous **c.** continuous **d.** discrete **e.** continuous **f.** continuous **4.5 a.** continuous **b.** discrete **c.** discrete **d.** discrete **e.** discrete **f.** continuous **4.7** 0, 1, 2, 3, ...; discrete **4.9** continuous; discrete **4.11** gender; IQ score **4.13** allergy to penicillin; blood pressure **4.15** table, graph, formula **4.17** no **4.19 a.** .25 **b.** .35 **c.** .8 **4.21 a.** .7 **b.** .3 **c.** 1 **d.** .2 **e.** .8

4.23 a. 3.8 **b.** 10.56 **c.** 3.25 **e.** no **f.** yes   **4.25 a.** $\mu_x = 1, \mu_y = 1$ **b.** distribution of x **c.** $\mu_x = 1, \sigma^2 = .6; \mu_y = 1, \sigma^2 = .2$ **4.27 a.** 1 **b.** .24
c. .39   **4.29 a.** BB, BG, GB, GG **b.** 1/4, 1/4, 1/4, 1/4, **c.** $P(0) = .25, P(1) = .5, P(2) = .25$ **d.** $P(0) = .222, P(1) = .513, P(2) = .265$
4.31 a. .23 **b.** .081 **c.** .77 **4.33 a.** MC, MS, MB, MO, ML, CS, CB, CO, CL, SB, SO, SL, BO, BL, OL **b.** equally likely, with $p = 1/15$
d. $P(0) = 6/15, P(1) = 8/15, P(2) = 1/15$ **e.** .667   **4.35** $-\$0.263$ **4.37 a.** $P(6) = .282, P(7) = .065, P(8) = .339, P(9) = .032,$
$P(10) = .008, P(11) = .274$ **b.** .274   **4.39** $0.25   **4.41 a.** 0, 1, 2 **b.** $P(0) = .625, P(1) = .250, P(2) = .125$

4.43 $p(x) = \binom{7}{x}.2^x.8^{7-x}$ $(x = 0, 1, 2, \ldots, 7)$ **4.45 a.** 15 **b.** 10 **c.** 1 **d.** 1 **e.** 4   **4.47 a.** $\mu = 12.5, \sigma^2 = 6.25, \sigma = 2.5$
b. $\mu = 16, \sigma^2 = 12.8, \sigma = 3.578$ **c.** $\mu = 60, \sigma^2 = 24, \sigma = 4.899$ **d.** $\mu = 63, \sigma^2 = 6.3, \sigma = 2.510$ **e.** $\mu = 48, \sigma^2 = 9.6, \sigma = 3.098$
f. $\mu = 40, \sigma^2 = 38.4, \sigma = 6.197$ **4.49 a.** 0 **b.** .998 **c.** .137   **4.51 a.** $p = .5$ **b.** $p < .5$ **c.** $p > .5$ **4.53 b.** 40 **c.** 24 **d.** (30.2, 49.8)
4.55 b. $n = 20, p = .8$ **c.** .174 **d.** .804 **e.** 16   **4.57 b.** .4752 **c.** .0769   **4.59 a.** .375 **b.** .5   **4.61 a.** .791 **b.** .056 **c.** rare event if $p = .1$
4.63 a. .1 **b.** .7 **c.** .4783 **d.** .0078 **e.** yes   **4.65 b.** $\mu = 2.4, \sigma = 1.47$ **c.** $p = .9, q = .1, n = 24, \mu = 21.6, \sigma = 1.47$ **4.69** standard
normal   **4.71 a.** .4772 **b.** .3413 **c.** .4987 **d.** .2190   **4.73 a.** 0 **b.** .8413 **c.** .8413 **d.** .1587 **e.** .6826 **f.** .9544 **g.** .6934 **h.** .6378   **4.75 a.** 1
b. -1 **c.** 0 **d.** -2.5 **e.** 3   **4.77 a.** 1.645 **b.** 1.96 **c.** -1.96 **d.** 1.28 **e.** 1.28   **4.79 a.** 30 **b.** 14.32 **c.** 40.24 **d.** 16.84 **e.** 19.76 **f.** 36.72
g. 48.64   **4.81 a.** $-.25$ **b.** .44 **c.** .0122   **4.83 a.** .1583 **b.** .2434 **c.** .0096 **d.** .9406 **e.** no; $P(x \le 25) \approx 0$ **4.85 a.** .8413 **b.** .7528
4.87 a. .4107 **b.** .1508 **c.** .9066 **d.** .0162 **e.** .8418   **4.89 a.** no; $P(x \le 9) = .7527$ **b.** no; $P(x \le 2) = .0139$ **4.91 a.** (62.26, 65.74);
(61.01, 66.99); (59.72, 68.28); (58.90, 69.10); (57.30, 70.70)   **4.93** .7019   **4.95 a.** $z_L = -.675, z_U = .675$ **b.** $-2.68, 2.68$ **c.** $-4.69, 4.69$
d. .0074, 0   **4.99 a.** .68 **b.** .95 **c.** .997   **4.101** plot c   **4.103 a.** not normal (skewed right) **b.** $Q_L = 37.5, Q_U = 152.8, s = 95.8$
c. IQR/s = 1.204   **4.105 a.** 7 **b.** 6.444 **c.** IQR/s = 1.09   **4.107** no; IQR/s = .82   **4.109 a.** histogram too peaked **b.** more than
95%   **4.111** both distributions approx. normal   **4.113** IQR/s = 1.3, histogram approx. normal   **4.115** lowest possible time (0 minutes) is less than 1 standard deviation below the mean   **4.119 a.** yes **b.** $\mu = 10, \sigma^2 = 6$ **c.** .726 **d.** .7291   **4.121 a.** .345; .3446 **b.** .115;
.1151 **c.** .924; .9224   **4.123 a.** .1788 **b.** .5236 **c.** .6950   **4.125 a.** 16.25 **b.** 3.49 **c.** 1.07 **d.** .1762   **4.127 a.** $\mu = 1,500, \sigma^2 = 1,275$
b. .0026 **c.** no   **4.129** ≈ 0 **4.131 a.** no **b.** yes **c.** yes   **4.133** no; $P(x > 110) = .0018$ **4.135 a.** 300 **b.** 800 **c.** 1   **4.139 c.** 1/16
4.141 c. .05 **d.** no
4.143 a.

M	1	1.5	2	2.5	3	3.5	4	4.5	5
$P(M)$	.04	.12	.17	.20	.20	.14	.08	.04	.01

4.147 equal   **4.149** $\bar{x}$ is normally distributed for large n **4.151 a.** $\mu = 100, \sigma = 5$ **b.** $\mu = 100, \sigma = 2$ **c.** $\mu = 100, \sigma = 1$
d. $\mu = 100, \sigma = 1.414$ **e.** $\mu = 100, \sigma = .447$ **f.** $\mu = 100, \sigma = .316$ **4.153 a.** $\mu = 2.9, \sigma^2 = 3.29, \sigma = 1.814$
c. $\mu_{\bar{x}} = 2.9, \sigma_{\bar{x}} = 1.283$ **4.155 a.** $\mu_{\bar{x}} = 30, \sigma_{\bar{x}} = 1.6$ **b.** approx. normal **c.** .8944 **d.** .0228 **e.** .1303 **f.** .9699
4.157 a. $\mu_{\bar{x}} = 320, \sigma_{\bar{x}} = 10$, approx. normal **b.** .1359 **c.** ≈ 0 **4.159 a.** 79 **b.** 2.3 **c.** approx. normal **d.** .43 **e.** .3336   **4.161** No,
$P(\bar{x} < 103) = .0107$ **4.163 a.** 10; .0002 **b.** Central Limit Theorem **c.** .0170   **4.165 a.** $\mu_{\bar{x}} = .53, \sigma_{\bar{x}} = .0273$, approx. normal **b.** .0336
c. after   **4.167 a.** .0034 **b.** $\mu > 6$ **4.169** hand rubbing: $P(\bar{x} < 30) = .2743$; hand washing: $P(\bar{x} < 30) = .0047$; sample used hand
rubbing   **4.171 a.** yes **b.** no **c.** no **d.** yes   **4.173** false   **4.175 a.** .192 **b.** .228 **c.** .772 **d.** .987 **e.** .960 **f.** 14; 4.2; 2.05 **g.** .975
4.177 a. .6915 **b.** .0228 **c.** .5328 **d.** .3085 **e.** .0 **f.** .9938   **4.179 a.** .3821 **b.** .5398 **c.** .0 **d.** .1395 **e.** .0045 **f.** .4602   **4.181 a.** yes **b.** .051
c. .757 **d.** .192 **e.** 3.678   **4.183 b.** .05 **c.** 10   **4.185** 83.15%   **4.187 a.** .9406 **b.** .9406 **c.** .1140   **4.189 a.** 106 **b.** 2.73 **c.** approx. normal
d. -2.20 **e.** .0136   **4.191** .642   **4.193 a.** .0735 **b.** .3651 **c.** 7.29   **4.195** ≈ 0 **4.197 a.** No, prob. ≈ 0 **b.** likely that $\mu > 4.59$
4.199 a. .3264 **b.** 1.881 **c.** valid   **4.201** ≈ 292 **4.203 a.** .05, .20, .50, .20, .05 **b.** z-scores **c.** identical

Chapter 5
5.5 yes   **5.7 a.** 1.645   **b.** 2.58   **c.** 1.96   **d.** 1.28   **5.9 a.** 28 ± 4.53 **b.** 102 ± 2.77 **c.** 15 ± 3.92 **d.** 4.05 ± 3.92 **e.** no
5.11 a. 83.2 ± 1.25 **c.** 83.2 ± 1.65 **d.** increases   **e.** yes   **5.13 a.** 19.3   **b.** 19.3 ± 3.44 **d.** random sample, large n
5.15 39 ± 1.55 **5.17 a.** $.36 \pm .0099$ **c.** first-year: $.303 \pm .0305$; landfast: $.362 \pm .0177$; multiyear: $.381 \pm .0101$ **5.19 a.** μ
b. no; apply Central Limit Theorem   **c.** (0.4786, 0.8167)   **d.** yes   **5.21 a.** $1.13 \pm .67$ **b.** yes   **5.23 a.** 19 ± 7.83 **b.** 7 ± 5.90
c. SAT-Math   **5.25 a.** males: 16.79 ± 2.35; females: 10.79 ± 1.67 **b.** .0975   **c.** males   **5.27** Central Limit Theorem no longer
applies; σ unknown   **5.29 a.** large: normal; small: t-distribution   **b.** large: normal; small: unknown   **5.31 a.** 2.228   **b.** 2.567
c. -3.707 **d.** -1.771 **5.33. a.** 5 ± 1.88 **b.** 5 ± 2.39 **c.** 5 ± 3.75 **d.** (a) $5 \pm .78$, (b) $5 \pm .94$, (c) 5 ± 1.28; width
decreases   **5.35. a.** μ = population mean trap spacing   **b.** 89.86   **c.** population of trap spacings has an unknown distribution
d. 89.86 ± 10.76 **e.** random sample; population of trap spacings is normally distributed   **5.37 a.** (652.76, 817.40)   **d.** yes
5.39 a. .009   **b.** 2.306   **c.** $.009 \pm .0037$ **5.41** $.11 \pm .08$ **5.43 a.** both untreated: 20.9 ± 1.06; male treated: 20.3 ± 1.25; female
treated: 22.9 ± 1.79; both treated: $18.6 \pm .79$ **b.** female treated   **5.45 a.** $1.43 \pm .15$ **c.** 90% of all similarly constructed intervals
contain μ **5.47 a.** 37.3 ± 7.70 **b.** One hour before: $\mu > 25.5$ **5.49** mean of sampling distribution of $\hat{p}$ is equal to p **5.51 a.** yes
b. $.64 \pm .07$ **5.53 a.** yes   **b.** no   **c.** no   **d.** no   **5.55 a.** all American adults   **b.** 1,000 adults surveyed   **c.** p = proportion of
adults who say Starbucks coffee is overpriced   **d.** $.73 \pm .03$ **5.57 a.** set of all gun ownership status (yes/no) values for all U.S.
adults   **b.** true percentage of all adults who own a gun   **c.** .26   **d.** $.26 \pm .02$ **5.59 a.** $.63 \pm .03$ **b.** yes   **5.61** $.64 \pm .05$
5.63 a. $.338 \pm .025$ **5.65 a.** no   **b.** $.009 \pm .010$ **5.67** 95% confident that proportion of health care workers with
latex allergy who suspect they have allergy is between .327 and .541   **5.69** true   **5.71** 519   **5.73 a.** 482   **b.** 214   **5.75** 34
5.77 a. $n = 16$: $W = .98$; $n = 25$: $W = .784$; $n = 49$: $W = .56$; $n = 100$: $W = .392$; $n = 400$: $W = .196$ **5.79** 21   **5.81 a.** small sample
b. 125   **5.83** 21   **5.85** 40   **5.87** 14,735   **5.89** $n = 1,041$ **5.91** 271   **5.93** chi-square   **5.95** $n-1$ **5.97 a.** (4.54, 8.81)   **b.** (.00024,
.00085)   **c.** (641.86, 1809.09)   **d.** (.95, 12.66)   **5.99** (4.27, 65.90)   **5.101 a.** (308.17, 537.93)   **b.** yes

5.103 a. $(8.6, 45.7)$ **b.** $(2.94, 6.76)$ **c.** random sample from a normal population **5.105** $(1.4785, 2.9533)$
5.107 $(43.76, 1200.99)$ **5.109 a.** μ **b.** μ **c.** p **d.** p **e.** p **f.** σ^2 **5.111 a.** $t = 2.086$ **b.** $z = 1.96$ **c.** $z = 1.96$ **d.** $z = 1.96$
e. neither t nor z **5.113 a.** $.57 \pm .05$ **b.** 2,358 **5.115** (1) p; (2) μ; (3) μ; (4) μ **5.117** $.219 \pm .024$ **5.119 a.** $.03$
b. $.03 \pm .01$ **5.121 a.** $.90$ **b.** $.05$ **c.** 259 **5.123 a.** $\bar{x} = 1.86, s = 1.195$ **b.** skewed right **c.** $1.86 \pm .45$ **e.** 90%
5.125 a. $1.07 \pm .24$ **c.** $(.145, .568)$ **d.** normal **e.** 129 **5.127 a.** $.660 \pm .029$ **b.** $.301 \pm .035$ **5.129 a.** $.044 \pm .162$
b. no evidence of species inbreeding **5.131 a.** $.81 \pm .41$ **b.** 140 **c.** $(1.27, 1.89)$ **5.133 a.** $.390 \pm .149$ **b.** $.683 \pm .142$
c. 41.405 ± 21.493 **d.** $.390 \pm .149$ **5.135 a.** $.094$ **b.** yes **c.** $.094 \pm .037$ **5.137 a.** 49.3 ± 8.6 **c.** population is normal **d.** 60
e. $(.75, 43.87)$ **5.139. a.** 196 **5.141 a.** yes **b.** missing measure of reliability **c.** 95% CI for μ: $.932 \pm .037$

Chapter 6

6.1 null; alternative **6.3** α **6.5** reject H_0 when H_0 is true; accept H_0 when H_0 is true; reject H_0 when H_0 is false; accept H_0
when H_0 is false **6.7** no **6.9 g.** (a) .025, (b) .05, (c) $\approx$.005, (d) $\approx$.10, (e) .10, (f) $\approx$ .01 **6.11 a.** H_0:$\mu = 2400, H_a$:$\mu > 2400$
b. the probability of concluding that the mean gain in fees is greater than \$2,400 when, in fact, it is equal to \$2,400 is .05
c. $z > 1.645$ **6.13** H_0: $p = .75$ **6.15** H_0: $p = .05, H_a$: $p < .05$ **6.17 a.** H_0: $\mu = 15, H_a$: $\mu < 15$ **b.** conclude mean mercury level is
less than 15 ppm when mean equals 15 ppm **c.** conclude mean mercury level equals 15 ppm when mean is less than 15 ppm
6.19 a. H_0: No intrusion occurs **b.** H_a: Intrusion occurs **c.** $\alpha = .001, \beta = .5$ **6.21** random sample, large n **6.23 a.** $z = 1.67$, reject H_0
b. $z = 1.67$, fail to reject H_0 **6.25 a.** $z > 1.875$ **b.** $\approx .03$ **6.27 a.** Type I: conclude true mean response for all New York City public
school children is not 3 when mean equals 3; Type II: conclude true mean response equals 3 when mean is not equal to 3
b. $z = -85.52$, reject H_0 **c.** $z = -85.52$, reject H_0 **6.29 a.** H_0: $\mu = 85, H_a$: $\mu \neq 85$ **b.** conclude that the true mean Mach rating score
is different from 85 when, in fact, it is equal to 85 **c.** probability of concluding that the true mean Mach rating score is different from
85 when, in fact, it is equal to 85 is .10 **d.** $|z| > 1.645$ **e.** $z = 12.80$ **f.** reject H_0 **g.** no **6.31** $z = 4.17$, reject H_0 **6.33 a.** $z = 4.03$,
reject H_0 **6.35 a** no **b.** $z = .61$, do not reject H_0 **d.** no **e.** $z = -.83$, do not reject H_0 **6.37 a.** $z = -3.72$, reject H_0 **6.39** small
p-values **6.41 a.** fail to reject H_0 **b.** reject H_0 **c.** reject H_0 **d.** fail to reject H_0 **e.** fail to reject H_0 **6.43** .0150 **6.45** p-value $= .9279$,
fail to reject H_0 **6.47 a.** fail to reject H_0 **b.** fail to reject H_0 **c.** reject H_0 **d.** fail to reject H_0 **6.49 a.** p-value ≈ 0 **b.** reject H_0
6.51 p-value $< .0001$; reject H_0 at $\alpha = .01$ **6.53 a.** $z = 26.15$, p-value ≈ 0; reject H_0 **b.** $z = -14.24$, p-value ≈ 0; reject H_0
6.55 b. reject H_0 at $\alpha = .05$ **c.** reject H_0 at $\alpha = .05$ **6.57** mound-shaped, symmetric; t-distribution is flatter than z-distribution
6.61 a. $|t| > 2.160$ **b.** $t > 2.5$ **c.** $t > 1.397$ **d.** $t < -2.718$ **e.** $|t| > 1.729$ **f.** $t < -2.353$ **6.63 a.** $t = -2.064$; fail to reject H_0
b. $t = -2.064$; fail to reject H_0 **c.** (a) $.05 < p$-value $< .10$; (b) $.10 < p$-value $< .20$ **6.65 a.** H_0:$\mu = 95, H_a$: $\mu \neq 95$ **b.** variation in
sample data not taken into account **c.** $t = -1.163$ **d.** p-value $= .289$ **e.** $\alpha = .10$; probability of concluding mean strap spacing
differs from 95 when, in fact, it is equal to 95 is .10 **f.** fail to reject H_0 **g.** random sample from a normal population of strap spacing
measurements **h.** yes **6.67** $t = -.32$; fail to reject H_0 **6.69** yes; $z = 3.23$, p-value $= .005$, reject H_0: $\mu = 15$ **6.71** $t = 3.725$,
p-value $= .0058$, reject H_0 **6.73** yes, $t = -2.53$ **6.75 a.** $t = .21$, p-value $= .84$, fail to reject H_0: $\mu = 0$ **b.** average of
positive and negative scores will tend to cancel **6.77** qualitative **6.79 a.** yes **b.** no **c.** no **d.** no **e.** no **6.81 a.** $z = -2.00$ **c.** reject H_0
d. .0228 **6.83 a.** $z = 1.13$, fail to reject H_0 **b.** .1292 **6.85 a.** p **b.** H_0: $p = .02, H_a$: $p > .02$ **c.** $z = 14.23$; $z > 1.645$
d. reject H_0 **e.** large n; yes **6.87** yes, $z = 1.74$ **6.89** yes; $z = 3.14$, reject H_0: $p = .70$ **6.91** no, $z = 1.52$ **6.93 a.** no, $z = 1.49$
(but inadequate sample size) **b.** p-value $\approx .07$ **6.95** 10 **6.97** random sample from a normal population **6.99** false
6.101 a. $\chi^2 < 6.26214$ or $\chi^2 > 27.4884$ **b.** $\chi^2 > 40.2894$ **c.** $\chi^2 > 21.0642$ **d.** $\chi^2 < 3.57056$ **e.** $\chi^2 < 1.63539$ or $\chi^2 > 12.5916$
f. $\chi^2 < 13.8484$ **6.103 a.** $\chi^2 = 182.16$, reject H_0 **6.105 a.** $\chi^2 < 20.7065$ or $\chi^2 > 66.7659$ **b.** $\chi^2 = 63.72$ **c.** do not reject H_0
6.107 a. σ^2, the variance of the population of rock bounces **b.** H_0: $\sigma^2 = 10, H_a$: $\sigma^2 \neq 10$ **c.** $\chi^2 = 20.12$ **d.** $\chi^2 > 21.0261$ or
$\chi^2 < 5.22603$ **e.** do not reject H_0 **f.** population of rock bounces is normally distributed **6.109** $\chi^2 = 187.90$, do not
reject H_0: $\sigma^2 = 225$ **6.111** $\chi^2 = 12.61$, fail to reject H_0 **6.113** no; $\chi^2 = 13.77$, fail to reject H_0 **6.115** population not normal
6.117 a. .063 **b.** .500 **c.** .004 **d.** .151; .1515 **e.** .212; .2119 **6.119** $S = 16$, p-value $= .115$, do not reject H_0 **6.121 a.** H_0: $\eta = 15$,
H_a: $\eta \neq 15$ **b.** data not normal **c.** $S = 14$ **d.** .0309; reject H_0 **6.123 a.** H_0: $\eta = 2.8, H_a$: $\eta > 2.8$ **b.** $S = 21$, $z = 2.01$ **c.** .0222 **d.** do
not reject H_0 **6.125 a.** invalid inference **b.** sign test of H_0: $\eta = 95$ vs. H_a: $\eta \neq 95$ **c.** $S = 5$ **d.** .454 **e.** do not reject H_0
6.127 no; $S = 10$, p-value $= .3145$ **6.129** $S = 8$, p-value $= .8204$, do not reject H_0 **6.131** alternative **6.133** null **6.135 a.** $z = -1.78$,
reject H_0 **b.** $z = -1.78$, fail to reject H_0 **c.** $.29 \pm .063$ **d.** $.29 \pm .083$ **e.** 549 **6.137 a.** $\chi^2 = 63.48$, reject H_0 **b.** $\chi^2 = 63.48$, reject H_0
6.139 a. H_0: $p = .45$ **b.** H_0: $\mu = 2.5$ **6.141 a.** $t = -3.46$, reject H_0 **b.** normal population **6.143** $s = 8$, p-value $= .0391$, reject H_0
6.145 a. H_0: $p = .5, H_a$: $p \neq .5$ **b.** $z = 5.99$ **c.** yes **d.** reject H_0 **6.147** yes, $z = 12.36$ **6.149 a.** H_0: $\mu = 16, H_a$: $\mu < 16$
b. $z = -4.31$, reject H_0 **6.151 a.** $z = .70$, fail to reject H_0 **b.** yes **6.153 a.** no, $z = 1.41$ **b.** small α **8.155** $z = 15.46$, reject H_0:
$p = .167$ **6.157 b.** .37, .24 **6.159 a.** yes, $z = -2.33$ **b.** .8925 **6.161 a.** no at $\alpha = .01$, $z = 1.85$ **b.** yes at $\alpha = .01$, $z = 3.10$

Chapter 7

7.1 normally distributed with mean $\mu_1 - \mu_2$ and standard deviation $\sqrt{\left(\dfrac{\sigma_1^2}{n_1} + \dfrac{\sigma_2^2}{n_2}\right)}$

7.3 a. no **b.** no **c.** no **d.** yes **e.** no **7.5** b **7.7 a.** 14; .4 **b.** 10; .3 **c.** 4; .5 **d.** yes **7.9 a.** .5989 **b.** $t = -2.39$, reject H_0,
c. $-1.24 \pm .98$ **7.11 a.** $(\mu_1 - \mu_2)$ **b.** H_0: $(\mu_1 - \mu_2) = 0, H_a$: $(\mu_1 - \mu_2) > 0$ **c.** yes; fail to reject H_0 **7.13 a.** H_0: $(\mu_1 - \mu_2) = 0$,
H_a: $(\mu_1 - \mu_2) \neq 0$ **b.** $t = .62$ **c.** $|t| > 1.684$ **d.** fail to reject H_0; supports theory **7.15 a.** $\mu_1 - \mu_2$ **b.** 12.5 ± 10.2 **7.17 a.** $(-.60, 7.95)$

b. independent random samples from normal populations with equal variances **7.19** $t = .57$, p-value $= .573$, fail to reject H_0: $(\mu_1 - \mu_2) = 0$ **7.21** $t = 1.08$, fail to reject H_0 **7.23** no, $t = .18$ **7.25 a.** $\mu_1 - \mu_2$ **b.** no **c.** $.2262$ **d.** fail to reject H_0 **7.27 a.** no standard deviations reported **b.** $s_1 = s_2 = 5$ **c.** $s_1 = s_2 = 6$ **7.29** before **7.31 a.** $t > 1.833$ **b.** $t > 1.328$ **c.** $t > 2.776$ **d.** $t > 2.896$ **7.33 a.** $\bar{x}_d = 2$, $s_d^2 = 2$ **b.** $\mu_d = \mu_1 - \mu_2$ **c.** 2 ± 1.484 **d.** $t = 3.46$, reject H_0 **7.35 a.** $z = 1.79$, fail to reject H_0 **b.** $.0734$ **7.37 a.** two measurements for each patient—before and after **c.** $\bar{x}_d = -7.63$, $s_d = 5.27$ **d.** -7.63 ± 1.53 **e.** yes; evidence that $\mu_{\text{before}} < \mu_{\text{after}}$ **7.39 a.** μ_d **b.** paired difference **c.** H_0: $\mu_d = 0$, H_a: $\mu_d > 0$ **d.** $t = 2.19$ **e.** reject H_0 **7.41** $t = 2.92$, fail to reject H_0; no evidence of a difference **7.43 a.** two scores for each twin pair **b.** 95% CI for μ_d: 1.95 ± 1.91; control group has larger mean **7.45** $t = 3.00$, reject H_0; after camera mean is smaller than before camera mean **7.47** yes; $t = -2.31$, reject H_0 **7.53** $n_1 = n_2 = 24$ **7.55** $n_1 = n_2 = 49$ **7.57** $n_1 = n_2 = 130$ **7.59** $n_1 = n_2 = 25$ **7.61** true **7.63 a.** $T_1 \le 41$, $T_1 \ge 71$ **b.** $T_1 \ge 50$ **c.** $T_1 \le 43$ **d.** $|z| > 1.96$ **7.65** yes, $z = -2.47$ **7.67 a.** $T_1 = 62.5$, reject H_0 **b.** yes, $T_1 = 62.5$ **7.69 a.** rank sum test **b.** H_0: $D_{\text{Low}} = D_{\text{High}}$, H_a: $D_{\text{Low}} > D_{\text{High}}$ **c.** $T_2 \le 41$ **d.** do not reject H_0 **7.71 a.** H_0: $D_{\text{Bulimic}} = D_{\text{Normal}}$, H_a: $D_{\text{Bulimic}} > D_{\text{Normal}}$ **c.** 174.5 **d.** 150.5 **e.** $z > 1.28$ **f.** $z = 1.72$, reject H_0 **7.73** yes, $z = 2.86$ **7.75 a.** rank sum test **b.** H_0: $D_{\text{CMC}} = D_{\text{FTF}}$, H_a: $D_{\text{CMC}} < D_{\text{FTF}}$ **c.** $z < -1.28$ **d.** $z = -.21$, fail to reject H_0 **7.77 a.** data not normal **b.** $z = -6.47$, reject H_0 **c.** $z = -.39$, do not reject H_0 **d.** $z = -1.35$, do not reject H_0 **7.79** data must be ranked **7.81 a.** H_0: $D_A = D_B$, H_a: $D_A > D_B$ **b.** $T_- = 3.5$, reject H_0 **7.83 a.** H_0: $D_A = D_B$, H_a: $D_A > D_B$ **b.** $z = 2.5$, reject H_0 **c.** $.0062$ **7.85 a.** $z = -4.638$ **b.** yes, p-value ≈ 0 **7.87 a.** H_0: $D_{\text{good}} = D_{\text{average}}$, H_a: $D_{\text{good}} \ne D_{\text{average}}$ **d.** $T = 3$ **e.** $T \le 4$ **f.** reject H_0 **7.89** $T_+ = 3.5$, reject H_0; program was effective **7.91** no, $T_- = 7$ **7.93** yes, $T_- = 50.5$ **7.95** independent random samples from treatment populations, or, randomly assign treatments to experimental units **7.97** True **7.99 a.** plot b **b.** 9; 14 **c.** 75; 75 **d.** 20; 144 **e.** 95 (78.95%); 219 (34.25%) **f.** MST $= 75$, MSE $= 2$, $F = 37.5$; MST $= 75$, MSE $= 14.4$, $F = 5.21$ **g.** reject H_0; reject H_0 **h.** both populations normal with equal variances **j.** plot a: df(T) $= 1$, df(E) $= 10$, df(Total) $= 11$, SST $= 75$, SSE $= 20$, SS(Total) $= 95$, MST $= 75$, MSE $= 2$, $F = 37.5$; plot b: df(T) $= 1$, df(E) $= 10$, df(Total) $= 11$, SST $= 75$, SSE $= 144$, SS(Total) $= 219$, MST $= 75$, MSE $= 14.4$, $F = 5.21$ **7.101 b.** not valid **7.103 a.** exp. units: coaches; dep. variable: 7-point rating; factor: division; treatments: I, II, III **b.** H_0: $\mu_I = \mu_{II} = \mu_{III}$ **c.** reject H_0; **7.105 a.** completely randomized **b.** treatments: 3, 6, 9, 12 robots; dep. variable: energy expended **c.** H_0: $\mu_3 = \mu_6 = \mu_9 = \mu_{12}$, H_a: At least 2 μ's differ **d.** reject H_0 **7.107 a.** TV viewers **b.** recall score **c.** program rating; V, S, neutral **d.** variances not taken into account **e.** $F = 20.45$, p-value $= .000$ **f.** reject H_0; mean recall scores differ among program groups **7.109 a.** completely randomized; honey dosage, DM dosage, no dosage **b.** $F = 17.51$, p-value $= .000$, reject H_0: $\mu_{\text{Honey}} = \mu_{\text{DM}} = \mu_{\text{Control}}$ **7.111** yes, $F = 7.25$ **7.113 a.** $\mu_1 - \mu_2$ **b.** $\mu_1 - \mu_2$ **c.** μ_1, μ_2, μ_3 **7.115 a.** $t = .78$, fail to reject H_0 **b.** 2.5 ± 8.99 **c.** $n_1 = n_2 = 225$ **7.117 a.** $3.9 \pm .31$ **b.** $z = 20.60$, reject H_0 **c.** $n_1 = n_2 = 346$ **7.119** Yes; $T_- = 1.5$ **c.** $n_1 = n_2 = 346$ **7.121** p-value $= .871$, fail to reject H_0: $\mu_{\text{no}} - \mu_{\text{yes}} = 0$ **7.123 a.** time needed **b.** climbers **c.** paired experiment **7.125 a.** $z = -2.64$, reject H_0 **b.** $z = .27$, fail to reject H_0 **7.127** yes; $(7.43, 13.57)$ **7.129 a.** H_0: $\mu_d = 0$, H_a: $\mu_d \ne 0$ **b.** $z = 2.08$, p-value $= .0376$ **c.** reject H_0 **7.131 a.** $t = -.46$, fail to reject H_0 **b.** $T_1 = 95.5$, fail to reject H_0 **7.133** $t = 1.77$, fail to reject H_0 at $\alpha = .01$ **7.135** $F = 3.95$, reject H_0 **7.137** $-.33 \pm .22$ **7.139** $T_- = 50$; fail to reject H_0 **7.141** use of creative ideas ($z = 8.85$); good use of job skills ($z = 4.76$)

Chapter 8

8.1 large, independent samples **8.3 a.** binomial **b.** normal **8.5 a.** $.07 \pm .067$ **b.** $.06 \pm .086$ **c.** $-.15 \pm .131$ **8.7** $z = 1.16$, fail to reject H_0 **8.9 a.** $.55$ **b.** $.70$ **c.** $-.15 \pm .03$ **d.** 90% confidence **e.** Dutch boys **8.11 a.** $.143$ **b.** $.049$ **c.** $z = 2.27$, fail to reject H_0 **d.** reject H_0 **8.13 a.** $z = 94.35$, reject H_0 **b.** $-.139 \pm .016$ **8.15** yes, $z = 11.05$ **8.17** yes; $z = 2.13$, reject H_0 at $\alpha = .05$ **8.19** $.48 \pm .144$; proportion greater for those who slept **8.21** Theory 1: $z = 2.79$, reject H_0, support theory; Theory 2: $z = -.30$, fail to reject H_0, support theory **8.25** $n_1 = n_2 = 520$ **8.27 a.** $n_1 = n_2 = 5051$ **b.** may be impractical to obtain such a large sample **c.** difference almost meaningless **8.29** $n_1 = 520$, $n_2 = 260$ **8.31** (1) n identical trials, (2) k possible outcomes to each trial, (3) probabilities of outcomes sum to 1, (4) probabilities remain the same from trial to trial, (5) trials are independent **8.33 a.** 18.3070 **b.** 29.7067 **c.** 23.5418 **d.** 79.4900 **8.35 a.** $\chi^2 > 5.99147$ **b.** $\chi^2 > 7.77944$ **c.** $\chi^2 > 11.3449$ **8.37 a.** no, $\chi^2 = 3.293$ **c.** $.288 \pm .062$ **8.39 a.** jaw habits; bruxism, clenching, bruxism and clenching, neither **c.** H_0: $p_1 = p_2 = p_3 = p_4 = .25$, H_a: At least one p_i differs from $.25$ **d.** 15 **e.** $\chi^2 = 25.73$ **f.** $\chi^2 > 7.81473$ **g.** reject H_0 **h.** $(.37, .63)$ **8.41 a.** Pottery type; burnished, monochrome, painted, other **b.** $p_1 = p_2 = p_3 = p_4 = .25$ **c.** H_0: $p_1 = p_2 = p_3 = p_4 = .25$ **d.** $\chi^2 = 436.59$ **e.** p-value ≈ 0, reject H_0 **8.43 a.** $P(\text{GG}) = P(\text{BG}) = P(\text{GB}) = P(\text{BB}) = 1/4$ **b.** $10{,}722$ **c.** 187.04 **d.** reject H_0 **e.** $E_{\text{GG}} = 10{,}205.2$, $E_{\text{BG}} = 10{,}715.6$, $E_{\text{GB}} = 10{,}715.6$, $E_{\text{BB}} = 11{,}251.7$; $\chi^2 = 64.95$; reject H_0 **8.45 a.** Answer 1: 450; Answer 2: 627; Answer 3: 219; Answer 4: 23 **b.** H_0: $p_1 = p_2 = p_3 = p_4 = .25$ **c.** 329.75 for each category **d.** 634.36 **e.** reject H_0 **f.** $\chi^2 = 39.29$, reject H_0 **8.47** $\chi^2 = 4.84$, do not reject H_0 **8.49 a.** yes, $\chi^2 = 360.48$ **b.** $(.165, .223)$ **8.51** column totals for one qualitative variable are known in advance **8.53** expected cell counts are all at least 5 **8.55 a.** H_0: Row & Column are independent, H_a: Row & Column are dependent **b.** $\chi^2 > 9.21034$

c.

	Column 1	Column 2	Column 3
Row 1	14.37	36.79	44.84
Row 2	10.63	27.21	33.16

d. $\chi^2 = 8.71$, fail to reject H_0 **8.57** $\chi^2 = 12.33$, reject H_0 **8.59 a.** $.078$ **b.** $.157$ **c.** possibly **d.** H_0: Presence of Dog & Gender are independent **e.** $\chi^2 = 2.25$, fail to reject H_0 **f.** $\chi^2 = .064$, fail to reject H_0 **8.61 a.** $.653$ **b.** $.555$ **c.** $.355$ **d.** yes **e.** H_0: Authenticity of news story & Tone are independent **f.** $\chi^2 = 10.427$ (p-value $= .005$), reject H_0 **8.63 a.** masculinity risk (high and low); event (violent and

avoided-violent) **b.** 1,507 newly incarcerated men **c.** 260.8 **d.** 118.2, 776.2, 351.8 **e.** $\chi^2 = 10.1$ **f.** reject H_0 **8.65** $\chi^2 = 23.46$, reject H_0
8.67 yes; $\chi^2 = 3.29$, fail to reject H_0 **8.69** yes, $\chi^2 = 7.38$, p-value $= .025$; no **8.71 a.** no
b.

SHSS:C	CAHS		
	Low	Medium	High
Low	32	14	2
Medium	11	14	6
High	6	16	29

c. yes **d.** $\chi^2 = 46.71$, reject H_0 **8.73 a.** $\chi^2 = 4.407$, reject H_0 **b.** no **c.** .0438 **d.** .0057; .0003 **e.** p-value $= .0498$, reject H_0
8.75 false **8.77 a.** $z = -2.04$, reject H_0 **b.** $-.10 \pm .096$ **c.** $n_1 = n_2 = 72{,}991$ **8.79 a.** yes, $\chi^2 = 54.14$ **b.** no **c.** yes
d.

	Col. 1	Col. 2	Col. 3	Totals
Row 1	.400	.222	.091	.200
Row 2	.200	.222	.636	.400
Row 3	.400	.556	.273	.400

8.81 a. Location (downtown, central city, suburban area) **b.** H_0: $p_1 = .40$, $p_2 = .30$, $p_3 = .30$ **c.** 45.2, 33.9, 33.9 **d.** 6.17 **e.** .046; reject H_0
8.83 a.

Genetic Trait	LLD		
	Yes	No	Total
Yes	21	15	36
No	150	306	456
Total	171	321	492

b. $\chi^2 = 9.52$, reject H_0 **8.85** $\chi^2 = 35.41$, reject H_0 **8.87 a.** .153, **b.** .215 **c.** $-.062 \pm .070$ **8.89** Internet: $\chi^2 = .512$,
do not reject H_0; party: $\chi^2 = 164.76$, reject H_0; military: $\chi^2 = 8.3$, reject H_0; views: $\chi^2 = 174.39$, reject H_0; race: $\chi^2 = 69.18$,
reject H_0; income: $\chi^2 = 16.39$, reject H_0; community: $\chi^2 = 17.62$, reject H_0 **8.91 a.** $\chi^2 = 17.16$, reject H_0 **b.** $.376 \pm .103$ **c.** yes
8.93 $\chi^2 = 6.17$, reject H_0 **8.95 a.** no, $\chi^2 = 7.39$ **b.** $(.456, .518)$ **8.97 a.** $\chi^2 = 164.90$, reject H_0 **b.** Gender: $\chi^2 = 6.80$, reject H_0;
Education: $\chi^2 = 6.59$, fail to reject H_0; Work: $\chi^2 = 11.69$, fail to reject H_0; Satisfaction: $\chi^2 = 30.42$, reject H_0 **8.99 a.** $\chi^2 = 9.65$
b. 11.0705 **c.** no **d.** $.05 < p$-value $< .10$ **8.101** $\chi^2 = 2.28$; insufficient evidence to reject the null hypothesis of independence

Chapter 9

9.7 $\beta_1 = 1/3$, $\beta_0 = 14/3$, $y = 14/3 + (1/3)x$ **9.11** difference between the observed and predicted **9.13** true
9.15 b. $\hat{y} = 7.10 - .78x$ **9.17 c.** $\hat{\beta}_1 = .918$, $\hat{\beta}_0 = .020$ **e.** -1 to 7 **9.19 a.** $y = \beta_0 + \beta_1 x + \varepsilon$ **b.** $\hat{y} = 19.393 - 8.036x$ **c.** y-intercept:
when concentration $= 0$, predicted wicking length $= 19.393$ mm; slope: for every 1-unit increase in concentration, wicking length
decreases 8.036 mm **9.21 c.** hoop pine **9.23 a.** positive **9.25 a.** $y = \beta_0 + \beta_1 x + \varepsilon$ **b.** $\hat{y} = 250.14 - .629x$ **e.** slope
9.27 a. $y = \beta_0 + \beta_1 x + \varepsilon$ **b.** positive **c.** $\hat{\beta}_1 = 210.8$: for every additional resonance, frequency increases 210.8; $\hat{\beta}_0 = 1469.4$: no practical interpretation **9.29 a.** $\hat{y} = 86.0 - .260x$ **b.** yes **c.** positive trend for female students; positive trend for male students
d. $\hat{y} = 39.3 + .493x$; for every 1-inch increase in height for females, ideal partner's height increases .493 inch **e.** $\hat{y} = 23.3 + .596x$; for
every 1-inch increase in height for males, ideal partner's height increases .596 inch **f.** yes **9.31** yes, $\hat{y} = 5.22 - .114x$; decrease by
.114 pound **9.35 a.** 57.5; 3.194 **b.** 257.5; 6.776 **c.** 9.288; 1.161 **9.37** 11.14: SSE $= 1.22$, $s^2 = .244$, $s = .494$; 11.17: SSE $= 5.134$,
$s^2 = 1.03$, $s = 1.01$ **9.39 a.** SSE $= 22.268$, $s^2 = 5.567$, $s = 2.3594$ **b.** $\approx 95\%$ of wicking length values fall within 4.72 mm of their
respective predicted values **9.41 a.** SSE $= 2760$, $s^2 = 307$, and $s = 17.51$ **9.43 a.** 5.36 **b.** 3.42 **c.** reading score
9.45 a. $\hat{y} = 23.3 + .596x$; $s = 2.06$ **b.** $\hat{y} = 39.3 + .493x$; $s = 2.32$ **c.** males **9.47** 0 **9.49** divide the value in half
9.51 a. 95%: 31 ± 1.13; 90%: $31 \pm .92$ **b.** 95%: 64 ± 4.28; 90%: 64 ± 3.53 **c.** 95%: $-.84 \pm .67$; 90%: $-.84 \pm .55$
9.53 b. $\hat{y} = 2.554 + .246x$ **d.** $t = .627$ **e.** fail to reject H_0 **f.** $.246 \pm 1.81$ **9.55 a.** negative linear trend **b.** $\hat{y} = 9{,}658.24 - 171.573x$; for
each 1-unit increase in search frequency, the total catch is estimated to decrease by 171.573 kg **c.** H_0: $\beta_1 = 0$, H_a: $\beta_1 < 0$
d. $.0402/2 = .0201$ **e.** reject H_0 **9.57 a.** $t = -6.42$, reject H_0 **b.** $-.305 \pm .135$ **9.59** $-.0023 \pm .0019$; 95% confident that change in
sweetness index for each 1-unit change in pectin is between $-.0042$ and $-.0004$ **9.61 a.** $y = \beta_0 + \beta_1 x + \varepsilon$ **b.** $\hat{y} = -8.524 + 1.665x$
d. yes, $t = 7.25$ **e.** $1.67 \pm .46$ **9.63 a.** $t = 6.29$, reject H_0 **b.** $.88 \pm .236$ **c.** no evidence that slope differs from 1
9.65 a. $\hat{\beta}_0 = .515$, $\hat{\beta}_1 = .000021$ **b.** yes **c.** very influential **d.** $\hat{\beta}_0 = .515$, $\hat{\beta}_1 = .000020$, p-value $= .332/2 = .166$, fail to reject H_0
9.67 true **9.69 a.** perfect positive linear **b.** perfect negative linear **c.** no linear **d.** strong positive linear **e.** weak negative linear
f. strong negative linear **9.71 a.** $r = .985$, $r^2 = .971$ **b.** $r = -.993$, $r^2 = .987$ **c.** $r = 0$, $r^2 = 0$ **d.** $r = 0$, $r^2 = 0$ **9.73** .877
9.75 a. 18% of sample variation in points scored can be explained by the linear model **b.** $-.424$ **9.77 a.** moderate positive linear
relationship; not significantly different from 0 at $\alpha = .05$ **c.** weak negative linear relationship; not significantly different from 0 at
$\alpha = .10$ **9.79 b.** .185 **9.81 a.** moderately strong negative linear relationship between the number of online courses taken and
weekly quiz grade **b.** yes, $t = -4.95$ **9.83 b.** piano: $r^2 = .1998$; bench: $r^2 = .0032$; motorbike: $r^2 = .3832$, armchair: $r^2 = .0864$;

teapot: $r^2 = .9006$ **c.** Reject H_0 for all objects except bench and armchair   **9.85 a.** $H_0: \beta_1 = 0$, $H_a: \beta_1 < 0$ **b.** reject H_0 **c.** no; be careful not to infer a causal relationship   **9.87** $r = .570$, $r^2 = .325$ **9.89** $E(y)$ represents mean of y for all experimental units with same x-value   **9.91** true   **9.93 a.** $\hat{y} = 1.375 + .875x$ **c.** 1.5 **d.** .1875 **e.** $3.56 \pm .33$ **f.** 4.88 ± 1.06 **9.95 c.** 4.65 ± 1.12 **d.** $2.28 \pm .63; -.414 \pm 1.717$ **9.97 a.** Find a prediction interval for y when $x = 10$ **b.** Find a confidence interval for $E(y)$ when $x = 10$ **9.99** (92.298, 125.104)   **9.101** run 1: 90% confident that for all runs with a pectin value of 220, mean sweetness index will fall between 5.65 and 5.84   **9.103 a.** (67.16, 76.53); 95% confident that ideal partner's height is between 67.16 and 76.53 in. when a female's height is 66 in. **b.** (58.37, 66.85); 95% confident that ideal partner's height is between 58.37 and 66.85 in. when a male's height is 66 in. **c.** males; 66 in. is outside range of male heights in sample   **9.105 a.** (2.955, 4.066) **b.** (1.020, 6.000) **c.** prediction interval; yes   **9.107 a.** Brand A: $3.35 \pm .59$; Brand B: $4.46 \pm .30$ **b.** Brand A: 3.35 ± 2.22; Brand B: 4.46 ± 1.12 **c.** $-.65 \pm 3.61$ **9.109** yes; $\hat{y} = 7.77 + .000113x$, $t = 4.04$, reject $H_0: \beta_1 = 0$ **9.111** random sample, continuous variables   **9.113 a.** $|r_s| > .648$ **b.** $r_s > .450$ **c.** $r_s < -.432$ **9.115 b.** $r_s = .745$, do not reject H_0 **c.** $.05 < p$-value $< .10$ **9.117 b.** $r_s = .9429$ **c.** do not reject H_0 **9.119 c.** .713 **d.** $|r_s| > .425$ **e.** reject H_0 **9.121 b.** perfect positive rank correlation **c.** actual relationship may be nonlinear   **9.123 a.** .714 **b.** reject H_0 **9.125 a.** $-.877$ **b.** $-.907$ **c.** reject H_0 **d.** reject H_0 **9.127 b.** invalid inference **c.** .507 **d.** yes   **9.129** $E(y) = \beta_0 + \beta_1 x$ **9.131** true   **9.133 b.** $\hat{y} = x; \hat{y} = 3$ **c.** $\hat{y} = x$ **d.** least squares line has the smallest SSE   **9.135 a.** $y = \beta_0 + \beta_1 x + \varepsilon$; negative **b.** yes **c.** no   **9.137 a.** positive **b.** yes **c.** $\hat{y} = -12.62 + .363x$ **e.** slope: for each additional hit per 1,000 at bats, estimate number of games won to increase by .363 **f.** $t = 1.47$, fail to reject $H_0: \beta_1 = 0$ **g.** .1535; $\approx$ 15 of sample variation in games won is explained by the linear model **h.** no   **9.139 a.** $y = \beta_0 + \beta_1 x + \varepsilon$ **b.** $\hat{y} = 175.70 - .8195x$ **e.** $t = -3.43$, reject H_0 **f.** $r_s = -.733$ **g.** do not reject H_0 **9.141 a.** $y = \beta_0 + \beta_1 x + \varepsilon$ **b.** 92% of sample variation in metal uptake is explained by the linear model   **9.143 a.** $\hat{y} = 560.1 + 63.3x$; for every 1-unit increase in JIF, estimated cost increases by \$63.30 **b.** \$908.50 **c.** 63.3 ± 232.7 **d.** $\hat{y} = 326.5 + 1.48x$; \$736.66; $1.48 \pm .81$ **e.** $\hat{y} = 338.9 + 197.21x$; \$846.73; 197.21 ± 177.31 **9.145 a.** yes **b.** $\hat{\beta}_0 = -3.05$, $\hat{\beta}_1 = .108$ **c.** $.t = 4.00$, reject H_0 **d.** $r = .756$, $r^2 = .572$ **e.** 1.09 **f.** yes   **9.147 a.** $\hat{\beta}_0 = -13.49$, $\hat{\beta}_1 = -.0528$ **b.** $-.0528 \pm .0178$; yes **c.** $r^2 = .854$ **d.** (.5987, 1.2653)   **9.149 a.** yes; positive **b.** $y = \beta_0 + \beta_1 x + \varepsilon$ **c.** $\hat{\beta}_0 = 20.13$, $\hat{\beta}_1 = .624$ **9.151 a.** $\hat{y} = 46.4x$ **b.** $\hat{y} = 478.44 + 45.15x$ **d.** no, $t = .91$ **9.153 a.** no; $r^2 = .748$ **b.** yes, $18(\hat{\beta}_1) = -.98$

Index

Photo Credits